Introduction to Material and Energy Balances

Introduction to Material and Energy Balances

G. V. Reklaitis
School of Chemical Engineering
Purdue University
West Lafayette, Indiana 47907

With contributions by
Daniel R. Schneider
Monsanto Company
St. Louis, Missouri

John Wiley & Sons
New York   Chichester   Brisbane   Toronto   Singapore

Copyright © 1983, by John Wiley & Sons, Inc.

All rights reserved. Published simultaneously in Canada.

Reproduction or translation of any part of
this work beyond that permitted by Sections
107 and 108 of the 1976 United States Copyright
Act without the permission of the copyright
owner is unlawful. Requests for permission
or further information should be addressed to
the Permissions Department, John Wiley & Sons.

Library of Congress Cataloging in Publication Data:

Reklaitis, G. V., 1942–
  Introduction to material and energy balances.

  Includes index.
  1. Chemical engineering—Mathematics.   2. Mathematical optimization.   3. Nonlinear programming.   I. Schneider, Daniel R.   II. Title.

TP149.R44   1983   620'.0042   82-23800
ISBN 0-471-04131-9

Printed in the United States of America

20 19 18 17 16 15 14 13 12

Printed and bound by Quinn - Woodbine, Inc.

*TO
JANINE*

PREFACE

Steady-state material and energy balance computations are the most frequent type of calculations carried out by all chemical engineers. Yet, in spite of their universality, this class of computations is quite neglected in the chemical engineering curriculum. Unlike the transport, thermodynamics, and reactor design computations, which are taught in a rigorous and systematic fashion, balance calculations are generally presented at a purely intuitive level, usually in the first chemical engineering course, and then only briefly revisited in the senior design course. In the senior course, it is usually assumed that balance calculations are well known to the students; hence, having little systematic knowledge to fall back on, the students bludgeon through the flowsheet calculations required in design projects by sheer persistence. Given the inadequate academic training, it is not surprising therefore that balance calculations continue to be one of the vague arts of the profession. The comprehensive treatment of this subject presented here is an attempt to fill this obvious gap in chemical engineering education.

The book has three specific goals:
1. To provide a thorough exposition of balance equation concepts.
2. To develop a framework for the analysis of flowsheet information and specifications.
3. To present systematic approaches for manual and computer-aided solution of balance problems.

The first goal is met with a detailed development of the structure, properties, and interrelationships of species and element balances as well as a comprehensive discussion of both the general energy balance equation and the thermochemical calculations required to evaluate the terms arising in that equation. For instance, the homogeneity of the material and energy balance equations is used to explain the concept of basis and to clarify which variables can be assigned values as basis. The species and element balances are developed using the algebraic view of reaction stoichiometry and the rate-of-reaction concept. The latter device considerably simplifies the formulation of the balance equations for multiple reactions. The former approach leads directly to the valuable notion of independent reactions, to constructions for generating such reactions, and to a simple rule for testing the equivalence of species and element balances.

The energy conservation equation development includes discussion of traditional concepts such as state variables, state functions, and units of measure. In addition, the difficult but central issue of interconversion between the enthalpy, phase distribution, temperature, pressure, and composition of a system is also introduced. For simplicity, this discussion is confined either to single species or to ideal vapor–liquid systems. Moreover, the interconversion computations are restricted to the use of the elementary dew point, bubble point, and isothermal flash calculations. Evaluation of the enthalpy function using thermodynamic tables, heat capacity correlations, and heats of transition are also covered in detail. The heat-of-reaction calculations are fully developed using generalized stoichiometry constructions, and the steady-state energy balance equation is formulated both in the heat-of-reaction and the total enthalpy form. Special attention is directed at the selection of problem reference state and at the use of enthalpy data with different reference values. In short, the selection of topics addresses all essential considerations that arise in the selection, formulation, and evaluation of balance equations.

The second goal, of providing a framework for the analysis of process information, is achieved by extensive use of a "degree-of-freedom" analysis. This is essentially an orderly way of analyzing the information and counting the variables and balance equations associated with each unit in the process as well as with the entire process as a whole. It provides a systematic way for determining whether the problem is underspecified, overspecified, or correctly specified. Such a systematic analysis is extremely important in design applications, in which, by definition, the problem is underspecified, as well as in process analysis applications, in which overspecified problems often must be dealt with—usually by deleting inaccurate or superfluous data. The degree-of-freedom analysis also provides criteria for selecting the location of the basis for the calculations, for determining the order in which manual unit calculations should be executed, for identifying whether the material balances can be solved independently of the energy balances, and for determining whether or not carrying unknown variables from unit to unit will be required. Together with the equation-solving and root-finding techniques included in the text, the degree-of-freedom analysis thus serves to fulfill the third goal, that of providing a logical procedure for solving balance equations by manual means.

In addition to manual calculation methods, computer-oriented procedures for solving balance problems are also analyzed. The widely used sequential modular strategy is discussed and is contrasted to the simultaneous or equation-oriented approach. The discussion of computer-oriented strategies using flowsheets modeled in terms of elementary material balance modules allows the key issues of process flowsheeting to be introduced in a straightforward manner. Thus, the selection of tear streams, treatment of nonstandard specifications (or control blocks), an introduction to elementary multivariable equation-solving and sequencing methods, as well as a comparative examination of the sequential and simultaneous strategies can all be given purely in terms of material balance problems. After the basic energy balance chapters, computer-oriented strategies are reconsidered, and it is shown how the elementary material balance modules can be extended to accommodate energy balances. Thus, the way is pointed to the more complex unit operation modules used in practical flowsheeting systems. The discussion of modules involving energy balances also suggests the need for computerized storage and use of thermochemical and other physical properties data. An introduction is therefore given to computerized physical properties estimation systems. However, the scope of the exposition is confined to the properties, data, and methods discussed in considering manual calculations. The text thus meets the third goal by giving an elementary but thorough presentation of the various approaches for solving full-scale flowsheet problems via the computer.

Finally, in order to illustrate the problem-solving methodology and to expose the student to realistic chemical engineering applications, the text includes a range of multiunit process examples. These examples encompass not only conventional chemical processes such as the ethylene oxide/glycol plant, the ammonia process, the acetic anhydride process, and the Claus process, but also novel processes currently under development, for instance, those for the conversion of coal to gases and liquids. These problems are described in adequate detail to be understandable within the limited engineering vocabulary of the sophomore engineering student.

The level of the presentation is aimed at the sophomore chemical engineering student and uses mathematical concepts familiar to him or her: derivatives of nonlinear functions and manipulations involving linear equations. All additional mathematical methodology required in the exposition, for example, root-finding methods and algebraic properties of linear equations, is presented as needed so that its utility and applicability are immediately demonstrated in the context of the discussion. The computer-oriented portions of the exposition contain no programming details so that emphasis can be placed on the underlying methodology rather than on code implementation. A FORTRAN program listing and a description of a rudimentary modular balance program are given in the solution manual for use at the instructor's option.

Although this book is intended for the first course in chemical engineering, its scope is sufficiently broad that the entire contents might not be covered in one course. Cuts can be instituted as follows. First, Section 1.3 can be assigned to the student as review. Chapters 2 and 3 should be discussed in their entirety. Section 4.3 can be covered selectively, and Section 5.2.2 can be omitted without loss of continuity. Chapter 6, 7, and 8 should be covered in detail. However, in Chapter 9, Section 9.2.2 can be omitted without loss of continuity. The book includes nearly 200 worked examples many of which involve multiunit flowsheets of some complexity. It is normally possible to discuss only about one third of these examples in class. Moreover, while it is desirable to explain the details of the simpler examples, for the larger problems it is more important to focus on the strategy of solution and to leave the details for reading outside of the lecture room. Finally, the more advanced material in Chapters 3 and 4 as well as the discussion and implementation of the computer-oriented approaches treated in Chapters 5 and 9 are suitable topics for the process design course either as review or as supplement. Thus, the text can both serve as a primary reference for the introductory chemical engineering course and as a supplementary reference for the design course.

This work is really the result of the confluence of many sources and influences the complete acknowledgment of which is quite impossible. The linear algebraic treatment of stoichiometry clearly is extracted from the classical work of Aris and Denbigh. Elements of the degree-of-freedom analysis have been used sporadically by many chemical engineers: whoever the originator, to him goes the credit. The comprehensive book *Material and Energy Balance Computations* by Ernest Henley and Edward Rosen, which assuredly was ahead of its time, has been an important factor influencing the choice of topics. The very idea of writing this book must be credited to Daniel R. Schneider, who, while a colleague at Purdue, was a partner in evolving the basic approach, in developing the overall outline of the book, and even in formulating some of the examples. The text itself has benefited materially from the critical comments offered by Dan Schneider and by Profs. Lowell B. Koppel, Robert G. Squires, and Roger E. Eckert. Important clarifications and suggestions were also contributed directly by Profs. Alden Emery and Ferhan Kayihan and indirectly by the numerous teaching assistants and even more numerous students whose questions and skepticism helped to reshape the manuscript over the past seven years. Finally, I am thankful for the supportive atmosphere created by my colleagues in the School of Chemical Engineering and for the patient, conscientious efforts of the clerical staff of the School in typing and processing a succession of drafts.

G. V. Reklaitis

CONTENTS

Introduction to Material and Energy Balances

CHAPTER
1

Introduction

The objective of this book is to develop the concepts and solution methods that are required to determine the distribution of material and energy flows in a chemical process. These so-called material and energy balance calculations are both the most elementary and the most frequently performed computations carried out by chemical engineers. They are as basic a tool of the chemical engineer as are ledger-book balancing methods to an accountant. This book will help the reader to develop a mastery of this basic tool both for use in manual calculations as well as for implementation on computers. The consideration of solution strategies appropriate for computer implementation is given special attention because the use of computers to perform "bread-and-butter" balance calculations is growing in the field and is clearly the way of the future. However, our study of computer-oriented strategies is at the conceptual level, stopping short of delving into coding details. Hence, prior mastery of any specific programming language is not a prerequisite.

   In this first chapter, we begin with a discussion of the role of chemical engineers in general and of the place of balance calculations both in the work of the professional engineer as well as in the curriculum required of B.S. Chemical Engineering graduates. In the following section, we review some of the basic concepts derived from chemistry, physics, and mathematics that are employed in balance calculations. The chapter concludes with a preview of the organization of the contents of the remainder of the text.

1.1 THE ROLE OF THE CHEMICAL ENGINEER

The two primary functions of the chemical engineer are to develop and design processes that convert raw materials and basic energy sources into desired chemical products or higher energy forms and to improve and operate existing processes so that they become as safe, reliable, efficient, and economical as possible. The design function involves the synthesis of appropriate sequences of chemical and physical transformation steps and the selection of the conditions under which these transformations are to take place given basic information about the chemical reactions

1

and physical properties of the materials to be processed. The responsibility of the chemical engineer–designer thus begins with the basic chemical and physical information developed by chemists on a laboratory scale and concludes with the specifications of equipment for a full-scale plant. These specifications are then implemented by mechanical and structural engineers in fabricating the actual mechanical devices and constructing the finished plant. The chemical engineer thus has the challenging job of translating a laboratory concept into a full-scale commercial plant.

The duties of the chemical engineer in an existing plant include identifying and correcting malfunctions in the process, devising improved operating schedules and procedures, finding ways of increasing plant safety or reliability, and selecting new operating conditions to accommodate changes in feed conditions, product requirements, or unit performance characteristics. Execution of these duties requires a knowledge of the chemical and physical operations that are embodied in the process, the ability to interpret plant operating data as well as to select the measurements that should be taken, and the skill to perform the necessary engineering calculations that will allow values of unaccessible process variables to be deduced or plant performance to be predicted. The responsibility of the process engineer is thus to translate basic operating information and objectives into concrete action that can be taken by the plant operators, maintenance staff, or structural and mechanical engineering crews.

The following two examples are intended to illustrate both the types of technical problems addressed by chemical engineers and the general structure of chemical processes themselves.

Example 1.1   Ethylene Glycol Process Design   It is known that the important chemical ethylene glycol, $C_2H_4(OH)_2$, widely used as automobile antifreeze, can be produced by reacting ethylene oxide, C_2H_4O, with water in the liquid phase. The ethylene oxide intermediate can in turn be obtained by partial oxidation of ethylene, C_2H_4. This gas-phase reaction can be made to proceed at moderate temperatures by using a silver catalyst. The formation of the oxide is however accompanied by the undesirable complete oxidation reaction in which only CO_2 and H_2O are products. Based on further laboratory information about the relative yields of C_2H_4O and CO_2 as well as data on the properties of the various chemical species, the job of chemical engineers would be to design a two-step process which uses ethylene, air, and water as primary feeds, as shown in Figure 1.1.

Solution   Based on an analysis of laboratory reactor data, the design group might conclude that in order to minimize the formation of CO_2, the concentrations of C_2H_4 and O_2 in the feed to the oxidation reactor should be kept low. The N_2 present in the air would conveniently act as dilutent for this purpose. In addition, the fraction of the C_2H_4 that is allowed to react would be kept small. As a result, however, the reactor product would be dilute in C_2H_4O and would contain unconverted C_2H_4 and O_2 as well as the species CO_2, H_2O, and N_2. Clearly it will

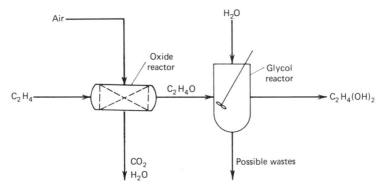

Figure 1.1   Two-stage conceptual glycol processes.

be necessary to devise means of separating these chemical species. The C_2H_4O must be recovered in relatively pure form to be used in the glycol formation reactor. The unconverted C_2H_4 and O_2 should be reclaimed for recycle to the oxide reactor, while the undesirable CO_2 and N_2 should be isolated and removed as waste products.

One possible separation scheme is shown in the process schematic given in Figure 1.2. The schematic consists of symbols representing process equipment units and of lines connecting these units representing physical pipes or ducts through which materials are transferred from unit to unit. These lines are commonly referred to as *streams* and are characterized by the flow rate of material, the composition of the species in the flowing mixture, temperature, and pressure. Schematics of the type of Figure 1.1 are commonly referred to as *flow diagrams* or *flowsheets* and will be used extensively throughout this book.

In the flowsheet of Figure 1.2, the stream leaving the oxide reactor is cooled by exchanging heat with the incoming feed to the reactor and is sent to an absorber unit. This unit separates C_2H_4O and a portion of the CO_2 from the remaining species in the product stream by absorption into cold water. This separation thus exploits the much higher solubility in water of C_2H_4O and CO_2 compared to the solubilities of the other species.

The solution from the absorber is fed to another separation device in which C_2H_4O and CO_2 are boiled off and the remaining water is recycled back to the absorber after being cooled using the cold absorber solution. The C_2H_4O and CO_2 are next separated by distillation at lower temperatures. The CO_2 and other light impurities are discarded as a waste stream while stream 12, the mixture of oxide and water, is transferred directly to the glycol reactor.

Returning to the gas stream leaving the oxide-absorber unit, a portion of that stream is split off and will be eliminated by incineration since there is no simple way of separating N_2 from the other gaseous species. (Such a separated waste stream is often called a *purge* stream.) The CO_2 in the remaining portion of the gas stream is removed by absorption into a solution of ethanolamine and water, which has a special affinity for CO_2. The resulting C_2H_4–O_2–N_2 mixture is com-

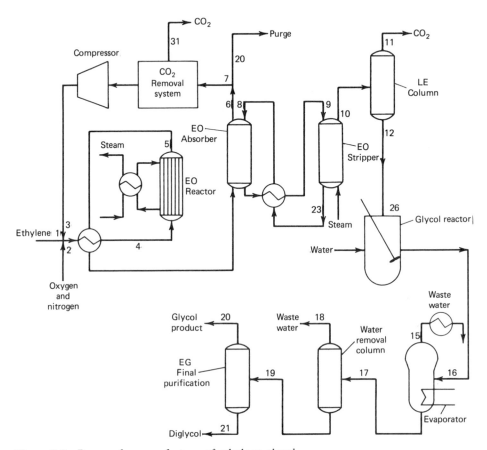

Figure 1.2 Process for manufacture of ethylene glycol.

pressed to make up for the pressure losses sustained in the previous processing of the gas stream, is mixed with fresh air and C_2H_4, and becomes the feed to the oxide reactor. The reader will note that in order to have devised the above separation scheme, knowledge of the relative solubilities of the species in water or ethanolamine solutions as well as information on the boiling points of C_2H_4O and CO_2 were indispensible.

As a result of the first part of the process, the intermediate C_2H_4O has been produced in a form suitable for reaction to ethylene glycol. As shown in Figure 1.2, the C_2H_4O solution is combined with a suitable amount of water in the glycol reactor. Since the oxide is quite reactive, there is no difficulty in achieving virtually 100% conversion. However, the concentration of C_2H_4O must be kept low to reduce the formation of diglycol, $(C_2H_4OH)_2O$, via a side reaction of ethylene oxide with the preferred product, the ethylene glycol, $C_2H_4(OH)_2$. Again, laboratory data are necessary to identify the C_2H_4O concentration as well as temperature which would be most advantageous. After reaction, the mixture of glycol, diglycol, and water must be separated to obtain a high-purity (say, 99%) glycol product. This is accomplished in several steps. First, an evaporator is used to boil off the bulk of the water. Next, a distillation unit is used to remove the residual water; and finally, a second distillation unit is used to separate glycol and diglycol. Having deduced the sequence of operations that are appropriate for this process, the design group will next proceed with detailed design calculations to determine the size and capacity of each of the individual equipment items as well as the composition, flow, temperature, and pressure of each of the streams flowing between the units. The design group will also perform some trade-off studies to examine the effects of changes in some of the design parameters. For instance, decreasing the ethylene oxide, C_2H_4O, concentrations in the glycol reactor feed will reduce the formation of diglycol, $(C_2H_4OH)_2O$, but will increase the amount of water that must be removed per unit of product formed in subsequent separation units. Thus, a balance must be found between loss of C_2H_4O to diglycol and water separation costs.

Another very important part of the design process is the determination of treatment methods for process waste streams, such as the water obtained in the glycol separation operations or the vent-gas stream produced in the light-ends columns. Finally, careful attention must be given to safety considerations such as the possibility of attaining explosive mixture proportions in the feed to the oxide reactor as well as the possibility of leaks of ethylene oxide, which is a very reactive substance. These considerations are no less critical in determining a good design than the selection of the main processing steps and their conditions.

The preceding example primarily illustrates the type of considerations involved in the development and design of a process. The following example presents issues which might have to be addressed by chemical engineers once the plant is built and operating.

Example 1.2   Ethylene Glycol Process Studies   The plant of Figure 1.2 is constructed, started up, and operates for a period of time. After some period of time, the following problems are posed to the plant process engineers:

(a)   The yield of ethylene oxide decreases gradually. What is causing the decrease?
(b)   A profitable market is found for diglycol. How can the operating conditions be altered to favor joint diglycol production?
(c)   The price of ethylene doubles. What can be done to improve ethylene utilization?
(d)   The demand for glycol declines because of reduced automotive sales. However, high-purity ethylene oxide is needed for polymerization applications. What can be done to recover a high-purity ethylene oxide intermediate product?

Solution   In case (a), plant tests might show that the silver catalyst is becoming less active because of fouling. The engineer would need to determine why this occurs and how to prevent fouling. Perhaps there are trace impurities introduced through the CO_2 removal system solution. Perhaps the reactor cooling system is malfunctioning and hot spots are occurring in the reactor. Suppose that as a result of measurements and possibly a few confirming laboratory-scale experiments, temperature control is found to be at fault. Then, modifications of the existing conditions or hardware will need to be proposed to correct the problem.

In case (b), the glycol-reactor feed concentration will need to be changed, as will the reactor operating temperature. These changes will affect the conditions in the downstream separation units. For instance, the evaporator will need to process a more concentrated solution thus requiring changes in heating rate and operating temperature. Moreover, depending upon the desired diglycol purity, an additional distillation unit may have to be added to the separation sequence. The new set of operating conditions will need to be carefully established and implemented. Equipment modifications may be necessary and a new column may have to be designed.

The improved ethylene utilization indicated in case (c) could be obtained in two ways: reduction in direct losses of ethylene and improved yields of products in each of the reactors. For instance, it might be desirable to develop a method of recovering C_2H_4 from the purge stream, say, by absorption into a heavier oil. Alternatively, since yield of C_2H_4O increases with a decrease in the fraction of C_2H_4 that is reacted in the oxide reactor, adjustment of the fraction reacted might be appropriate. Of course, this will increase the flow rate of the recycle stream and hence increase the load on the units processing the recycle flow. Perhaps an extra parallel compressor or extra heat exchange equipment may be desirable.

In the last case, case (d), three questions need to be considered.

1. The addition of separation units to purify a portion of the oxide solution which will then bypass the glycol production train, as shown in Figure 1.3.
2. The effects of reduced processing rates in the glycol production train. To what extent can the units be throttled down? Might it be necessary to only operate the glycol section periodically?
3. The possibility of increased production of ethylene oxide by scaling up the flows. What operations will prove to be the bottleneck?

In addressing these types of operational issues, the engineer will need to have insight into the process, to be able to anticipate trouble spots, to orchestrate the gathering of plant and laboratory data that guide the analysis process, and to perform the engineering calculations that will predict plant performance under the modified conditions.

Of course, chemical engineers are also involved in other important activities: research to further the understanding of the physical and chemical phenomena affecting new and existing processes; management of plant operating personnel; industrial sales and technical support for customer applications; and technical training of other chemical engineers. However, while these activities are both interesting and important, the core chemical engineering tasks remain process development and design as well as process engineering.

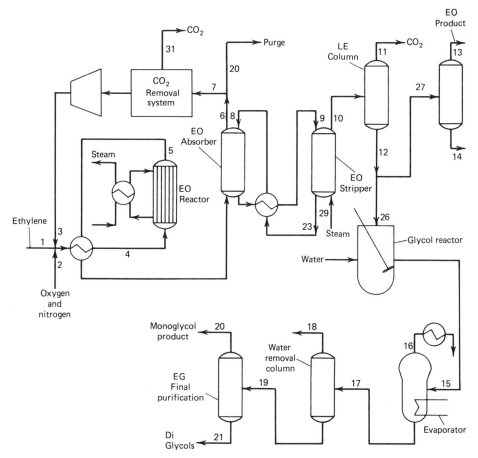

Figure 1.3 Process for manufacture of ethylene oxide and ethylene glycol.

1.2 THE ROLE OF BALANCE CALCULATIONS

Balance calculations are the computations based on the principles of conservation of mass and energy that serve to determine the flows, compositions, and temperatures of all streams in a flowsheet given selected or assumed information about the performance of some process equipment items or the properties of some streams. Since knowledge of the input streams and desired output streams to each process equipment item is essential information in the design of each such piece of equipment, it is clear that balance computations are of central importance in design. Similarly, since it is impractical or impossible to measure all streams in an existing process, balance calculations can be used to determine the flows and compositions of the unmeasured or unmeasurable streams in the process from known information about selected measured streams. Balance calculations thus have an important role in preliminary design, in final design, and in process operations. The following few examples illustrate the types of questions that can be answered using the methods discussed in this book.

Example 1.3   Space Station Life Support System[1]   In designing equipment for manned space missions, considerable attention must be devoted to the supply of air, water, and food as well as the disposal of respiratory and bodily wastes. While for short-term missions supply needs can be met from onboard caches and wastes can simply be stored or vented, for long-term missions recycle of waste products becomes important in order to minimize the need for large supply caches. The most important waste recycle systems involve recovery of respiratory CO_2 for reprocessing into O_2 and recovery and reuse of water from respiration and urine. Figure 1.4 shows the key elements of such a reprocessing system.

Food, which is represented as C_2H_2 because the carbon/hydrogen ratio in the average diet is about unity, is oxidized by the mission crew to produce CO_2 and H_2O using the O_2 in the cabin atmosphere.

The reclaimed water is electrolyzed to produce O_2 and H_2 gases in an electrolysis cell. The O_2 is returned for use in the metabolism of food, while the H_2 is used to reduce CO_2 to form methane, CH_4, and H_2O.

The CO_2 reduction reaction is known as the Sabatier reaction and can be carried out over a ruthenium catalyst in a tubular reactor. The products of reaction can easily be separated, with the water being recycled to the electrolysis cell while the CH_4 can be vented to space.

The foremost question to be answered in considering the system of Figure 1.4 is, assuming all reactions go to completion and all separations are perfect, which chemical species, O_2, H_2, or H_2O, must be stored to sustain the system. The next question is how much of that species must be provided per unit of food (C_2H_2) metabolized.

Solution   The material balance calculations discussed in Chapters 2 and 3 of this book can readily provide answers to both of these questions. Through such calculations, it can be shown that the system is H_2 deficient. Hence, a cache of H_2 must in principle be provided. An alternate and more convenient solution is to store the H_2 as water, H_2O, by using ordinary frozen foods. The use of water provides extra oxygen atoms, allowing some of the food carbon to be vented to space in the form of CO_2 and thus lowering the duty on the Sabatier reactor. The precise ratio of CO_2 to CH_4 which will be vented can, again, be computed using material balance calculations.

After these preliminary calculations, it is next necessary to select the equipment which will be used to reclaim H_2O and CO_2 from the space cabin atmosphere and

---

[1] P. J. Lunde, "Modeling, Simulation, and Operation of a Sabatier Reactor," Paper 56E, AICHE 74th National Meeting, New Orleans, March 1973.

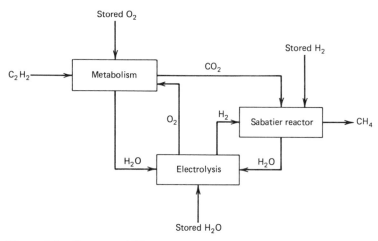

Figure 1.4 Conceptual life support system.

urine, to separate the Sabatier reactor products, and to carry out the electrolysis. Each of these real devices performs less than ideally, and the actual performance data must then be incorporated into final balance calculations which will yield the actual species flows for the life support system. Thus, material balance calculations prove useful both in the preliminary and in the final design stages of the system.

Example 1.4 Solar-Powered Chemical Heat Pump[2]

The system shown in Figure 1.5 and developed for the U.S. Department of Energy[2] uses solar energy, outside air, and a reversible chemical reaction to provide heating or cooling as well as hot water for homes. The reaction of methanol with anhydrous calcium chloride to produce the solid calcium chloride methanolate, $CaCl_2 \cdot 2CH_3OH$, is employed as follows.

Solar energy is used to heat a heat-transfer fluid to 130°C. The hot fluid is circulated through a bed of methanolate pellets to which it transfers heat, resulting in the decomposition of the methanolate. The released hot methanol vapor is cooled to form a liquid in the condenser unit. In the winter, this cooling is accomplished by heating inside air from, say, 20° to 40°C. In the summer, outside air can be used for this purpose provided it is at no more than 35°C. In either case, the liquid methanol is stored in a tank for use when needed in the second half of the process.

When either heating or cooling is desired, the valve to a bed of anhydrous calcium chloride, $CaCl_2$, will be opened allowing methanol vapor from the evaporator unit to be absorbed and reacted with the $CaCl_2$. As vapor is consumed, more liquid will be evaporated to replace it. The evaporation process requires heat, which is supplied in the summer by cooling inside air or in the winter by cooling outside air (provided it is above −15°C). The reaction of CH_3OH vapor with $CaCl_2$

---

[2] "Chemical Heat Pump Cools as well as Heats," *Chem. Eng. News*, 36–37 Anon., (Oct. 20, 1980).

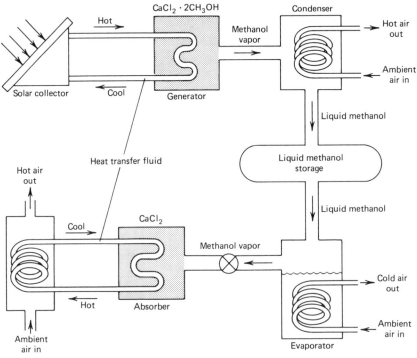

Note: Generator and absorber beds are exchanged on alternate cycles.

Figure 1.5 Solar powered chemical heat pump (reproduced from *Chemical & Engineering News*, Oct. 20, 1980).

releases heat, which is recovered from the bed using a heat-transfer fluid and is transferred either to a hot-water heater, to inside air, or to outside air depending upon requirements. The $CaCl_2$ bed and the reservoir of liquid methanol represent stored energy because, when they are allowed to combine, heating and/or cooling can be made to take place. A system of this type would use several beds of anhydrous $CaCl_2$ which would be used as generators or absorbers in alternating cycles.

Given a system of this type with fixed solar collector area, known average solar flux, and known outside air temperature, the answers to the following questions would be of interest.

1. If the inside air is available at 20°C and is heated to 40°C, how many cubic meters of air can be processed per hour?
2. How much liquid methanol could be accumulated over an average winter day? How much heat could be provided overnight?
3. If the inside air is available at 30°C and is cooled to 20°C, how many cubic meters of air can be processed per hour?
4. What is the pressure inside the evaporator when the inside air is cooled to 20°C?

Solution   These questions having to do with predicting the flows and conditions in an operating or proposed system can be readily answered using the energy balance calculations discussed in Chapters 6 through 8 of this book. Note that in this case, given the rate at which solar energy is delivered and certain measured stream properties (the temperatures), balance calculations would be used to determine the remaining unknown flows or properties.

The preceding two examples considered chemical engineering systems which involve relatively few streams and a relatively small number of distinct chemical species. In other process applications, such as the flowsheet of Figure 1.3, there may be many streams and the process may involve a substantial number of chemical species. The calculations required to determine all of the stream flows, compositions, and temperatures can become quite laborious and complex. Consequently, it is expedient to computerize the balance calculations using methods discussed in Chapters 5 and 9 of this text.

We conclude this section with a brief discussion of the place of material and energy balance calculations in the curriculum required of B.S. Chemical Engineers. Normally, the topics discussed in this book form the core of the first course in Chemical Engineering. The balance equations themselves are subsequently employed in the thermodynamics course, the reactor design course, the staged separations course, as well as the transfer operations courses. However, in these courses material and energy balances will normally only be developed for the individual units or operations studied in that particular course. The more extensive application of the subject matter of this text typically occurs in the process design course. In the design course, attention is again focused on entire process flowsheets of the type of Figure 1.3; and hence flowsheet balance calculations, particularly the methods discussed in Chapters 5 and 9, become quite important. Thus, material and energy balance concepts permeate the chemical engineering curriculum and find their heaviest utilization in the capstone design course. To compensate for the information loss which can occur because of the wide temporal separation between the first Chemical Engineering course and the capstone design course, the student may find it quite beneficial to review key chapters of this book, especially Chapters 5 and 9, prior to beginning the design course.

1.3 REVIEW OF BASIC CONCEPTS

In this section, we briefly summarize the elementary concepts drawn from chemistry, physics, and mathematics that underly balance computations. Specifically, we consider the conservation principles, elementary chemical stoichiometry, equation solving, and manipulation of dimensional quantities. The primary purpose of this review is to outline the prerequisites for the developments given in subsequent chapters. The reader who finds the discussion of this section too terse is strongly encouraged to refer to basic chemistry, physics, algebra, and calculus texts for expanded treatments of selected topics.

1.3.1 The Conservation Principles

One of the major accomplishments of the theory of relativity is the formulation of the principle that the total of the mass and energy of a system is conserved. This principle, which forms the basis of material and energy balance calculations, is of course a hypothesis since it has never been conclusively demonstrated. However, it is a very solid hypothesis since it has never been disproved experimentally.

A precise statement of the principle of conservation of mass and energy requires a careful definition of terms. First, the term *system* is understood to mean that bounded portion of the universe which is under study. The mass m of the system refers to the amount of matter at zero velocity relative to some selected reference point (sometimes called *rest mass*). The energy E of the system refers to energy in *all* possible forms. Finally, it is understood that a quantity is *conserved* if it can neither be created nor destroyed. Thus, all changes in the amount of the conserved quantity present in a system can be accounted for by simply measuring the tranfer of that quantity back and forth across the system boundaries.

Let $(d/dt)(m + E)_S$ denote the rate of change of the mass and energy of the system with time at a given point in time. Furthermore, let $(\dot{m} + \dot{E})_I$ and $(\dot{m} + \dot{E})_O$ denote the input and output rates of mass and energy into and out of the system. Then, the principle of mass and energy conservation reduces to the single statement

$$\frac{d}{dt}(m + E)_S = (\dot{m} + \dot{E})_I - (\dot{m} - \dot{E})_O \qquad (1.1)$$

In the absence of nuclear reactions or speeds approaching that of light, the extent of interconversion between mass and energy is negligible. Consequently, the single conservation equation can be separated into two statements:

$$\frac{d}{dt}m_s = \dot{m}_I - \dot{m}_O \qquad (1.2)$$

and

$$\frac{d}{dt}E_s = \dot{E}_I - \dot{E}_O \qquad (1.3)$$

The first is refered to as the *principle of conservation of mass*, and the second, as the *principle of conservation of energy*. These two principles were in fact postulated by early researchers in chemistry and mechanics and were considered to be independent laws. The theory of relativity showed that this is not the case in general. However, for most chemical engineering applications, excluding those involving nuclear reactors, the separation of the general conservation principle into two independent principles is an excellent and very convenient approximation. Equation (1.2) is the starting point for the material balance portion of this text; eq. (1.3) is the basis for the energy balance portion.

Special Cases   The conservation principle equations given above are written for the general case in which both sides of each equality are time-varying functions.

Such systems are said to be *dynamic*. In most of the applications to be considered in subsequent chapters, it is assumed that the system is at *steady state*, that is, that all properties of the system are invariant with time. For steady-state systems, it follows that the time derivatives dm_s/dt and dE_s/dt, often called the *accumulation terms*, are identically equal to zero. Thus, the conservation equations reduce to

$$\dot{m}_I = \dot{m}_O$$
$$\dot{E}_I = \dot{E}_O$$

or, the rate of mass transfer (energy transfer) into the system must be equal to the rate of mass transfer (energy transfer) out of the system. The difference between steady-state and dynamic systems is illustrated in the following example.

Example 1.5 Consider the system consisting of a barrel which has a capacity of 100 kg water, is empty at time $t = 0$, and is filled at the rate of 10 kg water per minute. At time $t = 10$ min, the barrel is full and commences to overflow. For t less than 10 min, water enters at $\dot{m}_I = 10$ kg/min, but $\dot{m}_O = 0$. Thus, from the mass conservation equation, eq. (1.2),

$$\frac{dm_s}{dt} = 10 - 0 = 10 \text{ kg/min}$$

Although the input and output mass transfer rates are time invariant, the accumulation term is nonzero, and therefore the system is *not* at steady state. The barrel is a *dynamic* system.

For times greater than 10 min, the barrel is filled to the rim, and hence the mass of water in the system is constant with time, that is, $dm_s/dt = 0$. Since the input rate remains constant at 10 kg/min, the mass conservation equation reduces to

$$\frac{dm_s}{dt} = 0 = 10 \text{ kg/min} - \dot{m}_O$$

or

$$\dot{m}_O = 10 \text{ kg/min}$$

The system has reached a *steady state*: all flows and properties are time invariant.

Systems can further be classified based on the occurrence of mass transfer across the system boundaries. An *open* system is one in which there is transfer of mass into or out of the system. A *closed* system is one in which there is no transfer of mass across the system boundaries. For a closed system, both m_I and m_O are equal to zero. Thus, eq. (1.2) reduces to the form

$$\frac{dm_s}{dt} = \dot{m}_I - \dot{m}_O = 0$$

or m_s is constant with time. Although a closed system must have constant mass, it can be either steady state or dynamic in terms of the properties having to do with energy. Thus, the full dynamic form of the energy conservation equation may be applicable. A system is said to be *isolated* if it is closed and, in addition, there is no energy transfer across the system boundaries. For an isolated system, $\dot{E}_I = \dot{E}_O = 0$ and eq. (1.3) reduces to the form,

$$\frac{d\dot{E}_s}{dt} = \dot{E}_I - \dot{E}_O = 0$$

Thus, an isolated system has both constant mass and constant energy level.

Example 1.6   Suppose at $t = 20$ min, the input flow of water to the barrel of Example 1.5 is shut off. Providing there are no leaks in the barrel, the output flow also becomes zero. Thus,

$$\frac{dm_s}{dt} = 0$$

and the system clearly is closed with constant mass of 100 kg. The system need not be isolated. If the sun shone upon it, the water in the barrel would warm up. If the outside air temperature would drop below freezing, the water in the barrel would eventually turn to ice. In both cases, the energy content of the system would change with time. For the system to become isolated, it would be necessary to cover the barrel and to insulate it so that no transfer of energy as heat or any other form could occur.

In subsequent chapters, we largely deal with open steady-state systems. Such systems typically represent the desired operating mode of large-scale chemical plants of the type of Figure 1.3.

1.3.2   Chemical Stoichiometry

In most applications of balance calculations, it is not sufficient to deal with mixtures on the basis of total mass alone. Particularly in systems involving chemical transformations it is necessary to focus on the individual chemical compounds or species of which the mixtures are composed. In this section, we summarize the basic facts from the atomic theory of matter and the notions of molecular formula, stoichiometric equations, and atomic and molecular weight which are prerequisites for applying the principle of conservation of mass to individual chemical species.

Molecules and Reactions   Under the atomic theory of matter, chemical compounds are composed of bonded aggregates, called *molecules*, consisting of atoms of one or more of the 103 known types of basic chemical building blocks called elements. Each molecule contains an integral number of atoms of its constituent elements and can thus in large part be characterized by the number and type of its atoms. This information is conveniently expressed in terms of a construction called a *molecular formula* which has the general form $A_a B_b C_c$, where each capital letter denotes the symbol for a specific element and the lower-case subscript indicates the number of atoms of that element per molecule of that chemical compound. A standard set of element symbols has been agreed upon and is summarized in Appendix 1. Using that symbol library and the molecular formula convention, the compound benzene, consisting of six carbon and six hydrogen atoms per molecule, is denoted C_6H_6.

The *chemical reaction* of two species to form one or more new product species is a process in which the reacting molecules are rearranged and their constituent elements are redistributed to result in the desired product species molecules. This process occurs in such a way as to preserve the identity of the atoms of the different elements. Thus, under the atomic theory of matter, the atoms of each type of elements are *conserved* during a chemical reaction.

Stoichiometric Equations   Since atoms are conserved and since the product molecules also contain integral numbers of atoms of the elements present in the reactant molecules, it follows that the reactant molecules must combine to form the product molecules in ratios that are integers or simple fractions. A compact way of expressing both the ratios in which specific compounds combine to form specific product compounds and the molecular formulas of the compounds themselves is the *stoichiometric reaction equation*.

If a molecules of compound with molecular formula A combine with b molecules of compound with molecular formula B to form c and d molecules of products C and D, respectively, then the stoichiometric equation for this reaction is

$$aA + bB \rightarrow cC + dD$$

The coefficients $a, b, c,$ and d are called *stoichiometric coefficients* of their respective species.

By convention, the direction of the arrow indicates the products of the irreversible reaction. A reversible reaction is denoted by a double arrow. For example, the stoichiometric equation for the reversible reaction of carbon monoxide and hydrogen to form methane plus water is

$$CO + 3H_2 \rightleftharpoons CH_4 + H_2O$$

Since atoms must be conserved during reaction, the stoichiometric coefficients appearing in the stoichiometric equation must result in the occurrence of the same number of atoms of any given element on each side of the reaction equation. A stoichiometric equation for which this is true is said to be *balanced*.

Example 1.7   The reaction

$$CO_2 + 4H_2 \rightleftharpoons CH_4 + 2H_2O$$

is balanced. There is a single C atom, two O atoms, and eight H atoms on each side of the equation.

The reaction

$$C_3H_6 + 4O_2 \rightarrow 3CO_2 + 3H_2O$$

is not balanced. There are eight O atoms in the reactant side and nine O atoms on the product side. The stoichiometric coefficient of O_2 should be changed to $\frac{9}{2}$ for the equation to represent a valid reaction.

The stoichiometric equation very clearly and compactly summarizes several important features of a system undergoing reaction.

1. Since the numbers of atoms of each type of element are unchanged, the mass of each type of element in the system is unchanged, that is, it is conserved.
2. Since the number of molecules of each type of chemical compound involved in the reaction will change, the mass of each reacting compound is *not* conserved.
3. Since species that do not appear in the stoichiometric equation are by definition unaffected by the reaction, the mass of each such inert species will be conserved.

These observations will prove quite central in our formulation of material balance equations.

Atomic and Molecular Weights   While the stoichiometric equation indicates the proportions in which molecules must be combined to form products, it is necessary to deal with the mass or mass flow rate of each reactant when actually preparing the reactant mixtures. To relate ratios of molecules to actual masses of each compound, it has proved convenient to use the concepts of atomic weight, molecular weight, and mole.

The *atomic weight* of an element is the relative mass of one atom of that element based on a standard but arbitrary scale which sets the mass of one atom of the carbon-12 isotope at exactly 12.

The *molecular weight* of a compound is the sum of the products of the atomic weight of each constituent element times the number of atoms of that element present in one molecule of the compound.

Both the atomic weight and the molecular weight are relative numbers, as they are based on the value of 12 assigned to carbon-12. The atomic weights of the elements are tabulated in Appendix 1.

Example 1.8   Calculate the molecular weight of C_3H_6 using Appendix 1.

Solution   The atomic weights of carbon and hydrogen are 12.0115 and 1.00797, respectively. The atomic weight of C is not 12 exactly because the standard atomic weights are determined using the mixture of isotopes of that element that occurs naturally. By definition, the molecular weight of C_3H_6 is equal to

$$3(12.01115) + 6(1.00797) = 42.08127$$

In practice, one would typically round this number to the nearest tenth.

A *gram mole*, or simply *mole*, of a substance is the amount of that substance that contains as many elementary entities as there are atoms in 12 g carbon-12. If the substance is an element, the elementary species will be atoms. If the substance is a compound, the entities will be molecules.

Since the gram mole and the molecular weight of a compound are both defined relative to carbon-12, the mass of one mole of a compound can be calculated as the product of the mass of one mole of carbon-12 times the molecular weight of the compound. That is,

$$\text{One gram mole} = 12 \text{ g carbon-12} \times \frac{\text{molecular weight of } x}{12}$$

$$= (\text{molecular weight of } x) \text{ grams}$$

where the number 12 in the denominator is the basis of the scale of atomic and molecular weights, the number assigned to carbon-12.

Because the gram mole is defined in terms of a specific amount of carbon-12 expressed in grams, one could equally define moles in terms of other measures of mass, for example, a kilogram mole or a milligram mole. For instance, 1 kgmol would be defined in terms of the number of atoms in 12 kg carbon-12. Thus, 1 kgmol would be equal to 10^3 gmol.

Example 1.9 Suppose the new unit of mass called the gold brick is defined as one gold brick = 1000/3 g.

(a) Calculate the mass in gold bricks of one gold-brick mole of H_2O.
(b) Calculate the number of gram moles per gold-brick mole.

Solution By definition, one gold-brick mole of water is the mass of water which contains as many molecules as there are atoms in 12 gold bricks of carbon-12. Since the molecular weight of H_2O is about 18, we have

$$\text{One gold-brick mole} = 12 \text{ gold bricks of carbon-12} \times \frac{18}{12}$$

$$= 18 \text{ gold bricks of } H_2O$$

Since 1 gold brick = 1000/3 g, it follows that one gold-brick mole = 18 gold bricks of $H_2O \times \frac{1000}{3}$ g/gold brick = 6000 g H_2O. But from the definition of the gram mole,

$$\text{One gram mole} = 18 \text{ g } H_2O$$

Therefore,

$$\text{One gold-brick mole} = \frac{6000 \text{ g } H_2O}{18 \text{ g } H_2O/\text{gmol}}$$

$$= \frac{1000}{3} \text{ gmol}$$

The above example shows that the definitions of the mole in terms of different mass units will lead to a mole unit of different total mass. Since the mass of a carbon-12 atom is a fixed quantity, changing the unit of mass on the reference mass of carbon-12 will change the number of atoms that are contained in one mole of carbon-12. The number of atoms in one gram mole of carbon-12 is known as Avogadro's number, A, and its most accurately known value[3] is $6.0220943 \times 10^{23} \pm 6.3 \times 10^{17}$. Since

$$\frac{12 \text{ g carbon} - 12/\text{gmol}}{\text{Mass of one carbon} - 12 \text{ atom}} = A \text{ atoms/gmol}$$

the number of elementary entities in 1 kgmol will be equal to $10^3 A$, in one gold-brick mole will be equal to 1000/3A, and so on. Fortunately, the calculations of the mole–mass equivalences use the relative atomic/molecular weights and do not explicitly require Avogadro's number or its multiples. Thus, knowledge of its precise numerical value is not essential for most calculations.

The key points to retain from the preceding discussion are the following. First, although the standard mole is the gram mole, mole units can be defined in terms of any desired unit of mass. Second, the mole-to-mass conversion factor for any compound in any mass unit will always be numerically equal to the molecular weight of the compound. For instance, for water there are 18 g per gmol, 18 kg/kgmol, 18 gold bricks/gold-brick mole, and so forth. For this reason, the molecular weight is often directly written as the mass–mole conversion factor with an assigned set of units, although strictly speaking it is dimensionless. We employ this loose practice in subsequent chapters because of its widespread usage in the field.

1.3.3 Equation-Solving Concepts

The application of the conservation laws to steady-state systems ultimately requires the solution of sets of balance equations for the values of unknown process flow rates. In general, the equations will be algebraic, may be linear or nonlinear, and their number may be quite large. Since the solution of such equation sets can be a formidable task, it is appropriate to devote attention to the study of efficient solution strategies and methods, as is done in subsequent chapters. For the present, we briefly recall some elementary equation-solving concepts which ought to be familiar to the reader.

Linear Equations A linear equation of the variables x_1 through x_N is a function of the form

$$a_1 x_1 + a_2 x_2 + \cdots + a_N x_N = b$$

where the coefficients a_i and the right-hand side b are known constants. Each term in the function contains a single variable, and each variable occurs to the first power. If the linear equation contains but a single unknown, say, x_1, then the unique solution to the equation can be obtained by mere division, that is, $x_1 = b/$

---

[3] Anon., "Metrology: A More Accurate Value of Avogadro's Number," *Science, 183*, 1037–1038 (Sept. 20, 1974).

a_1. If the linear equation contains more than one variable, then it will have an infinite number of solutions since we can always set all variables but one to arbitrary values and solve the resulting single-variable equation for the remaining unknown.

The process of setting $N - 1$ variables to fixed values is equivalent to augmenting the multivariable linear equation with $N - 1$ trivial equations of the form $x_i = c_i$. This is but a special instance of the well-known fact that a *system* of linear equations has a unique solution if and only if the system contains as many independent equations as there are unknowns. A set of linear equations is *independent* if and only if no one equation in the system can be obtained by adding together multiples of any of the remaining equations.

Example 1.10   The system of equations

$$x_1 + 0x_2 = 1$$
$$0x_1 + 1x_2 = 2$$

is *independent* because neither equation can be expressed as a multiple of the other.
   The system of equations

$$x_1 + 0x_2 = 1$$
$$0x_1 + 1x_2 = 2$$
$$1x_1 + 1x_2 = 3$$

is dependent because the third is equal to the sum of the first two.

The solution of systems of linear equations can be accomplished in two ways: by Cramer's rule or by variable elimination.

Cramer's Rule   Cramer's rule is a classical construction which expresses the solution of a system of linear equations in terms of ratios of determinants of the array of coefficients of the equations. In the case of a system of two equations in two unknowns,

$$a_{11}x_1 + a_{12}x_2 = b_1$$
$$a_{21}x_1 + a_{22}x_2 = b_2$$

The solution will be given by

$$x_1 = \frac{\det \begin{vmatrix} b_1 & a_{12} \\ b_2 & a_{22} \end{vmatrix}}{\det \begin{vmatrix} a_{11} & a_{12} \\ a_{21} & a_{22} \end{vmatrix}} = \frac{b_1 a_{22} - b_2 a_{12}}{a_{11} a_{22} - a_{12} a_{21}}$$

$$x_2 = \frac{\det \begin{vmatrix} a_{11} & b_1 \\ a_{21} & b_2 \end{vmatrix}}{\det \begin{vmatrix} a_{11} & a_{12} \\ a_{21} & a_{22} \end{vmatrix}} = \frac{b_2 a_{11} - b_1 a_{21}}{a_{11} a_{22} - a_{12} a_{21}}$$

The form of the solution extends in an obvious way to larger systems. In each case, the denominator of the ratios will consist of the determinant of the array of variable coefficients, while the numerator for the ith variable will consist of the determinant of the array formed when the ith column of the array of variable coefficients is replaced by a column of right-hand side constants. Unfortunately, the evaluation of determinants becomes quite cumbersome and error prone for systems larger than 2. Hence, the use of Cramer's rule is not recommended beyond the case $N = 2$.

Variable Elimination   The solution strategy for linear equation sets which is always applicable is successively to solve one of the equations of the set for one of the unknowns and to eliminate that variable from the remaining equations by substitution. The elimination process is continued until the system is reduced to a single equation in one unknown. This approach is considerably simplified if each of the original equations contains only a few of the unknowns.

Example 1.11   Solve the system of equations given below using elimination.

$$2x_1 + 2x_2 + 4x_3 = 5$$
$$x_1 \qquad + x_3 = 2$$
$$x_1 + 3x_2 \qquad = 7$$

Solution   Suppose we solve the second equation for x_3:

$$x_3 = 2 - x_1$$

and the third equation for x_2:

$$x_2 = \tfrac{1}{3}(7 - x_1)$$

The results can be substituted into the first equation to obtain a single equation in x_1. That is,

$$2x_1 + 2(\tfrac{1}{3}(7 - x_1)) + 4(2 - x_1) = 5$$

Collecting terms, we obtain

$$-\tfrac{8}{3}x_1 = -\tfrac{23}{3}$$

or

$$x_1 = \tfrac{23}{8}$$

The values of the remaining variables can then be obtained by back-substituting the value of x_1 in the expressions for x_2 and x_3. Thus, $x_2 = \tfrac{11}{8}$ and $x_3 = -\tfrac{7}{8}$. The solution can and always should be checked by substituting it into the three equations.

The variable elimination strategy is quite adequate for manual calculations involving perhaps up to five equations, particularly if each variable is not present in every equation. We use this approach extensively in Chapters 2 and 3 and select, whenever possible, the equations to be solved so that the above condition is met. For computer solution or manual solution of larger problems, formulations of the elimination strategy are available which operate on the array of detached equation coefficients. Such an approach is discussed in Chapter 4.

Nonlinear Equations   A nonlinear equation is quite simply any equation that is not linear. As such, a nonlinear equation can take on any of an unlimited number of different forms. For some of these forms, analytical solution formulas are available; hence, determination of the solution of a specific equation is a straightforward matter. For instance, the well-known analytical solution of the quadratic equation

$$f(x) = ax^2 + bx + c = 0$$

given by the formula

$$x = \frac{-b \pm (b^2 - 4ac)^{1/2}}{2a}$$

can easily be evaluated for any specific values of the constants a, b, and c. By contrast, the equally elementary equation

$$f(x) = x \exp(ax) - b = 0$$

has no analytical solution formula; hence, for any specific values of a and b, its solution requires the use of either graphical or numerical methods.

Graphical Solution   The graphical method for solving a nonlinear equation in a single variable involves the construction of a graph of the function $f(x)$. The construction of the graph of course requires that the function be evaluated at a suitable number of trial points and that a smooth curve be drawn through these trial points. The solution is then obtained as the intersection of the curve with the x axis. In principle, the accuracy of the solution can be improved as desired by repeating the graphical construction using more closely spaced trial points and a graph with a finer scale.

Example 1.12   Obtain an estimate of the solution of the function $f(x) = x \exp(x) - 5 = 0$ graphically.

Solution   Since at $x = 0$, $f(x)$ is negative, while at $x = 2$ it is positive, the solution must lie in the interval $0 \leqslant x \leqslant 2$. The function can be evaluated at a suitable number of trial points, say, 10 equidistant points in the interval. The resulting (x, y) pairs can then be used to construct the graph shown in Figure 1.6. From the graph, the solution can be estimated to be about 1.33, whereas the solution exact to seven decimals is 1.3267247. At $x = 1.33$, the function value is 0.0288, while at the more exact solution the function value is 3.05×10^{-7}.

Numerical Solution   Graphical solution is adequate in applications in which the function is easy to evaluate and a solution with two- or three-figure accuracy is acceptable. Repeated application of graphical constructions is generally inefficient

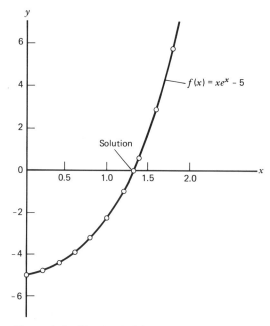

Figure 1.6   Graph of $f(x) = xe^x - 5$.

as a means of improving accuracy. If increased accuracy is desired or if the function is complex to evaluate, then *numerical* methods are much to be preferred. A *numerical method* or *algorithm* is a clearly defined series of logical and algebraic steps which, given an estimate x^{k-1} of the solution, will generate an improved estimate x^k by using information about the function at x^{k-1} as well as possibly at prior solution estimates. The most common type of function information used by numerical algorithms is the value of the function at the previous solution estimates.

The key requirement of a successful numerical method is that it successively improve the solution estimate. In this context, an improved solution estimate is simply one whose function value is closer to zero then that of the previous solution estimate. If the algorithm has this property, then there is some assurance that after repeated application of its steps, a sufficiently accurate solution estimate will be obtained. The algorithm is then said to have *converged*. Normally, an equation-solving algorithm will be terminated when the absolute value of the function at the current solution estimate is less than some specified accuracy parameter ε_1 (say, 10^{-4}). Sometimes, an algorithm might also be terminated if the relative difference between successive solution estimates becomes smaller than some specified tolerance ε_2 (say, 10^{-5}). That is, if x^k is the current estimate and x^{k-1} is the previous estimate of the solution, then one would terminate the application of the numerical method if

$$|f(x^k)| < \varepsilon_1$$

and/or

$$|x^k - x^{k-1}| < \varepsilon_2 |x^k|$$

The simplest of the numerical methods for the solution of the equation $f(x) = 0$ is the *interval halving* or bisection method. In this method, given two solution estimates x^L and x^R, one with a positive function value and the other with a negative function value, the next estimate z of the solution is obtained by setting

$$z = \frac{x^L + x^R}{2}$$

Then, if $f(z) < 0$, the new estimate z is used to replace the point x^L or x^R that has a negative function value. Alternatively, if $f(z) > 0$, the new estimate z is used to replace the point with positive function value. The calculations continue in this fashion until either $f(z)$ becomes sufficiently close to zero or the difference between the points x^L and x^R becomes sufficiently small. Observe that since the interval $x^L \leqslant x \leqslant x^R$ within which the solution of $f(x)$ must lie is reduced by one half with each trial point, the size of the interval after N trial points will be $(\overline{x}^R - \overline{x}^L)/2^N$, where $\overline{x}^R$ and $\overline{x}^L$ are the initial estimates with which the bisection method is started. As a result, the desired accuracy of the estimate of the solution x^* can also be controlled by choosing N appropriately. The complete interval halving algorithm is summarized below:

Given N, ε_1, and the bounds x^R and x^L such that $f(x^L) f(x^R) < 0$, set $n = 1$.

Step 1 Calculate $z = \dfrac{x^R + x^L}{2}$ and evaluate $f(z)$.

Step 2 If $|f(z)| \leqslant \varepsilon_1$, stop. The point z is the solution.

Step 3 If $f(z) f(x^R) \leqslant 0$, set $x^L = z$ and go to Step 5. Otherwise, continue.

Step 4 If $f(z) f(x^L) < 0$, set $x^R = z$ and continue.

Step 5 If $n = N$, stop. Otherwise, set $n = n + 1$ and continue with Step 1.

Example 1.13 Obtain a solution of $f(x) \equiv x \exp(x) - 5 = 0$ such that $|f(x)| \leqslant 10^{-2}$ starting with $x^L = 1.2$ and $x^R = 1.4$ using interval halving. As a conservative estimate, set $N = 10$.

Solution At x^L, $f(x^L) = -1.0159$; and at x^R, $f(x^R) = 0.67728$. Following Step 1, we obtain $z = (1.2 + 1.4)/2 = 1.3$ and $f(z) = -0.2299$. Since Step 2 is not satisfied, we check

$$f(z) f(x) = (-0.2299)(0.67728) < 0$$

Following Step 3, we set $x^L = z = 1.3$ and continue with Step 1. The results of the next six iterations are summarized below. At the seventh trial point, the requirement $|f(z)| \leqslant 10^{-2}$ is satisfied and, hence the calculations are terminated.

x^L	1.3	1.3	1.325	1.325	1.325	1.325
x^R	1.4	1.35	1.35	1.3375	1.33125	1.328125
$f(z)$	0.2075	-0.0151	0.0952	0.03981	0.01229	-0.00142

After seven iterations, the solution x^* has been bracketed within the interval $1.326563 \leqslant x^* \leqslant 1.328125$. The length of this interval is equal to the initial interval 1.4–1.2 divided by 2^7.

As evident from the example, the performance of the interval halving method is sure but rather slow. Hence, in Chapter 5 we study a broader selection of algorithms for the numerical solution of a single nonlinear equation.

Systems of Equations Solution of systems of nonlinear equations is considerably more difficult than the single equation case. As with linear equations, it is sometimes possible to use the variable elimination strategy to reduce the system to a single nonlinear equation. However, unlike in the linear case, the success of this strategy is not guaranteed since it may simply not be possible to explicitly solve any given nonlinear equation for a single unknown. For instance, the function in the previous example cannot be directly solved for x. In most cases, it is at best only possible to use the elimination strategy to reduce the system of equations to a smaller set; and when this is possible, it is useful to do so. Graphical solution of systems of nonlinear equations is normally not feasible; hence, in most cases numerical solution is the only alternative. In Section 5.2 we discuss such methods and their use with computers. The development of these methods assumes familiarity with the concepts of *derivative* and *partial derivative*. The reader unfamiliar with these concepts is strongly encouraged to consult the introductory chapters of any calculus text.

1.3.4   Dimensional Quantities and Their Manipulation

Scientific measurements and engineering calculations are normally performed using quantities whose magnitudes are expressed in terms of standard units of measure or dimensions. Thus, the mass of water in a barrel might be reported in kilograms or the length of a pipe in meters. While the association of dimensions with physical quantities is essential in removing ambiguities which can arise in communicating and using these quantities, it does require knowledge of the definitions of the standard units, the ability to convert between equivalent units, and the skill to manipulate dimensional quantities in a consistent fashion. In this section, we briefly review these topics.

Units of Measure A dimensional quantity is simply one that is defined by a magnitude and the name of a unit of measure. In order for the unit of measure to serve its function, its definition must be held in common by all users. Moreover, the definition should, if possible, be verifiable through standardized physical experiments. These considerations have led to the worldwide adoption of the International System of Units formalized in 1960 and last revised in 1971. The SI system has seven base units: the meter, kilogram, second, ampere, kelvin, mole, and candela. With the exception of the kilogram, these units are defined in terms of appropriate physical experiments.

The definitions of the base units of concern in balance calculations are the following.

The unit of *length*, the *meter* (m), is the length equal to 1,650,763.73 wavelengths in vacuum of the radiation corresponding to the transition between the levels $2p_{10}$ and $5d_5$ of the krypton-86 atom.

The unit of *mass*, the *kilogram* (kg), is the mass of a prototype held by the International Bureau of Weights and Measures in Paris.

The *second* (s) is the duration of 9,192,631,770 periods of the radiation corresponding to the transition between the two hyperfine levels of the ground state of the cesium-133 atom.

The *temperature* unit, the *kelvin* (K), is the fraction 1/273.16 of the thermodynamic temperature of the triple point of water.

The definition of the mole was given in Section 1.3.2, while the precise definitions of the remaining units can be found elsewhere.[4]

These SI units can be used with standard prefixes and their symbols to designate decimal submultiples. Preferred submultiples are

$$10^9 = \text{giga (G)}$$
$$10^6 = \text{mega (M)}$$
$$10^3 = \text{kilo (k)}$$
$$10^3 = \text{milli (m)}$$
$$10^6 = \text{micro } (\mu)$$
$$10^9 = \text{nano (n)}$$

Thus, 10^3 m can be called a kilometer and abbreviated km.

The base units can be used in combination to obtain derived units, such as the unit of force, the newton (N), defined as one kilogram meter per second squared.

Other various units of length, mass, time, or temperature which have come into use for historical reasons are by agreement defined in terms of the SI units. For instance, the hour (h) is defined as 3600 s; the ton (t), as 10^3 kg; the liter (l), as 10^{-3} m^3; and the centimeter (cm), as 10^{-2} m. Similarly, the unit of mass, the pound mass (lb_m), until recently widely used in the English-speaking world, is defined as 0.45359237 kg and the unit of length, the foot (ft), as 0.3048 m. Paralleling the definition of the mole given in Section 1.3.2, a molar unit called the pound mole (lbmol) is sometimes used in conjunction with mass quantities expressed in lb_m. Following the discussion given earlier, it is clear that 1 lbmol = 453.59237 gmol.

Manipulation of Dimensional Quantities   Since the units of a dimensional quantity are as important as the magnitude of that quantity, both should always be used in reporting. The units should also be explicitly shown for each dimensional quantity involved in a computation as an aid in deducing the units of the result and in verifying the consistency of the units involved in the intermediate steps of the calculation. If the units are shown explicitly along with the magnitudes of the quantities, then the unit symbols can be manipulated just like any other algebraic quantity. Specifically, the manipulation of units takes place via the following rules:

---

[4] *ASTM/IEEE Standard Metric Practice*, ASTM E380-75, Institute of Electrical and Electronic Engineers, New York, N.Y., Jan. 30, 1976.

1. The addition or subtraction of quantities all expressed in the same units yields a result expressed in those units. The addition or subtraction of quantities expressed in different units is obviously meaningless.
2. The multiplication or division of quantities expressed in arbitrary units yields a result which will have units given by the product or ratio of the units of those quantities.
3. The division of quantities in the same units gives as result a dimensionless quantity, that is, the units cancel.
4. The product of quantities expressed in the same units yields a result which has these units raised to a suitable exponent.

In short, the units associated with dimensional quantities follow the conventional rules of algebraic manipulation.

Example 1.14   Unit Manipulation Rules

(a) $10 \text{ kg/h} + 20 \text{ kg/h} = 30 \text{ kg/h}$.
(b) $10 \text{ kg/s} - 7200 \text{ kg/h}$ is not defined.
(c) $20 \text{ m} \times 10 \text{ m} \times 5 \text{ m} = 1000 \text{ m}^3$.
(d) $10 \text{ kg/s} \times 3600 \text{ s/h} = 36{,}000 \text{ kg} \cdot \text{s/s} \cdot \text{h} = 36{,}000 \text{ kg/h}$.

(e) $\dfrac{15 \text{ m/h} \times 4 \text{ kg/m}^3}{10 \text{ mol/m}^2 \times 25 \text{ kg/mol}} \times 50 \text{ h} = 12 \dfrac{\text{m}^3 \cdot \text{mol} \cdot \text{kg} \cdot \text{h}}{\text{m}^3 \cdot \text{mol} \cdot \text{kg} \cdot \text{h}} = 12$, where 12 is dimensionless.

Unit Conversion   In the above example, the difference $10 \text{ kg/s} - 7200 \text{ kg/h}$ can only be computed if the units of both quantities are the same. Since both quantities have dimensions of mass per unit time, they ought to be expressible in the same units. In particular, since by definition $1 \text{ h} = 3600 \text{ s}$, the rate of 7200 kg/h can be transformed to a rate expressed in kg/s by multiplying the former by the dimensionless ratio 1 h/3600 s, that is,

$$7200 \text{ kg/h} \times 1 \text{ h}/3600 \text{ s} = 2 \text{ kg/s}$$

In this manipulation, we have merely rearranged the definition of the hour to a dimensionless ratio and have applied the ratio in such a way as to allow the hour units to cancel out. The process of successive application of the definition of units to transform a quantity in one set of units to the equivalent quantity expressed in another desired set of units is called *unit conversion*. The process of unit conversion is clearly just a special instance of the application of the unit multiplication rule. In material balance computations, the unit conversions required will normally only involve direct interconversion between mass, mole, and time units. In energy balance applications, more involved unit conversions will be required, and these are given special attention in Section 6.4.

Example 1.15 Elementary Unit Conversion

(a) Convert a flow of 10 kg/s to a flow in tons per hour.
(b) Convert a flow of CO_2 of 88 kg/h to pound moles per hour.

Solution

(a) To convert seconds to hours and kilograms to tons, the definitions 1 h = 3600 s and 1 t = 10^3 kg are needed. Applying the definitions successively, we obtain

$$100 \text{ kg/s} \times 1 \text{ t}/10^3 \text{ kg} \times 3600 \text{ s/h} = 360 \text{ t/h}$$

(b) To convert mass to moles, we need the definition of the pound mole. Since the molecular weight of CO_2 is 44, it follows that

$$1 \text{ lbmol } CO_2 = 44 \text{ lb}_m$$

To use this mass-to-mole conversion factor, we first must convert mass in kg to mass in lb_m using the definition 1 lb_m = 0.45359237 kg. The combined result is

$$88 \text{ kg/h} \times 1 \text{ lb}_m/0.45359237 \text{ kg}$$

$$\times 1 \text{ lbmol } CO_2/44 \text{ lb}_m = 4.41 \text{ lbmol } CO_2/h$$

Composition Conversions   One special type of unit conversion which is required quite frequently in material balance computations is that of transforming one measure of the composition of a mixture to another measure. This need arises because composition can be expressed in various ways:

1. Molar concentration $\bar{c}$, moles of a component per volume of solution.
2. Mass concentration $\hat{c}$, mass of a component per volume of solution.
3. Mole fraction x, moles of one component per mole of mixture.
4. Mass fraction w, mass of one component per unit mass of mixture.
5. Solvent-free mole/mass fractions, moles or mass of a given species per mole of mixture exclusive of the solvent or some other designated component.

While composition can be measured and reported in various units, the principle of conservation of mass only applies to mass of species or to moles in nonreacting cases. Consequently, for material balance purposes it is ultimately necessary and preferable to deal with mole/mass fractions or, better yet, species mole/mass flows rather than with some of the other composition measures.

To carry out the conversions between composition measures, some or all of the following bulk properties of mixtures are required: density, specific or molar volume, and average molecular weight. These properties can simply be viewed as conversion factors which are specific to a given mixture.

The mass or molar *density* of a mixture, designated respectively as $\hat{\rho}$ and $\bar{\rho}$, is the mass or number of moles of mixture per unit volume of mixture.

The *specific volume* $\hat{V}$ or *molar volume* $\tilde{V}$ of a mixture is, respectively, the reciprocal of the mass or molar density.

The *average molecular weight* of a mixture, $\tilde{M}$, is simply the sum of the products of the mole fraction times the molecular weight of each of the mixture components.

If x_s indicates the mole fraction of species s and M_s is its molecular weight, then

$$\tilde{M} = \sum_s x_s M_s$$

where the sum is over all species s in the mixture. In general, the conversion of concentration measures to mole or mass fractions involves two steps: first, conversion of the unit volume of mixture to mass or mole quantities and second, if appropriate, conversion between the moles/mass of the species in question. Thus, for a given species s,

$$x_s = \tilde{c}_s \ (\text{mol/volume}) \ \tilde{V} \ (\text{volume/mol})$$

$$= \hat{c}_s \ (\text{mass/volume}) \ \tilde{V} \ (\text{volume/mol}) \ \frac{1}{M_s} \ (\text{mol/mass})$$

and

$$w_s = \tilde{c}_s \ (\text{mol/volume}) \ \hat{V} \ (\text{volume/mass}) \ M_s \ (\text{mass/mol})$$

$$= \tilde{c}_s \ (\text{mol/volume}) \ \tilde{V} \ (\text{volume/mol}) \ M_s \ (\text{mass/mol}) \ \frac{1}{\tilde{M}} \ (\text{mol/mass})$$

In the last equality, note that the molecular weight of species s, M_s, converts the moles of species s in the concentration to mass of species s. The average molecular weight $\tilde{M}$ converts the moles of mixture in the molar volume $\tilde{V}$ to mass of mixture.

Example 1.16 A solution of NaOH in water has a molarity of 2.0 and a density of 53 kgmol/m³. Calculate the mole fraction of NaOH and the mass density of the solution in tons per cubic meter.

Solution The molarity is a molar concentration defined in units of gmol/l. Thus, it is first necessary to convert liters to cubic meters and then cubic meters of solution to moles of solution. Thus,

$$x_{\text{NaOH}} = \frac{2.0 \text{ gmol NaOH}}{1} \times \frac{1 \text{ l}}{10^{-3} \text{ m}^3} \times \frac{1 \text{ kgmol}}{10^3 \text{ gmol}} \times \frac{1 \text{ m}^3}{53 \text{ kgmol}}$$

$$= 0.0377 \text{ kgmol NaOH/kgmol solution}$$

Since by definition 1 kgmol solution contains 0.0377 kgmol NaOH, the balance, or 0.9623 kgmol, must be water. The average molecular weight is therefore given by

$$\tilde{M} = x_{\text{NaOH}} M_{\text{NaOH}} + x_{\text{H}_2\text{O}} M_{\text{H}_2\text{O}}$$

$$= 0.0377(40) + 0.9623(18) = 18.83$$

The mass density will be equal to

$$\rho \ (kg/m^3) = \rho \ (kgmol/m^3) \ M \ (kg/kgmol)$$
$$= 53 \ kgmol/m^3 \times 18.83 \ kg/kgmol = 998.0 \ kg/m^3$$

Since $1 \ t = 10^3 \ kg$, this is equivalent to $0.998 \ t/m^3$.

Example 1.17 A mixture contains 10 g/l each of toluene and xylene in benzene. If the mixture density is $0.85 \ g/cm^3$, calculate the benzene free mass fraction of toluene.

Solution First the mass concentrations of toluene and xylene must be converted to mass fractions. Then, the benzene free mass fraction can be calculated from its definition. Thus,

$$w_{C_6H_5CH_3} = \frac{10 \ g \ toluene}{1} \times \frac{1 \ l}{10^{-3} \ m^3} \times \frac{cm^3}{0.85 \ g \ solution} \times \frac{(10^{-2} \ m)^3}{1 \ cm^3}$$
$$= \frac{10^{-2}}{0.85} \frac{g \ toluene}{g \ solution} = 0.01176$$

It is easy to verify that $w_{C_6H_5(CH_3)_2}$ will also be equal to 0.01176. Then, by definition, the benzene free mass fraction of toluene will be equal to

$$\frac{Mass \ toluene}{Mass \ toluene \ + \ mass \ xylene}$$

$$= \frac{10^{-2}/0.85 \ g \ toluene/g \ solution}{10^{-2}/0.85 \ g \ toluene/g \ solution \ + \ 10^{-2}/0.85 \ g \ xylene/g \ solution} = 0.5$$

Dimensional Equations Since the units of a dimensional quantity are merely labels expressing the scale in terms of which the magnitude of the quantity is to be interpreted, a dimensional quantity need not always be a constant. It can also be a variable. If an equation involves dimensional variables as well as some dimensional constants, then each of the terms of the equation becomes a dimensional quantity. The rules for manipulating dimensional quantities are equally applicable to such terms, and hence to equations, as they are to dimensional constants. In particular, rule 1 (p. 26) indicates that if an equation consists of a sum of terms, then each term must have the same units. Thus, the right-hand side constant of a linear equation must have the same units as each of the terms on the left-hand side of the equation. Moreover, the multiplication or division of factors or variable groupings must also follow rules 2 through 4. An equation which satisfies rules 1 through 4 is said to be *dimensionally homogeneous*. All dimensional equations must be dimensionally homogeneous.

Example 1.18 A mill pond has two inlets, one of which supplies water at 1 ft^3/min, and one outlet which services the mill, as shown in Figure 1.7. If the water flow to the mill is 7440 lb$_m$/h, what must be the additional inlet flow to the pond,

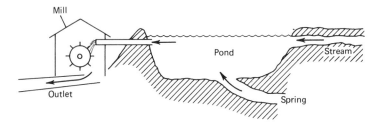

Figure 1.7 Schematic of Example 1.18.

assuming that the pond level is to remain constant? The density of water is 62 lb_m/ft^3.

Solution If the pond level and the flows are constant with time, then the system consisting of the water in the pond must be at steady state. The application of the mass conservation equation, eq. (1.2). Thus yields

$$\text{Flow water in} = \text{flow water out}$$

or

$$1 \text{ ft}^3/\text{min} + x = 7440 \text{ lb}_m/\text{h}$$

where x is the unknown additional water inlet flow. For this equation to be meaningful, the units of both quantities on the left-hand side must be the same. Thus, the unknown flow x must be defined in terms of ft^3/min. Moreover, the units of the right-hand side of the equation must be the same as those of the left-hand side. The flow rate of 7440 lb_m/h must therefore be converted to ft^3/min, that is,

$$7440 \text{ lb}_m/\text{h} \times \text{ft}^3/62 \text{ lb}_m \times 1 \text{ h}/60 \text{ min} = 2 \text{ ft}^3/\text{min}$$

Thus, the dimensionally homogeneous way of writing the above equation is

$$1 \text{ ft}^3/\text{min} + x \text{ (ft}^3/\text{min)} = 2 \text{ ft}^3/\text{min}$$

The solution is obviously $x = 1 \text{ ft}^3/\text{min}$.

Example 1.19 The density of liquid ethanol in g/cm^3 as a function of temperature T in degrees Kelvin has been correlated via the equation

$$\rho \text{ (g/cm}^3) = 1.032 - 5.392 \times 10^{-4}T - 8.712 \times 10^{-7}T^2$$

(a) What are the units of the three constants if the equation is dimensionally homogeneous?
(b) How would the correlation be altered to yield liquid density in gmol/l?

Solution (a) For the equation to be dimensionally homogeneous, all three terms on the right-hand side of the equation should have the same units as those of the density (g/cm^3). Therefore, the constant 1.032 should have units of g/cm³; the constant 5.392 × 10⁻⁴ should have the units of g/K · cm³; and the constant 8.712 × 10⁻⁷ should have the units of g/K² · cm³.

(b) Since the molecular weight of C_2H_5OH is 46, it follows that

$$\rho \ (gmol/l) = \rho \ (g/cm^3) \times \frac{10^3 \ cm^3}{1 \ l} \times \frac{gmol}{46 \ g}$$

$$= \rho \ (g/cm^3) \times \frac{10^3}{46} \ (gmol \cdot cm^3/g \cdot l)$$

Therefore, to convert the correlation to give density in gmol/l, all three constants in the correlation should be multiplied by $10^3/46$.

It should be noted that while each equation in an equation set must be dimensionally homogeneous, it certainly is not necessary that all equations in the set be expressed in the same units. In many cases, this is in fact impossible because the equations will have arisen from different physical laws (e.g., a mass conservation equation and an energy conservation equation) and hence may involve entirely different dimensional quantities. It is however critical that in each instance in which a given variable appears in the equations of the set that variable be expressed in the same units. Thus, if x represents the flow of water into the system in units of m³/h in one equation, then it must be similarly defined in all other equations.

This completes our review of basic concepts. We conclude the chapter with an overview of the organization of the remainder of the book.

1.4 PREVIEW OF SUBSEQUENT CHAPTERS

The chapters of this book can be divided into three parts. Chapters 2 through 5 deal with the application of the principle of conservation of mass. Chapters 6 through 8 deal with the application of the principle of conservation of energy. Finally, Chapter 9 discusses the solution of problems in which the two types of balances must be solved jointly.

With the exception of Chapter 5, each of the material balance chapters is organized with a dual purpose: to develop the particular structure and properties of a particular type of balance equation and to present analysis and solution strategies for treating increasingly more complex flowsheet problems. Thus, Chapter 2 develops the properties of species material balances for nonreacting systems, Chapter 3 discusses the properties of species material balances for reacting systems, while Chapter 4 presents the characteristics of material balance equations written on the individual chemical elements. Concurrently, as we proceed from Chapters 2 through 4, we advance from the solution of flowsheet problems with just a single processing unit and a few streams to flowsheets with many units and numerous streams. In Chapter 5, we cap the development of material balance concepts with a discussion of iterative and computer-oriented solution strategies.

The energy balance chapters differ from the material balance development principally in that considerable attention must be devoted to a study of the data, correlations, and subsidiary calculations necessary to evaluate the energy content terms in the balance equations. The solution of the energy balances themselves, by contrast, presents relatively few new difficulties. Chapter 6 reviews the separate terms which must be included in the general energy balance equation, introduces the new units and variables involved in energy balances, and develops the general form of the steady-state energy balance equation. Chapter 7 considers the type of thermodynamic information required to solve problems involving nonreacting systems and develops a framework for analyzing single-unit problems. Chapter 8 considers the reacting case, focusing on the additional thermodynamic data required and on the alternate forms of the energy balance equation that can be written for reacting systems. In addition, the multiunit analysis developed in the material balance part is extended to the energy balance case. Finally, Chapter 9, which in large part parallels Chapter 5, discusses iterative and computer-oriented strategies for solving the combined material and energy balances of complete process flowsheets. The reader who masters all of these topics can be confident of having learned the most important family of calculations performed by chemical engineers.

CHAPTER

2

Material Balances in Nonreacting Systems

In this chapter, we begin our study of balance calculations by considering the simplest type of system, namely, the steady-state open system in which neither chemical nor nuclear transformations are occurring. In spite of its relative simplicity, the nonreacting system is an extremely important case to consider because of its frequent occurrence in practice. Although one or more chemical reactors form the core of nearly all processing systems, the major portions of such plants, in terms of both the number of units involved and the total plant investment cost, are concerned with the physical transport, separation, and mixing of materials before and after reaction. Hence, the major portions of most chemical plants can be viewed as nonreacting systems.

The discussion of this chapter proceeds in three major phases. The first phase, contained in Section 2.1, concerns the formulation of material balance problems. We study what types of variables are involved in considering material balances, what the different forms are in which information about a particular system can be specified, and what kinds of equations can be constructed using the principle of conservation of mass in relating the process variables and specified information. Next, we analyze the single-unit case and determine a systematic way of either solving such problems or else recognizing that solution is not possible (Section 2.2). Finally, we extend the single-unit analysis to help us perform material balances on multiunit nonreacting systems (Section 2.3). In this last section, we focus especially on the strategy of solving problems because, without a plan of attack developed in advance, the larger-flowsheet material balancing problems can become quite unapproachable.

2.1 FORMULATING THE MATERIAL BALANCE PROBLEM

As defined in Chapter 1, an open steady-state system is a restricted portion of the universe in which no net accumulation of mass is observed over time and which receives matter from and discharges matter to its environment at a constant rate. Depending upon the level of aggregation of interest, the system can correspond to an entire chemical plant, a complete functional unit within the plant, or perhaps only an individual piece of equipment which is part of a unit. Yet, regardless of the internal complexities of the system, any open system can always be represented for material balance purposes as a box that has matter transferred in and out by means of several distinct material flows. Each of these input and output material flows will in general consist of a mixture of chemical compounds which must be identified as part of the definition of the system.

For example, the multistage desalination plant shown in Figure 2.1 is a complex process involving numerous pieces of equipment and considerable transfer of material between units. The input stream of brackish water is evaporated in stages, with the steam produced in each stage used as heating medium in the subsequent stage. The condensed vapor obtained at each stage is, after further recovery of heat, merged into a single pure-water product stream while the concentrated brine discharged from the last stage is disposed as a waste stream. In spite of its internal mechanical complexities, the composite plant can for material balance purposes be reduced to a single box, as shown in Figure 2.2, with one input and two output streams, the feed and waste streams consisting of salt and water and the product stream of water alone.

2.1.1 Material Balance Variables

The first steps in defining a material balance problem are thus to select the boundaries of the system; to identify all input and output streams, that is, all material flows which cross the system boundaries; and to identify the chemical species

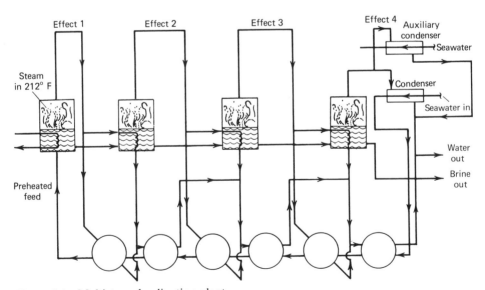

Figure 2.1 Multistage desalination plant.

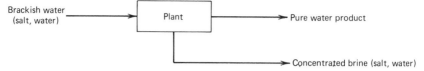

Figure 2.2   Input/output view of the desalination plant.

contained in each such stream. Next, in order to be able to accurately account for all materials entering and leaving the system, we will have to know the flow rates of each chemical species present in each stream. There are a number of different but equivalent ways of representing this information.

First of all, we can simply associate with each chemical species j in each stream the species flow rate N_j (moles of j per unit time) or F_j (mass of j per unit time). The total flow of the stream, either in moles N or mass units F, will then be given by summing the individual species flows over all species j that are present in the stream, as follows:

$$N = \sum_j N_j$$

$$F = \sum_j F_j$$

Hence, if the individual species flows are known, the total flow is a dependent variable.

An alternate way of representing a stream is to give its total flow, either in moles or in mass units, together with the *composition* of the stream. Compositions can of course be defined in a number of different ways. Two measures of composition of a species j which are most conveniently employed in material balance calculations are the *mass fraction* (weight fraction) w_j and the *mole fraction* x_j. From the definition, it follows that the mass fractions of all S species in a stream must sum to 1:

$$\sum_{j=1}^{S} w_j = 1 \tag{2.1}$$

Similarly, mole fractions of all S species in a stream must sum to 1:

$$\sum_{j=1}^{S} x_j = 1 \tag{2.2}$$

These two parallel measures for characterizing a stream, molar flow rates and mole fractions or mass rates and mass fractions, are of course completely equivalent. If the molecular weight M_j is known for each of the S species in the stream, then

$$N = \sum_{j=1}^{S} (w_j F / M_j) = F \sum_{j=1}^{S} (w_j / M_j) \tag{2.3}$$

and

$$x_j = (w_j F / M_j)/N = (w_j / M_j) / \sum_{j=1}^{S} (w_j / M_j) \tag{2.4}$$

The flow-composition and species flow forms of describing a stream are similarly equivalent and interconvertible. If the species flows are given, then the flow-composition variables can be calculated via

$$w_j = \frac{F_j}{F} \quad \text{or} \quad x_j = \frac{N_j}{N}$$

Thus, the choice of which of these several representations to use, mass or mole units, species flows or compositions, is determined largely by convenience. For instance, since it is often convenient to specify compositions of key species in process streams, one is naturally led to the use of the flow-composition variables. On the other hand, as we shall see in subsequent sections, the use of species flows sometimes leads to simpler balance equation formulations.

Example 2.1   Suppose a stream of 100 kg/h of brackish water consists of 0.05 mass fraction salt (NaCl) and 0.95 mass fraction water. Since, the molecular weight of the salt is approximately 58.5 kg/kgmol while that of water is 18 kg/kgmol, the corresponding molar rate will be

$$N = 0.05(100)/58.5 + 0.95(100)/18 = 5.363$$

The equivalent mole fractions will be

$$x_{salt} = \frac{0.05(100)/58.5}{5.363} = 0.01594$$

$$x_{H_2O} = \frac{0.95(100)/18}{5.363} = 0.98406$$

The equivalent species molar flows will be

$$N_{salt} = x_{salt}N = 0.0855 \text{ kgmol/h}$$

$$N_{H_2O} = x_{H_2O}N = 5.2778 \text{ kgmol/h}$$

Summarizing the preceding, regardless of which measure is used, the complete identification of a stream containing S species will require knowledge of the total flow rate and the S compositions or species flows. Note, however, that since the composition of each stream must satisfy the normalizing equations, eq. (2.1) or (2.2), it is actually sufficient if only $S - 1$ compositions are given. The remaining one can be calculated simply by difference:

$$x_S = 1 - \sum_{j=1}^{S-1} x_j$$

or

$$w_S = 1 - \sum_{j=1}^{S-1} w_S$$

We shall thus say that every stream containing S species has associated with it S *independent stream variables,* namely, the flow rate and any $S - 1$ compositions or species flows. These S variables must be known for complete identification of the stream. Similarly, since the sum of the species flows equals the total flow, only $S - 1$ species flows will be independent if the total flow is specified.

Example 2.2   Consider again the desalination plant of Figure 2.2. The number of independent stream variables associated with the plant is five: the flow rate and one composition fraction each for the feed stream and the waste stream and only the flow rate of the product stream because it involves only a single species. Certainly, it would be sufficient in Example 2.1 to simply state that the stream of 100 kg/h brackish water consists of 5% salt and the rest water. On the other hand, it would be clearly inadequate to specify both the salt fraction, $w_{salt} = 0.05$, and the water fraction, $w_{H_2O} = 0.95$, but not the flow rate.

2.1.2   MATERIAL BALANCE EQUATIONS AND THEIR PROPERTIES

The principle of conservation of mass and the atomic theory of matter provide assurance that in an open steady-state system the mass and number of molecules, hence also number of moles, of each chemical species will be conserved. Thus, in the nonreacting, steady-state case regardless of the type of process occurring within the system, the total mass or moles as well as the mass or moles of each chemical and each atomic species entering the system must be exactly equal to that leaving.

The conservation principle thus provides us with relationships which must be satisfied by the streams entering and leaving the system. More specifically, it provides us with a set of equations relating the stream variables associated with each stream of the process. We shall exploit these equations both to solve for all of the unknown stream variables when only some are given as well as to determine whether the values of the stream variables given in a particular application are consistent.

The following example illustrates the most common use of material balance computations, namely, to calculate values of the remaining unknown stream variables when only a portion of the stream variables are specified.

Example 2.3   Consider the steady-state desalination process of Figure 2.2. In the process, suppose that seawater containing 0.035 mass fraction salt is evaporated to produce 1000 lb/h pure water. Determine the throughput of seawater required if corrosion considerations limit the waste brine mass fraction to 0.07.

Solution   The solution of this problem is of course of prime importance to the design engineer because the seawater feed rate determines the capacities of the process evaporators, pumps, and transfer lines that will be required. The system and the associated stream variables expressed in mass units are shown in Figure 2.3.

As noted before, the process involves only five stream variables since the water mass fractions in the seawater and waste brine can be obtained from

$$w_{H_2O}^S = 1 - w_{salt}^S$$
$$w_{H_2O}^B = 1 - w_{salt}^B$$

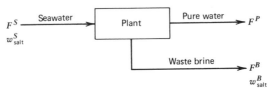

Figure 2.3 Flowsheet for Example 2.3, desalination plant.

The conservation law states that both total mass as well as the mass of each individual species are conserved. Thus, since total mass is conserved,

$$\text{Total mass in} = \text{total mass out}$$

or

$$F^S = F^P + F^B$$

Since salt is conserved,

$$\text{Total mass of salt in} = \text{total mass of salt out}$$
$$F^S w^S_{salt} = F^B w^B_{salt}$$

Finally, since water is likewise conserved,

$$F^S(1 - w^S_{salt}) = F^B(1 - w^B_{salt}) + F^P$$

Note, from the problem statement, that

$$F^P = 1000 \text{ kg/h}$$
$$w^S_{salt} = 0.035$$
$$w^B_{salt} = 0.07$$

With this information, the above equations become

$$F^S = 1000 + F^B$$
$$F^S(0.035) = F^B(0.07)$$
$$F^S(0.965) = F^B(0.93) + 1000$$

The second of these equations can be reduced to

$$F^S = 2F^B$$

which, when substituted into the first equation, yields $F^B = 1000$ and $F^S = 2000$ kg/h. The application of the conservation law thus allows us to solve for the required feed rate.

In the above example involving two species, water and salt, the conservation principle provided us with three material balance equations. Note, however, that the system of three equations obtained is *redundant* because if the salt balance equation is added to the water balance equation, the total mass balance equation is generated. The complete set of three conservation equations is said to be *dependent*; any two of the three are *independent* and can be used to generate the third. This observation is in fact a general property of balance equations.

Independent Balance Equations   In general, if the system involves S species, the conservation law will yield $S + 1$ material balance equations, one for each of the S species and one for the total mass. Of these $S + 1$ equations, only S will be *independent*; the $S + 1$st can always be generated from the others.

That this generalization must hold true regardless of the number of species or streams is easily seen from the following demonstration. Regardless of the number of streams involved, the conservation of species j requires that

$$\sum_{\substack{i=\text{input}\\ \text{stream}}} F^i w_j^i = \sum_{\substack{i=\text{output}\\ \text{stream}}} F^i w_j^i \tag{2.5}$$

The conservation of total mass requires that

$$\sum_{\substack{i=\text{input}\\ \text{stream}}} F^i = \sum_{\substack{i=\text{output}\\ \text{stream}}} F^i \tag{2.6}$$

Now, if all the species conservation equations are added together, we obtain

$$\sum_{\text{inputs}} F^i \sum_{j=1}^{S} w_j^i = \sum_{\text{outputs}} F^i \sum_{j=1}^{S} w_j^i \tag{2.7}$$

But the weight fractions must all sum to 1, namely,

$$\sum_{j=1}^{S} w_j^i = 1 \qquad \text{for all streams } i$$

Consequently, eq. (2.7) reduces just to eq. (2.6), the total mass balance. The argument can be repeated to generate any chosen species balance from the total balance and the $S - 1$ remaining species balance equations.

As a result of the dependence of the complete set of balance equations, we always have a choice of which S equations to use in any given application. The only criterion upon which to base this choice is ease of solution.

Homogeneity of the Balance Equations   A second important observation which can be made from Example 2.3 and the general species balance equation, eq. (2.5), is that the balance equations are *homogeneous* in the flow rates of the streams.

Example 2.4   In the previous example, the seawater throughput rate of the desalination plant was calculated assuming that a pure-water product rate of 1000 kg/h was required. Intuitively, it seems reasonable that if a product rate of 100 kg/h is chosen, then the throughput rate should proportionately be scaled down. That this is the case can be verified by repeating the calculations of Example 2.3.

The total mass balance will be

$$F^S = 100 + F^B$$

the salt balance,

$$F^S(0.035) = F^B (0.07)$$

and the water balance,

$$F^S(0.965) = F^B (0.93) + 100$$

The solution of any two of the above equations is

$$F^B = 100 \quad \text{and} \quad F^S = 200$$

In an analogous fashion, if the product rate is raised to 10,000 kg/h and the calculations are repeated, the seawater and brine flow rates are increased by a power of 10. This, too, can be easily verified by solving the balance equations.

A system of equations in which the values of a set of variables can be uniformly scaled such that the resulting values continue to satisfy the equations is said to be *homogeneous* in those variables. Formally stated, an equation $f(x, y) = 0$ in the two variables x and y is homogeneous in y if, given any solution $(\bar{x}, \bar{y})$, any constant times $\bar{y}$ is also a solution. The balance eqs. (2.5) and (2.6), involve only two types of variables, compositions and flows. It is easy to see that these equations are always homogeneous in the flows F^i because, if any set of flows $\bar{F}^i$ satisfies the balance equations and if α is any number, the flow rates $\alpha \bar{F}_i$ also satisfy the balance equations. This follows because

$$\sum (\alpha \bar{F}^i) w_j^i = \alpha \sum \bar{F}^i w_j^i$$

Thus, since α can be factored out of both sums in Eq. (2.5),

$$\alpha \left(\sum_{\substack{\text{input} \\ \text{streams}}} \bar{F}^i w_j^i \right) = \alpha \left(\sum_{\substack{\text{output} \\ \text{streams}}} \bar{F}^i w_j^i \right)$$

As a consequence of the homogeneity of the balance equations, we can take any solution and scale all of the flows by any factor and still be sure that the principle of conservation of mass is not violated.

The Basis Concept   As a further consequence of the homogeneity of the balance equations, if none of the stream flow rates is assigned a value in the problem statement, then any one of the stream flows can be assigned an arbitrary magnitude for purposes of the calculation. This is referred to as choosing a *basis* of the calculation.

Example 2.5   Consider again the desalination plant balance equations but suppose that the product rate is not specified. As before, the three balance equations are

$$F^S = F^P + F^B$$
$$F^S(0.035) = F^B(0.07)$$
$$F^S(0.965) = F^B(0.93) + F^P$$

Certainly, all three equations can be divided by the variable F^P to yield

$$(F^S/F^P) = 1 + (F^B/F^P)$$
$$(F^S/F^P)0.035 = (F^B/F^P)0.07$$
$$(F^S/F^P)0.965 = (F^B/F^P)0.93 + 1$$

The problem can apparently be solved in terms of the new variables,

$$f^S = (F^S/F^P) \qquad f^B = (F^B/F^P)$$

The solution is

$$f^B = 1 \qquad \text{and} \qquad f^S = 2$$

These variables, which represent flows of stream B and S per unit of stream P, thus remain fixed regardless of the particular choice of flow rate assigned to stream P. Consequently, if the product flow rate is not specified in the problem statement, there is no loss of generality in assigning F^P any arbitrary and convenient value for the calculation.

The same calculation can be carried out if the balance equations had been divided by any other flow rate. For instance, if F^B is used as divisor, the balances become

$$(F^S/F^B) = (F^P/F^B) + 1$$
$$(F^S/F^B)0.035 = 0.07$$
$$(F^S/F^B)(0.965) = 0.93 + (F^P/F^B)$$

The solution to this set is

$$(F^S/F^B) = 2 \qquad \text{and} \qquad (F^P/F^B) = 1$$

which is the same as obtained previously, since

$$f^S = \frac{(F^S/F^B)}{(F^P/F^B)} = \frac{2}{1} = 2$$

and

$$f^B = (F^B/F^P) = \frac{1}{(F^P/F^B)} = 1$$

Consequently, if the product flow rate had not been assigned a value in the problem formulation, either of the other streams B or S could be assigned a value as *basis* for the calculation.

Although the above analysis was carried out on a particular example, it should be apparent that it relies only on the fact that the balance equations are homogeneous in the stream flows. Hence, we can conclude that in general, if in a balance problem no stream flows are specified, the flow rate of any single stream in the system can be fixed at any convenient numerical value for purposes of solving the balance equations. The numerical value or *basis* of the calculation chosen should be clearly stated in reporting any partial solution to the problem since it serves to scale the magnitude of all other streams.

2.1.3 Material Balance Information

In the preceding sections, we have noted that a material balance problem consists of the following elements:

1. The selected system with its input and output streams.
2. The stream variables which describe the flow rate and composition of each stream.
3. The set of material balance equations of which at most S are independent, where S is the total number of different species that appear in the streams.
4. The basis selected for the computation.

In addition to these items, most material balance problems will also involve various specifications which are imposed on the system. These specifications serve to reduce the number of unknown stream variables and thus are important to the formulation of the problem. In general, the specified information can take a variety of forms, but commonly it will consist of either the direct assignment of values to stream variables or else the imposition of relationships between stream variables. In the examples we have considered up to this point, the given information was of the former type, that is, specified compositions or flow rates. The following several examples illustrate use of the latter type of specification.

Basically, three types of relations among stream variables commonly occur in material balance problems:

1. Fractional recoveries.
2. Composition relationships.
3. Flow ratios.

Regardless of the type that occurs in a given problem, these relations are merely treated as additional equations which can be used together with the independent material balances to solve for the unknown stream variables.

The next example illustrates a fractional recovery type of side condition.

Example 2.6 A feed stock available at the rate of 1000 mol/h and consisting of (all in mol %)

20% Propane (C_3)
30% Isobutane (i-C_4)
20% Isopentante (i-C_5)
30% Normal pentane (C_5)

is to be separated into two fractions by distillation. The distillate is to contain all of the propane fed to the unit and 80% of the isopentane fed to the unit and is to consist of 40% isobutane. The bottoms stream is to contain all the normal pentane fed to the unit. Calculate the complete distillate and bottoms analysis.

Solution The system, in this instance, consists of the distillation unit and three streams: feed, distillate, and bottoms. Assuming that each stream contains all four species, the system will involve a total of twelve stream variables, namely, the flowrate and three of the compositions for each of the three streams. Moreover, since the system involves four distinct species, it will be possible to write four independent material balances. The specified independent information consists of

1. Three independent feed compositions, for instance, 20% C_3, 30% i-C_4, and 20% i-C_5.
2. Two independent distillate compositions, 0% C_5 and 40% i-C_4.
3. One bottoms composition, 0% C_3.
4. The feed rate, 1000 mol/h.

The systems diagram, or flowsheet, listing the stream variables and the given direct stream information is shown in Figure 2.4. The mole fractions of i-C_5 are shown in parentheses to indicate that they are dependent stream variables not explicitly used in solution.

The material balance equations for the system are, first for the total moles,

$$N^M = N^D + N^B$$

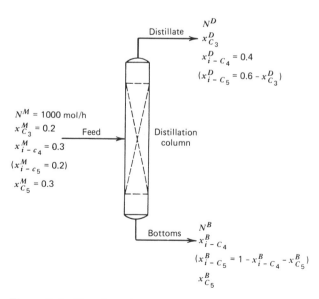

Figure 2.4 Flowsheet for Example 2.6, hydrocarbon distillation.

and for each species

C_3 balance: $\qquad 0.2N^M = N^D x_{C_3}^D$

i-C_4 balance: $\qquad 0.3N^M = 0.4N^D + x_{i\text{-}C_4}^B N^B$

i-C_5 balance: $\qquad 0.2N^M = N^D(1 - 0.4 - x_{C_3}^D) + N^B(1 - x_{i\text{-}C_4}^B - x_{C_5}^B)$

C_5 balance: $\qquad 0.3N^M = N^B x_{C_5}^B$

In addition to the specified flow and compositions, a recovery condition is imposed, namely, that 80% of the i-C_5 is recovered in the distillate. This condition is represented by the equation

$$0.8(0.2N^M) = N^D(1 - 0.4 - x_{C_3}^D)$$

The recovery condition thus is an indirect way of presenting information about the stream variables.

Since $N^M = 1000$, the recovery condition together with the C_3 balance yield

$$N^D = 600$$

and thus,

$$x_{C_3}^D = 0.333$$

The total mole balance, then, can be solved for

$$N^B = 400$$

and the i-C_4 balance,

$$x_{i\text{-}C_4}^B = 0.15$$

Finally, the C_5 balance can be used to calculate

$$x_{C_5}^B = 0.75$$

The recovery side condition is thus just another relation used together with the balance equations to solve the problem.

Composition relationships are probably the most common type of side condition imposed upon material balance problems. Typically, they take the form of the simple proportionality

$$x_j^i = Kx_j^k$$

between the compositions of a given species in two different streams. An equation of this type arises quite naturally from empirical observation of the distribution of a solute between two immiscible solvent streams or the relationship between the equilibrium mole fractions of a species in contacting liquid and vapor phases. It also occurs when a given stream is divided flow-wise into two or more smaller streams. In that case, the compositions of each of the branches must be the same, that is, the composition of species j in branch i must equal that of species j in branch k.

Of course, more general composition relationships can also occur. For instance, suppose a slurry consisting of a solid suspended in a solution is allowed to settle and a portion of the clear solution decanted. If there is no chemical adsorption of the solute on the solid particles, then clearly the composition of the decanted solution must be equal to that of the solution which remains entrained in the settled solids. This condition can, for each species j composing the solution, be written as

$$\frac{w_j^1}{1 - w_{solid}^1} = w_j^2$$

where superscript 1 indicates the settled slurry and 2 the decanted solution. The following example illustrates the occurrence of a composition relationship of this type.

Example 2.7   A crucial step in the production of aluminum from bauxite ore is the separation of alumina from the remaining mineral impurities in the ore. In the Bayer process, this is accomplished by treating bauxite with aqueous NaOH to produce $NaAlO_2$. Since $NaAlO_2$ is water soluble while the residual mineral constituents of bauxite are not, a separation can be achieved by allowing the minerals to settle out and decanting the aqueous solution of $NaAlO_2$ and unreacted NaOH. In order to further recover any $NaAlO_2$ entrained in the settled mineral solids, this "mud" is repeatedly washed with water and allowed to settle, and the wash water is decanted. Figure 2.5 shows one stage of this washing–settling process.

In this stage, a feed slurry consisting of 10% solids, 11% NaOH, 16% $NaAlO_2$, and the rest water is washed with a wash water stream containing 2% NaOH to yield a decanted solids-free solution containing 95% water and a settled mud containing 20% solids. How much $NaAlO_2$ is recovered in the decanted solution if slurry is fed at the rate of 1000 lb/h? The system with a set of independent stream variables is shown in Figure 2.6.

Solution   The material balance problem involves four streams containing a total of 13 stream variables: four each for streams 2 and 3, three for stream 4, and

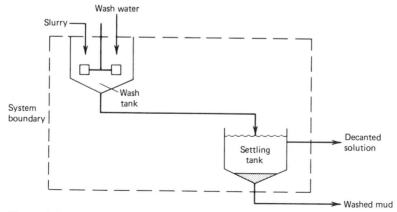

Figure 2.5   Schematic for Example 2.7, washing of bauxite ore.

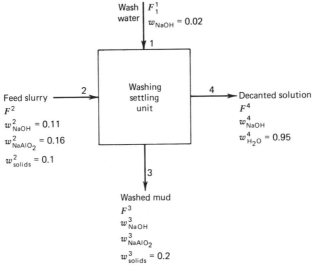

Figure 2.6   Flowsheet for Example 2.7, washing of bauxite ore.

two for stream 1. The system involves four distinct species; hence, at most four independent material balances can be written. These are

Total mass: $F^1 + F^2 = F^3 + F^4$

Solids: $0.1F^2 = 0.2F^3$

Water: $0.63F^2 + 0.98F^1 = (0.8 - w^3_{NaOH} - w^3_{NaAlO_2})F^3$

$$+ \ 0.95F^4$$

NaOH: $0.11F^2 + 0.02F^1 = w^3_{NaOH}F^3 + w^4_{NaOH}F^4$

NaAlO$_2$: $0.16F^2 = w^3_{NaAlO_2}F^3 + (0.05 - w^4_{NaOH})F^4$

Assuming that the solutes NaOH and NaAlO$_2$ have no tendency to adsorb to the settled solids, then the concentration of the solution entrained in stream 3 must be the same as that of stream 4. Hence,

$$\frac{w^3_{NaOH}}{1 - 0.2} = \frac{lb \ NaOH}{lb \ solids\text{-}free \ solution} = w^4_{NaOH}$$

and

$$\frac{w^3_{NaAlO_2}}{1 - 0.2} = w^4_{NaAlO_2} = 0.05 - w^4_{NaOH}$$

Note that a third composition relation involving water would be redundant, since

$$w^3_{NaOH} + w^3_{NaAlO_2} + w^3_{H_2O} = 0.8$$

and

$$w^4_{NaOH} + w^4_{NaAlO_2} + w^4_{H_2O} = 1.0$$

Any four of the above balance equations together with the two composition relations can now be used to solve for the required stream variables.

Note that the problem specification imposes a basis of 1000 lb/h slurry fed to the system. With this basis, it follows from the solids balance that $F^3 = 500$ lb/h. Then, if the two composition relations are added, the equation

$$\frac{1}{0.8} (w^3_{NaOH} + w^3_{NaAlO_2}) = 0.05$$

is obtained. This can be substituted into the water balance to yield

$$630 + 0.98F^1 = (0.8 - 0.04) 500 + 0.95F^4$$

Eliminating F^1 by using the total balance

$$F^1 = 500 - 1000 + F^4$$

the results $F^4 = 8000$ lb/h and $F^1 = 7500$ lb/h follow immediately. Finally, from the composition relations and the NaOH balance, we can calculate the compositions $w^3_{NaOH} = 0.02476$, $w^4_{NaOH} = 0.03095$, and $w^3_{NaAlO_2} = 0.01524$. Thus, of the 0.16(1000) $= 160$ lb/h $NaAlO_2$ fed to the system,

$$8000(0.05 - 0.03095) = 152.4 \text{ lb/h}$$

is recovered in the decanted solution.

Again, from an operational point of view, the composition relations amount to just another collection of equations which must be solved in conjunction with the material balance equations. However, as we shall observe in later chapters, composition relations are a major source of complication in the solution of flowsheet problems in which the streams internal to the system must also be calculated.

In the final example of this section, we consider the utilization of relations imposed between stream flow rates. Almost exclusively, flow side conditions take the form of simple ratios

$$\frac{F^i}{F^j} = \text{constant}$$

For instance, this type of relation is introduced quite naturally when the proportions in which the input streams to a process are to be added are specified. It is also quite useful to construct side conditions of this type when the flow rates of several streams in a process are specified. For instance, suppose M flows F^i in a flowsheet are specified. This specification can be replaced by $M - 1$ relations:

$$\frac{F^i}{F^r} = \text{constant}$$

$$= \left(\frac{\text{specified flow rate of stream } i}{\text{specified flow rate of stream } r}\right) \qquad i \neq r$$

In this way, we are free to choose any convenient basis for the calculations and at the same time we can be sure that the final scaled solution will satisfy all the specified flow rates.

Example 2.8 Recovery of a desired solute from a solution can sometimes be accomplished by using a second solvent which is immiscible with the solution but which preferentially dissolves the solute. This type of separation process is known as solvent extraction. In the system shown in Figure 2.7, benzene is separated from a refinery stream containing 70% (mass) benzene in a mixture of paraffin and napthene hydrocarbons by means of liquid SO_2. When 3 lb SO_2 is used per 1 lb process feed, a residual, or raffinate stream containing $\frac{1}{6}$ (mass fraction) SO_2 and the remainder benzene is obtained. The extract stream contains all of the non-benzene material, some SO_2, and about $\frac{1}{4}$ lb benzene per 1 lb nonbenzene hydrocarbons. Under those conditions, what is the percent recovery of benzene (pounds benzene in raffinate per pound benzene in the feed)?

Solution The extraction system involves four streams and three distinguished species: SO_2, benzene, and the lumped nonbenzene hydrocarbons. Thus, the number of independent balance equations will be three and the number of stream variables, eight (extract, three; feed, two; raffinate, two; solvent, one). The system with stream variables and given compositions is shown in Figure 2.8.
In addition to the compositions of the feed and raffinate streams, two relations are imposed on the stream variables: the ratio of solvent to feed,

$$\frac{F^2}{F^1} = 3$$

and the ratio of mass of benzene (B) to mass of nonbenzene (NB) material in the extract,

$$\frac{F^3 w_B^3}{F^3 w_{NB}^3} = \frac{w_B^3}{w_{NB}^3} = 0.25$$

These two relations, together with any three of the four material balance equations,

Total mass balance: $\qquad F^3 + F^4 = F^1 + F^2$

Benzene balance: $\qquad F^3 w_B^3 + 0.8333 F^4 = 0.7 F^1$

Nonbenzene balance: $\qquad F^3 w_{NB}^3 = 0.3 F^1$

SO_2 balance: $\qquad F^3 (1 - w_B^3 - w_{NB}^3) + 0.1667 F^4 = F^2$

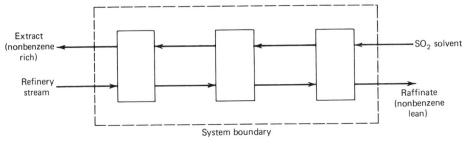

System boundary

Figure 2.7 Schematic for Example 2.8, SO_2 solvent extraction.

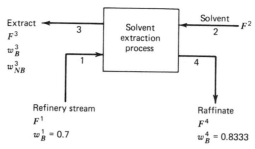

Figure 2.8 Flowsheet for Example 2.8, SO_2 solvent extraction.

can be used to solve for the unknown stream variables. Note that since no flow rate is specified, a basis for the calculation can be arbitrarily chosen; for instance, the choice $F^1 = 1000$ lb/h can be made. With that choice, the solvent-to-feed ratio gives $F^2 = 3000$ lb/h. The nonbenzene balance, together with the extract composition ratio, then yields

$$F^3 w_B^3 = 75$$

which, when substituted into the benzene balance, results in

$$75 + 0.8333F^4 = 700 \quad \text{or} \quad F^4 = 750 \text{ lb/h}$$

Next, from the total balance,

$$F^3 = F^1 + F^2 - F^4 = 3250 \text{ lb/h}$$

and finally,

$$w_B^3 = 75/3250 = 0.0231$$
$$w_{NB}^3 = 4(75/3250) = 0.09231$$

The benzene separation for the process is

$$100 \times \frac{\text{lb Benzene in raffinate}}{\text{lb Benzene in feed}} = \frac{625}{700} \times 100 = 89.3\%$$

2.2 ANALYSIS OF THE MATERIAL BALANCE PROBLEM

Our study of the application of the conservation law to open steady-state systems has focused on the individual components that make up the material balance problem: the stream variables, the balance equations, the directly specified information, and the imposed subsidiary relations. When dissected into these elements, the problem of calculating the material flows leaving or entering the system is revealed as simply one of solving a set of algebraic equations, which are usually linear, but may involve nonlinearities, for a certain number of unknown variables. As in any algebraic problem, two questions must be resolved prior to blindly submerging into the details of algebraic manipulation. First, we must decide whether in fact the equations we have assembled can yield a proper solution. If no solution is possible, then clearly we must return to reexamine the underlying physical system to determine whether any important features have been neglected, any false assumptions made, any incorrect or inaccurate information used, or, perhaps, how any important engineering design parameters left free for our choice can best be specified. Second, if the algebraic problem can be solved, then it behooves us to determine a strategy which will lead to an accurate solution in an efficient manner. Both of these questions are considered in the following two sections.

2.2.1   The Degree of Freedom

The determination of whether the algebraic model we have constructed will yield a physically realistic solution is in general a difficult matter. However, there is one simple index which can give a good indication of when the material balance problem is not likely to yield to solution. That index is the *degree of freedom* of the problem.

In simplest terms, we know from algebra that in order to solve a set of equations in, say, N unknowns, it is necessary that the set consist of N independent equations. If less than N independent equations are available, no solution is possible. If more than N equations are available, then one could choose any N for solution. However, there is always the risk because of errors or inconsistencies that the solution we obtain will depend upon which N equations we choose to use. Hence, the safest course is always to make sure that the number of variables and equations is in balance prior to attempting solution. The degree of freedom is simply an index which measures that balance. The attendant degree-of-freedom analysis is simply a systematic mechanism for counting all the variables, balance equations, and relations that are involved in the problem.

The degree of freedom of a system is defined as follows:

Degree of freedom = total number of independent stream variables
  − total number of independent balance equations
  − total number of specified independent stream variables
  − total number of subsidiary relations

If the degree of freedom is *positive*, then the problem is said to be *underspecified*, and it is not possible to solve for all the unknown stream variables. If the degree of freedom is *negative*, then the problem is *overspecified*, and redundant (possibly inconsistent) information must be discarded before a unique solution can be obtained. If the degree of freedom is *zero*, then the problem is *correctly specified*, that is, the number of unknown stream variables exactly balances the number of available equations. Each of these situations is illustrated in the following series of examples.

Example 2.9   Consider the distillation problem discussed in Example 2.6 and illustrated in Figure 2.4. As shown there, the total number of stream variables is 12 and is obtained by simply adding together the number of species present in each stream. The total number of independent balance equations is equal to the number of distinct species present in the system, in our case, four: C_3, i-C_4, i-C_5, and C_5. The number of specified compositions is six and, in addition, the steady-state feed rate is given. Finally a recovery condition is imposed on the system, yielding one subsidiary relation. Totaling all these in Figure 2.9, we calculate the degree of freedom as zero.

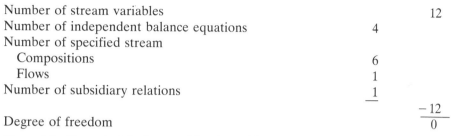

Number of stream variables		12
Number of independent balance equations	4	
Number of specified stream		
Compositions	6	
Flows	1	
Number of subsidiary relations	1	
		-12
Degree of freedom		$\overline{0}$

Figure 2.9 Example 2.9, degree-of-freedom table.

Hence, the problem is correctly specified and will yield a unique solution, as we verified in actually solving the problem.

Note that there is an arbitrariness in the counting of variables and specifications involving species not present in streams. For instance, in the preceding example, we could account for the absence of C_5 in the distillate stream and C_3 in the bottoms stream by simply counting only *three* stream variables with each of these streams. Thus, the problem would only have 10 stream variables. However, in this case the number of specified compositions would have to be reduced by two since the conditions $x_{C_5}^D = 0.0$ and $x_{C_3}^B = 0.0$ would be redundant. The resulting degree of freedom would be the same, namely,

Degree of freedom = 10 variables − 4 balance equations

− 4 compositions − 1 flow − 1 relation = 0

Either way of treating species which are known not to be present is correct. The choice is a matter of personal preference.

Example 2.10 Titanium dioxide, TiO_2, is a white hiding pigment manufactured in large quantities and used heavily in the paint and paper industries. In a new pigment plant that is to produce 4000 lb/h dry TiO_2 product, an intermediate stream consisting of TiO_2 precipitate suspended in an aqueous salt solution, is to be purified of salt so that the final product contains, on a water-free basis, at most 100 parts per million (1 ppm = mass fraction of 10^{-6}) of salt. The salt removal is to be accomplished by washing the precipitate with water. If the raw pigment stream contains 40% TiO_2, 20% salt, and the rest water (all mass %) and if the washed pigment is, upon settling, projected to consist of about 50% (mass) TiO_2 solids, what will the composition of the waste wash water stream be? The answer to that question clearly will be significant if this saline water is to be disposed of in a nearby river.

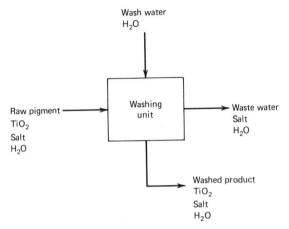

Figure 2.10   Flowsheet for Example 2.10, pigment washing.

Solution   The diagram of the system, its streams, and the components present in each stream are shown in Figure 2.10.

The number of stream variables will be equal to the total of the number of species present in each stream

Raw pigment	3
Wash water	1
Washed pigment	3
Waste water	2
	9

The number of distinct species present in the system are three: TiO_2, salt, and water. Hence, three independent balances are available. The number of specified compositions are three: two for the raw pigment and one for the washed product. Finally, the product flow rate on a dry basis is given (4000 lb/h) and a second relation is imposed, namely, that on a dry basis the product is to contain 100 ppm salt. From the summary given in Figure 2.11, the degree of freedom is 1; hence, the problem as formulated as *underspecified*.

Thus, we conclude that the problem cannot be solved for all unknown stream variables. To confirm this conclusion, let us nonetheless attempt solution. The

Number of stream variables	9
Number of independent balances	3
Number of specified stream compositions	3
Number of subsidiary relations	2
	−8
Degree of freedom	1

Figure 2.11   Example 2.10, degree-of-freedom table.

selected independent stream variables and stream information are summarized in Figure 2.12. In terms of these variables, a set of independent balance equations is

Total mass balance: $F^1 + F^2 = F^3 + F^4$

TiO_2 balance: $0.4F^1 \quad = 0.5F^3$

Salt balance: $0.2F^1 \quad = w_{salt}^3 F^3 + w_{salt}^4 F^4$

The fourth balance, that for water, is dependent. In addition to the balances we have the requirement that on a dry basis the washed pigment contain at most 100 ppm salt, or

$$\frac{\text{lb salt}}{\text{lb Dry product}} = \frac{w_{salt}^3 F^3}{F^3(0.5 + w_{salt}^3)}$$

$$= \frac{w_{salt}^3}{0.5 + w_{salt}^3} = 10^{-4}$$

and that the product flow rate on a dry basis be 4000 lb/h,

$$F^3(0.5 + w_{salt}^3) = 4000 \text{ lb/h}$$

The first equation yields $w_{salt}^3 = 5.0005 \times 10^{-5}$, and the second, $F^3 = 7999.2$ lb/h. From the TiO_2 balance,

$$F^1 = 9999 \text{ lb/h}$$

Thus, the salt balance is reduced to

$$1999.8 = 0.4 + w_{salt}^4 F^4$$

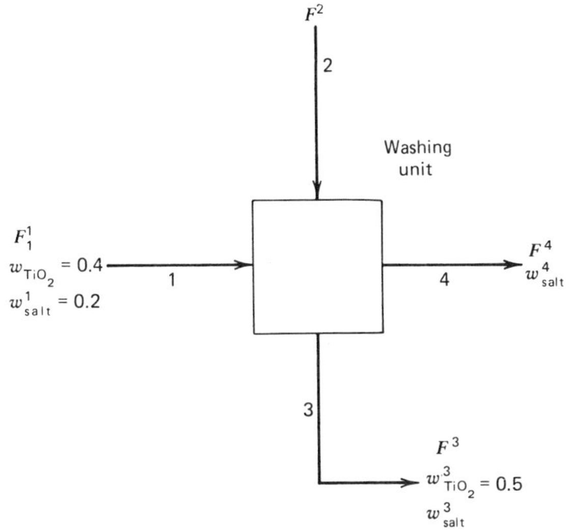

Figure 2.12 Flowsheet for Example 2.10, with specifications.

and the total balance to

$$9999 + F^2 = 7999.2 + F^4$$

The result is a system of two equations in the three unknowns F^2, F^4, and w_{salt}^4. The problem is underspecified by one variable, as the degree-of-freedom analysis indicated.

Clearly, what is missing in the problem specification is some indication of what the utilization of wash water ought to be. Typically, that will be a design variable which will be determined by the engineer based on process economics. As we noted in the Bayer process example (Example 2.7), washing is usually accomplished in continuous washer/settler units. Often, a number of such units are connected in series and operated countercurrently, that is, the feed slurry is introduced at one end and transferred from one unit to the next, while the fresh wash water is introduced at the other end and reused in each successive unit. Given an initial feed concentration, the final desired washed product concentration can be achieved by varying the number of units and the wash water flow rate. If more units are used, less water is required, and vice versa. The optimum design is obtained by balancing the cost of process water against the cost of buying and operating additional units.

Suppose that by conducting a design study of this type, it is determined that a wash water utilization of 6 lb H_2O/lb feed is optimum. With this additional specification, the degree of freedom of the problem is reduced to zero, that is, the problem becomes completely specified.

With $F^2/F^1 = 6$, we can complete the solution to obtain

$$F^2 = 6(9999) = 59{,}994 \text{ lb/h}$$

From the total balance,

$$F^4 = 61{,}993.8 \text{ lb/h}$$

and from the salt balance.

$$w_{salt}^4 = 1999.4/F^4 = 0.0323$$

From the above example, it is apparent that, in the case of an underspecified problem, only some of the unknown stream variables can be determined. The underspecification indicates that either some additional information needs to be obtained or else some additional free stream variables must be specified by design considerations in order to complete the solution of the balance problem.

Example 2.11   Upon completing the design of the washing system based upon the calculations carried out in Example 2.10, the engineer submits his project to the local regulatory agency for approval of a permit to divert 60,000 lb/h river water and discharge a stream of 62,000 lb/h water containing 3.23% salt. The agency rejects the permit request on the grounds that such a discharge would significantly affect the potability of the downstream river water (an acceptable limit for human consumption being 0.02%). The agency rules that at most a discharge of 30,000

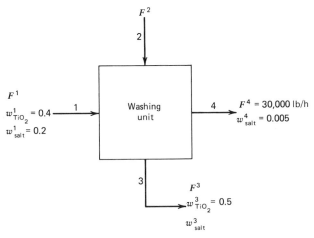

Figure 2.13 Flowsheet for Example 2.11, pigment washing.

lb/h with a salt content of 0.5% is allowable. The chastened engineer, who should have foreseen the pollution problem, returns to his systems diagram and enters the new information, as shown in Figure 2.13.

Since the product specifications must still be met, namely, a dry-basis salt composition of 0.01 % and a dry-basis product flow rate of 4000 lb/h, the reformulated problem will have nine stream variables, three balance equations, two relations as before, but now also four specified stream compositions and one specified flow rate. Hence, the degree of freedom, as shown in Figure 2.14, is −1.

The problem is *overspecified*. Again, let us confirm this conclusion by carrying out the computations.

The balance equations are:

Total balance: $F^1 + F^2 = F^3 + 30,000$

TiO_2 balance: $0.4F^1 = 0.5F^3$

Salt balance: $0.2F^1 = x_{salt}^3 F^3 + 0.005(30,000)$

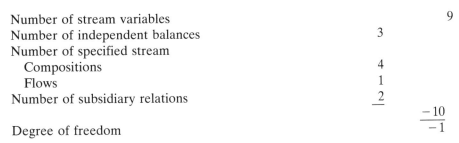

Number of stream variables		9
Number of independent balances	3	
Number of specified stream		
Compositions	4	
Flows	1	
Number of subsidiary relations	2	
		−10
Degree of freedom		−1

Figure 2.14 Example 2.11, degree-of-freedom table.

The two subsidiary relations again yield

$$x^3_{salt} = 5.0005 \times 10^{-5}$$

$$F^3 = 7999.2 \text{ lb/h}$$

From the TiO_2 balance, again,

$$F^1 = 9999 \text{ lb/h}$$

and from the total balance,

$$F^2 = 28,000.2 \text{ lb/h}$$

However, the salt balance is violated:

$$0.2(9999) = 1999.8 \neq 0.4 + 150 = 150.4$$

This contradiction is brought about because the superfluous information in this overspecified problem is *inconsistent*. This, in fact, is usually the case with overspecification: it is not possible to satisfy all of the imposed conditions. Although often overspecification indicates incorrect information, sometimes, as is the case in this example, overspecification indicates that the basic formulation is inadequate. For instance, in the above plant, the only way to meet all the output specifications is to devise a supplementary mechanism for removing salt from the system, either by reducing the salt content of the feed upstream of the washing unit or else by designing a waste water purification system. In the former case, by allowing the raw pigment salt content to be a variable, the problem becomes correctly specified and we can calculate the allowable feed salt concentration. In the latter case, a supplementary mechanism for salt removal introduces an additional output stream of salt into the problem, hence, increasing the number of stream variables by one. Again, the problem becomes correctly specified and we can calculate the rate at which salt must be removed.

The lesson to be retained at this point is that prior to attempting the solution of a material balance problem, you should first carefully determine how many unknowns there are in the problem, what the given information is, and how many independent material balances it will be possible to write. With these data, calculate the degree of freedom of the problem. If the problem is correctly specified, proceed with the solution. Otherwise, the problem must be studied further to determine what information can best be deleted if the problem is overspecified or else what additional information can be introduced if the problem is underspecified.

2.2.2 Solution Strategy

Once it is determined that the material balance problem is correctly specified, the pertinent equations and information can be assembled and the resulting algebraic problem can, at least in principle, be solved. As in any problem involving multiple equations in multiple unknowns, the ease with which a solution can be obtained depends upon the structure of the equations. However, unlike most problems involving the solution of multiple equations, in the material balance case we have some limited choice over that structure. That choice lies in selecting which set of S balance equations of the $S + 1$ possible equations we will use and in selecting the location of the basis for the calculation. The objective in making these choices is to achieve a sequential rather than a simultaneous solution of the balance equations and relations. Clearly, the work involved in solving a set of equations simultaneously, say, in the linear case by using Cramer's rule, is considerably more than that required if we can solve the equations one at a time, each for a single unknown variable. The strategy of solution is thus to order the calculations so that, if possible, no simultaneous solution is required.

Let us first consider the choice of balance equations. Since for a system involving S species it is possible to write $S + 1$ balance equations of which only S are independent, it is clear that we can choose among $S + 1$ different combinations involving S equations each. Which set should be chosen? A good rule of thumb is to choose for solution that set of balances which involves the fewest unknowns per equation. Such a set of equations will not only be the simplest but also the most likely to lend itself to sequential solution. The simplest way of recognizing which equations are likely to involve the fewest unknowns is to look for *species* that appear in the *fewest* streams and to then use their balance equations. For instance, in Example 2.10, TiO_2 appears in only two streams, while water appears in all four. Clearly, the TiO_2 balance

$$0.4F^1 = 0.5F^3$$

is a better candidate for solution than the water balance

$$0.4F^1 + F^2 = (0.5 - x_{salt}^3)F^3 + (1 - x_{salt}^4)F^4$$

As a second rule of thumb, the total balance should generally be employed because even though it involves all streams, it can never involve unknown compositions. For instance, the total mass balance

$$F^1 + F^2 = F^3 + F^4$$

is a simpler equation and involves fewer unknowns than the above water balance.

Finally, there is the matter of basis selection. Recall that because the balance equations are homogeneous, if no flows are specified in the problem then any one flow can be *assigned* a numerical value as a basis for the calculation. Actually, as a further consequence of homogeneity, even if a stream flow is assigned a numerical value as part of the system definition, this specification can be disregarded and another stream flow chosen as a basis providing the resulting solution is rescaled to match the specified stream flow. Thus, if by choosing a different stream flow as basis we can break up a possible simultaneous solution, the nature of the equations allows us to do so. We refer to this maneuver as *relocating the basis*. As shown in the following example, it often serves as a useful strategy for simplifying the solution of the balance equations.

Example 2.12 The conventional method for separating ethyl alcohol from an alcohol/water mixture is by distillation. However, this procedure can at best only produce a product 95% by volume alcohol because alcohol and water form a constant boiling mixture, called an azeotrope, of that composition. Simple distillation cannot eliminate the azeotrope. Instead, if a pure alcohol product is desired, benzene is added to the feed solution. The benzene itself forms an azeotrope with water but one which has a lower boiling point than the alcohol, and thus the alcohol can be purified.

Suppose it is desired to produce 1000 lb/h pure ethyl alcohol in this manner by distilling a feed mixture containing 60% water and 40% alcohol (in mass %). If the distillate composition (in mass %) is 75% benzene and 24% water, with the rest alcohol, how much benzene must be fed to the column?

Solution   In this problem, the system shown in Figure 2.15 consists of a single distillation unit and has two input streams, the alcohol/water feed and the benzene feed, and two output streams, the azeotropic distillate and the alcohol product.

The four input/output streams involve a total of seven stream variables: the flow rate and water composition of the feed; the flow rate of the benzene stream; the flow rate, water composition, and benzene composition of the distillate; and the flow rate of the pure alcohol product. The system processes three species and hence will yield three independent material balances. In addition, three compositions, one flow, and no subsidiary relations are specified. A degree-of-freedom calculation results in a degree of freedom of zero.

Number of stream variables		7
Number of independent balances	3	
Number of specified		
Compositions	3	
Flows	1	
		−7
Degree of freedom		0

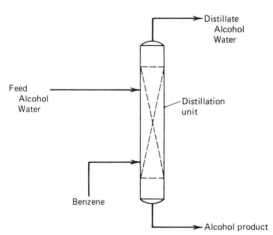

Figure 2.15   Schematic for Example 2.12, extractive distillation.

The problem is correctly specified and can be solved to yield a unique solution. The stream variables involved in the system, together with the given information, are shown in Figure 2.16. The four possible balance equations for the system are

Total mass balance: $F^M + F^B = F^D + F^A$

Benzene balance: $F^B = 0.75F^D$

Water balance: $0.6F^B = 0.24F^D$

Alcohol balance: $(1 - 0.6)F^M = (1 - 0.75 - 0.24)F^D + F^A$

Note that benzene and water occur in only two streams each, hence, they should be chosen for use. Since all compositions are known, the alcohol balance should be used in preference to the total balance, since it involves fewer variables. Next, note that although F^A has been specified at 1000 lb/h in the problem definition, a simultaneous solution of the balances can be avoided if instead we choose as basis the flow F^D. If F^D were assigned a value, then, from the benzene and water balances, F^B and F^M could be calculated directly. Hence, suppose we choose $F^D = 2000$ lb/h as basis. Then,

$$F^B = 0.75(2000) = 1500 \text{ lb/h}$$
$$F^M = (0.24/0.6)(2000) = 800 \text{ lb/h}$$

and from the alcohol balance,

$$F^A = 0.4(800) - 0.01(2000) = 300 \text{ lb/h}$$

The solution with basis $F^D = 2000$ lb/h is

$$(F^A, F^B, F^D, F^M) = (300, 1500, 2000, 800)$$

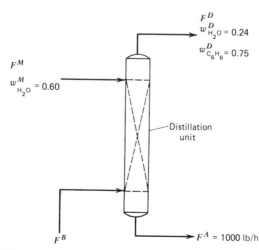

Figure 2.16 Flowsheet for Example 2.12, extractive distillation.

Since the balance equations are homogeneous in the flow rates, this solution can be scaled by any arbitrary factor and the result will still satisfy the balance equations. In particular, we can choose a factor $\frac{10}{3}$ which will then scale F^A to match the 1000 lb/h required in the problem statement. Hence, the final desired solution is

$$(F^A,\ F^B,\ F^D,\ F^M) = \tfrac{10}{3}(300,\ 1500,\ 2000,\ 800)$$
$$= (1000,\ 5000,\ 6667,\ 2667)\ \text{lb/h}$$

In summary, the preferred solution strategy is to seek for a sequential rather than a simultaneous solution of the material balance problem. The device of relocating the basis coupled with a judicious choice of balance equations can in many cases achieve that goal.

2.3 SYSTEMS INVOLVING MULTIPLE UNITS

In the preceding sections, we observed that regardless of the internal complexities of the system under consideration, the system could always be represented as a single box which communicates with the environment via several input and output streams. This corresponds to a macroscopic view of processing systems, since we always regard the system as a single unit within which the product materials are prepared. However, most chemical engineering systems typically consist of a sequence of processing steps, each of which occurs within a specially designed individual unit. Certainly, in order to be able to design such integrated multiunit systems, the chemical engineer must, in addition to the input and output streams of the entire plant, also know the flows and compositions of all of the internal streams connecting the various units composing the processing system. This requires a more detailed analysis of systems than our previous macroscopic view, since we now are concerned with the materials processing steps occurring within the system boundaries. However, it is by no means a microscopic view, since we do not delve into the mechanical details of the individual units but rather consider each to be an inscrutable "black" box.

In this section, we therefore consider the application of the conservation law to multiple-unit processes. As will become apparent, the analysis of such processes can be undertaken quite readily in terms of the logical framework developed for single unit or overall systems.

2.3.1 Independent Sets of Balance Equations

Consider the system shown in Figure 2.17 composed of two units and several input, output, and connecting streams. Suppose that each stream of this process involves the same S species. If unit I is viewed as a separate system, then, since it involves S species, it will be possible to relate the stream variables associated with unit I by means of S independent material balance equations. Similarly, if unit II is treated in isolation from the total process, then, since it too involves S species, it will again be possible to write S independent material balance equations relating the stream

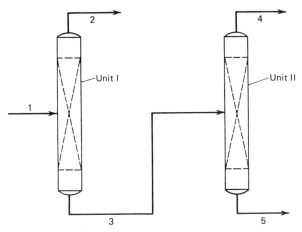

Figure 2.17   A two-unit process.

variables of streams 3, 4, and 5. Finally, if the entire two-unit process is viewed as a single unit, that is, if we draw the system boundaries as shown in Figure 2.18, so that only the streams entering and leaving the overall process are considered, then it will be possible to write an additional set of S independent balances. This set of balance equations will however only relate the stream variables associated with streams 1, 2, 4, and 5. The stream variables associated with the internal stream, stream 3, will not appear in this set of balance equations.

We refer to the set of balance equations written over the entire process as *overall* balances. We refer to the set of balances written for each unit separately as *unit* balances.

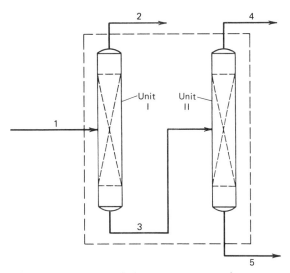

Figure 2.18   Overall balances on a two-unit process.

In the above two-unit system, the three points of view—unit I in isolation, unit II in isolation, and the overall process viewed as a single unit—thus provide us with two sets of unit balances plus a set of overall balances. Each separate set of material balance equations contains at most S independent equations; hence, the three sets provide us with a total of $3S$ equations from which to determine all stream variables. The question is whether all $3S$ equations can be used for solution, that is, whether they are all independent.

Recall that in the case of the single unit involving S components, a total of $S + 1$ material balances could be written: one material balance equation for each component plus the total mass balance. However, only S of these balance equations were independent; the $S + 1$st provided no additional information. The analogous situation develops for the sets of material balances that it is possible to write for multiunit processes. If a process contains M units each of which involve the same S components, then only M sets of balances will be independent. Thus, for the above two-unit process, although it is possible to develop three sets of balance equations (one for each unit separately and one for the overall process), only two sets will be independent. This is illustrated in the following example.

Example 2.13 Consider a separation train consisting of two distillation columns which are designed to separate a three component mixture of benzene, toluene, and xylene into three streams, each rich in one of the species. The system diagram listing the species present in each stream is given in Figure 2.19.

Given a feed rate of 1000 mol/h of a mixture consisting of 20% benzene, 30% toluene, and the rest xylene (all mole %), a bottoms product of 2.5% benzene and 35% toluene is obtained in the first unit and an overhead product of 8% benzene and 72% toluene in the second unit. Determine how much material will be processed by each unit and how this material will be divided among the outlet streams.

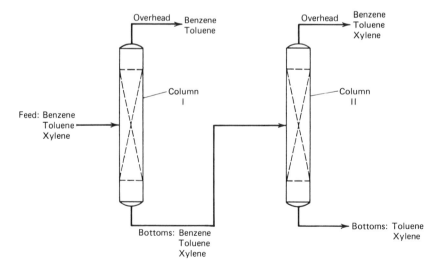

Figure 2.19 Schematic for Example 2.13, two-column separation train.

The process with its stream variables and specified compositions is shown in Figure 2.20.

Solution Consider just unit I. The unit processes three species: benzene, toluene, and xylene. Therefore, it is possible to write three independent species balances:

Benzene balances: $\qquad 200 = x_B^2 N^2 + 0.025N^3$

Toluene balances: $\qquad 300 = (1 - x_B^2)N^2 + 0.35N^3$

Xylene balances: $\qquad 500 = 0.625N^3$

The fourth balance, the total mole balance, will be dependent.

If unit II is considered in isolation, then, since it also involves the three components, it is possible to write its set of balances.

Benzene balance: $\quad 0.025N^3 = 0.08N^4$

Toluene balance: $\quad 0.35N^3 = 0.72N^4 + x_T^5 N^5$

Xylene balance: $\quad 0.625N^3 = 0.20N^4 + (1 - x_T^5)N^5$

Finally, if the separation train is considered as a single overall unit involving three species, three further balances can be written.

Benzene balance: $\quad 200 = x_B^2 N^2 + 0.08N^4$

Toluene balance: $\quad 300 = (1 - x_B^2)N^2 + 0.72N^4 + x_T^5 N^5$

Xylene balance: $\quad 500 = 0.20N^4 + (1 - x_T^5)N^5$

Note that if the benzene balances for units I and II are added, the above overall benzene balance equation is generated. Similarly, if the toluene balances

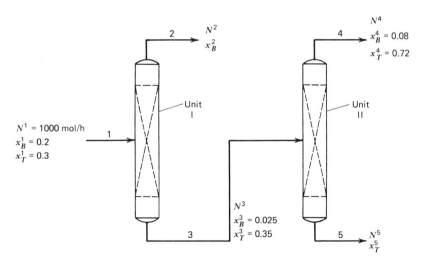

Figure 2.20 Flowsheet for Example 2.13.

for units I and II are added, the overall toluene balance is generated. Clearly, only six of the above nine equations are independent. Thus, the problem could be solved using either the unit I and unit II balances or else the overall balances together with a set of unit balances. The particular choice of balances to be used is dictated only by the ease with which the equations can be solved.

The conclusion reached via the two-unit example can be generalized to systems with M units each of which involves all S species. Each of the M units can be considered in isolation; hence, an independent set of S material balance equations can be written for each unit. Moreover, the whole multiunit process can be considered as a single unit; hence, an independent set of S overall material balance equations can also be written. However, since the component balances for the individual units can always be summed to yield the corresponding overall balances, only M of these $M + 1$ sets of material balances will be independent.

Of course, if the number of units is three or more, the number of possible sets of material balances is even larger than $M + 1$. For example, if a process consists of three units, as shown in Figure 2.21, then, in addition to the individual unit balances, various *combined* balances could also be written. For instance, units I and II could be considered as a combined single unit and a set of material balances written to describe this combination. Alternatively, combined balances could be formulated for units II and III considered as a single unit and even for I and III combined. Regardless of the type of combination balances used, however, if the process consists of M units each of which contains the S species, then only M sets of material balances will be independent.

Returning to the three-unit process, suppose the unit I balances, the combined unit I/unit II balances, and the overall balances are used. Consider the subsystem consisting of units I and II. Clearly, the balances around unit I and the combined unit I/unit II balances (which are the overall balances for the process consisting of just units I and II) provide the same information as the sets of unit balances around I and II. Moreover, the individual balances around I and II together with the

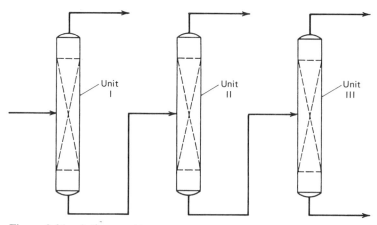

Figure 2.21   A three-unit process.

r erall balances around I, II, and III provide exactly the same information as individual balances around units I, II, and III. Thus, any combination of sets of unit and combined balances can always be reduced to an equivalent set of individual unit balances. If this equivalent set consists of exactly the M sets of individual balances for the M units in the process, then the set of combination balances is independent. In view of the large number of combination balances it is possible to generate by combining units in various ways and, especially, in view of the tedium involved in verifying whether the particular set of combined balances used is in fact independent, the use of combined balances is not advocated. Rather, a solution strategy relying only on individual unit balances or, if necessary, the overall balances will be used in this text.

The discussion of this section was carried out assuming that each of the multiple units in the system involved the same set of S species. Exactly analogous conclusions will be obtained if some of the units involve only a subset of the S species. Regardless of the number of species S_i which are processed by unit i, it is clear that S_i independent unit i balances can be written. The total number of independent balances that can thus be assembled for a process involving M units is the sum of the number of independent balances which can be written for each individual unit; that is,

$$\text{Maximum number of independent} \atop \text{material balances} = \sum_{i=1}^{M} S_i$$

If $S_i = S$ for all units i, then the maximum number of independent balances for the M unit process will be $M \times S$, a conclusion we reached earlier.

2.3.2 Degree-of-Freedom Analysis

Having established a rule for determining how many balance equations are available for solving for the unknown stream variables in a multiunit process, we now proceed to update the degree-of-freedom analysis which was presented for the single-unit case to accommodate the multiunit case. This updating is rather easily accomplished because the rules for counting stream variables and specified information remain unchanged. Thus, we define the degree of freedom of a multiunit system in a form analogous to the single-unit case as

Degree of freedom of the system = total number of stream variables
 - total number of independent material balances
 - total number of specified independent stream variables
 - total number of imposed subsidiary relations

As before, a degree of freedom of zero indicates that the system is completely specified and all unknown stream variables can be calculated. A degree of freedom which is positive would signal an underspecified problem, while a negative degree of freedom would signal that the problem is overspecified. If either of these cases occurred, the problem would have to be studied further to determine what specifications should be added or deleted.

Example 2.14   Consider again the distillation system described in Example 2.13. From Figure 2.20, it is clear that the total number of stream variables is 13: three each for streams 1, 3, and 4 and two each for streams 2 and 5. The maximum number of independent material balances is six, three each for units I and II. Finally, the total number of specified stream variables is seven: two independent compositions for each of streams 1, 3, and 4 and the flow rate of stream 1. Since no subsidiary conditions were imposed on the stream variables, the degree of freedom of the two-unit system is zero. The degree-of-freedom tabulation is summarized in Figure 2.22.

Referring to the balance equations written in Example 2.13, note that the three unit I balance equations involve just three unknowns: x_B^2, N^2, and N^3. The set of three unit II balances involves four unknown stream variables: x_T^5, N^3, N^4, and N^5; while the set of three overall balances contains five unknowns: x_B^2, x_T^5, N^2, N^4, and N^5. Since we have a choice of which two sets of balances to use, it is clearly best to solve the unit I balances first, followed by a solution of either one of the remaining two sets of balance equations.

Commencing with a solution of the unit I balances, we first solve the xylene balance for N^3 to obtain

$$N^3 = 500/0.625 = 800 \text{ mol/h}$$

This value is then substituted into the benzene and toluene balances to yield two simultaneous equations in N^2 and x_B^2. These are easily solved by adding them together, thus eliminating the product $x_B^2 N^2$. The result is

$$N^2 = 200 \text{ mol/h}$$

which, when substituted into the benzene balance, gives

$$x_B^2 = 0.9$$

To determine the molar flow and composition of the bottoms stream of unit II, either the overall or the unit II balances can be solved to obtain

$$N^5 = 550 \text{ mol/h}$$
$$N^4 = 250 \text{ mol/h}$$

and $x_T^5 = 9/11$

Number of stream variables		13
Number of independent balance equations		
Unit I	3	
Unit II	3	
Number of specified stream		
Compositions	6	
Flows	1	
Number of subsidiary relations	0	
		-13
Degree of freedom		0

Figure 2.22   Example 2.14, degree-of-freedom table.

All unknown stream variables have thus been calculated and the process is completely specified as expected.

Note that in solving the system of balance equations in the above example, the choice of which set of balances to solve first was made by counting the number of unknowns involved in each balance set. The unit I balances were chosen because they reduced to a system of three equations in three unknowns, that is, they had degree of freedom zero. This suggests that the selection and solution of the sets of balance equations for a multiunit system is facilitated if the degree-of-freedom calculation is carried out not only for the entire system but for each individual unit as well. Hence, for each multiunit process we construct a degree-of-freedom table which will contain a column for each unit as well as a composite column for the entire process.

Example 2.15 Let us construct a degree-of-freedom table for the distillation system considered in the previous examples. First, consider unit I. The feed to unit I has three species, hence three stream variables associated with it. The overhead stream from unit I has two, while the bottoms stream has three stream variables. Hence, unit I has eight stream variables associated with it. The unit processes three species; consequently, it is possible to write three independent material balances. Next we consider unit II. This unit is fed by the bottoms of unit I, which requires three variables. The bottoms and overheads require two and three stream variables, respectively. Hence, unit II has eight variables also and, since it, too, processes three species, it will have associated with it three independent material balances. Unit I has four independent compositions specified for the streams entering and leaving it. Unit II has four also. The feed to unit I has been specified as 1000 mol/ h. This is the basis for the computation. With this, the degree of freedom for each unit can be determined. The results are tabulated together with the degree-of-freedom calculation summarized in Figure 2.22 and shown in Figure 2.23.

The entire process is correctly specified, and since unit I is also completely specified, the unit I balances provide a convenient starting point for solution, as we observed in Example 2.14.

Several interesting features of the degree-of-freedom table are noteworthy. First of all, note that the number of balances for the process is simply the sum of the number of balances for the individual units. However, note that neither the

	Unit I	Unit II	Process
Number of stream variables	8	8	13
Number of independent balances	3	3.	6
Number of specified streams			
Compositions	4	4	6
Flows	1	0	1
Number of subsidiary relations	0	0	0
	−8	−7	−13
Degree of freedom	0	1	0

Figure 2.23 Example 2.15, degree-of-freedom table.

number of stream variables nor the specified information for the process can be obtained by adding the corresponding entries for the individual units. This comes about because, as can be seen from Figure 2.23, both the unit I and unit II tabulations will involve the stream variables and specified information associated with stream 3, the common stream. Thus, adding the corresponding entries for the individual units could in effect represent a double counting of the variables and specifications associated with stream 3. Thus, the total number of stream variables for the process should be calculated by totaling the number of stream variables associated with each unit and subtracting the number of variables associated with streams connecting adjacent units. The total number of items of information is best obtained by counting the specified data directly from the flowsheet.

Furthermore, it is important to note that although the degree of freedom of unit I is zero, this does not necessarily imply that the entire process is correctly specified. In the case of Example 2.15, the process is completely specified. However, if one of the specified compositions of stream 4 is deleted, then the degree of freedom of the entire plant becomes 1 (underspecified), while the degree of freedom of unit I remains at zero. Another possibility in the relationship between unit and process degrees of freedom is illustrated in the next example.

Example 2.16   In the four-stage evaporation system shown in Figure 2.24, a 50% by weight sugar solution is concentrated to 65% by evaporating an equal amount of water in each of the four stages. With a total input of 50,000 lb/h, a product stream of 35,000 lb/h is produced. Determine the compositions of the intermediate streams.

Solution   We proceed with the construction of a degree-of-freedom table. Note that each individual unit involves five stream variables and, since it processes two species, will have associated with it two independent balances. Specified information is available only for the feed and product streams; hence, it will be associated with units I and IV. Finally, the requirement that there be an equal amount evaporated at each stage translates to three flow relations: $F^2 = F^4$, $F^4 = F^6$, and $F^6 = F^8$. Since each of these relations is imposed on streams associated with two different units, they cannot directly be assigned to any one unit. However,

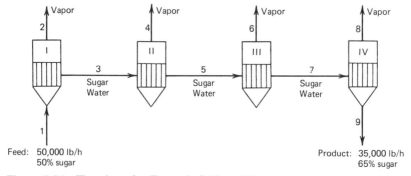

Figure 2.24   Flowsheet for Example 2.16, multistage evaporation.

	Unit I	Unit II	Unit III	Unit IV	Process
Number of stream variables	5	5	5	5	14
Number of material balances	2	2	2	2	8
Number of specified					
Compositions	1	—	—	1	2
Flows	1	—	—	1	2
Number of subsidiary relations	—	—	—	—	3
	−4	−2	−2	−4	−15
Degree of freedom	1	3	3	1	−1

Figure 2.25 Example 2.16, degree-of-freedom table.

they are available for the solution of the entire process. The resulting degree-of-freedom table is shown in Figure 2.25.

Note that the plant is overspecified, yet all of the individual unit balances are underspecified. The underspecification of an individual unit merely means that the balances for that unit cannot be solved independently of the remaining units.

In general, regardless of the degree of freedom of the plant, the degree of freedom of each of the individual units must always be greater than or equal to zero. If any individual unit is overspecified, the problem is incorrectly specified. A more extensive discussion of incorrect specifications will be taken up in later chapters.

2.3.3 Special Multiunit Configurations

Before proceeding with a discussion of the strategy of solving multiunit material balance problems, we digress to consider briefly two special multiunit process configurations which frequently occur in chemical engineering systems. These are the process with recycle and the process with bypass. Neither of these configurations should present any special problems for analysis—although the algebraic solution could become more complicated—if they are recognized as simply multiunit processes that involve a special type of unit.

Recycle Stream A recycle is simply a stream which is split off from the outlet of a unit and sent back as the input of an upstream unit. The unit with recycle may be viewed as a three unit process composed of a mixer, the inner unit, and a flow splitter, as shown in Figure 2.26. As with any multiunit process, balances can be

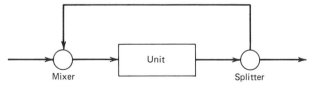

Figure 2.26 System with recycle.

written around each individual unit as well as an overall balance around the entire process. As before, only three of these four sets of balances will be independent.

The only unit in this system requiring any special attention is the splitter. A flow splitter is simply a device in which a given input stream is divided flow-wise into two or more smaller streams. Since the division is purely on the basis of flow, the compositions of all of the branches into which the main stream is split must be the same. As discussed in Section 2.1.3, this condition can be translated into a set of subsidiary relations

$$x_j^i = x_j^k \quad \text{or} \quad w_j^i = w_j^k$$

between the compositions of species j in branches i and k. If the stream being split contains S species, then relations of the above form will be imposed for each of the S species and between every pair of output branches. Since the sum of the composition fractions of any stream must equal 1, for any given pair of branches, only $S - 1$ relations of the above type will be independent. The remaining relation can always be generated from the other $S - 1$. That is, since for any stream i

$$\sum_{j=1}^{S} x_j^i = 1$$

and any stream k

$$\sum_{j=1}^{S} x_j^k = 1$$

the conditions

$$x_j^i = x_j^k \quad j = 1, \ldots, S - 1$$

imply that

$$x_S^i = x_S^k$$

From the above, we can conclude that if a stream containing S species is split into two branches, then $S - 1$ composition relations are automatically imposed. If we extend this to a flow split into N branches, then, since the composition of each pair of branches will be related by $S - 1$ equalities, there will be a total of $(N - 1)(S - 1)$ composition relations imposed on the compositions of the N branch streams.

These relations, which we call *splitter restrictions,* are intrinsic to the definition of a splitter and, hence, should automatically be included both in the degree-of-freedom analysis and in the solution of any problem which involves such a unit.

Example 2.17 Consider a flow splitter which divides a stream containing malt, hops, and water into three branches, as shown in Figure 2.27. Suppose the composition (in mass %) of the input stream is given as 20% malt, 10% hops, and the rest water. Further, the branch flows are regulated so that $F^2 = 2F^3$ and $F^3 = \frac{1}{3}F^4$. If $F^1 = 1000$ lb/h, what are the flow rates in each of the branches?

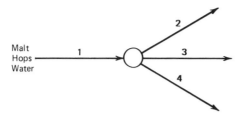

Figure 2.27 Flowsheet for Example 2.17, flow splitter.

Solution Since each of the four streams contains the three species, the system involves 12 stream variables and will yield three independent material balances. The problem, furthermore, has two specified compositions, a specified flow, and two relations defining the split. Finally, since a three-component stream is being split into three branches, the operation of the splitter imposes

$$(N - 1)(S - 1) = (3 - 1)(3 - 1) = 4$$

splitter restrictions. Summarizing this itemization in the degree-of-freedom table shown in Figure 2.28, we observe that the problem is correctly specified.

Of course, this fact can be observed by inspection. Our point, however, is that unless the splitter restrictions are properly accounted for, the degree-of-freedom calculation will be in error.

The solution to the problem is easily obtained by first using the total mass balance

$$1000 = F^2 + F^3 + F^4$$

together with the flow ratio relations

$$F^2 = 2F^3 \quad \text{and} \quad F^3 = \tfrac{1}{3}F^4$$

By substituting out F^2 and F^4,

$$1000 = 2F^3 + F^3 + 3F^3$$

Number of variables		12
Number of balances	3	
Number of specifications		
Compositions	2	
Flows	1	
Number of relations		
Flow ratios	2	
Splitter restrictions	4	
	—	-12
Degree of freedom		0

Figure 2.28 Flow splitter, degree-of-freedom analysis.

or

$$F^3 = \tfrac{1}{6} \times 10^3 \text{ lb/h}$$

and

$$F^2 = \tfrac{1}{3} \times 10^3 \text{ lb/h}$$
$$F^4 = \tfrac{1}{2} \times 10^3 \text{ lb/h}$$

The compositions of the branches then follow from the species balances

Malt: $\qquad 200 = w_M^2 F^2 + w_M^3 F^3 + w_M^4 F^4$

Hops: $\qquad 100 = w_H^2 F^2 + w_H^3 F^3 + w_H^4 F^4$

and the splitter restrictions

$$w_M^2 = w_M^3 \qquad w_M^3 = w_M^4$$
$$w_H^2 = w_H^3 \qquad w_H^3 = w_H^4$$

that is,

$$w_M^2 = w_M^3 = w_M^4 = 0.2$$
$$w_H^2 = w_H^3 = w_H^4 = 0.1$$

Although this result can be obtained by inspection, it should be noted that it is a consequence of the material balances and the splitter restrictions.

With the simple convention developed above for counting splitter restrictions, any process involving recycle streams can be readily analyzed.

Example 2.18 Staged distillation columns are devices which separate volatile materials by boiling off the more volatile components. In order for a clean separation to take place, these devices require that at least part of the vapor boiled off be condensed and returned to the column. This is necessary to ensure that both liquid and vapor phases coexist throughout the column.

Consider the recycle system consisting of a distillation column and its condenser shown in Figure 2.29. For material balance purposes, we can ignore the condenser since it merely produces a phase change in the stream. It does not affect mass and composition. Suppose the column is used to separate a three-component mixture consisting of 7% acetone, 61.9% acetic acid, and 31.1% acetic anhydride. The column is designed to yield a bottoms stream containing no acetone and a distillate containing 10% acetone and 88% acetic acid. If the column is operated so that 60% of the overhead is returned as reflux, calculate all flows assuming all compositions are in mol % and that 700 mol/h of distillate is to be produced.

Solution The system flowsheet indicating the species present in each stream and all specified compositions is shown in Figure 2.30 (on p. 74). We begin with the construction of the degree-of-freedom table shown in Figure 2.31.

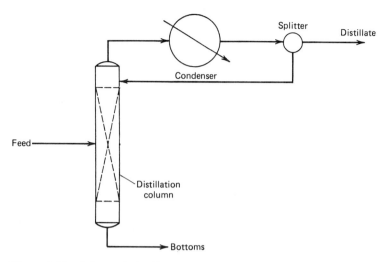

Figure 2.29 Schematic for Example 2.18, distillation column with reflux.

Note that, since the splitter has two output branches and involves three species, it will have associated with it

$$(M - 1)(S - 1) = (2 - 1)(3 - 1) = 2$$

splitter restrictions.

From the table, it is evident that the process is correctly specified. Moreover, since the degree of freedom of the splitter is zero, it is apparent that the solution

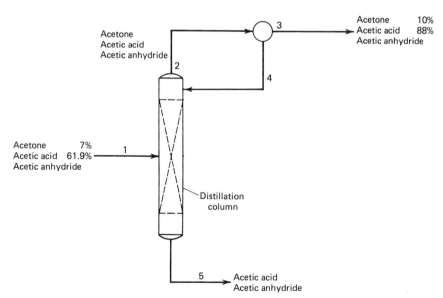

Figure 2.30 Flowsheet for Example 2.18, distillation column with reflux.

	Column	Splitter	Process
Number of stream variables	11	9	14
Number of balances	3	3	6
Number of specified			
Compositions	2	2	4
Flows	—	1	1
Number of relations			
Reflux ratio	1	1	1
Splitter restrictions	—	2	2
	-6	-9	-14
Degree of freedom	5	0	0

Figure 2.31   Example 2.18, degree-of-freedom table.

of the problem should be begun with the splitter balances. The splitter balance set consists of

Total mole balance: $N^2 = 700 + N^4$

Acetone balance: $x_A^2 N^2 = 0.1(700) + x_A^4 N^4$

Acetic acid balance: $x_{HAC}^2 N^2 = 0.88(700) + x_{HAC}^4 N^4$

The reflux condition can be expressed as

$$N^4 = 0.6 N^2$$

and the splitter restrictions yield

$$x_A^4 = x_A^3 = 0.1$$
$$x_{HAC}^4 = x_{HAC}^3 = 0.88$$

From the total mole balance and the reflux ratio, we obtain

$$N^2 = 1750 \text{ mol/h}$$

and thus

$$N^4 = 1050 \text{ mol/h}$$

From the splitter restrictions and the species balance, it then follows immediately that

$$x_A^2 = 0.1$$
$$x_{HAC}^2 = 0.88$$

The solution of the problem can now be completed by solving the column balances. These are

Total mole balance: $N^1 + N^4 = N^2 + N^5$

Acetone balance: $0.07 N^1 + 0.1 N^4 = 0.1 N^2$

Acetic acid balance: $0.619 N^1 + 0.88 N^4 = 0.88 N^2 + x_{HAC}^5 N^5$

With N^2 and N^4 known, the acetone balance yields

$$N^1 = 1000 \text{ mol/h}$$

Then, the total balance can be solved for N^5:

$$N^5 = 300 \text{ mol/h}$$

Finally, the acetic acid balance can be used to calculate the composition:

$$x_{HAC}^5(300) = 0.619(1000) - 0.88(700) = 3$$

or

$$x_{HAC}^5 = 0.01$$

Note again that, although the composition of stream 2 can be determined by inspection, it properly is a consequence of the splitter restrictions and the species material balances. If the degree-of-freedom analysis is to be carried out correctly, the appropriate number of splitter restrictions must always be included. With the simple rule for counting splitter restrictions given previously, the analysis of processes involving recycle streams is not significantly different from that required for any other multiunit process.

Bypass Stream A bypass stream is, as the name implies, a stream which is split off from another so as to avoid processing by a downstream unit. Again, as in the case of the recycle, this configuration is easily treated as simply a multiunit process consisting of a splitter, the unit, and a mixer, as shown in Fig. 2.32. The analysis of processes involving bypass streams is thus essentially the same as that carried out for recycle streams. In fact, the solution of the material balances for processes involving bypass is usually easier than that for processes involving recycles.

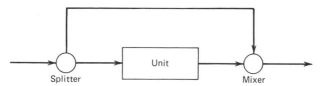

Figure 2.32 System with bypass stream.

Example 2.19 Fresh orange juice typically consists of about 12% (mass %) dissolved solids, largely sugars, in water. In order to reduce the cost of shipping, the juice is often concentrated prior to shipping and then reconstituted by adding water at the destination. Concentration must be carried out in specially designed, short residence-time evaporators operated at below atmospheric pressures in order to reduce the loss of volatile and thermally sensitive flavor and aroma components present in trace amounts. Since some loss of these components is nonetheless unavoidable, a widely accepted approach is to somewhat overconcentrate the juice and then add a small amount of fresh juice (called a cutback) to the concentrate

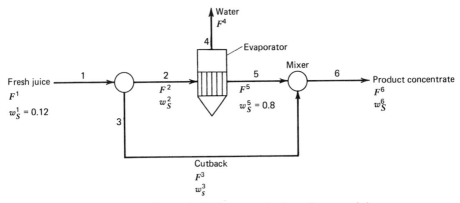

Figure 2.33 Flowsheet for Example 2.19, concentration of orange juice.

to produce a product of improved aroma and flavor. Suppose that 10% of the feed to such a process is used as cutback and that the evaporator is operated to produce an outlet concentrate containing 80% dissolved solids. If the process feed rate is 10,000 lb/h fresh juice, calculate the rate at which water must be evaporated and the composition of the final product.

Solution The flowsheet of the system with stream variables is shown in Figure 2.33. In the analysis of this problem, we will assume that the mass of the volatile trace components lost in evaporation is negligibly small. The degree-of-freedom table for the process is shown in Fig. 2.34.

The number of splitter restrictions in this instance is just one, since there are only two output branches involving two species.

The degree-of-freedom analysis indicates that the problem is correctly specified and that the problem solution should be initiated with the splitter balances.

The splitter balance equations are

Total mass balance: $10,000 = F^2 + F^3$

Solids balance: $0.12(10,000) = w_S^2 F^2 + w_S^3 F^3$

The splitter restriction is

$$w_S^2 = w_S^3$$

and the cutback fraction condition is

$$0.1 F^1 = F^3$$

From the total balance and the above conditions, we easily obtain

$$F^2 = 9000 \text{ lb/h}$$
$$F^3 = 1000 \text{ lb/h}$$

	Splitter	Evaporator	Mixer	Process
Number of variables	6	5	6	11
Number of balances	2	2	2	6
Number of specified				
Compositions	1	1	1	2
Flows	1	—	—	1
Number of relations				
Split ratio	1	—	—	1
Splitter restrictions	1			1
	-6	-3	-3	-11
Degree of freedom	0	2	3	0

Figure 2.34 Example 2.19, degree-of-freedom table.

Then, from the solids balance it is clear that

$$w_S^2 = w_S^3 = 0.12$$

Continuing with the evaporator balances, we have

Solids balance: $0.12(9000) = 0.8F^5$

Water balance: $0.88(9000) = F^4 + 0.2F^5$

These yield

$$F^5 = 1350 \text{ lb/h}$$
$$F^4 = 7650 \text{ lb/h}$$

Finally, the solution is completed by using the mixer balances:

Total balance: $F^5 + F^3 = F^6$

Solids balance: $0.8(1350) + 0.12(1000) = w_S^6 F^6$

From the former, we have

$$F^6 = 2350 \text{ lb/h}$$

and from the latter,

$$w_S^6 = 1200/2350 = 0.51$$

2.3.4 Strategy of Solution

As in the single-unit case, once it is determined that the material balance problem is correctly specified, we are left with the purely algebraic problem of solving a number of equations for an equal number of unknowns. Again, that solution process can be substantially simplified if we adopt the strategy of ordering the calculations so that, if possible, no simultaneous solution is required. However, because of the generally large number of equations and unknowns that arise in multiunit problems, it is expedient to implement the sequencing strategy at two levels. At the primary level, we seek to determine which sets of balance equations to solve and then, at the second level having made that decision, we apply the guidelines presented in Section 2.2.2 for selecting the equations to solve within the individual balance sets. In determining which balance sets to employ for solution and in what order, we basically have two choices to make: we have the choice of the location of the basis and we have the choice of which M balance equation sets from among the $M +$ 1 possible sets we shall employ.

Let us first consider the choice of balance set. Recall that for an M unit process it is possible to write M individual unit balances as well as a set of overall balances. Hence, we always have the option of substituting the set of overall balance equations for one of the unit balance sets. The criterion for whether or not to make this substitution is simply the degree of freedom. If the degree of freedom of the overall balance set is zero while none of the individual sets has degree of freedom zero, then the overall balances are clearly the preferred set with which to initiate problem solution. Consequently, in carrying out the degree-of-freedom analysis for the process, if it is evident that no unit has degree of freedom zero, then the column corresponding to the overall balances should be added to the table and the degree of freedom of the overall balances determined. As illustrated in the following example, the overall balances often are useful in initiating solution because in many process design applications both the feed composition as well as the composition, and possibly the flow rates, of the primary product streams will be specified. From this information the engineer will seek to calculate or assign values to most internal streams and process waste streams. Hence, the overall balances will tend to have more specified information associated with them and, thus, will be likely to have a lower degree of freedom than the individual units of the process.

Example 2.20   Consider again the solvent extraction process of Example 2.8. However, instead of drawing our systems boundary around the entire multiunit process as shown in Figure 2.7, suppose we seek to calculate the internal streams as well. As before, 1000 lb/h of a refinery stream containing 70% benzene in a mixture of nonbenzene hydrocarbons is extracted by means of SO_2 to yield a process raffinate stream containing $\frac{1}{6}$ (mass fraction) SO_2 and the remainder benzene. The process extract stream contains all the nonbenzene material, some SO_2, as well as $\frac{1}{4}$ lb benzene per 1 lb nonbenzene hydrocarbons. Suppose further that 3 lb SO_2 is used per 1 lb process feed, that the separation of benzene is 92% in unit I and 80% in unit II, and that the benzene compositions in the feed streams to units II and III are 86.25 and 95%, respectively. The process flowsheet with specified compositions is shown in Figure 2.35.

Solution   We begin with the construction of the degree-of-freedom table for the process. As shown in Figure 2.36, the process involves 18 stream variables, nine balance equations, five specified stream variables, and four subsidiary relations.

The degree of freedom of the process is zero; however, none of the individual units has degree of freedom zero. Hence, we proceed to examine the overall balance set. Recall that the overall balances correspond to a view of the process as a single

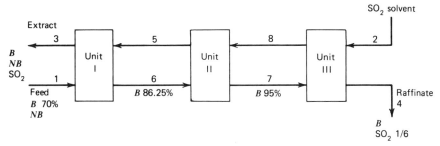

Figure 2.35 Flowsheet for Example 2.20, SO_2 solvent extraction.

unit and involve only the stream variables associated with the process input and output streams. In our case, the overall balances will involve streams 1, 2, 3, and 4, that is, eight stream variables, three balances, two compositions (those of streams 3 and 4), and two subsidiary relations (the SO_2/feed ratio and the benzene/non-benzene ratio in stream 3). Note that only those relations which involve the stream variables associated with streams 1, 2, 3, and 4 can be counted with the overall balances. Tabulating these numbers, we obtain

	Unit I	Unit II	Unit III	Process
Number of variables	10	10	8	18
Number of balances	3	3	3	9
Number of specified				
Compositions	2	2	2	4
Flows	1			1
Number of relations				
SO_2/feed ratio				1
lb benzene/lb nonbenzene in				
stream 3	1			1
80% separation in II		1		1
92% separation in I	1			1
	-8	-6	-5	-18
Degree of freedom	2	4	3	0

Figure 2.36 Example 2.20, degree-of-freedom table.

Number of variables	8
Number of balances	3
Number of specified	
Compositions	2
Flows	1
Number of relations	2
	-8
Degree of freedom	0

we note that the degree of freedom is zero. Hence, it is expedient to start the solution of the problem with the set of overall balances.

Solution of the overall balances simply involves repeating the calculations carried out in Example 2.8. The pertinent equations are

Total mass balance: $1000 + F^2 = F^3 + F^4$

Nonbenzene balance: $300 = w_{NB}^3 F^3$

Benzene balance: $700 = w_B^3 F^3 + \frac{5}{6} F^4$

and the side conditions

$$F^2/F^1 = 3$$
$$w_B^3/w_{NB}^3 = 0.25$$

The solution is obtained as in Example 2.8. The resulting stream flow rates are

$$F^2 = 3000 \text{ lb/h}$$
$$F^4 = 750 \text{ lb/h}$$
$$F^3 = 3250 \text{ lb/h}$$

and the composition is

$$w_B^3 = 0.0231$$
$$w_{NB}^3 = 0.09231$$

With stream 3 known, we can continue with the unit I balances

Total balance: $1000 + F^5 = 3250 + F^6$

Benzene balance: $700 + w_B^5 F^5 = 75 + 0.8625 F^6$

SO_2 balance: $w_{SO_2}^5 F^5 = 2875$

and with the separation conditions, that is, that 92% of the benzene fed to unit I is recovered in stream 6:

$$0.8625 F^6 = 0.92(w_B^5 F^5 + 700)$$

The benzene balance together with the separation relation yields

$$0.8625 F^6 = 0.92(75 + 0.8625 F^6)$$
$$0.069 F^6 = 69$$
$$F^6 = 1000 \text{ lb/h}$$

From the total balance,

$$F^5 = 3250 \text{ lb/h}$$

and from the separation condition,

$$w_B^5 = 0.0731$$

Finally, the SO_2 balance yields

$$w_{SO_2}^5 = 0.8846$$

It remains to solve the unit II balances. These are

Total balance: $\quad 1000 + F^8 = 3250 + F^7$

Benzene balance: $\quad 862.50 + w_B^8 F^8 = 0.95 F^7 + 237.5$

SO_2 balance: $\quad w_{SO_2}^8 F^8 = 2875$

and the separation condition

$$0.95 F^7 = 0.80(862.5 + w_B^8 F^8)$$
$$= 0.80(0.95 F^7 + 237.5)$$

This yields

$$F^7 = 1000 \text{ lb/h}$$

and, from the total balance,

$$F^8 = 3250$$

Finally, from the separation condition and the SO_2 balances, we obtain, respectively,

$$w_B^8 = 0.1$$

and

$$w_{SO_2}^8 = 0.8846$$

Thus, all intermediate streams have been determined. Note that since the overall balances were used, the unit III balance set is redundant.

The second item of choice in solving the multiunit balance problem is the location of the basis for the calculation. Recall from our discussion of the single unit case that even if a flow is specified in the problem formulation, the homogeneity of the balance equations allows us to specify any other flow as basis for the calculation providing that the resulting solution is rescaled to match the given flow rate. Since the collection of individual unit balances as well as the overall balances are likewise homogeneous, the strategy of basis relocation can again be employed. In the multiunit case, however, our primary goal in relocating the basis is to choose a basis which will insure that, if possible, the degree of freedom of one of the individual or the overall balance sets will be zero. Hence, we employ the degree of freedom table to determine with which balance set the basis ought to be associated. Having made that determination, we can then use the rules given in Section 2.2.2 to select which stream flow rate of the streams involved in the chosen balance set ought to be specified as basis. We thus again employ a two level approach:

first, we select the balance set, then we select the particular stream. Now, since the basis assignment only involves a single specification, it is clear that we ought to pick a balance set which has degree of freedom one, because with the basis specification the resulting degree of freedom of that balance set will be zero.

The simple device of basis relocation is illustrated in the next example.

Example 2.21   The four-unit separation train shown in Figure 2.37 has been designed to separate 1000 mol/h of a hydrocarbon feed containing 20% CH_4, 25%

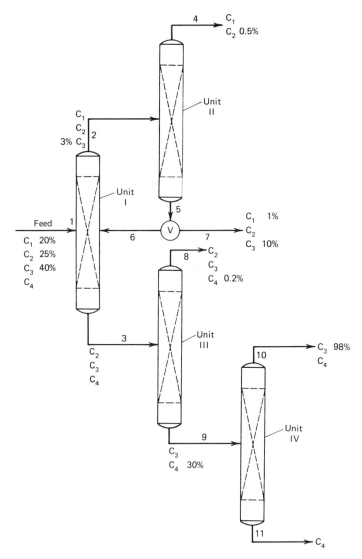

Figure 2.37   Flowsheet for Example 2.21, four-column hydrocarbon separation train.

C_2H_6, 40% C_3H_8, and the rest C_4H_{10} (all in mol %) into five fractions. Given the indicated compositions all in mol %, calculate all stream flow rates in the process assuming that the recycle to unit I is 50% of the bottoms from unit II.

Solution   The degree-of-freedom table including the column for the overall balances is shown in Figure 2.38.

In the construction of the table, note that unit V is a flow splitter which involves three species split into two branches; hence, two splitter restrictions are appropriate. Since unit V is a splitter, the two compositions of stream 7 can be carried around to streams 5 and 6 and hence can be used in solving the unit I as well as the unit II balances. In Figure 2.38, these carried compositions are indicated by (+2). Finally, the overall balances will involve only the stream variables associated with streams 1, 4, 7, 8, 10, and 11.

From Figure 2.38, it is evident that the process is correctly specified. However, neither the overall balance set nor any of the individual unit balances have degree of freedom zero. Observe that units II, IV, and V all have degree of freedom equal to 1 and none involves the specified flow rate. Consequently, if the specified flow rate of stream 1 is disregarded and instead a basis is selected with any of the streams associated with one of these three units, then that unit will have degree of freedom zero. In this manner, we will have generated a good candidate for the balance set with which to initiate solution. Suppose we choose our basis to be associated with unit V. In particular, let us select a basis of 1000 mol/h of stream 5.

From the splitter total mole balance

$$1000 = N^6 + N^7$$

and the given split ratio

$$N^6 = 0.5N^5 = 500 \text{ mol/h}$$

it follows that

$$N^7 = 500 \text{ mol/h}$$

	Unit I	Unit II	Unit III	Unit IV	Unit V	Overall Process	Balances
Number of variables	13	8	8	5	9	29	15
Number of balances	4	3	3	2	3	15	4
Number of specified							
Compositions	4 (+2)	2 (+2)	2	2	2	10	8
Flows	1	—	—	—	—	1	1
Number of relations							
Splitter restrictions	—	—	—	—	2	2	—
Recycle ratio	—	—	—	—	1	1	—
Degree of freedom	2	1	3	1	1	0	2

Figure 2.38   Example 2.21, degree-of-freedom table.

From the splitter restrictions and the species mole balances, it then follows immediately that

$$x^5_{C_1} = x^6_{C_1} = x^7_{C_1} = 0.01$$
$$x^5_{C_3} = x^6_{C_3} = x^7_{C_3} = 0.1$$

From the splitter restrictions and the species mole balances, it then follows immediately that

$$x^5_{C_1} = x^6_{C_1} = x^7_{C_1} = 0.01$$
$$x^5_{C_3} = x^6_{C_3} = x^7_{C_3} = 0.1$$

With this, we can continue with the unit II balances. These are

Total mole balance: $N^2 = N^4 + 1000$

C_3 mole balance: $0.03N^2 = 0.1(1000)$

C_2 mole balance: $x^2_{C_2}N^2 = 0.005N^4 + 0.89(1000)$

From the C_3 balance, we have

$$N^2 = 10^4/3 \text{ mol/h}$$

and from the total balance,

$$N^4 = \tfrac{7}{3} \times 10^3 \text{ mol/h}$$

The remaining equation, the C_2 balance, then yields

$$x^2_{C_2} = 0.2705$$

Continuing with the *unit I* balances, the C_1 balance yields $N^1 = 11,633$ mol/h, while the total balance can be solved to obtain $N^3 = 8800$ mol/h. The compositions of stream 3 then follow immediately from the remaining two balances. Thus, $x^3_{C_4} = 0.1983$ and $x^3_{C_3} = 0.5231$.

Next, the *unit III* balances are solved. The simultaneous solution of the total mole and C_4 balances yields

$$N^8 = 3003.4 \text{ mol/h}$$
$$N^9 = 5796.6 \text{ mol/h}$$

The unknown C_2 composition is then calculated from the C_2 balance:

$$x^8_{C_2} = 2451.8/N^8 = 0.8163$$

The solution of the problem is completed with the *unit IV* balances. The C_3 balance gives

$$N^{10} = 4140.4 \text{ mol/h}$$

and the total balance,

$$N^{11} = 1656.2 \text{ mol/h}$$

The complete solution, calculated in terms of the basis $N^5 = 1000$ mol/h, must now be rescaled to correspond to the given basis of $N^1 = 1000$ mol/h. Since the flow of stream 1 calculated in terms of the former basis is 11,633 mol/h, the desired solution can be obtained by scaling all calculated flows by the factor

$$10^3/11,633 = 8.596 \times 10^{-2}$$

The scaled results for the process are tabulated in the stream table shown in Figure 2.39.

The degree-of-freedom table for a multiunit process thus serves several important functions. It first of all tells us if the problem is correctly specified. Next, it indicates with which unit balance set the computations should be begun. If no unit has degree of freedom zero, the table will show whether or not the overall balances should be solved first. If no single balance set has degree of freedom zero, the table will further indicate whether a relocation of the basis will result in a set of balances with degree of freedom zero. Finally, if none of these devices yields a unit with degree of freedom zero, the table will, as discussed in Chapter 5, provide guidance in how a simultaneous solution is to be attempted.

2.4 SUMMARY

In this chapter, we examined in detail the elements involved in setting up, analyzing, and solving material balance problems. We found that process streams can be defined in terms of four sets of variables: total molar flow and species mole fractions, total mass flow and species mass fractions, individual species molar flows, and individual species mass flows. We then formulated the material balance equations for nonreacting systems and discussed their properties: homogeneity and independence. Next, we considered the various forms in which specifications can be imposed on a system: direct stream variable specification, fractional recovery relations, composition ratios, and flow ratios. Having assembled these elements of the material balance problem, we introduced the notion of degree of freedom of a problem and considered the consequence of having overspecified or underspecified problems. We concluded the analysis of single-unit systems by discussing the use of two solution strategy devices: basis relocation and balance equation selection.

In the second half of this chapter, we expanded the preceding analysis to include multiunit flowsheets. We observed that although in general it is possible to write a large number of independent balance equations, there is a maximum number of these which will be independent. That maximum can be determined simply by summing up the number of independent balance equations associated with each individual unit. Next, we introduced the alternate use of overall balance and observed that although various further combination balances are sometimes very effective, in general their use is attended with complications.

We then extended the degree-of-freedom analysis to include multiunit processes and considered several special multiunit configurations quite prevalent in

	Stream Number										
	1	2	3	4	5	6	7	8	9	10	11
Flow (mol/h × 10⁻³)	1.0	0.2865	0.7565	0.2006	0.0859	0.04298	0.04298	0.2582	0.4983	0.3559	0.1424
Composition (mole fraction)											
C₁	0.20	0.6995	—	0.995	0.01	0.01	0.01	—	—	—	—
C₂	0.25	0.2705	0.2786	0.005	0.89	0.89	0.89	0.8163	—	—	—
C₃	0.40	0.030	0.5231	—	0.10	0.10	0.10	0.1817	0.7	0.98	—
C₄	0.15	—	0.1983	—	—	—	—	0.002	0.3	0.02	1.0

Figure 2.39 Example 2.21, stream table.

flowsheets: the recycle and the bypass. We noted that these configurations are readily accommodated by means of a special process unit called the flow splitter. Finally, we concluded the chapter with examples illustrating the use of basis relocation and balance set selection which can often reduce a multiunit problem to a sequence of zero degree-of-freedom single-unit problems. At this stage, you should be able to formulate and solve most multiunit material balance problems that do not involve chemical reactions.

PROBLEMS

2.1 (a) A feed stream to a dryer is specified as 1000 lb_m/h. What is the rate in kg/min?

(b) The production rate of ammonia in a process is given as 10^5 lbmol/day. What is the equivalent rate in gmol/h?

2.2 A still produces a mixture of 90 mol % ethanol (C_2H_5OH) and the rest water.

(a) Calculate the mass fraction of ethanol.

(b) If the production rate is 1000 lbmol/h, calculate the equivalent rate in kg/min.

(c) For the production rate in (b), calculate the molar flows of the stream components in kgmol/h.

2.3 A stream consisting of:

$$\begin{array}{ll} H_2O & 0.4 \\ C_2H_5OH & 0.3 \\ CH_3OH & 0.1 \\ CH_3COOH & 0.2 \end{array}$$

all in weight fractions, is fed to a distillation column at the rate of 1000 lb_m/h. Convert these stream variables to:

(a) Molar species flows rates.

(b) Total molar flow and mole fractions.

(c) Mole fractions, water-free basis.

2.4 A solution containing Na_2S, $NaOH$, and Na_2CO_3 in water is called "white liquor" and is used in the paper industry to process wood pulp. Suppose that laboratory analysis indicates 50 g/l Na_2S, 250 g/l $NaOH$, and 100 g/l Na_2CO_3. If the solution density is 1.05 g/cm^3, calculate the species molar flows corresponding to a total stream flow of 1000 kgmol/h.

2.5 A process for producing methane from synthesis gas and steam is fed 6 kgmol/min of a gas consisting of 50% H_2, $33\frac{1}{3}$% CO, and the rest CH_4 (all on a mole basis) as well as 72 kg/min of steam (H_2O). The products are 3 kgmol/min of liquid water and 96 kg/min of a gas consisting of 25% CH_4,

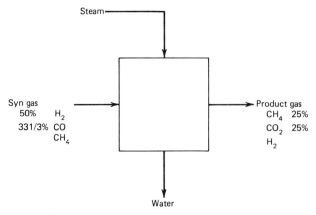

Figure P2.5

25% CO_2, and the rest H_2 (all on a mole basis) (see Figure P2.5). Determine to three significant figures whether:

(a) Total mass is conserved.
(b) Total moles are conserved.
(c) The moles of each type of atom are conserved.
(d) The mass of each type of chemical species is conserved.
(e) What do you conclude is happening in this process?

2.6 A process for producing methyl iodide, CH_3I, has the input and output streams indicated in Figure P2.6. The waste stream consists of 82.6% (by weight) HI and the rest H_2O, while the product stream consists of 81.6% (by weight) CH_3I and the rest CH_3OH. Determine to three significant figures whether:

(a) Total mass is conserved.
(b) Total number of moles is conserved.
(c) The number of moles of each type of atom is conserved.

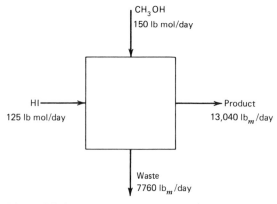

Figure P2.6

(d) The mass of each type of chemical species is conserved.
(e) What do you conclude is happening in the process?

2.7 In a sulfuric acid plant, 100 lbmol/h of a stream consisting of 90 mol % H_2SO_4 in water is mixed with 200 lbmol/h of a stream consisting of 95 mol % H_2SO in water and 200 lbmol/h of a stream consisting of 15 mol % SO_3 in N_2. The result is 480 lbmol/h of a mixed stream containing 170 lbmol/h N_2, 62.5 mol % H_2SO_4, no H_2O, and the rest SO_3. Determine by calculation whether:

(a) Total mass is conserved.
(b) Total number of moles is conserved.
(c) The number of moles of each type of atom is conserved.
(d) Mass of each species is conserved.
(e) If total mass did not balance, what is the most likely explanation?
(f) If total number of moles do not balance, what is the most likely explanation?

2.8 A gas containing 79.1% N_2, 1.7% O_2, and 19.2% SO_2 is mixed with another containing 50% SO_2, 6.53% O_2, and 43.47% N_2 to produce a product gas containing 21.45% SO_2, 2.05% O_2, and 76.50% N_2. All compositions are in mol %. Determine:

(a) How many independent stream variables there are in the problem.
(b) How many material balances can be written and how many of them will be independent.
(c) The ratio in which the streams should be mixed.

2.9 Acetone can be recovered from a carrier gas by dissolving it in a pure water stream in a unit called an absorber. In the flowsheet of Figure P2.9, 200 lb_m/h of a stream containing 20% acetone is treated with 1000 lb_m/h of a

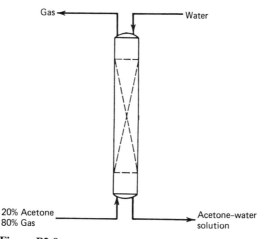

Figure P2.9

pure water stream to yield an acetone-free overhead gas and an acetone–water solution. Assume no carrier gas dissolves in the water.

(a) Determine how many independent stream variables and material balances there are in the problem.

(b) Write out all the material balance equations.

(c) Calculate all unknown stream variables.

2.10 An oil-free vegetable protein meal can be obtained from cottonseed by using hexane to extract the seed oil from cleaned seeds (see Figure P2.10). Given a raw cottonseed feed consisting of (by weight):

$$14\%\quad \text{cellulose material}$$
$$37\%\quad \text{meal}$$
$$49\%\quad \text{oil}$$

calculate the composition of the oil extract obtained when 3 lb_m hexane is used per 1 lb_m raw seeds.

2.11 In a distillation column, an equimolar mixture of ethanol, propanol, and butanol is separated into an overhead stream containing $66\frac{2}{3}\%$ ethanol and no butanol and a bottoms stream containing no ethanol (see Figure P2.11). Calculate the overhead and bottoms streams and compositions for a feed rate of 1000 mol/h.

2.12 The feed to a distillation column contains 36% benzene by weight, the remainder being toluene. The overhead distillate is to contain 52% benzene by weight, while the bottoms are to contain 5% benzene by weight. Calculate:

(a) The percentage of the benzene feed which is contained in the distillate.

(b) The percentage of the total feed which leaves as distillate.

2.13 One method of determining the volumetric flow rate of a turbulently flowing process stream is to inject small, metered amounts of some easily dispersed fluid and then to measure the concentration of this fluid in a sample of the mixed stream withdrawn a suitable distance downstream. Suppose a stream containing 95 mol % butane and 5 mol % O_2 is injected with 16.3 mol/h

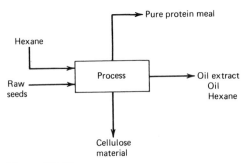

Figure P2.10

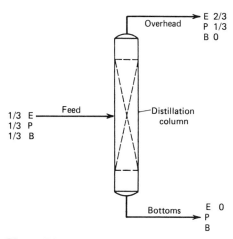

Figure P2.11

O_2. A downstream sample analyzes 10 mol % O_2. Calculate the flow rate of the process stream.

2.14 The spent acid from a nitrating process contains 43% H_2SO_4 and 36% HNO_3. This dilute acid is to be strengthened by the addition of concentrated sulfuric acid containing 91% H_2SO_4 and concentrated nitric acid containing 88% HNO_3. The product is to contain 41.8% H_2SO_4 and 40.0% HNO_3. Calculate the quantities of spent and concentrated acids that should be mixed to yield 100 lb_m of the fortified mixed acid.

2.15 In a plant four process streams are mixed to give 2000 lb_m/h of a single stream. The four inlet stream and exit product stream compositions are shown below.

	Inlet Streams				Exit Stream
	1	**2**	**3**	**4**	
H_2SO_4 (wt %)	80	0	30	10	40
HNO_3 (wt %)	0	80	10	10	27
H_2O (wt %)	16	20	60	72	31
Inerts (wt %)	4	0	0	8	2

Calculate the proportions in which the streams must be mixed if:

(a) 2 lb_m of stream 1 must always be used per 1 lb_m of stream 3 to yield the exit stream with the above composition.

(b) 2 lb_m of stream 1 is used per 1 lb_m of stream 3 and 3 lb_m of stream 2 is used per 1 lb_m of stream 4 to yield the same exit stream mixture.

(c) The exit stream must have no inerts and equal weight percents of the other stream components.

(d) The exit stream inert content can be arbitrary, but the remaining exit stream components must be present in equal weight percents.

2.16 A manufacturer mixes three alloys to yield 10,000 lb_m/h of a desired alloy. The alloy compositions (wt %) are given below:

	Feed Alloys			
Components	**1**	**2**	**3**	**Desired Alloy**
A	60	20	20	25
B	20	60	0	25
C	20	0	60	25
D	0	20	20	25

(a) Calculate the rates at which the three feed alloys should be supplied. Can all required conditions be met?

(b) Suppose a fourth feed alloy consisting of 20% B, 20% C, and 60% D could also be used. Calculate the rates for the four feed alloys.

(c) Suppose the desired alloy is only required to have 40% A and equal amounts of B and C (the D weight fraction is unspecified). In what proportions should the four alloys be mixed to satisfy this requirement, and what will the composition of the desired alloy be?

2.17 A slurry consisting of $CaCO_3$ precipitate in a solution of NaOH and H_2O is washed with an equal mass of a dilute solution of 5% (wt) NaOH in H_2O. The washed and settled slurry which is withdrawn from the unit contains 2 lb_m of solution per 1 lb_m of solid ($CaCO_3$). The clear solution withdrawn from the unit can be assumed to have the same concentration as the solution withdrawn with the solids (see Figure P2.17). If the feed slurry contains equal mass fractions of all components, calculate the concentration of the clear solution.

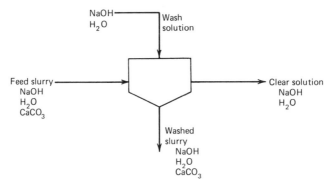

Figure P2.17

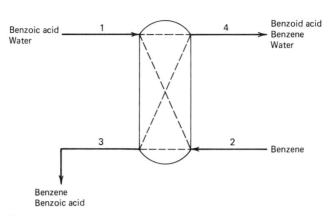

Figure P2.18

2.18 Benzoic acid can be extracted from a dilute water solution by contacting the solution with benzene in a single-stage extraction unit. The mixture will separate into two streams: one consisting of benzoic acid and benzene and another stream consisting of all three constituents as shown in the schematic of Figure P2.18.

Benzene is slightly soluble in water; thus, stream 4 will contain 0.07 kg benzene/kg water. Benzoic acid will distribute between streams 3 and 4 as follows:

$$\left(\frac{\text{Mass benzoic acid}}{\text{Mass benzene}}\right)_{(3)} = 4\left(\frac{\text{Mass benzoic acid}}{\text{Mass (benzene + water)}}\right)_{(4)}$$

The feed solution, stream 1, contains 2×10^{-2} kg acid/kg water and is fed at the rate of 10^4 kg/h:

(a) Show that the problem is underspecified.

(b) Suppose benzoic acid extracted into stream 3 is worth $1/kg and that fresh benzene (stream 2) costs 3¢/kg. Construct a graph of net profit vs. benzene flow and select the optimum benzene flow rate.

2.19 The feed to a two-column fractionating system is 30,000 lb_m/h of a mixture containing 50% benzene (B), 30% toluene (T), and 20% xylene (X). The feed is introduced into column I and results in an overload consisting of 95% benzene, 3% toluene, and 2% xylene. The bottoms from column I are fed to the second column, resulting in an overhead from column II containing 3% benzene, 95% toluene, and 2% xylene (see Figure P2.19). Assume that 52% of the feed appears as overhead in the first column and that 75% of the benzene fed to the second column appears as overhead, calculate the composition and flow of the bottoms stream from the second column.

2.20 The feed to a unit consisting of two columns contains 30% benzene (B), 55% toluene (T), and 15% xylene (X). The overhead stream from the first

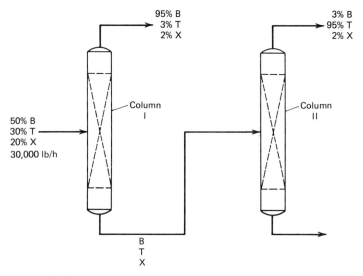

Figure P2.19

column is analyzed and contains 94.4% B, 4.54% T, and 1.06% X. The bottoms from the first column is fed to the second column. In this second column, it is planned that 92% of the original T charged to the unit shall be recovered in the overhead stream and that the T shall constitute 94.6% of the stream. It is further planned that 92.6% of the X charged to the unit shall be recovered in the bottoms from this column and that the X shall constitute 77.6% of that stream (see Figure P2.20). If these conditions are met, calculate:

(a) Analysis of each stream leaving the unit.

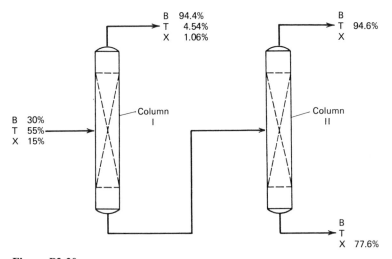

Figure P2.20

(b) Percentage recovery of benzene in the overhead stream from the first column.

2.21 A slurry consisting of TiO_2 precipitate in a salt water solution is to be washed in three stages as shown in the flowsheet of Figure P2.21. If the feed slurry consists of 1000 lb_m/h of 20% TiO_2, 30% salt, and the rest water, calculate the wash water rate to each stage. Assume that:
(a) 80% of the salt fed to each stage leaves in the waste solution.
(b) The stages are operated so that the slurry leaving contains one third solids.
(c) In each stage, the salt concentration in its waste solution is the same as the salt concentration of the solution entrained with the slurry leaving that stage.

2.22 Evaporators are typically used to concentrate solutions by boiling off some of the solvent. To economize on the energy input required, evaporation is often carried out in stages with each stage providing some of the evaporation duty. In the multistage evaporation shown in Figure P2.22, a 50% by weight sugar solution is concentrated to 65% by evaporating an equal amount of water in each of the four stages. With a total input of 50,000 lb_m/h, determine the concentrations of the intermediate streams.

2.23 A mixture of alcohol and ketone containing 40 wt % alcohol is washed with successive water washes to remove the alcohol. The wash water used contains 4 wt % alcohol. The water and ketone are mutually insoluble. The distri-

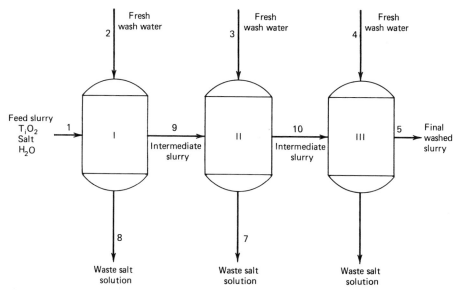

Figure P2.21

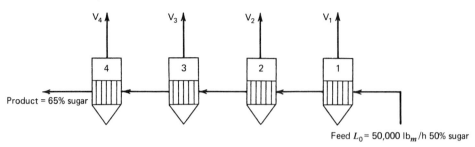

Figure P2.22

bution of alcohol (A) between the ketone (K) and the water streams (W) is given by:

$$\left(\frac{\text{Mass alcohol}}{\text{Mass ketone}}\right)_{(4)} = \frac{1}{4}\left(\frac{\text{Mass alcohol}}{\text{Mass water}}\right)_{(3)}$$

(See Figure P2.23A)

(a) If 150 lb of stream 2 is used per 200 lb feed mixture, calculate the composition of stream 4 after one wash.

(b) If the washing is repeated with 150 lb wash at each stage, how many wash stages are required to remove 98% of the original alcohol (see Figure P2.23B)?

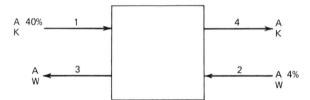

Figure P2.23A

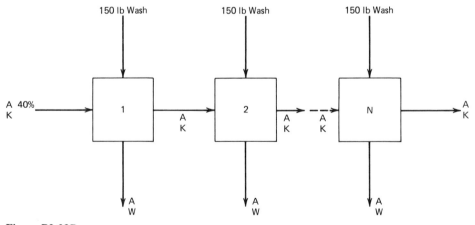

Figure P2.23B

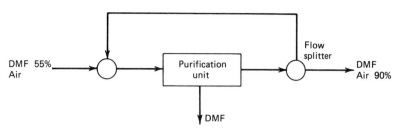

Figure P2.24

2.24 A purification system with recycle is used to recover the solvent DMF from a waste gas containing 55% DMF in air. The product is to have only 10% DMF (see Figure P2.24). Calculate the recycle fraction assuming that the purification unit can remove two thirds of the DMF in the combined feed to the unit.

2.25 A widely employed method of gas purification is to selectively absorb the undesirable constituents of the gas into a specifically selected liquid medium. The liquid medium is subsequently regenerated by chemical or heat treatment to release the absorbed material. In a particular installation, the purification system for the removal of sulfur compounds, designed to operate at a feed rate of up to 820 mol/h, is temporarily subjected to a feed rate of 1000 mol/h. Since the absorption system simply can accommodate only 82% of this flow, it is proposed that the overload be bypassed and that the exit H_2S concentration of the absorption system be reduced far enough so that the mixed exit stream contains only 1% H_2S and 0.3% COS on a mole basis (see Figure P2.25). Calculate all flows in the system. The feed stream consists of (mole basis) 15% CO_2, 5% H_2S, and 1.41% COS, with the remainder being CH_4.

2.26 Suppose the operation of the absorption system of Problem 2.25 is modified so that all of the COS and one mole of CO_2 per mole of H_2S are absorbed as shown in the schematic in Figure P2.26. The feed stream consists of (mole basis) 15% CO_2, 5% H_2S, and 1.41% COS, with the remainder being CH_4. Again, 18% of the feed bypasses the absorption system. Calculate all flows in the system.

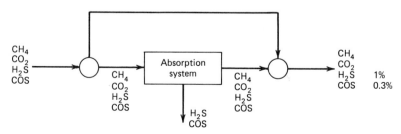

Figure P2.25

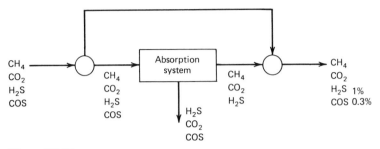

Figure P2.26

2.27 Instant coffee is produced via the flowsheet in Figure P2.27. Roasted, ground coffee is charged with hot water to a large percolator in which water-soluble material is extracted. The extract is spray dried to produce the product, and the wet grounds are partially dewatered prior to disposal as land fill or by incineration. For simplicity, the coffee feed is assumed to consist of solubles,

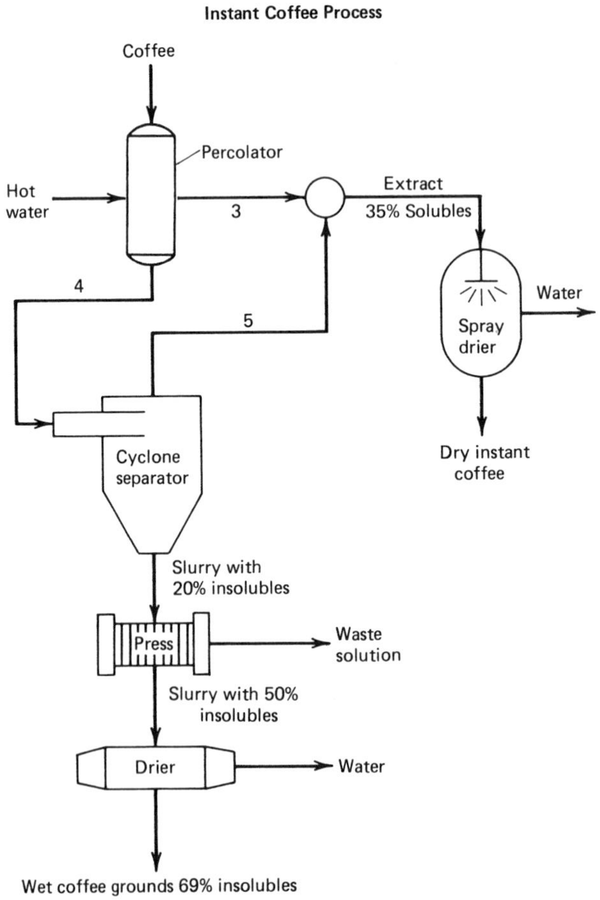

Instant Coffee Process

Coffee

Percolator

Hot
water

3

Extract
35% Solubles

Spray
drier

Water

4

5

Cyclone
separator

Dry instant
coffee

Slurry with
20% insolubles

Press

Waste
solution

Slurry with 50%
insolubles

Drier

Water

Wet coffee grounds 69% insolubles

Figure P2.27

insolubles, and no water. The standard charge consists of 1.2 lb water per lb coffee. As a reasonable approximation, it may be assumed that the solubles-to-water ratio in the two streams leaving the percolator is the same. Similarly for the separator and the press, but not for the drier.

(a) Given the above information and the indicated compositions, is the problem completely specified?

(b) Suppose we are not interested in streams 3, 4, and 5 so that the percolator, separator, and mixer can be lumped into a single "black box." Calculate the ratio of solubles recovered to solubles lost in the waste stream.

2.28 In the plant flowsheet considered in Problem 2.27, recovery of solubles as product is quite low. In an effort to improve the recovery, suppose the waste solution from the press is recycled back to the percolators. However, to reduce the possible release of bitter-tasting material during pressing, the dewatering rate is decreased so that a slurry of 40% insolubles is generated, as shown in the flowsheet in Figure P2.28. To handle the higher water content slurry, the drier operation is also adjusted to produce a 62.5% insolubles coffee grounds. Calculate the recovery ratio resulting from this modification. Assume that the feed coffee composition is 0% water and 32.7% insolubles and allow the feed water-to-coffee ratio to be variable. The solubles-to-water ratios in both streams leaving the press are the same.

2.29 Oilseed protein sources include soybean, cottonseed, peanut, sunflower, copra, rapeseed, sesame, safflower, castor, and flax. Commonly, the separation of the oil from the protein meal is performed by solvent extraction. The analysis of cottonseed is 4% hull, 10% linter, 37% meal, and 49% oil. During the extraction step, about 2 lb_m solvent, hexane, must be used per 1 lb_m clean seeds processed (see Figure P2.29). For each ton of raw seeds to be processed, determine the amount of oil and oil-free meal produced

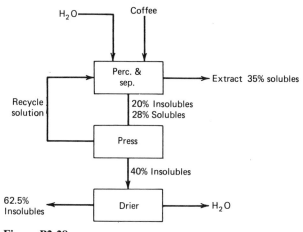

Figure P2.28

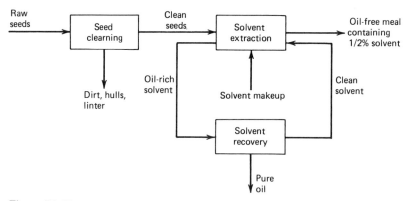

Figure P2.29

and the amount of hexane that must be recycled through the solvent extraction unit.

2.30 Consider the absorber–stripper system shown in Figure P2.30. In this system, a stream (1) containing 30% CO_2, 10% H_2S, and an inert gas (I) is scrubbed of H_2S and CO_2 by using a solvent to selectively absorb the H_2S and CO_2. The resulting stream (5) is fed to a flash unit in which the pressure is reduced, and as a result, some of the CO_2, H_2S, and some of the solvent are separated as an overhead stream (6). The remaining solution is split, with half being returned to the absorber and half sent to a stripping unit. In the stripping unit, the pressure is reduced further, resulting in an overhead stream (10) consisting of 30% solvent as well as unknown amounts of CO_2 and H_2S. The bottoms stream from the stripper (9), consisting of pure solvent, is recycled back to the absorber after mixing with some additional pure solvent which makes up for the solvent lost in the flash and stripper overheads. Suppose the absorber is operated so that the outlet overhead stream (2) from the unit contains no H_2S and only 1% CO_2. Suppose further that the stripper feed solution (8) contains 5% CO_2 and that the overhead from the flash (6) contains 20% solvent. Finally, the flash is operated so that 25% of the CO_2 and 15% of the H_2S in stream 1 are released in the overhead (6).

(a) Construct a degree-of-freedom table, determine the solution order, and solve the problem.

(b) Suppose that instead of specifying the split to be 50%, we specified the CO_2 composition in the stripper overhead to be 40%. How would this affect the degree of freedom of the process? How would this affect the degree of freedom of the *overall* balances?

2.31 As shown in the flow diagram, Figure P2.31, a countercurrent thickener consisting of three stages is used in a Kraft paper mill to wash a "white mud" consisting of 35% solids ($CaCO_3$) and 17% NaOH in water. Two wash streams are used, the first (stream 5) contains 4% suspended solids, 6% NaOH, and the rest H_2O; the second (stream 8) contains no solids, 2%

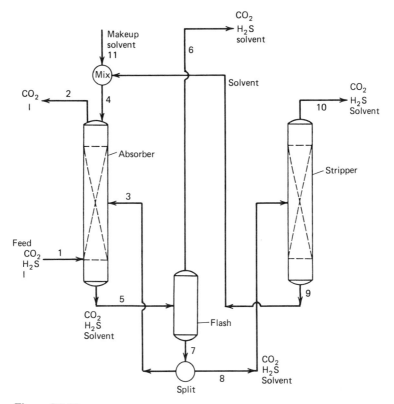

Figure P2.30

NaOH, and the rest H_2O. The clear liquids from stage I and stage II contain 0.5% suspended solids; the clear liquid from stage III contains 0.4% solids. The flow of wash liquid to stage II (stream 5) is 1.5 times the flow of the feed sludge (stream 1), and the washed sludge from stage II contains one third solids. The washed sludge from the third stage (stream 9) contains 32.5% solids and 2.5% NaOH, and the flow rates of streams 7 and 9 are equal. All compositions are given on a mass basis. Assume that in each stage the clear solution and the solution contained in the washed sludge have the same concentration.

(a) Construct a degree-of-freedom table and determine whether the problem is correctly specified.
(b) Suppose the degree-of-freedom table prior to selecting a basis shows the entries given below:

Mixer	I	II	III	Overall
6	4	4	1	2

Where should the basis be selected? Which balance set should be solved first?

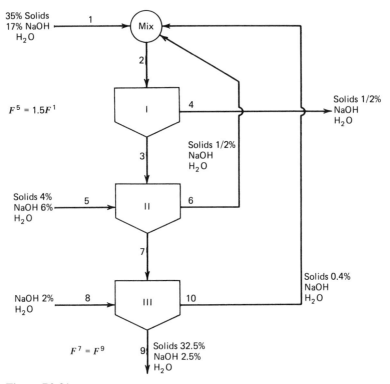

Figure P2.31

(c) Calculate the concentration of NaOH in stream 4. Explain your solution strategy and show all calculation steps.

(d) Suppose we assume that the solids content of the recycle streams (10 and 6) is negligible. How does this affect the degree of freedom of the problem?

(e) Suppose the specified flow ratio of streams 5 and 1 is replaced by the NaOH composition of stream 7 ($w_{NaOH}^7 = 0.10$). How does this affect the degree of freedom of the problem? How does this affect the solution of the problem?

2.32 In the process for making cellulose acetate from cotton linters, acetic acid, and acetic anhydride, a residual dilute acetic acid stream is produced. Process profitability requires that this dilute acid be reclaimed for further process use by purification and concentration. An accepted scheme for accomplishing this involves solvent extraction of acetic acid with ether. The mixture of dilute acid and ether separates into two phases: an ether phase rich in acetic acid and a water phase which contains only small amounts of residual acid, dissolved ether, and other impurities. The ether phase is then distilled to separate the solvent and the acetic acid. The resulting acetic acid is subjected to further distillation to reduce the water content and yield the final high-purity reactant. The either residual in the water phase is reclaimed in a

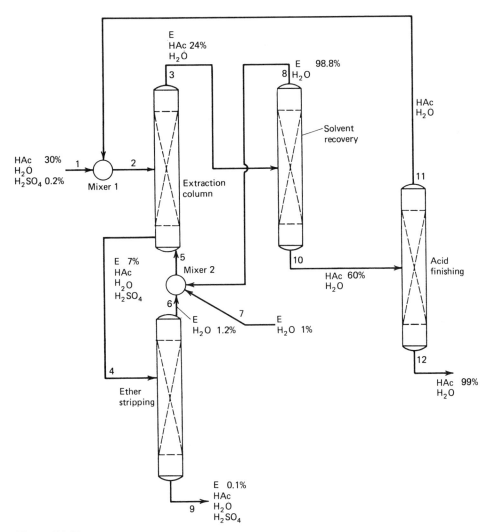

Figure P2.32

stripping column while the bottoms is disposed of as waste. In the process shown in the flow diagram, Figure P2.32, dilute acid consisting of 30% acetic acid, 0.2% H_2SO_4, and the rest water is treated to yield a reclaimed 99% pure acetic acid. Assume that 67.5% of the acetic acid fed to the finishing column is recovered as product and that 2.3 lb_m dilute acid is fed to the process per 1 lb_m dilute acid recycled. The composition given on the flowsheet are all in mass %.

(a) Carry out the degree-of-freedom analysis to show that the process is correctly specified.
(b) Determine where the basis ought to be located and with which unit the calculations ought to be initiated.

    (c)   Calculate the remaining composition of stream 3.

    (d)   How would you go about calculating all other streams?

2.33 Consider the system of Problem 2.32. Suppose that instead of specifying the ether composition of stream 4, the ether composition of stream 3 is specified. All other specifications remain as in Problem 2.32. How is the degree of freedom affected? How is the problem solution affected?

2.34 Consider the modification of Example 2.21 shown in Figure P2.34. The feed composition remains

$$C_1 \quad 20\%$$
$$C_2 \quad 25\%$$
$$C_3 \quad 40\%$$
$$\text{and } C_4 \quad 15\%$$

The stream leaving unit 2 and that leaving unit 6 are both split 50%.

    (a)   Is the problem correctly specified?

    (b)   How is the solution procedure used in Example 2.21 affected?

    (c)   Suppose instead of the input to unit 4 being specified (i.e., $C_4 = 30\%$), the input to unit 3 is set at $C_2 = 20\%$. What changes will be brought about?

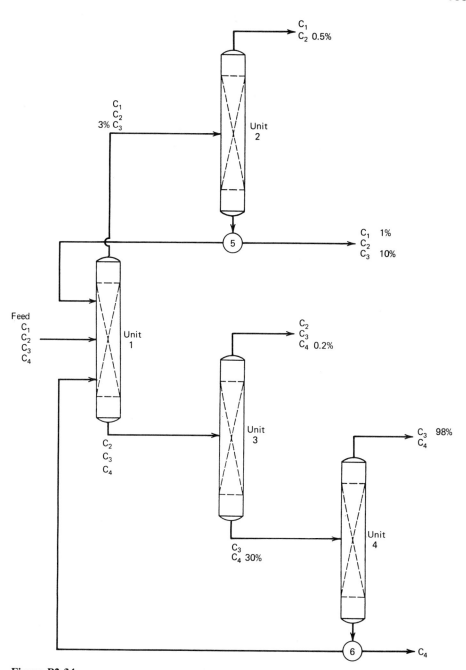

Figure P2.34

CHAPTER
3
Species Balances in Reacting Systems

The major hypothesis which underlies Chapter 2 is that chemical species are conserved when they undergo purely physical processes. This hypothesis is intuitive, reasonable, and readily verifiable empirically. It represents a level of sophistication in applied physics and mathematics which was attained by Western Civilization centuries, if not a millennium, ago. However, the formulation of species balance equations which could account for the changes taking place in a chemically reacting system requires a considerably higher degree of scientific sophistication. In particular, it requires the establishment of molecular formulas for chemical compounds, the understanding of chemical transformation in terms of the rearrangement of atoms, and the representation of chemical reactions in terms of the now familiar stoichiometric reaction equations. These concepts really only began to be formulated in the late 1700s. In fact, the first exposition of an atomic theory of chemistry is generally credited to the English chemist John Dalton in a work dated 1803.

Dalton showed that much of the existing experimental evidence on chemical reactions as well as many of the partially formulated explanations could be unified and rationalized if chemical substances could be assumed to consist of fixed ratios of atoms of different types of elements. He thus established a logical basis for the *law of constant proportionality*, which states that when given substances combine chemically to produce a given product, they always combine in the same ratios; and the *law of multiple proportions*, which states that when one substance combines with another in more than one porportion, these proportions bear a simple ratio to one another. These laws, combined with Avogadro's concept of molecule (1811) and the tacit assumption that atoms of each type of element must be conserved, led to the development of the symbolism, now familiar to every student of general chemistry, of expressing what happens in a chemical reaction in terms of a stoichiometric equation. The stoichiometric equation, of course, is essential to constructing species balances in the presence of chemical reactions. Thus, the reacting species balance has behind it a rich, considerable, but rather short scientific history.

In this chapter, we are concerned with both how to use the information contained in stoichiometric equations in constructing species balance equations as well as how to use the resulting balance equations themselves to solve material balance problems. We begin by studying the case in which only a single reaction occurs within the system (Section 3.1), then extend the results obtained to include multiple simultaneous chemical reactions (Section 3.2), and then conclude by developing a general algebraic approach for analyzing multiple chemical reactions (Section 3.3). Concurrently with this development we extend the degree-of-freedom analysis of the previous chapter to accommodate reacting systems and continue our further investigation of strategies for solving multiunit material balance problems.

3.1 SPECIES BALANCES WITH SINGLE CHEMICAL REACTION

As noted earlier, the principle of conservation of mass asserts that in an open steady-state system, the input and output mass (or molar) rates of each type of element in the system must balance regardless of whether chemical reactions are taking place. However, since we recognize that the chemical reaction process involves a rearrangement of atoms and molecules to form various different molecular compounds, it follows that the input and output mass (or molar) rates of each chemical species in a reacting system are not balanced. Rather, some chemical species will be depleted by reaction while others will be produced. Consequently, in the presence of chemical reaction, the species material balance equation

$$\text{Molar rate of species } s \text{ in} = \text{molar rate of species } s \text{ out}$$
$$N_s^{\text{in}} = N_s^{\text{out}}$$

is no longer valid.

We define the difference between the output and input rates, R_s, to be the molar production rate of species s:

$$R_s = N_s^{\text{out}} - N_s^{\text{in}}$$

or

$$R_s = \frac{F_s^{\text{out}} - F_s^{\text{in}}}{M_s}$$

where M_s is the molecular weight of species s.

Example 3.1 In a continuous process for synthesizing ammonia via the reaction $N_2 + 3H_2 \rightleftharpoons 2NH_3$, 40 mol/h H_2 and 12 mol/h N_2 are fed to the catalytic reactor to yield an output stream of 8 mol/h N_2, 28 mol/h H_2, and 8 mol/h NH_3. From the definition, the production rates of each species will be:

$$R_{NH_3} = N_{NH_3}^{\text{out}} - N_{NH_3}^{\text{in}} = 8 - 0 \quad = 8 \text{ mol/h}$$
$$R_{N_2} = N_{N_2}^{\text{out}} - N_{N_2}^{\text{in}} \quad = 8 - 12 \quad = -4 \text{ mol/h}$$
$$R_{H_2} = N_{H_2}^{\text{out}} - N_{H_2}^{\text{in}} \quad = 28 - 40 = -12 \text{ mol/h}$$

The reactants are depleted and accordingly have a negative production rate, while the product NH_3 has a positive production rate.

With the introduction of the species production rates, the species material balances in the presence of chemical reactions therefore become

$$N_s^{out} = N_s^{in} + R_s \qquad (3.1)$$

or

$$F_s^{out} = F_s^{in} + M_s R_s \qquad (3.2)$$

Apparently, one additional material balance variable, namely, the production rate, is required for each species. Note, however, that following Dalton's laws of proportionality for a given chemical reaction, the rates of production or depletion are not independent but rather must all be proportional to one another, with the constants of proportionality determined by the stoichiometric coefficients of the reaction in question. Hence, if one rate is given, all the others can be easily calculated.

Example 3.2 From the given reaction stoichiometry for ammonia synthesis,

$$N_2 + 3H_2 \rightleftharpoons 2NH_3$$

it follows that

$$\frac{R_{NH_3}}{-R_{N_2}} = \frac{2}{1} \qquad \frac{R_{H_2}}{R_{N_2}} = \frac{3}{1} \qquad \frac{R_{NH_3}}{-R_{H_2}} = \frac{2}{3}$$

Suppose the input rates of N_2, H_2, and NH_3 are specified as 12, 40, and 0 mol/h, respectively, and the rate of production of N_2 is given as -4 mol/h. Then

$$R_{NH_3} = 2(-R_{N_2}) = 8 \text{ mol/h}$$
$$R_{H_2} = 3(R_{N_2}) \quad = -12 \text{ mol/h}$$

Consequently, from the material balances,

$$N_{NH_3}^{out} = N_{NH_3}^{in} + R_{NH_3} = 0 + 8 \quad = 8 \text{ mol/h}$$
$$N_{N_2}^{out} = N_{N_2}^{in} + R_{N_2} \quad = 12 - 4 = 8 \text{ mol/h}$$
$$N_{H_2}^{out} = N_{H_2}^{in} + R_{H_2} \quad = 40 - 12 = 28 \text{ mol/h}$$

We conclude that in the presence of a single chemical reaction involving S chemical species, the production rate of any one single species suffices to determine the production rates of the remaining $S - 1$ species. Hence, the material balance equations will, in addition to the usual stream variables, effectively include one additional independent variable, namely, the production rate of a selected reference species.

3.1.1   The Rate of Reaction Concept

Although formulation of the balance equations in terms of the production rate of a selected reference species is a perfectly viable approach, it is much more desirable to define a measure of the production rate attained via a given reaction which is independent of the species selected as reference. This can be achieved in a natural way as shown in the following example.

Example 3.3   In Example 3.2, we observed that from the chemical reaction equation it follows that

$$\frac{R_{NH_3}}{-R_{N_2}} = \frac{2}{1} \qquad \frac{R_{H_2}}{R_{N_2}} = \frac{3}{1} \qquad \frac{R_{NH_3}}{-R_{H_2}} = \frac{2}{3}$$

These relations can conveniently be rewritten as

$$\frac{R_{NH_3}}{2} = \frac{R_{H_2}}{-3} = \frac{R_{N_2}}{-1}$$

In this form, it is quite apparent that the rate of production of NH_3 divided by its stoichiometric coefficient and the rate of depletion of the reactants N_2 and H_2 divided by their respective stoichiometric coefficient are constants which are the same for each species. This fact can be confirmed by forming these ratios for the rates calculated in Example 3.1. In particular, $R_{NH_3}/2 = 4$, $R_{H_2}/-3 = -12/-3 = 4$, and $R_{N_2}/-1 = 4$ mol/h. The common value of these ratios we henceforth call the rate of the chemical reaction.

The rate of reaction concept can be generalized by introducing a general notation for the stoichiometric coefficients of a chemical reaction. In particular, let σ_s represent the stoichiometric coefficient of species s in the chemical reaction with the convention that the coefficient is assigned a *negative* sign for reactants and a *positive* sign for products. Then, the *rate of reaction r* of any given chemical reaction is defined as

$$r = \frac{R_s}{\sigma_s} \qquad s = 1, \ldots, S$$

From the definition, the rate of production of any species s involved in the reaction can be obtained by multiplying the rate of reaction by the stoichiometric coefficient of that species. That is,

$$R_s = \sigma_s r \qquad s = 1, \ldots, S$$

Moreover, it follows that the species mole balance equation, eq. (3.1), can always be written as

$$N_s^{out} = N_s^{in} + \sigma_s r \qquad s = 1, \ldots, S \qquad (3.3)$$

and the species mass balance eq. (3.2) as

$$F_s^{out} = F_s^{in} + \sigma_s M_s r \qquad s = 1, \ldots, S \qquad (3.4)$$

In this general form, it is quite apparent that in the presence of a chemical reaction only one additional unknown variable, namely, the rate of reaction r, needs to be introduced in order to account for the changes taking place in the stream flow rates. Note further that the rate of reaction will have the units of mol/h and hence enters the balance equation, eq. (3.3) just like a species flow rate. Consequently, the balance equations are homogeneous not only in the flow rates, as before, but also in the rate of reaction.

Example 3.4   Consider again the conditions given in Example 3.2. That is, 12, 40, and 0 mol/h N_2, H_2, and NH_3, respectively, are fed to a reactor in which ammonia is produced via the reaction

$$N_2 + 3H_2 \rightleftharpoons 2NH_3$$

If the output rate of N_2 is given as 8 mol/h, calculate the output rates of the remaining species.

Solution   First, following the sign convention for reactants and products, note that

$$\sigma_{N_2} = -1 \qquad \sigma_{H_2} = -3 \qquad \sigma_{NH_3} = 2$$

Thus, the species mole balances, eqs. (3.3), for this system will be:

$$N_{N_2}^{out} = N_{N_2}^{in} + (-1)r = 12 - r = 8$$
$$N_{H_2}^{out} = N_{H_2}^{in} + (-3)r = 40 - 3r$$
$$N_{NH_3}^{out} = N_{NH_3}^{in} + 2r \;\; = 0 + 2r$$

From the first balance equation, it is evident that

$$r = 4 \text{ mol/h}$$

Consequently, the remaining two balances yield

$$N_{H_2}^{out} = 40 - 3(4) = 28 \text{ mol/h}$$
$$N_{NH_3}^{out} = 0 + 2(4) \;\; = 8 \text{ mol/h}$$

The above example, in addition to illustrating the use of the general balance equations, serves to bring out two points. First, it ought to be apparent that the numerical value of the rate of reaction, although independent of any species in the reaction, is dependent upon the numerical values which are assigned to the stoichiometric coefficients. Since all the stoichiometric coefficients in a given reaction can be multiplied or divided by any arbitrary factor without affecting the validity of the reaction stoichiometry, it follows that, when reporting values of the rate of a reaction, the precise form of the chemical reaction equation must also be indicated. Nonetheless, the simplicity obtained by using the rate of reaction concept more than offsets this slight inconvenience.

Example 3.5 Suppose the calculations of the previous example are repeated but the reaction equation is written as

$$\tfrac{1}{2}N_2 + \tfrac{3}{2}H_2 \rightleftharpoons NH_3$$

Solution In this case,

$$\sigma_{N_2} = -\tfrac{1}{2} \qquad \sigma_{H_2} = -\tfrac{3}{2} \qquad \sigma_{NH_3} = +1$$

Thus, the species balance equations become:

$$8 = N_{N_2}^{out} = 12 + \left(-\tfrac{1}{2}\right)r$$

$$N_{H_2}^{out} = 40 + \left(-\tfrac{3}{2}\right)r$$

$$N_{NH_3}^{out} = 0 + (+1)r$$

The N_2 balance now yields:

$$r = 8 \text{ mol/h}$$

which is double the rate calculated in the previous example. However, note that all the output rates remain unchanged, namely,

$$N_{H_2}^{out} = 40 - \left(\tfrac{3}{2}\right)8 = 28 \text{ mol/h}$$

$$N_{NH_3}^{out} = 0 + (1)8 = 8 \text{ mol/h}$$

Thus, although the numerical value of the rate of reaction depends upon the numerical values assigned to the stoichiometric coefficients, the output rates will remain unchanged since these depend only upon the ratios of the stoichiometric coefficients.

The second feature of the balance equations, eq. (3.3) or (3.4), which must be noted is that the rate of reaction r generally serves as an intermediate variable in the calculations. Thus, as in Examples 3.4 and 3.5, typically the input and output rates of some species s are given and the species balance is then used to calculate the rate of reaction. Then, having calculated r, and knowing either the input or the output rate for any other species, the unknown flow rate is calculated. In subsequent courses in the Chemical Engineering curriculum, it is shown that the rate of reaction can often be calculated from the temperature, pressure, composition, and flow patterns of the materials in the reactor, independently of the species balance equations. Such calculations will not be our concern. Rather, in keeping with our macroscopic view of process flowsheets, the rate of reaction will always either be calculated from one of the balance equations or else be specified indirectly via one of the traditional measures of the progress of a chemical reaction discussed in the following section.

3.1.2 The Limiting Reactant and Conversion

A familiar and commonly used measure of the progress of a chemical reaction is *fractional conversion*, or simply *conversion* of a species. As traditionally defined, the conversion of reactant s, denoted by X_s, is the fraction of that reactant which is depleted by reaction:

$$X_s = \frac{N_s^{in} - N_s^{out}}{N_s^{in}}$$

The specification of the conversion of a species amounts to the imposition of a *relationship* between the input and output flow rates of the species in question. This relationship can, together with the corresponding species balance equation, be used to calculate the rate of the reaction. Thus, since

$$N_s^{out} - N_s^{in} = \sigma_s r$$

and, from the definition of the conversion of species s,

$$N_s^{in} X_s = N_s^{in} - N_s^{out}$$

it follows upon substitution that

$$r = \frac{N_s^{in} X_s}{-\sigma_s} \tag{3.5}$$

Hence, given the conversion, the rate of reaction can always be calculated and the balance calculations completed using this rate.

Example 3.6 Modern processes for the production of nitric acid are based upon the oxidation of ammonia synthesized via the Haber reaction. The first step in the oxidation process is the reaction of NH_3 with O_2 over a platinum catalyst to produce nitric oxide. This reaction follows the stoichiometric equation

$$4NH_3 + 5O_2 \rightleftharpoons 4NO + 6H_2O$$

Under a given set of reactor conditions, 90% conversion of NH_3 is obtained with a feed of 40 mol/h NH_3 and 60 mol/h O_2. Calculate the output rates of all species from the reactor.

Solution Following our sign conventions,

$$\sigma_{NH_3} = -4 \qquad \sigma_{O_2} = -5 \qquad \sigma_{NO} = +4 \qquad \text{and} \qquad \sigma_{H_2O} = +6$$

Then, from eq. (3.5),

$$r = \frac{N_{NH_3}^{in} X_{NH_3}}{-\sigma_{NH_3}} = \frac{40(0.9)}{-(-4)} = 9 \text{ mol/h}$$

With this value of the rate of reaction, all output rates can be calculated from the species balance equations:

$$N_{NH_3}^{out} = N_{NH_3}^{in} - 4r = 40 - 4(9) = 4 \text{ mol/h}$$

$$N_{O_2}^{out} = N_{O_2}^{in} - 5r = 60 - 5(9) = 15 \text{ mol/h}$$

$$N_{NO}^{out} = N_{NO}^{in} + 4r = 0 + 4(9) = 36 \text{ mol/h}$$

$$N_{H_2O}^{out} = N_{H_2O}^{in} + 6r = 0 + 6(9) = 54 \text{ mol/h}$$

Note that since the fractional conversion is always given as a positive fraction, it follows that

$$N_s^{in} - N_s^{out} > 0$$

Consequently, the conversion is defined only for *reactants*. Moreover, since the conversion is defined as a relationship between the input and output flow rates of a species, it must always be referred to a particular reactant. If the conversion is stated without specifying a particular reactant, then it is by convention assumed that the conversion is to be referred to the limiting reactant. By definition, the *limiting reactant* is that reactant which is first depleted as the reaction is allowed to proceed.

Consider the species balance for reactant s,

$$N_s^{out} = N_s^{in} + \sigma_s r$$

Since s is a reactant, by convention $\sigma_s < 0$. Hence, as the reaction proceeds and r increases, a value of r will be reached at which $N_s^{out} = 0$. This value can be calculated as

$$r = \frac{N_s^{in}}{-\sigma_s}$$

Each reactant will have a characteristic value of r at which it will be depleted. Clearly, the smallest of these characteristic reaction rates will be the value of r at which the reaction will stop since some reactant will no longer be available. The reactant with the smallest ratio of $-N_s^{in}/\sigma_s$ will therefore be the limiting reactant. This simple formula thus provides a convenient criterion with which to identify the limiting reactant.

Example 3.7 Consider the reaction of Example 3.6 and suppose that 80% conversion is obtained with an equimolar mixture of ammonia and oxygen fed at the rate of 100 mol/h. Calculate the output rates of all species.

Solution The conversion is not referred to either reactant. Hence, it must refer to the limiting reactant. Since the feed is an equimolar mixture, the reactant input rates are

$$N_{NH_3}^{in} = 50 \text{ mol/h}$$

and

$$N_{O_2}^{in} = 50 \text{ mol/h}$$

Since

$$\frac{N_{NH_3}^{in}}{-\sigma_{NH_3}} = \frac{50}{4} > \frac{50}{5} = \frac{N_{O_2}^{in}}{-\sigma_{O_2}}$$

oxygen is the limiting reactant. From eq. (3.5), the rate of reaction is thus

$$r = \frac{X_{O_2}N_{O_2}^{in}}{-\sigma_{O_2}} = \frac{0.8(50)}{5} = 8 \text{ mol/h}$$

The species balance equations then yield the output rates

$$N_{NH_3}^{out} = 50 - 4(8) = 18 \text{ mol/h}$$

$$N_{O_2}^{out} = 50 - 5(8) = 10 \text{ mol/h}$$

$$N_{NO}^{out} = 0 + 4(8) = 32 \text{ mol/h}$$

$$N_{H_2O}^{out} = 0 + 6(8) = 48 \text{ mol/h}$$

If it had been assumed that the conversion given was that of ammonia, the output rates would have been 10, 0, 40, and 60 mol/h, respectively.

3.1.3 Degree-of-Freedom Analysis

In the preceding sections, we have determined what form the species material balance equations must take in the presence of a chemical reaction, what new material balance variables are introduced, and what new types of relations typically are specified. These elements can now be combined, as in the nonreacting case, to calculate the degree of freedom of the resulting algebraic problem which in turn can be used to determine a strategy of solution.

Consider first the species balance equations, eq. (3.3) or (3.4). If we adopt the convention that $\sigma_s = 0$ for any species which is present but is not involved in the reaction (i.e., is *inert*), then, as before, it follows that one balance equation can be written for each species present. Moreover, as was the case with nonreacting systems, all S individual species balances can be summed to yield the total balance, either in moles,

$$N^{out} = N^{in} + r \sum_{s=1}^{S} \sigma_s \tag{3.6}$$

or in mass units,

$$F^{\text{out}} = F^{\text{in}} + r \sum_{s=1}^{S} \sigma_s M_s \tag{3.7}$$

Paralleling the argument used in Section 2.1.2, it again follows that the total balance can be substituted for one of the species balances but that only S of these $S + 1$ equations will be independent. Hence, as before, the number of independent equations available for solution will be equal to the number of species.

The number of material balance variables will consist of the sum of the number of species associated with each input and output stream plus, in the presence of a chemical reaction, one additional new variable, the rate of reaction. The number of specified stream variables and relations will similarly be counted as before, with the exception that often a new relation, the conversion, may be imposed. Finally, since the species balance equations are homogeneous in the flows and the rate of reaction, if none is specified, any of the flows or the reaction rate can be assigned an arbitrary value as basis for the calculation.

In summary, the only noticeable modification in the degree-of-freedom accounting which needs to be made in the presence of a chemical reaction is to count one additional variable, the rate of reaction.

Example 3.8   The stoichiometric H_2–N_2 mixture (75% H_2–25% N_2) for synthesizing ammonia is made by mixing "producer" gas (78% N_2–20% CO–2% CO_2) and "water" gas (50% H_2–50% CO). The carbon monoxide, which acts as synthesis catalyst poison, is removed by reacting this gas mixture with steam to form carbon dioxide and hydrogen via the shift reaction

$$CO + H_2O \rightleftharpoons CO_2 + H_2$$

The CO_2 is subsequently removed by scrubbing with a suitable absorbant. Assuming that all compositions are in mol % and that just enough steam is added to completely convert all the CO, calculate the ratio in which the producer and water gas streams ought to be mixed.

Solution   The flowsheet of the system with the specified information is shown in Figure 3.1. The flowsheet involves nine stream variables plus the rate of the shift reaction. The system contains five species, the four reacting species represented in the reaction equation and the inert, N_2. Thus, it will be possible to write five independent material balance equations. In addition, four independent compositions are specified. Thus, if a basis is chosen for the calculations, the degree of freedom of the system is $10 - 5 - 4 - 1 = 0$. The problem is correctly specified. Note that the complete conversion of CO is accounted for by indicating no CO in the product stream.

With $\sigma_{N_2} = 0$, $\sigma_{CO} = -1$, $\sigma_{H_2O} = -1$, $\sigma_{CO_2} = +1$, and $\sigma_{H_2} = +1$, the species balance equations are

N_2 mole balance: $0.25N^5 = 0.78N^1$

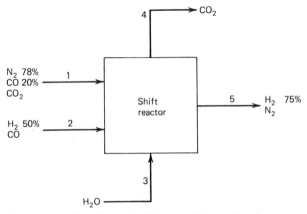

Figure 3.1 Flowsheet for Example 3.8, synthesis gas preparation.

CO mole balance: $\qquad 0 = 0.2N^1 + 0.5N^2 - r$

H_2O mole balance: $\qquad 0 = N^3 - r$

CO_2 mole balance: $\qquad N^4 = 0.02N^1 + r$

H_2 mole balance: $\quad 0.75N^5 = 0.5N^2 + r$

Choosing as basis 100 mol/h of stream 1, the N_2 balance immediately yields

$$N^5 = 312 \text{ mol/h}$$

Then, adding the CO and the H_2 balances, r is eliminated and

$$N^2 = 0.75(312) - 0.2(100) = 234 - 20 = 214 \text{ mol/h}$$

The remaining streams are determined by using the CO balance to calculate r:

$$r = 0.2(100) + 0.5(214) = 127 \text{ mol/h}$$

and the H_2O and CO_2 balances to calculate

$$N^3 = 127 \text{ mol/h} \quad \text{and} \quad N^4 = 2 + 127 = 129 \text{ mol/h}$$

Note that the reaction rate only serves as an intermediate variable which is used in the calculation of the remaining streams.

The preceding analysis can readily be extended to multiple-unit systems. As in the nonreacting case, a set of balances can be made around each individual unit as well as around the process as a whole. The main difference to be noted is that if one of the units in the process is a reactor, then, when overall balances are written, the overall process must be viewed as a reactor. In particular, an overall reaction rate must be accounted for as a variable and the reaction stoichiometry specified for the reactor used. These modifications are illustrated in the next example.

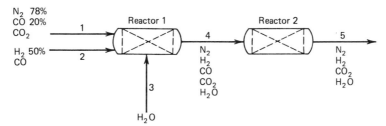

Figure 3.2 Flowsheet for Example 3.9, two-stage shift conversion reactor.

Example 3.9 Because of the essentially complete conversion of CO which must be achieved in order to avoid poisoning the synthesis catalyst, the shift conversion reaction of Example 3.8 is normally conducted in two separate reactor beds each of which contains a different type of catalyst. The first reactor, which does the bulk of the conversion, uses a cheaper catalyst; the second reactor uses a more expensive catalyst which can achieve nearly complete conversion of the remaining CO.

In the process shown in Figure 3.2, a mixture of producer and water gas with compositions as given in Example 3.8 is reacted with steam to produce a product stream containing H_2 and N_2 in a 3:1 ratio. If the steam flow rate is adjusted to be twice the total dry gas flow rate and if 80% conversion occurs in the first reactor stage, calculate the composition of the intermediate stream.

Solution In determining the degree of freedom for the system, observe that reactor 1 and reactor 2 both involve the shift reaction. Thus, each will have associated with it a rate of reaction; hence, the entire process itself must involve these two variables. However, in considering overall balances, the entire system is viewed as a single reactor in which the shift reaction occurs. Thus, the overall balances will have only a single rate of reaction associated with them. The remaining entries in the degree-of-freedom table, shown in Figure 3.3, are obtained in the usual fashion.

	Reactor 1	Reactor 2	Process	Overall Balances
Number of variables	12	10	17	11
Number of balances	5	5	10	5
Number of specified compositions	3	—	3	3
Number of relations				
Excess H_2O	1	—	1	1
Conversion	1	—	1	—
$H_2 : N_2$ ratio	—	1	1	1
Degree of freedom	2	4	1	1
Basis			-1	-1
			0	0

Figure 3.3 Example 3.9, degree-of-freedom table.

The process is completely specified, and it appears that solution should be initiated with the overall balances. Choosing as basis 100 mol/h of stream 1, the overall mole balances are:

N_2 balance: $\qquad N_{N_2}^5 = 0.78(100) = 78$ mol/h

CO balance: $\qquad 0 = 0.2(100) + 0.5N^2 - r$

H_2O balance: $\qquad N_{H_2O}^5 = N^3 - r$

CO_2 balance: $\qquad N_{CO_2}^5 = 0.02(100) + r$

H_2 balance: $\qquad N_{H_2}^5 = 0.5N^2 + r$

In addition, two relations apply: the H_2-to-N_2 ratio

$$N_{H_2}^5 = 3N_{N_2}^5 = 3(78) = 234 \text{ mol/h}$$

and the steam ratio

$$N^3 = 2(N^1 + N^2)$$

If the H_2 and CO balances are added to eliminate r, the result can be solved for N^2:

$$N^2 = 234 - 20 = 214 \text{ mol/h}$$

This result, when substituted into the CO balance, yields

$$r = 20 + 107 = 127 \text{ mol/h}$$

Next, N^3 is calculated from the steam ratio:

$$N^3 = 2(100 + 214) = 628 \text{ mol/h}$$

Finally, the CO_2 and H_2O balances give

$$N_{CO_2}^5 = 129 \text{ mol/h}$$

$$N_{H_2O}^5 = 628 - 127 = 501 \text{ mol/h}$$

This completes the solution of the overall balances. Either the reactor 1 or the reactor 2 balances could now be used to solve for the remaining unknown stream, stream 4. Note from the degree-of-freedom table that reactor 1 had a degree of freedom of 2. Consequently, if the flow rates of any two of the three streams 1, 2, or 3 were known (the third is fixed by the specified steam ratio), the reactor 1 balances would be completely specified. On the other hand, the degree of freedom of reactor 2 was 4. This indicates that if three variables associated with stream 5 were known (the fourth is already fixed by the N_2-to-H_2 ratio) and one additional variable associated with stream 4 were known, then the reactor 2 balances would be completely specified. As a result of the solution of the overall balances, the flow rates of streams 1, 2, 3, and 5 as well as the complete composition of stream 5 are known. Hence, the degree of freedom of reactor 2 will be reduced to 1, while that of reactor 1 will become zero. Clearly, the reactor 1 balances should be used to complete the solution of the problem.

Proceeding with these balances, note first that since the conversion is specified, eq. (3.5) can be used to calculate the rate of reaction, namely,

$$r = \frac{X_{CO} N_{CO}^{in}}{+1} = 0.8[0.2(100) + 0.5(214)] = 101.6 \text{ mol/h}$$

With r known, the flow rates of each species in stream 4 can be calculated from the species balances in a straightforward fashion:

N$_2$ balance: $N_{N_2}^4 = 0.78(100) = 78 \text{ mol/h}$

CO balance: $N_{CO}^4 = 127 - r = 25.4 \text{ mol/h}$

H$_2$O balance: $N_{H_2O}^4 = 628 - 101.6 = 526.4 \text{ mol/h}$

CO$_2$ balance: $N_{CO_2}^4 = 2 + r = 103.6 \text{ mol/h}$

H$_2$ balance: $N_{H_2}^4 = 107 + r = 208.6 \text{ mol/h}$

The composition of stream 4 is thus, in mole fractions,

$$(N_2, CO, H_2O, CO_2, H_2) = (.083, 0.027, 0.559, 0.110, 0.221)$$

Note in the above example that if a CO balance is performed around the second reactor, a rate of 25.4 mol/h is obtained. Hence, the rate calculated using the overall balances (127 mol/h) is just the sum of the rates of reaction of each separate reactor. This is the case in general because each species overall balance is just the sum over all units of the individual unit balances for that species. When the summation if performed, the intermediate stream terms will cancel out and thus only the process input/output streams and the sum of the unit reaction rates will remain. Thus, the overall reaction rate must always be equal to the sum of the unit reaction rates.

We conclude this section with one further multiunit example in which several of the units are reactors but in which the reactions occurring in the reactors are different.

Example 3.10 In a closed-cycle life support system described in Example 1.3, carbon dioxide and water from respiration and urine are reprocessed for reuse as shown in Figure 3.4. The food input represented as C_2H_2 is consumed via the reaction

$$C_2H_2 + \tfrac{5}{2}O_2 \rightarrow 2CO_2 + H_2O$$

The respiration products are separated by condensation of the H_2O, and the remaining waste gas containing an N_2-to-CO_2 ratio of about 1:100 is reacted to produce water via the reaction

$$CO_2 + 4H_2 \rightarrow CH_4 + 2H_2O$$

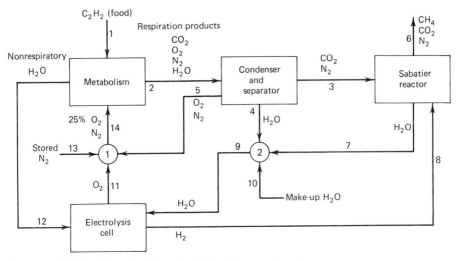

Figure 3.4   Flowsheet for Example 3.10, life support system.

The condensed H_2O is routed to a cell in which O_2 and H_2 are produced by electrolysis:

$$H_2O \rightarrow H_2 + \tfrac{1}{2}O_2$$

In order to maintain a "normal" cabin atmosphere, enough N_2 is mixed with the O_2 so as to obtain a cabin air feed containing 25% O_2. Since some of the N_2 is lost with the vent gas, N_2 must be added to the system from a cache.

Assuming that the organism requires 7.5 mol O_2 per 1 mol C_2H_2 for metabolism and that 10% of the H_2O produced by oxidation of food is recovered from urine, determine all flows and compositions in the system based on 1 mol/day C_2H_2.

Solution   As is evident from the problem statement, the system contains three reactors each involving a different reaction. Hence, the balances for each of these units will require the introduction of a rate-of-reaction variable. These three rate variables must therefore be counted in determining the degree of freedom of the process. The completed degree-of-freedom table for the life support system is shown in Figure 3.5.

In the table, the rate-of-reaction variables are separately denoted by a $+1$ to emphasize their distinction from the conventional stream variables.

As is evident from Figure 3.5, the process is completely specified. Since the degree of freedom of the metabolism is 1, the basis should be chosen at that unit and the balances associated with this unit solved first.

Using as basis 1 mol/day C_2H_2, the specified oxygen-to-feed ratio implies that 7.5 mol/day O_2 must be supplied. Thus, a cabin air feed of 30 mol/day is required. With this, the species balances for the metabolism unit can be written in terms of the stoichiometry of the reaction

$$C_2H_2 + \tfrac{5}{2}O_2 \rightarrow 2CO_2 + H_2O$$

	Metabolism	Condenser/ Separator	Electrolysis	Sabatier Reactor	Mixer 1	Mixer 2	Process
Number of variables	8 + 1	9	4 + 1	7 + 1	6	4	22 + 3
Number of balances	5	4	3	5	2	1	20
Number of specified compositions	1				1		1
Number of relations							
CO_2 : N_2 ratio		1		1			1
O_2 : C_2H_2 ratio	1						1
H_2O split	1						1
Degree of freedom	1	4	2	2	3	3	1
Basis	-1						-1
	0						0

Figure 3.5 Example 3.10, degree-of-freedom table.

These are:

N_2 balance:	$N^2_{N_2} = 0.75(30) = 22.5$ mol/day
O_2 balance:	$N^2_{O_2} = 7.5 - \frac{5}{2}r$
C_2H_2 balance:	$0 = 1 - r$
CO_2 balance:	$N^2_{CO_2} = 0 + 2r$
H_2O balance:	$N^2_{H_2O} + N^{12} = 0 + r$

From the C_2H_2 balance, it is evident that $r = 1$ and, hence, that $N^2_{O_2} = 5.0$ mol/day and $N^2_{CO_2} = 2$ mol/day. The assumption $N^{12} = 0.1(N^2_{H_2O} + N^{12})$ then implies via the H_2O balance that

$$N^{12} = 0.1 \text{ mol/day}$$
$$N^2_{H_2O} = 0.9 \text{ mol/day}$$

Stream 2 therefore consists of the following species flows:

$$(N^2_{CO_2}, N^2_{O_2}, N^2_{N_2}, N^2_{H_2O}) = (2, 5, 22.5, 0.9) \text{ mol/day}$$

With the completion of the metabolism balance calculations, the flow of stream 12 has been determined, and hence the degree of freedom of the electrolysis unit will have been reduced to 1. Similarly, the flow of stream 14 has been determined, so that the degree of freedom of mixer 1 will be reduced to 2. Finally, all four stream variables associated with stream 2 have been calculated; thus, the degree of freedom of the separator unit will be reduced to zero. These considerations indicate that the condenser/separator balances should be calculated next.

Since the separator does not involve reaction, these balances take the simpler and more familiar form employed in Chapter 2:

N_2 balance: $22.5 = N^3_{N_2} + N^5_{N_2}$

O_2 balance: $5 = N^5_{O_2}$

CO_2 balance: $2 = N^3_{CO_2}$

H_2O balance: $0.9 = N^4_{H_2O}$

Since the N_2-to-CO_2 ratio is specified to be 1:100, it immediately follows that $N^3_{N_2} = 0.02$ mol/day and, from the N_2 balance, that $N^5_{N_2} = 22.48$ mol/day.

At this juncture, the stream 3 flow rate has been determined (the ratio of N_2 to CO_2 was already fixed), thus the Sabatier reactor will have degree of freedom 1. The flow rate of stream 4 is known, which means that mixer 2 has degree of freedom 2. Finally, the degree of freedom of mixer 1, which had already been reduced to 2, becomes zero since both stream 5 variables have been calculated. It thus is apparent that the calculations should proceed with the mixer 1 balances:

N_2 balance: $N^{13} + 22.48 = 22.5$

O_2 balance: $N^{11} + 5 = 7.5$

Clearly,

$$N^{13} = 0.02 \text{ mol/day}$$
$$N^{11} = 2.5 \text{ mol/day}$$

With the flow rate of stream 11 known and that of stream 12 calculated previously, the degree of freedom of the electrolysis unit becomes zero. It becomes the next unit to be considered.

The reaction in the cell is

$$H_2O \rightarrow H_2 + \tfrac{1}{2}O_2$$

In terms of this stoichiometry, the species balances are:

H_2O balance: $0 = N^9 + 0.1 - r$

H_2 balance: $N^8 = 0 + r$

O_2 balance: $2.5 = 0 + \tfrac{1}{2}r$

Clearly, $r = N^8 = 5$ mol/day and $N^9 = 4.9$ mol/day.

Since the degree of freedom of mixer 2 is only reduced to 1, even with stream 9 known, while that of the Sabatier reactor becomes zero, the latter unit is calculated next. The reaction occurring in this unit is

$$CO_2 + 4H_2 \rightarrow CH_4 + 2H_2O$$

so that the species balances are

CO_2 balance: $N^6_{CO_2} = 2 - r$

H_2 balance: $0 = 5 - 4r$

CH_4 balance: $N^6_{CH_4} = 0 + r$

H_2O balance: $N^7 = 0 + 2r$

N_2 balance: $N^6_{N_2} = 0.02$

From the H_2 balance, $r = 1.25$ mol/day and, thus, $(N^6_{CO_2}, N^6_{CH_4}, N^6_{N_2}) = (0.75, 1.25, 0.02)$ mol/day.

The material balance calculations conclude with the mixer 2 balances to yield

$$N^{10} = N^9 - N^4 - N^7 = 4.9 - 0.9 - 2.5 = 1.5 \text{ mol/day}$$

Thus, 1.5 mol makeup water and 0.02 mol stored N_2 are required per 1 mol C_2H_2 metabolized.

In the preceding example, the order in which the unit balances were solved was determined by updating the degree of freedom of the individual process units after each set of unit balances was complete. This updating can be conveniently carried out by augmenting the degree-of-freedom table with one additional row for each unit balance. These additional rows can be used to record the new information that is obtained as each set of unit balances is calculated. In the previous example, the initial values of the unit degrees of freedom are given in the bottom row of Figure 3.5. After the metabolism balances are completed, the remaining units accrue new information as follows: condenser/separator, four (stream 2); electrolysis cell, one (stream 12); Sabatier reactor, none; mixer 1, one (flow of stream 14); and mixer 2, none. These items can be entered as shown below.

	Metabolism	Condenser/ Separator	Electrol- ysis	Sabatier Reactor	Mixer 1	Mixer 2
Initial degree of free- dom	0	4	2	2	3	3
Metabolism balance		−4	−1		−1	

Clearly, the degree of freedom of the condenser becomes zero, and its balances ought to be solved next. After these calculations are completed, the following new information count is obtained: electrolysis cell, none; Sabatier reactor, one (flow of stream 3); mixer 1, two (stream 5); and mixer 2, one (stream 4). The updated degree of freedom table reads:

	Metabolism	Condenser/ Separator	Electrol- ysis	Sabatier Reactor	Mixer 1	Mixer 2
Initial degree of free- dom	0	4	2	2	3	3
Metabolism balances		−4	−1		−1	
Cond./sep. balances				−1	−2	−1

Mixer 1 has been reduced to degree of freedom zero, and hence its balances are solved next. This process is continued as outlined in the solution of the example until all unit balances have been solved. The row-by-row entries of the composite table are summarized in Figure 3.6.

	Metabolism	Condenser/ Separator	Electrol- ysis	Sabatier Reactor	Mixer 1	Mixer 2
Initial degree of freedom	0	4	2	2	3	3
Metabolism balances		−4	−1		−1	
Cond./sep. balances				−1	−2	−1
Mixer 1 balances			−1			
Electrolysis balances				−1		−1
Sab. react. balances						−1

Figure 3.6 Example 3.10, degree-of-freedom updating table.

Although for small problems, much of this counting and updating can be carried out mentally, the above tabular arrangement does offer a systematic mechanism for developing the unit balance sequencing. Its use is recommended in all multiunit problems.

In concluding this section, note that overall balances were not required in the solution of Example 3.10. If they had to be used, then to be consistent with our earlier observations, the overall system would have to be viewed as a large reactor in which three different reactions were occurring simultaneously. In writing the overall balances, it thus seems intuitive that one ought to use three reaction rates within the species balance equations. On the other hand, recall that in writing the overall balances for example 3.9, only one reaction rate was used because the reaction occurring in both reactor units was the same. The questions which naturally arise are:

1. How many reaction rates is it proper to use in the presence of simultaneous reactions?
2. How does one incorporate multiple reaction rates into a species balance equation?

The answers to these questions are the subject of the remainder of this chapter.

3.2 SPECIES BALANCES WITH MULTIPLE CHEMICAL REACTIONS

Whenever a chemical reaction occurs within a system, it follows by definition that any given chemical species will either be depleted by the reaction or it will be produced. When multiple simultaneous reactions occur, then any given chemical species may be produced by some of the reactions but depleted by others. Hence, the observed species production or depletion rates will actually be the resultant of a number of competing reaction trends. Thus, to calculate the overall production rate of any species, it is necessary to determine the individual production rates of that species by each reaction and then to add these to derive the net rate. From this observation it follows that if the species production rate R_s is understood to represent the algebraic sum of the production rate of species s by all chemical reactions occurring in the system, then the defining relation

$$R_s = N_s^{out} - N_s^{in}$$

can again be employed to construct a species balance equation. For instance, given two chemical reactions and the production rates R'_s and R''_s of species s by these two reactions, then paralleling eq. (3.1), the species balance equation can be written as

$$N_s^{out} = N_s^{in} + R_s = N_s^{in} + R'_s + R''_s$$

Example 3.11   The reduction of magnetite ore, Fe_3O_4, to metallic iron can be carried out by reacting the ore with hydrogen gas. The reactions taking place are:

$$Fe_3O_4 + H_2 \rightarrow 3FeO + H_2O$$
$$FeO + H_2 \rightarrow Fe + H_2O$$

When 4 mol/h H_2 and 1 mol/h Fe_3O_4 are introduced into a reactor, a steady-state output of 0.1 mol/h magnetite and 2.5 mol/h Fe is obtained, along with other chemical species. Calculate the complete reactor output.

Solution   Since Fe_3O_4 only reacts via the first reaction, $R''_{Fe_3O_4} = 0$, and it follows that $R_{Fe_3O_4} = R'_{Fe_3O_4}$. Thus, from the definition of the rate of production,

$$R_{Fe_3O_4} = N_{Fe_3O_4}^{out} - N_{Fe_3O_4}^{in}$$
$$R'_{Fe_3O_4} = 0.1 - 1 = -0.9 \text{ mol/h}$$

Similarly, since Fe is only produced by the second reaction,

$$R_{Fe} = R''_{Fe}$$

and, again from the definition of the rate,

$$R_{Fe} = R''_{Fe} = 2.5 - 0 = 2.5 \text{ mol/h}$$

From the stoichiometry of the first reaction,

$$R'_{H_2} = R'_{Fe_3O_4} = -0.9 \text{ mol/h}$$
$$R'_{H_2O} = -R'_{Fe_3O_4} = 0.9 \text{ mol/h}$$
$$R'_{FeO} = -3R'_{Fe_3O_4} = 2.7 \text{ mol/h}$$

and from the stoichiometry of the second reaction,

$$R''_{H_2} = -R''_{Fe} = -2.5 \text{ mol/h}$$
$$R''_{H_2O} = +R''_{Fe} = 2.5 \text{ mol/h}$$
$$R''_{FeO} = -R''_{Fe} = -2.5 \text{ mol/h}$$

Note from the reaction equations that H_2 is depleted by both reactions, H_2O is produced by both reactions, while FeO is produced by the first and depleted by the second reactions. Thus, since the rate in the species balance represents the net rate of production by all reactions, it follows that

$$N_{H_2}^{out} = N_{H_2}^{in} + R_{H_2} = N_{H_2}^{in} + R'_{H_2} + R''_{H_2}$$

$$= 4 - 0.9 - 2.5 = 0.6 \text{ mol/h}$$

$$N_{H_2O}^{out} = N_{H_2O}^{in} + R_{H_2O} = N_{H_2O}^{in} + R'_{H_2O} + R''_{H_2O}$$

$$= 0 + 0.9 + 2.5 = 3.4 \text{ mol/h}$$

$$N_{FeO}^{out} = N_{FeO}^{in} + R_{FeO} = N_{FeO}^{in} + R'_{FeO} + R''_{FeO}$$

$$= 0 + 2.7 - 2.5 = 0.2 \text{ mol/h}$$

The output of the reactor thus consists of 0.6 mol/h H_2, 3.4 mol/h H_2O as well as Fe_3O_4, FeO, and Fe issuing at the rate of 0.1, 0.2, and 2.5 mol/h, respectively.

Although the calculations in Example 3.11 were carried out in terms of the production rates of two species, we recognize from the discussion of the single-reaction case that they could equally well have been carried out if a species-independent rate of reaction had been introduced for each individual reaction. This in fact leads to the very general and useful formulation of the balance equations that is given in the next section.

3.2.1 Generalized Stoichiometry

In Section 3.1.1, the symbol σ_s was introduced to denote the stoichiometric coefficient of species s in a given chemical reaction with the convention that $\sigma_s < 0$ for reactants and $\sigma_s > 0$ for products. In the presence of multiple chemical reactions, a given species may be involved in many of the reactions, and in each of these it will have assigned a specific stoichiometric coefficient. Hence, to differentiate between reactions, the symbol for the stoichiometric coefficient must be modified to include a second subscript to indicate the reaction in question.

Given a system with S species and R reactions, the stoichiometric coefficient of species s in reaction r will be denoted by σ_{sr}. Using this notation, by analogy to our previous definition of the rate of a reaction, the rate of the rth reaction of a set of R reactions will be given by

$$r_r = \frac{R_{sr}}{\sigma_{sr}} \qquad s = 1, \ldots, S$$

where R_{sr} is the production rate of species s by reaction r. From this it follows that the net production rate of a species s by the R chemical reactions will be given by the sum

$$R_s = \sum_{r=1}^{R} R_{sr} = \sum_{r=1}^{R} \sigma_{sr} r_r \qquad (3.8)$$

Equation (3.8) provides an explicit relation between the rates of reaction of the R chemical reactions occurring in the system and the rate of production of any selected species.

Example 3.12 Consider again the reaction system of the previous example:

$$Fe_3O_4 + H_2 \rightarrow 3FeO + H_2O$$

$$FeO + H_2 \rightarrow Fe + H_2O$$

Suppose the chemical species are indexed in the order

$$s = 1 \qquad Fe_3O_4$$
$$s = 2 \qquad FeO$$
$$s = 3 \qquad Fe$$
$$s = 4 \qquad H_2$$
$$s = 5 \qquad H_2O$$

and the reactions are indexed in the order in which they are written. Then, the stoichiometric coefficients of the first reaction will be

$$\sigma_{11} = -1 \qquad \sigma_{21} = 3 \qquad \sigma_{31} = 0 \qquad \sigma_{41} = -1 \qquad \sigma_{51} = 1$$

and those of the second reaction,

$$\sigma_{12} = 0 \qquad \sigma_{22} = -1 \qquad \sigma_{32} = 1 \qquad \sigma_{42} = -1 \qquad \sigma_{52} = 1$$

In terms of these coefficients and the rates of reaction, r_1 and r_2, of the two reactions, the rate of production of Fe_3O_4 and Fe will be given by

$$R_{Fe_3O_4} = R_1 = \sigma_{11}r_1 + \sigma_{12}r_2 = -r_1$$

$$R_{Fe} = R_3 = \sigma_{31}r_1 + \sigma_{32}r_2 = r_2$$

The rate of production of FeO, on the other hand, will involve both reaction rates:

$$R_{FeO} = R_2 = \sigma_{21}r_1 + \sigma_{22}r_2 = 3r_1 - r_2$$

The definition of the net species production rate, given by eq. (3.8), permits us to write a general mole balance equation for species s in the presence of R chemical reactions, namely,

$$N_s^{out} = N_s^{in} + \sum_{r=1}^{R} \sigma_{sr}r_r \qquad s = 1, \ldots, S$$

This equation can be formulated in mass units by incorporating the species molecular weights M_s to yield

$$F_s^{out} = F_s^{in} + M_s \sum_{r=1}^{R} \sigma_{sr}r_r \qquad s = 1, \ldots, S$$

It thus appears that in the presence of R simultaneous chemical reactions, S species balances can again be written, provided that a rate of reaction variable r_r is introduced for each of the chemical reactions. As was the case with just a single chemical reaction, the reaction rates again have the units of mol/h and the species balance equations again are homogeneous in both the flows and the rates of reaction.

Example 3.13 Chlorination of benzene produces a mixture of mono-, di-, tri-, and quadrosubstituted products via the reaction chain:

$$C_6H_6 + Cl_2 \rightarrow C_6H_5Cl + HCl$$

$$C_6H_5Cl + Cl_2 \rightarrow C_6H_4Cl_2 + HCl$$

$$C_6H_4Cl_2 + Cl_2 \rightarrow C_6H_3Cl_3 + HCl$$

$$C_6H_3Cl_3 + Cl_2 \rightarrow C_6H_2Cl_4 + HCl$$

The primary product in the chlorination is trichlorobenzene, which is sold as a dry cleaning agent, but of course concurrent production of the other chlorobenzenes is unavoidable. Suppose a 3.6:1 molar feed ratio of Cl_2 to benzene yields, on a HCl- and Cl_2-free basis, a product with the composition (all mol %)

C_6H_6	1%
Chlorobenzene	7%
Dichlorobenzene	12%
Trichlorobenzene	75%
Tetrachlorobenzene	5%

If 1000 mol/h benzene is charged to the reactor, calculate the mol/h HCl byproduct and $C_6H_3Cl_3$ primary product produced.

The system flowsheet with species and compositions is shown in Figure 3.7. Note that the system involves four simultaneous chemical reactions each with its own rate of reaction. These four variables together with the nine stream variables thus total 13 variables. Balances can be written on each of the seven reacting species, and four compositions and a stream ratio are specified. Hence, with the given basis, the degree of freedom of the system is

$$\text{Degree of freedom} = 13 - 7 - 4 - 1 - 1 = 0$$

Given a feed rate of 1000 mol/h C_6H_6, the feed ratio implies that $N^2 = 3600$ mol/h Cl_2. With this, the balance equation becomes:

Benzene balance:	$0.01N^4 = N^1 - r_1$
Chlorobenzene balance:	$0.07N^4 = 0 + r_1 - r_2$
Dichlorobenzene balance:	$0.12N^4 = 0 + r_2 - r_3$
Trichlorobenzene balance:	$0.75N^4 = 0 + r_3 - r_4$

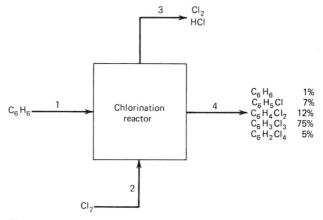

Figure 3.7 Flowsheet for Example 3.13, benzene chlorination.

Tetrachlorobenzene balance: $0.05N^4 = 0 + r_4$

Cl$_2$ balance: $N^3_{Cl_2} = N^2 - r_1 - r_2 - r_3 - r_4$

HCl balance: $N^3_{HCl} = 0 + r_1 + r_2 + r_3 + r_4$

The reaction rates in the above balances are indexed in the order in which the chemical reaction equations are given.

With N^2 known, the first five balances can be summed to yield

$$N^4 = N^1 = 1000 \text{ mol/h}$$

From this, the benzene balance gives

$$r_1 = 990 \text{ mol/h}$$

the chlorobenzene balance,

$$r_2 = 990 - 70 = 920 \text{ mol/h}$$

the dichlorobenzene balance,

$$r_3 = 920 - 120 = 800 \text{ mol/h}$$

the trichlorobenzene balance,

$$r_4 = 800 - 750 = 50 \text{ mol/h}$$

Consequently, it follows that

$$N^3_{Cl_2} = 3600 - 990 - 920 - 800 - 50 = 840 \text{ mol/h}$$

and

$$N^3_{HCl} = 0 + 2760 = 2760 \text{ mol/h}$$

Of the 1000 mol of stream 4, the trichloride amounts to 750 mol/h.

It should be apparent that, as in the single-reaction case, the numerical values of the rates of reaction will be dependent upon the magnitudes of the stoichiometric coefficients which are used in writing the reaction equations. However, the species input and output rates, which depend only on the ratios of these coefficients, will not be affected. The reader ought to verify this by repeating the solution of the previous example with the stoichiometric coefficients of one of the reactions multiplied by some nonzero constant.

In most cases, when R chemical reactions take place, the species balance equations will require the introduction of R rate-of-reaction variables. However, cases do arise in which this rule must be qualified. In particular, fewer than R reaction rates may be necessary in situations in which redundant chemical reactions are used in the formulation of the material balance problem.

Example 3.14 Isomers are chemical compounds that have the same molecular formula but different molecular structures. Because of the differences in structure, the various isomeric forms of a compound generally have different physical and chemical properties. If a particular isomer has a property which can be industrially exploited, then there is incentive for converting all the other isomers to that particular form. One example of this case is the isomer triplet consisting of 1-butene, C_4H_8, and the geometric isomers *cis*-2-butene and *trans*-2-butene. These three isomers can be interconverted over an alumina catalyst and the interconversion can be represented by the monomolecular isomerization reactions

$$\text{1-Butene} \rightleftharpoons \textit{cis}\text{-2-butene}$$

$$\textit{cis}\text{-2-Butene} \rightleftharpoons \textit{trans}\text{-2-butene}$$

$$\textit{trans}\text{-2-Butene} \rightleftharpoons \text{1-butene}$$

Suppose that under given conditions of temperature and pressure, 60% of a feed stream of pure 1-butene is converted to a product containing 25% *cis*-2-butene and unspecified compositions of the other two isomers. Calculate the unknown compositions. The process flowsheet is shown in Figure 3.8.

Solution Clearly, the system involves four stream variables and, since three reactions are given, three rates of reaction. Thus, there are seven variables. A balance can be made on each of the three species in the system and a conversion

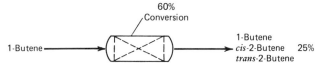

Figure 3.8 Flowsheet for Example 3.14, butene isomerization.

and composition are specified. Summarizing these items in the degree-of-freedom table shown below:

Number of variables		7
Number of balances	3	
Number of specified compositions	1	
Number of specified relations	1	
Basis	1	
		-6
Degree of freedom		1

it appears that, even if a basis is chosen, the problem is underspecified. This suggests that it is not possible to solve for all seven variables. Nonetheless, for purposes of illustration, let us attempt solution.

Choosing as basis 100 mol/h 1-butene feed, the specified conversion reduces, using the definition of conversion, to the relation

$$X_1 = 0.6 = \frac{N_1^{in} - N_1^{out}}{N_1^{in}} = \frac{100 - N_1^{out}}{100}$$

or

$$N_1^{out} = 40 \text{ mol/h}$$

The three species balance equations, written in terms of the rates of reaction of the three reactions given, are:

1-Butene balance: $N_1^{out} = 100 - r_1 + r_3 = 40$

cis-2-Butene balance: $N_{c-2}^{out} = 0 + r_1 - r_2$

trans-2-Butene balance: $N_{t-2}^{out} = 0 + r_2 - r_3$

Since the product stream consists of 25% cis-2-butene,

$$N_{c-2}^{out} = 0.25(N_1^{out} + N_{c-2}^{out} + N_{t-2}^{out})$$
$$= 0.25[(100 - r_1 + r_3) + (r_1 - r_2) + (r_2 - r_3)]$$
$$= 25 \text{ mol/h}$$

But this implies that

$$r_1 - r_2 = 25 \text{ mol/h}$$

while the 1-butene balance indicates that

$$r_1 - r_3 = 60 \text{ mol/h}$$

Thus,

$$N_{t-2}^{out} = r_2 - r_3 = (r_1 - r_3) - (r_1 - r_2)$$
$$= 60 - 25 = 35 \text{ mol/h}$$

The entire outlet stream therefore consists of

$$N^{out} = (N_1^{out}, N_{c-2}^{out}, N_{t-2}^{out}) = (40, 25, 35) \text{ mol/h}$$

Somehow, it was possible to calculate all the stream variables of interest without calculating all the rates of reaction. Since the reaction rates only serve as intermediates in the calculation of the stream variables, being unable to determine the rates is not a significant problem. The significant point is to be able to recognize this anomalous situation when it occurs so that the true degree of freedom of the system can be deduced.

To this end, note that the entire solution to the above problem could be carried out in terms of two quantities involving differences of reaction rates, namely, $r_1 - r_2$ and $r_1 - r_3$. This suggests that the balances could have been written in terms of only two reactions with reaction rates

$$r_1' = r_1 - r_3$$
$$r_2' = r_1 - r_2$$

Two reactions which might serve this purpose are:

$$\text{1-Butene} \rightarrow \textit{trans}\text{-2-butene}$$
$$\textit{trans}\text{-2-Butene} \rightarrow \textit{cis}\text{-2-butene}$$

With only two reactions and two reaction-rate variables, the above problem will have degree of freedom zero, indicating that all stream and rate variables can be calculated.

To verify this, let us write the species balance equations

1-Butene balance:	$N_1^{out} = 100 - r_1'$
cis-2-Butene balance:	$N_{c-2}^{out} = 0 + r_2'$
trans-2-Butene balance:	$N_{t-2}^{out} = 0 + r_1' - r_2'$

With a basis of 100 mol/h 1-butene and the specified conversion, the first balance equations yields $r_1' = 60$ mol/h. The *cis*-2-butene balance combined with the 25% composition specification then implies that $r_2' = 25$ mol/h. Thus, as before,

$$N_{t-2}^{out} = 0 + 60 \text{ mol/h} - 25 \text{ mol/h} = 35 \text{ mol/h}$$

Obviously, from the material balance point of view, the two- and three-reaction cases both predict the same stream flow rates. Since in the former case we can deal with fewer reactions and can calculate the reaction rates, it appears that solution using the smaller reaction set is preferable.

Two sets of chemical reactions are said to be *equivalent* if the values of the stream variables predicted by both sets are the same. A set of chemical reactions which can be reduced to an equivalent set containing fewer reactions is said to be *dependent*. As confirmed by the detailed analysis given in Section 3.3, a set of reactions can intuitively be recognized as being dependent if one of the chemical reactions can be obtained by adding together or subtracting multiples of the remaining reactions.

In the case of Example 3.14, the first isomer reaction

$$1\text{-Butene} \rightleftharpoons cis\text{-}2\text{-butene}$$

can clearly be obtained by adding the second and third reactions, and "canceling" the *trans*-2-butene which appears as both reactant and product in the resulting stoichiometric equation. The set of three isomerization reactions is thus dependent.

A dependent set of reactions will thus always include one or more reactions that are redundant, that is, can be deleted without affecting the solutions of the balance equations. On the other hand, a set of reactions is said to be *independent* if it is not possible to reduce the set to a smaller equivalent reaction set. For instance, the reactions

$$cis\text{-}2\text{-Butene} \rightleftharpoons trans\text{-}2\text{-butene}$$

$$trans\text{-}2\text{-Butene} \rightleftharpoons 1\text{-butene}$$

are an independent set of reactions because if either one is deleted, the remaining one is inadequate to describe the production of one of the isomer species.

A useful rule of thumb for recognizing independence of a set of reactions is that a set will be independent if each reaction in the set involves at least one chemical species which is not present in the remaining reactions. In general, the determination of whether a set of reactions is independent cannot be easily made without additional calculations. In Section 3.3, these calculations are detailed and a complete discussion of the concepts of dependence and independence of chemical reactions given. However, the reader may skip this analysis in the first reading of the text. It suffices at this point to recognize that in solving the material balances for a reacting system, only independent sets of chemical reactions need to be considered; hence, only as many reaction rate variables need to be introduced as there are independent reactions.

3.2.2 The Fractional Yield

From the balance equations, eq. (3.8), it is evident that the complete determination of the output of a reactor from given inputs requires that the rate of reaction of each of the R independent reactions be specified. This specification can be direct, by assigning values to the r_r's, or indirect, by imposing relationships from which the reaction rates can be deduced. As in the single-reaction case, the conversion of a reactant can again be used as a measure of the progress of chemical reactions. However, in general, in the presence of multiple reactions, the simple relationship, eq. (3.5), between conversion and reaction rate no longer is valid. Rather, since the reactant depletion rate may involve several reactions, eq. (3.5) becomes

$$X_s = \frac{-\sum_{r=1}^{R} \sigma_{sr} r_r}{N_s^{\text{in}}}$$

Hence, the conversion is no longer directly proportional to the rate of reaction but rather merely provides another relationship which can be used to calculate reaction rates, provided sufficient additional information is given.

Since the conversion is only defined for species which are depleted ($R_s < 0$), other supplementary measures of the progress of chemical reactions have been adopted by chemical engineers. For instance, terms such as selectivity, relative yield, overall yield, or fractional yield, each with widely varying definitions, are often used in technical publications. The most common of these is the specification of the fractional yield of a given product from a given reactant.

The *fractional yield* Y_{pq} of product p from reactant q is defined as the ratio of the net rate of production of product p to the rate of production which would be attained if the entire rate of depletion of reactant q could be allocated to produce p alone. In equation form,

$$Y_{pq} = \frac{R_p}{R_p^{\text{MAX}}}$$

where R_p^{MAX} is the maximum possible production rate of product p at the observed value of the rate of depletion of reactant q.

If product p is the desired output of the process, then the yield is a measure of the efficiency with which p is produced from reactant q. A yield near 100% indicates that most of the reactant is being converted to the desired product; a yield of less than 50% indicates that most of reactant q is wasted in undesirable side reactions.

Example 3.15 Polyglycols are produced by catalytic hydration of ethylene oxide followed by successive additions of the oxide to the resulting glycols. The chemical reactions describing this process are:

$$H_2O + C_2H_4O \rightarrow C_2H_4(OH)_2$$
$$C_2H_4(OH)_2 + C_2H_4O \rightarrow (C_2H_4OH)_2O$$
$$(C_2H_4OH)_2O + C_2H_4O \rightarrow (C_2H_3OH)_3(H_2O)$$

Suppose that when 100 mol/h ethylene oxide react completely in an excess of water, 10 mol/h mono-, 30 mol/h di-, and 10 mol/h triglycol are obtained. Calculate the yield of diglycol from the oxide.

Solution   Since no diglycol is fed to the reactor, the production rate of diglycol is simply 30 mol/h. If the entire 100 mol/h oxide could be used to produce diglycol alone, then the production rate of diglycol would be 50 mol/h. To verify this, let r_1, r_2, and r_3 be the reaction rates of the above three reactions in the order listed. Then, if the mono- and triglycol output rates are set to zero, we have

Oxide balance: $0 = 100 - r_1 - r_2 - r_3$

Mono- balance: $0 = 0 + r_1 - r_2$

Tri- balance: $0 = 0 + r_3$

From these balances, it follows that

$$r_3 = 0 \quad \text{and} \quad r_1 = r_2 = 50 \text{ mol/h}$$

Consequently, since the rate of production of diglycol is given by

$$R_{di} = r_2 - r_3$$

it follows that

$$R_{di}^{MAX} = 50 \text{ mol/h}$$

Thus, the yield of diglycol from oxide is

$$Y = \frac{30}{50} = 0.6 \quad \text{or} \quad 60\%$$

In many applications, it is found that the fractional yield is a function of the conversion of the principal reactant. For instance, the yield of preferred product may increase as conversion of the major reactant is reduced. Consequently, it is usual to specify both quantities when describing the performance of reactors in which multiple reactions occur.

Example 3.16 The ethylene oxide reactant used in glycol production is made by the partial oxidation of ethylene with an excess of air over a silver catalyst. The primary reaction is

$$2C_2H_4 + O_2 \rightarrow 2C_2H_4O$$

Unfortunately, some of the ethylene also undergoes complete oxidation to CO_2 and water via the reaction

$$C_2H_4 + 3O_2 \rightarrow 2CO_2 + 2H_2O$$

Suppose that with a feed containing 10% ethylene and an ethylene conversion of 25%, a yield of 80% oxide is obtained from this reactant. Calculate the composition of the reactor outlet stream.

Solution The flowsheet of the reactor with all species listed is given in Figure 3.9. Note that since each reaction involves a species not present in the other, the reactions are independent. The system involves nine stream variables and, since

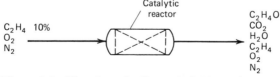

Figure 3.9 Flowsheet for Example 3.16, partial oxidation of ethylene.

there are two reactions occurring, two reaction rates. Thus, there are 11 variables. From the given reaction equations, it is clear that there are five distinct reacting species present. With the inert N_2, it will thus be possible to write six species balances. One composition is given, that of C_2H_4; the O_2-to-N_2 ratio must be that in air; and a yield and conversion are specified. In addition, since no flow is specified in the problem statement, we are free to choose a basis for the calculation. From the summary of these items shown in the degree-of-freedom table below, it is clear that the problem is completely specified.

Number of variables		9 + 2
Number of balances	6	
Number of specified compositions	1	
Number of specified relations	3	
Basis	1	-11
Degree of freedom		0

Suppose as basis for the calculation we choose 1000 mol/h total feed. Then, from the specified inlet composition,

$$N^{in}_{C_2H_4} = 0.1(1000) = 100 \text{ mol/h}$$

Of the remaining 900 mol/h reactor feed, 21% is O_2 and 79% is N_2. Hence,

$$N^{in}_{O_2} = 0.21(900) = 189 \text{ mol/h}$$

$$N^{in}_{N_2} = 900 - 189 = 711 \text{ mol/h}$$

Since the conversion is specified as 25%, it follows that

$$0.25 = \frac{N^{in}_{C_2H_4} - N^{out}_{C_2H_4}}{N^{in}_{C_2H_4}}$$

or

$$N^{out}_{C_2H_4} = 75 \text{ mol/h}$$

The yield specification corresponds to the relationship

$$0.8 = \frac{R_{C_2H_4O}}{R^{MAX}_{C_2H_4O}} = \frac{N^{out}_{C_2H_4O} - N^{in}_{C_2H_4O}}{R^{MAX}_{C_2H_4O}}$$

The maximum rate of production of C_2H_4O is attained if the entire 25 mol/h C_2H_4 which is converted is used to produce the oxide, that is, no CO_2 byproduct is produced. A CO_2 balance

$$0 = 0 + 2r_2$$

implies that $r_2 = 0$. Hence, from a C_2H_4 balance,

$$75 = 100 - 2r_1 - r_2 = 100 - 2r_1$$

it follows that

$$r_1 = 12.5 \text{ mol/h}$$

Thus,

$$R^{\text{MAX}}_{\text{C}_2\text{H}_4\text{O}} = 2r_1 = 25 \text{ mol/h}$$

The yield relationship thus reduces to

$$0.8 = \frac{N^{\text{out}}_{\text{C}_2\text{H}_4\text{O}} - 0}{25}$$

or

$$N^{\text{out}}_{\text{C}_2\text{H}_4\text{O}} = 20 \text{ mol/h}$$

Having extracted all the above information from the specified relations, the way has been prepared to write and solve the species balance equations. These are

C$_2$H$_4$O balance: $20 = 0 + 2r_1$

C$_2$H$_4$ balance: $75 = 100 - 2r_1 - r_2$

O$_2$ balance: $N^{\text{out}}_{\text{O}_2} = 189 - r_1 - 3r_2$

H$_2$O balance: $N^{\text{out}}_{\text{H}_2\text{O}} = 0 + 2r_2$

CO$_2$ balance: $N^{\text{out}}_{\text{CO}_2} = 0 + 2r_2$

N$_2$ balance: $N^{\text{out}}_{\text{N}_2} = 711$

From the C$_2$H$_4$O balance,

$$r_1 = 10 \text{ mol/h}$$

and from the C$_2$H$_4$ balance,

$$r_2 = 5 \text{ mol/h}$$

These reaction rates can then be used to immediately calculate the remaining species output rates.

3.2.3   Degree-of-Freedom Analysis

With the concepts introduced up to this point, we have the capability of writing general species balance equations in the presence of multiple chemical reactions. We have observed that when R independent reactions take place, the species balance equations will require the introduction of R new variables, the rates of each of the reactions, and that the balance equations will be homogeneous in these variables. We have also noted that the extent and efficiency with which the reactions proceed are commonly specified by means of the conversion and fractional yield concepts. In general, these concepts are not immediately translatable to explicit values of the rates of individual reactions, but rather result in relationships between the reaction rates and the stream variables which can be employed in solving the balance equations. As shown in Example 3.16, these elements of the material balance problem with multiple chemical reactions can be assembled to determine the degree of freedom of the individual reactor balances.

In this section, we consider the extension of the single-unit analysis to multiunit processes. This extension is straightforward except for three complications which can arise in calculating the degree of freedom and in deducing a strategy of solution for multiunit processes. These problems, all stemming from or arising in the use of overall balances, are

1. The possibility of dependence of reactions in the overall balances.
2. The possibility of exhausting the independent balances for a given species prior to completing all the unit balances.
3. The presence of chemical intermediates which appear neither in the process input nor in the process output streams.

The first of these complications is rather easily resolved by applying the conclusions derived in the single-unit case. If one or more of the units of a process are reactors, then, for overall balance purposes, the entire process is viewed as a single reactor in which all of the reactions are taking place simultaneously. As noted in the single-unit case, if this complete set of reactions is dependent, then it can be reduced to an equivalent independent set. Thus, when constructing the degree-of-freedom table, the number of reaction variables associated with the overall balances should be equal to the number of independent reactions.

Example 3.17   The gaseous reduction of magnetite, Fe_3O_4, discussed in Example 3.11, is industrially performed in a two-stage countercurrent system shown in Figure 3.10. The reactions taking place are

$$Fe_3O_4 + H_2 \rightarrow 3FeO + H_2O$$
$$FeO + H_2 \rightarrow Fe + H_2O$$

with the first reaction largely occurring in the first stage and the second reaction, in the second stage.

Pilot plant studies show that if 10 mol reducing gas, consisting of 33% H_2, 66% N_2, and 1% H_2O, enter stage 2 per 1 mol iron product leaving that stage, then a product consisting of 98 (mol)% Fe is obtained. When the resulting gas stream is passed up to the first stage, an intermediate partially reduced stream consisting of 2% Fe is obtained.

Assuming that these data apply to the full-scale process with recycle, calculate all stream compositions if 10% of the gas leaving stage 1 is purged and if the condenser is operated so that the recycle stream contains only 0.005 mole fraction water.

Solution   The system as given involves five units, two of which are reactors. From the stream compositions, it is clear that although both reduction reactions

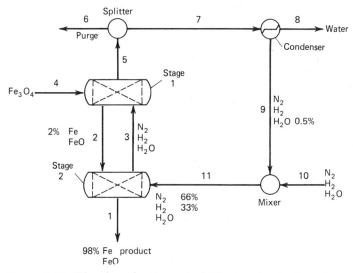

Figure 3.10 Flowsheet for Example 3.17, staged reduction of magnetite.

occur in stage 1, only the second reaction occurs in stage 2. Thus, stage 1 will require two rate variables and stage 2, only one rate variable. The entire process will consequently involve three reaction rate variables. On the other hand, if overall balances are to be considered, then the system is viewed as one single reactor in which all reactions take place simultaneously. Although strictly speaking a total of three reactions are occurring in the two reactor stages, this set of three reactions is obviously *dependent* because the second reaction in stage 1 is identical with the reaction which occurs in stage 2. Consequently, for purposes of the overall balances, only two rate-of-reaction variables will have to be introduced, one for each of the two independent reactions. The overall balances will consist of a species balance on each species appearing in the reaction equations (Fe_3O_4, FeO, Fe, H_2, and H_2O) and a balance on the inert, N_2; hence, a total of six balance equations. The remaining entries in the degree-of-freedom table can be established in the usual fashion and are summarized in Figure 3.11.

From the unit degrees of freedom, it is apparent that a basis should be chosen at stage 2 and the computations initiated with this unit. Choosing as basis 1000 mol/h of stream 11, it follows from the gas/product ratio that $N^1 = 100$ mol/h. Since the single reaction

$$FeO + H_2 \rightleftharpoons Fe + H_2O$$

occurs in stage 2, the unit balances are:

N_2 balance: $N^3_{N_2} = 0.66(1000) = 660$ mol/h

H_2 balance: $N^3_{H_2} = 0.33(1000) - r''_2$

H_2O balance: $N^3_{H_2O} = 0.01(1000) + r''_2$

	Mixer	Stage 2	Stage 1	Splitter	Condenser	Process	Overall Balances
Number of variables	9	10 + 1	9 + 2	9	7	30	10 + 2
Number of balances	3	5	6	3	3	20	6
Number of specified compositions	3	4	1	—	1	5	1
Number of relations							
Splitter restrictions	—	—	2	—		2	—
Purge fraction	—	—	1	—		1	—
Gas/product ratio	—	1	—	—	—	1	—
Degree of Freedom	3	1	4	3	3	1	5
Basis						−1	
						0	

Figure 3.11   Example 3.17, degree-of-freedom table.

Fe balance: $0.98(100) = 0.02N^2 + r_2''$

FeO balance: $0.02(100) = 0.98N^2 - r_2''$

where r_2'' is used to indicate the rate of the second of the two reactions occurring in stage 2.

The last two balances can be added to yield

$$N^2 = 100 \text{ mol/h}$$

which, when substituted into either balance, gives

$$r_2'' = 96 \text{ mol/h}$$

With the rate determined, the remaining balances can be immediately solved to obtain

$$N^3 = (N_{N_2}^3, N_{H_2}^3, N_{H_2O}^3) = (660, 234, 106) \text{ mol/h}$$

At this point, stream 3 and the flow of stream 2 have been calculated; hence, stage 1 will have degree of freedom zero. The flow of stream 11 has been fixed; hence, the degree of freedom of the mixer has been reduced to 2. The flow of stream 1 has been calculated; hence, the overall balances will have the degree of freedom 4. Clearly, the calculations ought to proceed with *stage 1*.

In stage 1, both reactions occur. If the reaction rates are indexed in the order in which the reactions are listed, the unit balances are

N_2 balance: $N_{N_2}^5 = 660 \text{ mol/h}$

H_2 balance: $N_{H_2}^5 = 234 - r_1' - r_2'$

H_2O balance: $N_{H_2O}^5 = 106 + r_1' + r_2'$

Fe balance: $0.02(100) = 0 + r_2'$

FeO balance: $\qquad 0.98(100) = 0 + 3r'_1 - r'_2$

Fe_3O_4 balance: $\qquad 0 = N^4 - r'_1$

From the Fe balance, $r'_2 = 2$ mol/h; and from the FeO balance, $r'_1 = 100/3$ mol/h. Thus, $N^4 = 100/3$ mol/h and

$$N^5 = (N^5_{N_2}, N^5_{H_2}, N^5_{H_2O}) = (660, 198.67, 141.33) \text{ mol/h}$$

With stream 5 calculated, the splitter degree of freedom has been reduced to zero. The splitter balances, combined with the specified 10% purge fraction, then yield

$$N^6 = (66, 19.867, 14.133) \text{ mol/h}$$

and

$$N^7 = (594, 178.8, 127.2) \text{ mol/h}$$

At this point, since streams 6 and 4 as well as the flow of stream 1 have been determined, the overall balances appear to have degree of freedom zero. Also, with stream 7 calculated, the condenser will have degree of freedom zero. Either set of balances could be solved next. Suppose the condenser is chosen. The balance equations are

N_2 balance: $\qquad N^9_{N_2} = 594$ mol/h

H_2 balance: $\qquad N^9_{H_2} = 178.8$ mol/h

H_2O balance: $\qquad N^9_{H_2O} + N^8 = 127.2$ mol/h

Since $N^9_{H_2O} = 0.005N^9$, it follows that $N^9_{H_2O} = 3.88$ mol/h and $N^8 = 123.32$ mol/h. Thus,

$$N^9 = (594, 178.8, 3.88) \text{ mol/h}$$

The calculations conclude with the mixer balances, which yield

$$N^{10} = (66, 151.2, 6.12) \text{ mol/h}$$

The degree-of-freedom updating summary for the problem is shown in Figure 3.12.

	Mixer	Stage 2	Stage 1	Splitter	Condenser	Overall Balances
Initial degree of freedom	3	0	4	3	3	5
Stage 2 balances	−1		−4			−1
Stage 1 balances				−3		−1
Splitter balances					−3	−3
Condenser balances	−2					

Figure 3.12 Example 3.17, degree-of-freedom updating table.

The preceding example, along with serving to illustrate the solution of a multiunit problem, also provides an instance of the second type of complication—that of dependent species balances.

As can be seen from Figure 3.12, after the completion of the splitter balances, both the condenser and the overall balances appeared to have net degree of freedom zero. In carrying out the solution, we chose to solve the condenser balances rather than the overall balances. Suppose we now choose the latter and attempt to solve them. The overall balance equations are:

N_2 balance: $N_{N_2}^6 = N_{N_2}^{10}$

H_2 balance: $N_{H_2}^6 = N_{H_2}^{10} - r_1 - r_2$

H_2O balance: $N_{H_2O}^6 + N^8 = N_{H_2O}^{10} + r_1 + r_2$

Fe balance: $0.98N^1 = 0 + r_2$

FeO balance: $0.02N^1 = 0 + 3r_1 - r_2$

Fe_3O_4 balance: $0 = N^4 - r_1$

where r_1 and r_2 are the overall rates of the two given independent reactions. In the calculations prior to this point, the following stream variables were determined:

$$N^1 = 100 \text{ mol/h}$$
$$N^4 = 100/3 \text{ mol/h}$$
$$N^6 = (N_{N_2}^6, N_{H_2}^6, N_{H_2O}^6) = (66, 19.867, 14.133) \text{ mol/h}$$

The overall balances had degree of freedom 5; hence, with these additional five calculated values, it ought to be possible to solve the above equations. Proceeding with the solution, we find that the Fe balance gives $r_2 = 98$ and the Fe_3O_4 balance, $r_1 = 100/3$. But note that with the rates and N^1 known, the FeO balance is redundant while the H_2O balance still contains two unknowns,

$$14.133 + N^8 = N_{H_2O}^{10} + (394/3)$$

Clearly, something is amiss with our calculation of the net degrees of freedom.

The paradox can be resolved if we first note that each of the six species involved in the process is not present in every unit. Thus, while it is possible to write all six species balances for some units, for others only three species balances can be written. To summarize which species balances can be written for what units, it is convenient to construct the species balance table shown in Figure 3.13. In this table, each column indicates which species are balanced in the given unit. The column corresponding to the entire process is simply the sum of each row of unit entries and indicates how many independent balances on a given species are available for solving the entire problem. In particular, we note that Fe_3O_4 is involved in only stage 2 and hence that only one independent Fe_3O_4 balance equation is available for solution. On the other hand, N_2 occurs in every unit; consequently, five independent N_2 balance equations can be used in solving the material balance problem. Note further that the last row of balance equation totals is identical to

Species	Number of Balances					
	Mixer	**Stage 1**	**Stage 2**	**Splitter**	**Condenser**	**Process**
Fe_3O_4			1			1
FeO		1	1			2
Fe		1	1			2
H_2	1	1	1	1	1	5
H_2O	1	1	1	1	1	5
N_2	1	1	1	1	1	5
Total number of balances	3	5	6	3	3	20

Figure 3.13   Example 3.17, species balance table.

the second row of the degree-of-freedom table given in Figure 3.11. Hence, the above tabulation can be viewed as just an expansion of the "number of balances" row of a degree-of-freedom table.

Now, in solving Example 3.17, after having completed the stage 2, stage 1, and splitter calculations, it is clear from the above species balance table that

1 Fe_3O_4 balance
2 FeO balances
2 Fe balances
3 H_2 balances
3 H_2O balances
3 N_2 balances

will have been expended. Comparing this to the total number of balances available for the entire process, it is clear that all of the independent Fe_3O_4, FeO, and Fe balances are at this point exhausted.

Recall from the discussion of Section 2.3.1 that an overall species balance is nothing more than the sum of all of the unit balances for that species. Consequently, if all of the individual unit species balances have already been employed in solution, then the overall species balance becomes redundant, i.e., algebraically dependent. In the case of Example 3.17, then, although potentially all six species balances could be written, only three of these, the H_2, the H_2O, and the N_2 balances, will actually be independent. After the completion of the stage 1, stage 2, and splitter balances, the overall Fe_3O_4, FeO, and Fe balances thus become redundant. Summarizing, we see that instead of having six independent equations available, as indicated in the degree-of-freedom table, actually at this stage of the calculations only three remain.

To confirm this note that the stage 1 Fe_3O_4, FeO, and Fe balances

Fe_3O_4: $0 = N^4 - r_1'$

FeO: $0.98N^2 = 0 + 3r_1' - r_2'$

Fe: $0.02N^2 = 0 + r_2'$

when added to the corresponding stage 2 balances

FeO: $0.02N^1 = 0.98N^2 - r_2''$

Fe: $0.98N^1 = 0.02N^2 + r_2''$

yield the following combined species balance equations:

Fe$_3$O$_4$: $0 = N^4 - r_1'$

FeO: $0.02N^1 = 0 + 3r_1' - r_2' - r_2''$

Fe: $0.98N^1 = 0 + r_2' + r_2''$

These are identical to the corresponding species overall balances written previously, provided that the following identification

$$r_1 = r_1'$$
$$r_2 = r_2' + r_2''$$

between the overall and individual unit reaction rates is made.

Since the individual unit reaction rates have already been calculated, the above relations further imply that the overall reaction rates are also known. Hence, rather than having ten stream variables and two reaction rates unknown, we find that at this stage of the calculations both reaction rates are known and that, with the previously calculated results, the unknown stream variables have been reduced to four.

Summarizing the calculation of the net degree of freedom of the overall balances, we find that we had a maximum of 12 variables. Of these, the product stream variables (two), the purge stream variables (three), the feed stream variables (one), and the two overall reaction rates are known. Hence, only four variables remain. Of the maximum number of six balance equations, only three independent balances remain. Consequently, the net degree of freedom of the overall balances is 1, which is in agreement with the fact that in attempting solution of the overall balances we ended up with one equation in two unknowns. Clearly, the decision to use the condenser balances rather than the overall balances in solving Example 3.17 was correct.

It should be emphasized that the above complication of redundant species balances can potentially arise whenever overall balances are introduced into the solution strategy. Consequently, if it is found that overall balances must be used in the solution of a material balance problem, it is recommended that a species balance tabulation such as that used in the preceding discussion be incorporated into the problem degree-of-freedom table.

We conclude this section with an example illustrating the third type of complication which can arise when overall balances are used with multiunit processes, namely, the problem of internal chemical intermediates. Frequently, in chemical processing the production of a desired product or separation of a mixture of chemicals can only be achieved via a sequence of reaction steps producing chemical intermediates which subsequently are consumed in later processing stages. When such a process is viewed from an overall balance point of view, these intermediate species do not appear in the process input and output streams and consequently can lead to an incorrect counting of variables. However, since all reactions occurring within the system must be considered in writing the overall balances, this variable-counting problem can be avoided if the reacting species for the overall balances are always counted directly from the assembled reaction stoichiometry.

Example 3.18 One route toward separating the uranium, U, and zirconium, Zr, contained in spent nuclear fuel involves the reaction of these elements with HCl. In the simplified flowsheet shown in Figure 3.14, 10 mol/h of a 90% Zr–10% U mixture is reacted with an HCl stream containing some water to produce metal chlorides via the reactions

$$U + 3HCl \rightarrow UCl_3 + \tfrac{3}{2}H_2$$

$$Zr + 4HCl \rightarrow ZrCl_4 + 2H_2$$

The U and Zr are completely converted to chlorides, provided the total HCl fed to the reactor is twice the amount required by the reaction stoichiometry. The UCl_3 produced is a solid and hence is readily removed from the remaining gaseous reaction products. The vapors from this first reactor are passed to a second reactor where the $ZrCl_4$ vapor is reacted with steam to produce solid ZrO_2 according to the reaction

$$ZrCl_4 + 2H_2O \rightarrow ZrO_2 + 4HCl$$

The reaction goes to completion and the ZrO_2 solid is separated from the gaseous reactor products.

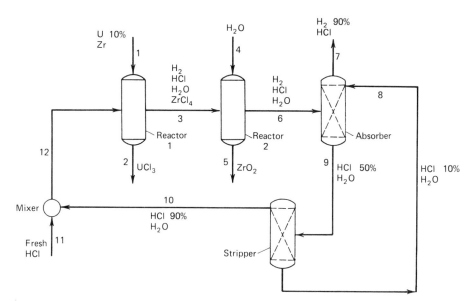

Figure 3.14 Flowsheet for Example 3.18, separation of uranium and zirconium.

The remainder of the process is concerned with the concentration of the residual HCl for recycle back to the first reactor. The gases leaving the second reactor are sent to an absorber which uses a 90% H_2O–10% HCl solution as a solvent to absorb HCl. This results in an absorber liquid containing 50% HCl and 50% H_2O and an overhead gas containing 90% H_2 and 10% HCl. The absorber liquid is sent to a stripper which boils off most of the HCl to produce a 90% HCl–10% H_2O recycle vapor. The remaining stripper bottoms is cooled and recycled to the absorber for use as solvent.

Calculate all flows and compositions, assuming all specified compositions are in mol %.

Solution   As stated, the problem contains two reactors. Of these, reactor 1, involves two reactions, while reactor 2 involves only one. For overall balance purposes, all three reactions must be considered as taking place simultaneously. Since each of the three reactions involves a chemical species not present in the others (U and Zr in the first two and ZrO_2 in the third), by our rule of thumb the three reactions are independent. Thus, three reaction rate variables must be associated with the overall balances. Note that one of the products of reactor 1, $ZrCl_4$, is a chemical intermediate which is completely converted in reactor 2 and thus appears in neither the process input streams 1, 4, and 11 nor in the process output streams 2, 5, and 7. Correct counting of the maximum number of overall species balances requires that we count this species along with those appearing in the process input and output streams. This is best accomplished by counting all of the species appearing in the reaction equations and adding to these any nonreacting species present in the process inputs or outputs. In our case, the reacting species are U, UCl_3, Zr, $ZrCl_4$, ZrO_2, H_2, HCl, and H_2O. Since there are no additional inert species, the overall balances will consist of eight species balance equations. The remaining entries are accumulated following the usual procedures and are summarized in Figure 3.15. Note that in constructing the table, the species balances were itemized separately anticipating the use of overall balances.

The problem is completely specified, and a choice exists of whether to proceed with the overall or the stripper balances. If the stripper balances are solved first, then, since the flows of streams 8 and 9 would be calculated, the degree of freedom of the absorber would be reduced to 1. Moreover, once the flow of stream 10 is calculated, the degree of freedom of the mixer would be reduced to 1. In either case, an impasse would be reached. Consequently, suppose the overall balances are solved first.

Since all three reactions must be included in the overall balances, let r_1 correspond to the rate of the uranium chlorination reaction, r_2 to the zirconium chlorination, and r_3 to the rate of the zirconium oxidation reaction. Choosing as basis of the calculation the suggested 10 mol/h of metal feed, the eight species balance equations thus are:

U balance: $0 = 1 - r_1$

Zr balance: $0 = 9 - r_2$

	Reactor 1	Reactor 2	Absorber	Stripper	Mixer	Process	Overall Balances
Number of variables	9+2	9+1	9	6	5	23+3	8+3
Number of balances							
U	1					1	1
UCl_3	1					1	1
Zr	1					1	1
$ZrCl_4$	1	1				2	1
ZrO_2		1				1	1
H_2	1	1	1			3	1
H_2O	1	1	1	1	1	5	1
HCl	1	1	1	1	1	5	
Total	7	5	3	2	2	19	8
Number of specified compositions	1	3	3	1		5	2
Number of relations							
HCl/feed ratio	1					1	
Degrees of freedom	2	5	3	1	2	1	1
Basis						−1	
						0	

Figure 3.15 Example 3.18, degree-of-freedom table.

UCl_3 balance:	$N^2 = 0 + r_1$
$ZrCl_4$ balance:	$0 = 0 + r_2 - r_3$
ZrO_2 balance:	$N^5 = 0 + r_3$
H_2 balance:	$0.9N^7 = 0 + \frac{3}{2}r_1 + 2r_2$
H_2O balance:	$0 = N^4 - 2r_3$
HCl balance:	$0.1N^7 = N^{11} - 3r_1 - 4r_2 + 4r_3$

The first two balance equations indicate that $r_1 = 1$ mol/h and $r_2 = 9$ mol/h. The $ZrCl_4$ balance then yields $r_3 = 9$ mol/h; and with the three rates determined, all stream flows can readily be obtained. In particular, from the UCl_3 balance, $N^2 = 1$ mol/h; from the ZrO_2 balance, $N^5 = 9$ mol/h; from the H_2 balance, $N^7 = 21.67$ mol/h; and from the HCl balance, $N^{11} = 5.167$ mol/h.

At this point, since the flows of streams 1 and 2 have been determined, it appears that the degree of freedom of reactor 1 has been reduced to zero. However, from Figure 3.15 it is evident that the single balance equation which is available for each of the species U, Zr, and UCl_3 has already been expended in writing the overall balances. Consequently, only four of the seven reactor 1 balance equations can be counted. In addition, since the overall reaction rates of the first two reactions have been calculated and are equivalent to the corresponding reactor 1 rates, the two reactor 1 rates are known. Hence, the true net degree of freedom of reactor 1 is 1, i.e., $2 - 2 + 3 - 2 = +1$. By a similar argument, it follows that the net degree of freedom of reactor 2 is 3. Since the flow of stream 7 has been calculated, the net degree of freedom of the absorber is 2; and since stream 11 is known, that of the mixer is 1. Apparently, we have reached an impasse in the sequencing. Note, however, that with stream 1 known, the condition that the HCl fed to reactor 1 be twice that required to react completely with the U and Zr metals fed yields

$$N_{HCl}^{12} = 2(3N_U^1 + 4N_{Zr}^1) = 2(3 + 36) = 78 \text{ mol/h}$$

With this, the degree of freedom of the mixer becomes zero. The mixer balances are:

HCL balance: $78 = 0.9N^{10} + 5.167$

H_2O balance: $N_{H_2O}^{12} = 0.1N^{10}$

Thus, $N^{10} = 80.926$ mol/h and $N_{H_2O}^{12} = 8.093$ mol/h.

With the flow of stream 12 known, the net degree of freedom of reactor 1 becomes zero; and with the flow of stream 10 known, that of the stripper similarly is reduced to zero. Thus, we arbitrarily choose to continue with the stripper balances. These are:

HCl balance: $72.833 + 0.1N^8 = 0.5N^9$

H_2O balance: $8.093 + 0.9N^8 = 0.5N^9$

Variable N^8 can be calculated by subtracting the two equations from one another. Thus, we obtain

$$64.740 = 0.8N^8 \quad \text{or} \quad N^8 = 80.926 \text{ mol/h}$$

Finally, by substitution, it follows that $N^9 = 161.852$ mol/h.

The net degree of freedom of the absorber is now reduced to zero. Again we have the choice of either using the reactor 1 or the absorber balances. Suppose we select the latter, since they can be solved by inspection to give

$$N^6 = (N_{H_2}^6, N_{H_2O}^6, N_{HCl}^6) = (19.5, 8.093, 75) \text{ mol/h}$$

The flows of streams 4 and 5, the rate of the third reaction, as well as stream 6 are now known. Moreover, the ZrO_2 balance has already been used up in the overall balances. Thus, the reactor 2 net degree of freedom is zero, i.e., $5 - 6 + 1 = 0$. Either of the reactor balances could be solved next. Proceeding with the reactor 2 balances, the result

$$N^3 = (N_{ZrCl_4}, N_{ZrO_2}, N_{H_2}, N_{H_2O}, N_{HCl})$$
$$= (9, 0, 19.5, 8.093, 39) \text{ mol/h}$$

is easily calculated. Note that the ZrO_2 balances is truly redundant because from the overall balances we already know that $r_2 = r_3 = 9$ mol/h. A summary of the degree-of-freedom updating is given in Figure 3.16.

In this section, we have examined in some detail the complications that can occur while updating the degree of freedom of the system and its units as the

	Reactor 1	Reactor 2	Absorber	Stripper	Mixer	Overall Balances
Initial degree of freedom	2	5	3	1	2	<u>0</u>
Overall balances						
Stream variables	−2	−2	−1		−1	
Reaction rates	−2	−1				
Redundant balances	+3	+1				
Reactant ratio condition					<u>−1</u>	
Mixer balances	<u>−1</u>			<u>−1</u>		
Stripper balances			<u>−2</u>			
Absorber balances		<u>−3</u>				

Figure 3.16   Example 3.18, degree-of-freedom updating table.

calculations proceed. The reader should note that although these complications can arise only if overall balances are employed, they also may arise whenever arbitrary combination balances are selected. It is partly because of these complexities that we recommended earlier that combination balances other than the overall balance be avoided.

3.3   THE ALGEBRA OF MULTIPLE CHEMICAL REACTIONS

In Section 3.2.1, a general formulation of the species material balance equations for systems involving multiple reactions was given in terms of species-independent rates of reaction and a doubly subscripted set of stoichiometric coefficients of the reactions. In analyzing some examples, we found that, although in most cases as many reaction rates were required as there were reactions taking place, there were instances in which fewer reaction rate variables and hence fewer reactions would suffice. We noted that these cases could be identified by establishing that one or more of the reactions given could actually be generated by adding or subtracting multiples of the remaining reactions. This rule of thumb was supplemented with an additional rule that a set of reactions could not be further reduced if each reaction in the set involved at least one species not present in the remaining reactions.

In this section, we discover that these observations are actually consequences of the algebraic properties of the array of stoichiometric coefficients σ_{sr} of the given reaction set. In particular, it will be demonstrated that the notions of dependence and independence of a reaction set that we discussed earlier are in fact equivalent to the linear dependence and independence of the vectors of stoichiometric coefficients of the reactions. We will confirm that in general if a reaction set is dependent, it can always be reduced to a smaller equivalent set. In addition, a procedure will be presented which can be used both for determining how many reactions must be contained in the reduced set and for selecting such a set. Finally, we shall learn how to use the stoichiometric array to deduce whether the specifications on a reactor are improper even though they are correct in number.

3.3.1 Linear Independence of Reactions

As a preliminary to the subsequent discussion, several elementary but essential algebraic concepts need to be established.

First, a set of vectors is said to be *linearly independent* if it is not possible to express any given vector as a linear combination of the remaining members of the set. A set of vectors is termed *linearly dependent* if it is not independent. In equation form, a collection of R vectors, each with S components, denoted by

$$\mathbf{x_r} = \begin{pmatrix} x_{1r} \\ x_{2r} \\ \vdots \\ x_{Sr} \end{pmatrix} \qquad r = 1, \dots, R$$

is linearly dependent if there can be found a set of constants α_r, $r = 1, \dots, R$, not all identically equal to zero, such that

$$\sum_{r=1}^{R} \alpha_r x_{sr} = 0 \qquad s = 1, \dots, S$$

If the only set of constants α_r which satisfies this condition is $\alpha_r = 0$, $r = 1, \dots$, R, then the set of vectors is linearly independent.

For instance, the vectors $\mathbf{x}_1 = \binom{2}{2}$ and $\mathbf{x}_2 = \binom{1}{1}$ are linearly dependent because there do exist constants $\alpha_1 = 1$ and $\alpha_2 = -2$ such that

$$(1)x_{11} + (-2)x_{12} = (1)(2) + (-2)(1) = 0$$
$$(1)x_{21} + (-2)x_{22} = (1)(2) + (-2)(1) = 0$$

On the other hand, the vectors

$$\mathbf{x}_1 = \begin{pmatrix} 1 \\ 0 \\ 0 \\ 0 \end{pmatrix} \qquad \mathbf{x}_2 = \begin{pmatrix} 0 \\ 1 \\ 0 \\ 1 \end{pmatrix} \qquad \mathbf{x}_3 = \begin{pmatrix} 0 \\ 0 \\ 1 \\ 1 \end{pmatrix}$$

are linearly independent since it is impossible to find constants α_1, α_2, and α_3, all not equal to zero, such that

$$\alpha_1 \begin{pmatrix} 1 \\ 0 \\ 0 \\ 0 \end{pmatrix} + \alpha_2 \begin{pmatrix} 0 \\ 1 \\ 0 \\ 1 \end{pmatrix} + \alpha_3 \begin{pmatrix} 0 \\ 0 \\ 1 \\ 1 \end{pmatrix} \equiv \begin{pmatrix} \alpha_1 \\ \alpha_2 \\ \alpha_3 \\ \alpha_2 + \alpha_3 \end{pmatrix} = \begin{pmatrix} 0 \\ 0 \\ 0 \\ 0 \end{pmatrix}$$

An additional concept which will prove to be useful is that of the *basis* of a set of vectors. A linearly independent subset of vectors selected from a dependent set of vectors is said to be a basis of the dependent set, if every vector in the parent set can be expressed as a linear combination of the members of the subset. By implication, every set of vectors always has a basis.

For instance, the set of vectors $x_1 = \binom{1}{0}$, $x_2 = \binom{0}{1}$, $x_3 = \binom{2}{1}$, and $x_4 = \binom{1}{1}$ is linearly dependent. The subset consisting of x_1 and x_2 is independent and spans the entire set since the other members of the set can be expressed in terms of these two vectors. In particular,

$$x_3 = 2x_1 + x_2 = 2\binom{1}{0} + \binom{0}{1} = \binom{2}{1}$$

$$x_4 = x_1 + x_2 = \binom{1}{0} + \binom{0}{1} = \binom{1}{1}$$

The importance of obtaining a basis of a set of vectors is that it allows us to work with a smaller set of vectors without really losing any information about the parent set.

With these preliminaries established, let us proceed to relate these concepts to the stoichiometry of multiple chemical reactions. Given a reacting system involving S species and R reactions, consider the set of S component vectors formed by collecting the stoichiometric coefficients associated with each reaction into a separate vector. The first property of this set of vectors that we demonstrate is the equivalence between the notion of linear dependence and independence of this set and the definition of independence and dependence of the set of reactions introduced previously.

Recall that a set of reactions was defined to be dependent if it could be reduced to a smaller set so that the stream variables, or more directly the species production rates, predicted by both sets were the same. A set of reactions was defined to be independent if it could not be reduced to a smaller equivalent set. Now suppose the given set of R reactions in S species is dependent and suppose that an equivalent set can be formed by deleting the Rth reaction. From eq. (3.8), the production rate of any given species s is given by

$$R_s = \sum_{r=1}^{R} \sigma_{sr} r_r \qquad s = 1, \ldots, S$$

Since the set formed by considering the first $R - 1$ reactions is equivalent to the complete set of reactions, it must also be true that

$$R_s = \sum_{r=1}^{R-1} \sigma_{sr} r_r' \qquad s = 1, \ldots, S$$

Since by definition the species production rates predicted by either set must be the same, we have

$$\sum_{r=1}^{R} \sigma_{sr} r_r = \sum_{r=1}^{R-1} \sigma_{sr} r_r' \qquad s = 1, \ldots, S$$

This equality is true for all choices of r_r, $r = 1, \ldots, R$; consequently, it must also hold true for some choice which has $r_R \neq 0$. But then we have

$$\sigma_{sR} = \frac{1}{r_R} \sum_{r=1}^{R-1} \sigma_{sr}(r_r' - r_r) \qquad s = 1, \ldots, S$$

that is, the stoichiometric coefficients of the Rth reaction can be expressed as linear combinations of the remaining $R - 1$ reactions. Therefore, the R vectors of stoichiometric coefficients are linearly dependent if the reactions are dependent.

The converse to the above statement is also true, that is, if stoichiometric coefficients of a set of reactions are linearly dependent, then the reactions are dependent. To verify this, suppose that the stoichiometric coefficients of a set of R chemical reactions are linearly dependent. Moreover, suppose that a subset consisting of the first R' reactions has linearly independent coefficients and that these reaction coefficients form a basis for the entire set of coefficients. Since the first R' vectors of stoichiometric coefficients are a basis of the remaining $R - R'$, there must exist constants α_{rd} such that

$$\sigma_{sd} = \sum_{r=1}^{R'} \alpha_{rd}\sigma_{sr} \qquad \begin{array}{l} d = R' + 1, \ldots, R \\ s = 1, \ldots, S \end{array} \qquad (3.9)$$

By eq. (3.8), the species production rates for these R reactions are given by

$$R_s = \sum_{r=1}^{R} \sigma_{sr}r_r \qquad s = 1, \ldots, S$$

Splitting this sum into two parts,

$$R_s = \sum_{r=1}^{R'} \sigma_{sr}r_r + \sum_{r=R'+1}^{R} \sigma_{sr}r_r \qquad (3.10)$$

and substituting eq. (3.9), it follows that

$$R_s = \sum_{r=1}^{R'} \sigma_{sr}r_r + \sum_{d=R'+1}^{R} \left(\sum_{r=1}^{R'} \alpha_{rd}\sigma_{sr} \right) r_d$$

where the outer index was changed to d for convenience.

Interchanging the order of summation,

$$R_s = \sum_{r=1}^{R'} \sigma_{sr}r_r + \sum_{r=1}^{R'} \sigma_{sr} \sum_{d=R'+1}^{R} \alpha_{rd}r_d$$

and combining the two sums, the result is

$$R_s = \sum_{r=1}^{R'} \sigma_{sr} \left(r_r + \sum_{d=R'+1}^{R} \alpha_{rd}r_d \right) \qquad s = 1, \ldots, S$$

Finally, let

$$r_r' = r_r + \sum_{d=R'+1}^{R} \alpha_{rd}r_d$$

Then,

$$R_s = \sum_{r=1}^{R'} \sigma_{sr}r_r' \qquad (3.11)$$

This implies that the subset of reactions with linearly independent stoichio-metric coefficients will yield the same species production rates as the complete set. Consequently, by definition, the complete set of reactions must have been dependent. This conclusion, combined with the converse statement established earlier, confirms that the two notions of dependence are identical. Moreover, as an obvious corollary, it is also true that the two notions of independence, that is, independence of reactions and linear independence of their stoichiometric coefficients, are also identical concepts.

Example 3.19 In Example 3.14, we concluded that the reaction system

$$1\text{-Butene} \rightleftharpoons cis\text{-2-butene}$$

$$cis\text{-2-Butene} \rightleftharpoons trans\text{-2-butene}$$

$$trans\text{-2-Butene} \rightleftharpoons 1\text{-butene}$$

was dependent because the same solution to the problem could be obtained if just the last two reactions were used. If the species are indexed in the order 1-butene, cis-2-butene, $trans$-2-butene, then the stoichiometric coefficients for each of these reactions are

$$\sigma_1 = \begin{pmatrix} \sigma_{11} \\ \sigma_{21} \\ \sigma_{31} \end{pmatrix} = \begin{pmatrix} -1 \\ +1 \\ 0 \end{pmatrix} \qquad \sigma_2 = \begin{pmatrix} \sigma_{12} \\ \sigma_{22} \\ \sigma_{32} \end{pmatrix} = \begin{pmatrix} 0 \\ -1 \\ +1 \end{pmatrix}$$

$$\sigma_3 = \begin{pmatrix} \sigma_{13} \\ \sigma_{23} \\ \sigma_{33} \end{pmatrix} = \begin{pmatrix} 1 \\ 0 \\ -1 \end{pmatrix}$$

Now note that

$$(1)\sigma_{11} + (1)\sigma_{12} + (1)\sigma_{13} = (-1) + 0 + (1) = 0$$

$$(1)\sigma_{21} + (1)\sigma_{22} + (1)\sigma_{23} = (+1) + (-1) + 0 = 0$$

$$(1)\sigma_{31} + (1)\sigma_{32} + (1)\sigma_{33} = 0 + 1 + (-1) = 0$$

Consequently, the stoichiometric coefficients of these reactions are linearly dependent.

The above demonstration of the identity between the concepts of dependence has several very important consequences.

1. It confirms our rule of thumb that a set of reactions is dependent if one reaction can be generated by adding or subtracting multiples of the remaining ones. This rule is clearly true since the stoichiometric coefficients are linearly dependent.
2. It confirms our rule of thumb that a set of reactions is independent if every reaction contains one species not present in the remaining reactions. If every vector of stoichiometric coefficients has a nonzero component where every

other vector has zero components, then the complete set of vectors of stoichiometric coefficients must be linearly independent.

3.   It allows us to use algebraic constructions for calculating a basis of a collection of vectors if our "simple" rule for identifying dependence becomes awkward to apply. One such standard calculation scheme, reviewed in the next section, will both identify dependence of a set of reactions and select an equivalent and independent subset.

4.   Equation (3.11) clearly establishes that in calculating the species production rates, it is always sufficient to use only an independent and equivalent subset of the entire set of reactions. The remaining reactions can be discarded from the material balance calculations without any loss of information.

3.3.2 Determination of Linear Independence

The rule of thumb that a set of reactions is independent if every reaction contains one species not present in the remaining reactions is convenient because it allows us to establish independence merely by inspecting the reaction equations. Unfortunately, in many instances this rule proves too restrictive because most species occur in more than one reaction. For these cases, a more general test must be applied in order to determine independence. In this section, such a test will be detailed. The test involves four elementary operations systematically carried out on an array consisting of the stoichiometric coefficients of the reactions. The goal of these operations is to reduce the array of stoichiometric coefficients to a form from which independence or dependence can be deduced by inspection.

To motivate the subsequent discussion, let us first consider the following example.

Example 3.20   Consider the three reactions occurring in Example 3.18:

$$U + 3HCl \rightarrow UCl_3 + \tfrac{3}{2}H_2$$

$$Zr + 4HCl \rightarrow ZrCl_4 + 2H_2$$

$$ZrCl_4 + 2H_2O \rightarrow ZrO_2 + 4HCl$$

By our rule of thumb, these reactions are independent because the species U, Zr, and ZrO_2 each occur in only one of the reactions. With the reactions ordered as they are written and the eight species sequenced in the order (U, Zr, ZrO_2, UCl_3, $ZrCl_4$, HCl, H_2, H_2O), the three vectors of stoichiometric coefficients will be

$$\sigma_1 = \begin{pmatrix} -1 \\ 0 \\ 0 \\ 1 \\ 0 \\ -3 \\ \tfrac{3}{2} \\ 0 \end{pmatrix} \quad \sigma_2 = \begin{pmatrix} 0 \\ -1 \\ 0 \\ 0 \\ 1 \\ -4 \\ 2 \\ 0 \end{pmatrix} \quad \sigma_3 = \begin{pmatrix} 0 \\ 0 \\ 1 \\ 0 \\ -1 \\ 4 \\ 0 \\ -2 \end{pmatrix}$$

The three vectors are clearly independent because even if just the first three components are considered, there are no possible nonzero values of α_1, α_2, and α_3 such that

$$(-1)\alpha_1 + (0)\ \alpha_2 + (0)\alpha_3 = 0$$
$$(0)\alpha_1 + (-1)\alpha_2 + (0)\alpha_3 = 0$$
$$(0)\alpha_1 + (0)\ \alpha_2 + (1)\alpha_3 = 0$$

If the stoichiometric vectors are assembled into an 8×3 array,

$$\begin{pmatrix} -1 & 0 & 0 \\ 0 & -1 & 0 \\ 0 & 0 & 1 \\ 1 & 0 & 0 \\ 0 & 1 & -1 \\ -3 & -4 & 4 \\ \frac{3}{2} & 2 & 0 \\ 0 & 0 & -2 \end{pmatrix}$$

then it is evident that the rule of thumb amounts to searching this array to find three rows (species) each of which have a single nonzero entry in a different column (reaction). In general, it may not be possible to find enough rows with this property. For those cases, we construct a scheme which will reduce the array to the form shown in the array above, that is, with the first R rows containing their only nonzero entries on the diagonal.

The four elementary operations upon which the array reduction procedure is based are:

1. Multiplying one of the columns of the array by an arbitrary constant.
2. Interchanging rows of the array.
3. Interchanging columns of the array.
4. Adding a multiple of one column to another column.

As is shown next, the new reaction set resulting from the execution of any of the elementary operations will yield the same species production rates as the original reaction set. Consequently, with each application of any of these operations, equivalent reaction sets will always be generated. Let us verify this property for each operation.

First of all, multiplying a column by a constant corresponds to scaling all the stoichiometric coefficients of a reaction by that constant. As shown previously, such scaling only causes the rate of the corresponding reaction to be scaled accordingly. Interchanging a row or column of the array simply corresponds to interchanging the order of the species or reactions as they are listed. The reaction set remains unchanged.

The fourth operation requires a closer examination. Consider a system in S species and R reactions with stoichiometric coefficients σ_{sr}. Suppose the stoichiometric coefficients of the kth reaction are replaced by a multiple α of the coefficients of the jth reaction added to those of the kth reaction:

$$\sigma_{sk}^* = \sigma_{sk} + \alpha\sigma_{sj} \qquad s = 1, \ldots, S$$

If these new coefficients, designated by asterisks, are used to calculate species production rates, we have

$$R_s^* = \sum_{r=1}^{R} \sigma_{sr}^* r_r^*$$

$$= \sum_{\substack{r=1 \\ r \neq j \\ r \neq k}}^{R} \sigma_{sr} r_r + (\sigma_{sk} + \alpha\sigma_{sj})r_k^* + \sigma_{sj} r_j^*$$

This expression can be rewritten to the form

$$R_s^* = \sum_{\substack{r=1 \\ r \neq j \\ r \neq k}}^{R} \sigma_{sr} r_r + \sigma_{sk} r_k^* + \sigma_{sj}(r_j^* + \alpha \hat{r}_k^*)$$

But with the change of variable

$$r_k = r_k^*$$
$$r_j = r_j^* + \alpha r_k^*$$

the expression for the species production rates reduces to

$$R_s^* = \sum_{r=1}^{R} \sigma_{sr} r_r = R_s$$

Hence, the new set of reactions will result in the same species production rates as the old set. In other words, the two sets are equivalent. All four elementary operations therefore only generate equivalent reaction sets.

Using only these four operations, an algorithm, which we shall call the *array reduction procedure*, can be formulated. The procedure is initiated with a given array of stoichiometric coefficients and consists of the following steps:

Step A   Take the first column and divide each entry in that column by the first element.

Step B   Add appropriate multiples of the first column to each remaining column so as to make the first entry in each of these columns equal to zero.

For each succeeding column, the same sequence of steps is repeated, that is, for the jth column:

Step A   Take the jth column and divide each entry in the jth column by the jth element in that column. If the jth element is zero, go to Step C.

Step B Add appropriate multiples of the jth column to each remaining column so that the jth entry in each of these columns is equal to zero. At the conclusion of these calculations, set $j = j + 1$, and go to Step A.

Step C If the jth element in the jth column is zero, interchange row j with any row below row j which has a nonzero entry in the jth position. Return to Step A. If no row has a nonzero entry in the jth column, interchange column j with any column to the right of j. Return to Step A.

The process terminates when all columns have been reduced or if all remaining columns are identically zero. The nonzero columns in the resulting array, say, there are R', will be linearly independent since the jth column will have 1 in the jth position and zeros for the remaining $R' - 1$ of the first R' entries. Since the entire algorithm consists of only a sequence of elementary operations, the reduced array of stoichiometric coefficients must be that of a reaction set which is equivalent to the original reaction set. Thus, if the reduced set has only R' linearly independent columns (the remaining $R - R'$ columns are identically equal to zero), then the original set of stoichiometric coefficients must have only R' linearly independent columns. The algorithm thus shows that the original set of R reactions really can be reduced to R' independent reactions. One particular set of R' independent reactions can be constructed by just using the nonzero columns of the reduced array.

The above reduction process is illustrated with the following example.

Example 3.21 The following reactions are postulated to occur when synthesis gas is made by steam reforming of methane:

$$CH_4 + CO_2 \rightleftharpoons 2CO + 2H_2$$
$$CO + H_2O \rightleftharpoons CO_2 + H_2$$
$$CH_4 + H_2O \rightleftharpoons CO + 3H_2$$
$$CH_4 + 2H_2O \rightleftharpoons CO_2 + 4H_2$$

If the synthesis gas is to be used for the catalytic production of methanol via the reaction

$$CO + 2H_2 \rightleftharpoons CH_3OH$$

then, to minimize by-product formation, it is desirable that the synthesis gas have a slight excess of hydrogen. Thus, suppose a reformer is operated to produce a synthesis gas with a $H_2:CO$ ratio of 2.2 from a feed containing 50% CH_4, 35% H_2O, and 15% CO_2. If 80% conversion of methane can be expected in the reformer, calculate the complete composition of the synthesis gas.

The system shown in Figure 3.17 in this case consists of the reformer, a single input stream, and a single output stream. Prior to commencing with the degree-of-freedom analysis, the specified reactions must be examined to determine the

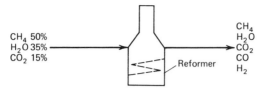

Figure 3.17 Flowsheet for Example 3.21.

number of independent reactions. If the reactions are sequenced in the order in which they are written above, the stoichiometric array becomes:

	1st Reaction	2nd Reaction	3rd Reaction	4th Reaction
CH_4	-1	0	-1	-1
CO_2	-1	1	0	1
CO	2	-1	1	0
H_2O	0	-1	-1	-2
H_2	2	1	3	4

Step A   Divide each entry in the first column by -1.

Step B   Add -1 times each entry of the new first column to each entry of the third column. Add -1 times each entry of the first column to each entry of the fourth column. The result of this operation is:

$$\begin{pmatrix} 1 & 0 & 0 & 0 \\ 1 & 1 & 1 & 2 \\ -2 & -1 & -1 & -2 \\ 0 & -1 & -1 & -2 \\ -2 & 1 & 1 & 2 \end{pmatrix}$$

Proceed with the second column.

Step A   Division by $+1$ leaves the array unchanged.

Step B   Add -1 times the second column to the third column. Add -2 times the second column to the fourth column. Add -1 times the second column to the first column. The result of these operations is the reduced array

$$\begin{pmatrix} 1 & 0 & 0 & 0 \\ 0 & 1 & 0 & 0 \\ -1 & -1 & 0 & 0 \\ 1 & -1 & 0 & 0 \\ -3 & 1 & 0 & 0 \end{pmatrix}$$

Since the remaining two columns are identically zero, the reduction process terminates. The above stoichiometric array, which represents a set of reactions that is equivalent to the original set, has only two independent columns. Hence, the original reaction set can have had at most two independent reactions. The two independent reactions given by the above reduced array are

$$CO + 3H_2 \rightleftharpoons CH_4 + H_2O$$
$$CO + H_2O \rightleftharpoons CO_2 + H_2$$

These correspond to just the third and second reactions of the original set of reactions. Of course, this choice of independent reactions is not the only one. For instance, the first two reactions of the original set will also form an independent set, as will be shown by repeating the calculation. The choice of which set to use is arbitrary. The important point is that a material balance problem stated in terms of the four original reactions can be solved using just two independent reactions.

With the number of independent reactions known, we can proceed with the degree-of-freedom analysis. From the flowsheet it is apparent that the system involves eight stream variables. Since there are two independent reactions, 10 variables need to be calculated. There are five species balances available for solution and two independent compositions, a conversion relation, and the $H_2 : CO$ ratio specified. Thus, with the choice of basis, the problem has degree of freedom zero.

If a basis of 100 mol/h CO in the product is assumed, then from the specified $H_2 : CO$ ratio, we immediately conclude that

$$N_{H_2}^{out} = 220 \text{ mol/h}$$

Suppose that the stoichiometric coefficients of the two independent reactions

$$CO + 3H_2 \rightleftharpoons CH_4 + H_2O$$
$$CO + H_2O \rightleftharpoons CO_2 + H_2$$

are used to construct the balance equations. Then, the species balances are:

CH$_4$ balance: $N_{CH_4}^{out} = 0.5N^{in} + r_1$

H$_2$O balance: $N_{H_2O}^{out} = 0.35N^{in} + r_1 - r_2$

CO$_2$ balance: $N_{CO_2}^{out} = 0.15N^{in} + r_2$

CO balance: $100 = 0 - r_1 - r_2$

H$_2$ balance: $220 = 0 - 3r_1 + r_2$

In addition, we know from the conversion relation that

$$N_{CH_4}^{out} = N_{CH_4}^{in}(1 - X_{CH_4}) = 0.5N^{in}(1 - 0.8) = 0.1N^{in}$$

Adding the CO and H_2 balances, we obtain immediately that

$$r_1 = 320/-4 = -80 \text{ mol/h}$$

Substituting this back into one of these balances,

$$r_2 = -20 \text{ mol/h}$$

can be calculated. Note that the negative reaction rates merely indicate that the two reactions actually proceed in the direction opposite to that given, that is, CH_4 is a product rather than a reactant. With r_1 known, the conversion relationship and the CH_4 balance yield

$$N^{in} = 200 \text{ mol/h}$$

With the reaction rates and the inlet flow known, the remaining outlet flows can be immediately calculated:

$$(N_{CH_4}^{out}, N_{H_2O}^{out}, N_{CO_2}^{out}, N_{CO}^{out}, N_{H_2}^{out}) = (20, 10, 10, 100, 220) \text{ mol/h}$$

The corresponding outlet compositions, in mole fractions, are (0.0556, 0.0278, 0.0278, 0.2778, 0.6111).

Suppose we repeat the solution of the problem using the first two reactions of the given reaction set, namely,

$$CH_4 + CO_2 \rightleftharpoons 2CO + 2H_2$$

$$CO + H_2O \rightleftharpoons CO_2 + H_2$$

Using the previous basis, $N_{CO}^{out} = 100$ mol/h and from the specified ratio, $N_{H_2}^{out} = 220$ mol/h. The material balances written in terms of the above stoichiometry are:

CH_4 balance:	$N_{CH_4}^{out} = 0.5N^{in} - r_1$
H_2O balance:	$N_{H_2O}^{out} = 0.35N^{in} - r_2$
CO_2 balance:	$N_{CO_2}^{out} = 0.15N^{in} - r_1 + r_2$
CO balance:	$N_{CO}^{out} = 0 + 2r_1 - r_2 = 100$ mol/h
H_2 balance:	$N_{H_2}^{out} = 0 + 2r_1 + r_2 = 220$ mol/h

Again adding the CO and H_2 balances, we obtain

$$r_1 = 80 \text{ mol/h}$$

and then, from the CO balance above,

$$r_2 = 60 \text{ mol/h}$$

From the conversion relationship, $N_{CH_4}^{out} = 0.1N^{in}$; therefore, the outlet stream vector in the previously given species order can be calculated as

$$N^{out} = (20, 10, 10, 100, 220) \text{ mol/h}$$

Since the reaction stoichiometry is different, the calculated reaction rates are different. However, since the two reaction sets used are equivalent, the outlet flows obtained are the same.

Note that although the rates of the two independent reactions could be determined, the material balances are insufficient to calculate the rates of the remaining two reactions. Hence, from a material balance point of view, we have no way of knowing which of the four reactions is the most significant or how much methane is consumed by the three reaction paths given. Detailed kinetic experiments must be conducted to answer such questions.

3.3.3 Independent Specifications

By means of the *array reduction procedure,* the set of chemical reactions postulated to occur in the system can be reduced to a linearly independent set. The significance of this reduction is that, if the material balance equations are written in terms of independent reactions, then, provided the problem is properly specified, it will be possible to calculate all the reaction rates as well as all the stream variables. The pivotal issue, of course, is what is a proper specification? Up to this point, we have implicitly assumed that proper specification merely means assigning values to the correct number of stream variables or imposing the correct number of subsidiary relations. However, even if the problem is expressed in terms of a set of independent reactions and even if the correct number of specifications is imposed, it is possible that the problem cannot be solved because the specifications were imposed on the wrong species. This possibility is illustrated in the following example.

Example 3.22 Suppose that a stream of methane ·is completely burned with oxygen to produce a product gas containing CO_2, CO, and H_2O and no residual CH_4 or O_2. The combustion of CH_4 can be described by means of the two reactions

$$CH_4 + 2O_2 \rightleftharpoons 2H_2O + CO_2$$

$$2CH_4 + 3O_2 \rightleftharpoons 4H_2O + 2CO$$

Note that these two reactions are independent because each involves a species not present in the other reaction (CO_2, CO). From the reactor flowsheet, Figure 3.18, it is clear that the unit will involve five stream variables. With the two reaction rates, this means the problem has seven variables. Since a balance can be written on each of the five species and since we are free to select a basis, the problem clearly has degree of freedom 1. Thus, solution ought to be possible with one additional specification. Using a basis of 1 mol/h CH_4, the balance equations are:

CH_4 balance: $0 = 1 - r_1 - 2r_2$

O_2 balance: $0 = N^2 - 2r_1 - 3r_2$

H_2O balance: $N^3_{H_2O} = 0 + 2r_1 + 4r_2$

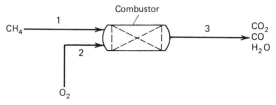

Figure 3.18 Flowsheet for Example 3.22.

CO$_2$ balance: $N^3_{CO_2} = 0 + r_1$

CO balance: $N^3_{CO} = 0 + 2r_2$

Now, suppose as the additional specification it is required that

$$N^3_{H_2O} = 2N^1 = 2 \text{ mol/h}$$

Then, the water balance reduces to

$$2 = 2r_1 + 4r_2$$

while the CH$_4$ balance is

$$1 = r_1 + 2r_2$$

The remaining balances each contain one unknown stream variable, in addition to the unknown rates. The water and CH$_4$ balances, therefore, need to be used to first calculate the rates. But, since these two equations are clearly just multiples of each other, simultaneous solution of them is not possible even though the correct number of specifications was imposed. Apparently, the specification is in some way dependent, that is, it leads to a dependent set of balance equations.

To investigate whether the numerical value of the $N^3_{H_2O}$-to-N^1 ratio has any influence, suppose the problem is resolved with the requirement that

$$N^3_{H_2O} = 3N^1 = 3 \text{ mol/h}$$

In that case, the water balance reduces to

$$3 = 2r_1 + 4r_2$$

while the CH$_4$ balance reduces to

$$1 = r_1 + 2r_2$$

This choice of specification leads to the inconsistency

$$3 = 2r_1 + 4r_2 \neq 2(r_1 + 2r_2) = 2$$

Since in both cases the specification is imposed on the exit water stream, it appears that the difficulty lies in having selected that particular species for specification. To confirm this suspicion, suppose the specification

$$N^3_{CO_2} = \tfrac{1}{2}N^1 = \tfrac{1}{2} \text{ mol/h}$$

is used. In this case, the CO$_2$ balance reduces to

$$\tfrac{1}{2} = r_1$$

which, when substituted into the CH$_4$ balance, yields

$$r_2 = \tfrac{1}{4}$$

With the rates determined, all flows can be calculated without further difficulty. The question of interest is, why was a specification on the H_2O species improper?

Recall that, for material balance purposes, the reaction rates serve purely as intermediate variables which, once determined, facilitate the calculation of the unknown stream variables. The values of the rates are usually not specified directly but are obtained by solving some of the material balance equations possibly in conjunction with other subsidiary relations. Typically, as in Examples 3.16 and 3.21, the subsidiary relations are used to deduce values of some of the stream variables. These are substituted into the species balance equation, and then R of the balance equations are used to calculate the R rates. Clearly, in order for the R independent reaction rates to be calculable, the R balance equations selected must contain no unknown stream variables and must be linearly independent. If these conditions are met and if the correct number of stream variables have been specified, then the remaining $S - R$ balance equations should suffice to calculate the remaining $S - R$ unknown stream variables. However, if the R selected balance equations are not linearly independent, the rates cannot be determined, and thus the remaining unknown stream variables will also remain indeterminate.

To make these verbal arguments more precise, we shall define a reacting species to be *completely specified* if all of its input and output flow rates are known. If a species is completely specified, then, since the left-hand side of its species balance equation

$$\Delta N_s \equiv N_s^{\text{out}} - N_s^{\text{in}} = \sum_{r=1}^{R} \sigma_{sr} r_r$$

will be a constant, the only remaining unknowns will be the reaction rates. From this, it is clear that in order to determine the R rates, R species will have to be completely specified. Moreover, to guarantee that the corresponding R species balances will suffice to calculate the R rates, this set of R balance equations,

$$\Delta N_s = \sum_{r=1}^{R} \sigma_{sr} r_r$$

must have linearly independent coefficient σ_{sr}.

We are thus led to the following definition: For species balances, a specification is *independent* if the rows of the stoichiometric matrix corresponding to the R completely specified species are linearly independent.

The determination of linear independence can usually be made by inspection; but, if need be, the *array reduction procedure* presented in the preceding section can be employed. In any case, for a reacting system, we can at this point conclude that for a specification to be *proper*, that is, for it to lead to a single and physically realistic solution, it is necessary that the specification be both independent and correct in number.

Example 3.23   The stoichiometric array for the reaction system given in the previous example is:

	1st Reaction	2nd Reaction
CH_4	-1	-2
O_2	-2	-3
H_2O	2	4
CO_2	1	0
CO	0	2

In the first two cases considered in Example 3.22, both the CH_4 and the H_2O species are completely specified. The rows of the array corresponding to these species are

$$(-1 \quad -2)$$

and

$$(\ 2 \quad\ 4)$$

These are clearly multiples of each other; hence, the complete specification of those two species is not independent. A solution cannot be obtained regardless of what numerical values are assigned to the input and output rates of these species. Depending upon the exact choice of values, either the two species balances will be dependent or inconsistent.

On the other hand, in the third case treated in Example 3.22, the CH_4 and CO_2 species are completely specified. The rows of the stoichiometric array corresponding to these species,

$$(-1 \quad -2)$$
$$(\ 1 \quad\ 0)$$

are clearly independent. Consequently, it was possible to obtain a solution to the balances.

3.4 SUMMARY

In this chapter, we have formulated the species material balance equations for systems involving single or multiple reactions. We have considered several different measures of the progress of a set of reactions and illustrated their use. We then incorporated these elements into an analysis of systems consisting of single or multiple reactor units. We showed how this analysis can be used to sequence the unit balances of a multiunit flowsheet and considered in some detail the precautions which must be taken when overall balances are used. Finally, we investigated some of the algebraic properties of reaction sets, determined how a set of dependent reactions could be identified, and showed how an independent subset could be isolated. We concluded by pinning down more precisely what is meant by a proper problem specification. At this stage, you should be confident of solving any material balance problem involving a known set of chemical reactions.

PROBLEMS

3.1 (a) Write a balanced chemical reaction for the reaction of 1 mol $C_8H_{12}S_2$ with O_2 to produce CO_2, H_2O, and SO_2.
(b) Calculate the rate of production of all species if 2 mol/h $C_8H_{12}S_2$ is reacted with a stoichiometric amount of O_2.
(c) Calculate the rate of reaction.

3.2 The combustion of C_3H_6 to CO_2 and H_2O can be described by either the reaction

$$C_3H_6 + \tfrac{9}{2}O_2 \rightarrow 3CO_2 + 3H_2O$$

or the reaction

$$2C_3H_6 + 9O_2 \rightarrow 6CO_2 + 6H_2O$$

Suppose 10 mol/h C_3H_6 is reacted with 50 mol/h O_2 with complete conversion of C_3H_6. Calculate the rates of reaction obtained with each reaction. Explain how the two rates are related and why.

3.3 Consider the reaction

$$3C_2H_5OH + 2Na_2Cr_2O_7 + 8H_2SO_4 \rightarrow 3CH_3COOH \\ + 2Cr_2(SO_4)_3 + 2Na_2SO_4 + 11H_2O$$

(a) If a reactor feed has the composition (mol %) of 20% C_2H_5OH, 20% $Na_2Cr_2O_7$, and the rest H_2SO_4, which is the limiting reactant?
(b) If a reactor is fed 230 kg/h C_2H_5OH, what feed rates of the other two reactants would be required to have a stoichiometric feed mixture?

3.4 A reactor is fed an equimolar mixture of chemicals A, B, and C to produce the product D via the reaction

$$A + 2B + \tfrac{3}{2}C \rightarrow 2D + E$$

If the conversion in the reactor is 50%, calculate the number of moles of D produced per mole of reactor feed.

3.5 A reaction with stoichiometric equation

$$A + 3B \rightarrow 2D$$

is carried out with 20% conversion of A. The reactor feed stream contains 25% A and 75% B by weight and has a rate of 1000 kg/h. If the molecular weight of A is 28 and that of B is 2,

(a) Calculate the molecular weight of D.

(b) Calculate the composition of the exit stream on a weight basis.

3.6 Chlorine dioxide gas is used in the paper industry to bleach pulp produced in a Kraft mill. The gas is produced by reacting sodium chlorate, sulfuric acid, and methanol in lead-lined reactors:

$$6NaClO_3 + 6H_2SO_4 + CH_3OH \rightarrow 6ClO_2 + 6NaHSO_4 + CO_2$$

Suppose 14 mol of an equimolar mixture of $NaClO_3$ and H_2SO_4 is added per 1 mol CH_3OH (see Figure P3.6).

(a) Determine the limiting reactant.

(b) Calculate the reactant flows required to produce 10 metric tons per hour of ClO_2, assuming 90% conversion is obtained.

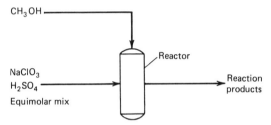

Figure P3.6

3.7 In the sulfuric acid industry, oleum is a term used for 100% acid containing free, unreacted SO_3 dissolved in the acid. A 20% oleum, for example, is 20 lb SO_3 in 80 lb 100% acid, per 100 lb mix. Oleum can also be designated as a percentage of sulfuric acid in excess of 100%. It is figured as the pounds of 100% acid which would be produced by adding enough water to 100 lb oleum to dissolve all the free SO_3. Using these definitions, calculate:

(a) Pounds of 25% oleum that can potentially be produced per 100 lb sulfur.

(b) Percentage of sulfuric acid corresponding to the 25% oleum.

3.8 Sodium hypochlorite is formed according to the reaction

$$2NaOH + Cl_2 \rightarrow NaOCl + NaCl + H_2O$$

in a continuous reactor by bubbling Cl_2 through a concentrated 40% (mass) NaOH solution. Suppose the solution of NaOH in H_2O is fed at 1000 kg/h and the Cl_2 gas, at 10 kgmol/h.

(a) Calculate the degree of freedom assuming the conversion is specified.

(b) Determine which is the limiting reactant.

(c) Calculate the outlet composition of the reactor assuming 100% conversion of the limiting reactant.

(d) Calculate the outlet composition of the reactor assuming 60% conversion of the limiting reactant.

3.9 An old process for producing hydrochloric acid involves heating a mixture of $NaHSO_4$ and $NaCl$ in a special furnace. Upon reaction, the residual Na_2SO_4 remains as a solid while the HCl is recovered as a gas. If the reaction follows the stoichiometry

$$NaHSO_4 + NaCl \rightarrow Na_2SO_4 + HCl$$

and the reactants are fed in a stoichiometric ratio, calculate the amount and composition of the residual solids. Assume that conversion is 95% complete and that $NaCl$ is fed at the rate of 5844 lb/day.

3.10 Superphosphate is produced by reacting calcium phosphate with sulfuric acid according to the reaction

$$Ca_3(PO_4)_2 + 2H_2SO_4 \rightarrow CaH_4(PO_4)_2 + 2CaSO_4$$

If 20,000 kg/day of raw calcium phosphate containing 14% inert impurities is reacted with 15,000 kg/day 92% H_2SO_4, determine the rate of production of superphosphate assuming the reaction is 95% complete. Which is the limiting reactant?

3.11 In a process for the catalytic hydration of ethylene to ethyl alcohol, only a fraction of the ethylene is converted. The product is condensed and removed after each pass through the converter and the unconverted gases are recycled. The condenser may be assumed to remove all the alcohol and the recycle gases will contain 6.5% (mol) water vapor. The conversion of ethylene per pass through the converter is 4.5%. The molar ratio of water to ethylene in the feed to the converter, after mixing the recycle gas with fresh feed, is 0.55. Calculate all streams in the process.

3.12 Acetic acid can be produced via the reaction

$$3C_2H_5OH + 2Na_2Cr_2O_7 + 8H_2SO_4$$
$$\rightarrow 3CH_3COOH + 2Cr_2(SO_4)_3 + 2Na_2SO_4 + 11H_2O$$

In the recycle system shown in Figure P3.12, 90% overall conversion of C_2H_5OH is obtained with a recycle flow equal to the feed rate of fresh C_2H_5OH. The feed rates of fresh H_2SO_4 and $Na_2Cr_2O_7$ are 20 and 10%,

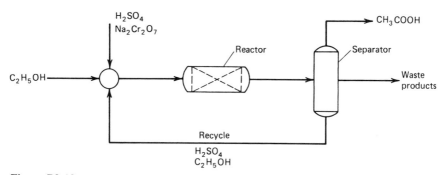

Figure P3.12

respectively, in excess of the stoichiometric amounts required for the fresh C_2H_5OH feed. If the recycle stream contains 94% H_2SO_4 and the rest C_2H_5OH, calculate the product flow and the conversion of C_2H_5OH in the reactor.

3.13 Perchloric acid can be produced via the flowsheet shown in Figure P3.13. The reaction follows the stoichiometry

$$Ba(ClO_4)_2 + H_2SO_4 \rightarrow BaSO_4 + 2HClO_4$$

If the H_2SO_4 fed to the reactor is 20% in excess of the stoichiometric amount required for reaction with the fresh feed of $Ba(ClO_4)_2$ and if 1000 lb/h of stream 1 is fed, calculate all unknown stream variables. Assume all compositions are in weight fractions.

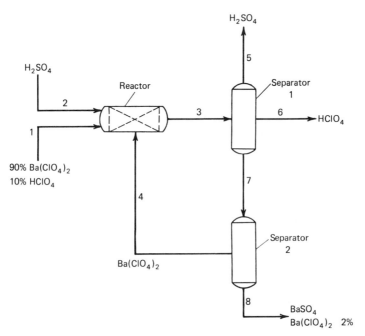

Figure P3.13

3.14 The reaction

$$2A + 5B \rightarrow 3C + 6D$$

is carried out in a reactor with 60% conversion of B. Most of the unreacted B is recovered in a separator and recycled to the reactor (Figure P3.14). The fresh feed to the reactor consists of A and B, with the fresh A being 30% in excess of the stoichiometric amount required to react with the fresh B (see Figure P3.14). If the overall conversion of B in the process is 95%, calculate the product and recycle flows required to produce 100 mol/h C.

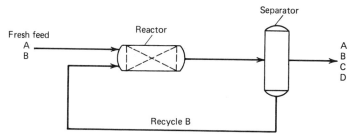

Figure P3.14

3.15 The solvent ethyl ether is made industrially by the dehydration of ethyl alcohol using a sulfuric acid as catalyst (Figure P3.15):

$$2C_2H_5OH \rightarrow (C_2H_5)_2O + H_2O$$

Assuming the recycle is one half of the feed rate to the process, the feed rate is 1000 kg/h alcohol solution, containing 85% (weight) alcohol, and the recycle alcohol solution will have the same composition as the feed, calculate

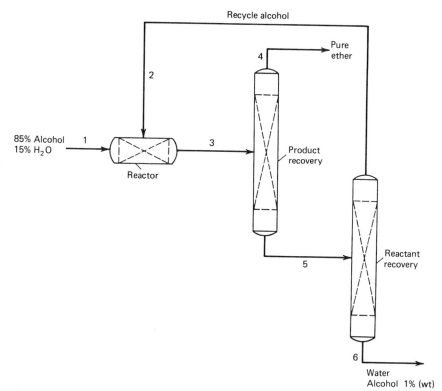

Figure P3.15

the production rate of ether, the loss of alcohol in stream 6, the conversion in the reactor, and the conversion for the process.

3.16 Hydrogen is used to reduce 1 ton/h Fe_2O_3 to metallic iron according to the reaction

$$Fe_2O_3 + 3H_2 \rightarrow 2Fe + 3H_2O$$

The water is condensed and the unreacted hydrogen is recycled (see Figure P3.16). Because the hydrogen in the fresh feed contains 1% CO_2 impurity, some of the unreacted hydrogen must be purged. Calculate the flow rate and the composition of the purge stream required to limit the CO_2 in the reactor feed to 3.5% if the ratio of recycle to fresh feed is 5 : 1 on a molar basis.

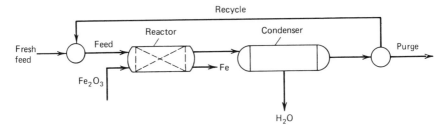

Figure P3.16

3.17 Methyl iodide can be produced by reacting hydriodic acid with an excess of methanol via the reaction

$$HI + CH_3OH \rightarrow CH_3I + H_2O$$

A typical process for large-scale production of methyl iodide is indicated in Figure P3.17. The process conditions are:

   1. The feed to the reactor contains 2 mol CH_3OH per 1 mol HI.
   2. 50% conversion of HI is obtained in the reactor.
   3. 90% of the H_2O entering the first separator leaves in stream 5.
   4. All compositions are on a molar basis.

How many moles of CH_3I are produced per mole of fresh HI feed?

3.18 A mixture containing 68.4% H_2, 22.6% N_2, and 9% CO_2 reacts according to the scheme

$$N_2 + 3H_2 \rightarrow 2NH_3$$
$$CO_2 + H_2 \rightarrow CO + H_2O$$

until the mixture contains 15% NH_3 and 5% H_2O. Calculate the mole fractions of N_2, H_2, CO_2, and CO.

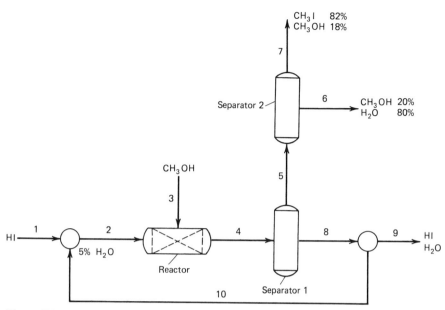

Figure P3.17

3.19 Acetaldehyde, CH_3CHO, can be produced by catalytic dehydrogenation of ethanol, C_2H_5OH, via the reaction

$$C_2H_5OH \rightarrow CH_3CHO + H_2$$

There is, however, a parallel reaction producing ethyl acetate, $CH_3COOC_2H_5$:

$$2C_2H_5OH \rightarrow CH_3COOC_2H_5 + 2H_2$$

Suppose that in a given reactor the conditions are adjusted so that a conversion of 95% ethanol is obtained with an 80% yield of acetaldehyde. Calculate the composition of the reactor product assuming the feed consists of pure ethanol.

3.20 Product C is produced from reactants A and B via the following three reactions:

$$2A + B \rightarrow 2D + E$$
$$A + D \rightarrow 2C + E$$
$$C + 2B \rightarrow 2F$$

With a feed ratio of 2 mol A per 1 mol B and 80% conversion of A, a product mixture containing 4 mol A per 1 mol B and 6 mol combined products C, D, E, and F per 1 mol combined residual reactants A and B is obtained (see Figure P3.20). Assuming the problem is correctly specified and using a feed of 200 mol/h A,

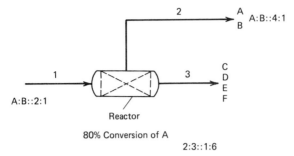

Figure P3.20

(a) Calculate the three reaction rates and the reactor outlet flows.
(b) Calculate the fractional yield of C from A.

3.21 Carbon disulfide is used in viscose rayon and cellóphane manufacture and in the production of carbon tetrachloride. In the generally preferred process, vaporized sulfur is reacted with methane according to the reactions

$$CH_4 + 4S \rightarrow CS_2 + 2H_2S$$

$$CH_4 + 2S \rightarrow CS_2 + 2H_2$$

$$CH_4 + 2H_2S \rightarrow CS_2 + 4H_2$$

For a feed containing 4 mol sulfur per 1 mol methane, calculate the composition of the product if 90% conversion of methane and 70% conversion of sulfur is obtained.

3.22 Hydrodealkylation is a process in which side chains, consisting of alkyl groups, are removed from aromatics by reaction with hydrogen to form the parent aromatic compound. For instance, toluene can be converted to benzene:

$$C_6H_5CH_3 + H_2 \rightarrow C_6H_6 + CH_4$$

Xylene can be converted to toluene:

$$C_6H_4(CH_3)_2 + H_2 \rightarrow C_6H_5CH_3 + CH_4$$

Pseudocumene and other C_9 hydrocarbons containing three CH_3 groups can be converted to xylenes:

$$C_6H_3(CH_3)_3 + H_2 \rightarrow C_6H_4(CH_3)_2 + CH_4$$

In a given application, a refinery reformate stream consisting of 5% benzene, 20% toluene, 35% xylene, and 40% C_9 hydrocarbons is reacted with hydrogen. If 5 mol H_2 is used per 1 mol feed, 80% conversion of toluene, 74% conversion of xylene, and 70% conversion of C_9 hydrocarbons are attained. The product stream is found to contain a small amount, 0.1% of biphenyl, indicating that the side reaction

$$2C_6H_6 \rightarrow C_6H_5C_6H_5 + H_2$$

occurs to some extent. Calculate the complete composition of the reactor outlet stream.

3.23 In the sulfuric acid industry, oleum is a term used for 100% acid containing free, unreacted SO_3 dissolved in the acid. A 20% oleum, for example, is 20 lb SO_3 in 80 lb 100% acid, per 100 lb mix. A contact sulfuric acid plant produces 10 tons 20% oleum and 40 tons 98% acid per day (see flowsheet, Figure P3.23). The oleum tower is fed with a portion of the acid from the 98% tower. The 98% tower is fed with 97% acid, obtained by diluting part of the 98% output. The gas fed to the oleum tower analyzes 10.8% SO_3. Calculate the amount of acid to be fed to each tower per day.

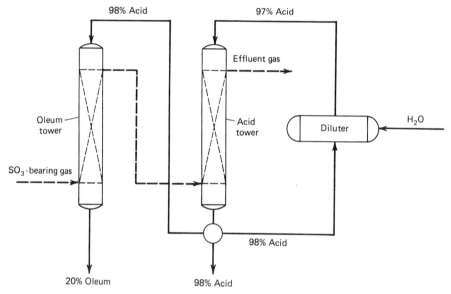

Figure P3.23

3.24 Recovery of Na_2CO_3 and its conversion to NaOH are the key elements of the Kraft process. In the simplified flowsheet shown in Figure P3.24, Na_2CO_3 is reacted with $Ca(OH)_2$ in the calciner via the reaction

$$Na_2CO_3 + Ca(OH)_2 \rightarrow 2NaOH + CaCO_3$$

The $CaCO_3$ is washed in a thickener and converted to CaO in a kiln via the reaction

$$CaCO_3 \rightarrow CaO + CO_2$$

The resulting lime (CaO) is hydrated in the slaker to again obtain $Ca(OH)_2$:

$$CaO + H_2O \rightarrow Ca(OH)_2$$

Use the compositions shown on the flowsheet and the additional specifications that

$$F^5 = 4F^3$$
$$F^1_{H_2O} = F^{10}$$

and assume that all reactions have 100% conversion. All compositions are in wt %.

(a) Show that the given reactions are independent.
(b) Construct a degree-of-freedom table.
(c) Deduce a calculation order which can be used to determine all streams. Give your reasoning in full detail.
(d) Solve the overall balances to the problem to determine the number of pounds $CaCO_3$ required per pound of Na_2CO_3 processed.

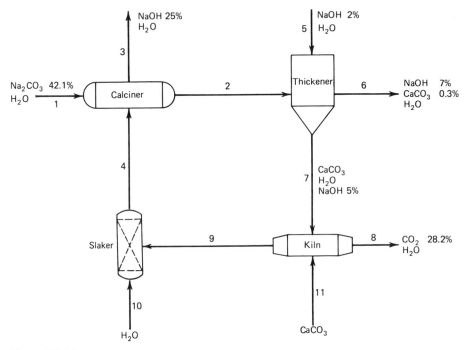

Figure P3.24

3.25 Product P is produced from reactant R according to the reaction

$$2R \rightarrow P + W$$

Unfortunately, both the reactant and the product P decompose to form by-product B according to the reactions

$$R \rightarrow B + W$$
$$P \rightarrow 2B + W$$

When a process feed consisting of one part inert I to 11 parts R is used and the recycle rate is adjusted to obtain a reactor feed R mole fraction of 0.85, a 50% conversion of R in the plant and an 80% yield of P from R for the plant are observed.

(a) Construct a degree-of-freedom table for the process shown in Figure P3.25. Is the process correctly specified?

(b) Suppose that with a new catalyst, only the first two reactions are observed. How does this affect the degree-of-freedom analysis of part (a)?

(c) Using the degree-of-freedom table for the conditions appropriate for part (b), deduce a calculation order which can be used to determine all streams.

(d) Solve the problem.

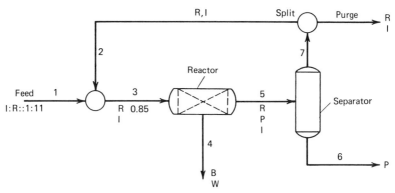

Figure P3.25

3.26 Sulfuric acid is produced by the successive oxidation of sulfur to SO_3 followed by reaction with H_2O. In the flowsheet shown in Figure P3.26, air and sulfur are first reacted in the sulfur burner to produce SO_2 via the reaction

$$S + O_2 \rightarrow SO_2$$

In this reaction, all of the sulfur is converted to SO_2. Assume air is 21 mol % oxygen and 79% N_2 and that 50% more oxygen is used than the stoichiometric amount required to convert the sulfur to SO_2. The gas stream from the sulfur burner is sent to the converter in which all of the SO_2 is oxidized to SO_3 with the aid of a catalyst. Next, the SO_3-containing gas stream is contacted with a stream of concentrated H_2SO_4 in the oleum tower. The SO_3 reacts with whatever water is present in the acid stream via the reaction

$$H_2O + SO_3 \rightarrow H_2SO_4$$

In addition, some of the SO_3 is dissolved in the pure H_2SO_4 to yield a product (oleum) consisting of 37.5 % SO_3 and the rest H_2SO_4. The gas stream leaving the oleum tower containing 12 mol % SO_3 is next contacted with a more dilute acid stream (80 % H_2SO_4 and the rest H_2O) in the acid tower. In this unit, all of the remaining SO_3 reacts to H_2SO_4. The dilute acid is prepared by recycling some of the concentrated acid and mixing it with water in the

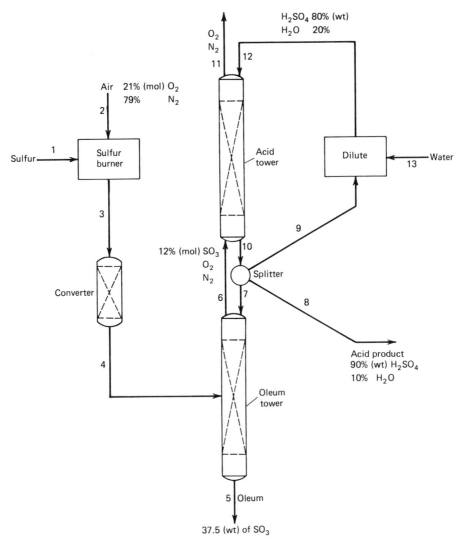

Figure P3.26

diluter. Calculate all flows in the process assuming that the plant is to produce 20,000 lb/day 90% H_2SO_4 product.

3.27 Benzene can be produced by the dealkylation of toluene following the reaction

$$C_6H_5CH_3 + H_2 \rightarrow C_6H_6 + CH_4$$

This catalytic reaction is, however, accompanied by the side reaction

$$2C_6H_5CH_3 + H_2 \rightarrow (C_6H_5)_2 + 2CH_4$$

producing the undesirable by-product diphenyl. Because of this side reaction, the conversion of toluene must be kept below 100% and a series of separations with recycle of unused reactants must be instituted. In the flowsheet shown in Figure P3.27, with a reactor feed (stream 3) containing 5 mol H_2 per 1 mol toluene, 75% conversion of toluene is attained.

(a) Assuming that the outlet stream 6 contains 5% benzene and 2% toluene, calculate the fractional yield of benzene in the reactor and the make-up H_2 requirements per mole toluene fed.

(b) Assuming that the outlet stream 6 is alternatively specified as containing 2% toluene and 58% CH_4, calculate the fractional yield of benzene.

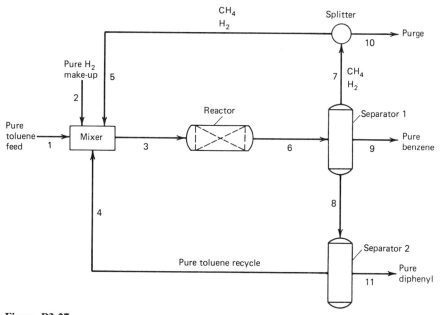

Figure P3.27

3.28 Product P is produced from reactant R according to the reaction

$$2R \rightarrow P + W$$

with side reactions

$$R \rightarrow B + W$$
$$P \rightarrow 2B + W$$

Only 50% conversion of R is achieved in the reactor (Figure P3.28) when a fresh feed containing 1 mol inerts I per 11 mol R is used. The unreacted R and inerts are separated from the reactor products and recycled. Some of the unreacted R and inerts are purged to limit the inerts level in the combined

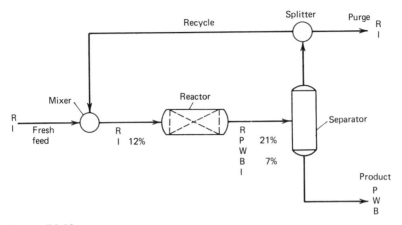

Figure P3.28

reactor feed to 12 (mol) %. If the reactor outlet stream analyzes 21% P and
7% B, on a molar basis, calculate all flows in the process for a fresh feed
rate of 100 mol/h.

3.29 Most modern processes for production of nitric acid are based on the sequential oxidation of ammonia to oxides of nitrogen followed by absorption
of these intermediates in water. In the flowsheet shown in Figure P3.29,
ammonia and air are mixed in a 1 : 10 molar ratio and reacted catalytically
in the first reactor stage. The two reactions occurring are the main reaction

$$4NH_3 + 5O_2 \rightarrow 4NO + 6H_2O$$

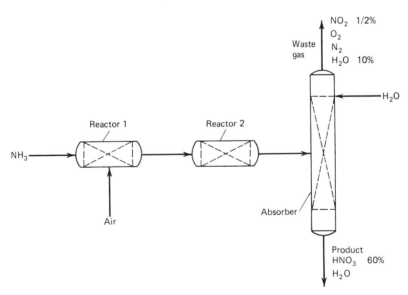

Figure P3.29

and the secondary reaction

$$2NH_3 + \tfrac{3}{2}O_2 \rightarrow N_2 + 3H_2O$$

All of the NH_3 is converted with a 95% selectivity for NO. In the second reactor stage, the NO produced is further oxidized to NO_2 via the reaction

$$NO + \tfrac{1}{2}O_2 \rightarrow NO_2$$

Finally, the reactor 2 effluent is further treated with water to produce the desired 60% HNO_3 product. The scrubber reaction is

$$2NO_2 + \tfrac{1}{2}O_2 + H_2O \rightarrow 2HNO_3$$

The waste gas from the process contains $\tfrac{1}{2}\%$ NO_2 and about 10% H_2O. The air composition may be taken as 21% O_2 and 79% N_2. Assuming all compositions are in mol %, calculate the composition of all streams in the process.

3.30 In a top-secret new process, HD_3A_4B is reacted with AD to produce primary products A_2B and A_3D_3 via the reaction

$$HD_3A_4B + AD \rightleftharpoons A_2B + A_3D_3 + HD$$

Unfortunately, there are two side reactions,

$$HD_3A_4B + 2\,AD \rightleftharpoons HAD_2 + A_2B + A_3D_3$$

$$HD + AD \rightleftharpoons HAD_2$$

which produce undesirable by-product HAD_2 (see Figure P3.30). It is desired to limit HAD_2 in the A_3D_3 product stream to 15% on a molar basis because it is difficult to separate HAD_2 from A_3D_3. The stream leaving the reactor contains 20% A_3D_3 on a molar basis. Calculate the production rates of products A_2B and A_3D_3 and the HD_3A_4B fresh feed rate if the feed rate of AD is 750 lbmol/h and the AD is completely reacted.

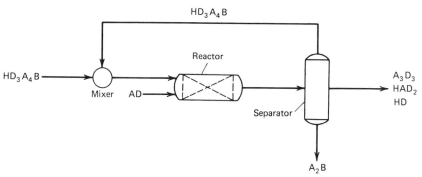

Figure P3.30

3.31 Calcium cyanamide can be manufactured from limestone ($CaCO_3$), coke, and nitrogen using the processing steps shown in the flowsheet (Figure P3.31).

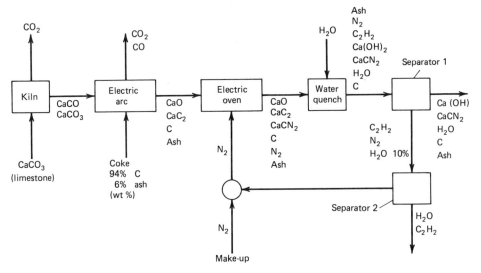

Figure P3.31

In the kiln, 90% of the $CaCO_3$ fed is thermally decomposed via the reaction

$$CaCO_3 \rightarrow CaO + CO_2$$

The CO_2 gas is removed and the remaining material is sent to an electric arc furnace where it is reacted with carbon to produce CaC_2:

$$CaO + 3C \rightarrow CaC_2 + CO$$

At this point, the remaining $CaCO_3$ is decomposed completely to yield additional CaO:

$$CaCO_3 \rightarrow CaO + CO_2$$

The coke feed rate is adjusted so that the amount of C fed is equal to the stoichiometric amount required for reacting with all of the available CaO. The CaO conversion in the arc furnace is not complete; only about 90% can be achieved. The CO_2 and CO gaseous by-products are separated, and the CaC_2 and CaO, carbon, and ash residuals are transferred to an electric oven in which the CaC_2 reacts with nitrogen to produce the cyanamide

$$CaC_2 + N_2 \rightarrow CaCN_2 + C$$

Conversion of CaC_2 is 80%, provided that an excess of N_2 is used (three times the stoichiometric requirement). The combined reaction products are then cooled by quenching with an excess of water. During the course of quenching, the remaining unreacted CaC_2 and CaO are completely converted to the hydroxide:

$$CaC_2 + 2H_2O \rightarrow Ca(OH)_2 + C_2H_2$$
$$CaO + H_2O \rightarrow Ca(OH)_2$$

The quenched stream is separated of its gaseous components—N_2, C_2H_2, and some water vapor. These gaseous components are further separated to yield a purified N_2 which is recycled back to the electric oven.

(a) Carry out the degree-of-freedom analysis to show that the process is underspecified.
(b) Carry out the balance calculations as far as possible. At what point do you run short of specified information? How much additional information would you need?

3.32 Ethylene oxide is made (Figure P3.32) by the partial oxidation of ethylene with oxygen using a siver catalyst:

$$2C_2H_4 + O_2 \rightarrow 2C_2H_4O$$

An undesirable side reaction also occurs:

$$C_2H_4 + 3O_2 \rightarrow 2CO_2 + 2H_2O$$

With a reactor inlet composition of 10% C_2H_4, 11% O_2, 1% CO_2, and the rest N_2 inert diluent, 25% conversion is observed and the reactor outlet stream is found to contain 2 mol CO_2 per 1 mol H_2O. The ethylene oxide is removed from the stream leaving the reactor by means of an absorber. The overhead gases from the absorber are found to contain 6 mol C_2H_4 per 1 mol CO_2. The C_2H_4O-containing liquid (4% C_2H_4O) from the absorber is sent to a steam stripper. The product, stream 12, analyzes 25% C_2H_4O. Part of the gases from the absorber are purged and the remainder is recycled to the reactor.

(a) Construct a degree-of-freedom table and show the process is correctly specified.
(b) Deduce a calculation order assuming all flows and compositions are to be calculated. Give explanations.
(c) Calculate the diluent N_2 required per mole of C_2H_4 fed to the process.
(d) Calculate the overall yield of C_2H_4O from C_2H_4 in the process. Calculate the fractional yield of C_2H_4O as it occurs in the reactor itself. Explain why these are different.

$$\text{Overall yield} = \frac{\text{moles } C_2H_4O \text{ actually produced}}{\begin{array}{c}\text{moles } C_2H_4O \text{ which could be}\\\text{produced if all } C_2H_4 \text{ fed was used to}\\\text{produce } C_2H_4O \text{ only}\end{array}}$$

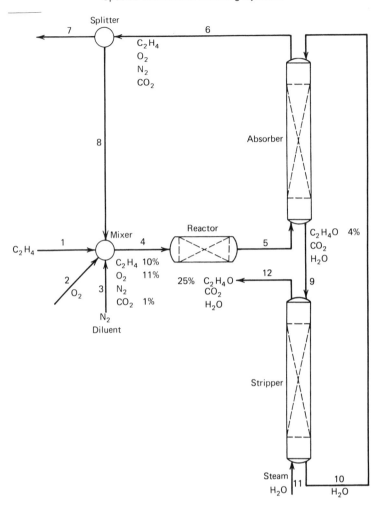

Figure P3.32

3.33 The important industrial chemical soda ash, Na_2CO_3, is produced from lime-stone, $CaCO_3$, salt, $NaCl$, and coke, C, using the Solvay process. In this ingenious process, the hypothetical chemical reaction

$$CaCO_3 + 2NaCl \rightarrow CaCl_2 + Na_2CO_3$$

which *does not* occur under industrially acceptable conditions, is carried out *indirectly* via a sequence of reactions involving the intermediate, ammonia. In the flowsheet shown in Fig. P3.33, coke (assumed to be 100% carbon) and limestone (assumed to be pure $CaCO_3$) are charged to a kiln in the ratio of 3 mol C to 4 mol $CaCO_3$. In the kiln, the carbon burns completely with air (21% O_2, 79% N_2):

$$C + O_2 \rightarrow CO_2$$

and provides the heat required to effect the thermal decomposition of $CaCO_3$:

$$CaCO_3 \rightarrow CaO + CO_2$$

Part of the flue gases from the kiln analyzing 36.75% CO_2 is purged and the remainder (64%) is sent to the carbonating unit. The solid residue from the kiln, assumed to consist entirely of CaO, is reacted with water in a unit called the slaker to produce a 35% calcium hydroxide solution, via the reaction

$$CaO + H_2O \rightarrow Ca(OH)_2$$

This hydroxide solution is reacted in the ammonia recovery unit with the ammonium chloride recycle from the carbonating unit to produce ammonia and by-product $CaCl_2$:

$$2NH_4Cl + Ca(OH)_2 \rightarrow 2NH_3 + CaCl_2 + 2H_2O$$

The recycle stream (stream 13) contains 4.7% dissolved CO_2, 7.65% NaCl, 61.54% H_2O, and the remainder the ammonium compounds. The $CaCl_2$ by-product stream also contains 80% H_2O and a small amount NH_4OH. Because

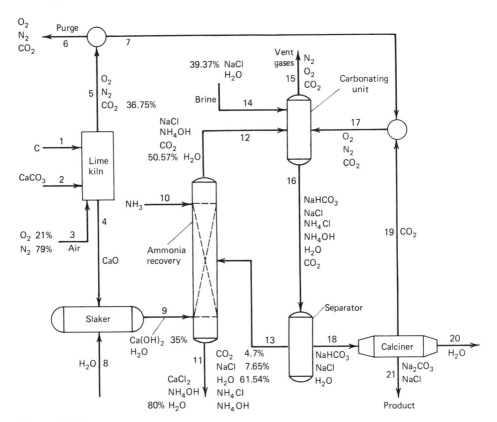

Figure P3.33

of this ammonia loss, some make-up NH_3 must be supplied to the process. Typical NH_3 consumption for the plant is 1 mol NH_3 per 17.5 mol Na_2CO_3 produced. The NH_3 introduced as make-up as well as the NH_3 produced by reaction immediately hydrolyze to ammonium hydroxide in the ammonia recovering unit:

$$NH_3 + H_2O \rightarrow NH_4OH$$

Stream 12, the concentrated (only 50.57% H_2O) ammonium hydroxide stream from the recovery unit, is next introduced into the carbonating unit in which the hydroxide reacts with NaCl, fed as a saturated brine solution containing 34.37% NaCl, and a CO_2-rich flue gas to produce $NaHCO_3$ via the reaction

$$NH_4OH + CO_2 + NaCl \rightarrow NaHCO_3 + NH_4Cl$$

The residual flue gases from the carbonating unit are vented while the product slurry, containing the $NaHCO_3$ precipitate, is sent to a separation subprocess. In the separator unit, the bicarbonate is filtered and washed in stages to produce a wet bicarbonate stream consisting of bicarbonate and some small amounts of NaCl. The residual liquid stream from the separation unit is recycled to the ammonia recovery unit while the wet bicarbonate stream is charged to a rotary furnace, called a calciner, in which bicarbonate is thermally decomposed to produce sodium carbonate:

$$2NaHCO_3 \rightarrow Na_2CO_3 + CO_2 + H_2O$$

The resulting solid product contains 2 mol NaCl per 100 mol sodium carbonate. The CO_2 and H_2O driven off in the calciner are cooled to condense out the H_2O. The CO_2 is mixed with the flue gases from the lime kiln and fed to carbonating unit. Assume all compositions are in mole fractions or %.

(a) Construct a degree-of-freedom table and show that the process is correctly specified.

(b) Outline the order in which the calculations ought to be carried out in order to determine all flows and compositions in the flowsheet.

(c) Calculate the moles of $CaCl_2$ solution produced (stream 11) per mole of brine feed (stream 14).

3.34 The important chemical intermediate acetaldehyde can be catalytically produced from partial oxidation of ethane. The primary reaction is

$$C_2H_6 + O_2 \rightarrow C_2H_4O + H_2O$$

However, there are a number of side reactions which also occur to a significant degree:

$$C_2H_6 + \tfrac{7}{2}O_2 \rightarrow 2CO_2 + 3H_2O$$
$$C_2H_6 + \tfrac{3}{2}O_2 \rightarrow CH_3OH + CO + H_2O$$

$$CH_3OH + \tfrac{1}{2}O_2 \rightarrow CH_2O + H_2O$$

$$2CO + 3H_2O \rightarrow C_2H_6 + \tfrac{5}{2}O_2$$

To reduce the formation of these various by-products, the reactor must be operated at low C_2H_6 conversion and high C_2H_6-to-O_2 ratios in the reactor feed. The process must, therefore, involve a high recycle rate and, because air is used as source of oxygen, must have a recycle purge stream to remove the inert N_2. To avoid the loss of valuable ethane in the purge, in the process flowsheet shown in Figure P3.34, the recycle stream is split into two equal parts. One part is subjected to a separation which will preferentially remove a stream of N_2, CO, and CO_2 for venting. The other half of the recycle stream is sent directly back to the reactor without treatment. Suppose under

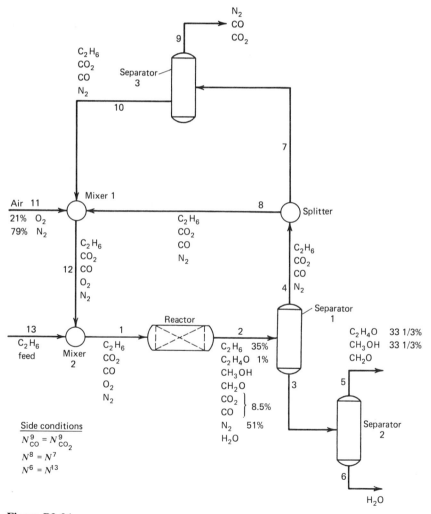

Figure P3.34

certain operating conditions the vent gas, stream 9, is found to contain equal molar amounts of CO and CO_2; the product stream (stream 5) is found to consist of $33\frac{1}{3}$ % C_2H_4O, $33\frac{1}{3}$ % CH_3OH, and $33\frac{1}{3}$ % CH_2O; and the reactor outlet stream is found to contain 35% C_2H_6, 51% N_2, 1% C_2H_4O, and 8.5% combined CO and CO_2. Also, it is found that 1 mol H_2O is formed in the process (stream 6) per 1 mol fresh C_2H_6 fed to the process (stream 13). All compositions and side conditions are in mole units.

(a)   Construct a degree-of-freedom table and show that the problem is correctly specified.

(b) . Choose the basis location and devise a unit-by-unit calculation order which ought to be followed if all streams in the flowsheet were to be calculated. Explain your reasoning in detail.

(c)   Calculate the *overall* yield of C_2H_4O from C_2H_6. The overall yield is defined as

$$\text{Overall yield} = \frac{\text{overall rate of } C_2H_4O \text{ production in the plant}}{\begin{array}{c}\text{overall rate of production of } C_2H_4O \text{ which could be}\\ \text{obtained if the } C_2H_6 \text{ converted was all used to}\\ \text{produce } C_2H_4O \text{ alone}\end{array}}$$

For this calculation, use as basis 12 mol/h of stream 5.

(d)   Suppose the reaction

$$2CO + O_2 \rightarrow 2CO_2$$

is added to the set given above. Does this reaction have any effect on the balance calculations? If so, what effect? Explain in detail.

CHAPTER
4

Element Balances

This chapter is concerned with the properties and use of the material balance equations that result when one applies the principle that the mass or moles of each type of element in a system must be conserved. We will find that this type of balance equation, which we call element balance, is indispensable if only an elemental analysis of the species in the system is available and is generally useful when the reaction stoichiometry is either unknown or very complex.

We begin in Section 4.1 by developing a general formulation for element balances in terms of the net output rates of each species and the coefficients of the atom matrix for the species present in the system. We shall observe that the element balance equations can be linearly dependent and learn to use a row reduction procedure to test for dependence which is analogous to the column reduction method of the previous chapter. Then, we extend our degree-of-freedom analysis to include element balances and analyze the complications which can arise from the use of overall element balances when solving multiunit problems.

In the next section, we study in detail the relationship between element and species balances to determine under what conditions one set might be preferable to the other. We derive a criterion which will allow us to recognize when the two types of balances are freely interchangeable. We conclude Section 4.2 by developing the concept of maximum number of independent reactions by means of which it is possible to explain why species balances are usually preferable if the reaction stoichiometry is given.

Having defined the roles of element and species balances for situations in which either could be used, we consider in Section 4.3 an area of applications in which element balances are often the only choice. This area of applications consists of material balance problems for systems involving the processing of fossil materials. We briefly review the origin and pertinent properties of fossil fuels, summarize the types of processes to which fossil fuels are most frequently subjected, for example, combustion, synthesis gas generation, and upgrading, and solve some material balance problems involving these processes.

Finally, in Section 4.4, a more advanced section which the reader may skip in the first reading, we study in detail the relationship between the atom and stoichiometric matrices of a reacting system. We devise a procedure for constructing a set of independent reactions from the atom matrix for the system if it is available. This construction is especially useful for problems in which the reaction stoichiometry is not completely defined. This section thus complements Section 4.2 and completes the analysis of the algebraic relationship between element and species balances.

Chapter 4 thus considers the construction, use, and properties of element balances, thoroughly examines the relationship between the two types of entity balances, and clarifies the conditions under which it is appropriate to use either type.

4.1 THE ELEMENT BALANCE EQUATIONS

Up to this point in our study of material balances, we have confined our selection of balanced entities to molecular species or convenient physical aggregates. We found that although the formulation of species balances for nonreacting systems was quite straightforward, for the reacting case, species balances required the introduction of a rate-of-production term into the balance equations. This was necessary because the entities which we chose to balance, namely, chemical species, actually were not conserved in reaction. Yet we know that in the absence of nuclear transformations, there do exist material entities which are conserved under both reacting and nonreacting conditions. These entities are the atoms of each type of element. Since the reaction process only involves the regrouping of atoms into different molecular aggregates, the number of atoms of each type of element in the system must remain unchanged. Hence, the number of moles of each element must be constant; and, since the atomic weights are fixed, it also follows that the mass of each type of element must be constant. We thus can conclude that for steady-state systems, we can write balances involving the mass or mole rates of every element present and that these balances require no correction for element production or depletion.

Example 4.1 Propylene, C_3H_6, can be produced by the catalytic dehydrogenation of propane, C_3H_8. Unfortunately, a number of side reactions occur which result in the concurrent production of lighter hydrocarbons as well as the deposition of carbon on the catalyst surface. Carbon deposition reduces the effectiveness of the catalyst so that it becomes necessary to periodically regenerate the catalyst by burning off the accumulated carbon. In a laboratory-scale reactor under given temperature, pressure, and residence time conditions, a feed of pure propane is converted into the following gaseous mixture of products (in mol %):

45% Propane
20% Propylene
 6% Ethane
 1% Ethylene
 3% Methane
25% Hydrogen

along with some evidence of carbon deposition. Assuming that the reactor can be scaled up to process 58.2 mol/day propane, calculate the rate of carbon deposition which can be expected.

Solution   From the species listed on the flowsheet (Figure 4.1), it is apparent that carbon and hydrogen are the only two types of elements occurring in the system. Following our discussion, it should be possible to write for each element an equation which would balance the molar input and output rates of that element as it is transported in each chemical species.

Using as basis 58.2 mol/day of stream 1, the molar input rate of hydrogen is just 8(58.2) mol/day since there are eight moles of hydrogen atoms in each mole of propane. The output rate of hydrogen is simply the sum over all species of the molar output rate of each chemical species times the number of moles of hydrogen atoms per mole of that species. Thus, the hydrogen output rate is

$$8(0.45N^2) + 6(0.2N^2) + 6(0.06N^2) + 4(0.01N^2)$$
$$+ 4(0.03N^2) + 2(0.25N^2) = 5.82N^2$$

Since there can be no production or depletion of hydrogen atoms, the input and output rates must be equal to each other, or

$$8(58.2) = 5.82N^2$$

Thus,

$$N^2 = 80 \text{ mol/day}$$

Similarly, the input rate of carbon atoms is 3(58.2) mol/day, while the output rate is the sum of the output rates of streams 2 and 3, or

$$N^3 + 3(0.45N^2) + 3(0.2N^2) + 2(0.06N^2)$$
$$+ 2(0.01N^2) + 0.03N^2 = N^3 + 2.12(80)$$

Again, since the input rate of carbon must equal the output rate, it follows that

$$3(58.2) = N^3 + 2.12(80)$$

or

$$N^3 = 5.0 \text{ mol/day carbon deposition}$$

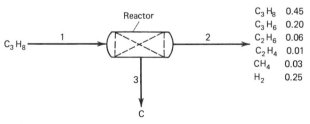

Figure 4.1   Flowsheet for Example 4.1, propane dehydrogenation.

In view of the number of reaction equations and hence reaction rates which would be required to describe the generation of the above product chemicals, it appears that the element balances provide a remarkably simple and straightforward alternative to species balances. On this basis, one might well be inclined to forgo the use of species balances completely and to solve all material balance problems with element balances. However, as we shall discover shortly, that would be ill considered for several reasons. First, even though species balances must incorporate reaction rate variables, it turns out that the net number of species balance equations available is never less than the number of element balances which can be written, and in many instances it is larger. Second, the element balance equations often are more difficult to solve because they involve more unknowns in each equation and hence are less likely to lend themselves to sequential solution. On the other hand, there is a substantial number of applications in which element balances must be used either because the reaction stoichiometry is not known or is very complex, or because the reactants are a complicated mixture of which only an elemental analysis is available. The former situation frequently occurs with reactions of hydrocarbons, the latter in the combustion of fossil fuels. Consequently, we will find that there is a place for both types of entity balances. In this chapter, we study the characteristics and use of element balances in order to better define the role of these balances in solving material balance problems.

4.1.1 The Atom Matrix and General Balance Equations

In order to elucidate the general properties of element balances, we begin by writing these equations in a general form. As in the species balance case, this requires the introduction of some notation and nomenclature. In particular, let us define the *net output rate of species s, $\overline{N}_s$,* for a system involving I input streams and J output streams as

$$\overline{N}_s = \sum_{j=1}^{J} N_s^j - \sum_{i=1}^{I} N_s^i$$

The net output rate of species s is thus the difference between the total output rate of species s through all output streams and the total input rate of species s through all input streams. Next, given a system involving S species composed of E elements, let α_{es} denote the number of atoms of element e in one molecule of species s. We shall refer to individual coefficients α_{es} as *atomic coefficients* and to the collection of such coefficients for a given system as composing the *atom matrix*.

In terms of this notation, the net molar outflow rate of element e with species s is $\alpha_{es}\overline{N}_s$. Therefore, the net molar outflow rate of element e with all species s, $s = 1, \ldots, S$, will be

$$\sum_{s=1}^{S} \alpha_{es}\overline{N}_s$$

Since the elements are conserved, this sum must be exactly equal to zero. Hence, we have constructed in general form the element balance equations for a system involving S species in E elements, namely,

$$\sum_{s=1}^{S} \alpha_{es}\overline{N}_s = 0 \qquad e = 1, \ldots, E$$

If we denote by A_e the atomic weight of element e, then the above molar balance equations can be expressed in mass units as

$$A_e \sum_{s=1}^{S} \alpha_{es}\overline{N}_s = 0 \qquad e = 1, \ldots, E$$

Summing the above element mass balance equations

$$\sum_{e=1}^{E} A_e \sum_{s=1}^{S} \alpha_{es}\overline{N}_s = 0$$

and noting that by definition the molecular weight of species s is

$$M_s = \sum_{e=1}^{E} A_s \alpha_{es}$$

we easily obtain

$$\sum_{s=1}^{S} M_s \overline{N}_s = 0 \qquad \text{or} \qquad \sum_{s=1}^{S} \overline{F} = 0$$

Verbally, this means that the element mass balances sum to yield the total mass balance. Thus, of the E element balances and the total mass balance, only at most E balance equations can be independent.

A further important property is that each element balance equation is homogeneous in the species flow rates. This follows because if $\overline{N}_s^*$, $s = 1, \ldots, S$ is a solution to the element balances, then, for any $\theta \neq 0$,

$$\sum_{s=1}^{S} \alpha_{es}(\theta\overline{N}_s^*) = \theta\left(\sum_{s=1}^{S} \alpha_{es}\overline{N}_s^*\right) = 0 \qquad e = 1, \ldots, E$$

Consequently, $\theta\overline{N}_s^*$ will also satisfy the element balance equations and the equations must be homogeneous by definition. Hence, paralleling our experience with species balances, given a system involving E elements, at most E balance equations can be written; and, if no flow is specified, any convenient basis for the calculation can always be selected.

Example 4.2 Formaldehyde, CH_2O, is produced industrially by partial oxidation of methanol, CH_3OH, with air over a silver catalyst. Under optimized reactor conditions, about 55% conversion of CH_3OH is attained with a feed consisting of 40% CH_3OH in air. Although CH_2O is the main product, some production of by-products such as CO, CO_2, and small amounts of formic acid, $HCOOH$, is unavoidable. Therefore, the gross reactor output is usually scrubbed to separate out a gas stream containing CO_2, CO, H_2, and N_2 and a liquid stream containing the unconverted CH_3OH, the products CH_2O and H_2O, as well as the undesirable by-product, $HCOOH$. Assuming that the liquid stream contains equal amounts of CH_2O and CH_3OH and ½% $HCOOH$, while the gas stream contains 7.5% H_2, calculate the complete compositions of both streams.

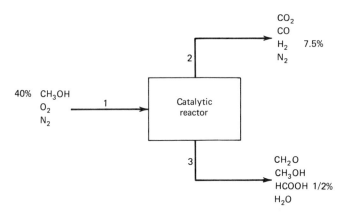

Figure 4.2   Flowsheet for Example 4.2, formaldehyde production.

Solution   From Figure 4.2 it is clear that the system involves four elements (O, H, C, N) and nine chemical species (O_2, H_2, CO, N_2, CO_2, CH_3OH, CH_2O, $HCOOH$, H_2O). If the elements and species are indexed in the order in which we have listed them, then the atom matrix is:

$$
\begin{array}{c c c c c c c c c c}
 & O_2 & H_2 & CO & N_2 & CO_2 & CH_3OH & CH_2O & HCOOH & H_2O \\
O & \left(\begin{array}{ccccccccc} 2 & 0 & 1 & 0 & 2 & 1 & 1 & 2 & 1 \\ 0 & 2 & 0 & 0 & 0 & 4 & 2 & 2 & 2 \\ 0 & 0 & 1 & 0 & 1 & 1 & 1 & 1 & 0 \\ 0 & 0 & 0 & 2 & 0 & 0 & 0 & 0 & 0 \end{array} \right) \\
H \\
C \\
N
\end{array}
$$

Suppose a basis of 1000 mol/day feed is chosen. Then, from the known CH_3OH composition and the fact that the O_2-to-N_2 ratio is that of air (21:79), we have

$$
N^1 = \begin{cases} 400 \text{ mol/day} & CH_3OH \\ 126 \text{ mol/day} & O_2 \\ 474 \text{ mol/day} & N_2 \end{cases}
$$

With the input rate of CH_3OH and its conversion known, the output rate of CH_3OH can be calculated as

$$
N^3_{CH_3OH} = 400(1 - 0.55) = 180 \text{ mol/day}
$$

Finally, since $N^3_{CH_3OH} = N^3_{CH_2O}$, it follows that

$$
N^3_{CH_2O} = 180 \text{ mol/day}
$$

With these results at hand, the species net outflow rates can be determined:

$$\overline{N}_{O_2} = 0 - 126 = -126$$

$$\overline{N}_{H_2} = 0.075N^2 - 0 = 0.075N^2$$

$$\overline{N}_{CO} = N^2_{CO} - 0 = N^2_{CO}$$

$$\overline{N}_{N_2} = N^2_{N_2} - 474 = N^2_{N_2} - 474$$

$$\overline{N}_{CO_2} = N^2_{CO_2} - 0 = N^2_{CO_2}$$

$$\overline{N}_{CH_3OH} = 180 - 400 = -220$$

$$\overline{N}_{CH_2O} = 180 - 0 = 180$$

$$\overline{N}_{HCOOH} = 0.005N^3 - 0 = 0.005N^3$$

$$\overline{N}_{H_2O} = N^3_{H_2O} - 0 = N^3_{H_2O}$$

The four element balances thus become

Oxygen balance: $2(-126) + 0(\overline{N}_{H_2}) + 1(N^2_{CO}) + 0(\overline{N}_{N_2}) + 2(N^2_{CO_2})$

$\qquad\qquad\qquad + 1(-220) + 1(180) + 2(0.005N^3) + 1(N^3_{H_2O}) = 0$

Hydrogen balance: $0(\overline{N}_{O_2}) + 2(0.075N^2) + 0(\overline{N}_{CO}) + 0(\overline{N}_{N_2}) + 0(\overline{N}_{CO_2})$

$\qquad\qquad\qquad + 4(-220) + 2(180) + 2(0.005N^3) + 2(N^3_{H_2O}) = 0$

Carbon balance: $0(\overline{N}_{O_2}) + 0(\overline{N}_{H_2}) + 1(N^2_{CO}) + 0(\overline{N}_{N_2}) + 1(N^2_{CO_2})$

$\qquad\qquad\qquad + 1(-220) + 1(180) + 1(0.005N^3) + 0(\overline{N}_{H_2O}) = 0$

Nitrogen balance: $0(\overline{N}_{O_2}) + 0(\overline{N}_{H_2}) + 0(\overline{N}_{CO}) + 2(\overline{N}_{N_2}) + 0(\overline{N}_{CO_2})$

$\qquad\qquad\qquad + 0(\overline{N}_{CH_3OH}) + 0(\overline{N}_{CH_2O}) + 0(\overline{N}_{HCOOH}) + 0(\overline{N}_{H_2O})$

$\qquad\qquad = 0$

Each of these balance equations is formed by summing the products of each coefficient of a given row of the atom matrix with the corresponding species net outflow rate. Note that since N_2 is an inert, the nitrogen balance indicates that the new outflow rate of N_2 should be zero, that is,

$$N^2_{N_2} - 474 = 0$$

which is what we would expect. The remaining three balances reduce to the form

$$N^2_{CO} + 2N^2_{CO_2} + 0.01N^3 + N^3_{H_2O} = 292 \qquad (4.1)$$

$$0.15N^2 + 0.01N^3 + 2N^3_{H_2O} = 520 \qquad (4.2)$$

$$N^2_{CO} + N^2_{CO_2} + 0.005N^3 = 40 \qquad (4.3)$$

In addition, the stream variables associated with streams 2 and 3 must, respectively, satisfy

$$N^2 = N^2_{CO_2} + N^2_{CO} + 0.075N^2 + 474$$

$$N^3 = 180 + 180 + 0.005N^3 + N^3_{H_2O}$$

or

$$0.925N^2 - 474 = N^2_{CO_2} + N^2_{CO} \tag{4.4}$$

$$0.995N^3 - 360 = N^3_{H_2O} \tag{4.5}$$

These two relations can be used to eliminate $N^3_{H_2O}$ and the quantity $(N^2_{CO_2} + N^2_{CO})$ from eqs. (4.2) and (4.3) to yield two simultaneous equations:

$$0.15N^2 + 2N^3 = 1240$$

$$0.925N^2 + 0.005N^3 = 514$$

The solution of these equations is

$$N^2 = 552.55 \text{ mol/day}$$

$$N^3 = 578.56 \text{ mol/day}$$

Substituting back into eq. (4.5), we obtain

$$N^3_{H_2O} = 215.67 \text{ mol/day}$$

These results, when used to simplify eqs. (4.1) and (4.4), yield two simultaneous equations:

$$N^2_{CO} + 2N^2_{CO_2} = 70.55$$

$$N^2_{CO} + N^2_{CO_2} = 37.11$$

The solution of this pair of equations is

$$N^2_{CO} = 3.67$$

$$N^2_{CO_2} = 33.44$$

The complete flows of stream 2 and the corresponding mole fractions are

$$N^2 = (N_{CO_2}, N_{CO}, N_{H_2}, N_{N_2}) = (33.44, 3.67, 41.44, 474) \text{ mol/day}$$

$$(x_{CO_2}, x_{CO}, x_{H_2}, x_{N_2}) = (0.0605, 0.0066, 0.0750, 0.8578)$$

For stream 3, the complete flows and corresponding mole fractions are

$$N^3 = (N_{CH_2O}, N_{CH_3OH}, N_{HCOOH}, N_{H_2O}) = (180, 180, 2.89, 215.67) \text{ mol/day}$$

$$(x_{CH_2O}, x_{CH_3OH}, x_{HCOOH}, x_{H_2O}) = (0.3111, 0.3111, 0.0050, 0.3728)$$

One property of element balances which can be noted in the above example and which generally holds true is that element balances are algebraically more complicated than species balances. That is, the element balances usually tend to involve more unknowns per balance equation. This comes about because each element balance involves every species net outflow rate, while the species balances, in addition to the reaction rates, involve only the net outflow rate of a single species.

A second property of element balances, which is illustrated in the next example, is that even though an element balance can be written for each element in the system, some of the balance equations can be redundant. Hence, although in general at most E independent balances can be written, sometimes fewer than E may actually be independent.

Example 4.3 Sulfuric acid can be prepared by absorbing SO_3 in water. Suppose a weak sulfuric acid recycle stream (20% H_2SO_4) and a water stream are used to absorb a stream of SO_3 received from a catalytic oxidizer (see Figure 4.3). In what proportions should the streams be mixed if a 50% H_2SO_4 solution is to be produced?

Solution The system involves three elements (S, O, H) and three species (SO_3, H_2O, H_2SO_4). Since all the compositions are known and a basis can be chosen, the three element balances ought to be sufficient to calculate the three unknown stream flow rates. If a basis of 100 mol of product is chosen, the element balances become

S balance: $1(0 - N^3) + 0(50 - N^1 - 0.8N^2) + 1(50 - 0.2N^2) = 0$

O balance: $3(0 - N^3) + 1(50 - N^1 - 0.8N^2) + 4(50 - 0.2N^2) = 0$

H balance: $0(0 - N^3) + 2(50 - N^1 - 0.8N^2) + 2(50 - 0.2N^2) = 0$

These balances simplify to

$$0.2N^2 + N^3 = 50$$

$$N^1 + 1.6N^2 + 3N^3 = 250$$

$$2N^1 + 2N^2 = 200$$

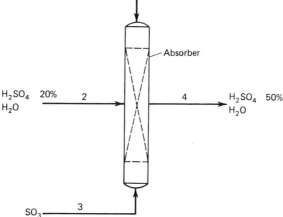

Figure 4.3 Flowsheet for Example 4.3, sulfuric acid production.

If three times the first equation is subtracted from the second, the result

$$N^1 + N^2 = 100$$

is obtained. But this is just the third equation multiplied by $\frac{1}{2}$. Consequently, the three balance equations are not independent and the problem actually is under-specified by one. The redundancy arises from the fact that oxygen occurs only in a fixed combination with hydrogen, as H_2O, and with sulfur, as SO_3. Since H_2SO_4 can be viewed as $(H_2O \cdot SO_3)$, the element balances amount to balancing on just the two groups of elements, H_2O and SO_3. This situation can occur frequently enough that in general the number of independent element balances should be determined as part of the problem analysis. Since in most instances the redundancy is not as transparent as is the case in the above example, a general algebraic approach such as that given in the next section is recommended.

4.1.2 The Algebra of Element Balances

From the form of the element balance equations

$$\sum_{s=1}^{S} \alpha_{es} \overline{N}_s = 0 \qquad s = 1, \ldots, S$$

it is apparent that the algebraic properties of element balances can be deduced from the corresponding properties of the atomic coefficients. In particular, since the coefficients of a given element balance equation are just the atomic coefficients of that element, questions concerning the dependence or independence of the set of element balances can be answered by considering the linear dependence/inde-pendence of the vectors of these coefficients. If we denote the row vector of atomic coefficients corresponding to element e by α_e, then it is clear that the set of element balance equations will be independent if and only if the set of row vectors α_e are linearly independent.

Example 4.4 The atom matrix for the system considered in Example 4.3 is:

$$\begin{array}{c} \\ S \\ O \\ H \end{array} \begin{array}{ccc} SO_3 & H_2O & H_2SO_4 \\ \begin{pmatrix} 1 & 0 & 1 \\ 3 & 1 & 4 \\ 0 & 2 & 2 \end{pmatrix} \end{array}$$

The three vectors formed by considering the rows of this array are

$$\alpha_S = (1, 0, 1)$$
$$\alpha_O = (3, 1, 4)$$
$$\alpha_H = (0, 2, 2)$$

Since

$$3\alpha_S + \tfrac{1}{2}\alpha_H = 3(1, 0, 1) + \tfrac{1}{2}(0, 2, 2) = (3, 1, 4) = \alpha_O$$

it is clear that the three vectors are dependent and, consequently, that the set of element balances formed with these coefficients will be dependent.

In Section 3.3.2, a procedure was presented for determining whether a set of column vectors is linearly dependent. This procedure was based on using four elementary column operations to reduce the array formed by the column vectors to a special structure. The reduced structure was such that dependence or independence could be determined by inspection. In particular, if the reduced array had one or more zero columns, the set of vectors was dependent and a span of the original set of vectors could be obtained by using the remaining nonzero vectors. The same procedure can be used to determine the corresponding facts about a set of row vectors of atomic coefficients providing the *array reduction* algorithm is carried out using *row* rather than *column* operations.

It can readily be shown that row operations do not affect the solution of the element balances. In other words, the solution of the balance equations formed by using the rows of the reduced array as coefficients will be the same as the solution obtained with the original set of element balances. For instance, row or column interchanges in the atom matrix merely amount to a reordering of elements and species, while multiplying a row by a constant merely corresponds to multiplying the element balance by a constant. The adding of one row to another amounts to carrying out the same operations on the element balances and clearly cannot affect the solution of these equations.

The equivalent *array reduction procedure* based on row operations, for the *i*th row, thus consists of the following steps:

Step A Divide each entry in the *i*th row by the *i*th element in that row. If the *i*th element is zero, go to Step C.

Step B Add appropriate multiples of the *i*th row to each remaining row so that the *i*th entry in each of these rows is equal to zero. At the conclusion of these calculations, set $i = i + 1$ and go to Step A.

Step C If the *i*th element of the *i*th row is zero, interchange column *i* with any column to the right of column *i* which has a nonzero entry in the *i*th position. Return to Step A. If no column has a nonzero *i*th entry, interchange row *i* with any row below the *i*th row.

As in the column case, the process terminates when all rows have been reduced or if all remaining rows are identically zero. If some of the rows are zero, then the original set of coefficients is dependent and the remaining nonzero rows constitute a basis of the original set. These calculations are illustrated with the following example.

Example 4.5 The solid fertilizer urea, $(NH_2)_2CO$, is produced by reacting ammonia with carbon dioxide. In a given plant, a mixture of NH_3 and CO_2 consisting of 33% CO_2 is fed to the reactor. The product stream is found to contain urea, water, ammonium carbamate, NH_2COONH_4, and unreacted CO_2 and NH_3 (Figure

4.4). Calculate the composition of the product stream if 99% conversion of CO_2 is obtained.

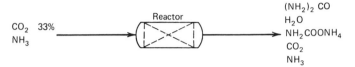

Figure 4.4   Flowsheet for Example 4.5, urea synthesis.

Solution   The process as shown involves four elements (C, O, H, N) and five chemical species (CO_2, H_2O, NH_3, $(NH_2)_2CO$, NH_2COONH_4). The atom matrix for this system, with elements and species ordered as listed, is:

$$
\begin{array}{c}
 \\
C \\
O \\
H \\
N
\end{array}
\begin{array}{ccccc}
CO_2 & H_2O \cdot & NH_3 & (NH_2)_2CO & NH_2COONH_4 \\
\left(\begin{array}{ccccc}
1 & 0 & 0 & 1 & 1 \\
2 & 1 & 0 & 1 & 2 \\
0 & 2 & 3 & 4 & 6 \\
0 & 0 & 1 & 2 & 2
\end{array}\right)
\end{array}
$$

The reduction calculations begin with the first row.

Step A   Divide the first row by 1.

Step B   Add -2 times each element of the first row to each corresponding element of the second row:

$$
\left(\begin{array}{ccccc}
1 & 0 & 0 & 1 & 1 \\
0 & 1 & 0 & -1 & 0 \\
0 & 2 & 3 & 4 & 6 \\
0 & 0 & 1 & 2 & 2
\end{array}\right)
$$

Since the first elements of the remaining two rows are already zero, Step B terminates. Proceed with the second row.

Step A   Divide the second row by 1.

Step B   Add -2 times each element of the second row to the corresponding element of the third row:

$$
\left(\begin{array}{ccccc}
1 & 0 & 0 & 1 & 1 \\
0 & 1 & 0 & -1 & 0 \\
0 & 0 & 3 & 6 & 6 \\
0 & 0 & 1 & 2 & 2
\end{array}\right)
$$

Since the second elements of the remaining two rows are already zero, Step B terminates. Proceed with the third row.

Step A   Divide each element of the third row by 3.

Step B   Add -1 times row three to each element of row 4.

$$\begin{pmatrix} 1 & 0 & 0 & 1 & 1 \\ 0 & 1 & 0 & -1 & 0 \\ 0 & 0 & 1 & 2 & 2 \\ 0 & 0 & 0 & 0 & 0 \end{pmatrix}$$

The third elements of the remaining rows are zero. Moreover, all the entries in the fourth row are zero. The algorithm terminates.

The above reduced array has a zero row, indicating that the complete set of four element balances is dependent. The first three rows are nonzero, indicating that the first three element balances will be independent and their use will be equivalent to using the complete set.

As stated, the problem has seven stream variables, a specified composition, a specified conversion, and a free choice of basis. Since only three independent element balances can be written, the problem is underspecified by one specification. Additional information is needed to complete the solution.

Although the above procedure allows us to determine how many independent element balances can be used in the solution of any given problem, a solution will actually be obtained only if the problem is properly specified. As was the case with species balances, proper specification implies not only that the correct number of stream variables be fixed and relations imposed, but also that these specifications be correctly distributed.

Example 4.6   Suppose a mixture of ethylene and butylene is hydrogenated to produce a product mixture consisting of ethylene, ethane, butylene, and butane. If the feed rates of C_2H_4 and H_2 are 10 and 7 mol/h, respectively, and if the output rates of C_2H_6, C_4H_8, and C_4H_{10} are 4, 5, and 3 mol, respectively, calculate the unknown feed rate of C_4H_8 and the unknown output rate of C_2H_4.

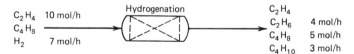

Figure 4.5   Flowsheet for Example 4.6, hydrogenation.

Solution   From Figure 4.5, it is apparent that the problem involves only two unknown species flow rates. Since the system involves two elements, H and C, the two element balances ought to be sufficient to calculate these two unknowns.

The atom matrix for the system is:

$$\begin{array}{ccccc} H_2 & C_2H_4 & C_2H_6 & C_4H_8 & C_4H_0 \end{array}$$

$$\begin{array}{c} H \\ C \end{array} \begin{pmatrix} 2 & 4 & 6 & 8 & 10 \\ 0 & 2 & 2 & 4 & 4 \end{pmatrix}$$

This array can be reduced to the form

$$
\begin{pmatrix}
1 & 0 & 1 & 0 & 1 \\
0 & 1 & 1 & 2 & 2
\end{pmatrix}
$$

indicating that the two element balances will be independent. The balances are:

Hydrogen balance: $2(0 - 7) + 4(N_{C_2H_4}^{out} - 10) + 6(4 - 0)$

$$+ 8(5 - N_{C_4H_8}^{in}) + 10(3 - 0) = 0$$

Carbon balance: $0(0 - 7) + 2(N_{C_2H_4}^{out} - 10) + 2(4 - 0)$

$$+ 4(5 - N_{C_4H_8}^{in}) + 4(3 - 0) = 0$$

When simplified to the form

$$8N_{C_4H_8}^{in} - 4N_{C_2H_4}^{out} = 40$$
$$4N_{C_4H_8}^{in} - 2N_{C_2H_4}^{out} = 20$$

it is obvious that the two equations are just multiples of each other. The problem cannot be solved because the specifications, although correct in number, are improperly distributed.

Note that the only two species in the preceding example which are not completely specified are C_2H_4 and C_4H_8. The two columns of the atom matrix corresponding to these two species are:

$$\begin{pmatrix} 4 \\ 2 \end{pmatrix} \quad \text{and} \quad \begin{pmatrix} 8 \\ 4 \end{pmatrix}$$

These two columns quite obviously are multiples of each other, that is, they are linearly dependent. We are thus led to a definition of an independent specification analogous to that given for species balances in Section 3.3.3, as follows:

For a system involving E independent element balances, a specification is *independent* if the columns of the atom matrix corresponding to the E species whose net outflow rates are unknown are linearly independent.

Note that if a system involves E independent element balances and has the correct number of specifications but has less than E unknown species net outflow rates, the specification cannot be proper. With less than E unknown net rates, both the input and the output rate of at least one species will be unknown. Since the element balances can only serve to calculate the net rate, this means that for that species the input and output rates cannot be determined separately. Thus, a proper specification necessarily must be both correct in number and be independent with exactly E unknown net species rates.

4.1.3 Degree-of-Freedom Analysis

As evident from the examples we have already considered, the degree-of-freedom analysis of single-unit systems in terms of element balances is remarkably uncomplicated and is the same for the reacting and nonreacting cases. The stream variables and relations are counted in the usual manner; the number of independent element balances is verified by means of the *array reduction procedure*; and an arbitrary basis is permitted since the element balance equations are homogeneous in the stream flow rates. If the degree of freedom is zero, then as a final check the independence of the specifications is verified.

The single-unit analysis is readily extended to multiple-unit problems. As in the case of species balances, the presence of multiple units introduces the possibility of using overall balances—only now the overall balances are on the elements. Construction of these balance equations is in the present instance less complicated than in the species case, because internal chemical species and their reactions need not be accounted for since there is no production term. Rather, only the elements and species actually appearing in the process input and output streams must be incorporated into the balances. Thus, once the input and output elements and species have been identified, their atom matrix can be constructed and a set of independent element balances determined using the *array reduction procedure* just as in the single-unit analysis. The only complication arises when the overall balances are substituted for one set of the individual unit balances and every element for which an overall balance is written does not appear in every individual unit. In that case, redundant element balances will be encountered which can confound the degree-of-freedom updating. The remedy is the same as in the species balance case. Namely, in constructing the degree-of-freedom table, we itemize which elements will be balanced in which units and determine how many independent balances are available in the process for each element. Then, in updating the degree of freedom of the units, we note when all the independent balances on a given element have been expended and correct the unit degrees of freedom accordingly. These considerations are illustrated in the following example.

Example 4.7   In view of the expected long-term energy shortages, considerable effort is being expended to develop processes for upgrading low-quality energy sources to high-quality fuels. The flowsheet shown in Figure 4.6 is an example of such a scheme. In it, high-sulfur coal, a fuel which otherwise could not be used because of air quality standards, is gasified to produce a clean synthetic natural gas (SNG).

Powdered coal and steam (0.8 lb H_2O per 1 lb coal) are charged to a reactor which at high temperature and pressure produces a complex mixture of gases including 10% (mole basis) methane. The heat for the reaction is provided by burning some of the charged coal with oxygen introduced directly into the reactor. The raw product gas is cooled, treated for removal of entrained ash, and sent to a second reactor, the shift converter. In this unit, the H_2-to-CO mole ratio of $0.56:1$ is increased by converting 62.5% of the CO. The shifted gas, still containing 10% CH_4, is next treated for removal of the acid gases, H_2S and CO_2. Finally, the residual gas, containing 49.2% H_2 on an H_2O-free basis, is sent to a catalytic reactor which produces additional methane. The final H_2O-free product gas contains 5% H_2, 0.1% CO, and the rest CH_4 (mole basis).

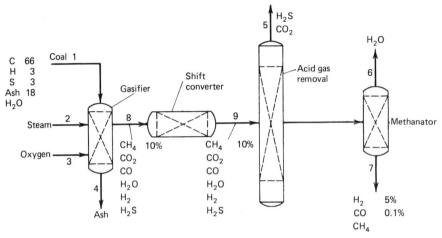

Figure 4.6 Flowsheet for Example 4.7, coal gasification.

Given a coal which analyzes (by weight)

66% C
3% H
3% S
18% Ash
10% H_2O

calculate all flows and compositions in the process based on a feed of 10,000 ton/day coal and assuming that 25% of the carbon fed ultimately appears in the product.

Solution From the given information, the process apparently involves four elements (H, C, S, O) and an inert aggregate consisting of inorganic minerals called ash. All units, except the methanator, involve the common species H_2, CH_4, H_2S, CO, CO_2, and H_2O, whose element matrix is:

$$\begin{array}{c c c c c c c} & H_2 & CH_4 & H_2S & CO & CO_2 & H_2O \\ H & \begin{pmatrix} 2 & 4 & 2 & 0 & 0 & 2 \\ C & 0 & 1 & 0 & 1 & 1 & 0 \\ S & 0 & 0 & 1 & 0 & 0 & 0 \\ O & 0 & 0 & 0 & 1 & 2 & 1 \end{pmatrix} \end{array}$$

It is easily shown that the four rows of this array as well as the three rows of the smaller array formed by deleting the S row and the H_2S and CO_2 columns each are independent. Consequently, the gasifier, shift converter, and acid gas removal unit each will have associated four independent element balances, while the methanator will have three. The gasifier will in addition have associated a balance on the ash aggregate. For purposes of the overall balances, the process input and output streams involve the same four elements, six species, as well as the ash. Consequently, the overall balances will also consist of the four independent element

	Gasifier	Shift Reactor	Acid Gas Removal	Methanator	Process	Overall Balances
Number of variables	14	12	12	8	30	14
Number of balances						
H	1	1	1	1	4	1
C	1	1	1	1	4	1
O	1	1	1	1	4	1
S	1	1	1	—	3	1
Ash	1	—	—	—	1	1
Total	5	4	4	3	16	5
Number of specified compositions	5	2	1	2	8	6
Number of relations						
Steam ratio	1				1	1
CO conversion		1			1	
H_2:CO ratio (shift)	1	1			1	
H_2:(CO + CH_4) ratio			1	1	1	
C efficiency					1	1
Degree of freedom	2	4	6	2	1	1
Basis					−1	
					0	

Figure 4.7 Example 4.7, degree-of-freedom table.

balances plus the ash balance. With the balances counted, the degree-of-freedom table for this flowsheet is readily constructed and is shown in Figure 4.7.

Evidently, the process is correctly specified and the calculations ought to be begun with the overall balances. Note that if the coal rate is selected as basis, then, since the coal composition is known, the balance species C, H, and S, which do not appear as such in any output stream, will be completely specified. In addition, with the feed stream known, the specified carbon efficiency relation and the known product composition will serve to determine the output rates of the species H_2, CO, and CH_4. Since these do not appear in any input stream, these species are also completely specified. Consequently, the only incompletely specified species are O_2, H_2O, H_2S, CO_2, and ash. These remain to be calculated using the five balance equations. The columns of the array formed by the atomic coefficients of these species,

$$\begin{array}{c} \\ O \\ H \\ S \\ C \end{array} \begin{array}{cccc} O_2 & H_2O & H_2S & CO_2 \\ \begin{pmatrix} 2 & 1 & 0 & 2 \\ 0 & 2 & 2 & 0 \\ 0 & 0 & 1 & 0 \\ 0 & 0 & 0 & 1 \end{pmatrix} \end{array}$$

are easily shown to be linearly independent. Hence, since the ash balance is clearly independent of the four element balances, the specifications associated with the overall balances are independent.

It is of interest to note that if the O_2 rather than the steam ratio had been specified, the specification would not have been proper. In that case, the only unspecified species would have been H_2O, H_2S, CO_2, and ash. These four balance entities are less than the number of available balance equations. The specification is improper because neither the input nor the output rate of H_2O is known. Hence, the determination of the net output rate of H_2O via the element balances would be inadequate to determine the input and output separately.

Proceeding with the solution of the overall balances, if a basis of 10,000 lb/h coal is chosen, then the steam ratio implies that $F^2 = 8000$ lb/h. The carbon efficiency relation indicates that

$$0.95N^7 = \frac{0.25(0.66 \times 10,000)}{12}$$

or

$$N^7 = 144.74 \text{ lbmol/h}$$

With these results, the ash balance reduces to

$$0.18(10,000) = F^4 = 1800 \text{ lb/h}$$

and the element balances are:

	Coal	O_2	H_2	CH_4	H_2S	CO	CO_2	H_2O	
H balance:	$(0 - 300)$	$+ 2(7.237 - 0)$	$2(0 - N^3)$	$+ 4(137.36 - 0)$	$+2(N^5_{H2S} - 0)$			$+2\left(N^6 - \dfrac{8000}{18}\right)$	$= 0$
C balance:	$\left(0 - \dfrac{6600}{12}\right)$			$+ (137.36 - 0)$		$+ (0.1447 - 0) +$	$(N^5_{CO2} - 0)$		$= 0$
O balance:						$+ (0.1447 - 0) + 2(N^5_{CO2} - 0) +$		$\left(N^6 - \dfrac{8000}{18}\right)$	$= 0$
S balance:	$\left(0 - \dfrac{300}{32}\right)$				$+ (N^5_{H2S} - 0)$				$= 0$

The carbon and sulfur balances can be solved to obtain

$$N^5_{CO_2} = 412.50 \text{ lbmol/h}$$

$$N^5_{H_2S} = 9.375 \text{ lbmol/h}$$

The hydrogen balance then yields

$$N^6 = 303.11 \text{ lbmol/h}$$

Finally, from the oxygen balance,

$$N^3 = 341.90 \text{ lbmol/h}$$

With the flows of streams 6 and 7 determined, the degree of freedom of the methanator balances has been reduced to zero. The degree of freedom of the gasifier appears to have been reduced to -1 since the flow of streams 1, 3, and 4 are now known (stream 2 is already accounted for by virtue of the steam-to-coal ratio). Note, however, that the process involves only a single ash balance and that this balance has already been expended as part of the overall balances. Thus, the degree of freedom of the gasifier is in effect zero. Since either the methanator or the gasifier balances can be solved next, let us proceed with the gasifier.

The gasifier balances are:

	Coal	O_2	H_2	CH_4	H_2S	CO	CO_2	H_2O	
H balance:	-300		$+2N_{H_2}^8$	$+4(0.1N^8)$	$+2N_{H_2S}^8$			$+2\left(N_{H_2O}^8 - \dfrac{8000}{18}\right)$	$=0$
C balance:	-550			$+0.1N^8$		$+N_{CO}^8 + N_{CO_2}^8$			$=0$
O balance:		$-2(341.90)$				$+N_{CO}^8 + 2(N_{CO_2}^8) +$		$\left(N_{H_2O}^8 - \dfrac{8000}{18}\right)$	$=0$
S balance:	-9.375				$+N_{H_2S}^8$				$=0$

In addition, the relation

$$N_{H_2}^8/N_{CO}^8 = 0.56$$

and the definition

$$0.9N^8 = N_{H_2}^8 + N_{CO}^8 + N_{CO_2}^8 + N_{H_2O}^8 + N_{H_2S}^8$$

apply.

Using the H_2-to-CO ratio, the H, C, and O balances reduce to

$$0.4N^8 + 1.12N_{CO}^8 + 2N_{H_2O}^8 = 1170.14$$

$$0.1N^8 + N_{CO}^8 + N_{CO_2}^8 = 550$$

$$N_{CO}^8 + N_{H_2O}^8 + 2N_{CO_2}^8 = 1128.24$$

Together with the definition of N^8,

$$0.9N^8 - 1.56N_{CO}^8 - N_{H_2O}^8 - N_{CO_2}^8 = 9.375$$

the above constitute a system of four linear equations in four unknowns. These can be solved by using the second equation to eliminate $N_{CO_2}^8$ from the third and fourth equations and using the first equation to eliminate $N_{H_2O}^8$ from the results. In this manner, the system of equations collapses to two equations:

$$0.4N^8 + 1.56N_{CO}^8 = 556.83$$

$$1.2N^8 \qquad\qquad = 1144.45$$

which are easily solved to yield

$$N^8 = 953.70 \text{ lbmol/h}$$

$$N_{CO}^8 = 112.40 \text{ lbmol/h}$$

Then, from the second of the original equations,

$$N_{CO_2}^8 = 342.23 \text{ lbmol/h}$$

and from the first,

$$N_{H_2O}^8 = 331.39 \text{ lbmol/h}$$

Therefore,

$$N^8 = (N_{CH_4}, N_{CO_2}, N_{CO}, N_{H_2O}, N_{H_2}, N_{H_2S})$$
$$= (95.37, 342.23, 112.4, 331.39, 62.94, 9.38) \text{ lbmol/h}$$

With the determination of stream 8, the shift converter balances will have been reduced to degree of freedom zero. Hence, these will be solved next. The balance equations are:

	H_2	CH_4	H_2S	CO	CO_2	H_2O	
H balance:	$2(N_{H_2}^9 - 62.94) +$	$4(0.1N^9 - 95.37) +$	$2(N_{H_2S}^9 - 9.38)$			$2(N_{H_2O}^9 - 331.39)$	$= 0$
C balance:		$(0.1N^9 - 95.37)$		$+ (N_{CO}^9 - 112.4) +$	$(N_{CO_2}^9 - 342.23)$		$= 0$
O balance:				$(N_{CO}^9 - 112.4) +$	$2(N_{CO_2}^9 - 342.23) +$	$(N_{H_2O}^9 - 331.39)$	$= 0$
S balance:			$(N_{H_2S}^9 - 9.38)$				$= 0$

In addition, a conversion relation is imposed:

$$0.625 = \frac{112.4 - N_{CO}^9}{112.4}$$

which reduces to $N_{CO}^9 = 42.15$ lbmol/h. From the sulfur balance, $N_{H_2S}^9 = N_{H_2S}^8 = 9.38$ lbmol/h. Using the C balance to eliminate the CH_4 term and the O balance to eliminate the H_2O term, the H balance simplifies to

$$2(N_{H_2}^9 - 62.94) - 8(N_{CO_2}^9 - 342.23) = +6(N_{CO}^9 - 112.4) = 6(-70.25)$$

or

$$2N_{H_2}^9 - 8N_{CO_2}^9 = -3033.5$$

Moreover, from the definition of N^9,

$$0.9N^9 = N_{H_2}^9 + N_{H_2S}^9 + N_{CO}^9 + N_{CO_2}^9 + N_{H_2O}^9$$

we obtain, using the same substitution, the additional equation

$$N_{H_2}^9 + 8N_{CO_2}^9 = 3433.1 \text{ lbmol/h}$$

If these two equations are added to eliminate $N_{CO_2}^9$, we obtain

$$N_{H_2}^9 = 133.20 \text{ lbmol/h}$$

and then, by substitution,

$$N_{CO_2}^9 = 412.49 \text{ lbmol/h}$$

Finally, when these results are used in the C and O balances, the remaining flows can be calculated:

$$N^9 = 953.70 \text{ lbmol/h} \quad\text{and}\quad N_{H_2O}^9 = 261.12 \text{ lbmol/h}$$

Stream 9 therefore consists of the flows

$$N^9 = (N_{CH_4}, N_{CO_2}, N_{CO}, N_{H_2O}, N_{H_2}, N_{H_2S})$$
$$= (95.37, 412.49, 42.15, 261.12, 133.2, 9.38) \text{ lbmol/h}$$

At this point, only the acid gas removal or the methanator balances remain to be solved. The former unit had degree of freedom 6. However, from the overall balances stream 5 was determined, reducing the degree of freedom to 4, while from the shift converter balances stream 9 was determined, thus reducing the degree of freedom to -1. But note that the three independent sulfur balances in the process have already been exhausted since they were used in the overall balances as well as the gasifier and shift converter balances. Thus, the degree of freedom of the acid gas removal unit is zero, as it should be. Proceeding with the balance equations for that unit, we have:

	H_2	CH_4	H_2S	CO	CO_2	H_2O
H balance:	$2(N_{H_2}^{10} - 133.2) +$	$4(N_{CH_4}^{10} - 95.37) +$	$2(0)$			$+ 2(N_{H_2O}^{10} - 261.12) = 0$
C balance:		$(N_{CH_4}^{10} - 95.37)$	$+$	$(N_{CO}^{10} - 42.15) +$	0	$= 0$
O balance:				$(N_{CO}^{10} - 42.15) + 2(0) +$		$(N_{H_2O}^{10} - 261.12) = 0$

and the relation

$$N_{H_2}^{10}/N_{CO}^{10} + N_{CH_4}^{10} + N_{H_2}^{10} = 0.492$$

Using the relation to eliminate $N_{H_2}^{10}$, the O balance to eliminate N_{H_2O}, and the C balance to eliminate N_{CO}^{10}, the H balance reduces to

$$6(N_{CH_4}^{10} - 95.37) = 0$$

or

$$N_{CH_4}^{10} = 95.37 \text{ lbmol/h}$$

The remaining species flows then follow immediately by back substitution. The result is

$$N^{10} = (N_{CH_4}, N_{CO}, N_{H_2O}, N_{H_2})$$
$$= (95.37, 42.15, 261.12, 133.2) \text{ lbmol/h}$$

All process streams have now been determined and can be scaled up to the required basis of 10,000 tons/day from the basis of 10,000 lb/h that was used in the calculations. The results of this scaling are summarized in the stream table shown in Figure 4.8.

The example brings out two points which deserve further comment. First, the example again confirms our previous observation that element balances generally are more likely to require simultaneous rather than the simpler sequential solution. In the present case, the gasifier, shift converter, and acid gas removal unit balances typically involved three or four unknown species net outflow rates per balance equation. The resulting set of simultaneous equations required extensive substitution to simplify the equations for solution.

	Stream Number							
	2	3	5	6	7	8	9	10
Total Flow (lbmol/day) $\times 10^{-4}$	88.888	68.380	84.375	60.622	28.948	190.74	190.74	106.37
Composition (mole fraction)								
CH_4	—	—	—	—	0.949	0.10	0.10	0.1793
CO	—	—	—	—	0.001	0.1179	0.0442	0.0793
H_2	—	—	—	—	0.05	0.0660	0.1397	0.2504
CO_2	—	—	0.9778	—	—	0.3588	0.4325	—
H_2O	1.0	—	—	1.0	—	0.3475	0.2738	0.4910
H_2S	—	—	0.0222	—	—	0.0098	0.0098	—
O_2	—	1.0	—	—	—	—	—	—

Figure 4.8 Example 4.7, stream table.

The degree-of-freedom updating summary is shown in Figure 4.9.

	Gasifier	Shift Reactor	Acid Gas Removal	Methanator	Overall Balances
Initial degree of freedom	1	4	6	2	<u>0</u>
Overall balances					
Stream variables	−2		−2	<u>−2</u>	
Redundant balances	<u>+1</u>				
Gasifier balances		<u>−4</u>			
Shift converter balances					
Stream variables			−5		
Redundant balances			<u>+1</u>		

Figure 4.9   Example 4.7, degree-of-freedom updating summary.

Second, a closer look at the acid gas removal balances suggests that the use of element balances for this nonreacting unit is unnecessarily complicated. Recall that after having calculated the output of the shift converter, the input stream to the acid gas removal unit is completely determined. In addition, one of the output streams, stream 5, is also known from the overall balance calculations. Clearly, if species balances were used, stream 10 could be calculated simply by difference without the tedious algebraic manipulations required in the element balance approach. Note further that if species balances are used, we actually have more balance equations to work with than if element balances are used. In particular, since the unit involves six species, six species balances could be written instead of, at most, the four balance equations available if elements are balanced. Since two more equations would be available, two specifications less would be required to solve the problem. The following question therefore arises: does this always happen and, if not, does it ever happen that element balances will prove more advantageous? These issues concerning the relationship between element and species balances comprise the subject matter of the next section.

4.2   THE RELATIONSHIP BETWEEN ELEMENT AND SPECIES BALANCES

In the previous section, we observed that element balances are appropriate if the reaction stoichiometry is not known or else, if only an elemental analysis of the reactants is available. In those cases, species balances cannot be written directly; hence, the use of element balances is the obvious alternative. However, in many cases the choice of which entity balance to employ is not as clear-cut. In particular, in the case of nonreacting systems in which all chemical species are identifiable, or in the case of reacting systems in which the reaction stoichiometry is known, either type of entity balance could in principle be used. In this section, we examine both of these two cases and derive a criterion which will allow us to recognize

when the two types of entity balances are interchangeable, as well as to determine
when one type is preferable to the other.

4.2.1 Nonreacting Systems

Regardless of whether or not chemical reactions are taking place in the system,
the element balances will always be

$$\sum_{s=1}^{S} \alpha_{es} \overline{N}_s = 0 \qquad e = 1, \ldots, E$$

where $\overline{N}_s$ is the *net* output rate of species s. In the nonreacting case, the corresponding species balances reduce simply to the conditions

$$\overline{N}_s = 0 \qquad s = 1, \ldots, S$$

Since both sets of equations are direct consequences of the principle of conservation
of mass, both must always be satisfied and both are always correct. The basis for
selecting one set over the other is thus not correctness but more simply which set
of balances will provide more equations for use in solution. The more equations
that are available, the less specified information will be required to solve the
problem and the less likely the possibility of encountering either inconsistent or
dependent specifications.

Now, if the maximum number of independent element balances, which we
designate as $\rho(\alpha)$, is exactly equal to the number of chemical species, S, then either
set of balances will provide the same number of equations and can be used interchangeably. But this need not always be the case, as is illustrated in the following
example.

Example 4.8 A gas stream containing 40% O_2, 40% H_2, and 20% H_2O (mol %)
is to be dried by cooling the stream and thus condensing out the water (see Figure
4.10). If 100 mol/h of a gas stream is to be processed, what is the rate at which
water will be condensed out and what is the composition of the dry gas?

Solution The process involves six stream variables, two specified independent
compositions, and a specified flow rate. Thus, if species balances are written on

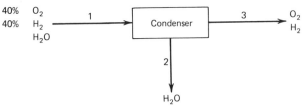

Figure 4.10 Flowsheet for Example 4.8, partial
condensation of a gas stream.

each of the three species, O_2, H_2, and H_2O, the problem is completely specified. The solution obtained by inspection will be

$$N^2 = 20 \text{ mol/h}$$

$$N^3_{O_2} = 40 \text{ mol/h}$$

$$N^3_{H_2} = 40 \text{ mol/h}$$

However, if element balances are used, then, since the problem involves only two elements, O and H, only two balance equations can be written:

O balance: $1(N^2 - 20) + 2(N^3_{O_2} - 40) = 0$

H balance: $2(N^2 - 20) + 2(N^3_{H_2} - 40) = 0$

These two element balances are insufficient to solve for the three unknown stream variables. From the element balance point of view, the problem appears underspecified. As we shall see in later sections, the explanation for this apparent paradox is that the element balances implicitly assume that all independent chemical reactions possible between the chemical species are actually taking place. In our case, there is a possible reaction between the three species.

$$\tfrac{1}{2}O_2 + H_2 \rightarrow H_2O$$

but there is no way of specifying within the element balances that this reaction is not taking place in the given system.

The above example clearly establishes that it is possible for $\rho(\alpha)$ to be less than S. The remaining question is, can $\rho(\alpha)$ be greater than S? The following simple argument demonstrates that it can not.

Recall that the independence of a set of element balances is determined by considering the row vectors α_e, $e = 1, \ldots, E$ and establishing whether or not these vectors are linearly independent. Now, given any collection of S component vectors, the maximum number that can be linearly independent is S. We can verify this fact constructively using our *array reduction procedure*. If in carrying out the row operations there are more rows than columns, then, once the first S rows are reduced to diagonal form, the remaining rows will be identically zero. Thus, it is always true that

$$\rho(\alpha) \leq S$$

Example 4.9   An aqueous solution of 20 mol % NaOH is to be prepared on a continuous basis by mixing pure NaOH and water (Figure 4.11). What is the addition rate of each required to prepare 100 mol/h solution?

Solution   The process involves four stream variables, a specified composition and flow, and, apparently, three element balances—those on Na, O, and H. Hence, from an element balance point of view, the problem appears overspecified. On the

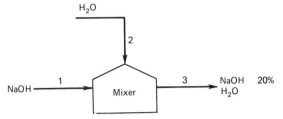

Figure 4.11 Flowsheet for Example 4.9, caustic preparation.

other hand, the problem appears correctly specified when analyzed in terms of species balances, since only two distinct species, NaOH and H_2O, are involved in the process.

If we proceed with the element balances, we obtain:

Na balance: $1(20 - N^1) = 0$

O balance: $1(20 - N^1) + 1(80 - N^2) = 0$

H balance: $1(20 - N^1) + 2(80 - N^2) = 0$

The first equation implies $N^1 = 20$, and then the second equation yields $N^2 = 80$. Clearly, the third is redundant.

The atom matrix

$$
\begin{array}{c}
 \\
Na \\
O \\
H
\end{array}
\begin{array}{cc}
NaOH & H_2O \\
1 & 0 \\
1 & 1 \\
1 & 2
\end{array}
$$

is easily reduced to the form

$$
\begin{pmatrix}
1 & 0 \\
0 & 1 \\
0 & 0
\end{pmatrix}
$$

indicating that only two of the element balances are independent.

The general conclusion that we have deduced is that in the nonreacting case, the number of independent element balances is always less than or equal to the number of species balances. Consequently, element balances should only be used if $\rho(\alpha) = S$. However, since the species balances will always be independent while the element balances may not be, it appears that the work of checking independence can be saved if species balances are always used in the nonreacting case. This general rule is further supported by the fact that species balances are always easier to solve because they always involve only a single net outflow rate while the element balances involve as many as there are species.

4.2.2 Reacting Systems

Given the same set of species, the element balances in the presence of chemical reactions are indistinguishable from their equivalents in the nonreacting case. However, the species balance equations do change to become

$$\overline{N}_s = \sum_{r=1}^{R} \sigma_{sr} r_r \qquad s = 1, \ldots, S$$

If the stoichiometric coefficients σ_{sr} of the R reactions are not known, then there appears to be no choice but to use element balances. However, if the stoichiometry of the reactions is known, we have a choice of which entity balance to use. Again, that choice ought to be dictated by the number of balance equations each can provide.

As in the nonreacting case, let us deduce the condition under which each set of entity balances will provide the same number of independent balance equations. Consider a chemically reacting system involving V stream variables and S species. Suppose that the stoichiometric coefficients of all reactions are known and that of the R given reactions, $\rho(\sigma)$ are independent. Following the degree-of-freedom analysis of chapter 3, the system will involve $V + \rho(\sigma)$ variables and S species balances. Hence, the system will have $V + \rho(\sigma) - S$ degrees of freedom.

Now let us repeat this analysis in terms of element balances. Suppose the system involves E elements and $\rho(\alpha)$ of the element balances are independent. Then, since the number of variables is V, the system will have degree of freedom $V - \rho(\alpha)$. In order for the degree of freedom of the problem to be the same regardless of which set of entity balances is employed, it follows that

$$V + \rho(\sigma) - S = V - \rho(\alpha)$$

or

$$\rho(\sigma) = S - \rho(\alpha)$$

If this condition is met, then the two sets of balances can be used interchangeably. If $\rho(\sigma) < S - \rho(\alpha)$, then the element balances will provide fewer equations than the species balances, and the latter should be used for solution. This is the case illustrated in the following example.

Example 4.10   Ozone is formed by the reaction of atomic oxygen with O_2 via the reaction

$$O + O_2 \rightarrow O_3$$

If 100 mol/h of a mixture containing 50% O and 50% O_2 is reacted with 50% conversion of the O_2, calculate the composition of the product.

Solution   If species balances are used, then, since the problem involves five stream variables and one rate of reaction, a composition, a conversion, and a flow rate, the three species balances ought to be adequate to calculate all flows and compositions. In particular, from the conversion specification,

$$N_{O_2}^{out} = N_{O_2}^{in} (1 - X_{O_2}) = 0.5(100)(1 - 0.5)$$
$$= 25 \text{ mol/h}$$

Then, from the O_2 balance,

$$r = 25$$

and from the remaining two balances,

$$N_O^{out} = 25 \text{ mol/h} \quad \text{and} \quad N_{O_3}^{out} = 25 \text{ mol/h}$$

However, if element balances are used, then, since only one element balance is available, the problem appears underspecified by one. Although the conversion relation still yields $N_{O_2}^{out} = 25$ mol/h, the O balance,

$$(N_O^{out} - 50) + 2(25 - 50) + 3(N_{O_3}^{out} - 0) = 0$$

is inadequate to solve for the two unknowns N_O^{out} and $N_{O_3}^{out}$.

The converse to the condition illustrated in the preceding example, namely, the situation

$$\rho(\sigma) > S - \rho(\alpha)$$

is a bit more complicated to analyze. We defer its discussion to Section 4.4. The result of the analysis given there, however, is that the above situation can *never* occur. In other words, regardless of the type of reacting problem and independent of the nature of the atom and stoichiometric arrays, it is always the case that

$$\rho(\sigma) \leqslant S - \rho(\alpha)$$

From this statement, it follows that

$$V + \rho(\sigma) - S \leqslant V - \rho(\alpha)$$

or, in words, the degree of freedom when analyzing by element balances is always equal to or greater than the degree of freedom when analyzing by species balances. Clearly, the entity balances can be used interchangeably only if the above inequality is satisfied as an equality. Otherwise, species balances should be used.

As in the nonreacting case, the above analysis suggests that testing of the equivalence condition can be avoided if we agree to always use species balances. However, unlike the nonreacting case, it does happen, especially in situations involving complex systems of reactions, that the element balances can sometimes be solved more easily than the species balances. Hence, it behooves us to maintain the element balance in our repertoire of problem solving tools.

Example 4.11 *N*-ethylaniline is produced industrially by vapor-phase catalytic reaction of aniline and ethanol. In a given plant, a reactant mixture of two thirds ethanol, C_2H_5OH, and one third aniline, $C_6H_5NH_2$, is reacted until two thirds of the aniline and one half of the ethanol are converted. The product is a mixture of aniline, ethanol, N-ethylaniline, $C_6H_5NHC_2H_5$, diethylaniline, $C_6H_5N(C_2H_5)_2$, water, and 1% ethylene, C_2H_4. The published research on this process indicates that the following set of reactions is generally accepted as taking place in the reactor:

$$C_6H_5NH_2 + C_2H_5OH \rightarrow C_6H_5NHC_2H_5 + H_2O$$

$$C_2H_5OH \rightarrow C_2H_4 + H_2O$$

$$C_6H_5NH_2 + C_2H_4 \rightarrow C_6H_5NHC_2H_5$$

$$C_6H_5NHC_2H_5 + C_2H_4 \rightarrow C_6H_5N(C_2H_5)_2$$

$$C_6H_5NHC_2H_5 + C_2H_5OH \rightarrow C_6H_5N(C_2H_5)_2 + H_2O$$

Calculate the composition of the reactor outlet stream.

Solution   Let us first solve the problem using species balances and then return to repeat the solution using element balances. Following Chapter 3, the first step in conducting the analysis in terms of the species balances is to determine the number of independent reactions. The stoichiometric array for the above five reactions in the order listed is:

$$
\begin{array}{lccccc}
 & (1) & (2) & (3) & (4) & (5) \\
C_6H_5NH_2 & -1 & 0 & -1 & 0 & 0 \\
C_2H_5OH & -1 & -1 & 0 & 0 & -1 \\
C_2H_4 & 0 & 1 & -1 & -1 & 0 \\
C_6H_5NHC_2H_5 & 1 & 0 & 1 & -1 & -1 \\
H_2O & 1 & 1 & 0 & 0 & 1 \\
C_6H_5N(C_2H_5)_2 & 0 & 0 & 0 & 1 & 1
\end{array}
$$

After carrying out the column operations, the following reduced array results:

$$
\begin{pmatrix}
1 & 0 & 0 & 0 & 0 \\
0 & 1 & 0 & 0 & 0 \\
0 & 0 & 1 & 0 & 0 \\
-2 & 1 & 1 & 0 & 0 \\
0 & -1 & 0 & 0 & 0 \\
1 & -1 & -1 & 0 & 0
\end{pmatrix}
$$

In the reduction, the third and fourth columns were interchanged. From the reduced array, it is clear that only three reactions are independent and that reactions 1, 2, and 4 can serve as an independent set.

As shown in Figure 4.12, the system consists of eight stream variables and involves two specified compositions and two conversions. Since the system involves

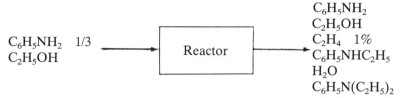

Figure 4.12   Flowsheet for Example 4.11, production of ethylaniline.

three independent reactions and six species, the degree of freedom of the reactor is, as shown in Figure 4.13, zero.

Proceeding with the solution of the problem, we select a basis of 9000 mol/h feed. Then, the conversion relations reduce to

Conversion of aniline: $\frac{2}{3} = \dfrac{r_1}{3000}$

Conversion of ethanol: $\frac{1}{2} = \dfrac{r_1 + r_2}{6000}$

Thus, $r_1 = 2000$ and $r_2 = 1000$ mol/h. With these calculated, the species balances are:

Aniline balance: $N_1^{out} = 3000 - r_1 = 1000$ mol/h

Ethanol balance: $N_2^{out} = 6000 - r_1 - r_2 = 3000$ mol/h

Ethylene balance: $N_3^{out} = 0 + 1000 - r_4 = 0.01N^{out}$

N-Ethylaniline balance: $N_4^{out} = 0 + 2000 - r_4$

Water balance: $N_5^{out} = 0 + 2000 + 1000 = 3000$ mol/h

Diethylaniline balance: $N_6^{out} = 0 + r_4$

In the balance equations, the subscripts on N^{out} refer to the species in the order in which they are listed in constructing the stoichiometric matrix.

Summing the above six equations, we obtain

$$N^{out} = 10,000 - r_4$$

Number of variables	8 + 3
Number of balances	6
Number of specified	
Compositions	2
Conversions	2
Basis	1
Degree of freedom	0

Figure 4.13   Example 4.11, degree-of-freedom table.

From the ethylene balance, it follows that

$$r_4 = 1000 - 0.01N^{out}$$

Substituting out r_4, we immediately have

$$N^{out} = \tfrac{1}{11} \times 10^5 \text{ mol/h}$$

Hence,

$$r_4 = \tfrac{1}{11} \times 10^4$$

The complete output stream thus consists of

$$N^{out} = (1000, 3000, \tfrac{10^3}{11}, \tfrac{12}{11} \times 10^3, 3000, \tfrac{10^4}{11}) \text{ mol/h}$$

Prior to repeating the calculation using element balances, we have to check whether the equivalence condition is satisfied. This requires the determination of the maximum number of independent element balances. The atom matrix for the system is:

	$C_6H_5NH_2$	C_2H_5OH	C_2H_4	$C_6H_5NHC_2H_5$	H_2O	$C_6H_5N(C_2H_5)_2$
N	1	0	0	1	0	1
O	0	1	0	0	1	0
C	6	2	2	8	0	10
H	7	6	4	11	2	15

After carrying out the row operations, the following reduced array is obtained:

$$\begin{pmatrix} 1 & 0 & 0 & 1 & 0 & 1 \\ 0 & 1 & 0 & 0 & 1 & 0 \\ 0 & 0 & 1 & 1 & -1 & 2 \\ 0 & 0 & 0 & 0 & 0 & 0 \end{pmatrix}$$

From this, it is apparent that only three element balances will be independent and that the balances on N, O, and C will constitute an independent set. Now, since the number of independent element balances, $\rho(\alpha)$, is 3, the number of independent reactions, $\rho(\sigma)$, is 3, and the number of species is 6, it follows that

$$3 = \rho(\sigma) = S - \rho(\alpha) = 6 - 3 = 3$$

Consequently, the problem analyzed in terms of elements will yield the same degree of freedom as in the species case, namely, zero, and solution using element balances may be carried out.

Since, from the conversion relationships,

$$N_1^{out} = 1000 \text{ mol/h} \quad \text{and} \quad N_2^{out} = 3000 \text{ mol/h}$$

the element balances become:

N balance:	$1(1000 - 3000)$			$+ 1(N_4^{out} - 0)$	$+ 1(N_6^{out} - 0)$	$= 0$
O balance:		$+ 1(3000 - 6000)$		$+ 1(N_5^{out} - 0)$		$= 0$
C balance:	$6(1000 - 3000)$	$+ 2(3000 - 6000)$	$+ 2(N_3^{out} - 0) + 8(N_4^{out} - 0)$		$+ 10(N_6^{out} - 0)$	$= 0$

In addition, from the given composition,

$$N_3^{out} = 0.01(N^{out})$$

$$= 0.01(4000 + N_3^{out} + N_4^{out} + N_5^{out} + N_6^{out})$$

Since the O balance yields

$$N_5^{out} = 3000 \text{ mol/h}$$

the problem reduces to the three equations

$$N_4^{out} + N_6^{out} = 2000$$

$$2N_3^{out} + 8N_4^{out} + 10N_6^{out} = 18{,}000$$

$$99N_3^{out} - N_4^{out} - N_6^{out} = 7000$$

Adding the first and third equations, it follows that $N_3^{out} = \frac{10^3}{11}$. Then, subtracting eight times the first from the second, we obtain $N_6^{out} = \frac{10^4}{11}$. Finally, substituting this value into the first equation, $N_4^{out} = \frac{12}{11} \times 10^3$ is recovered. The result is of course the same as that calculated using species balances.

4.2.3 The Maximum Number of Independent Reactions

The inequality relating the number of independent element balances and the number of independent reactions, in addition to serving as a criterion for deciding which entity balance ought to be used in any given situation, also provides an interesting insight into the number of chemical reactions that can be involved in the species balances for a given system. In particular, the inequality implies that for any system involving S species and E elements with specified atom matrix α, it is always true that the number of independent reactions, $\rho(\sigma)$, is never larger than $S - \rho(\alpha)$, regardless of the number of reactions making up the stoichiometric matrix, σ. Thus, $S - \rho(\alpha)$ must be the maximum number of independent reactions it is possible for a system with atom matrix α to have incorporated into the species balances. Since that maximum number depends only on S and α, it further follows that whenever we write element balances, we implicitly assume that $S - \rho(\alpha)$ reactions, that is, the maximum number, are actually taking place. Consequently, if fewer than the maximum number of reactions are specified in the problem formulation, because the others are for one reason or another not postulated to occur, then element balances should not be used. That is in effect what occurs in situations in which

$$\rho(\sigma) < S - \rho(\alpha)$$

On the other hand, if $\rho(\sigma) = S - \rho(\alpha)$, then, since either set of balances will be based on the same number of reactions, either set of balance equations can be used. These considerations are illustrated in the following example.

Example 4.12 Let us analyze the preceding four examples in the light of the above discussion. In Example 4.8, we examined a system involving the species O_2, H_2, and H_2O in which no reactions were postulated to occur. The atom matrix for this set of species is:

$$
\begin{array}{cc}
 & \begin{array}{ccc} O_2 & H_2 & H_2O \end{array} \\
\begin{array}{c} O \\ H \end{array} & \left(\begin{array}{ccc} 2 & 0 & 1 \\ 0 & 2 & 2 \end{array} \right)
\end{array}
$$

Clearly, the two rows of this array are independent. Thus, $\rho(\alpha) = 2$ and $S - \rho(\alpha) = 3 - 2 = 1$. In other words, the maximum number of independent reactions is one. For instance, $H_2 + \frac{1}{2}O_2 \rightarrow H_2O$ could be that reaction. Note, however, that since we assumed that no reactions are occurring, the actual number of independent reactions must be zero, that is, $\rho(\sigma) = 0$. Consequently,

$$
0 = \rho(\sigma) < S - \rho(\alpha) = 1
$$

that is, we have less than the maximum number of reactions actually occurring. Thus, element balances should not be used.

In Example 4.9, two species, NaOH and H_2O, occurred, and the associated atom matrix was found to have only two independent rows. Thus, $S - \rho(\alpha) = 2 - 2 = 0$, that is, the maximum number of independent reactions involving these two species is zero. Since no reaction was postulated to take place, we have

$$
\rho(\sigma) = S - \rho(\alpha) = 0
$$

and element and species balances could be used interchangeably.

In Example 4.10, we examined a system involving the species O, O_2, and O_3 which were subjected to the single reaction $O + O_2 \rightarrow O_3$. Certainly, since oxygen is the only element involved, $\rho(\alpha) = 1$ and, consequently, $S - \rho(\alpha) = 3 - 1 = 2$. The maximum number of reactions is two, but only one reaction is assumed to occur, namely, $\rho(\sigma) = 1$. Thus,

$$
\rho(\sigma) = 1 < S - \rho(\alpha) = 2
$$

and again element balances should not be used. In this case, the element balances will implicitly assume that a second reaction, such as $O_2 \rightarrow 2O$, is occurring, whereas no second reaction is postulated to occur in using the species balances.

Finally, in Example 4.11, three of the four rows of the atom matrix were found to be independent. Hence, since $S - \rho(\alpha) = 6 - 3 = 3$, the maximum number of independent reactions possible between the six species is three. Since three of the given five reactions were found to be independent, the maximum number of reactions is actually postulated to take place; thus, either set of entity balances can be used for problem solution.

The reader should be aware of the fact that the concept of maximum number of independent reactions is an algebraic result which has no relation to the number of reactions necessary to describe the mechanism of a chemical transformation. Any number of reactions may be assembled by a chemist to explain the intermediate and side reactions which take place in the course of proceeding from initial reactants to final products. The maximum number of reactions refers, in our context, to the maximum number of independent reaction rate variables and vectors of stoichiometric coefficients which can be introduced to relate species input flows to output flows in such a way that the principle of conservation of mass is satisfied.

4.3 APPLICATIONS INVOLVING CHEMICAL PROCESSING OF FOSSIL FUELS

Although, as we have concluded in the previous section, species balances are generally preferable to element balances, there is a substantial body of applications for which element balances are a very logical choice. These situations arise whenever the exact reaction stoichiometry is either unknown or very complex or when the reactants are a complicated mixture of which an analysis in terms of species is either inconvenient or very difficult to obtain. Such is the case whenever the system of interest involves the chemical processing of fossil fuels: natural gas, petroleum, oil shale, and coal. In this section, we briefly discuss the origin and properties of fossil fuels, summarize some of the standard ways of characterizing and analyzing them which are pertinent to material balances, and illustrate several typical instances of application of element balances to the solution of material balancing problems involving these raw materials.

4.3.1 Fossil Fuels and Their Constituents

Fossil fuels are hydrocarbon materials formed from the organic remains of plants and animals as a consequence of bacterial action and/or long-term compression and heating, generally under anaerobic conditions below the earth's crust. The principal types of fossil fuels are coal, oil shale, peat, petroleum, and natural gas. Although these alternate forms of fossil fuels have common constituents, namely, carbon, hydrogen, and varying amounts of nitrogen, sulfur, oxygen, and inorganic mineral matter, it is generally accepted that the processes involved in the formation of each type of fossil fuel are substantially different.

In the case of coal, the formation process begins with peat, a high-moisture-content material formed in swamps as a result of the decay of vegetable matter. At some point in the decomposition process, the layer of peat is covered up by water and subsequent layers of debris and sediment. The relatively rapid decomposition is arrested and a slower anaerobic change called coalification sets in. The geologic process of coalification is brought about by compression and heating and results in the successive reduction of the moisture and oxygen content of the solid through drainage and the evolution of carbon dioxide as well as some methane.

Depending upon the history and intensity of the temperature and pressure conditions to which the material is subjected, various types of coals are produced. Typical compositions of the common varieties of coal are shown in Figure 4.14.

It is generally believed that in the process of coalification, the fossil material progresses in a continuous fashion through each of the above varieties of coal. Hence, an anthracite coal will have existed first as peat, then as lignite, then bituminous, and finally in its anthracite form. The stages within the process of coalification of peat are referred to as the *rank* of the coal.

Oil shale, petroleum, and natural gas, on the other hand, are generally believed to have been derived from the organic remains of marine plants and animals incorporated into fine-grained sedimentary rocks. As these organic deposits were decomposed by the action of aerobic and anaerobic bacteria, the resulting products were either bound within the sedimentary rock, thus forming oil shale, or else migrated out of the sedimentary matrix, probably as a water emulsion, to collect in natural rock reservoirs of sand or limestone formations. Crude oils from various sites vary in composition but generally are composed of mixtures of paraffins, naphthenes, aromatics, possibly some olefins, and small amounts of organic compounds involving sulfur, nitrogen, and oxygen. Of these, the major constituents are paraffins, which are straight-chain hydrocarbons involving simple carbon–carbon bonds with general formula C_nH_{2n+2}, and naphthenes, which are cyclic analogs of the paraffins. The unsaturated cyclic hydrocarbons, the aromatics, and their acyclic analogs, the olefins, are generally present in much smaller amounts, if at all. Typically, crude oils will consist (by weight) of about 85% carbon, 11% hydrogen, and 4% impurities including organic compounds involving sulfur, nitrogen, and oxygen.

Natural gas is usually found above or near oil pools, trapped beneath a hard-cap rock and is composed of lighter hydrocarbon constituents of the parent petroleum. Typically, natural gas of the type supplied for the consumer use consists (on a mole basis) of 88% CH_4, 8% higher hydrocarbons, and the remainder H_2, N_2,

	Moisture Content in Raw State (%)	wt % on Dry Basis, Ash-Free		
		Carbon	Hydrogen	Oxygen
Wood	20	50	6	42.5
Peat	90	60	5.5	32.3
Brown coal	40–60	60–70	About 5	>25
Lignite	20–40	65–75	About 5	16–25
Subbituminous	10–20	75–80	4.5–5.5	12–21
Bituminous	10	75–90	4.5–5.5	5–20
Semibituminous	<5	90–92	4–4.5	4–5
Anthracite	<5	92–94	3–4	3–4

Figure 4.14 Composition of wood, peat, and various coals. (From *Fuel*, **23**, 70 (1944), reproduced with permission.)

and CO_2. The composition of hydrocarbons other than methane can vary depending upon how much of the hydrocarbons heavier than methane is removed at the well head.

4.3.2 Characterization of Fossil Fuels

Since most coals are not homogeneous substances but consist of a mixture of macromolecules, the only practical way of describing the composition of coals is to resort to an elemental analysis. The elemental or *ultimate* analysis of a coal involves a standardized chemical determination of the five major elements present in coal—carbon, hydrogen, oxygen, sulfur, and nitrogen—together with a lumped determination of inorganic, noncombustible residuals called ash. In the analysis, the weight percent of carbon and hydrogen is determined by collecting the products from the combustion of a standard sample. The nitrogen and sulfur are obtained by direct chemical analysis, and the weight percent of ash is found by heating a 1-g sample to 1340°F in the presence of oxygen and weighing the residual. The weight percent of oxygen is determined by difference and hence is the least reliable number in the ultimate analysis. Sometimes, the oxygen and hydrogen weight percents are reported as "combined water" and "net hydrogen." The combined water refers to the weight percent of the hypothetical water formed by adding two moles hydrogen per mole oxygen obtained in the analysis. The net hydrogen then consists of any hydrogen over and above that associated with the combined water. For precise calculation, the ash weight percent should be corrected to account for any iron pyrites, FeS_2, which will appear as Fe_2O_3 in the ash. Except for situations involving high-sulfur coals in which much of the sulfur is present as pyrites, this correction can be disregarded.

Although an ultimate analysis is necessary for any precise material balance calculations, this analysis is time consuming and tedious; hence, for purposes of routine plant operations, it is conventional to resort to a *proximate* analysis. The proximate analysis involves the determination of four lumped weight fractions: the moisture, the volatile matter, the ash content, and the fixed carbon of the coal sample. The moisture is determined by a standardized measurement of the weight loss during drying of a sample of the coal at about 110°C. The volatile matter in the coal is determined by measuring the weight loss of a 1-g sample of dry coal when it is heated for 7 min at 950°C. The ash weight fraction is determined as in the ultimate analysis, and the weight fraction of fixed carbon is calculated by difference. Often, the proximate analysis is supplemented by a separate determination of the weight fraction of sulfur because this sulfur is converted to the pollutant SO_2 when the coal is combusted. Typical ultimate and proximate analysis of some coals are given in Figure 4.15.

As noted in the previous section, petroleum is a complex mixture of several families of hydrocarbon compounds. Since an exact determination of the composition of every chemical species in this mixture is hopelessly difficult, an ultimate analysis is sometimes used and certainly will suffice for material balance calculation. However, as seen from the sample analysis given in Figure 4.16, the variation in

Coal Source	Type	Proximate Analysis (wt % Dry Basis)				Ultimate Analysis (wt % Dry Basis)					
		Moisture	volatile	Fixed Carbon	Ash	C	H	N	S	Ash	O
1. North Dakota Mercer	Lignite	31.70	41.60	46.61	11.49	63.63	4.29	0.72	1.22	11.49	18.65
2. Montana Rosebud	Subbituminous B	19.84	39.02	51.82	9.16	68.39	4.64	0.99	0.79	9.16	16.03
3. Illinois Knox No. 6	Subbituminous A	17.69	38.46	51.39	10.14	70.78	5.22	1.39	2.59	10.14	9.87
4. Illinois Saline No. 6	High volatile bituminous C	12.10	40.71	48.45	10.84	68.92	5.01	1.01	6.66	10.84	7.57
5. Indiana Owen Block 1	High volatile bituminous B	6.20	34.47	54.21	11.32	71.13	4.96	0.35	2.21	11.32	10.03
6. Kentucky Harlan C	High volatile bituminous A	1.50	36.16	61.77	2.07	82.86	5.49	1.49	0.44	2.07	7.66
7. Penna. Cambria L. Kittanning	Low volatile bituminous	0.64	18.62	74.37	7.01	82.90	4.46	0.86	1.09	7.01	3.68
8. Penna., Northumberland No. 8 (Leader)	Anthracite	1.25	7.09	84.59	8.32	83.67	3.56	0.55	1.05	8.32	2.85

Figure 4.15 Typical ultimate and proximate analysis of coals. (*Source.* Coal Conversion Systems Technical Data Book, U.S. DOE Report HCP/T2286-01, 1978, Section IA.50.1.)

Petroleum	Sp gr	At Temp. (°C)	C (%)	H (%)	N (%)	O (%)	S (%)
Pennsylvania pipeline	0.862	15	85.5	14.2			
Mecook, West Virginia	0.897	0	83.6	12.9		3.6	
Humbolt, Kansas	0.912		85.6	12.4			0.37
Healdton, Oklahoma			85.0	12.9			0.76
Coalinga, California	0.951	15	86.4	11.7	1.14		0.60
Beaumont, Texas	0.91		85.7	11.0	2.61		0.70
Mexico	0.97	15	83.0	11.0	1.7		4.30
Baku, U.S.S.R.	0.897		86.5	12.0		1.5	
Colombia, South America	0.948	20	85.62	11.91	0.54		

Figure 4.16 Typical ultimate analysis of several crude oils. (Reproduced from International Critical Tables II, National Academy Press, 1977.)

the carbon and hydrogen contents of crudes and their derivatives is remarkably small. Moreover, the variation in elemental weight percent gives a very poor indication of the physical properties, reacting properties, and utility of the petroleum stock as a fuel. Consequently, various additional approximate factors and lumped analyses have been adopted in the petroleum industry. These include a measure of the density of the stock, its flow characteristics, and a lumped analysis in terms of paraffin, olefins, naphthenes, and aromatics, referred to by the acronym PONA. For precise material balance calculation, nothing short of a complete species analysis or an ultimate analysis will suffice.

Natural gas is the least complicated of the conventional fossil fuels to characterize since it consists of relatively few chemical species—typically, methane, ethane, ethylene, propane, hydrogen, carbon dioxide, and nitrogen. Thus, the complete composition of natural gas and any produced synthetic gases is usually reported. These compositions are usually expressed in terms of mole or volume percent.

4.3.3 Major Types of Chemical Processing Operations for Fossil Fuels

The operations involving the direct chemical processing of fossil materials can be broadly classed into three categories: combustion, synthesis gas generation, and chemical upgrading. In this section, we briefly discuss the above three classes of processes and illustrate the use of element balances as a tool for solving flowsheet problems involving these processes.

Combustion The combustion of a fossil fuel with air or oxygen is an operation in which the fuel is oxidized to produce heat and partially or completely oxidized residual waste products. The primary waste product of combustion is a gas, consisting predominantly of CO_2, CO, H_2O, N_2, unreacted O_2, and any remaining unreacted gaseous fuel constituents. In addition, if the fuel is a solid or liquid, there may remain residual unreacted fuel and noncombustible inorganic material called ash.

The main operational concerns in combustion are: What rate of air or oxygen is or should be used with a given fuel feed rate? How efficiently is the fuel being used? And what is the concentration of any pollutants in the combustion waste gases? Answers to these questions are obtained by measuring one or more of the compositions of the flue gases, of the fuel, and/or of the uncombusted fuel residual and calculating the unknown unmeasured quantities by means of element or species balances. As we noted in Section 4.3.2, analysis of solid or liquid fossil fuels is usually given on an elemental basis, while that of gaseous fuels is often given in terms of species. Compositions of the combustion waste gases are usually measured and reported on a species basis because the products of combustion consist of relatively few species. A typical flue gas analysis will include the mole percent of CO_2, CO, O_2, possibly H_2 and CH_4, often SO_2 and H_2S, and N_2.

The standard experimental procedure for flue gas analysis, known as the *Orsat analysis,* involves the successive contacting of flue gas with solutions that preferentially absorb one or more of the constituents. After each stage of absorption, the volume of gas remaining is measured to determine by difference the volume of the absorbed constituent. For instance, a sample of gas is first bubbled through a 20% solution of KOH to remove CO_2. The reduction in volume of the gas sample at constant pressure can then be related to the mole fractions of CO_2 in the original sample. The remaining gas sample is then bubbled through a pyrogalic acid solution to remove O_2 and then through a cuprous chloride solution to remove CO. Further solutions can be used if appreciable amounts of CH_4, H_2, or SO_2 are suspected to be present.

Finally, the residual gas remaining is assumed to be N_2. Since each of the solutions used contain water, the sample always contains water vapors, and therefore the original water content of the flue gas cannot be measured. Thus, flue gas analyses are always reported on a water-free (i.e., dry) basis. Of course, in recent years, electronic automatic gas analyzers have become more and more prevalent. However, by convention, flue gas analyses are still always reported on a dry basis.

Once adequate data are available on fuel and flue gas composition, the next major quantity of interest in combustion is the relative amount of oxidizing agent introduced to react with the fuel. If too little oxidizer is introduced, the fuel will burn inefficiently, producing too much CO. Moreover, a considerable amount of it may exit completely unoxidized. On the other hand, if excessive amounts of oxidizer are used, the combustion may proceed too violently and too large a fraction of the heat generated by combustion will leave with the flue gases. As with any reaction, there is a calculatable stoichiometric ratio of oxidizer to fuel which corresponds to the oxidizer being present in precisely the required quantity for complete combustion of the fuel. Since in most combustion applications the oxidizing agent is O_2 gas, this exact stoichiometric amount in moles is known as the *theoretical oxygen.* It is calculated assuming that all combustible materials will oxidize to their most stable oxides. Thus, it is assumed that all carbon will oxidize to CO_2, all H to H_2O, and all S to SO_2. For purposes of the definition, N is usually assumed not to oxidize, and any O present in the fuel is deducted from the total amount of O_2 required to oxidize the other fuel constituents.

Since the chief source of oxygen for combustion is air, the theoretical oxygen is divided by the oxygen content of air to obtain the *theoretical air*. For purposes of this definition and in fact for most combustion calculations, air is usually assumed to contain 21 mol % O_2 and 79% N_2. The mol % N_2 thus includes the CO_2, argon, and other inert gases present in small amounts in air.

Example 4.13 Calculate the theoretical air for 100 lb of a dried coal with ultimate analysis C, 71.2%; H, 4.8%; S, 4.3%; O, 9.5%; ash, 10.2%.

Solution Based on this composition, the 100 lb coal will contain the following amounts of combustible materials:

Carbon: 71.2 lb or 5.933 lbatom
Hydrogen: 4.8 lb or 4.8 lbatom
Sulfur: 4.3 lb or 0.134 lbatom

Each mole of carbon requires 1 mol O_2 to form CO_2. Each mole of hydrogen requires ¼ mol O_2 to form H_2O; and each mole of sulfur requires 1 mol O_2 to form SO_2. Thus, the total number of moles of O_2 required is

$$5.933 + 1.2 + 0.134 = 7.267 \text{ mol}$$

From this must be deducted the moles of O_2 present in the coal, or

$$\tfrac{1}{2}(9.5/16) = 0.297 \text{ mol}$$

Thus, the theoretical oxygen is

$$7.267 - 0.297 = 6.970 \text{ mol/100 lb coal}$$

The theoretical air is therefore

$$\frac{\text{Theoretical oxygen}}{0.21} = 33.19 \text{ mol}$$

In most applications, more air than the theoretical amount is usually introduced for combustion. The extra air supplied above the theoretical amount is conventionally reported as *percent excess air*:

$$\% \text{ Excess air} = \frac{\text{moles of air actually supplied} - \text{theoretical air}}{\text{theoretical air}}$$

If the excess air is given, it can be used as an additional equation to be used in conjunction with the balance equation to complete the solution of the problem.

Example 4.14 A synthesis gas consisting of 0.4% CH_4, 52.8% H_2, 38.3% CO, 5.5% CO_2, 0.1% O_2, and 2.9% N_2 is burned with 10% excess air (see Figure 4.17). Calculate the composition of the flue gas assuming no CO is present.

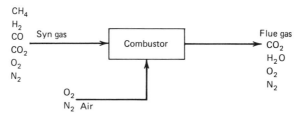

Figure 4.17   Flowsheet for Example 4.14, combustion of syn gas.

Solution   Since no CO is present, the flue gas must consist of CO_2, H_2O, O_2, and N_2.

The degree of freedom of the problem must therefore be:

Number of variables	12
Number of element balances (C, H, O, N)	4
Number of specifications	
Gas composition	5
Air composition	1
% Excess air	1
Basis	1
Degree of freedom	0

We begin the solution by selecting as basis 1000 mol/h synthesis gas and proceed with the calculation of the theoretical air for the fuel. The gas will contain

Carbon: $4 + 383 + 55 = 442$
Hydrogen: $4(4) + 2(528) = 1072$
Oxygen: $383 + 2(55) + 1(2) = 495$

The theoretical oxygen will therefore be

$$442 + \frac{1072}{4} - \frac{495}{2} = 462.5 \text{ mol/h}$$

and the theoretical air will be

$$\frac{462.5}{0.21} = 2202.4 \text{ mol/h}$$

The excess air relation by definition will be

$$\frac{\text{Actual air } - \text{ theoretical air}}{\text{Theoretical air}} = 0.1$$

or

$$\text{Actual air } = 1.1 \text{ theoretical air}$$
$$= 1.1(2202.4) = 2422.6 \text{ mol/h}$$

The air is thus fed to the combustor at the rate of

$$N_{N_2}^{in} = 1914 \text{ mol/h}$$

$$N_{O_2}^{in} = 508.8 \text{ mol/h}$$

With this, we can write the four element balances and determine the flue gas composition. The balances are:

N balance: $2(N_{N_2}^{out} - 1914 - 29) = 0$

H balance: $2(N_{H_2O}^{out}) + 2(-528) + 4(-4) = 0$

C balance: $(N_{CO_2}^{out} - 55) + (-383) + (-4) = 0$

O balance: $2(N_{O_2}^{out} - 1 - 508.8) + 2(N_{CO_2}^{out} - 55)$

$$+ (N_{H_2O}^{out}) + (-383) = 0$$

From the first three balances, it follows immediately that

$$N_{N_2}^{out} = 1943 \text{ mol/h} \qquad N_{H_2O}^{out} = 536 \text{ mol/h} \qquad N_{CO_2}^{out} = 442 \text{ mol/h}$$

Then, from the oxygen balance,

$$N_{O_2}^{out} = 46.3 \text{ mol/h}$$

The composition of the flue gas is therefore

$$(x_{CO_2}, x_{H_2O}, x_{O_2}, x_{N_2}) = (0.1490, 0.1806, 0.0156, 0.6548)$$

In most combustion calculations, the excess air is not known in advance. Rather, it is a quantity calculated once the balance calculations have been completed. A typical combustion calculation will involve a known Orsat analysis of the flue gas if the fuel is a solid, an averaged analysis of the combustion residue, and a partial analysis of the fuel. From this information, the complete fuel composition, the actual amount of air used, and the H_2O content of the flue gas can be calculated.

Synthesis Gas Generation   Synthesis gas is a gaseous mixture of predominantly CO and H_2 which can be used either as a fuel gas or as a feedstock for the synthesis of organic compounds. Synthesis gas generation is generally carried out by reacting a carbonaceous material with steam via the so-called steam–carbon reaction

$$C + H_2O \rightarrow CO + H_2$$

If the source of the carbon is a hydrogen-rich material such as natural gas, then a synthesis gas of higher hydrogen content is produced. In fact, steam reforming of methane,

$$CH_4 + H_2O \rightarrow CO + 3H_2$$

is a very common way of generating process hydrogen. If a material low in hydrogen content, such as coal, is used, then a gas rich in CO is produced which can either be used as a fuel or as a feedstock for the catalytic synthesis of alcohols and higher hydrocarbons. The use of the steam–carbon reaction to produce fuel gases began in the late 19th century and was quite prevalent up to the 1920s. Similarly, Fischer–Tropsch synthesis of alcohols from synthesis gas was practiced extensively in Europe, especially in Germany, until after World War II. Recently, there has been a resurgence of interest in both of these ways of utilizing the products of the steam–carbon reaction, especially in the production of fuel gas to be used in electric power generating stations.

One key characteristic of the carbon–steam reaction is that considerable quantities of heat are required to carry it out. This heat can be supplied externally, by reacting steam and carbon inside tubes which are heated in a furnace, as is done in steam reforming of methane, or directly by injecting oxygen into the reactor along with the steam and thus burning part of the feed carbon. An example of the former case is shown in Figure 4.18. The direct heating case we already encountered in Example 4.7 of this chapter. External heat supply has the advantage that the syn gas is kept separate from the furnace flue gases. Hence, air can be used as furnace fuel oxidant without concern that the large quantities of N_2 in the combustion air will dilute the syn gas product. Direct heating is advantageous because the reactor is simpler since it can be built without concern for separating syn gas from flue gases. However, direct heating requires the use of pure oxygen if N_2 dilution is a concern. The extraction of oxygen from air for this purpose is a well-known standard process, but it represents an additional undesirable expense.

Interconversion and Chemical Upgrading Chemical upgrading of fossil fuels is carried out for basically three reasons: first, to produce a fuel more convenient to transport, distribute, and use; second, to remove undesirable constituents in the raw fossil material; and, third, to form and recover desirable chemical feedstocks. The oldest process for upgrading fossil fuels is pyrolysis or thermal decomposition of carbonaceous material in the absence of air. Pyrolysis was first carried out on coal to produce methane-rich fuel gases, tars, and coke and has been used extensively in the petroleum industry to process crude oils. In pyrolysis, a number of processes take place: larger molecules are cracked into smaller molecules, smaller reactive molecules recombine or polymerize, and large polymers ultimately form tars and coke. The net process is one in which the molecular weight distribution of the feed material is changed to produce lighter hydrogen-rich materials and heavier carbon-rich materials. Depending upon the temperature and pressure conditions as well as the nature of any catalysts used, the distribution between stable lighter molecules, polymerized molecules, and tars or coke can be altered as desired. In the oil industry, various specialized thermal decomposition operations have been developed with the primary purpose of increasing the yield of hydrocarbons in the gasoline range from raw crude oil. Some of the more common pyrolysis operations, in the order of increasing severity of reaction conditions, are: thermal cracking, thermal (now more commonly catalytic) reforming, vis-breaking, and coking.

A second major category of processes for upgrading fossil materials involves treatment with hydrogen or hydrogenation. Coals typically have carbon-to-hydro-

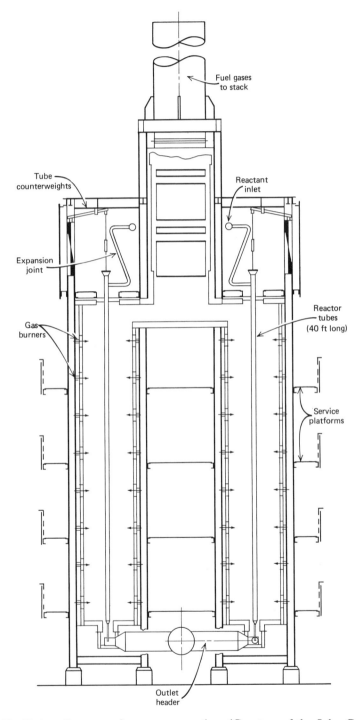

Figure 4.18   Twin-cell steam reformer cross section. (Courtesy of the Selas Corporation.)

gen ratios of about 1 : 0.9; hydrocarbon liquids have ratios of about 1 : 1.6; and light gases such as methane, of up to 1 : 4. Consequently, the process of converting coal to oil or gas can be viewed as essentially one of adding hydrogen to the already present carbon. This is in fact what is done in many coal conversion processes. For instance, several modern coal liquefaction processes involve reacting coal with hydrogen either directly or through an intermediate solvent. Similarly, the more advanced coal gasification processes treat coal directly with a hydrogen-rich gas so as to favor the production of CH_4 over that of synthesis gas.

A further objective in hydrogenation is to remove undesirable constituents such as nitrogen, oxygen, sulfur, and chlorine which are present as either organic or inorganic compounds in the fossil material. Removal of these elements is necessary since they will form air pollutants when the fossil fuel is ultimately burned or will act as chemical contaminants and catalyst poisons if the fossil material is subjected to further processing. Reaction with hydrogen generally results in the conversion of these elements to NH_3, H_2O, H_2S, and HCl which are gases that can be readily separated and disposed. Of course, since the conventional source of process hydrogen is the steam–carbon reaction, the hydrogenation of fossil material will always involve a substantial consumption of some of the fossil material itself. Thus, hydrogenation inherently imposes a conversion inefficiency. The various processes proposed differ primarily in their attempts to reduce this inherent waste of fossil material in the conversion processes.

We conclude this section with an example in which element balances are used to calculate the flows in a coal pyrolysis plant. The flowsheet which we shall examine is based on the COED process developed by the FMC Corporation and includes several stages of pyrolysis as well as a hydrotreating section in which the sulfur content of the raw coal oil is reduced by reaction with hydrogen.

Example 4.15   In the simplified process flowsheet shown in Figure 4.19, a dry pulverized high-sulfur bituminous coal with ultimate analysis of 72.5% carbon, 5% hydrogen, 9% oxygen, 3.5% sulfur, and 10% ash is contacted with a hot synthesis gas stream to produce a raw coal oil, a high methane content gas, and a devolatilized char. The high-methane-content product gas has a dry basis analysis (in mol %) of 11% CH_4, 18% CO, 25% CO_2, 42% H_2, and 4% H_2S, and the gas is found to contain 48 mol H_2O per 100 mol water-free gas. The raw coal oil has an ultimate analysis of (in wt %)

82% C
8% H
8% O
2% S

and about 150 lb raw oil is produced from 1000 lb dry coal. The devolatilized char from the first pyrolysis stage is transferred to a second stage in which it is gasified with steam and oxygen to produce the hot synthesis gas for the first stage containing (in mol %) 3% CH_4, 12% CO, 23% CO_2, 42% H_2O, 20% H_2, and a char consisting essentially of fixed carbon and ash. The raw-product gas from the first stage is sent

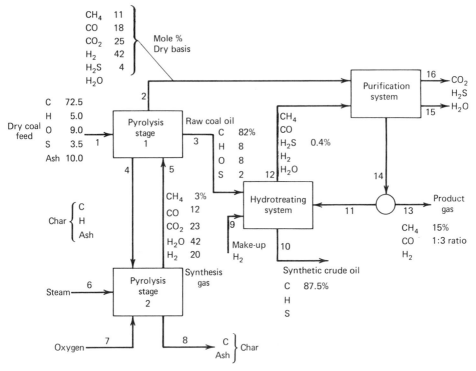

Figure 4.19 Flowsheet for Example 4.15, coal pyrolysis plant.

to the purification system where it is purified of H_2S, CO_2, and H_2O. The result is a product gas consisting of 15 mol % CH_4 and CO and H_2 in a 1 : 3 mole ratio suitable for methanation. This gas is split and the larger portion is recycled to the hydrotreating plant. In the hydrotreating plant, the raw coal oil is subjected to a mild hydrogenation using the recycle product gas as well as some make-up H_2 to produce a synthetic crude oil consisting of 87.5% C and only a very small amount of sulfur. Hydrotreating essentially removes oxygen and sulfur from the raw oil and somewhat increases the hydrogen content of the oil. The off-gases from hydrotreating, containing 4 mol CO and 15 mol H_2 per 1 mol H_2O and only 0.4 mol % H_2S, are sent to the purification system for removal of H_2S and H_2O. Calculate all flows in the process.

    Solution The process flow diagram with specified compositions is given in Figure 4.19. From the flowsheet, it is apparent that element balances will be required in both pyrolysis stages as well as in the hydrotreating system since these units involve streams specified in terms of an elemental analysis. By the same argument, the overall balances will also have to be written on the elements. The splitter and the purification system as nonreacting units can and should be solved using species balances. The complete degree-of-freedom table is shown in Figure 4.20.

	Pyrolysis Stage I	Pyrolysis Stage II	Hydro-system	Purif. System	Splitter	Process	Overall Balances
Number of variables	23	12	16	17	9	48	19
Number of balances	5	4	4	6	3	22	5
Specified compositions							
Coal	4					4	4
Raw oil	3		3			3	
Stage I gas	4			4		4	
Stage II gas	4	4				4	
Syn crude			1			1	1
Product gas			2	2	2	2	2
Stream 12			1	1		1	
Relations							
H_2O : gas ratio	1			1		1	
H_2 : H_2O and CO : H_2O in stream 12			2	2		2	
Oil : coal ratio	1					1	
Splitter restrictions					2	2	
Degree of freedom	1	4	3	1	2	1	7
Basis						-1	
						$\overline{0}$	

Figure 4.20   Example 4.15, degree-of-freedom table.

The problem is evidently correctly specified and, from the table, it is clear that the basis should be chosen with either the stage I pyrolysis unit or the purification system. Let us arbitrarily begin the problem solution with the stage I element balances and a basis of 1000 lb/h coal. From the specified oil yield, this choice of basis immediately implies that 150 lb/hr raw oil is produced.

With these flows determined, we can proceed to write the stage I element balances:

Ash balance: $F^4_{ash} = 0.1(1000) = 100$ lb/h

S balance: $-35 + 3 + (0.04N^2_{dry})32 = 0$

O balance: $-90 + 12 + 16\,(0.18N^2_{dry} - 0.12N^5)$

$+ \, 2(0.25N^2_{dry} - 0.23N^5)16$

$+ \, (N^2_{H_2O} - 0.42N^5)\,16 = 0$

C balance:

$$-725 + 123 + F_C^4 + 12\,(0.11N_{dry}^2 - 0.03N^5)$$
$$+ (0.18N_{dry}^2 - 0.12N^5)12 + 12\,(0.25N_{dry}^2 - 0.23N^5)$$
$$= 0$$

H balance:

$$-50 + 12 + F_H^4 + 4(0.11N_{dry}^2 - 0.03N^5)$$
$$+ 2\,(0.42N_{dry}^2 - 0.2N^5)$$
$$+ 2\,(0.04N_{dry}^2) + 2(N_{H_2O}^2 - 0.42N^5) = 0$$

From the sulfur balance, we obtain $N_{dry}^2 = 25$ lbmol/h, and from the H_2O-to-gas ratio, $N_{H_2O}^2 = 12$ lbmol/h. With these quantities known, the oxygen balance can be solved for

$$N^5 = 24.125 \text{ lbmol/h}$$

Next, the carbon balance yields $F_C^4 = 550.01$ lb/h; and, finally, the hydrogen balance can be solved for

$$F_H^4 = 12.81 \text{ lb/h}$$

As a result of the stage I balances, the flow of stream 2 is known and consequently the purification unit has degree of freedom zero. Also, with the flow of stream 5 and the composition and flow of stream 4 known, the degree of freedom of the stage II balances is reduced to zero. Either unit could be solved next. Suppose we arbitrarily choose to proceed with stage II. The stage II balances are:

Ash balance: $F_{ash}^8 = 100$ lb/h

C balance: $-550.01 + F_C^8 + 0.03(25.125)12 + 0.12(25.125)12$
$$+ 0.23(25.125)12 = 0$$

H balance: $-12.81 + 4(.03)25.125 + 2[0.42(25.125) - N^6]$
$$+ 2[0.2(25.125)] = 0$$

O balance: $-N^7(32) + 16[0.42(25.125) - N^6] + 16[0.12(25.125)]$
$$+ 32[0.23(25.125)] = 0$$

From the carbon balance, $F_C^8 = 440$ lb/h; and from the H balance, $N^6 = 10$ lbmol/h, or 180 lb/h. The O balance then yields $N^7 = 7.0625$ lbmol/h, or 226 lb/h. This completes the solution of the stage II balances.

At this point, the flows of streams 1, 6, and 7 and the flow as well as composition of stream 8 have been calculated. However, even with these five quantities determined, the overall balances are still underspecified. Since the degree of freedom of the hydrotreating unit has remained unchanged, it is clear that the purification unit balances ought to be solved next.

The purification system species balances are:

CO_2 balance: $6.25 = N_{CO_2}^{16}$

CH_4 balance: $2.75 + N_{CH_4}^{12} = 0.15N^{14}$

H$_2$ balance: $10.5 + N_{H_2}^{12} = 0.6375 N^{14}$

CO balance: $4.5 + N_{CO}^{12} = 0.2125 N^{14}$

H$_2$O balance: $12 + N_{H_2O}^{12} = N^{15}$

H$_2$S balance: $1 + 0.004 N^{12} = N_{H_2S}^{16}$

The side conditions imposed on stream 12 are

$$N_{CO}^{12}/N_{H_2O}^{12} = 4 \quad \text{and} \quad N_{H_2}^{12}/N_{H_2O}^{12} = 15$$

Substituting for N_{CO}^{12} in the CO balance and for $N_{H_2}^{12}$ in the H$_2$ balance, we obtain a pair of simultaneous equations:

$$4.5 + 4N_{H_2O}^{12} = 0.2125 N^{14}$$

$$10.5 + 15 N_{H_2O}^{12} = 0.6375 N^{14}$$

These can be easily solved to yield

$$N_{H_2O}^{12} = 1 \text{ lbmol/h}$$

$$N^{14} = 40 \text{ lbmol/h}$$

From the side conditions, it then follows that

$$N_{H_2}^{12} = 15 \text{ lbmol/h}$$

$$N_{CO}^{12} = 4 \text{ lbmol/h}$$

The CH$_4$ balance can next be solved for $N_{CH_4}^{12} = 3.25$ lbmol/h; and the H$_2$O balance, for $N^{15} = 13$ lbmol/h. Also, from the specified mole fraction of H$_2$S, it follows that $0.996 N_{H_2S}^{12} = 0.004 (23.25)$, or $N_{H_2S}^{12} = 0.09337$ lbmol/h. Thus, the H$_2$S balance will yield $N_{H_2S}^{16} = 1.09337$ lbmol/h.

As a result of these calculations, the flow and the one remaining composition of stream 12 have been determined. Consequently, the degree of freedom of the hydrotreating unit has been reduced to zero. Also, streams 15 and 16 have been calculated; and thus, with the previously calculated streams 1, 6, 7, and 8, it appears that the overall balances are overspecified by one. Note however that the ash balances have all been used up in the stage I and stage II calculations. Hence, of the five overall balance equations, only the four element balances remain. Therefore, the overall balances are in fact correctly specified. Again, we have a choice of which balance set to solve. Selecting the hydrotreating unit, we first construct the element balance equations:

O balance: $-0.08(150) + 16(4 - 0.2125 N^{11}) + 16(1) = 0$

S balance: $-0.02(150) + F_S^{10} + 32(0.09337) = 0$

C balance: $-0.82(150) + F_C^{10} + 12(4 - 0.2125 N^{11})$

$\qquad\qquad\qquad + 12(3.25 - 0.15 N^{11}) = 0$

H balance: $\quad -0.08(150) + F_H^{10} + 4(3.25 - 0.15N^{11})$

$\quad\quad\quad\quad\quad + 2(0.09337) + 2(1) + 2(15 - 0.6375N^{11} - N^9) = 0$

From the O balance, we can calculate N^{11} as

$$N^{11} = 20 \text{ lbmol/h}$$

and from the S balance,

$$F_S^{10} = 0.012 \text{ lb/h}$$

The carbon balance next yields

$$F_C^{10} = 123 \text{ lb/h}$$

Since the oil has a carbon content of 87.5%, we can now calculate the flow of stream 10:

$$F^{10} = 123/0.875 = 140.571 \text{ lb/h}$$

Therefore, by difference,

$$F_H^{10} = 140.57 - 123 - 0.012 = 17.56 \text{ lb/h}$$

Finally, from the H balance,

$$N^9 = 6.623 \text{ lbmol/h}$$

The calculations can be completed by using the total balance on the splitter to obtain

$$N^{13} = N^{14} - N^{11} = 40 - 20 = 20 \text{ lbmol/h}$$

Note that the hydrogen contained in the product gas,

$$0.6375(20) = 12.75 \text{ lbmol/h } H_2$$

is more than enough to supply the H_2 make-up required in the hydrotreating unit. Presumably a cryogenic separation will have to be carried out on the product gas to recover this amount of process hydrogen.

4.4 CONVERSION OF ELEMENT BALANCES TO SPECIES BALANCES

In Section 4.2, we considered the situation in which the species molecular formulas and the reaction stoichiometry were both known. We observed that although in such cases both element and species balances could always be constructed, the species balances were generally preferable. To identify when this occurred, we formulated, but only partially verified, a criterion involving properties of the stoichiometric and atomic arrays. We interpreted that criterion to correspond to checking whether the number of reactions which were postulated to occur actually was the same as the maximum number that could occur between the given set of species.

    In this section, we consider the case in which the reaction stoichiometry is either unknown or only partially known and show how species balances can still be constructed. We begin by developing a general relationship between the atom matrix for a given set of species and any possible reactions involving these species. From this relationship, it will be possible to establish the algebraic basis of the notion of maximum number of independent reactions and thus to complete the verification of the criterion given in Section 4.2.2. Then, we investigate a simple procedure for calculating a maximal set of reactions directly from the atom matrix and conclude by showing how these constructions can be used when solving problems in which the reaction stoichiometry is only partially known.

4.4.1 The Relation Between Atom Matrix and Reaction Stoichiometry

Consider a system involving S species and E elements with atom matrix α. Let σ_s, $s = 1, \ldots, S$ be the stoichiometric coefficients of any chemical reaction involving the S given species. For a chemical reaction to be properly balanced, it is necessary that the number of moles of each element present in the reactants be exactly equal to the number of moles of that element present in all the products. Since α_{es} is the number of moles of element e contained in 1 mol of species s, then $\alpha_{es}\sigma_s$ is the number of moles of element e in the number of moles of species s required by the reaction stoichiometry. In view of our sign convention for reactants and products, a balanced chemical reaction is therefore one for which

$$\sum_{s=1}^{S} \alpha_{es}\sigma_s = 0 \tag{4.6}$$

for each e, $e = 1, \ldots, E$.

Example 4.16 The chemical reaction

$$CH_3OH + \tfrac{3}{2}O_2 \rightarrow CO_2 + 2H_2O$$

is balanced because the number of moles of each type of element on either side of the reaction equation is the same. In our generalized notation, the atom matrix is:

$$
\begin{array}{cccc}
 & CH_3OH & O_2 & CO_2 & H_2O \\
C & 1 & 0 & 1 & 0 \\
H & 4 & 0 & 0 & 2 \\
O & 1 & 2 & 2 & 1
\end{array}
$$

With the species ordered as in the atom matrix, the vector of stoichiometric coefficients is

$$\sigma = \begin{pmatrix} -1 \\ -\tfrac{3}{2} \\ 1 \\ 2 \end{pmatrix}$$

Element Balances

Substituting these coefficients into eq. (4.6), we have for carbon

$$1(-1) + 0(-\tfrac{3}{2}) + 1(1) + 0(2) = 0$$

for hydrogen

$$4(-1) + 0(-\tfrac{3}{2}) + 0(1) + 2(2) = 0$$

and for oxygen

$$1(-1) + 2(-\tfrac{3}{2}) + 2(1) + 1(2) = 0$$

Consequently, our criterion for a balanced reaction appears to be correct.

Relations (4.6) must be satisfied by any chemical reaction involving the S species contained in the given atom matrix. If the atom matrix is known but the reaction stoichiometry is not, then these relations may be viewed as a set of linear equations in the unknown variables σ_s, $s = 1, \ldots, S$. Each solution σ^* satisfying the relations (4.6) will yield the stoichiometric coefficient of some reaction between the given S species. When viewed from this vantage point, the question of how many independent chemical reactions it is possible to write involving a specified set of chemical species reduces simply to the algebraic problem of determining how many linearly independent solutions it is possible for eq. (4.6) to exhibit.

Recall from algebra that a set of $\rho(\alpha)$ linearly independent equations in S unknowns will, provided the number of unknowns is greater than the number of independent equations, possess an infinite number of possible solutions. However, each and every member of this infinite family of solutions can be generated from a linear function of no more than $S - \rho(\alpha)$ parameters.

Example 4.17 Consider the two equations

$$x_1 - x_2 - x_3 + 2x_4 = 0$$
$$x_2 - x_3 + x_4 = 0$$

in the four unknowns x_1, x_2, x_3, and x_4. If we add the second equation to the first, we can solve for x_1 and x_2 in terms of x_3 and x_4:

$$x_1 = 2x_3 - 3x_4$$
$$x_2 = 1x_3 - 1x_4$$

If we add the trivial equations $x_3 = x_3$ and $x_4 = x_4$ to this set, then we can write all solutions $x^* = (x_1^*, x_2^*, x_3^*, x_4^*)$ of the original set of two equations in terms of two parameters x_3 and x_4. In other words, the infinite number of solutions which satisfy the two equations can all be generated via

$$x^* = \begin{pmatrix} x_1 \\ x_2 \\ x_3 \\ x_4 \end{pmatrix} = \begin{pmatrix} 2 \\ 1 \\ 1 \\ 0 \end{pmatrix} x_3 + \begin{pmatrix} -3 \\ -1 \\ 0 \\ 1 \end{pmatrix} x_4 \qquad (4.7)$$

where x_3 and x_4 are parameters which can be assigned to any arbitrary value.

In this example, we had two independent equations and four unknowns; thus, all solutions could be generated as a function of $S - \rho(\alpha) = 4 - 2 = 2$ parameters.

Another way of expressing the same fact is to say that the general solution of a set of $\rho(\alpha)$ independent linear equations in S unknowns can always be expressed as a linear combination of $S - \rho(\alpha)$ independent solutions. This can be verified in the above example by noting that since eq. (4.7) give the general solution, then the x^* obtained by setting $x_3 = 1$ and $x_4 = 0$,

$$x^* = \begin{pmatrix} 2 \\ 1 \\ 1 \\ 0 \end{pmatrix}$$

must be a solution. Similarly, the choice of $x_3 = 0$ and $x_4 = 1$ will likewise be a solution:

$$x^* = \begin{pmatrix} -3 \\ -1 \\ 0 \\ 1 \end{pmatrix}$$

These two solutions are clearly linearly independent because the two columns

$$\begin{pmatrix} 2 & -3 \\ 1 & -1 \\ 1 & 0 \\ 0 & 1 \end{pmatrix}$$

are linearly independent since each has a nonzero entry where the other does not. These two columns furthermore are just the columns of coefficients of the general solution, eq. (4.7). We have thus verified that the general solution can be expressed as a linear combination of $S - \rho(\alpha)$ (in our example, $4 - 2 = 2$) independent solutions. Now, if all solutions to a set of equations can be expressed as a linear combination of $S - \rho(\alpha)$ linearly independent solutions, then it follows from the definition of linear independence that there can be no more than $S - \rho(\alpha)$ linearly independent solutions of the given equation set.

Summarizing the above statements, we can say that given a set of $\rho(\alpha)$ linearly independent equations in S unknowns, that system of equations can possess no more than $S - \rho(\alpha)$ linearly independent solutions, and all solutions can be expressed as linear combinations of the selected set of independent solutions. Apply this conclusion to the system of E equations,

$$\sum_{s=1}^{S} \alpha_{es}\sigma_s = 0 \qquad e = 1, \ldots, E$$

in the S unknowns σ_s, $s = 1, \ldots, S$, we can conclude immediately that if $\rho(\alpha)$ of these equations are independent, then the system will involve at most $S - \rho(\alpha)$ independent solutions. Thus, the maximum number of algebraically independent reactions involving the S species with atom matrix α will be $S - \rho(\alpha)$. Moreover, given any arbitrary set of reactions with stoichiometric coefficient array σ, it follows that no more than $S - \rho(\alpha)$ of these can be independent, or

$$\rho(\sigma) \leq S - \rho(\alpha)$$

The fact that $\rho(\sigma)$ can never be greater than $S - \rho(\alpha)$ is therefore a direct consequence of the simple requirement that the chemical reactions, whatever they may be, must be balanced.

A further consequence of relations (4.6) is that, given the atom matrix α, we can use the array reduction procedure discussed previously to construct a maximal set of independent reactions when the actual reaction stoichiometry is partially or completely unknown. These constructed reactions can then be used to write species balances instead of the element balances we would otherwise have to write. The procedure for obtaining the set of reactions is described in the next section.

4.4.2 Construction of a Set of Independent Reactions

As illustrated in Example 4.17, given a system of linear equations which has more variables than equations, we can solve for some of the unknowns in terms of the remaining unknowns and thus obtain a general solution of the original equations in parametric form. In our discussion, we noted that, given a set of $\rho(\alpha)$ independent equations in S unknowns, the general solution will be expressed in terms of $S - \rho(\alpha)$ parameters and that the vectors of coefficients of these parametric equations will each be an independent solution of the original set of equations. The problem of determining a set of independent balanced chemical reactions involving S species with atom matrix α thus is the same as that of constructing a general solution in parametric form of the equation

$$\sum_{s=1}^{S} \alpha_{es}\sigma_s = 0 \qquad e = 1, \ldots, E$$

in the unknowns σ_s. In order to accomplish this, we need to devise a systematic procedure for solving for $\rho(\alpha)$ of the unknowns σ_s in terms of the remaining $S - \rho(\alpha)$ unknowns. This procedure is nothing more than a variant of the conventional way of solving a set of linear equations. One very common way of solving a set of linear equations is to add or subtract multiples of the equations from one another so as to cause each equation to contain only one variable. Once this form is achieved, the solution can be identified by inspection.

Example 4.18 Consider the system of equations

$$x_1 + 2x_2 - x_3 \;\; = 2 \tag{4.8}$$
$$2x_1 + 3x_2 \;\; = 1 \tag{4.9}$$
$$-x_1 + 3x_3 = 1 \tag{4.10}$$

We can begin to solve this set of equations by first eliminating variable x_1 from eqs. (4.9) and (4.10). For instance, if we subtract twice eq. (4.8) from eq. (4.9) and then add eq. (4.8) to eq. (4.10), we obtain

$$x_1 + 2x_2 - x_3 = 2 \tag{4.11}$$

$$0x_1 - x_2 + 2x_3 = -3 \tag{4.12}$$

$$0x_1 + 2x_2 + 2x_3 = 3 \tag{4.13}$$

Suppose we now eliminate variable x_2 from eqs. (4.11) and (4.13) by adding twice eq. (4.12) to eq. (4.11) and also twice eq. (4.12) to eq. (4.13). The result is

$$x_1 + 0x_2 + 3x_3 = -4 \tag{4.14}$$

$$0x_1 - x_2 + 2x_3 = -3 \tag{4.15}$$

$$0x_1 + 0x_2 + 6x_3 = -3 \tag{4.16}$$

Next, we can eliminate variable x_3 from eqs. (4.14) and (4.15) by subtracting $\frac{1}{2}$ times eq. (4.16) from eq. (4.14) and $\frac{1}{3}$ times eq. (4.16) from eq. (4.15):

$$x_1 + 0x_2 + 0x_3 = -\tfrac{5}{2} \tag{4.17}$$

$$0x_1 - x_2 + 0x_3 = -2 \tag{4.18}$$

$$0x_1 + 0x_2 + 6x_3 = -3 \tag{4.19}$$

We conclude by multiplying eq. (4.18) by -1 and eq. (4.19) by $\frac{1}{6}$ and have thus reduced the set of equations to a form in which we can read off our solution directly:

$$x_1 = -\tfrac{5}{2}$$

$$x_2 = 2$$

$$x_3 = -\tfrac{1}{2}$$

All the above operations could just as well have been carried out *without* manipulating the x_i's directly. Rather we could have written the original equations in detached coefficient form. The detached coefficient form of the equations is obtained by constructing an array which has one column for each variable, one column for the equation right-hand sides, and one row for each equation:

$$
\begin{array}{cccc}
x_1 & x_2 & x_3 & \text{RHS} \\
\end{array}
$$
$$
\begin{pmatrix}
1 & 2 & -1 & 2 \\
2 & 3 & 0 & 1 \\
-1 & 0 & 3 & 1
\end{pmatrix}
$$

Since each row represents the coefficients of one equation, the final result of the variable elimination procedure that we carried out previously can be represented as

$$
\begin{array}{cccc}
x_1 & x_2 & x_3 & \text{RHS} \\
\begin{pmatrix}
1 & 0 & 0 & -\frac{5}{2} \\
0 & 1 & 0 & 2 \\
0 & 0 & 1 & -\frac{1}{2}
\end{pmatrix}
\end{array}
$$

When viewed in detached coefficient form, the procedure that we carried out in eliminating variables is exactly the same as the *array reduction procedure* based on *row* operations that we discussed in Section 4.1.2.

The same process of solving linear equations by reducing the coefficients of the equations to diagonal form can also be employed if there are more unknowns than equations. In that case, the remaining nonreduced columns will simply correspond to the coefficients of those variables in terms of which the variables with reduced coefficients have been expressed.

Example 4.19 Consider the system of equations

$$2x_1 + 4x_2 - 2x_3 + 2x_4 = 0 \tag{4.20}$$

$$x_1 + \tfrac{3}{2}x_2 = 0 \tag{4.21}$$

$$3x_1 + 5x_2 - x_3 + x_4 = 0 \tag{4.22}$$

Let us first construct the detached coefficient array

$$
\begin{array}{ccccc}
x_1 & x_2 & x_3 & x_4 & \text{RHS} \\
\begin{pmatrix}
2 & 4 & -2 & 2 & 0 \\
1 & \frac{3}{2} & 0 & 0 & 0 \\
3 & 5 & -1 & 1 & 0
\end{pmatrix}
\end{array}
$$

We proceed in the reduction of this array by applying *array reduction procedure* based on row operations:

Step A Divide each entry of the first row by 2:

$$(1 \quad 2 \quad -1 \quad 1 \quad 0)$$

Step B Add -1 times this row to the second row and -3 times this row to the third row:

$$
\begin{pmatrix}
1 & 2 & -1 & 1 & 0 \\
0 & -\frac{1}{2} & 1 & -1 & 0 \\
0 & -1 & 2 & -2 & 0
\end{pmatrix}
$$

Step A Divide each entry of the second row by $-\frac{1}{2}$:

$$(0 \quad 1 \quad -2 \quad 2 \quad 0)$$

Step B Add -2 times this row to the first row and 1 times this row to the third row:

$$
\begin{pmatrix}
1 & 0 & 3 & -3 & 0 \\
0 & 1 & -2 & 2 & 0 \\
0 & 0 & 0 & 0 & 0
\end{pmatrix}
$$

The row reduction procedure terminates since the third row is identically zero. We conclude immediately that only two of the three equations are independent. Next, if we reconstruct the equations corresponding to the nonzero rows, we have

$$x_1 + 0x_2 + 3x_3 - 3x_4 = 0$$
$$0x_1 + x_2 - 2x_3 + 2x_4 = 0$$

or

$$x_1 = -3x_3 + 3x_4$$
$$x_2 = 2x_3 - 2x_4$$

In other words, we have reduced the system of three equations to two independent equations and, since these equations involved four unknowns, have solved for x_1 and x_2 in terms of the remaining two unknowns. If we now add the two trivial equations

$$x_3 = 1x_3 \tag{4.23}$$
$$x_4 = 1x_4 \tag{4.24}$$

we can write the general solution x^* of the above system of equations as

$$x^* = \begin{pmatrix} x_1 \\ x_2 \\ x_3 \\ x_4 \end{pmatrix} = \begin{pmatrix} -3 \\ 2 \\ 1 \\ 0 \end{pmatrix} x_3 + \begin{pmatrix} 3 \\ -2 \\ 0 \\ 1 \end{pmatrix} x_4$$

As we noted previously, the two columns of coefficients of this general solution are themselves linearly independent solutions of the original set of equations. Note, carefully, that these two columns were formed by reducing the original equation set and then adding on the two trivial eqs. (4.23) and (4.24) to the two nonreduced columns. In array form, this corresponds to taking the negative of the nonreduced part of the coefficient array:

$$-\begin{pmatrix} 3 & -3 \\ -2 & 2 \end{pmatrix}$$

and adding on a 2×2 array with 1's on the diagonal and zeros elsewhere:

$$\begin{pmatrix} -3 & 3 \\ 2 & -2 \\ 1 & 0 \\ 0 & 1 \end{pmatrix}$$

This construction will hold true in general, that is, the columns of linearly independent solutions will be formed by reducing the coefficient array using row operations and then adding the coefficients of the nonreduced columns a square array with 1's on the diagonal (often called an identity matrix) which will have the same number of columns as the number of nonreduced columns.

We can apply this procedure immediately to the problem of constructing a set of linearly independent stoichiometric reactions for a system with atom matrix α; that is, we first apply the *array reduction procedure* to reduce α to the usual form consisting of an identity matrix I, a block of nonreduced columns C, and one or more rows of zeros:

$$\begin{pmatrix} I & C \\ 0 & 0 \end{pmatrix}$$

Note that since I is square and has $\rho(\alpha)$ rows, C will have $\rho(\alpha)$ rows and $S - \rho(\alpha)$ columns. Then, we form the array composed of $-C$ and an identity matrix of size equal to the number of columns of C:

$$\begin{pmatrix} -C \\ I \end{pmatrix}$$

Based on our previous discussions, each of the $S - \rho(\alpha)$ columns of this array will consist of the stoichiometric coefficients of a linearly independent reaction involving the S species contained in α, that is, it will be a solution of

$$\sum_{s=1}^{S} \alpha_{es} \sigma_s = 0 \qquad e = 1, \ldots, E$$

In constructing the chemical reaction equations from the columns of the above array, the sign convention for products and reactants is understood to hold.

Example 4.20 Determine the maximum number and construct a set of independent chemical reactions involving the species H_2, CH_4, CO, CH_3OH, and H_2O.

Solution The atom matrix for this set of species is:

$$\begin{array}{c} \\ H \\ C \\ O \end{array} \begin{array}{ccccc} H_2 & CH_4 & CO & CH_3OH & H_2O \\ \left(\begin{array}{ccccc} 2 & 4 & 0 & 4 & 2 \\ 0 & 1 & 1 & 1 & 0 \\ 0 & 0 & 1 & 1 & 1 \end{array} \right) \end{array}$$

Applying the array reduction procedure to this matrix we obtain

$$\begin{pmatrix} 1 & 0 & 0 & 2 & 3 \\ 0 & 1 & 0 & 0 & -1 \\ 0 & 0 & 1 & 1 & 1 \end{pmatrix}$$

The three rows of the atom matrix are independent. Hence, $\rho(\alpha) = 3$; and since the system involves 5 chemical species, we have

$$S - \rho(\alpha) = 5 - 3 = 2$$

The maximum number of independent chemical reactions involving these five species is therefore two. Following the procedure outlined above, the stoichiometric coefficients of a pair of independent reactions will be obtained by taking the negative of the two nonreduced columns of the above reduced array and appending a 2×2 identity matrix:

$$\sigma = \begin{pmatrix} -2 & -3 \\ 0 & 1 \\ -1 & -1 \\ 1 & 0 \\ 0 & 1 \end{pmatrix}$$

Since in this stoichiometric array, the species will be ordered in the same sequence as in the atom matrix, the two reactions can be identified as

$$CO + 2H_2 \rightarrow CH_3OH$$

$$CO + 3H_2 \rightarrow CH_4 + H_2O$$

The two reactions generated are independent but are not unique. In other words, if the species had been ordered in a different sequence, then the reduction could have resulted in different numerical values for the nonreduced columns. However, whatever pair of independent reactions is generated, we can be sure that they are balanced and that the solution of any material balance problem calculated by using them will be exactly the same as that obtained using element balances.

4.4.3   Application to Systems with Incomplete Stoichiometry

The procedure for constructing a maximal set of independent reactions developed in the previous section can be employed whenever a situation is encountered in which the molecular formulas of all species are known but the reaction stoichiometry is not specified. The construction provides the option of using species balances where atom balances would otherwise have to be used. In most such cases, the choice between the two sets of entity balances is really a matter of convenience of solution. However, there are situations in which the use of a constructed set of reactions is preferable to using element balances. For instance, suppose it is desired to make overall balances on a process which involves several reactors, one or more of which has the stoichiometry unknown. In that case, either overall element balances must be used and the known chemical reactions disregarded; or else the above procedure is used to determine a set of reactions for the reactors with unknown stoichiometry and then overall species balances must be used with the complete set of reactions. As illustrated in the following example, overall balances using maximal sets of independent reactions, constructed for each reactor with unknown stoichiometry, may provide more independent balance equation, namely, a lower degree of freedom, then overall element balances.

Example 4.21   In an established process for producing acetic anhydride, $(CH_3CO)_2O$, pure acetone, $(CH_3)_2CO$, is partially decomposed in a direct-fired furnace to produce ketene, CH_2CO, along with other side products. The hot furnace products are quickly quenched with cold acetic acid, $C_2H_4O_2$. When sufficiently cooled, the ketene reacts completely with the acetic acid to produce the desired anhydride product via the reaction

$$CH_2CO + C_2H_4O_2 \rightarrow (CH_3CO)_2O$$

The resulting reactor outlet mixture is separated into a product stream containing 96% anhydride and 4% acetic acid, a pure acetone recycle stream, and a waste light-ends stream consisting of 70% CH_4, 15% CO, and the remainder C_2H_4 and H_2. The acetone recycle stream is usually quite large because the conversion of acetone in the furnace must be kept low in order to achieve a reasonable ketene yield. If in a given plant 4 mol acetone is recycled per 1 mol fresh acetone feed, calculate the composition of the furnace exit stream. The simplified flowsheet for the process is shown in Figure 4.21, where A indicates acetone, K indicates ketene, AA acetic anhydride, and HAc acetic acid.

Solution   As stated, the process involves two units in which reactions occur: the furnace and the quench reactor. Since no reaction stoichiometry is given for the furnace, it is clear that element balances ought to be used with that unit. A reaction is specified for the second reactor; consequently, species balances are appropriate for that unit. If overall balances are to be used, then, since the stoichiometry is only partially known, either overall element balances must be used or a set of reactions must be constructed for the furnace, in which case species balances ought to be used.

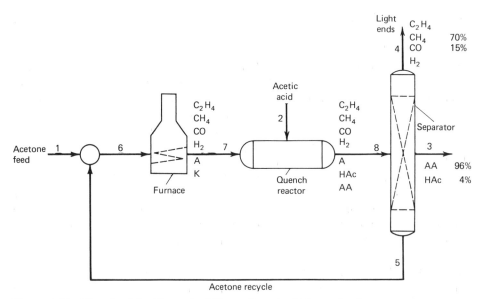

Figure 4.21   Flowsheet for Example 4.21, acetic anhydride production.

Let us begin by considering the atom matrix of the six species involved with the furnace:

$$
\begin{array}{c}
 \\
C \\
O \\
H
\end{array}
\begin{array}{cccccc}
CO & (CH_3)_2CO & CH_2CO & CH_4 & C_2H_4 & H_2 \\
\end{array}
\left(
\begin{array}{cccccc}
1 & 3 & 2 & 1 & 2 & 0 \\
1 & 1 & 1 & 0 & 0 & 0 \\
0 & 6 & 2 & 4 & 4 & 2
\end{array}
\right)
$$

As a result of the *array reduction procedure*, we obtain:

$$
\begin{array}{c}
 \\
C \\
O \\
H
\end{array}
\begin{array}{cccccc}
CO & (CH_3)_2CO & CH_2CO & CH_4 & C_2H_4 & H_2 \\
\end{array}
\left(
\begin{array}{cccccc}
1 & 0 & 0 & 0 & -2 & 1 \\
0 & 1 & 0 & 1 & 0 & 1 \\
0 & 0 & 1 & -1 & 2 & -2
\end{array}
\right)
$$

Thus, all three element balances will be independent and, since $\rho(\alpha) = 3$, $S - \rho(\alpha) = 6 - 3 = 3$. Consequently, three independent reactions can be written involving these six species. If species balances were to be employed, then the three reactions with stoichiometric coefficients formed from the last three columns of the above array, namely,

$$
\left(
\begin{array}{ccc}
0 & 2 & -1 \\
-1 & 0 & -1 \\
1 & -2 & 2 \\
1 & 0 & 0 \\
0 & 1 & 0 \\
0 & 0 & 1
\end{array}
\right)
$$

could be used. The three reactions that correspond to these coefficients are

$$(CH_3)_2CO \rightarrow CH_2CO + CH_4$$

$$2CH_2CO \rightarrow 2CO + C_2H_4$$

$$CO + (CH_3)_2CO \rightarrow 2CH_2CO + H_2$$

Note that while the first two reactions do occur as shown, the third does not. However, for material balance purposes, the exact form of the stoichiometry is not important. What is important is that the reaction involve the six chemical species and be independent of the other two reactions.

Next, if we consider the atom matrix associated with the overall element balances, we have:

$$
\begin{array}{c}
 \\
C \\
O \\
H
\end{array}
\begin{array}{cccccccc}
CO & (CH_3)_2CO & CH_4 & C_2H_4 & H_2 & (CH_3CO)_2O & C_2H_3OOH \\
\end{array}
\left(
\begin{array}{ccccccc}
1 & 3 & 1 & 2 & 0 & 4 & 2 \\
1 & 1 & 0 & 0 & 0 & 3 & 2 \\
0 & 6 & 4 & 4 & 2 & 6 & 4
\end{array}
\right)
$$

After carrying out the reduction, the following reduced array is obtained:

$$
\begin{pmatrix}
1 & 0 & 0 & -2 & 1 & 4 & 4 \\
0 & 1 & 0 & 2 & -1 & -1 & -2 \\
0 & 0 & 1 & -2 & 2 & 3 & 4
\end{pmatrix}
$$

This again indicates that all three element balances are independent. Note that ketene is not included in this array because it does not appear in the process input or output streams.

With this preliminary analysis concluded, we are ready to proceed with the degree-of-freedom analysis. Let us first construct the table assuming that species balances are employed with the furnace and the overall balances. In that case, the furnace will have three reactions and their reaction rates, the quench reactor will have a single reaction and its reaction rate, and the overall balances will have all four reactions and their reaction rates. Note that the overall species balances will consist of balances on each species appearing in the four chemical reactions—a total of eight balances. The complete degree-of-freedom table is shown in Figure 4.22.

The problem is correctly specified, and it appears that the problem solution could be initiated with either the mixer or the overall balances. Since solution of the mixer balances would still leave the furnace and the separator units with positive degrees of freedom, the calculations ought to be initiated with overall balances.

	Mixer	Furnace	Quench Reactor	Separator	Process	Overall Balances
Number of variables	3	7 + 3	14 + 1	14	23 + 4	8 + 4
Number of balances						
AA			1	1	2	1
HAc			1	1	2	1
A	1	1	1	1	4	1
K		1	1		2	1
C_2H_4		1	1	1	3	1
CH_4		1	1	1	3	1
CO		1	1	1	3	1
H_2		1	1	1	3	1
Total number of balances	1	6	8	7	22	8
Number of specified compositions	–	–	3	3	3	
Number of relations Feed/recycle ratio	1	–	–	–	1	–
Degree of freedom	1	4	7	4	1	1
Basis					−1	
					0	

Figure 4.22   Example 4.21, degree-of-freedom table, species balances.

On the other hand, if element balances are used with the furnace and the overall balances, then, while the furnace and process degrees of freedom remain unchanged, the overall balance column becomes:

Number of variables	8
Number of balances	3
Number of composition	3
Degrees of freedom	$\overline{2}$

Even with a basis selected, the overall balances will be underspecified. In this example, it is clearly advantageous to construct a set of reactions for the furnace and to use species balances throughout.

We therefore begin with the overall species balances constructed using the four reactions

$$(CH_3)_2CO \rightarrow CH_2CO + CH_4$$

$$2CH_2CO \rightarrow 2CO + C_2H_4$$

$$(CH_3)_2CO + CO \rightarrow 2CH_2CO + H_2$$

$$CH_2CO + C_2H_3OOH \rightarrow (CH_3CO)_2O$$

Using a basis of 100 mol/h of stream 3 and subscripting the reaction rates in the order in which the reactions are listed above, the following species balance equations can be written:

AA balance:	$96 = 0 + r_4$
HAc balance:	$4 = N^2 - r_4$
K balance:	$0 = 0 + r_1 - 2r_2 + 2r_3 - r_4$
A balance:	$0 = N^1 - r_1 - r_3$
CH_4 balance:	$0.7N^4 = 0 + r_1$
C_2H_4 balance:	$(0.15 - x_{H_2}^4)N^4 = 0 + r_2$
CO balance:	$0.15N^4 = 0 + 2r_2 - r_3$
H_2 balance:	$x_{H_2}^4 N^4 = 0 + r_3$

From the AA balance, it follows that $r_4 = 96$ mol/h and thus from the HAc balance, that $N^2 = 100$ mol/h. Adding the last three balance equations, the $x_{H_2}^4$ terms cancel, and we obtain

$$0.3N^4 = 3r_2 \quad \text{or} \quad r_2 = 0.1N^4$$

Substituting this into the CO balance to eliminate r_2, we have

$$r_3 = 0.05N^4$$

Now, since the CH_4 balance yields

$$r_1 = 0.7N^4$$

these three relationships for the rates can be substituted into the K balance, to give an equation in N^4 alone:

$$r_4 = 96 = 0.7N^4 - 2(0.1N^4) + 2(0.05N^4) = 0.6N^4$$

Consequently,

$$N^4 = 160 \text{ mol/h}$$

and

$$r_1 = 112 \text{ mol/h}$$
$$r_2 = 16 \text{ mol/h}$$
$$r_3 = 8 \text{ mol/h}$$

With these quantities known, the remaining flows and compositions are easily determined. The results are

$$N^4 = (N^4_{C_2H_4}, N^4_{CH_4}, N^4_{CO}, N^4_{H_2})$$
$$= (16, 112, 24, 8) \text{ mol/h}$$

with

$$x^4_{H_2} = 0.05$$
$$x^4_{C_2H_4} = 0.10$$
$$N^1 = 120 \text{ mol/h}$$

As a result of the overall balances, the mixer degree of freedom has been reduced to zero, while those of the remaining units continue to be positive. The mixer balance equation

$$120 + N^5 = N^6$$

together with the specified feed to recycle ratio

$$N^5/120 = 4$$

immediately yield

$$N^5 = 480 \text{ mol/h}$$
$$N^6 = 600 \text{ mol/h}$$

At this point, with all three outlet streams and the H_2 composition calculated, the separator degree of freedom has been reduced to zero. The two reactors continue to have positive degrees of freedom. Carrying out the separator balance calculations, we simply determine stream 8 by summing the species in the known exit streams. The result is

$$N^8 = (N^8_{AA}, N^8_{HAc}, N^8_A, N^8_{C_2H_4}, N^8_{CH_4}, N^8_{CO}, N^8_{H_2})$$
$$= (96, 4, 480, 16, 112, 24, 8) \text{ mol/h}$$

Finally, with streams 8 and 2 known, the degree of freedom of the quench reactor becomes -1. However, since at this point the HAc and AA balances have been exhausted and the rate of the fourth reaction has already been calculated via the overall balances, the degree of freedom properly becomes zero. With the rate known, the remaining six quench reactor balances are easily solved to yield

$$N^7 = (N_A^7, N_K^7, N_{C_2H_4}^7, N_{CH_4}^7, N_{CO}^7, N_{H_2}^7)$$
$$= (480, 96, 16, 112, 24, 8) \text{ mol/h}$$

The corresponding mole fractions are

$$x^7 = (0.6522, 0.1304, 0.0217, 0.1522, 0.0326, 0.0109)$$

The degree-of-freedom updating summary for this problem is shown in Figure 4.23.

Note that in the preceding example, although the degree of freedom of the process is the same regardless of which set of balances is used, the solution using the constructed reaction set led to the usual unit-by-unit solution. The solution using overall element balances would require the use of the partial solution strategies, which will be discussed in the next chapter.

	Mixer	Furnace	Quench Reactor	Separator	Process	Overall Balances
Initial degree of freedom	1	4	7	4	0	0
Overall balances	−1		−1	−3		
Mixer balances		−1		−1		
Separator balances						
Stream variables			−7			
Redundant balances			+2			
Reaction rate			−1			

Figure 4.23   Example 4.21, degree-of-freedom updating table.

4.5 SUMMARY

In this chapter, we developed a general formulation for balance equations written on individual elements. We found that the key algebraic properties of the element balance equations could be deduced from the atom matrix, in particular, that the maximum number of linearly independent element balances is equal to the maximum number of linearly independent rows in the atom matrix. Using the atom matrix, we further devised a simple criterion for determining when element and species balances are interchangeable. From this criterion, we could conclude that in general species balances are preferable since they are always independent and since they will always yield at least as many independent equations as will the element balances.

We next examined material balance applications involving fossil fuels. We briefly reviewed some general features of these materials and discussed typical processing operations: combustion, synthesis gas generation, and chemical up-grading. We found that element balances are the natural choice in such applications because often the composition of these materials can only be given as an elemental analysis and because chemical transformations involving fossil fuels often require large complex sets of stoichiometric equations. Finally, we introduced the concept of maximum number of independent reactions involving any set of species and devised a procedure for constructing a linearly independent set of reactions from the atom matrix. We now have completely defined the relationship between the atom and stoichiometric matrices and hence between the element and species balances. With this chapter, we thus conclude our study of the various types of balance equations and their properties. In the next chapter, we continue our discussion of material balance calculations by considering the manual and computer-aided solution of large complex flowsheet material balances problems.

PROBLEMS

4.1   Coal beneficiation is a physical separation process by means of which the sulfur content of coal is reduced (Figure P4.1). This process can remove sulfur which is present in the coal in the form of FeS (iron pyrites) but cannot remove the sulfur present as part of organic molecules. An additional benefit of beneficiation is that it will remove inorganic rock which might be mixed in with as-mined coal. However, the rock and FeS fraction always carries with it some good coal. In a given beneficiation application, a coal analyzing 5% sulfur (total) and 10% rock is cleaned of all pyritic sulfur and rock to yield a cleaned coal analyzing 2% sulfur. It is found that 10% of the "good" coal is lost in the beneficiation wastes. Determine the ratio of pyritic sulfur to total sulfur in the feed coal **(a)** using element balances; **(b)** using species balances.

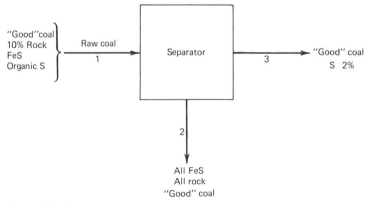

Figure P4.1

4.2 In a catalytic process for synthesizing a hydrocarbon-rich gas, a feed gas consisting of 0.8% CH_4, 7.6% CO_2, 35.2% CO, 51.9% H_2, 4.0% N_2, and 0.5% O_2 is converted to a gas analyzing 43.8% CO_2 and 0.2% CO (on a dry basis) and which is known to also contain CH_4, H_2, N_2, and H_2O. Calculate the composition of the product gas.

4.3 Wood can be converted to a raw synthesis gas consisting of CO and H_2 by thermal decomposition in the absence of air at about 1000°C (Figure P4.3). Smaller quantities of by-product organic chemicals are also produced. For instance, from 3300 lb of dry pine about

> 172 lb ethylene, C_2H_4
> 60 lb benzene, C_6H_6
> 25 lb acetylene, C_2H_2
> 17 lb propylene, C_3H_6
> 9 lb toluene, $C_6H_5CH_3$

The remaining product of pyrolysis is a fairly pure coke which can be assumed to be all carbon. If a pine which analyzes 50.3% C, 6.2% H, 43.5% O, and negligible amounts of ash is used as feed, calculate the pounds of synthesis gas, coke, and chemicals produced per ton of wood.

4.4 A waste stream of composition (mole basis) 25% A_3D_3, 15% HD, 10% A_2B, and the rest HAD_2 is burned with air. If the stable combustion products are H_2O, AO_2, DO_2, and BO, calculate the theoretical air requirements per mole of waste stream.

4.5 Two fuels are mixed and burned with air to produce a flue gas analyzing 7% CO_2, 1% CO, 7% O_2, and the rest N_2. If the fuel compositions are 80% CH_4 and 20% N_2 and 60% CH_4, 20% C_2H_6, and 20% N_2, calculate the ratio of the feed rates of the two fuels.

4.6 A natural gas containing 82.3% CH_4, 8.3% C_2H_6, 3.7% C_3H_8, and 5.7% N_2 is cracked at high temperature to produce a degraded gas and some coke residue. The cracked gas analyzes 60.2% H_2, 23.8% CH_4, 2.2% C_2H_6, 6.5% C_2H_4, 1.4% C_3H_6, 2.3% C_2H_2, and 3.6% N_2, and the coke may be considered to be pure carbon. Calculate the pounds of coke and the pound moles of cracked gas produced per 100 lbmol natural gas.

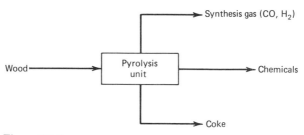

Figure P4.3

4.7 If a flue gas analyzes 1.0% CO, 2.6% O_2, 5.5% CO_2, and 90.9% N_2 on a dry basis when a fuel containing CH_4, CH_3OH, and N_2 is burned with 20% excess air, what is the composition of the fuel gas?

4.8 Methanol is catalytically oxidized with air (21% O_2, 79% N_2) to formaldehyde and water. Some of the formaldehyde is lost in a secondary reaction in which it is oxidized to formic acid. The effluent gas analyzes 6.6% HCHO, 0.4% HCOOH, 13.9% CH_3OH, 13.8% O_2, and 65.3% N_2 on a dry basis. Calculate the percent excess air used.

4.9 A gas containing 30% CS_2, 26% C_2H_6, 14% CH_4, 10% H_2, 10% N_2, 6% O_2, and 4% CO is burned with air. The stack gas contains 3% SO_2, 2.4% CO, and unknown amounts of CO_2, H_2O, O_2, and N_2. What percentage of excess air was used?

4.10 A fuel of unknown composition is burned with 20% excess air. The flue gas is analyzed and found to contain 8.4% CO_2, 1.2% CO, 4.2% O_2, and the rest N_2. If the fuel is known to consist of methane, ethane, and N_2, calculate its composition.

4.11 Two fuels are burned with air: a natural gas with composition 96% CH_4, 2% C_2H_2, and 2% CO_2 and a fuel oil with C-to-H ratio of 1:1.8 (Figure P4.11). The combined flue gases are found to contain 10% CO_2, 1% CO, 5% O_2, and 84% N_2 on a dry basis (H_2O is always present as product of combustion). Calculate the ratio in which the two fuels are consumed.

4.12 A petroleum gas mixture containing 65% propane, 25% propylene, and 10% butane (mol %) is burned with 40% excess air. All of the butane and propylene and 90% of the propane are consumed and no CO is found in the product gas. Calculate the stack gas composition on a dry basis assuming the air moisture content is negligible.

4.13 A stack gas containing 4.3% O_2, 8.0% H_2O, 8.0% CO_2, 8.6% CO, and 71.1% N_2 was obtained when a fuel containing C_2H_2, CO, and O_2 was burned

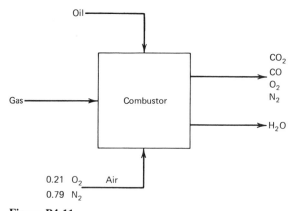

Figure P4.11

with air. Calculate **(a)** the composition of the fuel; **(b)** the percentage excess
air used.

4.14 The waste gases from a viscose rayon fiber plant analyze 35% CS_2, 10%
SO_2, and 55% H_2O. It has been suggested that these gases be disposed of
by burning with an excess of air. Calculate the minimum percent excess air
that must be used if the local air pollution regulations limit the SO_2 in the
stack gas to 2%. Assume that complete combustion of the carbon disulfide
is obtained.

4.15 A heavy fuel oil which analyzes (in wt %) C 84%, H 11.4%, N 1.4%, and
S 3.2% is to be burned in air.

(a) Calculate the theoretical air in moles required to burn 100 lb of the
oil.

(b) Under a given set of conditions, the stack gas obtained by burning this
oil with air is found to contain 4000 ppm (1 ppm $= 10^{-6}$ lb/1 lb) SO_2
on a dry basis. Calculate the percent excess air which was used.

4.16 A dry coal consisting of 3% ash, 1.5% N, 0.6% S, and unknown amounts
of C, H, and O is gasified to produce a fuel gas with dry-basis composition
23% H_2, 3.2% CH_4, 16.2% CO, 12% CO_2, 1% H_2S, and 44.6% N_2 and a
solid residue consisting of 10% carbon and the rest ash (Figure P4.16). If 1
kg steam is used per kilogram of coal fed, calculate the required air rate and
the moisture content of the fuel gas.

4.17 Acetic anhydride is produced from acetic acid by catalytic cracking (Figure
P4.17). In a given process, a feed consisting of 1 mol inerts per 50 mol acetic
acid produces a product analyzing 46% $(CH_3CO)_2O$, 50% H_2O, and 4%
CH_2CO. Calculate the composition of the purge stream.

4.18 A furnace is being operated by burning producer gas and air. The producer
gas analyzes 2% CH_4, 5% CO_2, 14% H_2, 20% CO, and 59% N_2. A sample

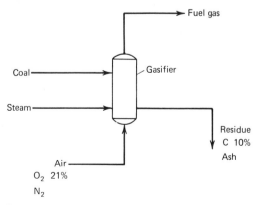

Figure P4.16

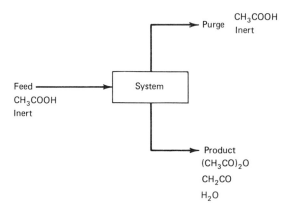

Figure P4.17

of flue gas obtained by water displacement analyzes 4.2% O_2, 11% CO_2, and 84.8% N_2.

(a) Show that the process is overspecified.

(b) Show that the data are inconsistent by showing that the solution obtained without using the carbon balance does not satisfy the carbon balance. Show that the data can be explained by assuming that some of the carbon dioxide in the flue gas dissolved in the water when the flue gas sample was obtained by water displacement.

(c) Show that 40% excess air was used.

Later, a sample of stack gas (H_2O, CO_2, O_2, and N_2) was found to contain only 9% CO_2 even though the furnace was being operated with the same producer gas and percentage excess air as before.

(d) Show that the process is overspecified.

(e) Show that the data are inconsistent by showing that the solution obtained without using the known percentage excess air is not consistent with the percentage excess air given. Show that the data can be explained by assuming that air leaks into the system between the furnace and the stack.

(f) Calculate the air leak-to-burner air ratio.

4.19 A plant utility gas can be produced by reacting powdered coal with steam and air (Figure P4.19). In a given application a coal with ultimate analysis 68% C, 6.5% H, 1.5% N, 1.2% S, 13.8% O, and 9.0% ash is converted to a gas with dry-basis analysis of 23% H_2, 3.2% CH_4, 16.2% CO, 13% CO_2 + H_2S, and 44.6% N_2. The conversion is carried out by using 0.95 lb steam per 1 lb coal. The air fed is found to contain 1.5 mol % H_2O and product gas 0.24 mol H_2O/mol dry gas. Calculate the fraction of the carbon fed which remains in the ash residue.

4.20 A furnace is being operated with a mixture of fuel gas and air. The fuel gas composition has been reliably measured as 5% CH_4, 2.5% C_2H_6, 20% CO,

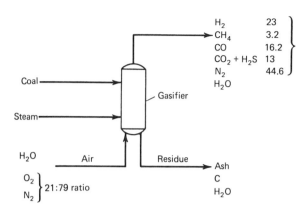

Figure P4.19

25% H_2, 2.5% CO_2, and the rest N_2. The flue gas is found to contain no measurable CO and 13% CO_2 on a dry basis. Measurements of O_2, N_2, and H_2O in the flue gas were not taken.

(a) Show that if the air is assumed to be dry, the problem is correctly specified and all flows can be calculated.

(b) Suppose the air was not dry. Instead, the combustion air could have contained from 5 to 10% water (mole basis). Determine the possible ranges of the flue gas composition and of the percent excess air.

(c) If the flue-gas CO_2 content were in error by $\pm \frac{1}{2}\%$ but the air water content was accurately known at 5%, what is the range of possible values of the percent excess air?

4.21 A process waste gas containing H_2S is burned with air (Figure P4.21). A sample of 1.285 mol gas leaving the furnace is analyzed and found to contain 0.1 mol CO_2 and 0.08 mol SO_2. Because of this high SO_2 content, the flue gas is scrubbed with H_2O. The resulting scrubbed gas has a dry-basis analysis of 1.0% CO, 7.5% CO_2, 2.6% O_2, and 88.9% N_2 and a water mole fraction of 1/11. The scrubber liquid consists of 2.5% CO_2, 8% SO_2, and the rest H_2O. If 10% excess air was used, calculate the fuel composition. All compositions are in mole fractions or percent.

4.22 Compound A_2BO can be made by partial oxidation of compound A_5B_3 via the reaction

$$A_5B_3 + O_2 \rightarrow 2(A_2BO) + AB$$

If the conditions are improperly controlled, the product A_2BO will further oxidize via the reaction

$$A_2BO + \tfrac{3}{2}O_2 \rightarrow 2AO + BO_2$$

If material balances were to be carried out in a reactor in which these reactions occurred, should element balances or species balances be used? Why?

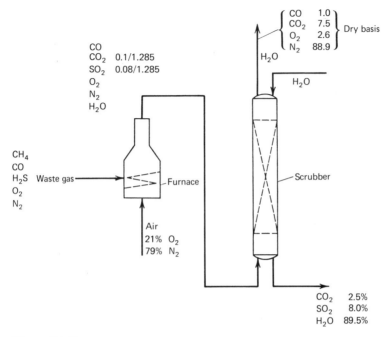

Figure P4.21

4.23 Formaldehyde can be produced industrially by partial oxidation with air over a silver catalyst (Figure P4.23). Suppose about $\frac{2}{3}$ mol formaldehyde is produced per 1 mol methanol reacted and that, in addition, some formic acid, H_2, CO, and CO_2 are produced. In order to ensure high utilization of the reactant, the product gas is scrubbed with H_2O to separate the off-gases. The residue is distilled to produce a pure recycle methanol. Assuming that the reactor inlet gas to the reactor is 35% (mol) methanol, calculate the moles of formaldehyde produced per mole of feed methanol, given the off-gas composition 20.2% H_2, 4.8% CO_2, 0.2% CO, 74.5% N_2, and 0.3% O_2.

4.24 A mixture of ammonia and air (21% O_2, 79% N_2) prepared by mixing 20 mol air per 1 mol ammonia is reacted to produce a complex mixture consisting

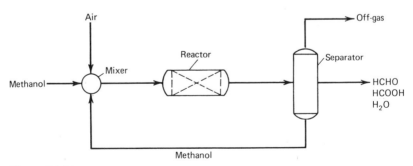

Figure P4.23

of H_2O, O_2, N_2, NH_3, NO, and NO_2 and containing 3% O_2 and 6% NO. Assuming that 80% conversion of ammonia is obtained, calculate the outlet compositions of the remaining species.

(a)   Assume that the reaction process follows the reactions

$$4NH_3 + 5O_2 \Rightarrow 4NO + 6H_2O$$
$$4NH_3 + 3O_2 \Rightarrow 2N_2 + 6H_2O$$
$$4NH_3 + 6NO \Rightarrow 5N_2 + 6H_2O$$
$$2NO + O_2 \Rightarrow 2NO_2$$
$$2NO \Rightarrow N_2 + O_2$$
$$N_2 + 2O_2 \rightleftharpoons 2NO_2$$

(b)   Assume that the reactions taking place are not known, so that element balances must be used.

4.25 What is the maximum number of independent chemical reactions that could take place between the five compounds given the atom matrix

$$\alpha = \begin{pmatrix} 1 & 0 & 1 & 2 & 0 \\ 1 & 0 & 0 & 0 & 1 \\ 3 & 2 & 1 & 4 & 2 \\ 0 & 1 & 0 & 1 & 0 \end{pmatrix}$$

involving four types of elements? Construct a set of independent reactions for this system.

4.26 When a feed of CO and H_2 is introduced into a catalytic reactor containing a nickel catalyst, a mixture consisting of CH_4, C_2H_6, C_2H_8, CH_3OH, C_2H_5OH, CO_2, H_2O, and some CO and H_2 is obtained.

(a)   What is the maximum number of chemical reactions which can take place among these chemical species?

(b)   Suppose the reactions

$$CO + 3H_2 \rightarrow CH_4 + H_2O$$
$$CO + H_2O \rightarrow CO_2 + H_2$$
$$CO + 2H_2 \rightarrow CH_3OH$$

are believed to take place. Construct enough additional reactions to form a set with maximum number.

4.27 In a recycle process, reactant HD_3A_4B is reacted with AD to produce primary products A_2B and A_3D_3 via the reaction

$$HD_3A_4B + AD \rightleftharpoons A_2B + A_3D_3 + HD$$

Two side reactions also take place:

$$HD_3A_4B + 2AD \rightleftharpoons HAD_2 + A_2B + A_3D_3$$
$$HD + AD \rightleftharpoons HAD_2$$

which produce undesirable product HAD_2 (Figure P4.27). It is desired to limit HAD_2 in the A_3D_3 product stream to 25% on a molar basis and HD to 12.5%. The stream leaving the reactor contains 20% A_3D_3 on a molar basis, and the process feed rate of AD is 750 mol/h.

(a) Determine whether the problem should be solved using either element or species balances.
(b) Construct a degree-of-freedom table, assuming element balances will be used for balances involving reactions.
(c) Solve the problem using element balances.
(d) Solve the problem using species balances. Compare the results of (c) and (d) and explain what has occurred.

4.28 A heavy fuel oil which analyzes (wt %) 84% C, 11.4% H, 1.4% N, and 3.2% S is desulfurized in the system shown in Figure P4.28. The desulfurization involves the catalytic reaction of the oil with a hydrogen-rich gas (44% H_2, 56% CH_4) to produce a product with 1 ppm nitrogen and 9 ppm sulfur. The reactor off-gases contain NH_3, H_2S, H_2, and CH_4. The H_2 and CH_4 are present in a 3:7 ratio, while H_2 and NH_3 are in a 4:1 ratio. This gas is subjected to two stages of purification: the first removes 90% of the H_2S and all of the NH_3; the second removes the remaining H_2S, a larger portion of CH_4, and a smaller amount of H_2. The second separation causes the process to lose some H_2. The recycle stream containing 40% H_2 (the rest is CH_4) is mixed with a fresh gas stream of 48% H_2 (the rest is CH_4) and then is returned to the reactor. Assume 60 lb oil is fed to the reactor per 1 lbmol gas. Note all gas compositions are in mol %. The oil compositions are wt %.

(a) Construct a degree-of-freedom table and show that the process is correctly specified.
(b) Develop a calculation order which is to be followed in order to calculate all unknown stream variables.
(c) Calculate the complete composition of the product oil.

4.29 Acetic anhydride, $(CH_3CO)_2O$, is produced from pure acetone, $(CH_3)_2CO$, by partially decomposition in a direct-fired furnace. Ketene, CH_2CO, along

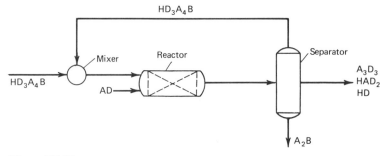

Figure P4.27

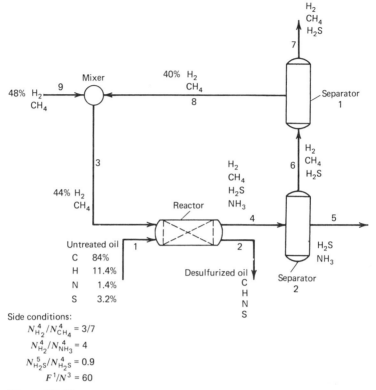

Side conditions:
$$N_{H_2}^4/N_{CH_4}^4 = 3/7$$
$$N_{H_2}^4/N_{NH_3}^4 = 4$$
$$N_{H_2S}^5/N_{H_2S}^4 = 0.9$$
$$F^1/N^3 = 60$$

Figure P4.28

with other side products are produced and are quickly quenched with cold acetic acid, C_2H_3OOH. When sufficiently cooled, the ketene reacts completely with the acetic acid to produce the desired anhydride product via the reaction

$$CH_2CO + C_2H_3OOH \rightarrow (CH_3CO)_2O$$

The resulting reactor outlet is first scrubbed with an acetic acid-rich recycle stream to produce a light-ends stream consisting of C_2H_4, CH_4, CO, and H_2. The absorber bottoms stream containing 60.75% acetone, 12.5% anhydride, and the rest acetic acid is next processed in an acetone column which produces a relatively pure acetone recycle stream and a bottoms product containing a small amount of acetone. This stream is further distilled in the anhydride column, resulting in an overhead stream consisting of 95% acetic acid. Assuming that the acetone conversion in the furnace is 20%, that the furnace exit gases contain 2 mol C_2H_4 and 12 mol ketene per 1 mol H_2, and that 1 mol acetic acid is used as quench for 6 mol acetone entering the furnace:

(a) Show that the process is correctly specified.
(b) Calculate all streams in the flowsheet (Figure P4.29).

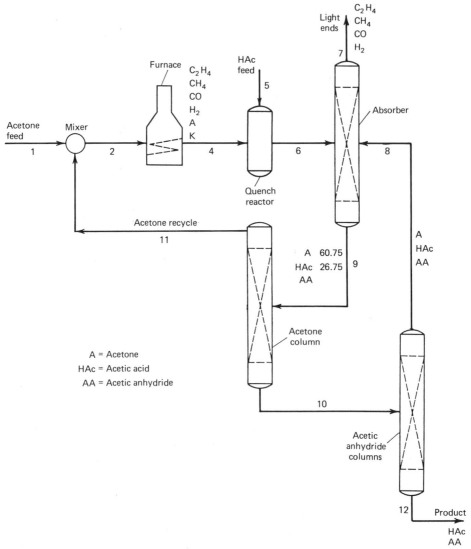

Figure P4.29

4.30 Two major undesirable constituents of the raw gas produced in coal gasification are H_2S and CO_2. These "acid" gases are typically removed by absorption into weakly alkaline solvents followed by regeneration of the absorbing solution, usually by heating. The resulting waste gas contains H_2S, CO_2, some H_2O, and small amounts of CH_4. Because of the high H_2S content, this gas cannot be directly vented to the atmosphere but rather must be further purified of this constituent. A very widely employed method is the Claus process for converting H_2S to elemental sulfur, a side benefit of which is the production of salable, high-purity sulfur.

In the flowsheet given in Figure P4.30, based on a design by the Foster-

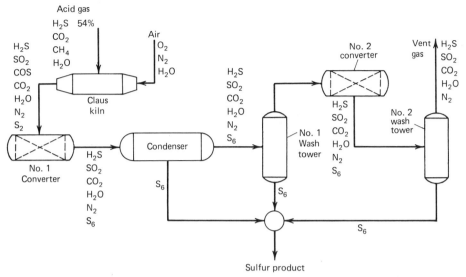

Figure P4.30

Wheeler Corporation, acid gas containing 54% H_2S is mixed with air and burned directly in a Claus kiln to produce a product analyzing (mol %) 4.6% H_2S, 4.6% SO_2, 4.6% COS, 10.0% CO_2, 24.0% H_2O, 3.7% S_2, and 48.5% N_2. The temperature in the kiln is carefully controlled so that SO_2 formation is reduced and the formation of elemental sulfur favored. The exit stream is subsequently cooled and reacted in a catalytic converter to further reduce the H_2S content to produce a product containing H_2S, SO_2, CO_2, H_2O, N_2, and 3.16% S_6. The sulfur vapor in this stream is reduced by passing through a cooler which condenses out about 68% of the sulfur vapor. The remaining stream is then further scrubbed of sulfur vapor by passing through a sulfur wash tower. The resulting gas, then, is passed through another catalytic converter the exit S_6 content of which is 0.51%. This gas is again scrubbed of S_6 to result in a waste gas consisting of CO_2, H_2O, N_2, and only a combined $\frac{3}{4}\%$ of H_2S and SO_2 which can be vented. Assume that the O_2-to-N_2 ratio in the humid outside air fed to the converter is 21:79.

(a) What fractions of the total sulfur products are obtained from the condenser, the No. 1 wash tower, and the No. 2 wash tower? The reactions occurring in the Claus kiln are:

$$2H_2S + O_2 \rightarrow S_2 + 2H_2O$$
$$H_2S + CO_2 \rightarrow COS + H_2O$$
$$\tfrac{1}{2}S_2 + O_2 \rightarrow SO_2$$
$$2H_2S + SO_2 \rightarrow \tfrac{3}{2}S_2 + 2H_2O$$
$$CH_4 + 2O_2 \rightarrow CO_2 + 2H_2O$$

(b) Suppose the specification of the No. 2 converter outlet stream is changed from 0.51% S_6 to $\frac{1}{2}\%$ H_2S. How does this affect the solution?

CHAPTER
5

Material Balances in Process Flowsheets

In the previous chapters, we have studied in detail the formulation and properties of material balances written on chemical species and elements both in reacting and nonreacting cases. Along with our analysis of the algebra of the balance equations themselves, we also considered the application of these equations to solving material balance problems involving multiple units. We examined the complications that can arise when the overall balance is introduced into the calculation process and developed a degree-of-freedom updating procedure which allowed us to determine the order in which the unit and/or overall balances are to be solved. In all the applications that we considered, it was always possible to solve multiunit problems as a sequence of zero-degree-of-freedom single-unit or overall balances. In general, this may not always be the case. In this chapter, we therefore discuss what to do if such sequencing is not possible. We consider approaches suitable for hand calculations and those which can be employed for computer calculations.

The approach for manual calculations involves carrying some unknown stream variables from unit to unit as the calculations proceed. This process may result in the need to solve one or more nonlinear equations in order to determine the values of the carried variables. The computer-oriented approaches we consider will involve two calculation strategies: the modular strategy, in which the general problem is solved by iterating on a unit-by-unit sequenced form of the problem; and the simultaneous solution strategy, in which all the unit material balances and specified relations are solved simultaneously. In the course of the discussion of these solution strategies, we will have need to review some numerical techniques for solving linear and nonlinear equations. The reader already familiar with these techniques can of course safely skip those portions of the chapter.

5.1 STRATEGY FOR MANUAL CALCULATIONS

The sequencing and calculation strategies we developed for multiunit problems have been motivated by the recognition that the easiest way to solve the set of balance equations is to solve them one at a time. Thus, we devised an accounting procedure based on the degree-of-freedom table which allowed us to select the order in which the units were to be individually calculated. We also learned to select a set of unit balances and the location of the basis such that the number of balance equations of a given unit which had to be solved simultaneously was minimized. We continue to follow this basic principle in the developments to be presented in this section. Even if solution of the multiunit problem cannot be broken down into a sequence of zero-degree-of-difficulty single-unit calculations, we shall continue to follow a unit-by-unit approach, with the exception that some of the unknown stream variables may have to be left undetermined until the entire problem solution has been completed.

5.1.1 Sequencing with Complete Solution

For convenience in the subsequent discussion, let us refer to the case in which the problem can be solved as a sequence of zero-degree-of-freedom single-unit balances as *sequencing with complete solution*. The following example reviews our procedure for determining the sequence of calculations and in addition illustrates the possibility of planning out this procedure prior to actually initiating any computations.

Example 5.1 The important chemical intermediate acetaldehyde can be catalytically produced from partial oxidation of ethane. The primary reaction is

$$C_2H_6 + O_2 \rightarrow C_2H_4O + H_2O$$

However, there are a number of side reactions which also occur to a significant degree:

$$C_2H_6 + \tfrac{7}{2}O_2 \rightarrow 2CO_2 + 3H_2O$$

$$C_2H_6 + \tfrac{3}{2}O_2 \rightarrow CH_3OH + CO + H_2O$$

$$CH_3OH + \tfrac{1}{2}O_2 \rightleftharpoons CH_2O + H_2O$$

$$2CO + 3H_2O \rightleftharpoons C_2H_6 + \tfrac{5}{2}O_2$$

To reduce the formation of these various by-products, the reactor must be operated at low C_2H_6 conversion and high C_2H_6-to-O_2 ratios in the reactor feed. The process must therefore involve a high recycle rate and, because air is used as source of oxygen, must have a recycle purge stream to remove the inert N_2. To avoid the loss of valuable ethane in the purge, in the process flowsheet shown in Figure 5.1, the recycle stream is split into two equal parts. One part is subjected to a separation which will preferentially remove a stream of N_2, CO, CO_2, and C_2H_6 for venting. The other half of the recycle stream is sent directly back to the reactor without treatment.

Under a given set of operating conditions with a reactor inlet stream consisting of 10 mol C_2H_6 per 1 mol CO and 0.9333 mol O_2 per 1 mol CO, the reactor outlet stream is found to contain 35% C_2H_6, 1% each of C_2H_4O, CH_3OH, and CH_2O, 4.25% CO, and 51% N_2 along with some CO_2 and H_2O. The outlet stream (10) from separator 3 is found to contain equimolar amounts of CO and CO_2. Assuming

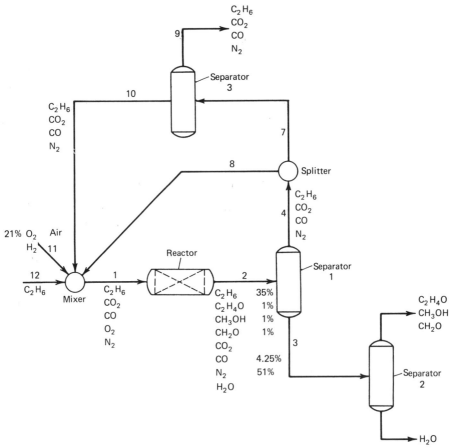

Figure 5.1   Flowsheet for Example 5.1, acetaldehyde plant.

that air contains 21% O_2 and 79% N_2, it is desired to calculate all streams in the flowsheet.

Following our usual approach, we begin by checking whether the given five reactions are in fact independent. The stoichiometric matrix with the reactions listed in the order in which they are given is:

	(1)	(2)	(3)	(4)	(5)
C_2H_4O	1	0	0	0	0
CO_2	0	2	0	0	0
CH_3OH	0	0	1	-1	0
CH_2O	0	0	0	1	0
CO	0	0	1	0	-2
H_2O	1	3	1	1	-3
O_2	-1	$-\frac{7}{2}$	$-\frac{3}{2}$	$-\frac{1}{2}$	$\frac{5}{2}$
C_2H_6	-1	-1	-1	0	1

Since the third column can be readily used to reduce the nonzero third element of the fourth column to zero, it is easy to recognize that all five reactions will be independent.

We next proceed to construct a degree-of-freedom table as shown in Figure 5.2. From the table, it is apparent that the problem is correctly specified and that the basis ought to be chosen at the reactor, since that unit will then have degree of freedom zero.

Suppose now that these calculations have been carried out. As a result of the reactor balance calculations, the remaining two unknowns in stream 2 will have been determined, as will the remaining three unknowns in stream 1. The latter comes about because, of the five stream variables associated with stream 1, two have in effect already been fixed in the problem statement through the two specified ratios. Since the degree of freedom of the mixer is still positive, that is, $7 - 3 = 4$, while that of separator 1 is zero, $2 - 2 = 0$, the calculations should be continued with the separator 1 balances. As a consequence of these unit balance calculations, streams 3 and 4 will be determined, and thus both the splitter and separator 2 will have degree of freedom zero. Suppose both of these unit balances are carried out. As a result of the separator 2 balances, streams 5 and 6 will be calculated but the degree of freedom of the overall balances will remain positive, that is, $6 - 3 - 1 = 2$. On the other hand, once the splitter balances are completed, stream 8 will have been determined and consequently the mixer degree of freedom will be reduced to zero. The separator 3 balances continue to remain positive. The mixer balances are thus carried out next. Finally, with both streams 7 and 10 determined, the separator 3 balances will have degree of freedom zero, $7 - 4 - 3 = 0$, and the solution can be completed with those balances.

Following our usual procedure, the degree-of-freedom updating can be summarized in an updating table as shown below in Figure 5.3.

Alternatively, the calculation sequencing can be shown in node-arc form in which the nodes indicate the balance sets and the directed arcs the sequence in which the calculations are carried out. The node-arc schematic for the preceding example is given in Figure 5.4. The dashed lines in this diagram indicate an alternate branching of the calculations, in our case, to the overall balances.

Note that we can in principle determine the entire calculation order without directly carrying out any numerical solution. Of course, since this approach is based purely on the data in the degree-of-freedom table, it will not deduce certain types of improper specifications. In general, those can only be observed by actually attempting the calculations.

5.1.2 Sequencing with Partial Solution

In Example 5.1, the problem could be solved via a unit balance sequencing with complete solution of each balance set. If the problem is modified by relocating one specification as given in the next example, it develops that this sequencing breaks down.

	Reactor	Sep 1	Sep 2	Sep 3	Splitter	Mixer	Process	Overall Balances
Number of variables	13 + 5	16	8	12	12	16	49	11 + 5
Number of balances	9	8	4	4	4	5	34	9
Number of specified compositions								
Stream 2	6	6					6	
Air						1	1	1
Number of specified relations								
$CO:CO_2$ ratio in stream 10				1		1	1	
$CO:O_2$ ratio in stream 1	1					1	1	
$CO:C_2H_6$ ratio in stream 1	1					1	1	
Splitter restrictions					3		3	
Split ratio					1		1	
Degree of freedom	1	2	4	7	4	7	1	6
Basis	−1						−1	
	0						0	

Figure 5.2 Degree-of-freedom table, Example 5.1.

	Reactor	Sep 1	Sep 2	Sep 3	Splitter	Mixer	Overall
Degree of freedom	0	2	4	7	4	7	6
Reactor balances		−2				−3	
Sep 1 balances			−4		−4		
Sep 2 balances							−4
Splitter balances				−4		−4	
Mixer balances				−3			−2
Redundant balances							+5
Known balances							−5

Figure 5.3   Degree-of-freedom updating table, Example 5.1.

Example 5.2   Suppose instead of specifying the O_2-to-CO ratio in the reactor inlet, the air-to-C_2H_6 feed ratio is given, for instance, 0.171 mol C_2H_6 in stream 12 per 1 mol air fed. The result of this relocation of information is to increase the degree of freedom of the reactor by 1 and decrease that of the overall balances by 1. If we retain the basis with the reactor, then the new unit degrees of freedom will be as shown below.

	Reactor	Sep 1	Sep 2	Sep 3	Splitter	Mixer	Process	Overall
Degree of freedom	1	2	4	7	4	7	0	5

Note that none of the units has zero degree of freedom, even though the process remains correctly specified. The fact that no unit has degree of freedom zero indicates that the balance set of any single unit cannot be completely solved for all the unknown stream variables. By way of illustration, suppose we solve the reactor balances using as basis 400 mol/h of reactor outlet stream 2. With the reactions indexed in the order in which they are listed in Example 5.1, the balance equations are:

C_2H_6: $140 = N^1_{C_2H_6} - r_1 - r_2 - r_3 + r_5$

C_2H_4O: $4 = 0 + r_1$

CH_3OH: $4 = 0 + r_3 - r_4$

CH_2O: $4 = 0 + r_4$

CO: $17 = N^1_{CO} + r_3 - 2r_5$

CO_2: $N^2_{CO_2} = N^1_{CO_2} + r_3 - 2r_5$

O_2: $0 = N^1_{O_2} - r_1 - \frac{7}{2}r_2 - \frac{3}{2}r_3 - \frac{1}{2}r_4 + \frac{5}{2}r_5$

H_2O: $N^2_{H_2O} = 0 + r_1 + 3r_2 + r_3 + r_4 - 3r_5$

N_2: $204 = N^1_{N_2}$

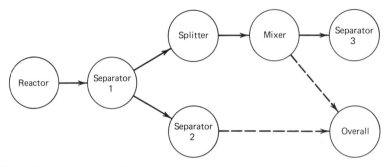

Figure 5.4   Calculation order, Example 5.1.

In addition, the definition of N^2 yields

$$N^2 = 400 = 373 + N^2_{CO_2} + N^2_{H_2O}$$

and the specified C_2H_6-to-CO ratio

$$N^1_{C_2H_6} = 10N^1_{CO}$$

The C_2H_4O, CH_3OH, and CH_2O balances can be immediately solved to obtain the rates

$$r_1 = 4$$
$$r_3 = 8$$
$$r_4 = 4$$

With these rates known, the CO and C_2H_6 balances reduce to

$$N^1_{CO} = 9 + 2r_5$$
$$N^1_{C_2H_6} = 152 + r_2 - r_5$$

From the side condition, we then can obtain a relation between r_2 and r_5:

$$152 + r_2 - r_5 = 10(9 + 2r_5)$$

or

$$r_2 = 21r_5 - 62$$

Using the definition of N^2, we can then express $N^2_{H_2O}$ and $N^2_{CO_2}$ in terms of r_5 alone:

$$N^2_{H_2O} = 60r_5 - 170$$
$$N^2_{CO_2} = 197 - 60r_5$$

With this, all the reactor input species flows can similarly be expressed as a function of r_5:

$$N^1 = \begin{cases} N^1_{C_2H_6} = 90 + 20r_5 \\ N^1_{CO_2} = 321 - 102r_5 \\ N^1_{CO} = 9 + 2r_5 \\ N^1_{N_2} = 204 \\ N^1_{O_2} = 71r_5 - 199 \end{cases}$$

Thus, the degree of freedom being equal to 1 indicates that all but one unknown variable can be calculated, or, as we see above, all stream variables can be expressed as functions of a single unknown. Although this situation is not as convenient as when all the variables can be determined, it nonetheless does offer some help in the solution of the problem. Note in particular that, as a result of the *partial* solution of the reactor balances, the two unknowns in stream 2, namely, the total flow and one composition, have in effect been reduced to one unknown, r_5. Similarly, three unknowns in stream 1 have also been reduced to a single unknown, the parameter r_5. This reduction in the number of unknowns in streams 1 and 2 serves to reduce the degree of freedom of the adjacent units, the mixer and separator 1. In particular, the degree of freedom of the mixer becomes $7 - 3 = 4$, while that of separator 1 becomes 1. Since separator 1 involves a clean component separation, the species flows of streams 4 and 3 will all be given in terms of r_5. In each case, this represents a reduction from four unknown stream variables to one unknown r_5. Hence, the degrees of freedom of the splitter and of separator 2 are reduced from 4 to 1. If the splitter balances are carried out, then stream 8 can be expressed in terms of r_5, since

$$N^8_{C_2H_6} = \tfrac{1}{2}N^4_{C_2H_6} \quad \text{and} \quad N^4_{C_2H_6} = N^2_{C_2H_6} = 140$$

$$N^8_{CO_2} = \tfrac{1}{2}N^4_{CO_2} \quad \text{and} \quad N^4_{CO_2} = N^2_{CO_2} = 197 - 60r_5$$

$$N^8_{CO} = \tfrac{1}{2}N^4_{CO} \quad \text{and} \quad N^4_{CO} = N^2_{CO} = 17$$

$$N^8_{N_2} = \tfrac{1}{2}N^4_{N_2} \quad \text{and} \quad N^4_{N_2} = N^2_{N_2} = 204$$

The degree of freedom of the mixer is therefore reduced by 3 and thus appears to be 1:

$$\text{Degree of freedom} = 7 - 3\,(\text{stream 1}) - 3\,(\text{stream 8}) = 1$$

Note, however, that both stream 1 and stream 8 are expressed in terms of the same unknown r_5. Consequently, the actual degree of freedom of the mixer is *zero*, that is, the mixer balances can be solved not only for the unknown feed and recycle streams (10, 11, 12) but also for the unknown r_5 carried over from the reactor balances. The subsequent calculations will proceed as before. Therefore, the complete calculation flow diagram can be given as shown in Figure 5.5.

In the flow diagram, the numerals 1 on the arcs are intended to indicate that an unknown variable has to be carried from, for example, the reactor balances to

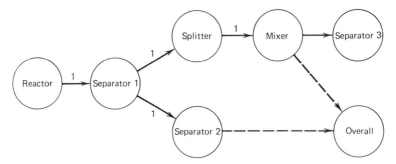

Figure 5.5   Calculation order, Example 5.2.

the separator 1 balances. If no arc numbers are shown, no variables need to be carried to subsequent units, and hence the preceding unit must have been completely solved.

The degree-of-freedom updating table which we constructed for Example 5.1 can be assembled for Example 5.2 as well. Since in a calculation involving partial solution, all of the stream variables are not determined, rather the number of unknown stream variables is merely reduced, care must be taken to ensure that the correct number of unknowns is deducted in each row of the table. Moreover, since it is generally difficult to determine in advance which carried variables will end up in what streams, the net degree of freedom calculated as part of the updating may sometimes be conservative. This was the case with the mixer which apparently had degree of freedom 1 but, since the same unknown was involved in two streams, actually had degree of freedom zero. The degree-of-freedom updating table for Example 5.2 is shown in Figure 5.6.

As illustrated by the preceding example, in cases in which partial solution is required, the sequencing rules are essentially the same. At the conclusion of the calculations for each set of unit balances, the next unit to be solved will be that which has the lowest degree of freedom. If that degree of freedom is zero, then the unit can be solved completely. If the degree of freedom is positive, then only

	Reaction	Sep 1	Sep 2	Sep 3	Splitter	Mix	Overall
Degree of freedom	1	2	4	7	4	7	5
Reactor balances		-1				-3	
Sep 1 balances			-3		-3		
Sep 2 balances							-3
Splitter balances				-3		-3	
Same variable in streams 3 and 1						-1	
Mixer				-3			-1
r_5 determined				-1			-1
Sep 3							

Figure 5.6   Degree-of-freedom updating table, Example 5.2.

partial solution is possible and the number of variables, which will have to be retained and carried over to the next unit selected, will be equal to the degree of freedom.

A noteworthy feature of the previous example is that the final determination of the carried unknown r_5 involved the solution of a single linear equation. This need not always be the case, as demonstrated in the following example.

Example 5.3 In an ammonia plant, a feed gas consisting of 74% H_2, 24.5% N_2, 1.2% CH_4, and 0.3% argon is catalytically reacted to produce NH_3. The products of reaction are refrigerated to separate out 75% of the NH_3 product per pass. The remaining process stream is recycled back to the reactor. In order to stabilize the buildup of the inerts CH_4 and A in the process, part of the recycle gas is purged. Suppose that the purge rate is adjusted so that the combined reactor feed contains 18% CH_4 and that 65% of the N_2 is converted per pass. Calculate all flows in the process. A simplified flowsheet for the process is given in Figure 5.7.

Solution The construction of the degree-of-freedom table for this example is quite straightforward since the process only involves a single reaction. The results are shown in Figure 5.8. Quite apparently the problem is correctly specified. However, none of the units has degree of freedom zero, hence, sequencing with partial solution will be necessary. Since the overall balances have the lowest degree of freedom, we select a basis with this set of balances and shall need to express the solution to these balances in terms of two unknowns. With the completion of this balance set, the five species flows of stream 6 will be expressed in terms of two unknowns and stream 4 will more than likely remain unknown. Thus, the splitter balances will have degree of freedom 3, namely, the original 6 minus 3, since the number of unknowns in stream 6 are reduced from five to two. A degree of freedom 3 indicates that one additional variable will have to be carried at the conclusion of

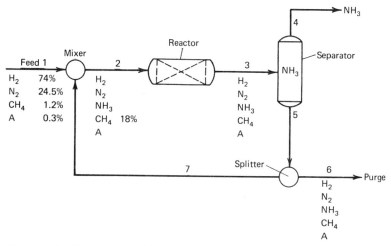

Figure 5.7 Flowsheet for Example 5.3, NH_3 synthesis loop.

	Mixer	Reactor	NH₃ Sep	Splitter	Process	Overall
Number of variables	14	10 + 1	11	15	31	10 + 1
Number of balances	5	5	5	5	20	5
Number of specified compositions						
Feed gas	3				3	3
CH₄ in stream 2	1	1			1	
Number of relations						
Conversion		1			1	
NH₃ recovery			1		1	
Splitter restrictions				4	4	
Degree of freedom	5	4	5	6	1	3
Basis					−1	
					0	

Figure 5.8 Degree-of-freedom table, Example 5.3.

the splitter balances. Thus, the species flows of streams 5 and 7 will be expressed in terms of three unknowns. The mixer degree of freedom thus will be 2, since it had been 4 after the overall balances; while that of the separator will be 2, since the same three variables will be involved in expressing the flows of streams 4 and 5. After these two balances are completed, only the reactor balances are left. But since the overall balances were used, these are no longer independent. Since the problem has degree of freedom zero, the conversion relation which remains to be used should suffice to calculate whatever unknowns remain.

ı he calculation sequencing order is shown in Figure 5.9.

Let us therefore select a basis of, say, 1000 mol/h of stream 1 and commence with the overall balances:

A   : $N_A^6 = 3$

CH₄ : $N_{CH_4}^6 = 12$

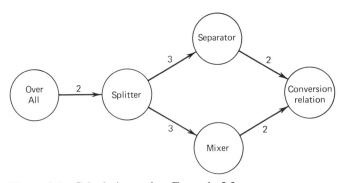

Figure 5.9 Calculation order, Example 5.3.

N_2 : $N_{N_2}^6 = 245 - r$

H_2 : $N_{H_2}^6 = 740 - 3r$

NH_3 : $N_{NH_3}^4 + N_{NH_3}^6 = 0 + 2r$

For convenience, let us use r and $N_{NH_3}^4$ as variables to be carried; then, the species flows of stream 6 will be

$$N_A^6 = 3$$
$$N_{CH_4}^6 = 12$$
$$N_{N_2}^6 = 245 - r$$
$$N_{H_2}^6 = 740 - 3r$$
$$N_{NH_3}^6 = 2r - N_{NH_3}^4$$

Continuing with the splitter balances, it is convenient to introduce as third variable the ratio

$$X = N^7/N^6$$

In terms of this variable, the splitter balances will result with stream 7 simply having the species flows given for stream 6 all multiplied by X, while the species flows of stream 5 will be those of stream 6 multiplied by $(1 + X)$.

Next, the mixer balances simply become:

A   : $N_A^2 = 3X + 3$

CH_4 : $N_{CH_4}^2 = 12X + 12$

N_2 : $N_{N_2}^2 = (245 - r)X + 245$

H_2 : $N_{H_2}^2 = (740 - 3r)X + 740$

NH_3 : $N_{NH_3}^2 = (2r - N_{NH_3}^4)X$

Summing up these balance equations, we have

$$N^2 = 1000(X + 1) - 2rX - N_{NH_3}^4 X$$

The known CH_4 composition therefore yields the nonlinear equation

$$12X + 12 = 0.18N^2 = 0.18[1000(X + 1) - 2rX - N_{NH_3}^4 X]$$

or

$$\frac{1 + X}{X} = \frac{0.18}{168}(2r + N_{NH_3}^4) \tag{5.1}$$

In principle, this equation could be used to eliminate X from the expressions for the stream 2 species flows so that these are expressed only in terms of r and $N_{NH_3}^4$, that is, two parameters.

We next proceed to the NH_3 separator balances:

A : $\quad N_A^3 = 3(1 + X)$

CH_4 : $\quad N_{CH_4}^3 = 12(1 + X)$

N_2 : $\quad N_{N_2}^3 = (245 - r)(1 + X)$

H_2 : $\quad N_{H_2}^3 = (740 - 3r)(1 + X)$

NH_3 : $\quad N_{NH_3}^3 = (2r - N_{NH_3}^4)(1 + X) + N_{NH_3}^4$

The specified recovery relation requires that

$$N_{NH_3}^4 = 0.75 N_{NH_3}^3$$
$$= 0.75[(2r - N_{NH_3}^4)(1 + X) + N_{NH_3}^4]$$

or

$$N_{NH_3}^4 = \frac{1.5r(1 + X)}{(1 + 0.75X)} \tag{5.2}$$

Again, this relation could be used to eliminate $N_{NH_3}^4$ from the equations for the species flow rates of stream 3 to express them in terms of only two variables.

Finally, we have the reactor conversion relation

$$0.65 = \frac{N_{N_2}^2 - N_{N_2}^3}{N_{N_2}^2} = \frac{245(X + 1) - rX - (245)(X + 1)}{245(X + 1) - rX}$$

or, upon rearrangement,

$$r = \frac{159.25(X + 1)}{(1 + 0.65X)} \tag{5.3}$$

The above three nonlinear equations clearly will suffice to determine the values of the three variables r, X, and $N_{NH_3}^4$. One approach to solving them is to use eq. (5.3) to eliminate r and eq. (5.2) to eliminate $N_{NH_3}^4$ from eq. (5.1). The result is

$$\frac{(1 + X)}{X} = \frac{0.18}{168} \frac{318.5(1 + X)}{(1 + 0.65X)} + \frac{318.5(1 - X)^2 \, 0.75}{(1 + 0.65X)(1 + 0.75X)}$$

Assuming that $(1 + X)$ is not zero, we can divide both sides of this equation by $(1 + X)$ and, after rearranging, obtain the quadratic equation

$$4.095X^2 - 134.8725X - 168 = 0$$

This can be solved using the formula for the roots of a quadratic equation. Thus,

$$X = \frac{134.8725 \pm [(134.8725)^2 + 4(168)(4.095)]^{1/2}}{2(4.095)}$$

Since only the positive root is significant, the solution is $X = 34.14$. With X

	Stream Number				
Species	**2**	**3**	**5**	**6**	**7**
A	105.42	105.42	105.42	3	102.42
CH_4	421.68	421.68	421.68	12	409.68
N_2	371.32	130.02	130.02	3.7	126.32
H_2	1289.65	565.76	565.76	16.1	549.65
NH_3	154.65	637.25	159.18	4.53	154.65

Figure 5.10   Stream table for Example 5.3.

calculated, eq. (5.3) can be used to calculate $r = 241.30$ mol/h and eq. (5.2) to calculate $N_{NH_3}^4 = 478.07$ mol/h.

Finally, we can use these three variables to calculate the stream flow rates which we had obtained in parametric form. The stream flows are given in Figure 5 10 in units of (mol/h).

Note that in general the equations to be solved need not have been quadratic. For instance, if we had not eliminated the factor $(1 + X)$, the resulting equation would have been a cubic, that is,

$$(1 + X)(4.095X^2 - 134.8725X - 168) = 0$$

which when multiplied out becomes

$$4.095X^3 - 130.78X^2 - 302.87X - 168.0 = 0$$

In order to solve that equation, it is necessary to use an iterative numerical procedure. In the next section, we discuss several common methods for determining the solution of a nonlinear equation.

5.1.3   Root-Finding Methods

As shown in Example 5.3, the use of the partial solution strategy can lead to the generation of a nonlinear equation which may have to be solved numerically. In this section, we discuss the several types of numerical equation-solving or root-finding methods which are in wide use. These are:

1. The method of successive substitution.
2. The secant method of Wegstein.
3. Newton's method and its secant approximation.

Method of Successive Substitution   The first, most popular, but not necessarily the most effective of these methods is the method of successive substitution. The basic idea behind this technique is to use the equation whose root is being sought to directly generate the next estimate for the trial-and-error calculation. Specifically, the equation $f(x) = 0$ is rearranged to the form

$$x = g(x)$$

In this form, the function $g(x)$ can be used to provide the next estimate of the root of $f(x)$. The iteration procedure is started with a guessed value of the root and is terminated when successive estimates of the root do not change significantly.

Formally stated, the algorithm proceeds as follows:

Step 0   Rearrange the function $f(x)$ so that it is in the form $x = g(x)$. Select an initial estimate x^0 and a suitable accuracy/termination criterion $\varepsilon > 0$. For trial k, $k = 1, 2, 3 \ldots$.

Step 1   Calculate

$$x^{k+1} = g(x^k)$$

Step 2   Test for convergence. If $|x^{k+1} - x^k| \leq \varepsilon|x^{k+1}|$, terminate the iterations. The root of $f(x)$ is x^{k+1}. If $|x^{k+1} - x^k| > \varepsilon|x^{k+1}|$, set $k = k + 1$ and go to Step 1.

Example 5.4   Solve the equation

$$f(x) = x^2 - 5x + 4 = 0$$

starting at the point $x^0 = 0$ to an accuracy of $\varepsilon = 10^{-3}$ using the method of successive substitution.

Solution   Suppose $f(x)$ is rearranged to the form

$$x = \frac{x^2 + 4}{5} \equiv g(x)$$

Step 1   At $x^0 = 0$, $g(x^0) = (0 + 4)/5 = 0.8 = x^1$

Step 2   Since x^0 and x^1 are substantially different, that is,

$$|0.8 - 0| > 10^{-3}|0.8|$$

we cannot terminate. Step 1 is repeated.

$$g(x^1) = \frac{(0.8)^2 + 4}{5} = 0.928 = x^2$$

The iterations continue in this fashion until the iterate $x^7 = 0.9997$ is reached, at which point the criterion $|0.9997 - 0.9993| < 10^{-3}(0.9997)$ is satisfied. The intermediate results are tabulated below:

x	0	0.8	0.928	0.972	0.989	0.996	0.998	0.9993
$g(x)$	0.8	0.928	0.972	0.989	0.996	0.998	0.9993	0.9997

Unfortunately, the method of successive substitution can and frequently does *diverge* rather than converge, as illustrated in the next example.

Example 5.5 Consider the equation

$$f(x) = x - 2(1 - x)^3 = 0$$

rearranged to the form

$$x = 2(1 - x)^3 \equiv g_1(x)$$

Solution The first few iterations starting at the point $x^0 = \frac{1}{2}$ soon begin to diverge:

x	0.5	0.25	0.844	0.00759	1.9548
$g_1(x)$	0.25	0.844	0.00759	1.9548	−1.7408

On the other hand, if $f(x)$ is rearranged to the form

$$x = 1 - \left(\frac{x}{2}\right)^{1/3} \equiv g_2(x)$$

then, starting at $x^0 = \frac{1}{2}$, the iterations will converge to the root:

x	0.5	0.3700	0.4302	0.4008	0.4148	0.4081	0.4113
$g_2(x)$	0.3700	0.4302	0.4008	0.4148	0.4081	0.4113	0.4100

Obviously, the nature of the function $g(x)$ directly influences the convergence properties of the method. In fact, it can be verified by simple examples shown graphically in Figure 5.11 that it is really the slope of the curve $g(x)$ in the region between the initial given x^0 and the root x^* which determines convergence. It can be shown formally that if the absolute value of the slope of $g(x)$, namely, $\left|\dfrac{dg}{dx}\right|$, is less than 1 in the region between x^0 and x^*, then the iterations will converge. If it is greater than 1, the iterations may diverge.

We can verify that this rule is satisfied by using Example 5.5. For the function $g_1(x)$ we have

$$\frac{dg_1}{dx} = -6(1 - x)^2$$

Therefore, at $x^* = 0.410245$,

$$\frac{dg_1}{dx} = -2.087 \qquad \text{or} \qquad \left|\frac{dg_1}{dx}\right| > 1$$

On the other hand, for the function $g_2(x)$,

$$\frac{dg_2}{dx} = -\tfrac{1}{3}(\tfrac{1}{2})\left(\frac{x}{2}\right)^{-\frac{2}{3}}$$

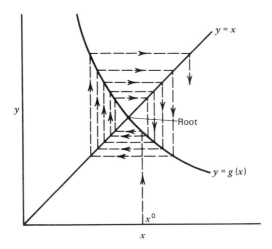

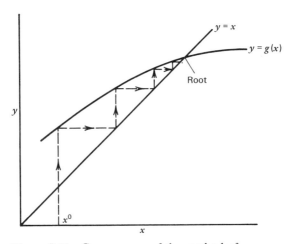

Figure 5.11 Convergence of the method of
successive substitution.

and $\left|\dfrac{dg_2}{dx}\right|_{x^*} = 0.479 < 1$, indicating convergence.

Since the root is not known in advance, this convergence result is not very useful
in guiding the selection of $g(x)$. Consequently, another class of methods has been
devised which actually attempts to use an approximation to the slope of $g(x)$. We
shall discuss one widely used representative of this class.

The Method of Wegstein   A more sophisticated root-finding method is one which
attempts to find the root of the equation

$$x = g(x)$$

by approximating $g(x)$ by a linear function whose slope is calculated at two trial points,

$$g(x) - g(x^1) = (\text{Slope})(x - x^1)$$

where

$$\text{Slope} = \frac{g(x^1) - g(x^0)}{x^1 - x^0} \equiv \left(1 - \frac{1}{\theta}\right)$$

Since we are seeking a point x such that

$$x = g(x)$$

the estimating equation becomes

$$x - g(x^1) = \left(\frac{\theta - 1}{\theta}\right)(x - x^1)$$

or, solving for x,

$$x = (1 - \theta)x^1 + \theta g(x^1)$$

This is the iteration formula of Wegstein. Graphically, it is depicted in Figure 5.12.

It has been found empirically that if the parameter θ is allowed to take on large absolute values, then the convergence of the method can be retarded. Consequently, it is frequently found expedient to reset θ to an upper positive limit α if it is calculated to be larger than α and to reset it to a lower negative value $-\beta$ if it is calculated to be large and negative. The bounds α and $-\beta$ are selected empirically; for instance, 5 and 10 might be typical values. The version of this algorithm which uses this type of bounding is sometimes called the *bounded Wegstein method*.

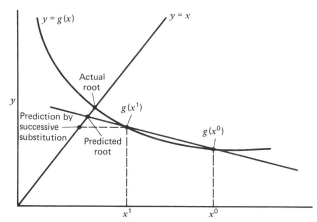

Figure 5.12   Graphical interpretation of Wegstein iteration formula.

The overall algorithm can be summarized as follows:

Step 0 Rearrange the function $f(x)$ so that it is in the form $x = g(x)$. Select an initial estimate x^0 and a suitable accuracy parameter $\varepsilon > 0$ and the bounds α and β. Calculate $x^1 = g(x^0)$ and $g(x^1)$. For $k = 1, 2, \ldots$.

Step 1 Calculate the slope:

$$\text{Slope} = \frac{g(x^k) - g(x^{k-1})}{x^k - x^{k-1}}$$

Set

$$\theta = \frac{1}{1 - \text{Slope}}$$

Step 2 If $\theta > \alpha$, set $\theta = \alpha$; if $\theta < -\beta$, set $\theta = -\beta$. Calculate

$$x^{k+1} = (1 - \theta)x^k + \theta g(x^k)$$

Step 3 Test for convergence. If $|x^{k+1} - x^k| \leq \varepsilon|x^{k+1}|$, terminate the iterations. Otherwise, set $k = k + 1$ and go to Step 1.

Example 5.6 Let us apply Wegstein's method to the function of Example 5.5, that is, $x = 2(1 - x)^3$ with $x^0 = \frac{1}{2}$ and $\varepsilon = 10^{-3}$. Step 0 of the iterations requires the calculation of

$$x^1 = g(x^0) = 0.25$$

and

$$g(x^1) = 0.844$$

In Step 1, we then calculate the slope and θ:

$$\text{Slope} = \frac{0.844 - 0.25}{0.25 - 0.5} = -2.376$$

$$\theta = \frac{1}{1 - \text{Slope}} = 0.2962$$

In Step 2, we check that θ is within bounds and then use Wegstein's formula to calculate the next estimate of the root:

$$x^2 = 0.7038(0.25) + 0.2962(0.844) = 0.4259$$

The iterations clearly have not yet converged; hence, we continue with Step 1, the calculation of the updated slope:

$$\text{Slope} = \frac{0.3783 - 0.844}{0.4259 - 0.25} = -2.6473$$

$$\theta = 0.2742$$

The new iterate calculated in Step 2 is

$$x^3 = 0.7258(0.4259) + 0.2742(0.3783) = 0.4128$$

The calculations continue for two more iterations before the termination criterion is met. The two iterations are

$$x^4 = 0.4102$$
$$x^5 = 0.41026$$

Note that the method of successive substitution diverged with this function.

Both the method of successive substitution and Wegstein's method require that it be possible to rearrange the equation $f(x) = 0$ to the form $x = g(x)$. Yet this is not always possible either because of algebraic complexity or because $f(x)$ cannot be directly expressed in terms of x.

Recall that the *interval halving* method discussed in Section 1.3.3 was suitable for solving this form of the root finding problem. The key characteristic of this method is that given an initial bracket of the root it will retain a bracket. Hence, it is guaranteed to converge. Unfortunately, its rate of convergence is rather slow: it uniformly reduces the remaining interval by one half with each trial. The root-finding method we consider next directly solves the equation $f(x) = 0$ and has a much improved rate of convergence, but at the price of requiring values of the derivative of $f(x)$.

Newton's Method   This very important technique solves the equation $f(x) = 0$ by constructing a local linear approximation to $f(x)$ by using the derivative of $f(x)$. In particular, given a point x^0, $f(x)$ can be approximated by

$$f(x) \simeq f(x^0) + \left.\frac{df}{dx}\right|_{x=x^0} (x - x^0)$$

Since we are seeking a root of $f(x)$, $f(x)$ is set equal to zero in the approximating function, that is,

$$0 = f(x^0) + \left.\frac{df}{dx}\right|_{x=x^0} (x - x^0)$$

and the equation is solved for x:

$$x = x^0 - \left(\left.\frac{df}{dx}\right|_{x=x^0}\right)^{-1} f(x^0)$$

Since the derivative is the best local approximation to the slope of $f(x)$, the resulting iteration formula can be expected to exhibit better rate of convergence than the simpler root-finding methods. This, in fact, is usually the case.

The steps of the method can be formulated as follows, given x^0 and an accuracy parameter $\varepsilon > 0$, for $k = 1, 2, 3, \ldots$:

Step 1   Calculate

$$\left.\frac{df}{dx}\right|_{x=x^{k-1}} = f'(x^{k-1})$$

Step 2   Calculate

$$x^k = x^{k-1} - (f'(x^{k-1}))^{-1} f(x^{k-1})$$

Step 3   If $|f(x^k)| \leqslant \varepsilon$, terminate. Otherwise, go to Step 1.

Example 5.7   Let us apply Newton's method to the function $f(x) = x^2 - 5x + 4 = 0$ of Example 5.4, starting at the point $x = 0$ with $\varepsilon = 10^{-3}$. Note that $\dfrac{df}{dx} = 2x - 5$.

Step 1

$$\left.\frac{df}{dx}\right|_{x=0} = -5$$

Step 2

$$x^1 = 0 - (-5)^{-1}(4) = 0.8$$

This was the same result obtained with successive substitution.

Step 1

$$\left.\frac{df}{dx}\right|_{x=0.8} = -3.4$$

Step 2

$$x^2 = 0.8 - (-3.4)^{-1}(0.64) = 0.9882$$

This is considerably better than the result with successive substitution, which was 0.928.

Step 1

$$\left.\frac{df}{dx}\right|_{x=0.9882} = -3.024$$

Step 2

$$x^3 = 0.9882 - (-3.024)^{-1}(0.03554) = 0.99995$$

This solution meets the termination criterion

$$f(x^3) = 0.00014 < 0.001$$

and the method terminates.

As can be seen from just this example, Newton's method has a superior rate of convergence. However, it has the disadvantages of being rather sensitive to the selection of initial guess and, even more important, of requiring exact derivatives of the functions. In many engineering applications this can be a serious drawback.

To avoid the need for derivatives, a difference approximation to the derivative of $f(x)$ is often used. This approximation is similar to the slope approximation used in Wegstein's method. Given two points x^1 and x^2 which are estimates, but not necessarily brackets, of the root, we approximate the derivative of $f(x)$ by

$$\frac{df}{dx} \approx \frac{f(x^2) - f(x^1)}{x^2 - x^1}$$

This estimate of the derivative is then used in Step 2 of Newton's method to obtain a new point. This point is used to replace one of the two "old" points, the one with the largest absolute value of $f(x)$. Alternatively, if x^1 and x^2 establish a bracket on the root, then, as in the interval-halving method, the retained old point should be the one which allows the bracket to be retained. Whichever rule is used, the resulting pair of points is used to repeat Step 1, that is, calculate a new estimate of the derivative. The process continues until the Step 3 termination check of Newton's method is satisfied. This approximation to Newton's method is called the *secant method*. As might be expected, the convenience of not having to compute the function derivative is purchased at the price of somewhat slower convergence to the root.

Let us conclude this section by carrying out the numerical solution of the cubic equation obtained in Example 5.3 by using the secant method.

Example 5.8   The equation to be solved is

$$f(x) = 4.095x^3 - 130.78x^2 - 302.87x - 168 = 0$$

With the secant method, $f(x)$ need not be rearranged. However, two initial estimates are required. Suppose we use the two points $x^1 = 25$ and $x^2 = 45$. Then, since $f(x^1) = -25,491.4$ and $f(x^2) = 94,535.2$, we can proceed with Step 1.

Step 1

$$\frac{df}{dx} \approx \frac{f(x^2) - f(x^1)}{x^2 - x^1} = 6001.3$$

Step 2

$$x^3 = x^2 - \left(\frac{df}{dx}\right)^{-1} f(x^2) = 45 - \frac{94,535.2}{6001.3} = 29.25$$

Step 3

$$f(x^3) = -18,443.2$$

The point is still far from the root. Hence, we reject the worst of the two previous points (x^2) and continue with Step 1.

Step 1

$$\frac{df}{dx} \approx \frac{f(x^3) - f(x^1)}{x^3 - x^1} = 1659.3$$

Step 2

$$x^4 = x^3 - \left(\frac{df}{dx}\right)^{-1} f(x^3) = 40.36$$

Step 3

$$f(x^4) = 43{,}824.2$$

We reject the worst of x^1 and x^3, and Step 1 is repeated with x^3 and x^4.

The iterations continue for several more cycles to obtain $x^5 = 32.54$, $x^6 = 34.784$, $x^7 = 34.08$, and $x^8 = 34.136$. The final point is in close agreement with previous results.

5.2 STRATEGY FOR MACHINE COMPUTATIONS

As illustrated in Example 5.3, the manual solution of material balance problems involving only a few units can become algebraically complex if sequencing with partial solution is required. For larger problems, the process of carrying variables and algebraic manipulation of balance equations to express stream flows in terms of the selected variables can consume many man-days of work, and thus computer-aided calculation methods become attractive. In this section, we discuss the essential features of two approaches to computerized material balance calculations: the sequential modular strategy and the simultaneous solution strategy. Our discussion will focus on the key ideas and numerical methods involved and will not delve into the details of the coding required to implement these calculation strategies. Efficient programs implementing both approaches are available, but the discussion of such programs is beyond the scope of this text.

5.2.1 The Sequential Solution Strategy

In studying the manual solution of material balance problems, we found it desirable to carry out the calculations on a unit-by-unit basis. This was advantageous for three reasons:

1. It systematized writing the balance equations because each block of equations involved only the few streams associated with that unit.
2. It simplified the analysis of the problem and the determination of a calculation sequence because each unit involved a relatively small number of balances, stream variables, subsidiary relations, and specifications which could be considered as a block.
3. It simplified the solution once the sequence was determined because only a small number of equations had to be manipulated at one time.

This natural blocking of the flowsheet balances into individual unit balances and specification sets also proves to be a very convenient way of organizing computerized calculations.

In constructing a computer program which will solve a balance problem, the same three steps must be executed that arise in manual calculations:

1. The balance equations must be assembled and coded.
2. The specifications must be inserted into the code, analyzed to determine if the problem is correctly specified, and a calculation order determined.
3. The actual numerical solution must be carried out.

To use the computer to maximum advantage, it is first of all desirable that the balance equations can be generated automatically, without having to specifically code each balance equation for every species and every unit. This can be done conveniently if we recognize that, for material balance purposes, any process unit can be represented as either a mixer, a flow splitter, a component separator, a reactor, or some combination of these four. The calculations involved in each of these four elementary modules can be coded in general form into special-purpose subroutines. Thus, whenever the material balances associated with one of these four types of unit need to be solved, the corresponding subroutine is called and executed by the computer without the need for the engineer to laboriously code the individual balance equations. This is a key idea in computerizing a sequential, unit-by-unit balance calculation scheme.

However, there are penalties which must be paid for this convenience in setting up the equations. First, the allowable specifications that can be imposed relating the streams associated with a unit must be severely restricted. Second, for every module, certain streams must always be completely known while others will always be calculated, and these roles cannot be interchanged. This comes about because, in writing program statements, known quantities must always appear in the right-hand side of the equality while unknown quantities must appear on the left. The role of known and unknown quantities cannot be interchanged without rewriting the program statement. The form of the unit modules thus imposes rigid constraints on the kinds of specifications that are acceptable and the order in which the streams must be calculated. We discuss these features in more detail in the next section.

The Elementary Material Balance Modules The four elementary modules of general interest in balance calculations are the mixer, the separator, the splitter, and the stoichiometric reactor. Let us consider each of these in turn.

Mixer The stream mixer, shown in Figure 5.13(a), is simply a device which accepts several input streams and aggregates them into a single output stream. The species balance equations for the mixer will simply consist of

$$N_s^{\text{out}} = \Sigma N_s^i \qquad s = 1, \ldots, S$$

where the sum is over all input streams i. If there are I input streams, then of the $S(I + 1)$ species flows, $S \times I$ must be specified in order to calculate the remaining S. The choice of the unknown stream is arbitrary, but by convention the output stream is usually taken to be the unknown stream while the inputs must all be known. The stream calculation order is therefore in the forward direction, from inputs to output.

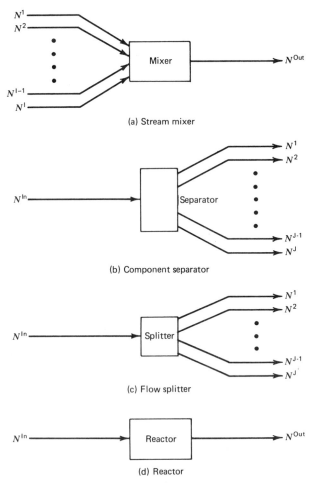

(a) Stream mixer

(b) Component separator

(c) Flow splitter

(d) Reactor

Figure 5.13   Elementary modules.

Separator   The component separator, shown in Figure 5.13(b), is a device which separates a feed stream into two or more output streams. If there are J output streams, then the species balance equations are

$$N_s^{\text{in}} = \sum_{j=1}^{J} N_s^j \qquad s = 1, \ldots, S$$

If the input stream is assumed to be known, then in order to calculate the J output streams, $S(J - 1)$ specifications must be imposed upon the unit. In general, these might take various forms: compositions, composition ratios, flow ratios, or some combination of the three. Because of the complexities involved in accommodating such a variety of specifications, it is conventional to require that all specifications be given as split fractions. The split fraction of component s in stream

j is defined by the ratio

$$t_s^j = \frac{N_s^j}{N_s^{in}}$$

Note that, by virtue of the material balances,

$$N_s^{in} = \sum_{j=1}^{J} t_s^j N_s^{in} = N_s^{in} \sum_{j=1}^{J} t_s^j$$

or

$$\sum_{j=1}^{J} t_s^j = 1$$

Thus, if $J - 1$ split fractions for a given species are specified, the Jth fraction can be calculated by difference. If the input stream and split fractions are known, then the separator balances simply reduce to the simple equations

$$N_s^j = t_s^j N_s^{in} \qquad \begin{aligned} s &= 1, \ldots, S \\ j &= 1, \ldots, J \end{aligned}$$

Again, because of this selection of known variables, the stream calculation order is in the forward direction, from input to output.

Splitter The flow splitter is a special separator, which we first encountered in Chapter 2. It has the property that the composition of any species is the same in all output streams. It is easy to verify that the equivalent of these splitter restrictions expressed in terms of the split fraction concept is that

$$t_s^j = t^j \qquad s = 1, \ldots, S \text{ for each } j = 1, \ldots, J$$

In other words, the split fractions associated with any given stream are all identical. With this modification, the splitter balance equations reduce to

$$N_s^j = t^j N_s^{in} \qquad \begin{aligned} s &= 1, \ldots, S \\ j &= 1, \ldots, J \end{aligned}$$

As in the case of the general separator, the split fractions and the input species flows are assumed known, and from these the output streams are calculated.

Reactor The reactor is a device in which selected chemical species are converted into selected product species in amounts governed by known stoichiometric coefficients and rates of reaction. For simplicity, it is conventional to formulate the reactor module balances in terms of a single input and a single output stream. Thus, the balances become, as in Chapter 3,

$$N_s^{out} = N_s^{in} + \sum_{r=1}^{R} \sigma_{sr} r_r \qquad s = 1, \ldots, J$$

The quantities assumed to be known in the reactor module are the input stream, the stoichiometric coefficients, and the rates of reaction. Usually, dimensionless quantities such as rates per mole of feed or conversions of key reactants and yields or selectivities of key products are specified instead of the rates, which are dimensional quantities. Frequently, only a single reaction is allowed per reactor module. Multiple reactions thus must be transformed to a series of reactions in series. Regardless of these details, the form of the module is such that the module balance will yield the outlet species flows given the inlet species flows. Thus, the calculation order is again in the direction of flow.

As noted previously, this fixed repertoire of modules can be used to represent the units in any arbitrary flowsheet, if necessary, by using the elementary modules in combinations. This feature is illustrated in Example 5.9.

Example 5.9 Nitric acid can be manufactured by oxidizing ammonia with air according to the reaction

$$NH_3 + 2O_2 \rightarrow HNO_3 + H_2O$$

and absorbing the HNO_3 in water. In the simplified conceptual flowsheet shown in Figure 5.14, the production of HNO_3 is integrated with a plant for the production of the reagent NH_3 from N_2, CH_4, and H_2O.

The NH_3 production facility consists of three basic units: (1) a steam reforming plant in which 3 mol steam is reacted with 1 mol CH_4 to produce a synthesis gas containing equimolar amounts of CO and CO_2; (2) a shift reaction/gas clean-up unit in which undesirable CO is eliminated and additional hydrogen is produced via the reaction

$$CO + H_2O \rightarrow CO_2 + H_2$$

to result in a H_2/N_2 gas containing the stoichiometric 25% N_2; and (3) the ammonia synthesis plant in which NH_3 is produced by direct reaction

$$N_2 + 3H_3 \rightarrow 2NH_3$$

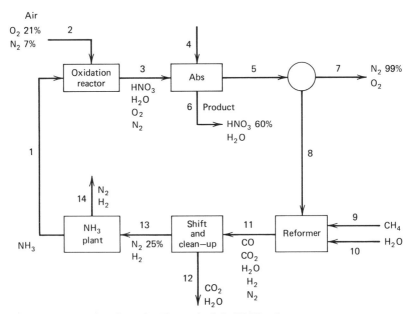

Figure 5.14 Flowsheet for Example 5.9, HNO_3 plant.

Since some of the synthesis gas is lost in purging and NH_3 separation, the conversion in the NH_3 synthesis plant may be taken at 90%.

The resulting NH_3 becomes the feed to the oxidation unit and then to the absorber. The final HNO_3 acid product will contain about 60% HNO_3. The residual gases from the absorber are split and part is purged while the remainder is sent to the reformer. The purge stream consists of 99% N_2 and 1% O_2. Formulate the problem in terms of elementary modules.

Solution When represented in terms of the four elementary modules, the process takes the form shown in Figure 5.15. In this flowsheet, the boundaries of the original units are indicated by dotted lines. Note that except for the splitter, all of the original units require more than one elementary module for their representation.

The Unconstrained Material Balance Problem and its Properties The use of these elementary modules to represent process flowsheets requires that all of the problem specifications be given in a very specific form. These specifications, which we call *natural specifications*, consist of:

1. The species flow rates in all process input streams.
2. The split fractions t_s^j for every species and for every output stream of all splitters and separators.
3. The reaction stoichiometry and reaction rates, or some equivalent measure of progress of reaction, for all reactions in all reactors.

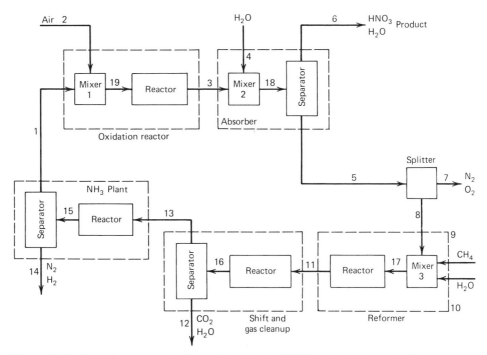

Figure 5.15 Modular representation of flowsheet of HNO_3 plant, Example 5.9.

A problem in which all the natural specifications are imposed is called an *unconstrained material balance problem*. The unconstrained material balance problem has three significant properties: it involves only linear equations in the species flows, it always has degrees of freedom zero, and the order of the unit calculations always follows the direction of the process stream flows. The first property is obvious since all the module equations reduce to constant-coefficient linear equations.

The second property is easily verified for one- and two-module flowsheets. For example, given an unconstrained problem consisting of a mixer and a reactor in series, since all mixer input streams are known, the mixer output stream can be calculated from the balances. Then, since the reaction rates and stoichiometry are assumed known and since the mixer output stream is the reactor input stream, the reactor output stream can be calculated from the reactor species balances. For the case of an unconstrained material balance problem with an arbitrary number of modules, the proof that the degree of freedom is zero must be carried out by induction. That is, we assume that the degree of freedom is zero for a flowsheet with $N - 1$ modules and show that after adding an Nth naturally specified module, the degree of freedom remains zero. Rather than carrying out that exercise, we shall instead verify that it is true by considering a specific example.

Example 5.10 Consider the acetaldehyde plant flowsheet given in Figure 5.1. The flowsheet is already in the form of elementary modules, therefore it does not have to be reformulated. Now, instead of the specifications given in Example 5.1, suppose all the natural parameters are specified, that is, all split fractions, process input streams, and reaction rates or equivalents are specified. This means that streams 11 and 12 to the mixer and the five reaction rates for the reactor are specified. Recall that the number of independent split fractions for a separator is equal to the number of species in the feed times one minus the number of output streams. For a splitter, the number of split fractions is simply equal to the number of output streams minus one, since the split fractions for any given output stream are all identical. Thus, assuming for simplicity that all nine species are present in every stream except for the process feed streams, then separators 1, 2, and 3 will each have nine specified split fractions and the splitter will have one. Given these specifications, we can construct a degree-of-freedom table for the flowsheet as shown in Figure 5.16. Note that in counting the number of variables and balances, it was assumed that all streams except streams 11 and 12 contained all nine species. From the degree-of-freedom table, it is evident that the process is completely specified. Thus, the example verifies that an unconstrained material balance problem always has degree of freedom zero.

The third property of the unconstrained material balance problem is that the unit-by-unit calculation order is always in the direction of flow. This property is a direct consequence of the particular calculation direction selected for the modules themselves. As we have formulated them, each elementary module requires that its input stream or streams be known at the time the module calculations are to be executed. It follows that the calculation order must always proceed in the

	Reactor	Sep 1	Sep 2	Sep 3	Splitter	Mixer	Process
Number of variables	18 + 5	27	27	27	27	30	98
Number of balances	9	9	9	9	9	9	54
Specified flows							
Stream 11						2	2
Stream 12						1	1
Specified reaction rates	5						5
Specified split fractions							
Sep 1		9					9
Sep 2			9				9
Sep 3				9			9
Splitter					1		1
Splitter restrictions					8		8
Degree of freedom	9	9	9	9	9	18	0

Figure 5.16 Degree-of-freedom table for Example 5.10.

direction of flow indicated in the flowsheet. Only the choice of the module at which the calculations are to be initiated is left free. This can be seen from Example 5.10. Note that in the degree-of-freedom table for that example, Figure 5.16, the degree of freedom of every unit is equal to the total number of species contained in those unit input streams that are internal process streams. Thus, if stream 1 were known, then the reactor balances could be solved completely. Next, the separator 1 balances could be calculated; then, the separator 2 and splitter balances, followed by the separator 3; and finally the mixer balances. If the calculations were initiated with any other unit, with the exception of separator 2, which is a terminal unit, the same direction of calculations would be employed.

If possible, the choice of the module at which the calculations are initiated should fall to a unit which has all of its input streams known. If no such unit is available, as in the above example, some stream must be selected as iteration vector, and the flowsheet calculations will be initiated with the unit to which the iteration stream is an input. In the former case, the material balance problem can be solved in a single-calculation pass through the flowsheet. In the latter case, an iterative calculation employing a multidimensional root-finding method to improve successive estimates of the iteration stream must be employed to converge the flowsheet balances. The following two examples examine each of these cases.

Example 5.11 Consider the separation train shown in Figure 5.17. Suppose the problem is unconstrained and hence has degree of freedom zero. Since the feed streams to separator 1 and the mixer as well as the split fractions of all separators and splitters are known, the problem can be solved in a completely sequential fashion. Beginning with separator 1, the flows of streams 2 and 3 are calculated. Then, the separator 2 calculations yield streams 4, 5, and 6 and the splitter calculations, stream 8. Since stream 14 is assumed known, the mixer will yield stream 9, and the separator 3 balances will determine streams 10 and 11. Finally, the

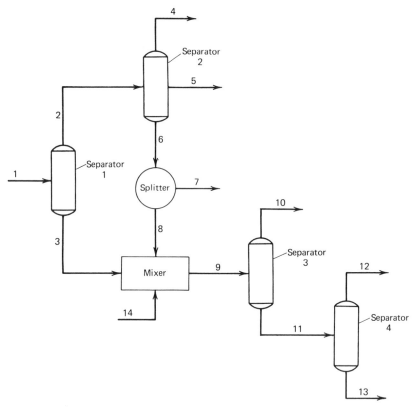

Figure 5.17 Flowsheet for Example 5.11, separation train.

separator 4 calculations will give streams 12 and 13 to complete the flowsheet balances. Thus, all streams are determined in a single-calculation pass through the flowsheet in the direction of flow.

Example 5.12 Consider the simplified benzene chlorination process given in Figure 5.18. In the process, a process stream containing benzene is reacted with a chlorine feed stream to produce mono- and dichlorobenzenes via the reactions

$$C_6H_6 + Cl_2 \rightarrow C_6H_5Cl + HCl$$
$$C_6H_5Cl + Cl_2 \rightarrow C_6H_4Cl_2 + HCl$$

The products of reaction are cooled to separate out a gas stream consisting of HCl and Cl_2 and a liquid stream containing benzene and its chlorinated products as well as a small amount of dissolved Cl_2. The gas stream is scrubbed with fresh benzene to yield a purified HCl gas stream and a liquid-bottoms stream which is sent to the reactor.

A portion of the liquid stream from the condenser is recycled to the reactor, while the remainder is sent to a series of separation units. The first separator recovers benzene for recycle to the reactor, while the second separates the two primary products C_6H_5Cl and $C_6H_4Cl_2$.

Solution   Note that the flowsheet given in Figure 5.18 is not in elementary module form; hence, it must be reformulated. The resulting flowsheet is shown in Figure 5.19. In this flowsheet, the reactor is replaced by a mixer and a reactor,

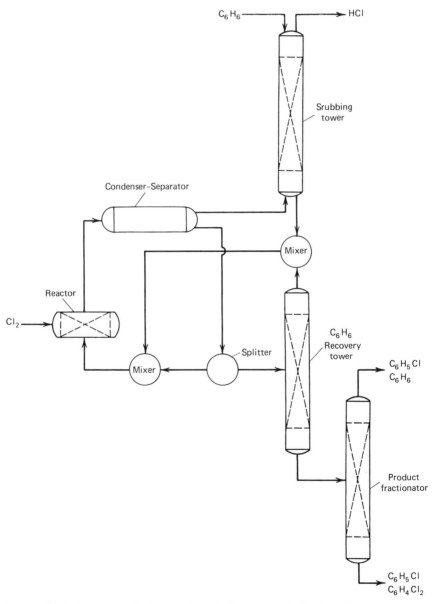

Figure 5.18   Flowsheet for Example 5.12, benzene chlorination plant.

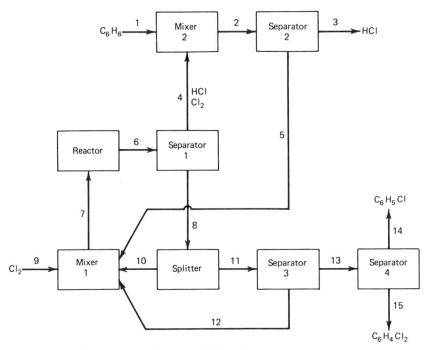

Figure 5.19   Flowsheet for Example 5.12 in elementary module form.

while the scrubbing tower is replaced by a mixer and a separator. Let us assume that this problem is unconstrained, that is, all feed streams, reaction rates, and split fractions are known. Thus, the degree of freedom is zero. Now, observe that since none of the separator input streams is known, the calculations cannot be initiated with any of these units. Moreover, since the reactor input stream is not known and input streams 5 and 12 to mixer 1 and stream 4 to mixer 2 are not known, the calculations can also not be initiated with these units. Clearly, the solution of this problem will require iteration on one or more streams.

However, whichever streams are used as iteration streams, the direction of calculations will always follow the material flows. For instance, if streams 2 and 8 are used, then, given values for stream 8, the splitter balances will yield values for streams 10 and 11. Next, the separator 3 balances will yield streams 12 and 13. Given values for stream 2, the separator 2 balances will yield streams 3 and 5. Then, with streams 5, 10, and 12 known, the mixer 1 balances can be solved for stream 7, followed by the reactor balances which will yield stream 6. Finally, the separator 1 balances will result in a new value of stream 8 and the mixer 2 balances, in a new value of stream 2.

Tearing and Converging of Streams   Given that iteration on one or more internal process streams is likely to be required for many if not most flowsheets, four issues must be resolved:

1.   How many streams will require iteration?

2. Which streams should be selected for iteration?

3. In what order should the tear streams be updated?

4. What numerical scheme should be used to update the successive values of the iterated streams?

Answers to the first two questions are, in the general case, a subject of continuing research. However, for the unconstrained material balance problem, it is possible to determine exactly the minimum number of streams which must be iterated upon, or *torn,* assuming that all streams contain the same number of species. We shall summarize the details of such a procedure in this section. The second issue, that of selecting which stream or streams it is best or most efficient to tear, is a complicated one. In general, one would like to select the tear streams so that the iterations would converge as quickly as possible. However, it is difficult to predict without actually solving the problem whether a given selection will lead to rapid convergence. We shall develop a simple selection rule which has proved to be satisfactory in practice but cannot be guaranteed to be foolproof.

A rigorous resolution of the third question is similarly beyond the scope of this text. We shall make use of a simplified procedure which will determine the order in which individual tear streams are to be processed. In this procedure, selection of tear streams which are to be simultaneously processed is accomplished by exhaustive enumeration. This simple approach proves satisfactory for all but the very large and complex flowsheets.

Finally, the last issue, that of selecting a suitable numerical iteration method, is the least complicated since development of suitable techniques follows quite closely the root-finding methods we have discussed in Section 5.1.3.

We begin our discussion of stream tearing or iteration by considering the problem of determining how many streams must be torn. Recall first of all that stream tearing must be introduced because of the presence of recycle loops in the flowsheet. Thus, it can be expected that the number of streams which must be torn is a function of the number of major recycle loops in the flowsheet. This being the case, it appears that what is needed is a way of identifying recycle loops. In our representation of flowsheets using only the four primitive modules, the only way to close a recycle loop is with a mixer. If we focus on the latter unit, then, since every recycle must be terminated with a mixer, there can never be more recycle loops than there are mixers. Thus we arrive at a very simple rule:

The maximum number of streams which have to be torn is given by the number of mixers in the flowsheet.

For instance, the flowsheet shown in Figure 5.19 contains two mixers; hence, at most two streams will have to be torn. Of course, this upper bound on the number of torn streams is conservative because not all mixers serve to close recycle streams. Some mixers merely serve to introduce external feed streams into the process or to join several process streams without involving recycle. For instance, the flowsheet given in Figure 5.15 contains three mixers. Yet, the flowsheet clearly consists of only one major recycle loop; tearing any one internal stream will suffice to initiate a series of unit-by-unit calculations which will determine all the other streams. Similarly, the flowsheet shown in Figure 5.17 has one mixer. Yet we observed that the elementary modules could be solved sequentially beginning with Separator 1 without any tearing. These examples indicate that our rule for predicting the number of streams which must be torn needs to be modified so as to obtain a more precise estimate. In particular, we need to devise a way of distinguishing between two types of mixers:

> *Essential Mixers*: Those which actually close major recycle loops.
> *Nonessential Mixers*: Those which only serve to introduce process feed streams or to join nonrecycling internal streams.

This functional classification of mixers can be obtained by following upstream the path taken by the material in each input stream of any given mixer. Since reactors, separators, and splitters each have only a single input stream, the upstream tracing of any mixer input stream will follow a unique path until either the output stream of a mixer is encountered or a process input stream is encountered. We call that path a *mixer branch*. If the mixer output stream that is encountered is that of the mixer at which the tracing was initiated, then clearly a recycle loop has been identified and the mixer must be *essential*.

On the other hand, if the mixer output stream encountered is different from that of the mixer whose inputs are being traced, then two situations can occur depending upon whether the encountered mixer is essential or not. If it is essential, and if all the other input streams lead either to other essential mixers or to process input streams, then the mixer whose inputs are being traced can be classified as *nonessential*. If the encountered mixer output stream is that of a nonessential mixer, then the tracing can be continued by following the input streams of the encountered mixer. Since that mixer is nonessential, all of its input streams must lead to output streams of other essential mixers or process input streams. If any of these essential mixer output streams are output streams of the original mixer being traced, then the traced mixer must be essential. Otherwise if all input traces lead to different mixer output streams or process input streams, the traced mixer must be nonessential.

The above input path tracing technique, although clearly defined at each step, is circular in that mixers are classified on the basis of their connections to other essential or nonessential mixers. Yet, we have not indicated how the classification procedure is to be initiated.

To make the classification procedure more precise, we define a *branch source* stream to be the first mixer output or process input stream encountered in tracing a mixer branch. Each mixer branch is thus terminated by a single-branch source stream. Since every mixer input has a unique branch associated with it, every mixer will have a unique set of branch source streams associated with it. Thus, given any arbitrary flowsheet expressed in elementary module form, one can construct a table of mixers, mixer output streams, and their associated branch source streams. For instance, the flowsheet shown in Figure 5.19 has two mixers with output streams 2 and 7. Mixer 2 has two branch source streams, process input stream 1 and mixer output stream 7. Similarly, mixer 1 has four branch source streams: process input stream 9, mixer output stream 2 (obtained by tracing stream 5), mixer output stream 7 (obtained by tracing the branch consisting of streams 10, 8, 6, and 7), and, again, mixer output stream 7 (obtained by tracing the branch: 12, 11, 8, 6, and 7). The branch source stream table for a flowsheet will be the starting point of the procedure to be outlined below. For simplicity in exposition, we assume that all source streams that are process input streams are deleted from the table.

Mixer-Labeling Procedure

Step 0 Initially label all mixers as essential and eligible.

Step 1 Scan the branch source stream table to find an eligible mixer which has a single source stream. If there are none, go to Step 4. Otherwise, continue with Step 2.

Step 2 If the output stream of the selected mixer does not appear in the branch source stream list for this mixer, label the mixer as nonessential and go to Step 3. Otherwise, label the mixer as ineligible for further consideration, and go to Step 1.

Step 3 Search the branch source stream table and replace each occurrence of the mixer output stream whose mixer was labeled as nonessential with the list of branch source streams associated with that nonessential mixer. Label the nonessential mixer as ineligible and go to Step 1.

Step 4 Search the table to find the eligible mixer with fewest source streams. If no eligible mixers remain, go to Step 5. Otherwise, continue with Step 2.

Step 5 The number of required tear streams is equal to the number of mixers labeled as essential.

All ties which occur in this procedure can be broken arbitrarily.

Let us illustrate the above procedure by applying it to the flowsheets given in Figures 5.1, 5.15, 5.17, and 5.19

Example 5.13 (a) The flowsheet in Figure 5.1 has a single mixer. The branch source stream table associated with it will consist of process input streams 11 and 12, stream 1 (obtained via the branch 10, 7, 4, 2, and 1), and stream 1 again (obtained via branch 8, 4, 2, and 1). For purposes of the labeling procedure, we ignore streams 11 and 12. Thus, the branch source stream list for mixer 1 consists only of stream 1. Step 1 thus directs us to Step 2. Since the output stream of mixer 1 appears in its source stream list, mixer 1 becomes ineligible for further consideration and remains labeled as essential. We continue with Step 1. There are no more mixers, hence we continue with Step 4, which in turn directs the search to Step 5. We thus have one essential mixer.

(b) The flowsheet in Figure 5.15 contains three mixers. Mixer 1 has two source streams—process input 2 and mixer output 17. Mixer 2 has two source streams—process input 4 and mixer output 19. Mixer 3 has three source streams—process inputs 9 and 10 and mixer output 18. If we delete process input streams, then all mixers have only a single source stream. Suppose we arbitrarily begin Step 1 with mixer 1. From Step 2, since the output stream for mixer 1, stream 19, does not appear as source stream for that mixer, mixer 1 is labeled nonessential and we continue with Step 3. Step 3 indicates that each occurrence of stream 19 in the source stream should be replaced by stream 17, the mixer 1 source stream. Thus, we obtain:

Mixer	Mixer Output Stream	Source Streams
2	18	17
3	17	18

Mixer 1 is labeled ineligible and we continue with Step 1. Suppose mixer 2 is next selected, from Step 2 this mixer is labeled nonessential, and in Step 3 the source stream of mixer 3 is replaced by stream 17. Mixer 2 is next labeled ineligible and the procedure returns to Step 1. Only mixer 3 remains, and in Step 2 it is labeled essential and ineligible since its output stream is also its source stream. The procedure terminates since the list of eligible mixers is exhausted. From Step 5, the flowsheet requires only one tear stream.

(c) The flowsheet of Figure 5.17 contains only a single mixer. The source stream list for that mixer consists only of streams 1 and 14, both of which are process input streams. Thus, this flowsheet has no essential mixers and does not require tear stream iterations.

(d) Finally, the flowsheet given in Figure 5.19 contains two mixers. We have already noted that the branch source stream lists, excluding process input streams, are:

Mixer	Mixer Output Stream	Source Streams
1	7	2, 7
2	2	7

Mixer 2 will be selected in Step 1 since it has only a single source stream. Since stream 2 is not a source stream for this mixer, mixer 2 is labeled nonessential in Step 2. In Step 3, stream 2 is replaced by stream 7 as source stream for mixer 1. Mixer 1 now has only a single source stream. Moreover, that source stream is also the output of that unit. Therefore, in Step 2, mixer 1 is labeled essential and ineligible. The procedure terminates because there are no more eligible mixers. The flowsheet can be solved using just one tear stream. To confirm this, suppose we tear stream 6. The Separator 1 calculation will give streams 4 and 8; the mixer 2 calculation will yield stream 2 and the splitter calculation, streams 10 and 11. The Separator 2 calculation will give 5 and the Separator 3 calculation will give

12. With streams 5, 10, and 12 known, the mixer can be solved to yield stream 7, followed by the reactor calculation which will result in a new value for stream 6. Thus, all streams can be calculated with just one torn or iterated stream.

It should be noted that the labeling as to essential and nonessential mixers obtained by the search procedure is not necessarily unique. For instance, in the case of Figure 5.15, if the mixers had been traced in the reverse order, mixer 1 would have been labeled as essential and the other two as nonessential. However, the total number of essential mixers will always be the same, independent of the ordering of the mixers.

The above search procedure thus satisfactorily resolves the first issue of interest: determining the minimum number of streams to tear. The second issue, selection of specific streams for tearing, can, for most practical purposes, be settled by choosing as tear streams the output streams of the essential mixers. Mixers are the only units that can have more than one input stream. Since summing several streams tends to dampen the variations in the individual streams, it is reasonable to expect that the mixer output streams will serve as reasonably stable iteration variables. Of course, there is no way of guaranteeing this in general; hence, in specific applications tear streams with better convergence properties can sometimes be determined by trial and error.

The third issue, that of determining the order in which tear streams are to be iterated upon, arises whenever the flowsheet involves more than one essential mixer. If more than one essential mixer is present, then either all tear streams must be updated at the same time or else they can be subdivided into groups which can be sequentially updated. In general, subdivision of tear streams is desirable because the computational effort required to solve nonlinear equations increases as an exponential function of the number of variables. Reduction of a large problem into a series of problems, each with fewer variables, is therefore computationally efficient. Accomplishment of such a subdivision requires that the tear streams be both subdivided or partitioned into simultaneously soluble groups and that these groups be sequenced in the order in which they are to be converged. Partitioning and sequencing of equation sets in general and tear streams in particular is the subject of continuing research. The interested reader is directed to the book by Westerberg et al.[1] for a more detailed discussion. For present purposes, we shall develop a simple partitioning and sequencing technique which makes use of the reduced branch stream table which we obtain upon applying our mixer-labeling procedure to identify a set of essential mixers. Before outlining that procedure, we first consider an example which will help motivate the constructions.

Example 5.14 Determine the tear streams and calculation order for the flowsheets given in Figure 5.20.

---

[1] A. W. Westerberg, H. P. Hutchison, R. L. Motard, and P. Winter, *Process Flowsheeting*, Cambridge University Press, London, 1979.

Solution   Consider first the flowsheet in Figure 5.20(a). The branch source stream list for the two mixers will be:

Mixer	Mixer Output Stream	Source Streams
1	2	1, 2
2	4	3, 2, 4

Both mixers are essential since each has its mixer output stream as source stream. Note that, since stream 1 is a process input stream, mixer 1 involves only a single mixer output stream, stream 2; hence, that tear stream can be converged independently of the other tear stream, stream 4. Since stream 5 will be known only after tear stream 2 has converged, stream 4 can be converged only after tear stream 2 is known. The branch source stream list for mixer 2 indicates that fact,

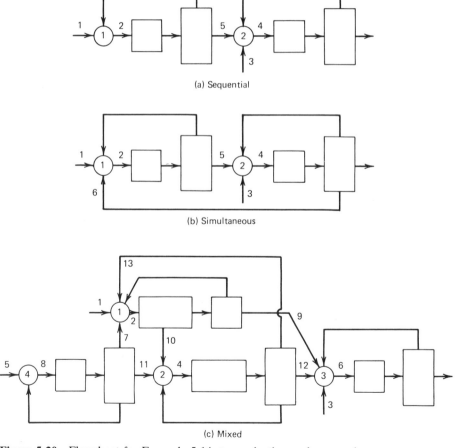

(a) Sequential

(b) Simultaneous

(c) Mixed

Figure 5.20   Flowsheet for Example 5.14, tear selection and sequencing.

namely, it shows that stream 4 can be traced to stream 2 and to itself. Thus, stream 2 must be converged before stream 4. The flowsheet of Figure 5.18(a) consists of sequential tear streams.

The flowsheet of Figure 5.20(b) differs from Figure 5.20(a) only in stream 6. The revised branch source stream list will be:

Mixer	Mixer Output Stream	Source Streams
1	2	1, 2, 4
2	4	3, 2, 4

Again, both mixers are essential since each has its output stream as source stream. In addition, however, each mixer also has the other tear stream as source stream. The presence of stream 6 has coupled the tear streams so that they can no longer be sequenced. In order to converge stream 2, stream 6 and hence stream 4 must be known; to converge stream 4, stream 5 and hence stream 2 must be known. The two tear streams must be converged simultaneously, and we can clearly recognize this fact from the source stream list.

Finally, consider the flowsheet of Figure 5.20(c). This flowsheet consists of four mixers with the following source stream lists, in which process input streams have been deleted:

Mixer	Mixer Output Stream	Source Streams
1	2	2, 4, 8
2	4	2, 4, 8
3	6	2, 4, 6
4	8	8

Again, all four mixers are essential since the output stream of each is also a source stream. Note that mixer 4 has only stream 8 as source stream. Thus, this tear stream can be converged independently. However, since stream 8 is a source stream for mixers 1 and 2, stream 8 must clearly be determined before attempting to converge the mixer 1 and 2 output streams. With stream 8 determined, streams 2 and 4 must be converged simultaneously since both streams are source streams for mixers 1 and 2. Next, observe that stream 6 is a source stream for mixer 3 only but that mixer 3 also has streams 2 and 4 as source streams. This suggests that stream 6 should be converged last after streams 2 and 4 have been determined. The tear streams should therefore be partitioned into three groups—stream 8 by itself, streams 2 and 4, and stream 6 by itself—and their convergence should be sequenced in that order.

From the preceding example, it is evident that all of the information essential to determining a partition and sequence for the tear streams in a flowsheet can be extracted from the source stream list of the essential mixers. If an essential mixer has only a single source stream, then it can be converged separately prior to the remaining tear streams. If a source stream only was associated with a single mixer but that mixer involved other source streams as well, then that stream can be converged separately but only after all the other source streams associated with that mixer are determined. These two cases can be readily recognized if we construct a source stream table which consists of a row for every essential mixer and a column for every source stream. The table will contain an entry 1 in row I, column J, if the Jth source stream is a source stream for the Ith essential mixer, and 0 otherwise. For instance, the source stream table for Figure 5.20(c) will be:

Mixer	Source Stream			
	2	**4**	**6**	**8**
1	1	1	0	1
2	1	1	0	1
3	1	1	1	0
4	0	0	0	1

The case of the tear stream to be solved first can be recognized by locating a row of the source stream table which contains only a single entry (assuming that process input streams are excluded from consideration). The tear stream to be solved last can be recognized by locating a column of the source stream table which contains only a single entry.

In general, this process of identifying singleton tear streams suitable for early or late convergence can be repeated. Each time an identification is made, a tear stream is eliminated from the list of those yet to be sequenced and others can become candidates for selection. At the conclusion of this identification of singleton tear streams, the remaining mixers will each involve multiple source streams and each stream will be a source stream for multiple mixers. These must be sequenced into simultaneous blocks based on the source streams they have in common. For most flowsheets, this can be done easily by examination of all possible ways of grouping the remaining tear streams in pairs, triplets, and so on. Exhaustive enumeration of groupings is feasible because, in most cases, flowsheets rarely involve more than a total of 10 essential mixers. For systematic approaches suitable for larger problems, the reader is again directed to Westerberg et al.[1] In summary, our tear stream sequencing procedure will consist of the following simple steps.

Tear Stream Sequencing   Given a flowsheet with M essential mixers and its source stream table, set $I = 1$.

Step 1   Select any row of the array which contains a single entry. If none can be found, go to Step 3.

Step 2   Label the column corresponding to that entry I and delete the row and column from further consideration. Set $I = I + 1$ and go to Step 1.

Step 3   Select any column of the array which contains a single entry. If none can be found, go to Step 5.

Step 4   Label the column corresponding to that entry $-I$ and delete the row and column from further consideration. Set $I = I + 1$ and go to Step 1.

Step 5   Determine the blocks of simultaneous tear streams by enumeration. If none can be found, go to Step 7.

Step 6   Label the columns involved in the block I, delete the corresponding rows and columns, set $I = I + 1$, and go to Step 1.

Step 7   All remaining tear streams must be determined simultaneously. The procedure terminates.

The tear streams are to be ordered beginning first with $I = 1$ and proceeding in increasing label magnitude until all positive I are exhausted. Then, continue with the negative label largest in absolute magnitude and proceeding until $I = -1$ is reached. All tear streams with the same label must be converged simultaneously.

Example 5.15   Determine the tear streams and the order in which they are to be converged for the flowsheet shown in Figure 5.21.

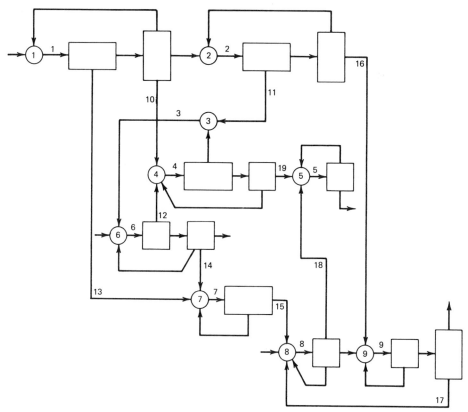

Figure 5.21   Flowsheet for Example 5.15, tear selection and sequencing.

Solution   The flowsheet has nine mixers. The source stream list for the nine mixers, exclusive of process input streams, is readily constructed using simple path tracing.

Mixer	Mixer Output Stream	Source Streams								
1	1	1								
2	2	1	2							
3	3		2	4						
4	4	1		4		6				
5	5			4	5			8		
6	6				3	6				
7	7	1				6	7			
8	8						7	8	9	
9	9		2					8	9	

Of the nine mixers, all are essential except mixer 3.

The source stream array which results when mixer 3 is eliminated is shown in Figure 5.22(a). In Step 1 of the sequencing procedure, row 1 is selected, column 1 is labeled 1, and the row and column are deleted. As a result of eliminating column 1, row 2 only has a single entry. Hence, column 2 is labeled 2 and the corresponding row and column are deleted. At this point, each row has more than one entry. We proceed with Step 3. The fourth column, corresponding to source stream 5, has a single entry; hence, that column is labelled -3. At this point, as seen from Figure 5.20(b), no remaining rows or columns have single entries. We arrive at Step 5, which directs us to search for simultaneous tear streams. We begin the enumeration by considering pairs of streams. The pair consisting of mixers 4 and 6 clearly can be treated as a block. We label the corresponding two columns 4 and delete them from further consideration. We continue with Step 1, and from Figure 5.22(c) it is evident that the mixer 7 row now has only a single entry. The corresponding column is labeled 5 and the row and column are deleted. No singleton rows and columns remain. We return to Step 5 and select the remaining pair to be treated simultaneously. Streams 7 and 8 are both labeled 6, and the procedure terminates

Reconstructing the sequence, stream 1 will be first, followed by stream 2. Next, streams 4 and 6 are determined as a pair, followed by stream 7, and then the pair 8 and 9. Only one stream with negative label remains, namely, stream 5, and it will be converged last. We can verify this sequence by referring to the flowsheet, Figure 5.21. Clearly, streams 1 and 2 are sequential and can be done separately. With 1 and 2 converged, streams 10 and 11 will be known. Mixers 4 and 6 are linked via streams 3 and 12, hence their output streams must be converged simultaneously. With stream 6 converged, stream 14 is known; and from previous calculations, stream 13 is known, hence 7 can be converged by itself. With streams 2 and 7 converged, streams 15 and 16 will be known, and thus streams 8 and 9 can

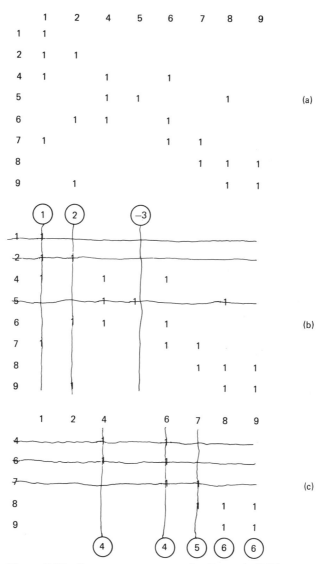

Figure 5.22 Source stream arrays for Example 5.15.

be converged. They must be handled simultaneously because of the coupling through stream 17. Finally, with streams 18 and 19 known, stream 5 can be converged. In this fashion, the entire flowsheet will have been calculated.

In concluding our discussion of sequencing, it is important to note that tear streams which are identified as requiring simultaneous solution can in fact be treated one at a time by *nesting* the flowsheet calculations. For instance, the flowsheet shown in Figure 5.20(b) has two tear streams, 2 and 4, which require simultaneous convergence. From an initial estimate of stream 4, an initial estimate of stream 6 can be obtained by merely executing the two downstream modules. With that estimate of stream 6, stream 2 can be converged. With the resulting value of stream 5, stream 4 can be converged to yield a new estimate of 6. Now, stream 2 must again be reconverged, and the process continues until both 2 and 4 finally are converged. Nesting of the convergence of simultaneous tear streams is the standard technique used to handle such tear streams in the sequential modular approach. Nesting proves to be convenient because it reduces the calculations to the sequential processing of one tear stream at a time. However, this convenience is obtained at the price of having to perform the nested calculations repeatedly.

Finally, we are ready to consider the fourth issue of concern in this section, namely, choice of the numerical scheme to be used to speed convergence of the values of the tear streams. In general, the problem of iterating on the tear streams is equivalent to that of finding a solution to the set of equations

$$x_n = g_n(x_1, x_2, \ldots, x_n) \qquad n = 1, \ldots, N$$

In this formula, each variable x_n can be identified as a species flow of one of the tear streams and each function g_n as the result of the calculations involved in going through the flowsheet to obtain a new value of x_n. In the case of the unconstrained material balance problem, the functions g_n are known to be linear because each individual module involves only linear equations. However, since in the modular or sequential strategy of computation the individual module calculations are carried out within the module subroutines, the explicit expressions for the functions $g_n(x)$ are not available for use in the iterations. Consequently, it is necessary to resort to numerical equation-solving techniques which do not explicitly require the functions or their derivatives. The two most commonly used techniques are direct extensions of the root-finding methods for single variable functions which we discussed in Section 5.1.3, namely, successive substitution and Wegstein's method. The multidimensional extension of Newton's methods will also be discussed, but it is generally not used with the modular approach.

Multidimensional Successive Substitution   In order to carry out successive substitution with a set of equations, it is necessary, as in the single-equation case, to rewrite these equations so that each is in the form

$$x_n = g_n(x_1, x_2, \ldots, x_n)$$

In other words, each equation has one of the variables isolated on the left-hand side and each variable appears on the left-hand side of exactly one equation. In flowsheet material balance calculations involving iterations on tear streams, this form of the equations occurs naturally without special rearrangement. In any case, assuming that the equations are in the above form, the iteration scheme parallels exactly the single-dimensional calculations. Starting with an assumed trial value of the x_n, labeled x^0, each equation $g_n(x)$ is evaluated at x^0 to yield a new value of the corresponding variable x_n. The next vector of x_n values then serves to initiate the next round of evaluations of the $g_n(x)$ functions, and so on, until successive estimates of the x_n do not change significantly.

The algorithm can be formally stated as follows:

Step 0   Select an initial estimate $x_n^0, n = 1, \ldots, N$, and a suitable accuracy criterion $\varepsilon > 0$. For trial $k, k = 1, 2, 3, \ldots$.

Step 1 For $n = 1, \ldots, N$, calculate

$$x_n^{(k+1)} = g_n(x^{(k)})$$

Step 2 Test for convergence. If $|x_n^{(k+1)} - x_n^{(k)}| \leq \varepsilon |x_n^{(k+1)}|$, for each n, $n = 1$, $\ldots, N$, terminate the iterations. The solution to the simultaneous equations is $x_n^{(k+1)}$, $n = 1, \ldots, N$. Otherwise, set $k = k + 1$ and go to Step 1.

Example 5.16 Find a solution to the equations

$$x_1 = 0.2x_1^2 + 0.1x_2 + 0.7$$
$$x_2 = 2(x_1 + 3x_2)^{-1} + 0.5$$

starting at the point $x^0 = (\frac{1}{2}, 2)$ to an accuracy $\varepsilon = 10^{-2}$.

Step 1 $g_1(x^0) = 0.2(0.5)^2 + 0.1(2) + 0.7 = 0.95 = x_1^{(1)}$

$$g_2(x^0) = 2(6.5)^{-1} + 0.5 = 0.8077 = x_2^{(1)}$$

Step 2 Since each component of x is substantially different from the starting values, we proceed with Step 1.

Step 1

$$g_1(x^{(1)}) = 0.2(0.95)^2 + 0.1(0.8077) + 0.7$$
$$= 0.9613 = x_1^{(2)}$$
$$g_2(x^{(1)}) = 2[0.95 + 3(0.8077)]^{-1} + 0.5$$
$$= 1.0929 = x_2^{(2)}$$

The iterations continue in this fashion until the iterate $x^{(6)} = (0.9993, 1.0015)$ is reached. At that point, both coordinates satisfy the criterion

$$|0.9993 - 0.9991| \leq 10^{-2}(0.9993)$$
$$|1.0015 - 0.9963| \leq 10^{-2}(1.0015)$$

From the intermediate results tabulated below, it is evident that the first component satisfies the termination criterion at the fourth iterations; however, the second component does not. Hence, the calculations must continue until both satisfy the test.

n	0	1	2	3	4	5	6
x_1	0.5	0.95	0.9613	0.9941	0.9948	0.9991	0.9993
x_2	2	0.8077	1.0929	0.9717	1.0116	0.9963	1.0015

As in the single-variable case, the method of successive substitution may fail to converge or may converge only very slowly. Moreover, convergence will be influenced by both the choice of starting trial value of x and the form of the functions $g_n(x)$.

Example 5.17   Consider the following modification of the equations of the previous example:

$$x_1 = 0.4x_1^2 + 0.1x_2 + 0.5$$
$$x_2 = 4(x_1 + 3x_2)^{-1}$$

Starting at the same initial point $x^0 = (0.5, 2)$, the first six iterations yield the sequence of values tabulated below.

n	0	1	2	3	4	5	6
x_1	0.5	0.8	0.8175	0.9185	0.9122	0.9594	0.9531
x_2	2	0.6154	1.5116	0.7473	1.2656	0.8494	1.1403

Quite obviously, convergence to the solution $x^* = (1, 1)$ is considerably slower than in Example 5.16. Suppose the iterations are repeated with the starting point $x^0 = (2, \frac{1}{2})$. As shown by the iteration summary given below, the successive values diverge.

n	0	1	2	3	4	5
x_1	2	2.15	2.4633	2.9988	4.138	7.5731
x_2	0.5	1.1429	0.7170	0.8669	0.7144	0.6322

Convergence tests similar to those given in the single-variable case can be derived for the multivariable case. In particular, it can be shown that if the sum of the absolute values of the partial derivatives of each function $g_n(x)$ is less than one at each iterate between the initial point x^0 and the solution x^*, then the iteration will converge. Unfortunately, this type of test is not particularly useful since the solution is not known in advance. Moreover, when this method is applied to iterately solve for the values of torn streams, then the form of the functions $g_n(x)$ is fixed by the selection of tear stream locations. Thus, if a problem fails to converge, two remedies can be employed; either restarting the iterations with another starting estimate of the tear streams or selecting different tear streams. The latter can usually be achieved by selecting a different indexing order for the flowsheet mixers and repeating the mixer-labeling procedures.

Multidimensional Wegstein   The single-variable version of Wegstein's method was developed by using a linear approximation to the function $g(x)$. The slope of this linear function corresponded to the derivative of $g(x)$ and was calculated by using the values of $g(x)$ at two previous iteration points. The same type of construction can be employed to develop a multidimensional extension. However, in this case, since the $g_n(x)$ are functions of several variables, the linear approximations should also be functions of several variables. Thus, given a point $x^{(1)}$, the linear approximation to each $g_n(x)$ should be given by

$$g_n(x) - g_n(x^{(1)}) = S_{n1}(x_1 - x_1^{(1)}) + S_{n2}(x_2 - x_2^{(1)}) + \cdots + S_{nN}(x_n - x_n^{(1)})$$

In the above expression, each slope S_{nj} should in principle be estimated by an approximation to the partial derivative of $g_n(x)$ with respect to x_j, that is,

$$S_{nj} = \frac{g_n(x^{(j)}) - g_n(x^{(1)})}{\Delta x_j} \tag{5.4}$$

where $x^{(j)}$ is a point such that

$$x_i^{(j)} = \begin{cases} x_j^{(1)} + \Delta x_j & \text{if } i = j \\ x_j^{(1)} & \text{if } i \neq j \end{cases} \tag{5.5}$$

and Δx_j is some selected increment in the jth variable. However, this type of approximation would require that the functions $g_n(x)$ be evaluated at N different points before the coefficients S_{nj} could all be calculated. To avoid this computational burden, a further simplification can be made; namely, that $g_n(x)$ can be approximated by the linear function

$$g_n(x) - g_n(x^{(1)}) = S_n(x_n - x_n^{(1)})$$

This is equivalent to assuming that for each function $g_n(x)$, the only significant slope is the one with respect to x_n; all others can be neglected. With this simplifying assumption, the calculation of the S_n values can be carried out using just one additional point:

$$S_n = \frac{g_n(x^{(1)}) - g_n(x^{(0)})}{x_n^{(1)} - x_n^{(0)}}$$

The algorithm which results from this approximation is the exact parallel of· the single-variable version and is given below:

Step 0   Select an initial estimate $x^{(0)}$, a suitable accuracy parameter $\varepsilon > 0$, and bounds α and β. Calculate $x_n^{(1)} = g_n(x^{(0)})$ and $g_n(x^{(1)})$, $n = 1, \ldots, N$.

For $k = 1, 2, 3 \ldots$,

Step 1   Calculate the slopes

$$S_n = \frac{g_n(x^{(k)}) - g_n(x^{(k-1)})}{x_n^{(k)} - x_n^{(k-1)}} \qquad n = 1, \ldots, N$$

Step 2   For $n = 1, \ldots, N$, let $\theta_n = 1/(1 - S_n)$. If $\theta_n > \alpha$, set $\theta_n = \alpha$; if $\theta_n < -\beta$, set $\theta_n = -\beta$. Calculate

$$x_n^{(k+1)} = (1 - \theta_n)x_n^{(k)} + \theta_n g_n(x^{(k)})$$

Step 3 Test for convergence. If $|x_n^{(k+1)} - x_n^{(k)}| \leq \varepsilon|x_n^{(k+1)}|$, for each n, then terminate the iteration. Otherwise, set $k = k + 1$ and go to Step 1.

Example 5.18 Let us apply Wegstein's method to the equations used in Example 5.16. These were

$$x_1 = 0.2x_1^2 + 0.1x_2 + 0.7$$
$$x_2 = 2(x_1 + 3x_2)^{-1} + 0.5$$

with $x^{(0)} = (\frac{1}{2}, 2)$, $\varepsilon = 10^{-2}$, and $\alpha = -\beta = 10$.

Step 0 Calculate

$$x_n^{(1)} = g_n(x^{(0)}) \quad \text{and} \quad g_n(x^{(1)}) \quad n = 1, \dots, N$$

as in the method of successive substitution.

Step 1 Calculate the slopes

$$S_1 = \frac{g_1(x^{(1)}) - g_1(x^{(0)})}{x_1^{(1)} - x_1^{(0)}}$$

$$= \frac{0.9613 - 0.95}{0.95 - 0.5} = 0.0251$$

$$S_2 = \frac{g_2(x^{(1)}) - g_2(x^{(0)})}{x_2^{(1)} - x_2^{(0)}}$$

$$= \frac{1.0929 - 0.8077}{0.8077 - 2} = -0.2392$$

Step 2 Calculate $\theta_1 = 1/(1 - S_1) = 1.0257$ and $\theta_2 = 1/(1 + 0.2392) = 0.8070$. Both θ values are within bounds. Thus,

$$x_1^{(2)} = (1 - 1.0257)(0.95) + 1.0257(0.9613)$$
$$= 0.9616$$
$$x_2^{(2)} = (1 - 0.8070)(0.8077) + 0.8070(1.0929)$$
$$= 1.0378$$

Step 3 The iterates clearly have not converged.

Step 1 Recalculate the slopes:

$$S_1 = \frac{0.9887 - 0.9613}{0.9616 - 0.95} = 2.3621$$

$$S_2 = \frac{0.9908 - 1.0929}{1.0378 - 0.8077} = -0.4437$$

Step 2 Since $\theta_1 = 1/(1 - 2.3621) = -0.7342$ and $\theta_2 = 1/(1 + 0.4436) = 0.6927$, then

$$x_1^{(3)} = (1 + 0.7342)(0.9616) + (-0.7342)(0.9887)$$

$$= 0.9417$$

$$x_2^{(3)} = (1 - 0.6926)(1.0378) + 0.6926(0.9908)$$

$$= 1.0052$$

The iterations continue as summarized below:

n	0	1	2	3	4	5	6
x_1	0.5	0.95	0.9616	0.9417	1.0209	1.0013	1.0009
x_2	2.0	0.8077	1.0378	1.0052	1.0053	1.0052	1.0049

Six iterations are required before the termination criterion is met. In this instance, convergence is not much better than that observed with successive substitution in Example 5.16. Convergence appears to have been retarded in the third iteration, in which a decrease rather than an increase in x_1 is calculated.

In the worst case in which the θ parameters always must be reset to their bounds, Wegstein's method will reduce to a scaled form of successive substitution. Hence, the worst case rate of convergence will be essentially that of *successive substitution*. And the same convergence conditions that are applicable to the latter method will apply to Wegstein's method. Furthermore, Wegstein's method may encounter difficulties if the slope S_n does not dominate the slopes associated with the other variables which are neglected in deriving the method. In practice, a check of the dominance assumption requires either calculation of the S_{nj} or evaluation of all of the partial derivatives of the functions $g_n(x)$. If $1 > |S_{nn}| >> |S_{nj}|\ j \neq n$ $n = 1, \ldots, N$, or, more directly, if $1 > |\partial g_n(x)/\partial x_n| >> |\partial g_n(x)/\partial x_j|\ j \neq n,\ n = 1, \ldots,$ N, over the range of x of interest, then Wegstein's method will converge successfully. In spite of the shortcomings of this technique, it remains the iterative algorithm used most commonly in conjunction with the sequential strategy for computerized material balancing.

For the sake of completeness, we conclude this section with a brief discussion of the multidimensional extension of Newton's method applied to equations of the form $x = g(x)$.

The Newton–Raphson Method A set of equations in the form $x_n = g_n(x)$, $n = 1, \ldots, N$ can, as in the development of Wegstein's method, be approximated by a linear approximation constructed at a point $x^{(1)}$. If the partial derivatives of the $g_n(x)$ functions can be readily computed, then a good linear approximation will be given by

$$g_n(x) - g_n(x^{(1)}) = \sum_{j=1}^{N} \left(\frac{\partial g_n}{\partial x_j}\right)_{x=x^{(1)}} (x_j - x_j^{(1)}) \qquad n = 1, \ldots, N$$

Since these approximating functions are to be used to obtain a solution satisfying $x_n = g_n(x)$, $n = 1, \ldots, N$, the term $g_n(x)$ in the above expression can be replaced by x_n. If the resulting expression is rearranged by grouping the terms in the unknowns x_j, the following equation is obtained:

$$\left[1 - \left(\frac{\partial g_n}{\partial x_n}\right)_{x = x^{(1)}}\right] x_n - \sum_{\substack{j=1 \\ j \neq n}}^{N} \left(\frac{\partial g_n}{\partial x_j}\right)_{x = x^{(1)}} x_j$$

$$= g_n(x^{(1)}) - \sum_{j=1}^{N} \left(\frac{\partial g_n}{\partial x_j}\right)_{x = x^{(1)}} x_j^{(1)} \qquad n = 1, \ldots, N \quad (5.6)$$

Equations (5.6) are a set of linear equations in the unknowns x_n which can be solved using the row reduction procedure discussed in Section 4.4.2. The solution obtained can then be used to calculate new values of the partial derivatives and to formulate a new set of linear equations which are then solved to obtain a new estimate of the x_n values.

The complete algorithm can be summarized in the following steps:

Step 0 Select $x^{(0)}$ and $\varepsilon > 0$.

For $k = 1, 2, 3, \ldots$,

Step 1 Calculate

$$\left(\frac{\partial g_n}{\partial x_j}\right)_{x = x^{(k)}} \qquad \begin{array}{l} n = 1, \ldots, N \\[4pt] j = 1, \ldots, N \end{array}$$

and set up eqs. (5.6).

Step 2 Solve eqs. (5.6) to obtain a solution $x^{(k+1)}$.

Step 3 Check for termination. If $|g_n(x^{(k)}) - x_n^{(k+1)}| \leq \varepsilon$ for each $n = 1, 2, \ldots, N$, then the iterations have converged. Otherwise, set $k = k + 1$ and go to Step 1.

Example 5.19 Solve the set of equations

$$x_1 = 0.25x_2^2 + 0.75 \equiv g_1(x)$$
$$x_2 = 3(2x_1 + x_2)^{-1} \equiv g_2(x)$$

using Newton's method starting at the point $(2, 2)$ and using the accuracy criterion $\varepsilon = 5 \times 10^{-3}$.

The partial derivatives of the functions are

$$\left(\frac{\partial g_1}{\partial x_1}, \frac{\partial g_1}{\partial x_2}\right) = (0, 0.5x_2)$$

$$\left(\frac{\partial g_2}{\partial x_1}, \frac{\partial g_2}{\partial x_2}\right) = (-6(2x_1 + x_2)^{-2}, -3(2x_1 + x_2)^{-2})$$

Step 1 At $x^{(0)} = (2, 2)$, the partial derivative values are

$$\left(\frac{\partial g_1}{\partial x_1}, \frac{\partial g_1}{\partial x_2}\right) = (0, 1)$$

$$\left(\frac{\partial g_2}{\partial x_1}, \frac{\partial g_2}{\partial x_2}\right) = \left(-\tfrac{1}{6}, -\tfrac{1}{12}\right)$$

and the function values are

$$g_1(x^{(0)}) = 1.75$$
$$g_2(x^{(0)}) = 0.5$$

Equations (5.6) thus take the form

$$(1 - 0)x_1 - (1)x_2 = 1.75 - 0(2) - 1(2) = -0.25$$
$$\tfrac{1}{6}x_1 + (1 + \tfrac{1}{12})x_2 = 0.5 + \tfrac{1}{6}(2) + \tfrac{1}{12}(2) = 1.0$$

or, in array form,

$$\begin{pmatrix} 1 & -1 \\ \tfrac{1}{6} & \tfrac{13}{12} \end{pmatrix} \begin{pmatrix} x_1 \\ x_2 \end{pmatrix} = \begin{pmatrix} -0.25 \\ 1.0 \end{pmatrix}$$

Step 2 Solution of these linear equations yields

$$x_1^{(1)} = 0.5833$$
$$x_2^{(1)} = 0.8333$$

Convergence has clearly not been attained; hence, another iteration must be initiated.

Step 1 At $x^{(1)}$,

$$\left(\frac{\partial g_1}{\partial x_1}, \frac{\partial g_1}{\partial x_1}\right) = (0, 0.4166)$$

$$\left(\frac{\partial g_2}{\partial x_1}, \frac{\partial g_2}{\partial x_2}\right) = (-1.5, -0.75)$$

and

$$g_1(x^{(1)}) = 0.9236$$
$$g_2(x^{(1)}) = 1.50$$

The array form of eqs. (5.6) becomes

$$\begin{pmatrix} 1 & -0.4166 \\ 1.5 & 1.75 \end{pmatrix} \begin{pmatrix} x_1 \\ x_2 \end{pmatrix} = \begin{pmatrix} 0.5764 \\ 3.000 \end{pmatrix}$$

Step 2 Solution of the equations yields

$$x_1^{(2)} = 0.9510$$
$$x_2^{(2)} = 0.8991$$

Repeating the iterations for one more cycle results in

$$x_1^{(3)} = 0.9967$$
$$x_2^{(2)} = 0.9982$$

Since $g_1(x^{(3)}) = 0.9991$ and $g_2(x^{(3)}) = 1.0028$, the convergence criteria

$$|0.9991 - 0.9967| \leqslant 5 \times 10^{-3}$$
$$|1.0028 - 0.9982| \leqslant 5 \times 10^{-3}$$

are clearly met.

The same problem when solved using Wegstein's method requires 13 iterations to converge to the same tolerance. A sample of the iterations is shown below.

n	0	1	2	3	5	8	11	13
x_1	2	1.75	0.8125	0.8734	0.8592	0.9764	1.0058	0.9974
x_2	2	0.5	0.7143	1.2824	1.0884	0.9817	0.9902	1.0046

The Wegstein iterations converge slowly because the assumption concerning the dominance of the S_n terms is not satisfied. For instance, at $x^{(0)} = (2, 2)$,

$$\left|\frac{\partial g_1}{\partial x_1}\right| = 0 < \left|\frac{\partial g_1}{\partial x_2}\right| = 1$$

and

$$\left|\frac{\partial g_2}{\partial x_2}\right| = \tfrac{1}{12} < \left|\frac{\partial g_2}{\partial x_1}\right| = \tfrac{1}{6}$$

and similarly at $x^* = (1, 1)$.

Newton's method is generally superior to the other two algorithms. However, it can be used only if either the partial derivatives can be evaluated or if they can be approximated by differences via eqs. (5.4) and (5.5). The latter approach is costly in terms of the number of times the functions $g_n(x)$ need to be evaluated. Since in applications involving iteration on tear streams a set of function evaluations corresponds to a complete calculation cycle through the flowsheet, it is clear why Newton's method is generally not used despite its superior convergence property. *The CONVER Module* In a flowsheet represented in terms of elementary modules, the direction of the streams also corresponds to the direction of the calculation.

As we noted earlier, the calculations will be initiated at specific locations, the tear streams, and will proceed downstream until the tear streams are again encountered. At that point, the iterative procedures discussed in the previous section come into play to predict better estimates of the tear streams to be used in the next cycle of calculations. Since the flowsheet doubles to indicate the order of module calculations, it is convenient to insert into the flow diagram, at the tear stream locations, a special calculation module which will execute the numerical equation-solving procedures required to update the tear stream values. This special purpose subroutine, or module, which we shall call CONVER, will thus contain one or more of the iterative procedures discussed previously. This is the mechanism by means of which these procedures are typically executed in modular material balancing computer programs.

The external structure of module CONVER is quite similar to that of the other four elementary modules, that is, it has a single input and a single output stream by means of which it is connected to the calculation flow diagram, or flowsheet. The output stream of the CONVER module is the tear stream, while the input stream is a fictitious stream representing the values of the tear stream obtained as the result of a cycle of calculations through the flowsheet. For instance, in Example 5.12, mixer 1 is the only essential mixer, and thus stream 7 is the tear stream. The execution of a predictive calculation to update stream 7 will be represented by inserting a fictitious stream, stream 16, and the CONVER module into the flowsheet between stream 7 and mixer 1, as shown in Figure 5.23. The position of the CONVER module serves to mark the point in the flowsheet at which the sequence of flowsheet calculations is to be initiated. That is, the flowsheet calculations will commence with a trial value of stream 7, and the elementary module calculations will be carried out in the direction of flow until all inputs to mixer 1 have been calculated. At that time, the mixer 1 calculation will be made to give stream 16. Finally, CONVER will be executed to predict the value of stream 7 to be used for the next cycle of computations. If successive substitution is used, then CONVER will simply transfer the values of stream 16 into stream 7.

On the other hand, if Wegstein's method is used, then CONVER will use the current value of streams 7 and 16 and the values of streams 7 and 16 saved from the previous cycle to predict the new value of stream 7. In the notation used in defining Wegstein's method, the current value of stream 7 can be identified as $x^{(k)}$ and the current value of stream 16, as $g(x^{(k)})$. The values of streams 7 and 16 saved from the previous cycle of iterations will correspond to $x^{(k-1)}$ and $g(x^{(k-1)})$, respectively.

If Newton's method is to be used in CONVER, then each prediction of the new value of stream 7 will require that N cycles of calculations through the flowsheet be carried out while incrementing the species flows in stream 7 one at a time. These intermediate results must all be stored for use in the Newton calculation. The CONVER subroutine thus is the fifth module which might be made available with any modular material balancing program.

The next example illustrates the formulation and solution of a material balance problem using the modular concept in its entirety.

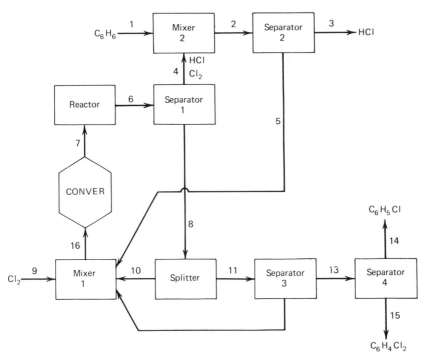

Figure 5.23   Flowsheet 5.19 with tear stream iteration module inserted.

Example 5.20   Acetic anhydride can be produced by thermal cracking of acetic acid to ketene, which is then reacted with additional acetic acid to form the anhydride. The desired reaction in the cracking furnace is

$$CH_3COOH \rightarrow CH_2CO + H_2O$$

But this reaction is accompanied by a side reaction,

$$CH_3COOH \rightarrow CH_4 + CO_2$$

Experimental studies have shown that, for a particular cracking temperature and pressure, fractional yield is a direct function of conversion of acetic acid (HAc).

For instance, at 700°C, the following experimental data have been reported.[2]

Fractional yield CH_2CO	0.03	0.037	0.0522	0.062	0.067	0.0722	0.08	0.09	0.101	0.11
Conversion HAc		0.3	0.4	0.6	0.7	0.75	0.8	0.85	0.9	0.95	1.0

The product stream from the cracking furnace is cooled rapidly, or quenched, and contacted with glacial acetic acid in a quench tower in which both absorption of ketene and the reaction

$$CH_2CO + CH_3COOH \rightarrow (CH_3CO)_2O$$

occur. The liquid stream leaving the quench tower consists of a mixture of two phases: an organic phase containing acetic acid, anhydride, and any unreacted ketene; and an aqueous phase consisting primarily of H_2O and acetic acid. This two-phase mixture is separated in a decanter. The organic phase is next sent to the acid recovery tower, the overhead of which is recycled to the cracking furnace. The bottoms stream is the desired anhydride product. The aqueous phase from the decanter contains a considerable amount of acetic acid which is recovered in the acid drying tower. This concentrated acid is also recycled to the cracking reactor.

In the flowsheet shown in Figure 5.24, the quench unit is operated so that all

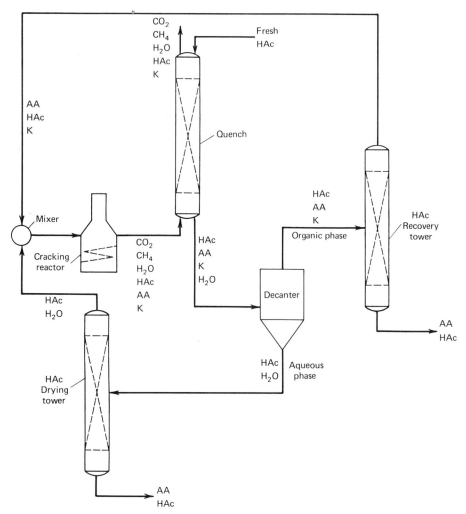

Figure 5.24 Flowsheet for Example 5.20, acetic anhydride plant.

of the CO_2 and CH_4, 5% of the H_2O, and 1% each of the unreacted acetic acid and ketene appear in the vent gas. Furthermore, the organic phase leaving the decanter contains 5% of the acetic acid and all of the ketene and anhydride fed to that unit. The acid recovery tower is operated so that 99% of the acetic acid, 1% of the anhydride, and all of the ketene fed appear in the overhead stream. The acid drying tower will operate so that 6% of the acetic acid and 96% of the H_2O fed to the unit will appear in the bottoms stream.

Calculate all streams in the process assuming 85% conversion of acetic acid in the cracking reactor, 95% conversion of ketene in the quench, and a process feed of 100 lbmol acetic acid per hour.

Solution   The first step in using the sequential strategy to solve this problem is to reformulate the flowsheet in terms of the four elementary modules. The only unit which requires reformulation is the quench unit. This unit involves mixing of two feed streams, reaction of ketene, and separation of the liquid and gas product streams. The flowsheet in elementary module form is shown in Figure 5.25.

The next step in analyzing the flowsheet is to determine if it has degree of freedom zero, or, equivalently, if all of the natural specifications are given. First, the stoichiometry of all three reactions is given as well as the conversion of the key reactants, acetic acid and ketene. From the given experimental data and the known 85% HAc conversion, the fractional yield of CH_2CO is known. Thus, in effect all of the natural reactor specifications have been assigned. All of the split fractions are known from the problem statement and are summarized in the table given below.

	Split Fractions (mol in overhead/mol in feed)			
	Sep1	**Sep2**	**Sep3**	**Sep4**
Acetic acid	0.01	0.05	0.99	0.94
Ketene	0.01	1.0	1.0	0.0
H_2O	0.05	0.0	0.0	0.04
CH_4	1.0	0.0	0.0	0.0
CO_2	1.0	0.0	0.0	0.0
Anhydride	0.0	1.0	0.01	0.0

Finally, the feed rate of acetic acid is also specified. Since all the natural specifications are known, the problem must have degree of freedom zero and is an unconstrained material balance problem.

The third analysis step is to determine if stream tearing and converging will be necessary. It is apparent from Figure 5.25 that the flowsheet contains two mixers. However, to verify whether both are essential, it is necessary to employ the mixer-labeling procedure. Suppose we apply that procedure to the mixers in the order in which they are indexed in the flowsheet.

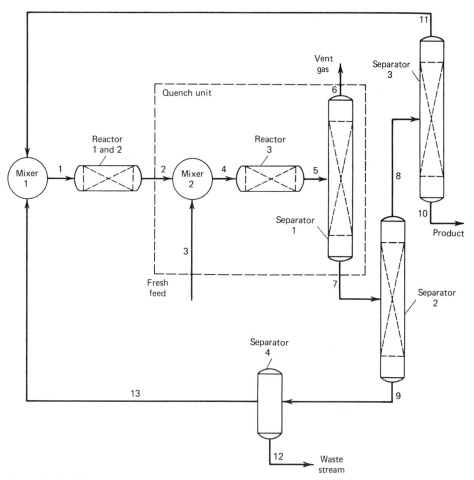

Figure 5.25 Flowsheet for Example 5.20 in elementary module form.

The branch source stream lists for the two mixers, exclusive of process input streams, are:

Mixer	Mixer Output Stream	Source Streams
1	1	4
2	4	1

If we begin the procedure with mixer 1, it will immediately be labeled nonessential. Stream 1 will be replaced by stream 4 in the source stream list for mixer 2. Mixer 2 remains essential because its source stream is also its output stream. Thus, only one tear stream is required, and stream 4 can be used for that purpose.

With the tear stream located, the flowsheet is next revised to show the location of the CONVER module. The revised flowsheet, Figure 5.26, indicates the order which is to be followed in executing the elementary modules in a computer program. Thus, the computations commence with a trial value of stream 4; proceed with the reactor 3, separator 1, and separator 2 calculations; branch to do the separator 3 and separator 4 calculations; and finish with the mixer 1, reactors 1 and 2, and mixer 2 calculations. The result is stream 14. If CONVER uses successive substitution, then stream 4 is set equal to stream 14 and another cycle is begun.

If CONVER uses Wegstein's method, then a new value of stream 4 is calculated which, in general, is different from stream 14. In either case, the CONVER subroutine will contain a termination test which checks whether the iterations have converged. If not, further cycles of computation are initiated until convergence is

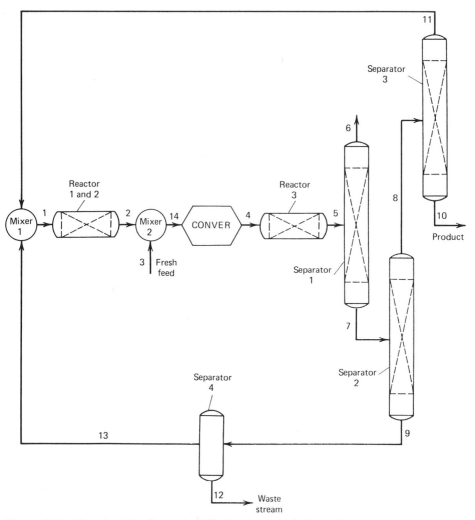

Figure 5.26 Flowsheet for Example 5.20 showing calculation order.

achieved. A summary of the successive values of stream 4 generated using Wegstein's method is given in Figure 5.27 and the complete stream table of the final solution, in Figure 5.28. The calculations converged to an accuracy of 1×10^{-5} in 16 iterations from the selected initial value of stream 4.

By contrast, suppose the solutions had been carried out manually. Then, from the degree-of-freedom table given in Figure 5.29 and constructed by referring to the flowsheet in Figure 5.24, a sequential solution, in which three variables would have to be carried along, would be required. The calculation order is shown in Figure 5.30.

Iteration No.	Species Flows (lbmol/h)					
	1	2	3	4	5	6
Acetic acid	50	100.35	113.71	115.25	115.16	115.16
Ketene	50	2.63	6.35	7.36	7.22	7.22
Water	50	2.06	6.30	7.25	7.13	7.13
Methane	50	1.82	71.40	79.39	78.94	79.01
Dioxide	0	1.82	71.40	79.39	78.94	79.01
Anhydride	0	0.48	0.03	0.06	0.08	0.06
Total flow	200	109.17	269.14	288.69	287.47	287.61

Figure 5.27 Wegstein iteration sequence, stream 4.

	Cracking Reactor	Quench	Decanter	HAc Recovery	HAc Drying	Mixer	Process	Overall
Number of stream variables	10 + 2	16 + 1	9	8	6	9	37	10 + 3
Number of balances	6	6	4	3	2	4	25	6
Specifications								
Conversions	1	1					2	
Selectivity	1						1	
Split fractions								
Quench		3					3	
Decanter			1				1	
HAc recovery				2			2	
HAc drying					2		2	
Degree of freedom	4	7	4	3	2	5	1	7
Basis					−1		−1	
					1		0	

Figure 5.29 Degree-of-freedom table, Example 5.20.

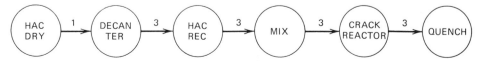

Figure 5.30 Calculation order, Example 5.20.

Stream No.	1	2	3	4	5	6	7	8	9	10	11	12	13	14	15
Total flow (lbmol/h)	101.75	187.63	100.0	287.63	280.76	159.47	121.29	12.66	108.64	6.92	5.73	12.63	96.01	287.63	108.62
Composition (mole fraction)															
CH_3COOH	0.9931	0.0808	1.000	0.4004	0.3857	0.0068	0.8839	0.4236	0.9375	0.0077	0.9255	0.4840	0.9972	0.4004	0.8670
CH_2CO	0.0035	0.0385	0	0.0251	0.0013	0.0000	0.0030	0.0283	0	0	0.0624	0	0	0.0251	0.0666
H_2O	0.0027	0.0381	0	0.0248	0.0254	0.0022	0.0559	0	0.0625	0	0	0.5160	0.0028	0.0248	0.0658
CH_4	0	0.4211	0	0.2747	0.2814	0.4955	0	0	0	0	0	0	0	0.2747	0
CO_2	0	0.4211	0	0.2747	0.2814	0.4955	0	0	0	0.9923	0.0121	0	0	0.2747	0
$(CH_3CO)_2O$	0.0007	0.0004	0	0.0002	0.0247	0	0.0572	0.5481	0	0	0	0	0	0.0002	0.0006

Figure 5.28 Final stream table for Example 5.20.

The Constrained Material Balance Problem The developments presented in the previous sections all rest upon the assumption that the process flowsheet can be expressed in terms of elementary modules and that all process input streams, reaction stoichiometry and rates, as well as all split fractions are specified. Yet, as can be seen from many of the examples discussed in Chapters 2 through 4, flowsheet specifications can take on many different forms. For instance, stream compositions, composition ratios, species flows, and ratios of species flows of internal process streams might well be specified. Since such alternate specifications are readily accommodated in manual calculations, it is reasonable to expect that the modular computer-aided strategy ought to be able to accommodate them as well. In this section, we find that, while the sequential modular strategy can in principle accommodate all of these various specifications, in practice the presence of such specifications increases the computational burden considerably.

For convenience of reference, let us introduce two definitions. A *constraint specification* is any condition imposed on the species flows which is not one of the natural specifications. For example, the assignment of a value to the composition of a species in a stream is a constraint specification. A *constrained* material balance problem is one in which at least one specification is a constraint.

A constrained problem may be viewed as an unconstrained problem in which one or more of the natural specifications are omitted and are replaced by constraints. Since unconstrained material balance problems always have degree of freedom zero, it follows that any correctly specified constrained problem must always have exactly as many independent constraint specifications as there are missing natural specifications. This simple observation allows the degree-of-freedom analysis for a flowsheet expressed in elementary modules to be simplified considerably. Instead of counting variables, balances, and specifications individually, it suffices to count the number of given constraint specifications and the missing natural specifications. The degree of freedom of the problem will simply be the difference; and, as before, if the difference is zero, the problem is correctly specified.

As we noted earlier, counting of the natural specifications is straightforward:

1. For every reactor module, there must be as many reaction rates or their equivalents specified as there are independent reactions.
2. For every separator module, there must be a split fraction specified for every species. Hence, if the number of output streams is J, there must be $(J - 1)S$ split fractions specified.
3. For every splitter, there must be $J - 1$ stream split fractions, since the split fractions associated with any given stream must by definition all be the same.
4. All process input streams must be specified.

Thus, the main effort in applying the above simple degree-of-freedom calculation rule is in reformulating the flowsheet into elementary module form.

Example 5.21 Consider the acetaldehyde plant discussed in Example 5.1. The flowsheet given in Figure 5.1 is already in elementary module form. Note that a number of constraint specifications are given in the problem statement. Hence, the problem is clearly constrained. To determine its degree of freedom, let us first count the number of natural specifications missing for each unit;

> Splitter: None missing since the split ratio is given.
> Separator 1: None missing since each component appears in only one output stream.
> Separator 2: None missing since each component appears in only one output stream.
> Separator 3: All four split fractions are unspecified.
> Mixer: One ideal specification, the flow of stream 12, is missing. The flow of stream 11 can be chosen as basis; and, since the composition is known, the species flows are known.
> Reactor: All five reaction rates are missing.

The total number of missing natural specifications is the sum of these and is equal to 10. The constraint specifications given are the two species ratios in stream 1, the seven compositions in stream 2 (including the implied specification that $N_{O_2}^2 = 0$), and the CO-to-CO$_2$ ratio in stream 10. The total number of constraints is thus 10, and the degree of freedom is 10 minus 10, or zero.

The above procedure is clearly simpler than the detailed accounting method which we employed for manual calculations, and thus it should be used in preference whenever a problem has been cast into elementary module form.

Having determined the degree of freedom of a constrained problem, we next need to deduce a way of solving it using the given elementary modules. Since for a correctly specified constrained problem the numbers of missing natural and given constraint specifications are equal, the constrained problem can be decomposed into an unconstrained problem with a certain number of unknown natural parameters and an equal number of side conditions. If the unknown natural parameters were assigned values, then the problem would be unconstrained and could be solved as such. However, the choice of natural parameter values which can be made is not free. Rather, the parameter values must be such that the constraint specifications are also satisfied. This suggests an iterative procedure in which the unknown natural parameters are assigned values, the corresponding unconstrained problem is solved, and then the constraint specification equations are used to predict improved values for the natural parameters. The process would continue until a set of natural parameters is obtained at which the constraint specifications are met within suitable tolerances. This of course amounts to an implicit simultaneous solution of the constraint specifications for the unknown natural parameters. Before developing a general solution for this problem, let us examine a small example.

Example 5.22 Consider a separator which divides a known stream containing acetic acid and water into an acetic acid-rich and a water-rich stream. The feed contains 60 mol/h acetic acid and 40 mol/h water. The separator will recover 90% of the acetic acid in the overhead, and the bottoms stream is to contain 80% (mol) of water.

The problem is clearly constrained since the split fraction, t, of H_2O is not given; instead, the bottoms composition is specified. This specification can be written as

$$0.2N_{H_2O}^B = 0.8N_{HAC}^B$$

The balance equations for the bottoms stream written in the separator module have the form

$$N_{H_2O}^B = t(40)$$

$$N_{HAC}^B = 0.1(60)$$

Now, if the latter equations could be substituted into the constraint specification, we would obtain an equation for t:

$$t(40) = 4[0.1(60)]$$

which could be directly solved for the unknown split fraction t. With the split fraction known, we could return to the separator module and calculate both the bottoms and the overhead streams. However, since we are using the modular approach, the balance equations for the bottoms stream are not available explicitly. All that can be done is to guess a value of t, to have the separator module calculate the resulting bottoms stream, and then to verify whether the equation

$$0.8N_{HAC}^B - 0.2N_{H_2O}^B = 0$$

is satisfied. If the equation is positive, then $N_{H_2O}^B$ is too large and t should be reduced. If the equation is negative, $N_{H_2O}^B$ is too small and t should be increased.

We can recognize this process as nothing more than a trial-and-error procedure for solving an equation of the form

$$f(t) = 0$$

where the dependence on t is implicit rather than explicit.

Let us apply one of the root-finding methods of Section 5.1.3 to solve for the appropriate value of t. Since our function

$$f(t) = 0.8N_{HAC}^B - 0.2N_{H_2O}^B = 0$$

is not explicit in t, one of the methods which do not require rearrangement to the form $t = g(t)$ is more convenient. Accordingly, let us use the secant method, with the initial estimates $t = \frac{1}{2}$ and $t = \frac{3}{4}$.

At the first point,

$$N_{H_2O}^B = t(40) = 20$$

$$N_{HAC}^B = 0.1(60) = 6$$

Therefore,

$$f(\tfrac{1}{2}) = 0.8(6) - 0.2(20) = 0.8$$

Similarly, at $t = \tfrac{3}{4}$,

$$f(\tfrac{3}{4}) = -1.2$$

Step 1

$$\frac{df}{dt} \approx \frac{f(\tfrac{3}{4}) - f(\tfrac{1}{2})}{0.75 - 0.5} = \frac{-1.2 - 0.8}{.25} = -8$$

Step 2

$$t^{(3)} = t^{(2)} - \left(\frac{df}{dt}\right)^{-1} f(t^2) = 0.75 - (-8)^{-1}(-1.2) = 0.6$$

Step 3 At $t^{(3)} = 0.6$,

$$N_{H_2O}^B = 0.6(40) = 24$$
$$N_{HAC}^B = 0.1(60) = 6$$

and therefore,

$$f(0.6) = 0.8(6) - 0.2(24) = 0$$

Since the root of the equation has been found, the iterations terminate with 0.6 as the correct water split fraction.

Note that the iterations converge rapidly to the exact root because the dependence of f on t is actually linear. Consequently, the derivative of f is a constant and the secant approximation is exact.

The preceding example suggests a general calculation procedure in which constrained problems are solved as a sequence of unconstrained problems. After each unconstrained problem solution, the unknown natural parameters will be updated, in general, by applying a multidimensional root-finding method to the constraint specification relations.

Let the functions $f_n(x)$, $n = 1, \ldots, N$, represent N constraint specifications and the variable x represent the N unknown natural parameters. The calculation procedure for constrained problems can then be summarized in the form of a flow chart as given in Figure 5.31. Note that this solution procedure consists of a double loop of iterations since the unconstrained material balance problem itself usually requires convergence calculations. In the next example, this procedure is illustrated in detail.

Example 5.23   A simple process, in which chemical A is converted into chemical B, consists of a mixer, a reactor, and a separator as shown in Figure 5.32(a). The product separator is operated so that 80% of the A and 40% of the B in the reactor

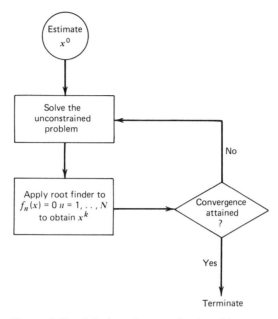

Figure 5.31 Solution of constrained problems.

outlet stream are recycled. The reactor operating conditions are adjusted so that the ratio of B to A in the reactor feed stream is 1 : 5 when a process feed of pure A is used. Assuming that the reaction stoichiometry is simply

$$A \rightarrow B$$

calculate the reactor conversion required to achieve these process conditions.

Solution We first note that if the flow of stream 1 is chosen to be 100 mol/h of A, then the only natural specification which is not given is the conversion. However, since the species ratio in stream 2 is specified, the problem has degree of freedom zero. Moreover, the problem is constrained and will have to be solved by iterating on the unknown conversion of A in the reactor. At each iteration, the corresponding unconstrained problem will have to be solved; hence, we need to check whether that calculation will itself require iteration. The flowsheet clearly has only a single mixer; and, since stream 5 can be traced back to the mixer via streams 3 and 2, the mixer is essential. Thus, solution of the problem will require iteration on stream 2. The calculation order will be as shown in Figure 5.32(b), with the convergence module inserted between the mixer and the reactor. Commencing with an estimate of stream 2 and of the conversion of A in the reactor, the reactor module will calculate stream 3. The separator module will yield stream 5 and the mixer stream 6. At that point, the convergence module will generate a new estimate of stream 2 and the cycle will continue until convergence is achieved.

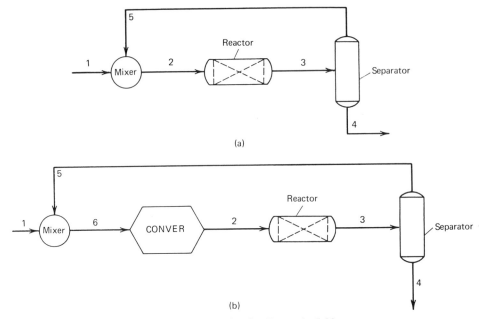

(a)

(b)

Figure 5.32 Flowsheet and calculation order for Example 5.23.

Let us carry out one cycle of such calculations. Recall that the reactor module equations will simply be

$$N_A^3 = (1 - X)N_A^2$$
$$N_B^3 = N_B^2 + XN_A^2$$

and the separator equations for stream 5,

$$N_A^5 = 0.8N_A^3$$
$$N_B^5 = 0.4N_B^3$$

The mixer will simply add stream 5 to the feed stream, that is,

$$N_A^6 = N_A^5 + 100$$
$$N_B^6 = N_B^5 + 0$$

Suppose we choose as initial estimates $X = 1$ and $N_A^2 = 100$, $N_B^2 = 0$. Then, the reactor module yields

$$N_A^3 = 0$$
$$N_B^3 = 100$$

and the separator module,

$$N_A^5 = 0$$
$$N_B^5 = 40$$

Finally, the mixer results are

$$N_A^6 = 100$$

$$N_B^6 = 40$$

If CONVER employs successive substitution, then

$$N_A^2 = N_A^6 = 100$$

$$N_B^2 = N_B^6 = 40$$

Repeating with the next cycle of calculations, we obtain

$$N_A^3 = 0 \qquad N_A^5 = 0 \qquad N_A^6 = 100$$

$$N_B^3 = 140 \qquad N_B^5 = 56 \qquad N_B^6 = 56$$

Convergence has still not been obtained since N_B^2 has changed from 40 to 56. If the iterations are repeated, the results tabulated below are obtained:

n	0	1	2	3	4	5	6	7	8	9
N_B^2	0	40	56	62.4	64.96	65.98	66.39	66.56	66.62	66.65

The iterations converge to $N_B^2 = 66.67$ mol/h.

The unconstrained problem has now been solved for the assumed value of the conversion, $X = 1$. At that solution, the constraint

$$\frac{N_B^2}{N_A^2} = \frac{1}{5}$$

is not satisfied, that is,

$$\frac{N_B^2}{N_A^2} = \frac{66.67}{100} = 0.6667$$

Consequently, we now must apply a root-finding technique to the constraint expressed in the form

$$f(X) = N_B^2 - 0.2N_A^2$$

to determine a new estimate of the conversion X with which to resolve the unconstrained problem.

If the secant method is employed, then the recycle calculation should be repeated with another estimate of X. With $X^{(2)} = 0.2$, the converged values of stream 2 will be

$$N_A^2 = 277.78$$

$$N_B^2 = 37.04$$

and the constraint value will be

$$f(0.2) = -18.52$$

The secant formula will yield the new conversion estimate

$$X^{(3)} = 0.427$$

The inner iteration on stream 2 followed by the outer iteration on the conversion is continued until a selected termination criterion on X is met. The outer iteration results on X are summarized in tabular form below:

n	3	4	5	6	7
X	0.4273	0.3231	0.3042	0.2998	0.3000
N_A^2	184.56	218.10	225.55	227.36	227.27
N_B^2	52.57	46.98	45.74	45.44	45.45

The converged value of the conversion is thus 0.3, and with this value the required B-to-A ratio is met.

While the above iterative strategy for accommodating constraint specifications within the modular balancing approach is reasonably successful when the number of such specifications is small, it becomes increasingly more prone to failure as this number increases. The difficulties arise from three sources:

1. Increased likelihood that the inner problem iterations will fail to converge.
2. Increased likelihood that the outer iterations on the constraints will fail to converge, especially if these conditions are nonlinear.
3. Substantially increased computation times, even if the inner and outer iterations do converge.

These difficulties can in part be mitigated by iterating on the tear streams and on the constraints simultaneously and by using more sophisticated equation solving methods. These more advanced modular strategies are the subject of continuing research.

5.2.2 The Simultaneous Solution Strategy

As suggested in the preceding discussion, if the flowsheet involves both a large number of constraint specifications and requires iteration on a sizable number of tear streams, then the sequential solution strategy can become rather inefficient. It becomes advantageous to employ the computer to solve the material balances for the entire flowsheet simultaneously. In this section, we consider the advantages and disadvantages of simultaneous solution and illustrate the basic concepts involved with several examples.

Linear Material Balance Problems Simultaneous machine solution of material balance problems can be especially attractive if all of the constraint specifications are linear in the species flows. In those cases, the balance equations and specifications set will simply comprise a large set of linear equations. Linear equations can be solved quite efficiently, without convergence cycles, using variants of the row reduction procedure discussed in Section 4.4.2. Moreover, computer programs which carry out these calculations are commonly available as part of the program library on most computer installations. The disadvantages of the simultaneous approach are that it can lead to large computer memory requirements, unless special storage schemes are employed, and that, unless suitable balance equation generation programs are available, the engineer is faced with the tedious task of explicitly coding each equation and side condition.

Example 5.24 Consider the flowsheet problem solved iteratively in Example 5.23. The balance equations for the three units will simply consist of:

Mixer balances:

$$N_A^2 = N_A^5 + 100 \tag{5.8}$$

$$N_B^2 = N_B^5 \tag{5.9}$$

Reactor balances:

$$N_A^3 = N_A^2 - r \tag{5.10}$$

$$N_B^3 = N_B^2 + r \tag{5.11}$$

Separator balances:

$$N_A^3 = N_A^4 + N_A^5 \tag{5.12}$$

$$N_B^3 = N_A^4 + N_A^5 \tag{5.13}$$

In addition to the balance equations themselves, the species flows are related via the split fraction specifications

$$N_A^5 = 0.8 N_A^3 \tag{5.14}$$

$$N_B^5 = 0.4 N_B^3 \tag{5.15}$$

and the composition relation

$$N_A^2 = 5 N_B^2 \tag{5.16}$$

This is a set of nine linear equations, involving eight species flows and the unknown reaction rate, which can be solved directly using a row reduction procedure. We shall illustrate the calculations involved by solving the 7×7 system obtained when the equations containing N_A^4 and N_B^4, that is, the separator balances, are deleted. The seven equations in detached coefficient form are

	N_A^2	N_B^2	N_A^5	N_B^5	N_A^3	N_B^3	r	RHS
(5.8)	1	0	−1	0	0	0	0	100
(5.9)	0	1	0	−1	0	0	0	0
(5.10)	−1	0	0	0	1	0	1	0
(5.11)	0	−1	0	0	0	1	−1	0
(5.14)	0	0	1	0	−0.8	0	0	0
(5.15)	0	0	0	1	0	−0.4	0	0
(5.16)	1	−5	0	0	0	0	0	0

The column headed RHS corresponds to the constant right-hand sides of the equations. The reader unfamiliar with the reformulation of linear equations to detached coefficient array form is advised to review Example 4.18 of Section 4.4.2. Solution of the equations in detached form is accomplished using the reduction procedure also detailed in that section.

Step A of the row reduction procedure consists of dividing the first row by 1. Step B involves addition of multiples of the first row to the remaining rows so that the first entries of the resulting rows become zero; that is, we add the first row to the third row and −1 times the first row to the last row to obtain

$$
\begin{pmatrix}
1 & 0 & -1 & 0 & 0 & 0 & 0 & 100 \\
0 & 1 & 0 & -1 & 0 & 0 & 0 & 0 \\
0 & 0 & -1 & 0 & 1 & 0 & 1 & 100 \\
0 & -1 & 0 & 0 & 0 & 1 & -1 & 0 \\
0 & 0 & 1 & 0 & -0.8 & 0 & 0 & 0 \\
0 & 0 & 0 & 1 & 0 & -0.4 & 0 & 0 \\
0 & -5 & 1 & 0 & 0 & 0 & 0 & -100
\end{pmatrix}
$$

Continuing with the second row, Step A consists of dividing each element of the second row by the second element. Then, in Step B, the second row is added to the fourth and five times the second is added to the last to obtain

$$
\begin{pmatrix}
1 & 0 & -1 & 0 & 0 & 0 & 0 & 100 \\
0 & 1 & 0 & -1 & 0 & 0 & 0 & 0 \\
0 & 0 & -1 & 0 & 1 & 0 & 1 & 100 \\
0 & 0 & 0 & -1 & 0 & 1 & -1 & 0 \\
0 & 0 & 1 & 0 & -0.8 & 0 & 0 & 0 \\
0 & 0 & 0 & 1 & 0 & -0.4 & 0 & 0 \\
0 & 0 & 1 & -5 & 0 & 0 & 0 & -100
\end{pmatrix}
$$

The reduction procedure is repeated until all rows are reduced, that is, until the diagonal form shown below is achieved.

N_A^2	N_B^2	N_A^5	N_B^5	N_A^3	N_B^3	r	RHS
1	0	0	0	0	0	0	227.273
0	1	0	0	0	0	0	45.455
0	0	1	0	0	0	0	127.273
0	0	0	1	0	0	0	45.455
0	0	0	0	1	0	0	159.091
0	0	0	0	0	1	0	113.636
0	0	0	0	0	0	1	68.182

The solution obtained is of course the same as that calculated iteratively in Example 5.22, namely,

$$N_A^2 = 227.273 \quad \text{and} \quad N_B^2 = 45.455$$

Since the rate for the elementary reaction is $r = 68.182$, the conversion of A is, as before,

$$X = \frac{r}{N_A^2} = \frac{68.182}{227.273} = 0.300$$

The example clearly demonstrates the simplicity of the simultaneous approach in the linear case. Instead of two nested iterative loops, the calculations can be performed exactly and without concerns about convergence. A further advantage is that no restrictions need to be imposed on the type of specifications that are allowed, provided that the relations are linear in the species flows. Subsidiary specifications simply appear as additional rows in the detached coefficient array and require no special handling.

A major disadvantage however is that, even for problems small enough for manual calculations, the number of simultaneous equations and hence the coefficient array can become quite large. For instance, Example 5.1 of this chapter would require the solution of 48 simultaneous linear equations. Yet this is a relatively small problem. A full-scale process problem could very easily involve over 1000 equations. Although the simultaneous solution of that many linear equations can be carried out routinely on the computer, the large amount of storage required for the associated coefficient array is a significant limitation.

Recall that the structure of the material balance equation sets is such that a species flow of any given stream will never appear in more than two balance equations: a balance equation associated with the unit for which that stream is an output and a balance equation associated with the unit for which the stream is an input. The only other appearances of a given species flow will be in equations arising from specifications. For instance, from the detached coefficient array of the equation set given for Example 5.24, it is apparent that every species flow appears no more than twice as a nonzero entry in the array. For large equation sets, this means that the overwhelming majority of array entries will be zeros. With a 500-equation problem, if every column contains only three nonzero entries, the array will consist of 497/500 or 99.4% zeros. Since a 500 × 500 array contains 250,000 entries, it is clear that the only viable way to use the simultaneous approach is to employ storage schemes which only save the nonzero entries and parameters identifying their positions in the array. Assuming 0.6% nonzeros and two position parameters for each nonzero, this means that a 500-equation problem can be stored using just 0.006 × 250,000 × 3 = 4500 entries. Of course, implementation of such storage schemes and of equation-solving routines that can operate on arrays stored in this fashion is not a simple task. At the present time, the only generally available material balance program which uses sparse storage organization and includes the feature of automated balance equation generation is the SYMBOL Program.[3]

Nonlinear Material Balance Problems   The key assumption underlying the simultaneous approach is that all balance equations and specification relationships are linear equations in the species flows. Yet, we know that while the balance equations will always be linear, the specifications need not always be linear. For instance, in Example 5.3 of this chapter, we encountered a quadratic equation. In this section, we discuss how to accommodate nonlinearities in the simultaneous solution approach.

In Section 2.1.3, we noted that there are basically three types of specifications which commonly occur in material balance problems, namely, specifications of

1.   Fractional recoveries or split fractions.
2.   Composition relationships.
3.   Flow ratios.

The first and last of these can always be expressed as linear relationships between species flows. The second type of specifications can however lead to nonlinear equations. The most common form of composition relationship is the simple proportionality

$$x_s^i = K x_s^j$$

For instance, when $K = 1$ and i and j are outlet streams of a splitter, we can recognize this composition relationship as the splitter restrictions. Since compositions are simply ratios of species flows, the above equations can equivalently be written as

$$\frac{N_s^i}{\sum_s N_s^i} = K \frac{N_s^j}{\sum_s N_s^j}$$

In general, the species s on either side of the equality might differ, and the sums

[3] *SYMBOL User Manual*, Computer Aided Design Center, Cambridge, England, 1973.

in the denominators might be over only some rather than all species. But, regardless of these details, with K specified, the resulting equation is nonlinear in the species flows. Moreover, if we multiply across by the denominators of the equation, we can observe that in fact the equation is quadratic. Note however that the nonlinearity can sometimes be avoided. For instance, if the ratio of the denominators is known or if the right-hand side of the expression is known, then the above condition reduces to a linear equation. A typical example might be a flow splitter in which the split ratios of the output streams are not known but the compositions are known. In that case, the splitter restrictions reduce to just linear flow ratios. However, in general, if the ratio of compositions of species in two different streams is specified or if the split ratio between two outlet streams of a flow splitter is not known, the material balance problem will involve nonlinear specification equations. In those cases, the simultaneous solution of the balance equations and specifications can no longer be carried out directly by using linear equation-solving methods. Rather, since nonlinear equations are involved, multivariable root-finding methods must be used.

In view of the fact that the major portion of the simultaneous equations to be solved will be linear, an advantageous iterative approach is to employ the Newton–Raphson method. Recall that this technique simply involves construction of a linear approximation of each nonlinear function using the partial derivatives of that function and then solution of the resulting set of linear equations to obtain a new estimate of the solution of the simultaneous equations. Clearly, any of the equations which are linear to begin with need not be linearized; rather, they can be appended directly to the linearized equations and the combined set solved simultaneously. Thus, in the presence of nonlinear specifications, the simultaneous approach requires the solution of a sequence of linear equation sets until convergence to the final solution is attained.

In order to define this solution procedure more precisely, we need to introduce the following notation. Let the variables x_n, $n = 1, \ldots, N$, indicate the unknown flows of all species in all streams. Then, let $\sum_{n=1}^{N} a_{in} x_n = b_i$, $i = 1, \ldots, L$, indicate the set of linear balance equations and specifications, and $f_l(x) = 0$, $l = 1, \ldots, N - L$, indicate the set of nonlinear specifications. Given a starting estimate x^0 of *all* species flows, the simultaneous solution will proceed in the following steps.

Step 1 Construct the linear approximations to each function $f_l(x)$, $l = 1, \ldots, N - L$, at the current iterate $x^{(k)}$:

$$\sum_{n=1}^{N} \left(\frac{\partial f_l}{\partial x_n}\right)_{x=x^{(k)}} x_n = \sum_{n=1}^{N} \left(\frac{\partial f_l}{\partial x_n}\right)_{x=x^{(k)}} x_n^0 - f_l(x^0) \tag{5.17}$$

The development of this equation parallels exactly the development of eq. (5.6) if we substitute $f_l(x)$ for $x_n = g_n(x)$.

Step 2   Solve the combined set of linear eqs. (5.17) and the linear equations

$$\sum_{n=1}^{N} a_{in}x_n = b_i$$

and denote the solution by x^{k+1}.

Step 3   Check for termination. If $|f_l(x^{k+1})| \leq \varepsilon$ for all $l = 1, \ldots, N - L$, then terminate the iterations. Otherwise, set $k = k + 1$ and continue with Step 1.

Note that this calculation procedure requires an initial estimate of all species flows. Although good initial estimates can sometimes be obtained by inspection or prior experience with the problem, the safest approach is to select reasonable values of species flows involved in the nonlinear specifications, one species flow per specification, and then to solve the remaining set of linear equations to obtain estimates of the other species flows in the flowsheet. This approach will minimize the number of flows which need to be estimated initially and will result in consistent values of the remaining flows.

We illustrate each step of the above calculation procedure by solving a small example.

Example 5.25   A process for the production of sodium hydroxide contains an intermediate stream consisting of $CaCO_3$ precipitate slurried in a solution of $NaOH$ and H_2O. This slurry is washed with water in three stages so as to reduce the $NaOH$ concentration in the slurry to a sufficiently low level. The flowsheet for this countercurrent washing process is shown in Figure 5.33. The washing stages can be assumed to operate so that the slurry leaving each wash stage will contain 2 lb solution per 1 lb $CaCO_3$ solid. Also, the concentrations of the solutions in the pair of streams leaving each stage can be assumed to be equal. If the slurry fed to the first stage contains 10% $NaOH$, 30% $CaCO_3$, and 60% H_2O, calculate the pounds of wash water required per pound of feed to achieve a 1% $NaOH$ concentration in the solution leaving in stream 4.

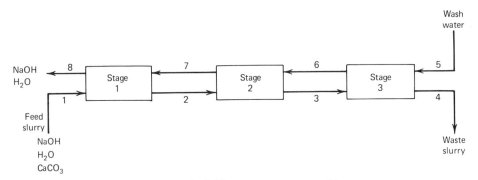

Figure 5.33   Flowsheet for Example 5.25, countercurrent washing.

As confirmed by the degree-of-freedom table shown in Figure 5.34, the process is correctly specified, provided a basis is selected. If we choose as basis 1000 lb/h stream 1, then from the degree-of-freedom table it is apparent that the problem will consist of:

Nine linear species balance equations.
Three linear relations arising from the solution/solid ratios.
One linear relation arising from the specified NaOH solution concentration in stream 4.
One linear relation arising from the specified concentration ratio at stage 3.
Two nonlinear relations arising from the concentration ratios at stages 1 and 2.

	Stage 1	Stage 2	Stage 3	Process
Number of stream variables	10	10	9	19
Number of balance equations	3	3	3	9
Number of specified compositions				
Stream 1	2			2
Stream 4			1	1
Number of specified relations				
Solid/solution ratio	1	1	1	3
Concentration ratio	1	1	1	3
Degree of freedom	3	5	3	1
Basis				-1
				$\overline{0}$

Figure 5.34 Degree-of-freedom table for Example 5.25.

Thus, we can expect that simultaneous solution will involve a system of 14 linear and two nonlinear equations. Let us proceed to assemble these equations. For convenience, we index the species in the following order: $NaOH = 1$, $H_2O = 2$, and $CaCO_3 = 3$.

Stage 1 Balance equations:

$$100 + F_1^7 = F_1^2 + F_1^8$$
$$600 + F_2^7 = F_2^2 + F_2^8$$
$$300 = F_3^2$$

Specified relations:

$$F_1^2 + F_2^2 = 2F_3^2$$
$$\frac{F_1^8}{F_1^8 + F_2^8} = \frac{F_1^2}{F_1^2 + F_2^2}$$

Stage 2   Balance equations:

$$F_1^2 + F_1^6 = F_1^3 + F_1^7$$
$$F_2^2 + F_2^6 = F_2^3 + F_2^7$$
$$F_3^2 = F_3^3$$

Specified relations:

$$F_1^3 + F_2^3 = 2F_3^3$$
$$\frac{F_1^7}{F_1^7 + F_2^7} = \frac{F_1^3}{F_1^3 + F_2^3}$$

Stage 3   Balance equations:

$$F_1^3 + 0 = F_1^4 + F_1^6$$
$$F_2^3 + F_2^5 = F_2^4 + F_2^6$$
$$F_3^3 = F_3^4$$

Specified relations:

$$F_1^4 + F_2^4 = 2F_3^4$$
$$\frac{F_1^4}{F_1^4 + F_2^4} = 0.01$$
$$\frac{F_1^4}{F_1^4 + F_2^4} = \frac{F_1^6}{F_1^6 + F_2^6}$$

Note that since the concentration of the stream 4 solution is known, the composition ratio can be reduced to a linear equation. If the trivial $CaCO_3$ balances are excluded, the resulting set of simultaneous equations will consist of the linear balance equations,

$$
\begin{array}{llllllll}
F_1^2 & & & & - F_1^7 & + F_1^8 & & = 100 \\
& F_2^2 & & & - F_2^7 & + F_2^8 & & = 600 \\
- F_1^2 & + F_1^3 & & - F_1^6 & + F_1^7 & & & = 0 \\
& - F_2^2 & + F_2^3 & - F_2^6 & + F_2^7 & & & = 0 \\
& & - F_1^3 + F_1^4 & + F_1^6 & & & & = 0 \\
& & - F_2^3 & + F_2^4 - F_2^5 & + F_2^6 & & & = 0
\end{array}
$$

the linear specifications,

$$
\begin{array}{ll}
F_1^2 + F_2^2 & = 600 \\
F_1^3 + F_2^3 & = 600 \\
F_1^4 + F_2^4 & = 600 \\
0.99F_1^4 - 0.01F_2^4 & = 0 \\
0.99F_1^6 - 0.01F_2^6 & = 0
\end{array}
$$

and the nonlinear specifications,

$$\frac{F_1^8}{F_1^8 + F_2^8} = \frac{F_1^2}{F_1^2 + F_2^2} \qquad \text{or} \qquad F_1^8(F_1^2 + F_2^2) = F_1^2(F_1^8 + F_2^8)$$

$$\frac{F_1^7}{F_1^7 + F_2^7} = \frac{F_1^3}{F_1^3 + F_2^3} \qquad \text{or} \qquad F_1^7(F_1^3 + F_2^3) = F_1^3(F_1^7 + F_2^7)$$

Hence, we have 11 linear and two quadratic equations which must be solved simultaneously. Following the calculation sequence outlined previously, we must begin with an estimated trial solution at which the linearizations to the two nonlinear constraints can be constructed. Since there are two nonlinear conditions, we need to estimate two species flows; and, since these two conditions are associated with stages 1 and 2, we should choose values for streams associated with those units. Note that the composition of the solution entering with the slurry is 14.3% NaOH. The composition of the wash solution obtained certainly will not be higher; suppose it is 10% NaOH. If nearly all the NaOH fed to the process appears in stream 8, this means that F_2^8 will be about 1000 lb/h. Furthermore, since the initial solution composition of 14.3% NaOH must be reduced to 1% in three stages, the reduction in concentration achieved in each stage must be at least $\frac{1}{2}$, that is, $\frac{1}{7} \times \frac{1}{2} \times \frac{1}{2} \times \frac{1}{2}$ $= \frac{1}{56} > \frac{1}{100}$.

Suppose we assume that one half of the NaOH is removed in the first stage, namely, 50 lb/h. Then, if nearly all of the NaOH fed is contained in stream 8, stream 7 must contain 100–50 or about 50 lb/h of NaOH. Thus, $F_1^7 = 50$ lb/h. We thus have estimated two species flows, one involved in the first nonlinear relation and the other in the second nonlinear relation. Given these two initial estimates, the remaining balance equations and specification are *linear* and can be solved using row reduction, as in Example 5.24, to yield the initial estimates of all process species flows, summarized below.

	Stream Number						
	2	**3**	**4**	**5**	**6**	**7**	**8**
NaOH (lb/h)	56	15.94	6	—	9.94	50	94
H$_2$O (lb/h)	544	584.06	594	994	984.06	944	1000

We are now ready for Step 1 of the calculation: construction of the linear approximations to the functions

$$f_1 = F_1^8(F_1^2 + F_2^2) - F_1^2(F_1^8 + F_2^8) = 0$$
$$f_2 = F_1^7(F_1^3 + F_2^3) - F_1^3(F_1^7 + F_2^7) = 0$$

Note that

$$\frac{\partial f_1}{\partial F_1^8} = F_2^2 \qquad \frac{\partial f_1}{\partial F_2^8} = -F_1^2 \qquad \frac{\partial f_1}{\partial F_1^2} = -F_2^8 \qquad \frac{\partial f_1}{\partial F_2^2} = F_1^8$$

Consequently, at the initially estimated set of flows,

$$\frac{\partial f_1}{\partial F_1^8} = 544 \qquad \frac{\partial f_1}{\partial F_2^8} = -56 \qquad \frac{\partial f_1}{\partial F_1^2} = -1000 \qquad \frac{\partial f_1}{\partial F_2^2} = 94$$

From eq. (5.17), we thus have

$$-1000F_1^2 + 94F_2^2 + 544F_1^8 - 56F_2^8$$
$$= -1000(56) + 94(544) + 544(94) - 56(1000) - (-4864) = -4864 \quad (5.18)$$

Similarly, since

$$\frac{\partial f_2}{\partial F_1^7} = F_2^3 \qquad \frac{\partial f_2}{\partial F_2^7} = -F_1^3 \qquad \frac{\partial f_2}{\partial F_2^3} = -F_2^7 \qquad \frac{\partial f_2}{\partial F_2^3} = F_1^7$$

upon substitution into eq. (5.18), we obtain the linearized equation

$$-944F_1^3 + 50F_2^3 + 584.06F_1^7 - 15.94F_2^7 = 14{,}155.64 \quad (5.19)$$

The resulting set of linear equations in detached coefficient form is given below.

F_1^2	F_2^2	F_1^3	F_2^3	F_1^4	F_2^4	F_2^5	F_1^6	F_2^6	F_1^7	F_2^7	F_1^8	F_2^8	RHS
1									-1		$+1$		100
	1									-1		1	600
-1		$+1$					-1		$+1$				0
	-1		$+1$					-1		$+1$			0
		-1		$+1$			$+1$						0
			-1		$+1$	-1		$+1$					0
1	$+1$												600
		1	$+1$										600
				1	$+1$								0
				0.99	-0.01								0
						0.99	-0.01						0
-10^3	$+94$										$+544$	-56	-4864
		-944	$+50$						584.06	-15.94			14,155.64

The first six rows in the array correspond to the linear balance equations. The next five rows correspond to the five linear specifications. Finally, the last two rows are eqs. (5.18) and (5.19).

Step 2 of the iterative procedure requires that this set of linear equations be solved to yield a new estimate of the species flows. The solution is given below.

	Stream Number						
	2	3	4	5	6	7	8
NaOH (lb/h)	41.16	17.97	6	—	11.97	35.16	94
H_2O (lb/h)	558.84	582.03	594	1196.99	1185.02	1161.83	1202.99

Step 3 requires a check for termination. Evaluating the nonlinear specifications at the above solution, we obtain

$$f_1 = 3015.89 \qquad \text{and} \qquad f_2 = -413.91$$

The iterations have not terminated since the deviations from zero are quite substantial. Hence, proceeding with Step 1, the nonlinear specifications are linearized at the current point. The result is

$$-1202.94 F_1^2 + 94 F_2^2 + 558.84 F_1^8 - 41.16 F_2^8 = 3020.01$$

and

$$-1161.83 F_1^3 + 35.16 F_2^3 + 582.03 F_1^7 - 17.97 F_2^7 = -413.91$$

The updated linearizations replace the old linearization in the detached coefficient array and the solution of the linear equations is repeated, as required by Step 2. The result is:

	Stream Number						
	2	**3**	**4**	**5**	**6**	**7**	**8**
NaOH (lb/h)	42.85	18.17	6	—	12.17	36.85	94
H_2O (lb/h)	557.15	581.83	594	1216.99	1204.82	1180.14	1222.99

At this point,

$$f_1 = -32.90 \qquad \text{and} \qquad f_2 = -2.708$$

If these values are not close enough to satisfy the Step 3 termination test, the iteration cycle can be repeated. The result of the final iteration is:

	Stream Number						
	2	**3**	**4**	**5**	**6**	**7**	**8**
NaOH (lb/h)	42.84	18.166	6	—	12.166	36.84	94
H_2O (lb/h)	557.16	581.834	594	1216.64	1204.47	1179.80	1222.64

The solution of nonlinear material balance problems as a sequence of simultaneous linearized subproblems has the obvious advantages that any and all types of nonlinear specifications can readily be accommodated. All that the engineer needs to supply are the nonlinear functions, their partial derivatives, and initial estimates of one distinct species flow for each nonlinear specification. The disadvantages are similar to those discussed for the linear simultaneous strategy, namely, large storage requirements and tedium of coding all balance equations and specification relations. These disadvantages can largely be overcome but require the development of sophisticated large-scale material balancing computer programs.

A disadvantage which cannot be overcome even if sparse array storage schemes and equation generation routines are employed is the inefficiency inherent in the repeated solution of the combined balance and specification equation sets. Only a portion of the equation set is altered in each calculation cycle—in Example 5.25, only two out of 13 equations. Yet, the entire set must be resolved each time. Recently, a decomposed solution strategy has been proposed which avoids this inefficiency by separating the solution of the balance equations from the solution of the specification equations.[4] The details of this approach are beyond the scope of the present treatment.

5.3 SUMMARY

In this chapter, we have discussed manual and computer-oriented solution strategies for solving flowsheet material balance problems. We found that manual solution of larger flowsheet problems can become very tedious if the specified information is inopportunely distributed. The partial solution strategy involving carrying unknown variables can accommodate most types of specifications. However, in the presence of nonlinear specifications, iterative equation-solving methods must be employed.

The computer-oriented strategies we discussed differ by the amount of problem structure that the methods exploit. The simultaneous strategy uses very little structure and is very flexible in accommodating any type of specification but requires solution of very large sets of simultaneous linear and nonlinear equations. The sequential modular strategy uses the natural structure of the flowsheet to sequence calculations but generally requires iteration on one or more tear streams. Moreover, it is restrictive in the types of specifications which can be imposed since constraint specifications require additional outer loops of iterative calculations.

At this point, the reader should have a good grasp of the fundamentals of each of these solution strategies and their relative advantages and disadvantages. With this chapter, we conclude our discussion of flowsheet material balance problems. In the next chapter, we commence with a similar study of energy balances.

PROBLEMS

5.1 Acetic anhydride can be produced from acetic acid by catalytic cracking. In the conceptual process shown in Figure P5.1, the acetic acid is reacted, the products are separated, and the unreacted acetic acid as well as inerts are recycled. The fresh feed consists of 1 mol inerts per 50 mol acetic acid, and the product analyzes 46% $(CH_3CO)_2O$, 50% H_2O, and 4% CH_2CO (molar basis).

---

[4] M. K. Sood, G. V. Reklaitis, and J. M. Woods, "Solution of Material Balances for Flowsheets Modeled with Elementary Modules: The Unconstrained Case" and "The Constrained Case." *AICHEJ*, **25**, 209 and 220 (1979).

(a) Determine the degree of freedom and a calculation order.

(b) Calculate the fraction of the acetic acid fed which must be purged to prevent the inerts from accumulating.

(c) Calculate the conversion of acetic acid per pass.

5.2 Consider the variation of Example 5.1 shown in Figure P5.2. The same five reactions occur; and, as before, to avoid loss of valuable ethane in the purge, the recycle stream is split into two equal parts. One part is subjected to a separation which will preferentially remove a stream of N_2, CO, and CO_2 for venting. The other half of the recycle stream is sent directly back to the reactor without treatment. Suppose under certain operating conditions the vent gas, stream 9, is found to contain equal molar amounts of CO and CO_2; the product stream (stream 5) is found to consist of $33\frac{2}{3}\%$ C_2H_4O, $33\frac{1}{3}\%$ CH_3OH, and $33\frac{1}{3}\%$ CH_2O and the reactor outlet stream is found to contain 35% C_2H_6, 51% N_2, 1% C_2H_4O, and 8.5% combined CO and CO_2. Also, it is found that 1 mol H_2O is formed in the process (stream 6) per 1 mol fresh C_2H_6 fed to the process (stream 13). All compositions and side conditions are in molar quantities.

(a) Show that the process is correctly specified.

(b) Devise a unit-by-unit calculation order.

(c) Calculate the overall yield of C_2H_4O from C_2H_6.

(d) Repeat the analysis assuming that the additional reaction

$$2CO + O_2 \rightarrow 2CO_2$$

is added to the set describing the reactor behavior. Explain what effect this reaction has on the calculations.

5.3 A variant of the recycle process for producing perchloric acid is given in Figure P5.3. The reaction

$$Ba(ClO_4)_2 + H_2SO_4 \rightarrow BaSO_4 + 2HClO_4$$

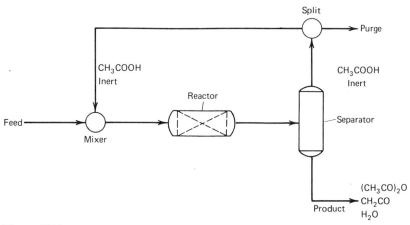

Figure P5.1

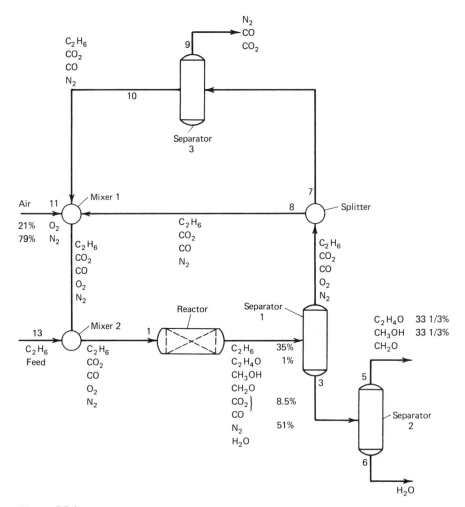

Figure P5.2

proceeds in the reactor with 80% conversion of $Ba(ClO_4)_2$. The ratio of number of moles H_2SO_4 to number of moles $Ba(ClO_4)_2$ in the combined feed to the reactor is 1:1.2. Assuming that the process feed consists of 90% (wt) $Ba(ClO_4)_2$ and the rest $HClO_4$ and that the recycle contains only $Ba(ClO_4)_2$, calculate all flows in the process.

5.4 In a modification of the process for preparing methyl iodide given in Figure P5.4, 2000 lb/day HI is added to an excess of CH_3OH.

(a) If the product and waste streams have the indicated wt % compositions and the reaction

$$HI + CH_3OH \rightleftharpoons CH_3I + H_2O$$

is 30% complete, what is the amount of HI recycled?

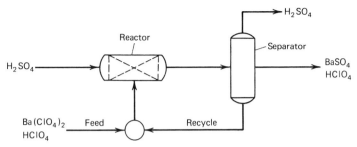

Figure P5.3

(b) Suppose the conversion can be increased to 65%, how does this affect the solution?

5.5 Product P is produced from reactant R according to the reaction

$$2R \rightarrow P + W$$

Unfortunately, because both reactant and product decompose to form by-product B according to the reactions

$$R \rightarrow B + W$$
$$P \rightarrow 2B + W$$

only 50% conversion of R is achieved in the reactor, and the fresh feed contains 1 mol inerts I per 11 mol R. The unreacted R and inerts are separated from the products and recycled. Some of the unreacted R and inerts must be purged to limit the inerts in the reactor feed to 12% on a molar basis (see Figure P5.5). For a product stream produced at the rate of 1000 lbmol/h and analyzing 38% product P on a molar basis, calculate the following (see Figure P5.5):

(a) The composition of the recycle stream on a molar basis.
(b) The fraction of the recycle which is purged [purge/(recycle + purge)].
(c) The fresh feed rate in lbmol/h.
(d) The composition of the product stream on a molar basis.
(e) The fraction of reactant R which reacts via the reaction $R \rightarrow B + W$.

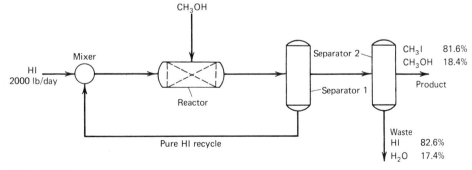

Figure P5.4

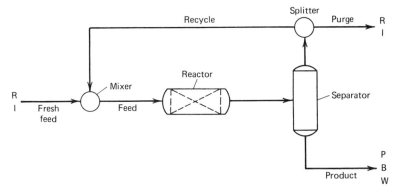

Figure P5.5

5.6   Recovery of Na_2CO_3 and its conversion to NaOH are the key elements of the Kraft process. In the simplified flowsheet shown in Figure P5.6, Na_2CO_3 is reacted with $Ca(OH)_2$ in the calciner via the reaction

$$Na_2CO_3 + Ca(OH)_2 \rightarrow 2NaOH + CaCO_3$$

The $CaCO_3$ is washed in a thickener and converted to CaO in a kiln via the reaction

$$CaCO_3 + H_2O \rightarrow Ca(OH)_2$$

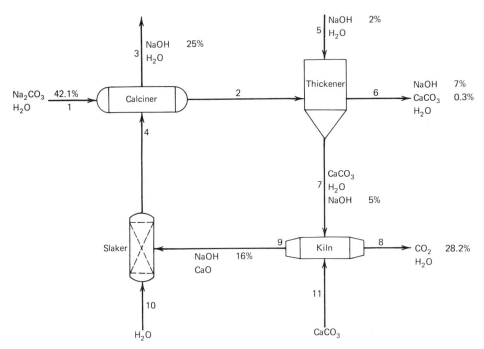

Figure P5.6

The resulting lime (CaO) is hydrated in the slaker to again obtain $Ca(OH)_2$:

$$CaO + H_2O \rightarrow Ca(OH)_2$$

Use the compositions shown on the flowsheet, the additional specifications that

$$F_{H_2O}^1 = F^{10}$$

and assume that all reactions have 100% conversion.

(a) Show that the problem is correctly specified and deduce a calculation order.

(b) Solve for all streams.

5.7 Iodine can be obtained by treating seaweed, which contains NaI, with H_2SO_4 and MnO_2. The reaction, which proceeds with 80% conversion of NaI, is

$$2NaI + MnO_2 + 2H_2SO_4 \rightarrow Na_2SO_4 + MnSO_4 + 2H_2O + I_2$$

Suppose the make-up MnO_2 rate is adjusted so that the total MnO_2 entering the reactor is in stoichiometric proportion to the entering NaI, while the make-up H_2SO_4 flow is adjusted so that the total H_2SO_4 is in 20% excess. The separator removes all reaction products and inerts from the reactants and the reactants are recycled. Suppose the seaweed contains 5% NaI, 30% H_2O, and the rest inerts (all wt %). The product iodine consists of 54% I_2 and the rest H_2O (see Figure P5.7).

(a) Show that the problem is correctly specified and outline a calculation order.

(b) Calculate the composition of the recycle stream and the tons of I_2 produced per ton of seaweed fed.

5.8 Consider the absorber–stripper system shown in Figure P5.8. In this system, a stream (1) containing 30% CO_2, 10% H_2S, and an inert gas is scrubbed of H_2S and CO_2 by using a solvent to selectively absorb the H_2S and CO_2. The resulting stream (5) is fed to a flash unit in which the pressure is reduced,

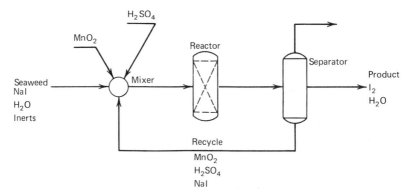

Figure P5.7

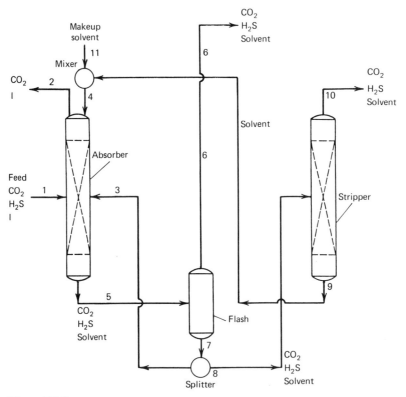

Figure P5.8

and as a result, some of the CO_2, H_2S, and some of the solvent are separated
as an overhead stream (6). The flash unit separation proceeds so that $x^6_{CO_2}$
$= 13.33x^7_{CO_2}$ and $x^6_{H_2S} = 6.9x^7_{H_2S}$. In the stripping unit, the pressure is reduced
further, resulting in an overhead stream (10) consisting of 30% solvent as
well as unknown amounts of CO_2 and H_2S. The bottoms stream from the
stripper (stream 9) consisting of pure solvent is recycled back to the absorber
after mixing with some additional pure solvent which makes up for the solvent
lost in the flash and stripper overheads. Suppose the absorber is operated
so that the outlet overhead stream (2) from the unit contains no H_2S and
only 1% CO_2. Suppose further that the stripper feed solution (stream 8)
contains 5% CO_2 and that the overhead from the flash (stream 6) contains
20% solvent. Construct a degree-of-freedom table, determine the solution
order, and solve the problem:

(a) Assume that stream 7 is split in half in the splitter.
(b) Repeat the solution assuming instead of (a) that 25% of the CO_2 in
      stream 1 is released in stream 6.

5.9 In the gasification process shown in Figure P5.9, a char of known composition
is reacted with steam to produce a synthesis gas. The heat for driving this

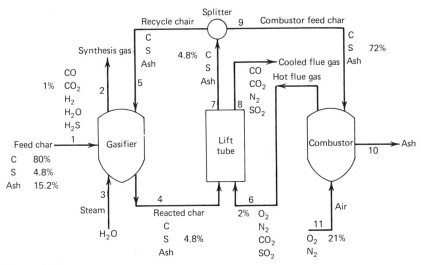

Figure P5.9

reaction is supplied by a recirculating char stream. Heat is supplied to the recirculating char stream by burning part of the char in a combustor and using the resulting hot flue gas to heat up the recirculating char in a lift tube reactor. The lift tube reactions are known to be

$$C + O_2 \rightarrow CO_2$$

$$C + CO_2 \rightarrow 2CO$$

$$S + O_2 \rightarrow SO_2$$

In the flowsheet, all gas compositions are on a mole basis and all solid compositions on a mass basis. The following additional specifications apply:

$$F^3 = \tfrac{1}{2}F^1$$

$$N^8_{CO} = \tfrac{1}{5}N^8_{CO_2}$$

$$F^5 = 15F^1$$

(a) Construct a degree-of-freedom table for the problem and show that the problem is correctly specified.

(b) Determine a calculation order for the process explaining your choice in detail.

(c) Solve the combustor balances and determine the percent excess air that was used.

(d) Calculate the ratio in which the char stream is split in the splitter.

5.10 Ethylene oxide is made by the partial oxidation of ethylene with an excess of air using a silver catalyst (Figure P5.10):

$$2C_2H_4 + O_2 \rightarrow 2C_2H_4O$$

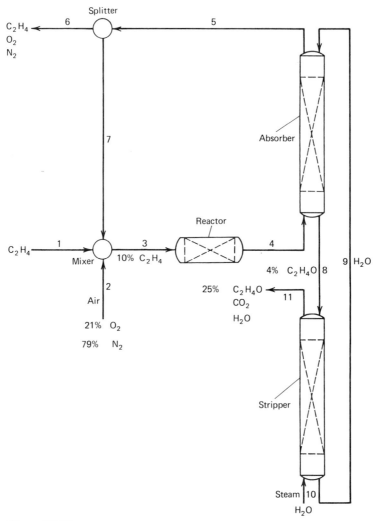

Figure P5.10

25% conversion was obtained with a feed containing 10% ethylene. Unfortunately, some of the ethylene is oxidized completely to carbon dioxide and water:

$$C_2H_4 + 3O_2 \rightarrow 2CO_2 + 2H_2O$$

The selectivity (moles C_2H_4 converted to C_2H_4O/total moles C_2H_4 reacted) is 80%. The ethylene oxide is removed from the stream leaving the reactor in the absorber. 20% of the gases leaving the absorber are purged, the remainder being recycled. The absorbate analyzes 4% ethylene oxide. The ethylene oxide is recovered from the absorbate by stripping with steam. The product analyzes 25% C_2H_4O. All compositions are on a molar basis.

(a) Show that the problem is correctly specified and outline a calculation order for manual solution.

(b) How many moles of steam are required to produce one mole of ethylene oxide?

(c) What is the composition of the purge stream?

5.11 Low-sulfur synthetic fuels can be produced from high-sulfur bituminous coal by pyrolysis, or thermal decomposition, of the coal at high temperatures. In the process flowsheet shown in Figure P5.11, a dry pulverized coal with ultimate analysis of (in wt %) 72.5% C, 5% H, 9% O, 3.5% S, and 10.0% ash is contacted with a hot synthesis gas stream to produce a raw coal oil, a high methane content gas, and a devolatilized char. The high-methane-content product gas has a dry-basis analysis (in mol %) of 11% CH_4, 18% CO, 25% CO_2, 42% H_2, and 4% H_2S, and the gas is found to contain 43 mol H_2O per 100 mol water-free gas. The raw coal oil has an ultimate analysis of (in wt %) 82% C, 8% H, 8% O, and 2% S; and about 150 lb raw oil is produced from 1000 lb dry coal. The devolatilized char from the first pyrolysis stage is transferred to a second stage in which it is gasified with steam and oxygen to produce the hot synthesis gas for the first stage containing (in mol

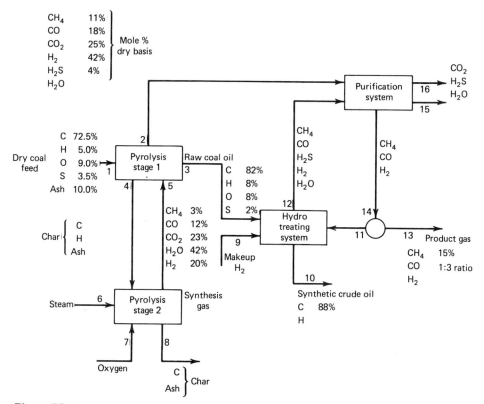

Figure P5.11

%) 3% CH_4, 12% CO, 23% CO_2, 42% H_2O, and 20% H_2 and a char consisting essentially of fixed carbon and ash. The raw-product gas from the first stage is sent to the purification system where it is purified of H_2S, CO_2, and H_2O. The result is a product gas consisting of 15 mol % CH_4 and CO and H_2 in a 1:3 mole ratio suitable for methanation. This gas is split, and the larger portion is recycled to the hydrotreating plant. In the hydrotreating plant, the raw coal oil is subjected to a mild hydrogenation using the recycle product gas as well as some makeup H_2 to produce a synthetic crude oil consisting essentially of 88% (wt) C and 12% H_2. Hydrotreating essentially removes oxygen and sulfur from the raw oil and somewhat increases the hydrogen content of the oil. However, the total amount of carbon in the oil remains unchanged, that is, $N_C^3 = N_C^{10}$. The off-gases from hydrotreating are sent to the purification system for H_2S and H_2O removal. Assume that $\frac{2}{3}$ mol recycle gas is fed to the hydrotreating unit per pound of raw oil.

(a) Construct a degree-of-freedom table and show that the process is correctly specified.

(b) Give a calculation order which ought to be followed if all streams had to be calculated. Indicate how many variables, if any, may have to be carried from unit to unit.

(c) Calculate the moles of makeup H_2 required per 1000 lb synthetic crude oil.

5.12   In an established process for producing acetic anhydride $(CH_3CO)_2O$, pure acetone $(CH_3)_2CO$ is partially decomposed in a direct-fired furnace to produce ketene, CH_2CO, along with other side products. The hot furnace products are quickly quenched with cold acetic acid, C_2H_3OOH. When sufficiently cooled, the ketene reacts completely with the acetic acid to produce the desired anhydride product via the reaction

$$CH_2CO + C_2H_3OOH \rightarrow (CH_3CO)_2O$$

The resulting reactor outlet is first scrubbed with an acetic acid-rich recycle stream to produce a light-ends stream consisting of C_2H_4, CH_4, CO, and H_2. The absorber bottoms stream containing 60.75% acetone, 12.5% anhydride, and the rest acetic acid is next processed in an acetone column which produces a relatively pure acetone recycle stream and a bottoms product containing 1.875% acetone (see Figure P5.12). Suppose the acetone conversion in the furnace is 20% and that the furnace exit gases contain 2 mol C_2H_4 and 12 mol ketene per 1 mol H_2. Calculate all streams in the flowsheet, assuming that all compositions are in mol % and that

(a) Stream 10 is further distilled in the anhydride column resulting in a product stream consisting of 96% acetic anhydride and no acetone.

(b) Stream 10 is distilled so that

$$x_A^{12} = 0.25x_A^8$$
$$x_{AA}^{12} = 50x_{AA}^8$$

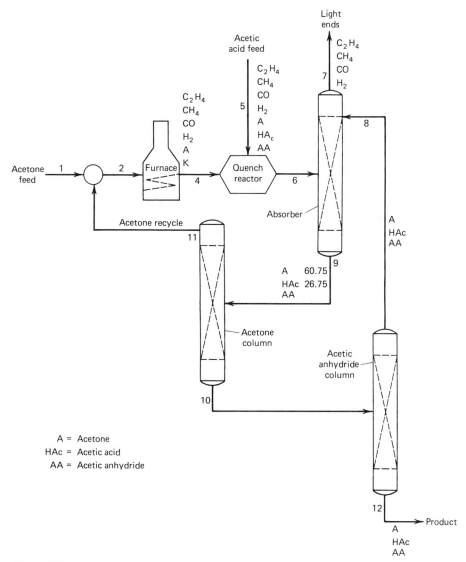

Figure P5.12

5.13 Consider the countercurrent washing process of Example 5.25. Deduce a unit-by-unit calculation order and show that the direct manual solution can be reduced to the iterative solution of a nonlinear equation in the NaOH flow in stream 8 of Figure 5.33. Solve the problem using a root-finding method assuming that the NaOH concentration in stream 4 is to be reduced to $\frac{1}{4}\%$.

5.14 Suppose that the specifications of Problem 5.2 are altered as follows:

1. The division of stream 4 into streams 7 and 8 is left unspecified.

2. Instead of specifying the joint mole fraction of CO and CO_2 in stream 2 to be 8.5%, the mole fraction of CO is set at 4.25%.
3. The ratio of C_2H_6 to CO in the reactor inlet is required to be 10 to 1.

Confirm that the problem remains correctly specified. Deduce a calculation order and show that up to three variables will need to be carried along in order to solve for all flows in the process.

5.15 Consider the following alternate set of specifications for the flowsheet of Problem 5.10:

1. Conversion of C_2H_4, 25%.
2. Selectivity for C_2H_4O, 80%.
3. Moles O_2/moles C_2H_4 at reactor inlet, 1.1.
4. Moles CO_2 per mole CO_2 free stream at reactor inlet, 0.015.
5. Ratio of CO_2 in stream 5 to CO_2 in stream 8, 5:1.
6. C_2H_4O mole fraction in stream 8, 0.045.
7. Ratio C_2H_4O in recycle (stream 9) to C_2H_4O in product (stream 11), 1:8.
8. Moles stream 10/mol stream 8, 0.125.
9. Air composition, 21% O_2.

Note that streams 3, 5, and 7 all contain CO_2.

(a) Show that the problem is correctly specified and determine a calculation order.
(b) Carry out the partial solution to obtain a nonlinear function in the recycle fraction N^7/N^5. Show that this equation has two solutions, solve the equation, and evaluate the remaining flows for these two solutions.
(c) Determine algebraically an expression for the molar flow of CO_2 at the inlet to the reactor as a function of the recycle fraction N^7/N^5. Plot the molar flow of CO_2 vs. (N^7/N^5) and use this plot to deduce over what ranges of values of specification 4 above the problem has multiple feasible solutions.

5.16 Determine the root of the equation

$$f(x) = x^3 - 3x^2 + 2x = 0$$

to three significant figures from the initial estimate $x = \frac{1}{4}$ using

(a) Secant method.
(b) Newton's method.

5.17 Determine the root of the equation

$$f(x) = 3x\,e^{-x} - 1 = 0$$

using interval halving and the initial estimates

$$x^L = 1.0 \qquad x^R = 2.0$$

Use ten trials.

5.18 Find the solution of the equation

$$x = e^{-x}$$

to three significant figures using **(a)** successive substitution and **(b)** Wegstein's method with bounds ± 5. Select your own starting point.

5.19 The function $f(x) = x^3 - 2x - 5 = 0$ has a root in the interval $\frac{3}{2} \leqslant x \leqslant 3$. Construct two iteration functions for use with the method of successive substitution, one that will lead to convergence and one that will not. Explain why and confirm with sample calculations.

5.20 The pressure, temperature, and volume per mole CO_2 are related via the semiempirical equation

$$P = \frac{82.06T}{V - 29.7} - \frac{6.377 \times 10^7}{T^{1/2}(V + 29.7)V}$$

where P is in atm, T is in °K, and V is in $cm^3/gmol$. Given that P is 40 atm and V is 700 $cm^3/gmol$, use a root finding technique to determine the temperature to three significant figures. (Hint: use $PV = 82.06T$ to obtain an initial estimate.)

5.21 Solve the system of equations

$$f_1(x) = x_1 x_2 - \frac{1}{2}x_2^2 = 0$$
$$f_2(x) = x_2^2 - 2x_1 x_2 + x_1^2 - 2 = 0$$

to three significant figures from the initial estimate $x^0 = (\frac{1}{2}, 1)$ using Wegstein's method and the bounds ± 10. Test your iteration functions at the solution $x = (\sqrt{2}, 2\sqrt{2})$.

5.22 Solve the system of equations

$$f_1(x) = 4x_1^3 - 27x_1 x_2^2 + 25 = 0$$
$$f_2(x) = 4x_1^2 x_2 - 3x_2^3 - 1 = 0$$

to three significant figures using the method of your choice. The solution lies in the positive quadrant.

5.23 Solve the system of equations

$$f_1(x) = -\frac{1}{2}x_1 + \frac{1}{3}(x_2^3 + x_3^3) + 0.3 = 0$$
$$f_2(x) = \frac{1}{2}x_1 x_2 - \frac{1}{2}x_2 + \frac{1}{5}x_3^5 = 0$$
$$f_3(x) = \frac{1}{3}x_2^3 - \frac{1}{4}x_3 + 0.04 = 0$$

from the initial estimate $x^{(0)} = (0, 0, 0)$.

(a) Use successive substitution.
(b) Use the Newton–Raphson method.

(c)   Check whether your choice of iteration function in (a) meets the convergence test at the solution $x^* = (0.60273, 1.056 \times 10^{-4}, 0.1600)$.

5.24   Consider the flowsheet shown in Figure P5.24 expressed in terms of elementary modules. Deduce the number of essential mixers, identify the tear streams, and specify a calculation order for the tear streams.

5.25   Figure P5.25 represents a COED coal pyrolysis flowsheet modeled in terms of elementary modules. Show that this flowsheet will require four tear streams. Identify a set of tear streams and deduce the order in which they ought to be processed.

5.26   For the flowsheet shown in Figure P5.26, which represents a conceptional SRC process, deduce the minimum number of tear streams, select which these should be, and specify the order in which they ought to be processed. Note that the hexagons labeled A and B represent streams which should be connected.

5.27   Use a suitable computer program to solve the flowsheet of the acetic anhydride plant of Example 5.20 for the following two cases:

(a)   The quench unit is operated so that no H_2O, acetic acid, or ketene appear in the vent gas.

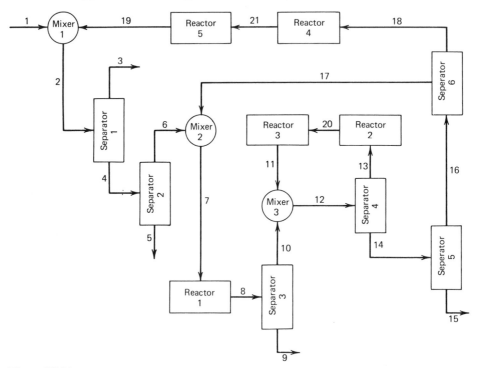

Figure P5.24

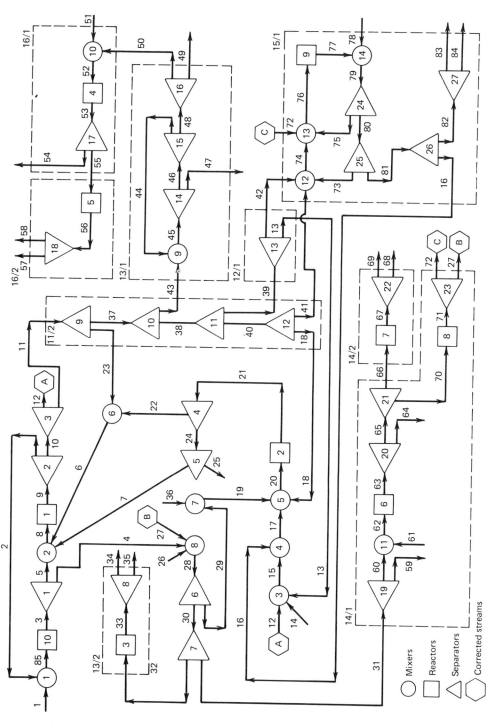

Figure P5.25

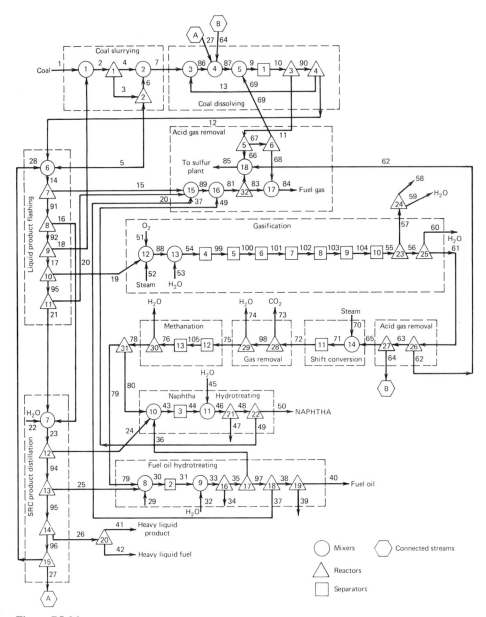

Figure P5.26

(b) The acid recovery tower is to be operated so that the anhydride product is 99% pure, and the acid drying tower is to be operated so that there is only a 5% loss of acetic acid from the entire process. (The split fractions of acetic acid in these two towers are therefore to be determined.)

5.28 A complete plant for the production of ethylene oxide and ethylene glycol is to be designed. As shown in the flowsheet (Figure P5.28), the plant will consist of four primary sections: the C_2H_4 partial oxidation loop, the ethylene oxide purification section, the hydration reactor, and the glycol purification section. Detailed process information for each separate unit is given as follows:

1. *Oxidation Reactor*

   Primary reaction: $\quad 2C_2H_4 + O_2 \rightarrow 2C_2H_4O$

   Secondary reaction: $\quad C_2H_4 + 3O_2 \rightarrow 2CO_2 + 2H_2O$

   With a reactor feed stream of 5% C_2H_4, 10% O_2, and 0.25% CO_2, the conversion of C_2H_4 is 40%, and the selectivity of C_2H_4O is 70%.

   NOTE: Selectivity = (moles C_2H_4 converted to C_2H_4O)/(total moles C_2H_4 converted).

2. *Absorber* All of the C_2H_4O and most of the CO_2 is absorbed from stream 4 and leaves in stream 8. 20% of stream 5 is purged and leaves the process as stream 6.
3. *Stripper* Stream 8, which in addition to some CO_2 and H_2O contains 4% C_2H_4O, is stripped with steam (stream 11) to drive off all of the CO_2 and C_2H_4O (together with

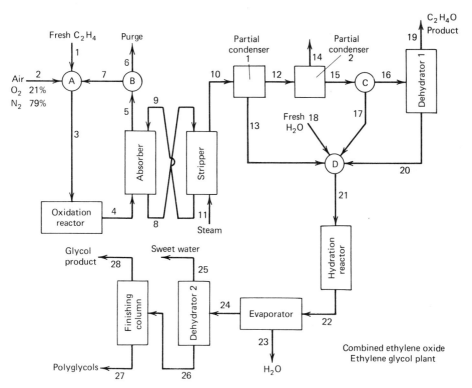

Figure P5.28

some H_2O) as the overhead (stream 10). The remaining H_2O is recycled to the absorber via stream 9.

4. *Partial Condensers* The gas stream (stream 10) leaving the stripper and containing 30% C_2H_4O is cooled in PCON 1 to condense out 50% of the water. The remaining gas is cooled even further in PCON 2 in which all of the H_2O and most of the C_2H_4O is liquified. The vent gas from PCON 2 contains essentially no H_2O, all of the CO_2, and a small amount of C_2H_4O (0.4%). The C_2H_4O/H_2O solution leaving the unit is split at point C into two streams, one of which will be further purified to produce the C_2H_4O product and the other which is sent ahead to the hydration reactor to produce ethylene glyc~¹

5. *Dehydrator 1* Stream 16 is separated into a C_2H_4O product analyzing 99.5% C_2H_4O and a bottoms stream (20) containing 40% C_2H_4O.

6. *Hydration Reactor* Streams 17 and 20 containing C_2H_4O and H_2O together with the condensed water from PCON 1 are mixed with additional fresh water to produce a feed to the hydration reactor containing a mole ratio of C_2H_4O to H_2O of 1:20. In the reactor, all the oxide is converted via the primary reaction

$$C_2H_4O + H_2O \rightarrow C_2H_4(OH)_2$$

as well as the secondary reaction

$$C_2H_4O + C_2H_4(OH)_2 \rightarrow (C_2H_4OH)_2O$$

The formation of diglycol via the second reaction is not desirable; hence reaction conditions are adjusted so that a 90% selectivity for $C_2H_4(OH)_2$ (ethylene glycol) is achieved.

7. *Separation Train* Since the ethylene glycol product must be 99.0% pure, the product stream leaving the reactor (stream 22) which contains a large amount of water must undergo several stages of purification. The first step is to remove enough water by evaporation to reduce the H_2O mol % in stream 24 to 75%. The remaining H_2O is removed by distillation in dehydrator 2. The overhead from this unit will contain 2% glycol and the rest H_2O. The bottoms (stream 26) will contain only about 0.5% H_2O and will be further purified in the finishing column to produce a bottoms stream containing 1% glycol (the remainder diglycol). The product stream, stream 28, contains some H_2O and diglycol impurities but is 99.0% pure. The plant is to produce 20×10^6 lb/yr ethylene oxide product and 25×10^6 lb/yr glycol product.

(a) Carry out a degree-of-freedom analysis to show that the problem is correctly specified.

(b) Determine a unit-by-unit calculation order.

(c) Construct an equivalent material balance flowsheet formulated in terms of elementary modules. Itemize the constraints. Check that the problem is correctly specified.

(d) Determine the tear streams, their sequence, and the appropriate sequence for the constraint iterations.

(e) Solve the problem using a suitable computer program.

5.29 A cracking gas consisting of light hydrocarbons, butylene, butanes, and heavy hydrocarbons with composition 61.3% light hydrocarbons, 5.2% C_4H_8, 6.4% $n\text{-}C_4H_{10}$, 4% $i\text{-}C_4H_{10}$, and 23.1% heavy hydrocarbons (all in wt %) is to be processed to produce C_8H_{18} as primary product. As shown in Figure P5.29, the fresh feed plus recycle lighter hydrocarbons are distilled to separate off the heavy and light hydrocarbons to yield a stream consisting only of C_4H_8, $n\text{-}C_4H_{10}$, and $i\text{-}C_4H_{10}$. The $i\text{-}C_4H_{10}$ and C_4H_8 are reacted to produce C_8H_{18} in the alkylation reactor with stoichiometry

$$C_4H_8 + i\text{-}C_4H_{10} \rightarrow C_8H_{18}$$

and 100% conversion of C_4H_8. The products of alkylation are separated to yield pure C_8H_{18} as bottoms stream. The overhead is further distilled to yield a $n\text{-}C_4H_{10}$ bottoms and an overhead containing light hydrocarbons and $i\text{-}C_4H_{10}$. These are separated perfectly in the final column. The $n\text{-}C_4H_{10}$ separated in column 4 is isomerized to produce $i\text{-}C_4H_{10}$. Two reactions occur in this reactor:

$$n\text{-}C_4H_{10} \rightarrow i\text{-}C_4H_{10}$$

$$n\text{-}C_4H_{10} \rightarrow \text{light hydrocarbons}$$

With 50% conversion of $n\text{-}C_4H_{10}$, a fractional yield of 96% $i\text{-}C_4H_{10}$ is ob-

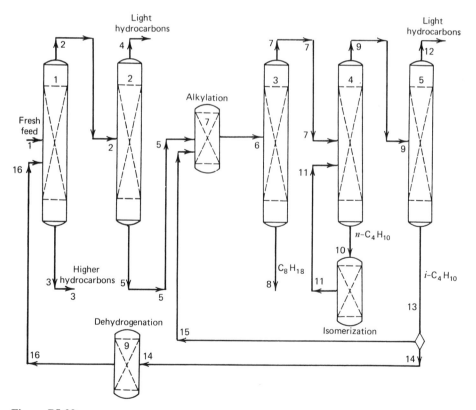

Figure P5.29

tained. The i-C_4H_{10} stream from column 5 is split, with 23% sent to a de-hydrogenation reactor in which the reactions

$$i\text{-}C_4H_{10} \rightarrow C_4H_8 + \tfrac{2}{58} \text{ (light hydrocarbons)}$$

$$i\text{-}C_4H_{10} \rightarrow \text{light hydrocarbons}$$

occur with 31% conversion of i-C_4H_{10} and a fractional yield of 84% C_4H_8. The molecular weights of the species are: light hydrocarbon, 58; C_4H_8, 56; n-C_4H_{10}, 58; i-C_4H_{10}, 58; C_8H_{18}, 114; heavy hydrocarbon, 200. The feed of cracking gas is 1000 lb/h.

(a) Construct a degree-of-freedom table to show the process is correctly specified.

(b) Develop a calculation sequence for hand calculations.

(c) Determine the tear streams which should be used for computer calculation.

(d) Solve the problem as posed above using a suitable computer program.

(e) Suppose the fractional yield of C_4H_8 in the dehydrogenation reactor is unknown but instead the specification is imposed that $w_{i\text{-}C_4H_{10}}^{16} = 0.69 w_{i\text{-}C_4H_{10}}^{14}$. Solve the problem using a suitable computer program.

5.30 A noxious solute must be removed from a waste water stream before it can be discharged to a nearby river. An idle four-stage extraction unit is available which could be used to extract some of the solute with an organic solvent. The resulting solute-rich organic solvent could be treated to recover the solvent, thus allowing the solvent to be recycled. The solvent recovery operation is limited to 1250 lb solvent per hour. The solute content of the waste water is 0.04 lb/lb water, the waste water flow is 10,000 lb/h, and the maximum allowable effluent solute content is 0.005 lb solute per 1 lb water. It is necessary to determine whether the four-stage extractor will prove adequate. The solute distributes between the solvent and water phases leaving each extraction stage as follows:

$$\frac{\text{lb solute}}{\text{lb solvent}} = 20\left(\frac{\text{lb solute}}{\text{lb water plus dissolved solvent}}\right)$$

The organic solvent is slightly soluble in water to the extent lb solvent/lb water $= 0.07$ (see Figure P5.30).

(a) Show that the problem is correctly specified if the solute content of the treated waste water is assumed unknown.

(b) Write the set of equations which must be solved if the simultaneous solution strategy is used.

(c) If x_i = lb solute/lb (water and dissolved solvent) leaving stage i and y_i = lb solute/lb solvent leaving stage i, show that the problem can be reduced to the simultaneous solution of a set of four equations for the four unknown x_i values.

(d) Solve the simultaneous set of equations.

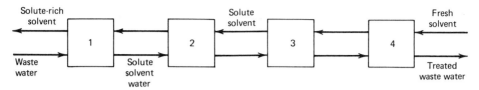

Solute-rich solvent | Solute solvent | Fresh solvent

1   2   3   4

Waste water | Solute solvent water | Treated waste water

Figure P5.30

5.31 In the recycle system shown in Figure P5.31, a feed of 1000 mol/h consisting of one third A and two thirds B is mixed with a recycle stream and reacted following the stoichiometry

$$A + B \rightarrow D$$

In the reactor, 20% of the entering A is converted to product. The resulting stream is separated so that the recycle stream contains 80% of the A, 90% of the B, and 10% of the D fed to the separator.

(a) Show that the process is correctly specified.
(b) Determine a calculation order for manual solution.
(c) Show that the problem is an unconstrained material balance problem.
(d) Solve the problem using the sequential modular strategy with successive substitution and the initial estimate of a zero recycle flow.

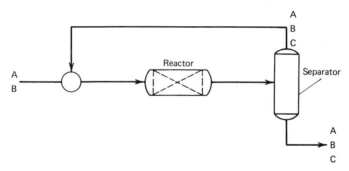

Figure P5.31

5.32 The two-stage countercurrent separation system shown in Figure P5.32 is used to process a feed of 1000 lb/h consisting of 30% A, 40% B, and 30% C. Both separators are operated so that 80% of the A, 90% of the B, and 10% of the C of the mixture fed to those units appear in the overhead stream (streams 5 and 4, respectively). Forty percent of the feed to splitter 1 is recycled and 60% of the feed to splitter 2 is recycled.

(a) Show that the problem is correctly specified.
(b) Determine a calculation order for manual solution.
(c) Determine how many essential mixers are present in the flowsheet.
(d) If the sequential strategy were used, which streams could be selected as tear streams and in what order should they be sequenced?

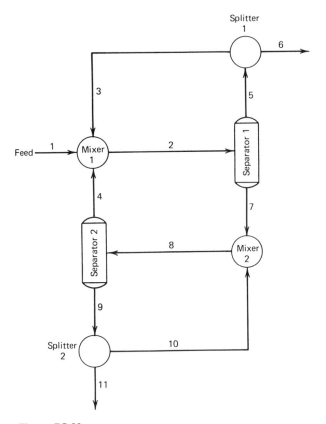

Figure P5.32

5.33 Nitric acid is formed by the combustion of ammonia, which in turn is formed by the reaction of nitrogen and hydrogen. In the plant shown on the accompanying flowsheet (Figure P5.33), 220 lbmol/h hydrogen is fed (stream 1), combined with a pure nitrogen stream (stream 2) and recycle gases, and reacted according to the equation

$$N_2 + 3H_2 \rightarrow 2NH_3$$

Conversion of hydrogen in the reactor is 0.25. All of the ammonia is removed in the condenser. 5% of the gas stream (5) is purged in the splitter before the rest is recycled to the reactor. Air at a rate of 1700 lbmol/h is fed to the burner, where all of the ammonia is reacted. Two reactions occur:

$$NH_3 + 2O_2 \rightarrow HNO_3 + H_2O$$
$$2NH_3 + \tfrac{5}{2}O_2 \rightarrow 2NO + 3H_2O$$

The yield of HNO_3 is 87%. Water at a rate of 110 lbmol/h is fed to the absorber, where the nitric acid is completely absorbed. The remaining gases

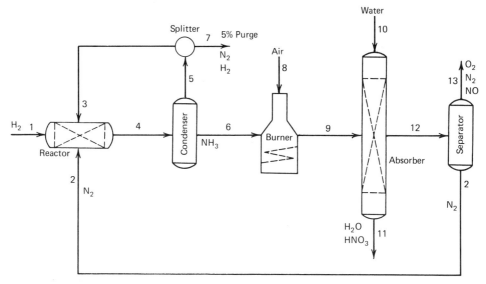

Figure P5.33

enter a separator where all of the O_2 and NO and 93% of the N_2 are removed overhead, and pure N_2 (stream 2) is recycled.

(a) Construct a degree-of-freedom table to show that the process is correctly specified.

(b) Develop a calculation sequence for hand calculations (but do not perform the calculations).

(c) Determine the tear streams which should be used for computer calculation.

(d) Determine all flow rates and compositions by computer calculation.

(e) Suppose the conversion of hydrogen in the ammonia reactor is not known, but instead it is known that the flow rate of the product, stream 11, is 340 lbmol/h. Calculate all stream compositions and flow rates and the conversion of hydrogen in the ammonia reactor, using a computer program.

5.34 Consider the following variation of Problem 5.29. Suppose the fractional yield of butane is still 84% but the conversion of isobutane in the dehydrogenation reactor is unknown. Instead, it is known that the concentration of isobutane in stream 5 (see Figure P5.29) is 3.5 mol %.

(a) Reformulate the flowsheet in elementary module form.

(b) Identify the unknown specifications and the constraints.

(c) Use a suitable computer program to solve the constrained problem.

5.35 Consider the process discussed in Example 5.12. Suppose two thirds of stream 8 (Figure 5.19) is recycled to the reactor (stream 10). The remainder of stream 8 is sent to separator 3 in which 96% of the unreacted benzene and all the chlorine are removed and also recycled to the reactor. In the final separator, 99.5% of the monochlorobenzene and all the benzene appear in stream 14, while all the dichlorobenzene appears in stream 15. With process feeds of 25 kgmol/h benzene and 25 kgmol/h chlorine, reactor outlet concentrations of 37% mono- and 3.3% dichlorobenzene are obtained (mole basis). Solve the problem using a suitable computer program, structuring the iteration loops so that the constraint iterations constitute the inner loop and the recycle iteration, the outer loop.

5.36 A hydrocarbon stream is to be stripped of gasoline range components via absorption in decane and subsequent flash separations to recover the absorbed species, as shown in Figure P5.36. The process feeds consist of cycle and casing head gas fed at 16,000 and 12,000 lb/h, respectively, and the absorber lean oil fed at 18,000 lb/h. The species separation fractions of all separators are known; however, the split of raw gasoline from stream 20 is not specified. Instead, the flow rate of raw gasoline is required to be 1500 lb/h. Given the data tabulated below:

	Module Number					
	2	**3**	**7**	**9**	**10**	**11**
	(Stream 4)	**(Stream 9)**	**(Stream 15)**	**(Stream 18)**	**(Stream 20)**	**(Stream 23)**
Component split fraction						
C1	1.0	0.8553	0.9863	1.0	0.0164	0
C2	1.0	0.6474	0.1657	0.9981	0.0648	0.9847
C3	1.0	0.3922	0.0543	0.9402	0.1238	0.8067
C4	1.0	0.0131	0.0179	0.6853	0.3492	0.7174
C5	1.0	0.0021	0.0089	0.6582	0.4206	0.7059
C6	1.0	0	0.0002	0.5666	0.6792	0.6139
C7	1.0	0	0	0.4839	0.8559	0.5238
C8	1.0	0	0	0.3890	0.9382	0.4496
C9	1.0	0	0	0.2866	0.9759	0.3752
C10	1.0	0	0.0007	0.0035	1.0	0.0841

	Stream 1	**Stream 2**	**Stream 6**
Feed flow (lb/h $\times$ 10^{-3})			
C1	13.76	10.2	0
C2	0.48	0.42	0
C3	0.48	0.42	0
C4	0.32	0.3	0
C5	0.32	0.14	0
C6	0.16	0.12	0
C7	0.16	0.12	0
C8	0.16	0.12	0
C9	0.16	0.06	0
C10	0	0	18

Split fractions: unit 4 to stream 11, 0.75; unit 6 to stream 7, 0.7; unit 8 to stream 21, unspecified.

(a) Identify the tear streams.
(b) Deduce a calculation order for tear streams (if more than one).
(c) Use a suitable computer program to solve the problem taking into account the constraint on the raw gasoline flow.

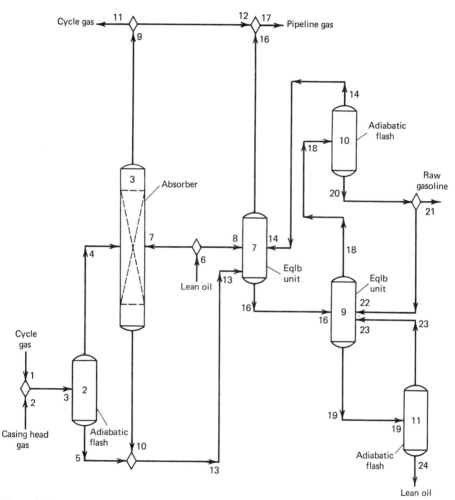

Figure P5.36

CHAPTER
6

Introduction to Energy Balances

When we first considered the law of conservation of matter, we observed that this "law" is nothing more than a convenient simplification. Actually, as proposed by the theory of relativity, neither mass nor energy is conserved independently; rather, their total is conserved. However, the interconversion of these two forms is only known to occur during nuclear reactions or conditions approaching the speed of light, conditions which are not normally encountered in chemical engineering practice. Hence, for most cases of engineering significance, we can claim both that mass of matter is conserved and that the total energy in a system is conserved. The latter statement is referred to as the principle of conservation of energy. This principle or law is the foundation of the discussion of this and of the next several chapters.

Our first concern in this chapter is to translate the general conservation law statement into equations which can be used to balance or calculate quantities of energy in process systems and streams. As was the case in the previous chapters, we shall need to examine the structure and properties of the *energy balance equations* and to learn how to manipulate them to solve problems. We shall find that although energy balance problems are easier to deal with because there are fewer equations to manipulate, the equations themselves are inherently more complex. Specifically, for each unit, we will write only a single equation expressing the fact that the total amount of energy is conserved. However, complications arise in formulating this single equation because energy can exist or be transferred in various forms and because the distribution of energy among these various forms can be readily changed by interconversion. For example, we know from physics that when a weight in an elevated position is dropped, the energy it possessed because of its elevation is converted to energy of motion. Although the total energy of the system, in this instance the weight, is conserved, the relative amount present in the two forms changes. Thus, although we can write only a single energy balance equation, that equation will need to include several different terms to account for the various forms in which energy can exist and the different mechanisms of energy transfer.

A further complication with energy balances is the following fact. In order to relate energy content to physically measurable variables, such as temperature, pressure, phase composition, and density, we need to use supplementary functions that quantify these relationships. In the material balance case, the variables directly involved in the balance equations, namely, flows and compositions, were the quantities of ultimate physical significance. In the energy balance case, the physically significant variables such as composition, phase, and temperature must first be converted to energy content. The energy contents of the various streams can then be balanced, and finally the calculated energy content of the previously unknown streams must be translated back to physically significant variables. These relations and transformations require additional data and supplementary calculations, some of which we introduce in Chapters 7 and 8. However, our treatment of this complex and very important subject will be quite limited. The detailed discussion of the relation between energy content and temperature, pressure, and phase composition is the main thrust of courses in chemical engineering thermodynamics. The emphasis in this text will primarily be on the manipulation and use of the balance equations themselves, rather than on the details of the energy content functions.

In this introductory chapter, we begin that discussion by first examining the various forms of energy and their interconversion. Next, we review the systems of units in terms of which energy and energy balance variables are commonly measured. Finally, we formulate the general energy balance equation and examine several special cases of it.

6.1 DEFINITION OF TERMS

In order to simplify the subsequent developments, it is expedient to first review commonly employed terms.

A *system* is understood to be the restricted part of the universe under consideration. The complement of the system we shall call the *environment* or *surroundings.* Thus, if a reactor and its contents is the system under study, then the environment will be understood to constitute all of the universe except the reactor and its contents (Figure 6.1).

A system can be either *closed,* one whose boundaries are *not* crossed by matter, or *open,* one whose boundaries are pervious to matter. Thus, a closed system is one of constant mass, but the converse need not be true.

In order to be able to quantitatively describe a system, it is necessary to deal with its measurable properties. For purposes of this text, we shall exclusively be concerned with macroscopic properties, that is, we treat matter as a continuum; hence the properties in which we are interested are those which measure bulk characteristics of matter. Typical examples of macroscopic properties of concern to us are temperature, pressure, specific volume or density, etc. These macroscopic properties are generally divided into two classes. Those properties that are additive or are proportional to the size of the system are called *extensive* properties. Typical examples are volume, mass, and energy content. Those properties that are inde-

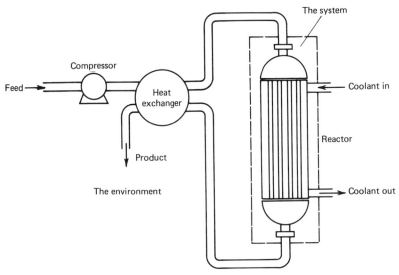

Figure 6.1   A system and its environment.

pendent of system size and thus will be unchanged if the system is subdivided are called *intensive* properties. Typical examples of these are temperature, pressure, mole fraction, and specific volume.

A system is said to be in a given *state* if all of its macroscopic properties have given separate fixed values. A system is said to be in a *steady state* if at each point in the system the values of its properties at that point do not vary with time. If point property values are time varying, then the system is said to be *dynamic* or unsteady state. All of the material balance problems that we have dealt with in the previous four chapters involved steady-state open systems. For instance, we considered processes such as the reactor shown in Figure 6.1 in which all flow rates were constant with time and in which there was no net accumulation of mass in the system. For a system to be steady state from the energy balance point of view, we require in addition that the temperature and pressure at each point in the reactor do not vary with time. Most of the situations we consider in the remainder of this text are steady-state systems.

However, in this chapter, to clarify the development of some key concepts, we have occasion to consider closed systems which do undergo changes in state, hence can be viewed as dynamic. But our concern is with relating the final and initial states of the system rather than with describing the time trajectory of this change. For instance, a batch reactor containing a mixture of chemical A and B at a certain temperature and pressure (State 1) is a closed system. As the reaction proceeds, the composition, temperature, and pressure of the system changes until state 2 is reached. The batch reactor is a dynamic system. However, we are not concerned with describing the concentration–time derivative dx_A/dt, rather, we are concerned with the changes in the system energy content and changes in the distribution of the various energy forms in going from state 1 to state 2.

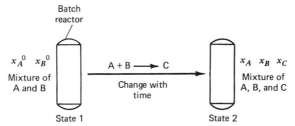

Figure 6.2   Batch reactor system.

The sequence or path of changes of state a system undergoes in going from an initial state to a final state is called a *process*. A process may take place under various conditions: constant temperature (isothermal), constant pressure (isobaric), or constant volume (isochoric), to name a few. For instance, the process which the batch reactor of Figure 6.2 undergoes is a constant-volume process. If care were taken to fix, by external means, the temperature in the reactor, the reaction could proceed isothermally as well. In this text, we learn to account for changes in state that arise by means of these specialized processes.

To assist us in accounting for the changes in state that occur during various types of processes, we make use of special functions called *state functions*. State functions are quantitative relationships between the intensive properties of a system. For example, the relationship between the specific volume, temperature, and pressure of a gas is a state function or equation of state. Such functions have the special feature that their value is dependent only on the system properties at a given state and are independent of the path or process taken to arrive at that state. Thus, if h is a state function, and if h_1 and h_2 are values of this state function for a given system in states 1 and 2, then the difference $h_2 - h_1$ will be the same regardless of the process the system undergoes in proceeding from state 1 to state 2. Moreover, if the system undergoes a *cyclic* process which returns it to state 1, that is, state 2 = state 1, then the difference $h_2 - h_1$ will be equal to zero.

Finally, an important concept, central to energy balance calculations, is that of *equilibrium*. A system is in thermodynamic equilibrium if its state does not change over the time frame of interest and any small change in its state can be easily reversed to return it to its initial state. Thus, for a system to be at equilibrium, it is not only necessary that it be at steady state but also that the system have no spontaneous inclination to change its state. The simplest examples of equilibrium are those of mechanical equilibrium. For example, the ball at rest at the bottom of the trough shown in Figure 6.3(a) meets the definition of equilibrium because both its state is unchanging with time and it has no tendency to be displaced from that state. If the ball were displaced slightly from its rest position, it would spontaneously tend to return to that position, although, without the damping effects of friction, it will in principle oscillate indefinitely about the equilibrium state. As a further illustration, consider a gas contained in a cylinder which is closed at one end with a movable, leak-proof, and frictionless piston, Figure 6.3(b). Suppose that gas is allowed to reach the same temperature as its surrounding and that

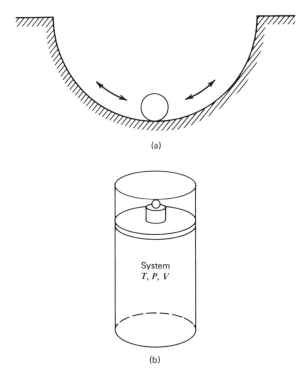

Figure 6.3 Simple equilibrium systems. (**a**) System at state of mechanical equilibrium. (**b**) Gas at state of thermodynamic equilibrium.

weights are added to the piston until the piston is freely supported by the gas. The system consisting of the gas is in a state of equilibrium. Its temperature, pressure, and specific volume will remain unchanged with time. Moreover, after a momentary compression caused by a small and momentary displacement of the piston, the gas will tend to return to its original pressure and specific volume.

The state of a system in equilibrium is always described in terms of intensive properties: temperature, pressure, mole fractions, specific volume, elevation, and velocity of the system. This is because the mass or extent of the system has no influence on its tendency to change its state. For example, all balls under the conditions shown in Figure 6.3(a) will be at equilibrium whether they have a mass of 0.1 kg or 0.001 kg. If one half of the gas in the cylinder is removed, the system of Figure 6.3(b) will attain a state of equilibrium with the same temperature, pressure, and specific volume as the original mass of gas. Because equilibrium states are describable exclusively in terms of intensive properties, the state functions employed to relate or predict equilibrium state properties always involve only intensive properties. For instance, the equation of state of a gas involves temperature, pressure, and specific volume. For the balance of this text, whenever a system undergoes a change in state, it will be assumed that the initial and final states are always equilibrium states.

With these definitions established, we can now turn to a discussion of the forms of energy and of one of the more important state functions, namely, the internal energy of a system.

6.2 FORMS OF ENERGY ASSOCIATED WITH MASS

Suppose that a system in a specified state is subjected to some process which causes its state to change. Since energy cannot be created or destroyed, it must be true in all cases that

$$\begin{pmatrix} \text{Input of energy} \\ \text{to the system} \\ \text{from the surroundings} \end{pmatrix} = \begin{pmatrix} \text{output of energy} \\ \text{to the surroundings} \\ \text{from the system} \end{pmatrix} + \begin{pmatrix} \text{accumulation of} \\ \text{energy within} \\ \text{the system} \end{pmatrix}$$

In other words, all changes in the energy inventory of the system between the initial and final states must be accounted for by energy interchanges between the system and its surroundings. In order to be able to convert this qualitative statement to a useful balance equation, we need both to specify the manner in which these energy interchanges can take place and to enumerate the forms in which the energy inventory can be maintained. In this section, we discuss the latter question. In Section 6.3 we consider the former issue, namely, the mechanisms by means of which energy can be transferred between the system and its surroundings.

Potential Energy   Conceptually, the simplest form of stored energy is the energy an object possesses by virtue of its relative position in a uniform gravitational field. If an object of mass m is at rest at elevation z relative to some datum plane in the earth's gravitational field of strength g (sometimes called the acceleration due to gravity), the *potential energy* of the object as a result of its position is given by

$$PE = mgz$$

Note that the potential energy is proportional to the mass of the object and hence is an extensive property. Furthermore, note that if z is positive, the potential energy is positive, and thus the object must have a higher energy inventory level than it would have were it at the datum plane. This stored energy can readily be transmitted to another object. For example, if the elevated object were connected via a pulley, as shown in Figure 6.4, to another object of infinitesimally smaller mass, it could be used to raise the second object to the level z while itself returning to the datum plane.

Kinetic Energy   The *kinetic energy* is the form of energy a system or object possesses relative to its state at rest by virtue of its bulk movement at a constant velocity v. Specifically, it is given by the familiar formula

$$KE = \tfrac{1}{2}mv^2$$

The velocity v in the above expression refers to the relative motion of the center

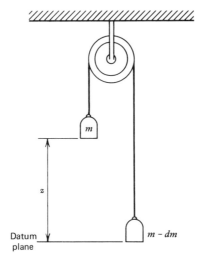

Figure 6.4 Potential energy example.

of gravity of the object. It does not include molecular and atomic translational or rotational motion. It only includes motion at the macroscopic level. Note that, as is the case with potential energy, the kinetic energy is proportional to the mass of the system and hence is an extensive property. Note, further, that this stored form of energy can be converted to other energy forms, such as potential energy.

Electric and Magnetic Field Energies   Just as a system possesses energy by virtue of its position in a gravitational field, it may also posess stored energy by virtue of electric and magnetic fields. Electrical potential energy $(PE)_e$ and magnetic potential energy $(PE)_m$ contributions normally are not of significance in chemical engineering applications.

Internal Energy   The internal energy is the stored energy a system possesses by virtue of the atomic and molecular energy of the matter of which it is constituted. In the case of monoatomic gases, the internal energy consists mainly of the kinetic energy associated with the individual atoms. In the case of molecules containing two or more atoms, the internal energy will include vibrational bond energies as well as molecular rotational energies; while in the case of denser gases, liquids, and solids it will further encompass energy due to molecular forces. In view of the complexity of the atomic and molecular energy components that the initial energy concept includes, the internal energy is sometimes alternatively defined as the form in which energy is stored when simple systems at rest undergo processes which involve no change in position in a gravitational field, no change in bulk velocity, as well as no energy changes resulting from electromagnetic field effects.

The internal energy is an extensive property since it is proportional to the mass of the system in question. It is a complex function of the state of the system, and hence it is not possible to write an explicit formula relating it to the state variables. However, it is a state function, that is, it is uniquely defined by the state of the system; and, as we shall learn in Chapter 7, it can be calculated using various subsidiary relations derived from experimental data. Typically, the internal energy is denoted by the symbol U.

We can summarize the discussion of this section with the following brief statement. The total accumulation of energy in a system in going from one state to another is given by the sum

$$\begin{array}{c}\text{Accumulation of}\\\text{energy in the system}\end{array} = \Delta U + \Delta KE + \Delta(PE)_g + \Delta(PE)_e + \Delta(PE)_m$$

In the next section, we consider the details of the input and output terms of our energy balance equation. To be specific, we consider the forms in which energy can be exchanged between a system and its surroundings.

6.3 Forms of Energy in Transition

Energy can be exchanged between a system and its surroundings in essentially four ways:

1. By transfer of mass.
2. By performing work.
3. By transfer of heat.
4. By field effects.

The first of these is most easily visualized. It is clear that the energy inventory of a system can be changed by the transfer of mass because at the very least that mass will have associated with it a certain amount of internal energy. Thus, if a unit mass of water is added to the system consisting of a pool of water, the internal energy of the system must increase by at least the internal energy of the unit mass added. The latter three modes of energy transfer are a little less obvious and in fact have been subject to considerable misinterpretation in the past. Consequently, we discuss these in more detail.

Work   In mechanics, work is defined as the product of a force times the distance through which it acts. Thus,

$$W = F\,\Delta z$$

if the force F is constant, or

$$W = \int_{z_1}^{z_2} F\,dz$$

if the force is variable over the distance z_1 to z_2. By its very definition, work indicates a transition or motion and hence can be recognized as a form of energy in transition. When the system does work on its surroundings, it transfers to it a certain quantity of energy. Similarly, when the surroundings do work on a system, the energy level of the system rises. For instance, when you compress a spring, you exert a force through a distance and thus perform work. That work raises the energy inventory of the spring. This increase in energy very definitely represents an increase in the ability of the spring itself to do work. When the spring is released, it in turn is able to perform work on the surroundings, which may be, say, the steel ball in a pinball game.

While in every case work can always be reduced to a force moving through a distance, it can be classified into various types, for example, work of compression, rotating shaft work, and electrical work. In chemical engineering applications, the most common type of work is the work in compressing or expanding fluids.

Consider a system consisting of a fluid of volume V confined in a horizontal cylinder fitted with a piston of cross-sectional area A, as shown in Figure 6.5. Suppose that an external force is applied to the piston against which the fluid expands by exerting a pressure or force per unit area, P, upon the piston. Thus, the work done by the system in expanding against the applied external force is

$$W_{syst} = +(PA)\,dz$$

Note that, by convention, work done by the system is assigned a positive sign and work done on the system is assigned a negative sign. Assuming that the cylinder is of uniform cross-sectional area,

$$dz = \frac{dV}{A}$$

Thus,

$$W_{syst} = +P\,dV$$

If the system is expanded from initial volume V_1 to final volume V_2 at constant pressure,

$$W_{syst} = +P\,\Delta V = +P(V_2 - V_1)$$

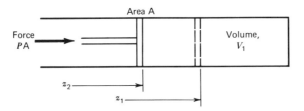

Figure 6.5 Compressive work.

However, if the pressure varies, as it normally will if the system is closed, then calculation of the work done by the system will require the integration

$$W_{syst} = + \int_{V_1}^{V_2} P \, dV$$

While this equation was derived for the special case of a fluid in a cylinder, the result is quite independent of geometry and is in fact a general expression which can be used to evaluate the work done by a fluid in expansion or compression. It is also important to note that the work done by the system in expanding is not necessarily equal to the work done by the surroundings in resisting this expansion. The actual work done by the surroundings will depend upon the nature of the process, the efficiency of the piston, and so on. In engineering calculations, we generally determine the work of compression or expansion of fluids by performing the fluid side calculation and then relate it to the actual work which must be done by the surroundings by means of efficiency factors.

Heat   When a system at a given temperature is placed in an environment which is at a higher temperature, the system temperature rises while that of the environment may decrease. What happens? As we understand it now, some of the energy of the higher-temperature environment is transferred to the lower-temperature system by some mechanism such as convection or conduction. The quantity of energy transferred is called *heat*. Thus, heat is a form of energy in transition: it is energy transferred by virtue of temperature differences. By convention, heat added to the system is positive and heat transferred from the system to the surroundings is negative.

This is a much more difficult concept to visualize and formalize than work. Consequently, it has in the past been subject to considerable misunderstanding. In the earlier days of science, heat was thought to be a weightless fluid which could pass from one object to another. A hot object was thought to contain more of this fluid than a cold object. The temperature of an object was thus understood as a measure of its quantity of caloric fluid. This concept of heat was finally disproved by Count Rumford (Ben Thompson) via the famous observation, made while supervising the boring of common barrels, that an infinite amount of caloric fluid or heat could be generated by friction. Thus, heat was recognized as a form of energy in transition instead of as a quantity stored in a system, while temperature was recognized as an intensive property of a system, a potential which is used as a measure of the driving force for heat transfer.

Electromagnetic Radiation   In addition to heat and work, energy can also be transferred by means of electromagnetic radiation. A traveling wave of electromagnetic radiation impinging upon a system will cause an energy transfer to that system. The mechanism for this transfer is called *radiation*. For most cases of process importance, the only significant form of radiation is that which involves electromagnetic waves in the infrared range and that which occurs by virtue of high temperature. This infrared radiation is normally included as one of the mechanisms of heat transfer. Thus, for purposes of this text, we only consider energy transfer via heat, work, and mass transfer.

6.4 SYSTEMS OF UNITS

In the International System of units (SI), now adopted worldwide, there are three fundamental units: the unit of mass, the *kilogram* (kg); the unit of length, the *meter* (m); and the unit of time, the *second* (s). In the alternate system of units now only employed in the United States (the American Engineering System (AE)), the fundamental units are the unit of length, the *foot* (ft); the unit of mass, the *pound mass* (lb_m); the unit of time, the *second* (s); and a fourth unit, the unit of force, the *pound force* (lb_f). The foot is defined to be exactly equal to 0.3048 m and the pound mass is exactly equal to 0.45359237 kg. The pound force is defined to be the force exerted upon one pound mass of material by gravity under conditions where the acceleration due to gravity, g, has the value 32.1740 ft/s². These two units however are not strictly independent since from Newton's second law of motion, force and mass are related via the equation $F = ma$. To allow this equation to hold under the above definition of pound force, it is necessary to introduce a conversion factor, g_c. Thus, in the American Engineering System, the above equation must be written $F = ma/g_c$.

The numerical value of g_c can be obtained by substituting the defining quantities for the pound force into this modified equation. Thus,

$$1 \; lb_f = \frac{1 \; lb_m \times 32.1740 \; \text{ft/s}^2}{g_c}$$

or

$$g_c = \frac{32.1740 \; lb_m \cdot \text{ft}}{lb_f \cdot \text{s}^2}$$

Whenever calculations involving mechanical energy terms are performed in American Engineering units, this factor must always be introduced. For example, the potential energy term must be written

$$(PE)_g = \frac{mgz}{g_c}$$

and the kinetic energy term must be written

$$KE = \frac{1}{2} \frac{mv^2}{g_c}$$

The Energy Units   Even though energy can be stored and transferred in various forms, it seems reasonable and appropriate that each of these forms should be measured in terms of a single common energy unit. In the SI system, that basic energy unit is the joule. The *joule* is a derived unit which is defined to be equal to one kilogram square meter per square second. In terms of the SI force unit, the newton (N), the joule is equal to one newton meter. Thus,

$$1 \text{ J} = 1 \text{ N} \cdot \text{m} = 1 \text{ kg} \cdot \text{m}^2/\text{s}^2$$

Unfortunately, because of the separate development of the heat, work, and electric energy concepts, specialized units have been devised and are in use for these quantities. The heat unit, the *kilocalorie* (kcal), initially was defined as the heat which must be supplied to achieve a unit temperature elevation in a kilogram of water. Because of disagreements as to the precise definition of this unit, two forms eventually came into use: the thermochemical kilocalorie and the International Steam Tables kilocalorie. Both of these units are now defined in terms of the joule:

$$1 \text{ thermochemical kcal} = 4.184 \times 10^3 \text{ J (exactly)}$$
$$1 \text{ International Tables kcal} = 4.1868 \times 10^3 \text{ J (exactly)}$$

In electric power applications, the unit of energy in common use is the kilowatt hour (kWh). This unit is exactly equal to 3.6×10^6 J. Although acceptable in the SI system, its use is not preferred.

The basic heat unit in the American Engineering System is the *British thermal unit* (Btu), while the unit of mechanical energy or work is the foot pound force (ft · lb$_f$). Both of these units are now defined in terms of the joule. However, as in the case of the kilocalorie, two definitions of Btu are in use: the thermochemical and the International Steam Tables Btu. Their definitions in terms of the joule are

$$1 \text{ thermochemical Btu} = 1.05435 \times 10^3 \text{ J}$$
$$1 \text{ International Tables Btu} = 1.055056 \times 10^3 \text{ J}$$

The definition of the ft · lb$_f$ is

$$1 \text{ ft} \cdot \text{lb}_f = 1.355818 \text{ J}$$

Although use of these alternate energy units will gradually diminish to be replaced by the joule, much of the data published in handbooks and engineering reference works is given in terms of these units. Hence, it is advisable that the reader develop a working familiarity with them.

The Temperature Scales   The fundamental temperature unit in the SI system is the thermodynamic temperature unit, the degree Kelvin (K). This temperature scale is used to define another commonly employed metric temperature scale, the Celsius scale. Temperatures in degrees Celsius (°C) can be related to temperatures in Kelvin via

$$t(°\text{C}) = T - 273.15 \text{ K}$$

where T is temperature in K. Note that the temperature interval of the degree Celsius is numerically exactly equal to one Kelvin.

The temperature unit in the American Engineering System is the degree Rankine (°R). By definition, the temperature interval of one degree Rankine is exactly equal to 5/9 of the temperature interval measured in Kelvin.

An alternate temperature scale in frequent everyday use in the United States is the Fahrenheit temperature scale. The conversion of temperatures in degrees Fahrenheit (°F) to Rankine temperatures is defined via the relation

$$t(°F) = T (°R) - 459.67°R$$

Note that the temperature interval of one degree Fahrenheit is exactly equal to the temperature interval of one degree Rankine.

The Pressure Units The second thermodynamically very important quantity is the pressure, or force per unit area. The basic SI pressure unit is the *pascal,* abbreviated Pa. It is a derived unit which is defined to be one newton per square meter. Because of its numerically very small value (the atmospheric pressure at sea level is about 1.01×10^5 Pa), the unit is often used with the prefix kilo, where 1 kPa = 10^3 Pa. A further frequently used multiple is the *bar,* which is defined to be equal to 10^5 Pa. The bar is used because it is numerically very close to the older, non-SI pressure unit, the standard *atmosphere.* By definition, 1 atmosphere is exactly equal to 1.01325×10^5 Pa, hence it is equal to 1.01325 bar.

The pressure unit in the AE system is the *pound force per square inch absolute,* abbreviated psia. Numerically, the standard atmosphere is equal to about 14.696 psia. The designation absolute is appended to this unit to contrast it from the alternate AE pressure unit *pound force per square inch gauge* (indicated psig), which is pressure measured relative to standard atmospheric pressure. This term has been in long-term engineering use because pressure-measuring devices or gauges frequently indicate pressure as the difference between the surrounding pressure of the atmosphere and the pressure in the vessel or container being monitored. Thus,

$$P \text{ (psia)} = P \text{ (psig)} + 14.696 \text{ psia}$$

Since the psia is a derived unit defined in terms of the pound force and the inch, it is related to the SI pressure unit, the pascal, via the definition of the independent mass, length, and time units. One psia is approximately equal to 6.894757×10^3 Pa.

Pressure is sometimes also reported in terms of a height of a column of standard fluid such as mercury [mm Hg (0°C)] or pure water [ft H_2O (39.2°F)]. These units arose because pressure measuring devices (manometers) were constructed using uniform internal diameter tubes filled with liquid. Since the force exerted on an area A at the base of a column of liquid of height z of density ρ units of mass per unit volume is given by

$$\text{Force} = PA = zA\rho$$

it follows that $P = z\rho$. Thus, the pressure at the base of the column of liquid is proportional to z, with the density as constant of proportionality. However, since the density of liquids varies with temperature, for precise measurements of pressure it is necessary to specify the temperature of the liquid when the measurement is taken. Thus, various temperature standards have entered usage, for instance, 0°C for Hg, 39.2F for H_2O. Under SI conventions, these various heights of liquid-based units are now defined in terms of the pascal as follows (approximately):

$$1 \text{ mm Hg}(0°C) = 0.133322 \text{ kPa}$$

$$1 \text{ in. } H_2O \ (60°F) = 0.24814 \text{ kPa}$$

$$1 \text{ ft } H_2O \ (39.2°F) = 2.98898 \text{ kPa}$$

In general, the use of these pressure units is discouraged under SI conventions.

Conversion of Units   Although unit conversion tables are listed in various engineering reference books, it is important to be able to carry out unit conversions quickly and systematically, and it is useful to remember a few of the common conversion factors. The preferred approach to follow in converting from one set of derived units to another set of derived units is as follows:

1. Convert all derived units in the original set to the major units in that system of units.
2. Convert the major units in that system to the major units in the desired system of units using the defining relations.
3. Convert these major units into the required derived units in the desired system.

In carrying out these conversions, it is also recommended that the units be explicitly written out along with their numerical conversion factors. In this fashion the conversion calculation can be checked by making sure that all intermediate units cancel out algebraically.

Example 6.1   Convert 1 kcal/kg $\cdot$ K to its equivalent in terms of Btu/lb$_m$ $\cdot$ °F.

Solution   The kilocalorie is a derived unit; thus, we first convert it to the fundamental unit, the joule. Next, we convert the joule to the Btu using the basic definition of the Btu in terms of the joule, convert the kg to lb$_m$ using the definition of pound mass, and convert degrees Kelvin to degrees Rankine to the derived unit, degrees Fahrenheit. In equation form,

$$1\frac{\text{kcal}}{\text{kg} \cdot \text{K}} \times \frac{4.184 \times 10^3 \text{ J}}{1 \text{ kcal}} \times \frac{1 \text{ Btu}}{1.05435 \times 10^3 \text{ J}} \times \frac{0.45359237 \text{ kg}}{1 \text{ lb}_m} \times \frac{1 \text{ K}}{1.8°R} \times \frac{1°R}{1°F}$$

$$= \frac{4.184 \times 10^3}{1.05435 \times 10^3} \times \frac{0.45354237}{1.8} \ \frac{\text{kcal} \times \text{J} \times \text{Btu} \times \text{kg} \times \text{K} \times °R}{\text{kg} \times \text{K} \times \text{kcal} \times \text{J} \times \text{lb}_m \times °R \times °F}$$

Note that the units cancel algebraically to yield

$$1.00000 \text{ Btu/lb}_m \cdot °F$$

While only six significant figures are shown in the result, the conversion is in fact exactly 1.0; that is,

$$1 \text{ kcal/kg} \cdot \text{K} = 1 \text{ Btu/lb}_m \cdot °F$$

Note also that because of the 1.8 conversion factor between K and °R, it is also true that

$$1 \text{ kcal/kg} = 1.8 \text{ Btu/lb}_m$$

We find these two groups of units occurring frequently in our discussions in Chapter 7.

Example 6.2 Convert 100 psia to the SI pressure unit, the pascal.

Solution To convert psia, we first need to convert the derived length unit, the inch, to the fundamental AE unit, the foot. Next, to convert the AE force unit to the independent mass, length, and time units, we need to introduce the g_c conversion factor. Then, we convert lb_m to kg using the definition of lb_m and convert ft to m using its definition. Finally, we convert the fundamental SI units to the derived SI unit, the pascal:

$$100 \text{ psia} = \frac{100 \text{ lb}_f}{\text{in.}^2} \times \left(\frac{12 \text{ in.}}{\text{ft}}\right)^2 \times 32.174 \frac{\text{lb}_m \cdot \text{ft}}{\text{lb}_f \cdot \text{s}^2}$$

$$\times \frac{1 \text{ ft}}{0.3048 \text{ m}} \times 0.4535937 \frac{\text{kg}}{\text{lb}_m}$$

$$= 6.894757 \times 10^5 \frac{\text{kg}}{\text{m} \cdot \text{s}^2}$$

$$= 6.894757 \times 10^5 \frac{\text{kg}}{\text{m} \cdot \text{s}^2} \times \frac{1 \text{ N}}{(\text{kg} \cdot \text{m})/\text{s}^2} \times \frac{\text{Pa}}{\text{N/m}^2}$$

$$= 6.894757 \times 10^5 \text{ Pa}$$

Thus,

$$100 \text{ psia} = 689.4757 \text{ kPa} = 0.6894757 \text{ MPa}$$

Example 6.3 Convert 1000°F to temperature in degrees Celsius.

Solution The degree Fahrenheit is an alternate temperature unit in the AE system, while the degree Celsius is an alternate temperature unit in the SI system. Thus, we first convert °F to °R, then use the definition of °R in terms of K, and finally convert K to °C:

$$1000°F = (1000 + 459.67)°R \times \frac{1\text{K}}{1.8°R}$$

$$= (1459.67/1.8 - 273.15)\text{K} \times \frac{1°\text{C}}{1\text{K}}$$

$$= 537.78°C$$

Table 6.1   Useful Approximate Conversion Factors

To convert from	To	Multiply by
atm	kPa	101.3
atm	psia	14.7
Btu	ft · lb$_f$	778
Btu	J	1055
Btu	kcal	0.252
ft^3	m^3	0.283
lb$_m$	kg	0.454
lb$_m$/ft^3	kg/m^3	16.0

Note that if we carry out this conversion in general form, we can obtain the useful and *exact* direct conversion relationship

$$t\ (°C) = \frac{5}{9}[t(°F) - 32]$$

Further useful but *approximate* conversion factors are summarized in Table 6.1. These factors may be used for quick calculations which do not require more than three-figure precision. For more accurate calculations, the exact conversion factors should be determined.

6.5   THE LAW OF CONSERVATION OF ENERGY

In the preceding sections, we discussed the forms in which energy could be stored and the mechanisms by means of which energy can be exchanged between a system and its surroundings. In this section, we assemble these various types of energy terms into a general energy balance equation. We consider closed systems first because this allows us to initially avoid the complications arising from the transfer of energy in and out of a system due to mass inputs and outputs.

6.5.1   The Basic Closed-System Balance Equation

If we assume that energy transfer by means of electromagnetic radiation is negligible, then the energy inventory E of a closed system can be altered in only two ways: by heat transfer and by performing work. Consider a system in a state with initial energy inventory E_1. Suppose that the system does work on its surroundings and let W represent that work. Similarly, suppose that the system exchanges heat with the surroundings and let Q represent the amount of heat transferred to the system. As a result of this interchange of heat and work, the system will achieve a new state with energy inventory E_2.

Since energy is a conserved quantity, the input of energy to the system must equal the output of energy from the system plus the change in the inventory of energy in the system. Thus,

$$Q = W + (E_2 - E_1)$$

where $E_2 - E_1$ represents the difference in the energy inventory of the system between the final and initial states. The rearranged equation

$$E_2 - E_1 = Q - W$$

is the statement of the principle of conservation of energy for closed systems.

In this equation, the sign convention is adopted that Q is positive if heat is added to the system and negative if heat is removed from the system. Also, W is positive if the system performs work on the surroundings and negative if the surroundings perform work on the system. This convention arises from the initial use of this law in the context of steam engines. In steam engines, heat is added to the system by burning a fuel under a boiler. High-pressure steam is raised in the boiler, and the steam is expanded in pistons to do work on the surroundings, namely, drive a machine of some type. (The reader will find that the steam engine is actually a good device for remembering this sign convention.)

To verify that the above equation and the sign convention are consistent with each other, consider the case in which no heat is transferred between system and surroundings and in which work is performed on the system. Thus, $Q = 0$ and $W < 0$. From the first law, $\Delta E = 0 - W = -W$. Since W is negative, ΔE is positive, indicating that the energy of the system is increased. For example, if the system is a spring and we (the surroundings) compress it (perform work on the system), the energy of this spring system is increased.

Similarly, if heat is added to a system but no work is done, then, since $Q > 0$ and $W = 0$, $\Delta E = Q - 0 > 0$ and the energy of the system is increased. For example, if the system is a kettle of water and heat is added to it by means of a burner, the energy of the system is increased. Thus, the statement of the first law and the sign convention we adopted are consistent with our intuitive understanding and everyday experience with energy, heat, and work.

An important limitation of the first law is that it only provides us with an equation by means of which we can calculate *energy differences* between two states of a system. It does not provide us with a way of calculating the absolute or total energy of a system in a given state. In fact, science has not yet been able to measure or calculate the total energy of even the simplest systems. For calculation purposes, we therefore have to select an arbitrary reference value for the energy of the system in an arbitrarily selected reference state. For example, we could define the reference energy of a 1-l bottle of 3.2 beer at 15°C and 2×10^5 Pa to be equal to zero. Then, the first law can be used to calculate the energy of the system at any other state in terms of the heat and work required to reach that state. Thus,

$$E = E^0 + Q - W$$

As an illustration, if we need to calculate the energy of the liter bottle of beer when it is warmed to 25°C, we would calculate Q and then would add Q to the reference energy.

6.5.2 Superimposition of Energy Forms

The first law provides us with an equation for calculating changes in the total energy of a system. As we noted earlier, that total energy can be maintained in many different forms. Let us now examine how the expressions for these different forms can be incorporated into the first law balance equation.

Potential Energy   Suppose a mass m is raised from elevation z_1 to elevation z_2 in a uniform gravitational field of strength g. If the mass is the system, then the work done by the system on the surroundings will be

$$W = -mg(z_2 - z_1)$$

If this work is done without adding any heat to the system, then

$$E_2 - E_1 = -W = mg(z_2 - z_1)$$

If $z_2 - z_1 > 0$, then the energy of the system is increased and that increase is stored in the form of potential energy. Thus,

$$\Delta E = \Delta PE$$

Kinetic Energy   Consider a mass m moving along a straight and level path with velocity v_1. Suppose that the mass is accelerated from v_1 to v_2 by the application of an outside force. The work performed on the system will be

$$W = -\int_{x_1}^{x_2} F\, dx = -\int_{x_1}^{x_2} \left(m\frac{dv}{dt} \right) dx$$

But $v = dx/dt$ or $dx = v\, dt$. Thus,

$$W = -m\int_{t_1}^{t_2} \frac{dv}{dt} v\, dt$$

$$= -m\int_{t_1}^{t_2} \frac{d}{dt} \left(\tfrac{1}{2}v^2 \right) dt$$

$$= -\tfrac{1}{2}mv^2 \Big|_{t_1}^{t_2}$$

$$= -\tfrac{1}{2}mv_2^2 - \left(\tfrac{1}{2}mv_1^2 \right)$$

If this work is done without adding any heat to the system, then

$$E_2 - E_1 = -W = \tfrac{1}{2}mv_2^2 - \tfrac{1}{2}mv_1^2$$

If $v_2 > v_1$, then the energy of the system is increased and that increase is stored in the form of kinetic energy. Thus,

$$\Delta E = \Delta KE$$

Internal Energy Next, consider a simple system at rest. Suppose subsequently its state is changed to another state wherein it remains at rest with no change in elevation but, let us say, the volume, temperature, and pressure are different. The heat and work transferred to achieve this change of state must all appear as stored internal energy. Thus,

$$E_2 - E_1 = Q - W = U_2 - U_1$$

Finally, consider a system initially in some state 1 with elevation z_1 and velocity v_1 and state property values T_1, P_1, and V_1. Suppose the system is subjected to each of the above three simple processes simultaneously. That is, work and heat are transferred so that the state variables become T_2, P_2, and V_2; the elevation becomes z_2; and the velocity becomes v_2. Then, it seems reasonable that the overall change in the energy of the system be given by

$$E_2 - E_1 = (U_2 - U_1) + (\tfrac{1}{2}mv_2^2 - \tfrac{1}{2}mv_1^2) + (mgz_2 - mgz_1)$$

We can write the total energy change in this separable way because it is known from experience that the components of the overall process are not *coupled*. This means that raising the system in a gravitational field and accelerating the system will not change its temperature, pressure, and volume. We say that the processes can be *superimposed*. Thus, the total change in the energy of the system can be represented as a sum of changes each associated with a single simple process which affects a single energy form.

The expanded form of the first law balance equation for closed systems becomes

$$\Delta U + \Delta KE + \Delta PE = Q - W$$

Various specialized forms of this balance equation can be obtained by simply deleting terms from the balance equation. To take an example from classical mechanics, one is often concerned with *isolated systems*, that is, closed systems which exchange neither heat nor work with their surroundings. During any process involving such a system, it follows by definition

$$Q = W = 0$$

Consequently,

$$\Delta E = Q - W = 0$$

or

$$\Delta U + \Delta PE + \Delta KE = 0$$

Suppose, furthermore, that the isolated system undergoes a process such that only its elevation and velocity are affected but the remaining properties are unchanged. Then,

$$\Delta PE + \Delta KE = 0$$

This equation expresses the statement that the mechanical energy in an isolated system is conserved. This is the earliest form in which the law of conservation of energy was first postulated. It only applies to simple engineering systems in which frictional effects are negligible.

Example 6.4 A 1-kg slug of a stream of water leaves the bottom outlet of a vertical pipe with a velocity of 2 m/s and falls 100 m to a pool below. What is the velocity as it hits the pool?

Solution The initial and final states are, as shown in Figure 6.6, the end of the pipe and the surface of the pool. The system consists of the 1-kg mass of water. Assuming that the slug of water is an isolated system and assuming no internal energy changes, then

$$mg(z_2 - z_1) + \tfrac{1}{2}m(v_2^2 - v_1^2) = 0$$

or

$$
\begin{aligned}
v_2^2 &= 2g(z_1 - z_2) + v_1^2 \\
&= 2(9.80665 \ m/s^2)(100 \ m - 0) + (2 \ m/s)^2 \\
&= 1965.33 \ m^2/s^2
\end{aligned}
$$

The solution is

$$v_2 = 44.33 \ m/s$$

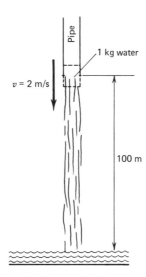

Figure 6.6 System for Example 6.4.

6.5.3 The General Conservation Equation

In this section, we derive the general form of the energy balance equation which is the quantitative statement of the principle of conservation of energy, or the first law of thermodynamics. For notational simplicity, we derive the equation by considering a flow system with a single inlet and a single outlet stream. The extension to multiple inlets and outlets then follows immediately. This general conservation equation subsumes the closed-system form that we have considered in the previous section and also subsumes the open steady-state system form which is employed in most flowsheet calculations. We shall discover that all of these special cases can be obtained by merely deleting the appropriate terms from the general balance equation.

Consider the arbitrary flow system shown in Figure 6.7 at some point t in time. The system has an inlet channel at an elevation z_1 relative to some datum plane and a single outlet channel at an elevation z_2. At time t, the center of mass of the system is located at elevation z and is moving with bulk velocity v. Mass is entering the system through the inlet channel at a mass flow rate dm_1/dt and is leaving the system at a mass flow rate dm_2/dt, and at time t the total mass of the system is m. The inlet fluid has an internal energy per unit mass $\hat{U}_1$, a temperature T_1, velocity v_1, pressure P_1, and volume per unit mass, $\hat{V}_1$. Similarly, the outlet fluid has unit

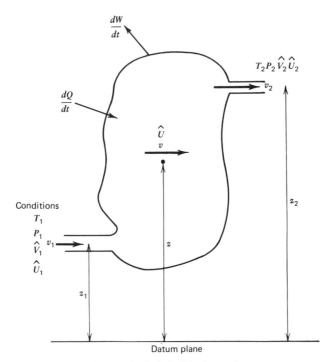

Figure 6.7   General single-inlet/single-outlet system.

internal energy $\hat{U}_2$, temperature T_2, velocity v_2, pressure P_2, and specific volume $\hat{V}_2$. At time t, the system has specific internal energy $\hat{U}$, heat is being added to the system at the rate dQ/dt, and work is being performed by the system on the surroundings at the rate dW/dt.

Since the mass entering the system at time t with rate dm_1/dt has associated with it internal, kinetic, and potential energies, the rate of energy input to the system by virtue of the flow of input mass will be given by

$$\left(\hat{U}_1 + gz_1 + \tfrac{1}{2}v_1^2\right)\frac{dm_1}{dt}$$

Similarly, the rate of energy removed from the system as a consequence of the output flow of mass at rate dm_2/dt will be

$$\left(\hat{U}_2 + gz_2 + \tfrac{1}{2}v_2^2\right)\frac{dm_2}{dt}$$

Finally, the total inventory of energy within the system at time t will be equal to the sum of its internal, kinetic, and potential energies, or

$$\left(\hat{U} + gz + \tfrac{1}{2}v^2\right)m$$

For any conserved quantity, the general accounting principle which must always hold is that the rate of input of the conserved quantity to the system must be equal to the rate of output of the conserved quantity from the system plus the rate of accumulation of the conserved quantity within the system. If energy is conserved, then it must be true that

$$\left(\begin{array}{c}\text{Rate of}\\\text{energy input}\end{array}\right) - \left(\begin{array}{c}\text{rate of}\\\text{energy output}\end{array}\right) = \text{rate of energy accumulation} \quad (6.1)$$

At time t, the rate of energy accumulation in the system of Figure 6.7 is simply equal to the rate of change of the energy inventory with respect to time. Thus,

$$\text{Rate of energy accumulation} = \frac{d}{dt}\left[\left(\hat{U} + gz + \tfrac{1}{2}v^2\right)m\right]$$

The rates of energy input and output will consist of two components: the input/output energy flows associated with mass crossing the system boundaries and the net transfer which occurs as heat and work. Since the heat transferred to the system is an input and the work done by the systems is an output, it follows that

$$\text{Rate of energy input} = \left(\hat{U}_1 + gz_1 + \tfrac{1}{2}v_1^2\right)\frac{dm_1}{dt} + \frac{dQ}{dt}$$

and

$$\text{Rate of energy output} = \left(\hat{U}_2 + gz_2 + \tfrac{1}{2}v_2^2\right)\frac{dm_2}{dt} + \frac{dW}{dt}$$

Substituting these expressions into the general balance expression, eq. (6.1), we obtain the energy conservation equation

$$(\hat{U}_1 + gz_1 + \tfrac{1}{2}v_1^2)\frac{dm_1}{dt} - (\hat{U}_2 + gz_2 + \tfrac{1}{2}v_2^2)\frac{dm_2}{dt} + \frac{dQ}{dt} - \frac{dW}{dt}$$

$$= \frac{d}{dt}[(\hat{U} + gz + \tfrac{1}{2}v^2)m] \quad (6.2)$$

While this equation was derived assuming a single inlet and a single outlet channel or stream, it is clear that additional channels or streams can be accommodated by merely including the energy contribution

$$(\hat{U} + gz + \frac{1}{2}v^2)\frac{dm}{dt}$$

associated with each stream into the balance equation with a plus sign for inputs and a minus sign for outputs. Thus, for multiple inlets with subscripts j and outlets with subscripts k, we can write

$$\sum_j (\hat{U} + gz + \tfrac{1}{2}v^2)_j\frac{dm_j}{dt} - \sum_k (\hat{U} + gz + \tfrac{1}{2}v^2)_k\frac{dm_k}{dt} + \frac{dQ}{dt} - \frac{dW}{dt}$$

$$= \frac{d}{dt}[(\hat{U} + gz + \tfrac{1}{2}v^2)m] \quad (6.3)$$

This general energy balance equation subsumes the special cases we have considered in the previous sections and serves as starting point for the generation of balance equations for all other special cases. Let us consider the following three special cases: the open steady-state system, the closed system, and the isolated system.

Steady-State Systems   By definition, a steady-state system is one whose point property values are invariant with time. Hence, it is a system in which all flow rates are constant with time and in which there is no net accumulation of mass or energy. Consequently, the accumulation term

$$\frac{d}{dt}[(\hat{U} + gz + \tfrac{1}{2}v^2)m]$$

must be equal to zero and the mass flows dm_j/dt and dm_k/dt can be replaced by the steady-state flow rates F^j and F^k. Thus, for steady-state open systems, we have the simplified energy balance equation

$$\sum_k (\hat{U} + gz + \tfrac{1}{2}v^2)_k F^k - \sum_j (\hat{U} + gz + \tfrac{1}{2}v^2)_j F^j = \frac{dQ}{dt} - \frac{dW}{dt} \quad (6.4)$$

This is the basic balance equation which we employ in our flowsheet calculations in the next several chapters.

Closed Systems Since a closed system can have no input or output flow, dm_j/dt and dm_k/dt must be zero for all j and k. The balance thus reduces to

$$\frac{dQ}{dt} - \frac{dW}{dt} = \frac{d}{dt}\left[(\hat{U} + gz + \tfrac{1}{2}v^2)m\right]$$

If we integrate this equation from initial time t^1 to final time t^2, corresponding to the initial and final states of the system, then we obtain the result

$$Q - W = \left[(U + gz + \tfrac{1}{2}v^2)m\right]_2 - \left[(\hat{U} + gz + \tfrac{1}{2}v^2)m\right]_1$$

where Q and W represent the total heat and work transferred in going from state 1 to state 2. Since a closed system must be at constant mass, it follows that $m_1 = m_2 = m$, and thus we recover the closed system balance

$$Q - W = \left[\Delta\hat{U} + g\Delta z + \tfrac{1}{2}(v_2^2 - v_1^2)\right]m$$

The isolated system case then follows by merely setting both Q and W to zero.

6.5.4 The Enthalpy Function

For flow systems which normally are of concern to chemical engineers, the work term dW/dt which arises in the general conservation equation will consist of two contributions: the flow work performed by the system as a result of material entering and leaving the system and all other work occurring within the system. Since the flow work can be directly associated with the separate input and output streams, it is convenient to separate out this work component and to combine it with the individual stream energy terms. This rearrangement of terms leads to the definition of a convenient special state function called the *enthalpy*, which is used to replace the internal energy function in flowsheet calculations.

Consider again the system of Figure 6.7. At time t, the input flow dm_1/dt enters the system at pressure P_1 and with specific volume $\hat{V}_1$. The surroundings, thus, perform work on the system to inject a volumetric flow $\hat{V}_1(dm_1/dt)$ into the system by exerting a pressure P_1. Specifically, the rate of work done by the surroundings will be equal to

$$P_1\hat{V}_1 \frac{dm_1}{dt}$$

Hence, the work done by the system will be

$$-P_1\hat{V}_1 \frac{dm_1}{dt}$$

Similarly, the work performed by the system at the outlet is equal to the work of expelling a volumetric flow $\hat{V}_2(dm_2/dt)$ against a constant pressure P_2. Thus, the work done by the system at the outlet is given by

$$P_2\hat{V}_2 \frac{dm_2}{dt}$$

Thus, for the single inlet/single outlet system, the total work done by the system will be

$$\frac{dW}{dt} = P_2 \hat{V}_2 \frac{dm_2}{dt} - P_1 \hat{V}_1 \frac{dm_1}{dt} + \frac{dW'}{dt}$$

where dW'/dt denotes all system work excluding the flow work. Substituting this expression into energy balance equation, eq. (6.2), of the previous section and grouping the flow work terms with the corresponding stream terms, we obtain the result

$$(\hat{U}_1 + P_1 \hat{V}_1 + gz_1 + \tfrac{1}{2}v_1^2)\frac{dm_1}{dt} - (\hat{U}_2 + P_2 \hat{V}_2 + gz_2 + \tfrac{1}{2}v_2^2)\frac{dm_2}{dt}$$

$$+ \frac{dQ}{dt} - \frac{dW'}{dt} = \frac{d}{dt}[(\hat{U} + gz + \tfrac{1}{2}v^2)m] \quad (6.5)$$

This analysis can of course be repeated in the multiple stream case with the result that a $P\hat{V}$ term is introduced for each stream. The grouping $\hat{U} + P\hat{V}$ which is thus naturally generated for each stream is called the *enthalpy function* and is usually denoted by the letter H. Thus, $\hat{H} = \hat{U} + P\hat{V}$ on a mass basis and $\tilde{H} = \tilde{U} + P\tilde{V}$ on a mole basis.

With this new energy function, eq. (6.3) becomes

$$\sum_j (\hat{H} + gz + \tfrac{1}{2}v^2)_j \frac{dm_j}{dt} - \sum_k (\hat{H} + gz + \tfrac{1}{2}v^2)_k \frac{dm_k}{dt} + \frac{dQ}{dt} - \frac{dW'}{dt}$$

$$= \frac{d}{dt}[(\hat{U} + gz + \tfrac{1}{2}v^2)m] \quad (6.6)$$

Note that in the special case of the steady-state system, the above balance equation reduces to a form which does not explicitly contain the internal energy function, namely,

$$\sum_k (\hat{H} + gz + \tfrac{1}{2}v^2)_k F^k - \sum_j (\hat{H} + gz + \tfrac{1}{2}v^2)_j F^j = \frac{dQ}{dt} - \frac{dW'}{dt} \quad (6.7)$$

Consequently, in steady-state applications, which are our primary concern in subsequent chapters, it will be the enthalpy rather than the internal energy function which will prove to be most useful. Note that by using the enthalpy function, the flow work is automatically taken into account with the result that in the steady-state case there is no need to ever deal with the combined work dW/dt explicitly. Hence, in subsequent usage of eq. (6.7), we drop the prime on the W with the understanding that the term dW/dt represents all work except the flow work.

Finally, the reader should note that the enthalpy is a *state* function, as is the internal energy U and the total energy E. As is the case for both E and U, absolute values of H are not known. Instead, an arbitrary reference value is assigned to H at some abitrary reference state, and a special form of the constant-pressure closed-system balance equation is used to calculate values of H at other states for tabulation purposes. A primitive form of an experimental apparatus which might be used for measuring the enthalpy of a gas, for instance, might be a cylinder equipped with a frictionless and leakproof piston which exerts a constant pressure P_1 on the gas contained within the cylinder. The gas is initially at some temperature T_1 and occupies a volume V_1. Suppose P_1, T_1, and $\hat{V}_1$ are selected as defining the reference state and suppose that a measured quantity of heat Q is transferred to the gas to raise its temperature to T_2 and its volume to $\hat{V}_2$ while the piston maintains the pressure at P_1. The gas constitutes a closed system, hence the conservation equation

$$Q - W = [\Delta \hat{U} + g\Delta z + \tfrac{1}{2}(v_2^2 - v_1^2)]m$$

is applicable.

Although the center of gravity of the gas will change as the volume expands, assume that the piston is sufficiently large in area so that the change in potential energy is negligible. Also, it is clear, since the system is at rest, that kinetic energy changes are zero. Thus, the balance equation reduces to

$$\Delta U = Q - W$$

The work done by the system in expanding V_1 to V_2 against the constant pressure P_1 will be

$$W = P_1(\hat{V}_2 - \hat{V}_1)m$$

Therefore,

$$m\,\Delta \hat{U} = Q - P_1(\hat{V}_2 - \hat{V}_1)m$$

or

$$Q = [(\hat{U}_2 - \hat{U}_1) + P_1(\hat{V}_2 - \hat{V}_1)]m$$
$$= [(\hat{U}_2 + P_1\hat{V}_2) - (\hat{U}_1 + P_1\hat{V}_1)]m$$

Using the definition of the enthalpy function, $\hat{H} = \hat{U} + P\hat{V}$, the equation reduces to

$$Q = (\hat{H}_2 - \hat{H}_1)m$$

Therefore, the enthalpy at the state T_2, P_1, $\hat{V}_2$ can be calculated using

$$\hat{H}_2 = \hat{H}_1 + \frac{Q}{m}$$

where $\hat{H}_1$ will be the chosen reference state enthalpy value. In this manner, by adjusting P and Q and measuring T, complete tabulations of H can be developed. In Chapter 7, we have occasion to examine and to use such tabulations for solving energy balance problems.

6.6 SUMMARY

In this chapter, we discussed the basic concepts and definitions necessary for the development of the energy balance equation. These included a study of the various forms in which energy can be stored and transferred. We reviewed the two main systems of units now in use and discussed the calculations required for conversion from one system of units to another. We then developed the formulation of the energy balance equation for closed systems and studied some of its properties. Next, we derived the general energy balance equation and showed that it subsumes both the closed and the open system cases. Finally, we introduced the concept of the enthalpy function. This state function was developed as a convenient way of grouping the internal energy and flow work terms. Use of the enthalpy function was shown to be particularly convenient in the case of open steady-state systems. In the next chapter, we investigate the various forms in which enthalpy data are tabulated or can be calculated and we learn how to solve simple process energy balance problems.

PROBLEMS

6.1 Which of the following properties are intensive and which are extensive: **(a)** concentration (kgmol/cm^3), **(b)** flow (mol/h), **(c)** velocity (ft/s), **(d)** specific volume (ft^3/lb$_m$), **(e)** pressure (lb$_f$/in.^{2}), **(f)** specific enthalpy (J/kg), **(g)** elevation (m), **(h)** mole fraction (mol/mol), **(i)** heat transfer rate (kcal/h), and **(j)** temperature (°F)?

6.2 Convert a pressure of 30 N/m^2 to **(a)** bar, **(b)** psia, and **(c)** atmospheres.

6.3 Convert a molar flux of 100 kgmol/m$^2 \cdot$ h to a mass flux in (lb$_m$/in.$^2 \cdot$ s) assuming the average molecular weight of the stream is 100 g/mol.

6.4 The specific internal energy of a closed system is 1200 kJ/kg, its specific volume is 0.01 ft^3/lb$_m$, and its pressure is 10 atm. Calculate the specific enthalpy in Btu/lb$_m$.

6.5 The pressure, volume, and temperature of an ideal gas at one set of conditions (denoted by subscript 1) can be related to corresponding values of pressure, volume, and temperature at another set of conditions (denoted by subscript 2) via the equation

$$\frac{P_1 V_1}{T_1} = \frac{P_2 V_2}{T_2}$$

Given $P_1 = 10$ atm, $V_1 = 100$ ℓ, $T_1 = 100$°C, $P_2 = 1000$ psig, and $T_2 = 100$°F, calculate the volume in in.^{3} at condition 2.

6.6 Which of the following quantities of work is largest in magnitude: 10^5 ft $\cdot$ lb$_f$, 10^6 Btu, 10^4 kJ, 50 kWh, 5×10^3 kcal, or 10^{10} kg $\cdot$ cm^2/min^2?

6.7 Calculate the value of the gas constant R in hp $\cdot$ h/lbmol $\cdot$ °R given that $R = 8.31343$ J/(mol $\cdot$ K). Use only exact conversion factors (i.e., definitions), if possible.

6.8 The value of a heat transfer coefficient is 200 Btu/h · ft^2 · °F. Calculate its value in SI units (J/s · m^2 · K).

6.9 A syrup has a viscosity of 500 centipoises (1 centipoise = 10^{-2} g/cm · s). Calculate its viscosity in lb$_m$/ft ·s.

6.10 A gas has a density of 0.005 lbmol/ft^3 and a molecular weight of 100 lb/ lbmol. Calculate its density in g/cm^3.

6.11 The normal human body temperature is taken to be 98.6°F. What is it in °C?

6.12 The temperature on the beach reads 40°C. What is the temperature in °F?

6.13 The heat capacity of a gas in J/mol · K is given by

$$C_p = 25 + 0.1T + 0.001T^2 - 10^{-6}T^3 + 10^{-9}T^4$$

where T is in K. Develop a heat capacity equation for C_p in Btu/lbmol · °R where the temperature is expressed in °R.

6.14 At what temperature, in °C, would a thermometer calibrated in K read the same as a thermometer calibrated in °F?

6.15 Originally, the Fahrenheit temperature scale was defined in terms of the ice point of water (32°F) and the steam point of water (212°F). Suppose that an inaccurate mercury-in-glass thermometer reads 35°F at the ice point and 210°F at the steam point.

    (a) What temperature would a correctly calibrated thermometer read when the inaccurate thermometer reads 70°F?

    (b) Is the inaccurate thermometer accurate at any temperature?

    (c) Determine the temperature range on the inaccurate thermometer for which the temperature read is within 1°F of that which would be read on a correctly calibrated thermometer.

6.16 The vapor pressure equation for a substance is

$$\ln p = 10 - \frac{6000}{t + 400}$$

where p is in psia and t is in °F.

    (a) Determine the constants A, B, and C of an equation in which pressure is in kPa and temperature is in K:

$$\ln(p \text{ (kPa)}) = A - \frac{B}{T \text{ (°K)} + C}$$

    (b) Verify your conversion by evaluating both equations at the given point $p = 100$ psia and $T = 212$°F.

6.17 Convert the coefficients of the Antoine equation for toluene given in Appendix 4 in SI units to obtain an equivalent equation in AE units (psia, °F).

6.18 A closed system has work performed on it at the rate of 100 hp and heat removed from it at the rate of 10^6 Btu/h. Calculate the rate of change of the energy inventory of the system in kilowatts.

6.19 A closed system with mass of 10 kg is subjected to the following changes:

1. Elevation change from 50 ft to 1000 ft.
2. Velocity change from 10 m/s to 100 m/s.
3. Change in internal energy from 1000 kJ/kg to 2000 kJ/kg.
4. Heat transferred to the system: 10^4 kcal.
5. Work performed by the system: 50 kWh.

Do these changes satisfy the closed-system energy balance equation?

6.20 A gas is transported in a horizontal pipeline at a flow of 10,000 lb_m/min. If it is moving at 10 ft/s, what rate of work (in horsepower) must be performed on the gas to increase its velocity to 200 ft/s at the same temperature and pressure (1 hp $= 550$ ft·lb_f/s)?

6.21 A 10-kg missile is flying vertically downward at 10 m/s at an elevation of 1000 m above the ground when it is accelerated by a burst from its rocket engine which imparts 10^5 J of work. What will be the missile velocity 10 m above the ground?

6.22 A liquid flowing at the rate of 10^4 U.S. gal/min which has a density of 70 lb_m/ft^3 and a bulk velocity of 5 ft/s is pumped from an elevation of 10 m to an elevation of 1000 m. If the velocity at the new elevation is 10 ft/s, what rate of work in (kJ/h) must be performed (1 U.S. gal $= 3.7854 \times 10^{-3}$ m^3)?

6.23 The specific enthalpy of a system of mass 15 kg at a selected reference pressure and temperature is taken to be 0 kJ/kg. Suppose that the state of the system is changed at constant pressure by adding 900 kcal of heat.

(a) Calculate the specific enthalpy at the new state.
(b) Suppose that the reference specific enthalpy value is defined to be 55 kJ/kg, calculate the specific enthalpy at the new state.

CHAPTER
7

Energy Balances for Nonreaction Systems

In order to use the open- and closed-system energy balance equations derived in the previous chapter, it is necessary to provide values of the enthalpy and internal energy functions which appear in these equations, given the macroscopic properties of the system in its final and initial states. For instance, in order to calculate Q for a constant-pressure process using the balance equation $Q = H_2 - H_1$, we must be able to evaluate H at the final and initial states. On the other hand, if H_1 and Q are known, the balance equation can be used to calculate H_2. However, in order to use that value of the enthalpy to deduce one or more of the unknown macroscopic properties of the final state, we must be able to invert the enthalpy function. Thus, solution of energy balance problems requires that we know the answers to the following questions:

1. What are the macroscopic properties of which H and U are functions?
2. In what forms are the H and U functions commonly available and how are they used?
3. How are the interconversion calculations between H and U and the macroscopic properties performed?

   In this chapter, we consider the first two questions in detail, restricting our attention to nonreacting systems. Our study of the third issue will of necessity be rather limited. The interconversion between temperature, pressure, phase composition, enthalpy, and internal energy is in general a very complex and difficult subject. It requires consideration of various equations of state, physical phase equilibrium, and other topics which are the subject matter of thermodynamics and are quite beyond the scope of this text. In this text and in particular in this chapter, we restrict our attention to simple systems for which interconversion is straightforward. We shall in general assume that:

1. The phases which are present in a system are always known.

2. The effects of pressure on H and U are negligible.
3. All mixtures are ideal.

These simplifying assumptions, which are satisfied in many engineering applications, will allow us to focus attention on the analysis of energy balance problems and on the development of solution strategies which will allow us to calculate the primary variables of concern, namely, T, Q, and W, as efficiently as possible.

7.1 CHARACTERIZATION OF THE STATE OF A SYSTEM

The enthalpy and internal energy functions are functions only of the variables defining the state of the system. In order to determine which and how many of the macroscopic properties of a system must be specified to unambiguously fix the enthalpy or internal energy of the system relative to some reference condition, we must first clarify how many properties must be specified to fix the state of a system. More precisely, we must clarify how many intensive properties must be specified to fix the equilibrium state of a system.

7.1.1 The Phase Rule

In general, the primary intensive properties used to characterize the equilibrium state of a system are the temperature, the pressure, and the composition of each of the phases. However, not all of these properties need to be specified in order to define the equilibrium state of a system. Rather, there is a precise number of system properties which, when specified, automatically defines the equilibrium state and thus fixes the values of the remaining intensive properties. That number of properties is called the thermodynamic *degree of freedom* of the system, and it depends upon the number of chemical components and the number of phases present in the system. The precise relationship between the degree of freedom D, the number of components C, and the number of phases ϕ is called the phase rule and is given by

$$D = C - \phi + 2$$

Correct application of this rule requires careful attention to the definitions of the three key quantities: phase, component, and degree of freedom.

The *phase* is an aggregate of matter which is homogeneous with respect to all its macroscopic properties: not only density, temperature, pressure, and composition, but also refractive index, dielectric constant, and so on. If the phase consists of a mixture of chemical species, then the mixing must occur down to the molecular level. Common phase examples are gas, liquid, and solid, and the most common multiphase system will involve a gas and a liquid phase. However, systems involving several liquid and solid phases do occur in applications of engineering significance.

For nonreacting systems, a chemical *component* as used in the context of the phase rule is the same as a molecular species. Thus, in the nonreacting case, $C = S$, where S is the number of species as used in the context of species material balances. In the reacting case, the number of components is equal to the number of species minus the number of independent chemical reactions, $\rho(\sigma)$, occurring between these species. Thus,

$$C = S - \rho(\sigma)$$

The presence of reactions thus serves to reduce the thermodynamic degree of freedom. This parallels the effective reduction in the number of independent species balances which occurs when solving material balance problems for reacting systems.

The degree-of-freedom concept as used in the phase rule is restricted to the primary intensive properties: temperature, specific volume, pressure, and composition of each of the phases. If compositions of the C components are expressed in terms of normalized measures such as mass or mole fractions, then, as is the case in material balancing, only $C - 1$ of these compositions are considered independent. Furthermore, the phase rule excludes properties such as elevation in a gravitational field, velocity, or other properties associated with mechanical or electromagnetic field effects.

The phase rule can be derived from basic thermodynamic principles, but such derivation requires concepts beyond the scope of this text.

7.1.2 Single-Component Systems

For single-component nonreacting systems, the phase rule reduces to

$$D = 3 - \phi$$

Since the degree of freedom cannot be negative, the phase rule indicates that the pure component can exist in at most three phases in equilibrium. The unique point at which a pure substance will be present in three equilibrium phases is called the *triple point*. For instance, the triple point of water at which ice, liquid water, and water vapor exist in equilibrium occurs at 273.16 K (by definition) and 611 Pa.

If the system consists of two phases, then it has one degree of freedom: either the temperature or the pressure may, for example, be specified but not both. A typical example of a two-phase single-species system is liquid water and its vapor. At any given pressure, there is a unique temperature at which water will boil. Similarly, at any given temperature, there is a pressure at which the two phases will coexist at equilibrium. The locus of temperature and pressure conditions at which vapor and liquid will be in equilibrium is called a *vapor pressure curve* (see Figure 7.1). The vapor pressure curve extends from the triple point involving vapor and liquid to the highest temperature and pressure at which vapor and liquid can exist at equilibrium. The latter point is called the *critical point*. For water, this point occurs at the *critical temperature* of 647.3K and the *critical pressure* of 22.109 $\times$ 10^3 kPa. Critical point conditions are used as the basis for many pure component property correlations and thus are key characteristics of a species. Values for a number of chemicals are given in Appendix 5. Vapor pressure curves in the form of an empirical equation (called an Antoine equation) of the type

$$\ln p = A - \frac{B}{T + C}$$

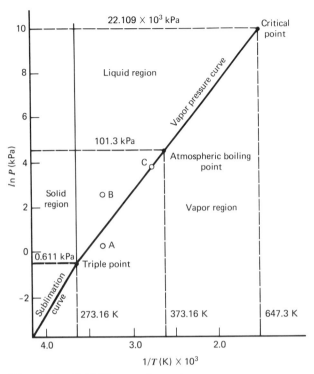

Figure 7.1 *P–T* Diagram for water.

are reported for various common chemicals in references such as the *Chemical Engineer's Handbook*.[1] A tabulation of Antoine equation constants is given in Appendix 4. These equations are obtained by choosing values of *A*, *B*, and *C* so as to fit experimentally measured equilibrium data over suitable temperature and pressure ranges. If a vapor pressure equation is available, interconversion between equilibrium *T* and *P* is easily carried out by rearranging the equation to calculate one property given the other.

Example 7.1 The vapor pressure of acetone is given by

$$\ln p = 16.7321 - \frac{2975.95}{T - 34.5228}$$

where *p* is expressed in mmHg and *T* in K. A vapor–liquid mixture of acetone is found to exist in equilibrium at 100 mm Hg. What is the equilibrium temperature?

Solution This is a two-phase, single-component system. Hence, its thermodynamic degree of freedom is 1. Thus, specification of the pressure defines the

---

[1] R. H. Perry and C. H. Chilton, Eds. *Chemical Engineers Handbook*, 5th ed., McGraw-Hill, New York, 1973.

equilibrium state. The equilibrium temperature can be calculated from the vapor pressure equation

$$\frac{2975.95}{T - 34.5228} = 16.7321 - \ln 100 = 12.12693$$

Therefore, $T = 245.40 + 34.5228 = 279.92$ K.

If the pure substance exists in a single phase, then, according to the phase rule, it has two degrees of freedom. Hence, any two of temperature, pressure, and molar or specific volume must be specified to define an equilibrium state, that is, to fix the third property. In the case of gas- or liquid-phase systems, this relationship between the equilibrium temperature, pressure, and molar volume is most commonly expressed in terms of an *equation of state*. The simplest equation of state is that of an *ideal gas*, namely,

$$P\tilde{V} = RT$$

where R is a proportionality constant, called the gas constant, which has the value 8.3143 J/mol · K, or 10.73 ft^3 · psia/lbmol · °R. As required by the phase rule, when any two of P, T, or $\tilde{V}$ are specified, the equation of state can be used to calculate the third property. Thus, in the ideal gas case, the interconversion between the three intensive equilibrium properties is quite straightforward.

Other more complex equations of state have been developed to accommodate conditions under which substantial deviations from the ideal gas law are encountered. For example, the Redlich–Kwong equation

$$P = \frac{RT}{\tilde{V} - b} - \frac{a}{T^{1/2}\tilde{V}(\tilde{V} + b)}$$

where a and b are empirical constants characteristic of a component is frequently used in precise engineering calculations. While more accurate than the ideal gas equation, this equation makes interconversion between P, T, and $\tilde{V}$ more difficult. Equations of state are also available for the liquid state. For a detailed discussion of equations of state and their use, the reader is referred to any of a number of chemical engineering thermodynamics texts.[2]

In the preceding discussion, we have considered how many properties must be specified to define the equilibrium state of a single-component system *given* the number of phases which are present in the system. A converse problem is that of determining how many properties must be specified in order to identify what phase or phases actually are present. For a single-component system, the phase rule indicates that this component can be present in as many as three phases and as few as one, since

$$\phi = 3 - D \qquad \text{and} \qquad D \geqslant 0$$

---

[2] K. E. Bett, J. S. Rowlinson, and G. Saville, *Thermodynamics for Chemical Engineers*, MIT Press, Cambridge, MA, 1975.

In order to identify which phases are present, we need to be able to specify the situation with the largest degree of freedom, namely, the one-phase case. Thus, two properties need to be specified. For instance, if P and T are known, we can first check whether this condition corresponds to one of the known triple points of the component. If not, three phases cannot be present. Next, if a two-phase equilibrium relation, for example, a vapor pressure curve, is available, we can check whether the P and T pair satisfies this relation. If not, then these two phases are not present in equilibrium. However, the individual phases may be present. For instance, if the specified pressure is below the equilibrium vapor pressure at the specified temperature, the component must exist in the gas phase. If the specified pressure is above the vapor pressure, the component must be in the liquid phase.

Example 7.2   Consider the P–T diagram for water shown in Figure 7.1. Suppose a system containing only water is in equilibrium at

1. 300 K and 1.3 kPa.
2. 300 K and 13 kPa.
3. 350 K and 41.69 kPa.

Identify the phases present.

Solution   Condition 1 corresponds to point A in Figure 7.1 and lies entirely in the vapor region. The pressure is below the equilibrium vapor pressure at 300°K.

Condition 2 corresponds to point B in Figure 7.1, which lies in the liquid region. The pressure is above the vapor pressure at 300 K.

Condition 3 corresponds to point C and lies on the vapor pressure curve. Both vapor and liquid phases will be present under these conditions. If we had known that both phases were present, the temperature or the pressure would have sufficed to specify the state of the system.

7.1.3 Multicomponent, Multiphase Systems

As the number of components and phases increases, the specification and determination of equilibrium states become increasingly complex. In general, in addition to specifying T, P, or $\tilde{V}$, it is also necessary to specify phase compositions. For example, for a two-component gas mixture, the phase rule indicates a degree of freedom of 3:

$$D = 2 - 1 + 2 = 3$$

Thus, if the temperature and pressure are specified and the mixture of the two gases is assumed to behave as an ideal gas, then the mixture molar volume can be calculated via

$$\tilde{V} = \frac{RT}{P}$$

If, in addition, the mole fraction x of one of the gases is specified, then the state of the system is completely specified.

If the two-component system has two phases present, then, since

$$D = 2 - 2 + 2 = 2$$

specification of any two of T, P, the mole fraction of a component in one phase, and the mole fraction of a component in the second phase will define the equilibrium state. For instance, if for a gas–liquid mixture the temperature and pressure are specified, there will be a unique pair of vapor and liquid compositions which will exist in equilibrium with each other. For gas–liquid mixtures, this relationship between the mole fractions of a component in the two phases is frequently expressed in terms of empirically determined "K" factors. The "K" factors are defined as

$$K_i = \frac{y_i}{x_i}$$

where y_i is the mole fraction of species i in the vapor phase and x_i is the mole fraction of species i in the liquid phase. This type of equilibrium data is commonly employed in distillation and other vapor–liquid separation calculations. For some simple mixtures, the K_i are functions only of T and P:

$$K_i = K_i(T, P)$$

For instance, for mixtures of components of similar molecular structure, K_i can be approximated quite well by the ratio of the vapor pressure of species i at the given temperature to the total system pressure. That is,

$$K_i = \frac{p_i(T)}{P}$$

This is usually called *Raoult's law*. In two-component cases in which Raoult's law is applicable, when T and P are specified, the K factors for each of the components will be fixed. Thus, the equations

$$K_1 = \frac{y_1}{x_i}$$

$$K_2 = \frac{y_2}{x_i}$$

can be solved for the unique composition, as predicted by the phase rule. To confirm this, rewrite the definition of K_1 as

$$K_1 = \frac{y_1}{x_1} = \frac{1 - y_2}{1 - x_2}$$

and substitute the definition

$$K_2 x_2 = y_2$$

to obtain

$$K_1(1 - x_2) = 1 - K_2 x_2$$

Solving for x_2, we have

$$x_2 = \frac{K_1 - 1}{K_1 - K_2} \tag{7.1}$$

or x_2 is expressed only in terms of the K values. The remaining composition can then be calculated via the equations

$$x_1 = 1 - x_2$$
$$y_1 = K_1 x_1$$
$$y_2 = K_2 x_2$$

Thus, given T and P, all phase compositions of the two-component, two-phase system can be determined.

Alternatively, if P and y_1 are specified, then again, from the phase rule, the state of the two-component, two-phase system is fixed. However, to determine the equilibrium T and remaining compositions, an iterative calculation will be required. In particular, since y_2 can be calculated from $1 - y_1$ and since by definition

$$x_1 + x_2 = 1$$

we obtain upon substituting the definitions of the K factors the equation

$$\frac{y_1}{K_1(T, P)} + \frac{y_2}{K_2(T, P)} = 1$$

In this equation, P and the y values are known, however, determination of the equilibrium T will require a trial-and-error calculation using the tabulated or correlated values of the K_i as functions of temperature.

In some complex mixtures, the K_i are functions also of the liquid compositions, that is,

$$K_i = K_i(T, P, x_1, x_2)$$

In such cases, the calculation of the equilibrium values of the remaining intensive variables becomes even more involved. Thus, once a system has two or more components and involves more than one phase, the interconversion between equilibrium properties in general becomes fairly complicated.

Dew and Bubble Points The problem of identifying the phases present in a system involving two or more components is similarly complicated. In the case of a two-component system, the phase rule indicates that these components can exist in as many as four or as few as one equilibrium phases. Therefore, to identify which phases are present, three properties must be known: T, P, and one composition. Suppose, for instance, that a mixture of two components containing mole fraction z_1 of the first component is allowed to equilibrate at fixed T and P conditions. If the mixture is to exist as a single gas phase, the specified temperature T must be above the *dew point* temperature for a gaseous mixture of composition z_1 at pressure P. The dew point temperature is the temperature at which the first liquid drop would form when the temperature of a mixture of vapors is slowly decreased at a specified constant pressure. If the mixture is to exist as a single liquid phase, then the specified temperature T must be below the *bubble point* for a liquid mixture of composition z_1 at pressure P. The *bubble point* temperature is the temperature at which the first vapor bubble would form when the temperature of a liquid mixture is slowly increased at a specified constant pressure. If the specified temperature is between the dew point temperature T_D and the bubble point temperature T_B, the mixture will exist as a gas–liquid phase mixture. Thus, identification of these possibilities will require determination of T_B and T_D; moreover, if T is between T_D and T_B, additional calculations must be performed to actually determine the gas–liquid phase compositions.

If the K value relationships are available for the two components in question, then the dew point determination proceeds as follows. By definition of the dew point, we assume that the mixture is all vapor with composition $y_1 = z_1$ and $y_2 = 1 - z_1$, and we seek to calculate the composition of the liquid which will be in equilibrium with this vapor at pressure P. This is equivalent to the two-component, two-phase case discussed previously in which the vapor composition and pressure are known and the liquid composition and temperature must be calculated. Thus, the mixture dew point temperature is found by finding T such that for fixed P and z_1 the relation

$$\frac{z_1}{K_1(T, P)} + \frac{1 - z_1}{K_2(T, P)} = 1 \tag{7.2}$$

is satisfied.

Similarly, to derive the bubble point temperature, we assume the mixture is all liquid with composition $x_1 = z_1$ and $x_2 = 1 - z_1$, and we seek to calculate the composition of the vapor which would be in equilibrium with this liquid at the specified pressure P. Thus, we seek y_1, y_2, and T such that

$$y_1 + y_2 = 1$$

and

$$y_1 = K_1(T, P)x_1$$
$$y_2 = K_2(T, P)x_2$$

Upon substitution, we obtain the equation

$$z_1 K_1(T, P) + (1 - z_1)K_2(T, P) = 1 \tag{7.3}$$

which must be solved by trial-and-error selection of T. The above dew and bubble point temperature calculations, although derived in terms of two components, can be immediately extended to more components by simply adding a term for each additional component in eqs. (7.2) and (7.3).

Example 7.3   An equimolar mixture of benzene and toluene is stored at 1000 mm Hg (133.22 kPa). Below what temperature will the mixture exist as a single liquid phase? Above what temperature will it exist as a single gas phase?

The vapor pressure curve for the pure components can be approximated using the Antoine equation:

$$\text{Benzene:} \quad \ln\left[p \text{ (mm Hg)}\right] = 16.1753 - \frac{2948.78}{T - 44.5633}$$

$$\text{Toluene:} \quad \ln\left[p \text{ (mm Hg)}\right] = 16.2665 - \frac{3242.38}{T - 47.1806}$$

where T is in K. Assume Raoult's law is sufficiently accurate for this system.

Solution   The K values for benzene and toluene will be given by

$$K_i = \frac{p_i}{P}$$

Thus,

$$\ln K_i = \ln p_i - \ln P = \ln p_i - 6.9078$$

Consequently, for benzene,

$$\ln K_1 = 9.2675 - \frac{2948.78}{T - 44.5633}$$

and for toluene,

$$\ln K_2 = 9.3587 - \frac{3242.38}{T - 47.1806}$$

To carry out the bubble point calculation, we need to find the value of T which satisfies the equation

$$f(T) = \tfrac{1}{2}K_1(T, 1000 \text{ mm Hg}) + \tfrac{1}{2}K_2(T, 1000 \text{ mm Hg}) - 1 = 0$$

Since this is a nonlinear root-finding problem, we can employ Newton's method, as discussed in Section 5.1.3. To evaluate the iteration formula

$$T = T^0 - \left(\frac{df}{dT}\bigg|_{T=T^0}\right)^{-1} f(T^0)$$

we need the derivative df/dT. This can be calculated directly from the formula for $\ln K$. If

$$\ln K = A - \frac{B}{T - C}$$

then

$$\frac{d}{dT}(\ln K) = \frac{+B}{(T - C)^2} = \frac{1}{K}\frac{dK}{dT}$$

Thus,

$$\frac{dK}{dT} = \frac{+KB}{(T - C)^2}$$

and the complete derivative of f will be given by

$$\frac{df}{dT} = \frac{1}{2}\frac{dK_1}{dT} + \frac{1}{2}\frac{dK_2}{dT} = \frac{1}{2}\left(\frac{K_1B_1}{(T - C_1)^2} + \frac{K_2B_2}{(T - C_2)^2}\right)$$

To begin the iteration, we need to supply an initial temperature estimate. Suppose we select $T^0 = 373$ K. Then,

$$f(373 \text{ K}) = -0.05587$$

$$\frac{df}{dT}\bigg|_{T=373\text{K}} = 0.02669$$

The new estimate of T becomes

$$T^1 = 373 \text{ K} - (0.02669)^{-1}(-0.05587) = 375.09 \text{ K}$$

At this temperature,

$$f(375.09 \text{ K}) = 0.001315$$

If we seek an accuracy of $\varepsilon = 10^{-4}$, then the iterations must be continued. The new estimate of T becomes

$$T^2 = 375.09 \text{ K} - (0.02796)^{-1}(0.001315) = 375.05 \text{ K}$$

At this temperature,

$$f(375.05 \text{ K}) = 6.5 \times 10^{-7}$$

which is within the accuracy of 10^{-4} required. Thus, the mixture bubble point temperature is 375.05 K, or 101.90°C.

To calculate the dew point, the above procedure must be repeated with

$$f(T) = \frac{0.5}{K_1(T)} + \frac{0.5}{K_2(T)} - 1 = 0$$

In this case,

$$\frac{df}{dT} = -0.5\left[\frac{B_1}{K_1(T - C_1)^2} + \frac{B_2}{K_2(T - C_2)^2}\right]$$

At an initial estimate of $T^0 = 383$ K,

$$f(383 \text{ K}) = -0.04031$$

$$\frac{df}{dT}\bigg|_{T=T^0} = -0.02673$$

Thus,

$$T^1 = 383 - (-0.02673)^{-1}(-0.04031) = 381.49 \text{ K}$$

At this temperature,

$$f(381.49 \text{ K}) = 0.00105$$

which does not meet our accuracy requirements. Thus, we continue with the iterations, to obtain $T^2 = 381.53$ K. At this temperature,

$$f(T^2) = 0.000090$$

which is within our accuracy requirement of 10^{-4}. The dew point of the mixture is found to be 381.53 K, or 108.38°C.

Hence, for $T \leq 375.05$ K, the mixture will exist as a liquid phase; while for $T \geq 381.53$ K, the mixture will exist as a gas phase.

As noted earlier, if the specified mixture temperature falls between the bubble and dew point temperatures at the given pressure, then the mixture will be present in equilibrium gas–liquid phases. For a system with two components present in two phases, the thermodynamic degree of freedom is 2, that is,

$$D = 2 - 2 + 2 = 2$$

Thus, with the temperature and pressure fixed, the two-phase equilibrium state is unequivocally defined. In particular, with T and P fixed, if K factor equations are available, then the system of equations

$$K_1(T, P) = \frac{x_1}{y_1} = \frac{1 - x_2}{1 - y_2}$$

$$K_2(T, P) = \frac{x_2}{y_2}$$

can be directly solved for the phase compositions. As was previously shown, eq. (7.1), the result is

$$x_2 = \frac{K_1 - 1}{K_1 - K_2}$$

and

$$x_1 = 1 - x_2$$

The y_i values can then be calculated directly from the definitions of the K values.

Example 7.4   Calculate the phase compositions of the mixture of Example 7.3 if the mixture is at 378.15 K and 133.22 kPa.

Solution   Using the expressions for the K_i given in Example 7.3 and evaluating the K_i at 378.15 K (105°C), we have

$$K_1 = 1.5340$$

$$K_2 = 0.6453$$

Using eq. (7.1), the result is

$$x_2 = \frac{1.5340 - 1}{1.5340 - 0.6453} = 0.6009$$

and

$$x_1 = 1 - x_2 = 0.3991$$

The vapor-phase compositions can immediately be calculated from the K_1 relation

$$y_1 = K_1 x_1 = 0.6122$$

$$y_2 = 1 - y_1 = 0.3878$$

Note that the species distribute between the vapor and liquid phases, with the former being richer in benzene and the latter richer in toluene. Note further that the phase compositions are *independent* of the bulk composition of the mixture. The bulk compositions were not used at all in this calculation. Specification of the bulk composition z_1 will only serve to fix the division of the bulk mass between the vapor and liquid phases.

Isothermal Flash Calculation   The above calculation of the phase compositions of a gas–liquid mixture of two components cannot be directly extended to more than two components. If more than two components are present, then a calculation known as an *isothermal flash calculation* must be performed. In general, as was the case with dew and bubble point calculations, an iterative numerical solution will be required.

Consider a system in two phases and S number of species and with T and P specified. The thermodynamic degree of freedom of this system will be

$$D = S - 2 + 2 = S$$

With T and P given, this is reduced to $S - 2$. Thus, if $S - 2$ of the bulk compositions of the mixture are given, the state of the system is thermodynamically fixed. If the composition of $S - 1$ species is specified, then the problem will also be fully specified from the point of view of the material balances.

To that end, consider the system shown in Figure 7.2. Given 1 mol/h of a mixture with bulk compositions z_i, $i = 1, \ldots, S$, at specified T and P, we seek to determine the distribution of this mixture into V mol/h of vapor of composition y_i, $i = 1, \ldots, S$, and L mol/h of liquid of composition x_i, $i = 1, \ldots, S$. Since, T, P, and the z_i are known, the only remaining unknowns are the vapor and liquid streams. Thus, we have $2S$ independent unknowns. There are S material balance

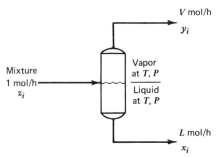

Figure 7.2   Isothermal flash unit.

equations available and S K-value specification equations. Thus, the constrained material balance problem is correctly specified.

The balance equations are

$$z_i = Vy_i + Lx_i \qquad i = 1, \ldots, S$$

and the K value equations are

$$y_i = K_i x_i \qquad i = 1, \ldots, S$$

Depending upon whether we use the K value equations to substitute out the y_i or the x_i variables, two forms of the balance equations can be obtained:

$$z_i = (VK_i + L)x_i \qquad (7.4)$$

$$z_i = (VK_i + L)\frac{y_i}{K_i}$$

If we use the expressions in terms of the x_i, solve for the x_i, and add them together, we obtain

$$\sum_{i=1}^{S} x_i = 1 = \sum_{i=1}^{S} \frac{z_i}{VK_i + L} \qquad (7.5)$$

Similarly, if we use the reduced balances expressed in terms of the y_i, solve for the y_i, and add the results, we have

$$\sum_{i=1}^{S} y_i = 1 = \sum_{i=1}^{S} \frac{K_i z_i}{VK_i + L} \qquad (7.6)$$

If we next subtract eq. (7.6) from eq. (7.5), term for term, the ones on the left-hand sides cancel and the result is

$$\sum_{i=1}^{S} \frac{z_i - K_i z_i}{VK_i + L} = 0$$

Since $L = 1 - V$, we obtain the reduced equation

$$\sum_{i=1}^{S} \frac{z_i(1 - K_i)}{1 + V(K_i - 1)} = 0 \qquad (7.7)$$

The only unknown in this equation is the vapor flow V, since the z_i are specified and the K_i are known because T and P are specified. Thus, the problem is reduced to solving eq. (7.7) for V, where V must take on a value between 0 and 1. Once V is determined, L can be calculated from the overall balance

$$L = 1 - V$$

Then, the x_i can be recovered from eqs. (7.4) and the y_i from the K value equations. Note that when $S > 2$, eq. (7.7) is a nonlinear equation in V. Note further that since V is based on a unit feed, it can be viewed as a flow ratio, moles vapor per mole feed.

Example 7.5   Repeat the calculation of Example 7.4 using the isothermal flash equation.

   Solution   As previously evaluated, at 378.15 K (105°C),

$$K_1 = 1.5340$$

$$K_2 = 0.6453$$

Since $z_1 = z_2 = \frac{1}{2}$, eq. (7.7) becomes

$$\frac{\frac{1}{2}(-0.5340)}{1 + V(0.5340)} + \frac{\frac{1}{2}(0.3547)}{1 + V(-0.3547)} = 0$$

Since only two terms in V are present, this equation can be directly rearranged to yield

$$1 - 0.3547V = \frac{0.3547}{0.5340}(1 + 0.5340V)$$

$$2(0.3547)V = 1 - \frac{0.3547}{0.5340} = 0.3358$$

$$V = 0.4733$$

From eq. (7.4),

$$x_1 = \frac{z_1}{1 + V(K_1 - 1)}$$

$$= \frac{0.5}{(1 + 0.4733(0.5340)} = 0.3991$$

which agrees with the previous solution.

7.2   ENERGY BALANCES USING TABULAR THERMODYNAMIC DATA

In the previous section, we discussed the use of the phase rule to determine how many state variables must be specified to uniquely define the state of a system. Since the enthalpy and internal energy functions are functions of the state variables only, their values will also be uniquely defined once the appropriate number of state variables, as dictated by the phase rule, is specified. In the case of single-component systems, we noted that no more than two and as few as none of the state variables had to be specified depending upon the number of phases present in the system. Consequently, for single components, the specific enthalpy and specific internal energy functions can be expressed as functions of at most two of any of the three state variables P, T, and $\hat{V}$. By convention, the specific enthalpy function for a single pure component is expressed in terms of T and P, while the pure-component specific internal energy function is expressed in terms of T and $\hat{V}$. Thus we write

$$\hat{H} = \hat{H}(T, P)$$

and

$$\hat{U} = \hat{U}(T, \hat{V})$$

Of course, in the region of T and P over which the pure component exists as a single phase, both T and P will need to be specified to determine the enthalpy. However, in regions in which two phases will exist in equilibrium, T and P cannot be independently specified, and hence the specific enthalpy function collapses to effectively become a function of a single variable.

For purposes of energy balance calculations, it would of course be most convenient if we had the pure-component specific enthalpy functions available as explicit algebraic functions of T and P, in the single-phase case, and, say, T in the two-phase case so that we could evaluate them directly. Presumably, we would need an enthalpy function for each of the possible single phases, solid, liquid, and vapor, as well as for the two-phase regions. Such functions are available for some substances such as water, ammonia, air, and standard refrigerants. However, they are typically quite complicated, contain many terms, and are awkward to evaluate for hand calculations. For example, the best available specific enthalpy function for water vapor or steam is an algebraic function involving some 60 constants determined by fitting experimental data. Moreover, to use such functions, we need to know which phase is present. Hence, we would need to have available additional relations such as a vapor pressure equation as well as equations of state which would allow us to carry out P, T, $\hat{V}$ interconversions. Therefore, for manual calculations, it has proved convenient to represent the pure-component specific enthalpy and internal energy functions in tabular form. Tabular representations can be fairly compactly constructed because the state functions will involve at most two independent variables which can be arranged in rows and columns. The tabular representation also allows convenient insertion of information which identifies phase transition boundaries and which permits state variable interconversion via equation-of-state data. The resulting tables thus can conveniently contain all of the necessary thermodynamic information for a pure-component species and hence are called *thermodynamic tables*.

In this section, we discuss how such tables are organized and what are the subsidiary calculations required for their use. We also illustrate how to use these tables to perform closed- and open-system energy balances. To avoid redundancy, we restrict our attention to a single tabulation, namely, to the thermodynamic tables for water. However, the reader should appreciate that the concepts involved are the same regardless of the pure component of concern.

7.2.1 The Steam Tables

The most complete and precisely known tabulations of the enthalpy and internal energy functions are those for water. This came about not only because water is such a ubiquitous substance but also because water, especially in its vapor form, has been the prime working fluid for devices such as steam engines and steam turbines. Precise design of these devices requires accurate values of these state functions as well as of the equation of state. Extensive and very detailed tabulation of the thermodynamic properties of liquid water and its vapor form, commonly called steam, can be found in the compilation due to Keenan and Keyes.[3] We use an abbreviated tabulation given in Appendix 8.

Before proceeding to discuss the contents of these tables, let us briefly review some of the terminology used in steam calculations. Recall that water can exist in solid, liquid, and vapor or steam phases and that under certain ranges of condition these phases will exist in equilibrium. At a given pressure, liquid water below its boiling or *saturation* temperature is referred to as *unsaturated liquid*; liquid water at its boiling temperature, as *saturated liquid*; the vapor produced at the boiling point, as *saturated steam*; and vapor heated beyond the boiling temperature, as *superheated steam*. Referring to the *P–T* diagram for water shown in Figure 7.1, unsaturated liquid is liquid in the region above the vapor pressure curve, saturated water and steam are the equilibrium liquid and vapor under conditions represented by the vapor pressure curve, and superheated steam is vapor in the region below and to the right of the vapor pressure curve. Superheated steam is usually characterized by a quantity called *degree of superheat*. The degree of superheat is equal to the number of temperature degrees that the steam is above its *saturation* or boiling temperature at the given pressure.

The steam table given in Appendix 8 consists of three parts: a saturated steam temperature table, a saturated steam pressure table, and a superheated steam table. All three tables list values of the specific volume, which is the reciprocal of the density, the specific internal energy, and the specific enthalpy. The saturated steam tables list the properties of liquid water and steam when they are at saturated conditions, that is, in equilibrium with each other. Since for a pure-component system present in two phases the degree of freedom is 1, the saturated steam tables give values of specific volume, enthalpy, and internal energy as a function of a single independent variable. The first table uses temperature as independent variable; the second table uses pressure as independent variable. Note that the saturated steam tables, in addition to listing the enthalpy of water and steam, also give a third intermediate column entitled "Evaporation." This column is simply

---

[3] J. H. Keenan and F. G. Keyes, *Thermodynamic Properties of Steam*, Wiley, New York, 1936.

the difference between the specific enthalpy of steam and liquid water and is called the *enthalpy of vaporization*, or the latent heat. Finally, note that at 0.01°C and 0.00611 bar (10^5 Pa), the internal energy of saturated liquid water is given as zero. This is the reference state for these steam tables, and the value of zero for the specific internal energy is the assigned reference value.

The third portion of the steam tables, the superheated steam table, gives the specific enthalpy $\hat{H}$, specific internal energy $\hat{U}$, and specific volume $\hat{V}$ of steam as a function of pressure and temperature. Since the thermodynamic degree of free-dom of a single-component, single-phase system is 2, both state variables must be specified to define the values of the other state variable, $\hat{V}$, and the values of the state functions $\hat{H}$ and $\hat{U}$. For convenience, the table also lists the saturated tem-perature below the pressure and includes the saturated steam and water properties in the second and third column. Finally, the table also gives the critical pressure, 221.2 bar, and critical temperature, 374.15°C, at which the distinction between steam and liquid water properties disappears.

Example 7.6

(a) Evaluate the enthalpy of 1 kg saturated liquid water at 80°C and determine the saturation pressure.

(b) Evaluate the enthalpy of 10 kg saturated vapor at 1.0 bar and determine its saturation specific volume.

(c) Evaluate the internal energy of 1 kg saturated vapor at a specific volume of 2.20 m³/kg.

(d) Evaluate the enthalpy of 100 kg steam at 20 bar and 500°C and calculate the degrees of superheat.

(e) Calculate the specific enthalpy of liquid water at the triple point.

Solution

(a) To determine saturation properties at a given temperature, it is most con-venient to use the saturated steam temperature table. At 80°C, the saturation pressure is 0.4736 bar and the specific enthalpy of liquid water is 334.9 kJ/kg. The enthalpy of 1 kg is thus 334.9 kJ.

(b) To determine saturation properties at a given pressure, it is appropriate to use the pressure table. At 1.0 bar, the saturation specific enthalpy of steam is 2675.4 kJ/kg and the specific volume is 1.694 m³/kg. Thus, the enthalpy of 10 kg steam is 26754 kJ.

(c) Although the saturated steam tables are arranged by either temperature or pressure, the specific volume is a perfectly acceptable independent state variable. To determine the other saturation properties given the saturated steam specific volume, we simply scan the steam specific volume column of either of the tables to locate the entry 2.20. In this case, the value 2.20 for $\hat{V}$ of steam corresponds to 92°C in the temperature table. The internal energy of saturated vapor under those conditions is 2496 kJ/kg. The internal energy of 1 kg is thus 2496 kJ.

(d) At 20 bar, we can confirm from the saturated steam pressure table that the saturation temperature is 212.4°C. The 500°C steam is thus superheated with $500 - 212.4 = 287.6$°C of superheat. To determine the specific enthalpy, we must use the superheated steam tables. At 20 bar and 500°C, the specific enthalpy is 3467 kJ/kg. Hence, the enthalpy of 100 kg is 346,700 kJ.

(e) The steam tables are given with the triple point as reference state and with the reference value of exactly zero for the specific internal energy. Since the internal energy and enthalpy are related via the definition

$$\hat{H} = \hat{U} + P\hat{V}$$

the value of $\hat{H}$ at the reference state can be calculated as

$$\hat{H}^0 = \hat{U}^0 + P^0\hat{V}^0$$

$$= 0.000 + 611 \frac{N}{m^2} \times 10^{-3} \frac{m^3}{kg} = 0.611 \frac{N \cdot m}{kg}$$

$$= 0.611 \times 10^{-3} \text{ kJ/kg}$$

The specific enthalpy of saturated water at the triple point is listed as $+0.0$ in the tables.

Interpolation   In the previous examples, the specified values of the state variables corresponded exactly to tabulated entries given in the tables. In general, this will not be the case. Rather, the specified values will fall somewhere between the tabulated entries. Thus, interpolation will usually be required to estimate the properties at this intermediate value. For instance, the pressure table gives properties at 100 and 105 bar; but the specific enthalpy of steam at 103.0 bar may be required. Clearly, the desired enthalpy will be somewhere between 2727.7 and 2718.7 kJ/kg. An adequate approximation for engineering calculations is to use *linear interpolation* between adjacent tabular values.

The general interpolation problem is that given the values of a function $y(x)$ at x^U and at x^L, denoted by y^U and y^L, respectively, estimate the value of $y(x)$ at an intermediate point x^i. If a simple proportional interpolation is used, then

$$\frac{y^i - y^L}{y^U - y^L} = \frac{x^i - x^L}{x^U - x^L}$$

or

$$y^i = y^L + \frac{y^U - y^L}{x^U - x^L}(x^i - x^L)$$

The above is of course recognizable as a linear equation for y^i in terms of x^i. Note that if several properties must be interpolated for the same independent variable x, then the above equation must be applied to each property separately.

Example 7.7 Determine the specific enthalpy and volume for steam at 5 bar and 320°C.

Solution At 5 bar, from the pressure table, the saturation temperature is 151.8°C. The steam is thus superheated. The superheated steam table contains entries for 300 and 350°C. To estimate $\hat{V}$ and $\hat{H}$ at 320°C, interpolation will be necessary. Interpolating first on the specific volume, we have

$$\hat{V}(320) = \hat{V}(300) + \frac{\hat{V}(350) - \hat{V}(300)}{350 - 300}(320 - 300)$$

$$= 0.522 + \frac{0.571 - 0.522}{50}20$$

$$= 0.542 \text{ m}^3/\text{kg}$$

Similarly, interpolating on the enthalpy,

$$\hat{H}(320) = \hat{H}(300) + \frac{\hat{H}(350) - \hat{H}(300)}{350 - 300}(320 - 300)$$

$$= 3065 + \frac{3168 - 3065}{50}20$$

$$= 3106.2 \text{ kJ/kg}$$

Quality of Steam In steam calculations, it is often necessary to deal with mixtures of steam and liquid water. For instance, the steam may have been used as a heating medium and some of it may have condensed. The resulting mixture will thus consist of saturated steam and entrained saturated liquid water slugs or droplets. The mass fraction of vapor in a mixture of liquid and steam, designated by X, is by convention called the *quality* of the steam. By definition, when $X = 1$, the steam is all saturated vapor; when $X = 0$, the steam is all saturated liquid. A steam with quality intermediate to these two extremes is said to be a *wet* steam.

Since wet steam will consist of saturated phases, the saturated steam tables can be used to calculate the specific enthalpy, internal energy, and volume of the vapor–liquid mixture. If the quality of the steam is given, then the mixture properties can be calculated as a linear combination of the individual saturated phase properties. Thus,

$$\hat{H}_{\text{mix}} = \hat{H}_V X + \hat{H}_L(1 - X)$$
$$\hat{U}_{\text{mix}} = \hat{U}_V X + \hat{U}_L(1 - X)$$
$$\hat{V}_{\text{mix}} = \hat{V}_V X + \hat{V}_L(1 - X)$$

Conversely, given one of the mixture properties, the quality can be calculated from the above relations by simply solving for X.

Example 7.8 Determine the pressure, specific volume, enthalpy, and internal energy of a wet steam at 230°C with quality 0.4.

Solution Since the steam has quality between 0 and 1, both phases must be present, and it can be assumed that these phases will be in equilibrium. Thus, from the phase rule, one macroscopic property suffices to define the state of the system. From the saturated steam pressure table, we find that at 230°C the saturation pressure is 28 bar. The saturated phase properties are

$\hat{V}$		$\hat{U}$		$\hat{H}$	
Water	**Steam**	**Water**	**Steam**	**Water**	**Steam**
0.001209	0.0714	987.1	2602.1	990.5	2802.0

Using the mixture relations, it follows that

$$\hat{H}_{\text{mix}} = 2802(0.4) + 990.5(0.6) \qquad = 1715.1 \text{ kJ/kg}$$

$$\hat{U}_{\text{mix}} = 2602.1(0.4) + 987.1(0.6) \qquad = 1633.1 \text{ kJ/kg}$$

$$\hat{V}_{\text{mix}} = 0.0714(0.4) + 0.001209(0.6) = 0.0293 \text{ m}^3/\text{kg}$$

Example 7.9 A pressure vessel with known volume of 0.3 m³ contains 10 kg water at a measured pressure of 60 bar. Calculate the temperature, quality, and specific enthalpy of the steam.

Solution The bulk specific volume of the vessel contents is $\hat{V}_{\text{mix}} = 0.3$ m³/10 kg $= 0.03$ m³/kg. At 60 bar, the saturated specific volumes of liquid water and steam are 0.001319 and 0.0324, respectively. Since the bulk specific volume lies between these two values, the steam is wet. The quality of the steam can be calculated from

$$\hat{V}_{\text{mix}} = \hat{V}_V X + \hat{V}_L (1 - X)$$

Thus,

$$0.03 = 0.0324 X + 0.001319(1 - X)$$

or

$$X = 0.923$$

The steam is 92.3% vapor and 7.7% liquid. The specific enthalpy of the mixture is

$$\hat{H}_{\text{mix}} = 2785 X + 1213.7(1 - X)$$

$$= 2785(0.923) + 1213.7(0.077) = 2664 \text{ kJ/kg}$$

7.2.2 Energy Balance Applications: Closed Systems

Given an extensive tabulation of thermodynamic properties such as that for steam, the solution of closed-system energy balance problems is generally quite straightforward. As shown in Section 6.5.3, the energy balance reduces to the form

$$m(\hat{U}_2 - \hat{U}_1) + mg(z_2 - z_1) + \tfrac{1}{2}m(v_2^2 - v_1^2) = Q - W$$

Frequently, the potential and kinetic energy terms will be either zero or negligible, leading to the simplification

$$m(\hat{U}_2 - \hat{U}_1) = Q - W$$

Moreover, in most closed-system applications of chemical engineering interest, the work term will arise due to expansion or compression. Hence, the balance reduces further to

$$m(\hat{U}_2 - \hat{U}_1) = Q - \int P \, dV$$

Finally, if the pressure is constant, then $\int P \, dV = P \, \Delta V$ and the enthalpy form $m(\hat{H}_2 - \hat{H}_1) = Q$ can be conveniently employed.

Given the simple form of the equations, the effort in solving closed-system energy balance problems is reduced to determining the appropriate values of the $\hat{U}$ and $\hat{H}$ function to insert into the balance equations. We illustrate several typical applications in the next three examples.

Example 7.10   One kilogram steam at a pressure of 1 bar, contained in a cylinder of cross-sectional area 1.69 m², freely supports a movable, leakproof cover of fixed weight, as shown in Figure 7.3. The cylinder is heated externally to raise the temperature of the steam from 100°C to 300°C. Assuming no heat loss to the environment, calculate the amount of heat required for this process.

Solution   The system of concern consists of the steam contained in the cylinder. With no steam leaks during the process, the system is clearly closed. As the steam expands upon heating, it will simply raise the cylinder cover and thus a constant pressure will be maintained. Therefore, the enthalpy form of the closed-system energy balance equation is most convenient to use. That is,

$$\Delta H + \Delta PE + \Delta KE = Q$$

Since the system is at rest, the kinetic energy term will be zero. There will however be a change in the elevation of the center of mass of the system as the steam expands. That change in elevation will probably be small, nonetheless, we shall retain the potential energy term in the balance equation.

From the superheated steam table, at 100°C and 1 bar, the specific enthalpy of steam is 2676 kJ/kg and the specific volume is 1.69 m³/kg; while at 300°C and 1 bar, the specific enthalpy and volume are 3074 kJ/kg and 2.64 m³/kg. Applying the above balance equation, we calculate that

$$Q = m\hat{H}_2 - m\hat{H}_1 + mg(z_2 - z_1)$$

Since the elevation of the center of mass of the cylinder is given by $z = L/2 =$

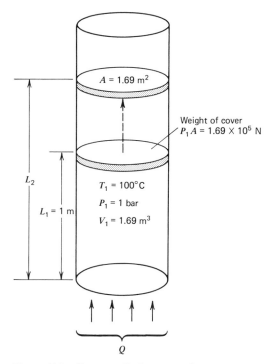

Figure 7.3   Steam cylinder example.

$\frac{1}{2}(V/A)$, we have

$$Q = m(\hat{H}_2 - \hat{H}_1) + \tfrac{1}{2}mg\,\frac{V_2 - V_1}{A}$$

$$= 1\text{ kg}(3074 - 2676)\,\frac{\text{kJ}}{\text{kg}} + \frac{1\text{ kg}}{2}\,\frac{9.8\text{ m}}{\text{s}^2}\,\frac{(2.64 - 1.69)\text{ m}^3}{1.69\text{ m}^2}$$

$$= 398\text{ kJ} + \frac{5.51\text{ J}}{2} = 398.0028\text{ kJ}$$

Clearly, the potential energy term is quite insignificant, as usually is in cases involving heating and cooling.

During this process, the system does perform work of expansion. That work can be calculated using the equation

$$W = \int P\,d\hat{V}$$

which, since P is constant, reduces to

$$W = P\,\Delta\hat{V} = 1 \times 10^5\,\frac{\text{N}}{\text{m}^2}\,(2.64 - 1.69)\frac{\text{m}^3}{\text{kg}} = 95\text{ kJ/kg}$$

Thus, of the total amount of energy transferred to the system as heat (398 kJ), about one fourth (95 kJ) is in turn consumed by the system in work of expansion. As a result, the internal energy of the system is increased by only the difference $Q - W$, that is,

$$\Delta U = Q - W = 398 - 95 = 303 \text{ kJ}$$

This calculation can be checked by directly using the values of U given in the tables.

Example 7.11 One kilogram steam at 100°C and 1 bar is contained in a pressure vessel. How much heat is required to raise the temperature of the steam in the vessel to 300°C, and what is the final pressure?

Solution Assuming no loss of steam from the pressure vessel, the system consisting of the steam is a closed system. Since the vessel walls are rigid, the process will occur at constant volume. Thus, assuming no loss of heat to the environment and since the center of mass of the system will not change, the energy balance equation reduces to

$$\Delta E = \Delta U = Q - W$$

Since the process occurs at constant volume, there is no work of expansion done on or by the system. Thus, $W = \int P \, dV = 0$ and the balance reduces further to

$$\Delta U = m_2 \hat{U}_2 - m_1 \hat{U}_1 = Q$$

At 100°C and 1 bar, the specific volume of the steam is 1.69 m³/kg. Since the system is closed and the volume is constant, the final specific volume must be the same:

$$\hat{V}_2 = 1.69 \text{ m}^3/\text{kg}$$

Clearly, at 300°C, the steam will continue to be superheated. Consequently, the two conditions suffice to fix the final state. Note that at 1 bar and 300°C, the specific volume is 2.64 m³/kg, while at 5 bar and 300°C, the specific volume is 0.522 m³/kg. The final state is somewhere between these values, that is, interpolation is required. Thus,

$$P_2 = 1 \text{ bar} + \frac{(5 - 1) \text{ bar } (1.69 - 2.64) \text{ m}^3/\text{kg}}{(0.522 - 2.64) \text{ m}^3/\text{kg}}$$

$$= 2.79 \text{ bar}$$

Similarly, the internal energy at the final state will be

$$\hat{U}_2 = 2811 \text{ kJ/kg} + \frac{(2803 - 2811) \text{ kJ/kg}}{(0.522 - 2.64) \text{ m}^3/\text{kg}} (1.69 - 2.64) \text{ m}^3/\text{kg}$$

$$= 2807.4 \text{ kJ/kg}$$

Therefore,

$$Q = 1 \text{ kg}(2807.4 - 2507.0)\text{kJ/kg} = 300.4 \text{ kJ/kg} = \Delta U$$

It is interesting to note that Q is larger in the case of the constant-pressure process (Example 7.10) than in the present constant-volume process. This occurs because in the constant-P case, in addition to raising the temperature from 100°C to 300°C, it is also necessary for the system to consume energy to perform work of expansion. In the constant-volume process, the system did not have to perform work of expansion.

Note further that the ΔU in both cases is very nearly the same, indicating that the effect of pressure on the internal energy function is small.

Finally, note that in this case we have used the steam tables to calculate the state variable P given the values of the other two state variables T and $\hat{V}$. This calculation is equivalent to evaluating the equation of state for steam, except that in the present case the equation of state is contained in the steam tables in tabular form.

Absolute and Relative Energy Values The reader should note that the $\hat{H}$ and $\hat{U}$ values contained in the energy balance equations are actually the absolute enthalpy and internal energy values. However, the enthalpy and internal energy values listed in the steam tables are only *relative* values. That is, they are values relative to an *assumed* triple point reference internal energy value of zero. The relation between the two is, in the case of the internal energy function, given by

$$\hat{U}(T, P) = (\hat{U}(T, P) - \hat{U}(T^0, P^0)) + \hat{U}(T^0, P^0)$$
$$= \hat{U}_r(T, P) + \hat{U}(T^0, P^0)$$

where $\hat{U}(T, P)$ is the absolute internal energy and $\hat{U}_r(T, P) = \hat{U}(T, P) - \hat{U}(T^0, P^0)$ is the internal energy relative to the reference state (T^0, P^0).

We can use the relative enthalpies for balance purposes because the internal energy of the reference state simply cancels in the balance equations. That is, if $\hat{U} = \hat{U}_r + \hat{U}^0$ is substituted into the first law equation, we obtain

$$Q - W = m(\hat{U}_2 - \hat{U}_1) = m[(\hat{U}_{2r} + \hat{U}^0) - (\hat{U}_{1r} + \hat{U}^0)]$$
$$= m(\hat{U}_{2r} - \hat{U}_{1r})$$

Thus, the balance equations can be written in terms of either absolute or relative internal energies. Since absolute internal energies are not available to us, we always use relative internal energies and typically delete the subscript r.

However, the reader should always take care that all of the internal energy or enthalpy values used in his balance calculations are referred to the *same* reference state. For instance, if the internal energy for the initial state is obtained from one source while the internal energy for the final state is obtained from another tabulation which may have a higher T and P range, it is important to check that both sources use the same zero reference state. If they do not, the enthalpy values should be corrected to the same zero reference state. The correction is carried out as follows. Suppose one table lists internal energies relative to reference state a, while another table gives internal energies for the same substance relative to reference state b. To convert enthalpies relative to a, $(U - U^a)$, to enthalpies relative to b,

$(U - U^b)$, we simply note that

$$(U - U^b) = (U - U^a) - (U^b - U^a)$$

That is, we correct the enthalpy relative to a by subtracting from it the enthalpy of reference state b relative to reference state a. The reference enthalpy correction calculation is illustrated in the next example.

Example 7.12   Calculate the heat required to raise the temperature of 1 kg steam at 885 bar from 500°C to 816°C, given that the specific enthalpy at the latter temperature is 1130 kJ/kg relative to saturated steam at 1.01325 bar (1 atm).

Solution   Clearly, at constant pressure, $Q = \Delta\hat{H} \cdot 1$ kg. The enthalpy at the initial condition, by interpolating from the superheated steam tables, is

$$\hat{H}_1 = 2723 \text{ kJ/kg} - \frac{407 \text{ kJ/kg}}{500 \text{ bar}} (385 \text{ bar}) = 2409.6 \text{ kJ/kg}$$

This enthalpy is relative to liquid water at the triple point. To convert the given relative enthalpy of 1130 kJ/kg at the final state to an enthalpy relative to the triple point, we use the correction formula

$$(\hat{H}_2 - \hat{H}(Tr)) = (\hat{H}_2 - \hat{H}(1 \text{ atm})) - [\hat{H}(Tr) - \hat{H}(1 \text{ atm})]$$

The quantity $[\hat{H}(1 \text{ atm}) - \hat{H}(Tr)]$ is just the enthalpy of steam at 1 atm relative to the triple point. From the saturated steam pressure table, this quantity is 2676.0 kJ/kg. Thus,

$$\hat{H}_2 - \hat{H}(Tr) = 1130 \text{ kJ/kg} - (-2676.0 \text{ kJ/kg}) = 3806 \text{ kJ/kg}$$

Finally, $Q = m(\hat{H}_2 - \hat{H}_1) = 1$ kg$(3806 - 2409.6)$ kJ/kg = 1196.4 kJ

Use of the uncorrected relative enthalpy would of course have led to a considerable error in the calculation of Q. In the present instance, it would lead to a negative Q.

7.2.3   Energy Balance Applications: Open Systems

As derived in Section 6.5, the general form of the energy balance equation for an open system involving multiple inlet and outlet streams is

$$\frac{dQ}{dt} - \frac{dW}{dt} = \sum_{\substack{\text{all} \\ \text{outlets}}} F^k(\hat{H} + gz + \tfrac{1}{2}v^2)_k - \sum_{\substack{\text{all} \\ \text{inlets}}} F^j(\hat{H} + gz + \tfrac{1}{2}v^2)_j$$

The internal energy form of the open-system equation is not normally used because it requires separate evaluation of the $P\hat{V}$ terms. The primary concerns in applying this equation to specific problems are to

1. Clearly define the system boundaries.
2. Carefully consider which terms can be neglected in problem solution.

3. Check that all heat and work contributions are included.
4. Verify that all enthalpy data sources for a given component use the same reference state.

We illustrate these concerns with three examples, all of which use steam table enthalpy data.

Example 7.13 Steam at 200°C and 7 bar enters a horizontal nozzle at a steady velocity of 60 m/s. The steam exits at a velocity of 600 m/s and a decreased pressure of 1.4 bar. Calculate the temperature and quality of the exit steam.

Solution A nozzle is a device whose function it is to increase the kinetic energy of a fluid by decreasing the pressure. The process is normally assumed to occur adiabatically because the heat loss or gain through the nozzle walls is usually negligible compared to the other energy terms.

The system under consideration, shown in Figure 7.4, is an open system since steam continuously enters and leaves the system boundaries. Moreover, the system clearly has a single inlet and a single outlet. The applicable energy balance equation is therefore

$$\frac{dQ}{dt} - \frac{dW}{dt} = F^2(\hat{H}_2 + gz_2 + \tfrac{1}{2}v_2^2) - F^1(\hat{H}_1 + gz_1 + \tfrac{1}{2}v_1^2)$$

By assumption, the system is undergoing an adiabatic process; thus, $dQ/dt = 0$. Also, since the nozzle has rigid walls, no work of expansion or compression is done by the system; hence, $dW/dt = 0$. Furthermore, since the inlet and outlet streams are at the same level, there is no change in elevation, $z_1 = z_2$. Finally, from a simple mass balance, $F^2 = F^1 = F$. The energy balance equation is thus reduced to

$$F[(\hat{H}_2 + \tfrac{1}{2}v_2^2) - (\hat{H}_1 + \tfrac{1}{2}v_1^2)] = 0$$

Dividing out by the mass flow and solving for $\hat{H}_2$, we obtain

$$\hat{H}_2 = \hat{H}_1 + \tfrac{1}{2}(v_1^2 - v_2^2)$$

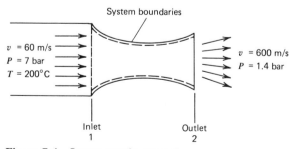

Figure 7.4 Steam nozzle example.

At 200°C and 7 bar, the steam is superheated and has an interpolated specific enthalpy of 2843.8 kJ/kg. Since the two velocities are known, we can calculate $\hat{H}_2$:

$$\hat{H}_2 = 2843.8 \text{ kJ/kg} + \tfrac{1}{2}(60^2 - 600^2) \frac{\text{m}^2}{\text{s}}(\text{kg/kg})$$

$$= 2843.8 \text{ kJ/kg} + (-178.2 \text{ kJ/kg}) = 2665.6 \text{ kJ/kg}$$

To determine the temperature of the exit steam, given that its exit pressure is 1.4 bar, we check the saturated steam table to determine whether the calculated enthalpy exceeds the saturated steam enthalpy. The saturated steam enthalpy at 1.4 bar is 2690.3 kJ/kg; thus, the exit steam must be wet and its temperature must be equal to the saturation temperature at 1.4 bar, namely, 109.3°C. The quality is easily calculated from

$$2665.6 = 2690.3X + 458.4(1 - X)$$

The solution is $X = 0.989$, that is, 0.011 of the steam is condensed.

Note that even with the very high outlet velocity of 600 m/s, the kinetic energy term is only about 6% of the inlet enthalpy term. Normally, in chemical process applications, stream velocities are less than 50 m/s, in which case the kinetic energy term would amount to 2.5 kJ/kg, or less than 0.1% of the enthalpy term.

Example 7.14   In the system shown in Figure 7.5, a stream consisting of saturated steam at 145 psia with a bulk velocity of 100 ft/s is passed through a superheater

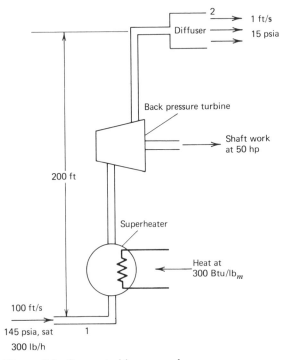

Figure 7.5   Steam turbine example.

which transfers heat at the rate of 300 Btu/lb$_m$ to the stream. The superheated steam is next expanded through a back-pressure turbine to develop 50 hp of shaft work and finally exits through a diffuser at 15 psia and a velocity of 1 ft/s. The elevation change between the inlet and the outlet of the system is 200 ft. Calculate the outlet temperature and steam quality, assuming negligible pressure drop due to friction and inlet steam flow of 300 lb/h.

Solution The superheater is simply a device used to heat the steam by external medium. The back-pressure turbine can be viewed as a rotary fan run in reverse. That is, rather than supplying power to the fan so that it can propel the gas, the pressurized gas is directed through the fan blades to turn the fan and its shaft. The rotating shaft can then be used to do useful work. The diffuser is simply the reverse of a nozzle: it decreases the kinetic energy of a fluid by increasing the pressure.

The system in this case consists of the steam flow between the inlet, labeled 1 in Figure 7.5, and the outlet from the diffuser, labeled 2. It is clearly a single-inlet, single-outlet open system. The system performs work at the rate of 50 hp and has heat added to it at the rate of 300 Btu/lb$_m$ or 9×10^4 Btu/h. The steam undergoes both an elevation change and a change in velocity. Consequently, the complete form of the energy balance must be used, namely,

$$\frac{dQ}{dt} - \frac{dW}{dt} = F^2 \left(\hat{H}_2 + \frac{g}{g_c} z_2 + \tfrac{1}{2} \frac{v_2^2}{g_c} \right) - F^1 \left(\hat{H}_1 + \frac{g}{g_c} z_1 + \tfrac{1}{2} \frac{v_1^2}{g_c} \right)$$

Note that if the problem is solved in the AE system of units, the conversion factor g_c must be included in the mechanical energy terms of the balance equation.

We begin by converting the work term into units of Btu/h. To do this, we must convert hp to the primary (ft $\cdot$ lb$_f$/s) units, then convert the mechanical energy units to Btu, and finally adjust the time units. We do this using the approximate conversion factors of Table 6.2. Thus,

$$\frac{dW}{dt} = 50 \text{ hp} \times 550 \frac{\text{ft} \cdot \text{lb}_f}{\text{s} \cdot \text{hp}} \times \frac{1 \text{ Btu}}{778 \text{ ft} \cdot \text{lb}_f} \times 3600 \text{ s/h}$$

$$= 1.2725 \times 10^5 \text{ Btu/h}$$

Since from a simple mass balance $F^1 = F^2$, the energy balance reduces to

$$(9 \times 10^4 - 1.2725 \times 10^5) \frac{\text{Btu}}{\text{h}}$$

$$= 300 \frac{\text{lb}}{\text{h}} \left[(\hat{H}_2 - \hat{H}_1) + \frac{g}{g_c} \times 200 \text{ ft} + \tfrac{1}{2} g_c^{-1} (1 - 10^4) \frac{\text{ft}^2}{\text{s}^2} \right]$$

To evaluate the specific enthalpy at inlet conditions, we must convert 145 psia to SI units. The equivalent SI pressure is 1×10^6 kPa, or 10 bar. Saturated steam at 10 bar has a specific enthalpy of 2776.2 kJ/kg. In terms of Btu/lb$_m$, the specific enthalpy is

$$\hat{H}_1 = 2776.2 \text{ kJ/kg} \times \frac{1 \text{ Btu}}{1.055 \text{ kJ}} \times \frac{0.454 \text{ kg}}{1 \text{ lb}_m} = 1194.7 \text{ Btu/lb}_m$$

The energy balance now becomes

$$\frac{-3.725 \times 10^4}{300} \frac{Btu}{lb_m} = \hat{H}_2 - 1194.7 \frac{Btu}{lb_m} + \frac{32.174 \, ft/s^2}{32.174 \, ft \cdot lb_m/s^2 \cdot lb_f} \times \frac{200 \, ft}{778 \, ft \cdot lb_f/Btu}$$

$$+ \tfrac{1}{2}(1 - 10^4)\frac{ft^2}{s} \times \frac{1 \, s^2 \cdot lb_f}{32.174 \, ft \cdot lb_m} \times \frac{1 \, Btu}{778 \, ft \cdot lb_f}$$

$$= \hat{H}_2 - 1194.7 + 0.257 - 0.200 \frac{Btu}{lb_m}$$

Note that both of the mechanical energy terms require the conversion of work units to the Btu energy unit. The calculated value of $\hat{H}_2$ becomes

$$\hat{H}_2 = 1070.5 \, Btu/lb_m$$

or, in SI units,

$$\hat{H}_2 = 2487.5 \, kJ/kg$$

At 15 psia, or 1.03 bar, the specific enthalpy of saturated steam is larger than this value; hence the steam is wet. The interpolated saturated steam and liquid specific enthalpies are 2676.7 and 421.4 kJ/kg. The quality is thus obtained from the equation

$$2487.5 = 2676.7X + 421.4(1 - X)$$

and is

$$X = 0.916$$

The exit temperature, again interpolated, is 100.5°C, or 212.9°F.

The reader should again note that the magnitudes of the energy term arising from a 200-ft elevation difference (0.257 Btu/lb$_m$) and of the kinetic energy term due to a velocity reduction from 100 ft/s to 1 ft/s (0.2 Btu/lb$_m$) are quite negligible compared to the magnitudes of the ΔH term (-124.2 Btu/lb$_m$), the work term (424.2 Btu/lb$_m$), and the heat term (300 Btu/lb$_m$).

The observation made in the previous two examples, that the magnitudes of the potential and kinetic energy terms are relatively small, actually applies to most chemical process applications. In addition, in most chemical process applications, the units involved are fixed vessels of constant volume and, with the exception of pumps, turbines, and compressors, do not involve significant work inputs or outputs. Thus, for most steady-state process energy balance applications, $dW/dt = 0$; consequently, the open steady-state energy balance equation can for such applications be reduced to

$$\frac{dQ}{dt} = \sum_j F^j\hat{H}^j - \sum_k F^k\hat{H}^k$$

This equation basically states that all changes in the energy content of the process streams occur as the result of heat transfer to or from the system. It reflects the fact that heat transfer to and from streams and units is the most common operation in chemical processing.

By assuming a time basis of, say, 1 h, the above equation is sometimes cast into the pseudo-closed system form

$$Q = \Delta H$$

The form of this equation is of course identical to that derived for closed constant-pressure systems. In both cases, the mechanical energy and work terms are excluded. However, in the open-system case, the process must proceed at constant pressure, and the balance equation will include the work of expansion or compression. In the open-system case, the inlet and outlet stream pressures need not be the same.

Example 7.15   Steam is used to preheat 300 kg/h of process water at 5 bar from 50°C to 150°C using the double pipe heat exchanger shown in Figure 7.6. The steam is available at 10 bar and is saturated. The condensate is returned as saturated liquid. Calculate the required steam flow.

Solution   The double-pipe heat exchanger is simply a device consisting of two concentric pipes with water flowing in the inner pipe, steam flowing in the outer pipe, and a layer of insulation on the outside. The flow of condensate out of the annular region is regulated by a device called a steam trap which allows only condensate but no vapor to escape.

Because of the inner wall between the two fluids, the water and steam do not mix. Instead, heat is transferred from the steam side to the water side through the metal tube wall. From the energy balance point of view, the double-pipe exchanger can be represented by the schematic shown in Figure 7.7. That is, the system can

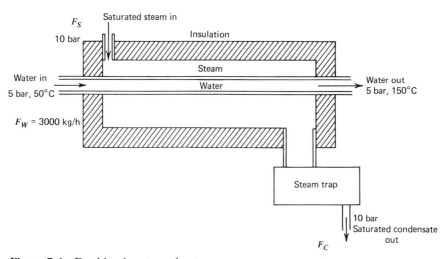

Figure 7.6   Double-pipe steam heater.

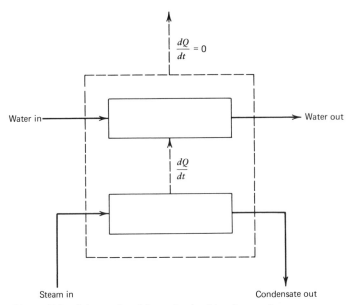

Figure 7.7   Schematic of flows in double-pipe exchanger.

be viewed as two separate tubes with water in one and steam in the other, with transfer of heat but not mass between the two streams. If the system boundary is drawn as shown in the figure, then that internal heat transfer is enclosed within the boundaries, and the only heat transfer to the environment is the heat loss through the insulation. Assuming that the insulation is adequate, the heat loss to the environment will be negligible. Thus, for energy balance purposes, the exchanger can be viewed as a multiple inlet/multiple outlet device which operates adiabatically. If in addition we assume that kinetic energy effects are negligible and note that there is neither external work nor a change in potential energy, the energy balance for the system becomes

$$0 = \sum_{\text{outlets}} F^k \hat{H}^k - \sum_{\text{inlets}} F^j \hat{H}^j$$

or specifically,

$$0 = F_w^{\text{out}} H_w \ (150°\text{C}, 5 \ \text{bar}) + F_c^{\text{out}} H_c \ (10 \ \text{bar})$$
$$- F_w^{\text{in}} H_w \ (50°\text{C}, 5 \ \text{bar}) - F_s^{\text{in}} H_s \ (10 \ \text{bar})$$

Since the two streams are not in contact, from simple mass balances,

$$F_c^{\text{out}} = F_s^{\text{in}} \quad \text{and} \quad F_w^{\text{in}} = F_w^{\text{out}}$$

Therefore,

$$F_s(H_s \ (10 \ \text{bar}) - H_c(\ 10 \ \text{bar})) = F_w[H_w \ (150°\text{C}, 5 \ \text{bar}) - H_w \ (50°\text{C}, 5 \ \text{bar})]$$

At 10 bar, the saturated steam and water enthalpies are 2776.2 and 762.6 kJ/kg, respectively. At 5 bar and 50°C, the water specific enthalpy is 209.7 kJ/kg; while at 150°C, it is 632.2 kJ/kg. Substituting in these values and solving for the unknown steam flow rate, we obtain

$$F_s = \frac{3000 \text{ kg/h } (632.2 - 209.7) \text{ kJ/kg}}{(2776.2 - 762.6) \text{ kJ/kg}} = 629.5 \text{ kg/h}$$

Note that a separate energy balance on the water side would result in the equation

$$dQ/dt = F_w(632.2 - 209.7) = 1.2675 \times 10^6 \text{ kJ/h}$$

whereas a separate balance on the steam side yields

$$dQ/dt = F_s(762.6 - 2776.2) = -1.2675 \times 10^6 \text{ kJ/h}$$

As expected, the heat added to the water side is equal to the heat given up by the steam side. This is of course taken into account in the combined balance equation.

For energy balance purposes, the actual heat transferred between the streams is not significant in itself. Thus, we generally use only the combined balance equation form. However, for design purposes, the individual stream balances will be important because the size of the heat exchanger will in large part be governed by the rate of heat transfer required.

In this section, we have illustrated the use of thermodynamic tables to carry out closed- and open-system energy balance calculations. We have for simplicity confined our examples to applications involving water. However, comparable tabulation of enthalpy and internal energy data are available for other common bulk chemicals. For example, the *Chemical Engineers Handbook*[4] gives abbreviated specific enthalpy and volume tables for acetylene, air, ammonia, argon, several comon light hydrocarbons, carbon dioxide, carbon monoxide, hydrogen, helium, mercury, oxygen, sulfur dioxide, and several commercial refrigerant fluids. Fairly extensive tabulation of enthalpy and specific volume data for hydrocarbon compounds are given in the API *Technical Data Book*.[5] Some of these tabulations are also given in graphic form as enthalpy–temperature or enthalpy–pressure diagrams. Graphs allow a more compact but less accurate presentation of wider data range. See, for example, Figure 7.8, which is an enthalpy–pressure plot for propylene. Note that secondary lines for temperature and specific volume are superimposed in this diagram. Graphic presentation of the data of course reduces the need for interpolation but at a substantial reduction in accuracy.

The principles and methodology we have described are unchanged, regardless of the chemical species involved in the system or the form, graphic or tabular, in which the thermodynamic data are available. Consequently, we leave further ex-

---

[4] R. C. Perry and C. H. Chitton, Eds., *Chemical Engineer's Handbook,* 5th ed., McGraw-Hill, New York, 1973.
[5] *Technical Data Book–Petroleum Refining,* 3rd ed., American Petroleum Institute, Washington, D.C., 1976.

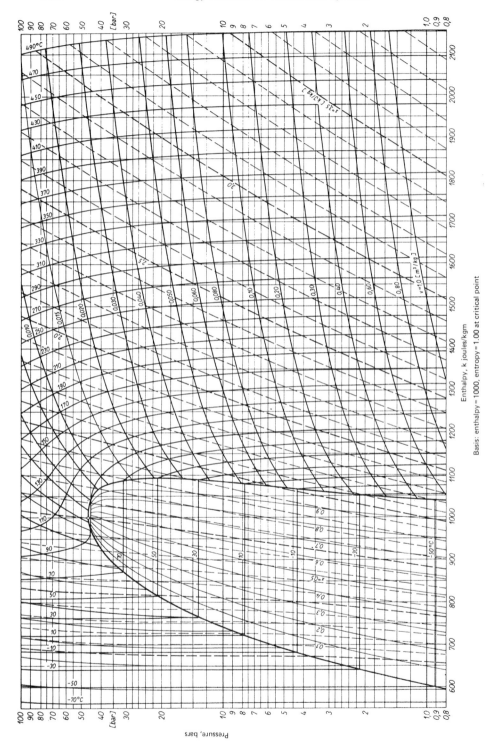

Enthalpy, k joules/kgm

Basis: enthalpy = 1000, entropy = 1.00 at critical point

Figure 7.8 Enthalpy-pressure plot for propylene. *Source:* K. Stephan and G. Scherer, *Chem. Ing. Tech.*, **33**, 417 (1961). Reproduced with permission from Verlag Chemie GmbH.

ercises involving such applications to the reader. Instead, in the next section we consider alternate, more compact but sometimes less accurate forms of enthalpy data.

7.3 ENERGY BALANCES WITHOUT COMPLETE THERMODYNAMIC TABLES

Extensive tabulations of enthalpies and internal energies such as those developed for steam are not available for most pure substances. In the absence of detailed tabulations, the enthalpy of a substance in a given state relative to some reference state must be determined by subdividing the relative enthalpy difference into several separate components: enthalpy differences associated with temperature changes at constant pressure and phase, enthalpy differences associated with pressure changes at constant temperature and phase, and enthalpy differences associated with phase changes at constant temperature and pressure. The contribution of each of these components, once evaluated, can then be combined to calculate the final overall relative enthalpy. For instance, suppose we wish to calculate the relative enthalpy of a single-component system at T_3, P_3, and phase b relative to a reference state at T_1, P_1, and phase a. Suppose the phase transition occurs at T_2 and P_2. Since the enthalpy is a state function, the difference between the final and initial state enthalpies is independent of the temperature and pressure path used to connect these states. Thus, we can formulate the following expression for the relative enthalpy by adding and subtracting like terms:

$$H_b(T_3, P_3) - H_a(T_1, P_1) = H_b(T_3, P_3) - H_b(T_2, P_3) + H_b(T_2, P_3)$$
$$- H_b(T_2, P_2) + H_b(T_2, P_2) - H_a(T_2, P_2) + H_a(T_2, P_2) - H_a(T_1, P_2)$$
$$+ H_a(T_1, P_2) - H_a(T_1, P_1)$$

If we group these terms pairwise, we can identify the following individual contributions:

$H_b(T_3, P_3) - H_b(T_2, P_3)$ = enthalpy difference due to change in temperature at constant pressure and phase

$H_b(T_2, P_3) - H_b(T_2, P_2)$ = enthalpy correction due to change in pressure at constant T and phase

$H_b(T_2, P_2) - H_a(T_2, P_2)$ = enthalpy difference arising from change in phase at constant T and P

$H_a(T_2, P_2) - H_a(T_1, P_2)$ = enthalpy difference due to change in temperature at constant P and phase

$H_a(T_1, P_2) - H_a(T_1, P_1)$ = enthalpy correction due to a change in pressure at constant T and phase

For many substances, adequate thermodynamic information is available so that these individual enthalpy changes can be evaluated. The overall enthalpy change can then be calculated by summing these contributions. In this section, we consider the form in which this information is available and learn how to carry out enthalpy balance calculations using this type of enthalpy information. In particular, we shall find that enthalpy changes at constant pressure and phase can be estimated using relations known as heat capacity functions. We learn how to use standard phase transition enthalpies and how to transform these transition enthalpies to other temperature and pressure conditions.

However, we shall not consider enthalpy corrections due to changes in pressure. Evaluation of enthalpy pressure corrections involves the use of real equations of state and thermodynamic concepts beyond the scope of this text. Fortunately, under normal T and P conditions for most substances, especially liquids and solids, the effect of pressure on enthalpy as well as internal energy is small. Recall, for instance, in Example 7.11, that the difference in internal energy between superheated steam at 1 bar and at nearly 3 bar was only 3.6 kJ/kg. By contrast, the difference in internal energy at 1 bar with a change of only 1°C in temperature is about 2 kJ/kg. Thus, the effect of temperature change will usually dominate the pressure effect by a considerable margin. Consequently, for purposes of this text, we shall neglect pressure corrections except in those cases when complete thermodynamic tables are available.

In this section, we also consider the calculation of enthalpies of mixtures of pure components. As we observed in our discussion of material balance problems, processes normally involve streams which contain multiple components. For many applications, it is adequate to assume that the enthalpy or internal energy of a stream can be calculated as a sum of the pure-component enthalpies or internal energies weighted by their mass or mole fractions:

$$\hat{H}_{\text{mix}}(T, P) = \sum_j w_j \hat{H}_j(T, P)$$

This is, in fact, how we calculated the enthalpy of wet steam in the previous section. However, for more complex mixtures, for instance, solutions of water and some acids, there are enthalpy corrections which must be added to the weighted sum to account for the enthalpy changes associated with the dissolution process. We discuss these corrections in Section 7.3.3.

7.3.1 Heat Capacities

In this section, we consider the definition and use of a quantity that allows us to calculate enthalpy and internal energy differences associated with processes in which there is no phase change and in which temperature is the independent variable. Suppose a closed system consisting of a pure component in a given phase and state (T, P) undergoes a small displacement in its state to $(T + dT, P + dP)$. Since, by convention, the enthalpy is a function of T and P, we can from calculus write that the differential enthalpy change dH will be given by

$$dH = \left(\frac{\partial H}{\partial T}\right)_p dT + \left(\frac{\partial H}{\partial P}\right)_T dP$$

If the change in state occurs at constant pressure, then $dP = 0$ and the above equation will reduce to

$$Q = dH = \left(\frac{\partial H}{\partial T}\right)_p dT$$

If it were possible to evaluate or approximate the partial derivative, then the above expression could be integrated to calculate the enthalpy difference or heat transferred as the result of a larger change in temperature from T_1 to T_2 at constant pressure. That is,

$$Q = \Delta \tilde{H} = \tilde{H}_2 - \tilde{H}_1 = \int_{T_1}^{T_2} \left(\frac{\partial \tilde{H}}{\partial T}\right)_p dT$$

Because of its utility in calculation Q, the quantity $(\partial H/\partial T)_p$ is called the *heat capacity at constant pressure* and is commonly denoted by C_p. Using this definition, the constant-pressure energy balance can for small changes in T be expressed as

$$Q = C_p \, dT$$

From this equation, the (molar) heat capacity at constant pressure can be interpreted as the quantity of heat which must be transferred per mole of the pure component to raise its temperature by one temperature unit. The heat capacity can also be defined per unit mass of pure component, in which case we use the symbol $\hat{C}_p$. It is interesting to note that one of the early definitions of the AE heat unit, the Btu, was that 1 Btu is the amount of heat that had to be transferred to 1 lb_m of pure liquid water at 1 atm and a specified temperature to raise the temperature of the water by 1°F. Under this definition of the Btu and using the above constant pressure energy balance, it follows that

$$1 \text{ Btu} = \hat{C}_p \times 1 \text{ lb}_m \times 1°F$$

or that the heat capacity per unit mass of liquid water under the specified temperature and pressure conditions is

$$\hat{C}_p(\text{liq. } H_2O) = 1.0 \text{ Btu/lb}_m \cdot °F$$

For this particular combination of dimensions, it happens that, as shown in Example 6.1, the heat capacity expressed in terms of the kcal unit will also be numerically the same. That is,

$$\hat{C}_p(\text{liq. } H_2O) = 1.0 \text{ kcal/kg} \cdot K$$

The definitions of heat and energy units in terms of the heating of water have since been abandoned as too imprecise. Instead, the definitions in terms of the joule are used as given in Section 6.4.

The heat capacity of water and of all substances is in general a function of temperature and pressure, since the enthalpy function from which C_p is derived is itself dependent on temperature and pressure. Thus, C_p data should in principle be reported in two-dimensional tables such as those for the superheated steam enthalpy. However, it is conventional to tabulate or correlate C_p values at a fixed reference pressure as a function of temperature only and, as we shall soon see, to correct for pressure effects via separate calculations.

Following the same arguments as used for developing the definition of C_p from the enthalpy function, we can also develop a heat capacity function for the internal energy function. Since by convention the internal energy is a function of T and $\tilde{V}$, we can for differential changes in T and $\tilde{V}$ at constant phase write

$$d\tilde{U} = \left(\frac{\partial \tilde{U}}{\partial T}\right)_{\tilde{V}} dT + \left(\frac{\partial \tilde{U}}{\partial V}\right)_{T} d\tilde{V}$$

For a closed system undergoing a constant volume process, $d\tilde{V} = 0$, and the differential balance can be integrated to yield

$$Q = \int_{T_1}^{T_2} \left(\frac{\partial U}{\partial T}\right)_{\tilde{V}} dT$$

The quantity $(\partial U/\partial T)_{\tilde{V}}$ is called the *heat capacity at constant volume* and is usually denoted by C_v. The heat capacity at constant volume has the same functional dependence as the internal energy, hence it is in general a function of temperature and specific volume. As in the case of C_p, the heat capacity at constant volume is usually tabulated or correlated at a chosen reference $\tilde{V}$ as a function of T only.

Before proceeding with a discussion of the form and use of heat capacity data, we digress briefly to examine the relationship between the two heat capacities which can be elucidated by considering an abstraction known as the *ideal gas*.

The Relationship Between C_p and C_v A theoretically interesting and practically significant model of real gas behavior can be obtained by considering an abstraction called the *ideal gas*. By definition, an ideal gas is one which satisfies the equation of state

$$P\tilde{V} = RT$$

where R is the gas constant with value 8.3143 J/mol · K. It can be shown that the ideal gas has the property that its internal energy is a function of temperature only, namely, $U = U(T)$. Therefore, since by definition $\tilde{H} = \tilde{U} + P\tilde{V}$, it follows upon substitution of the equation of state that

$$\tilde{H} = \tilde{U}(T) + P\tilde{V} = \tilde{U}(T) + RT$$

Consequently, the enthalpy of an ideal gas is also a function of temperature only. Moreover, if the resulting expression is differentiated with respect to temperature, the relation

$$\frac{d\tilde{H}}{dT} = \frac{d\tilde{U}}{dT} + R$$

is obtained. But these derivatives can be recognized as the heat capacities at constant pressure and constant volume, respectively, for the ideal gas. Thus, from the heat capacity definitions, it follows that

$$C_p = C_v + R$$

For an ideal gas, therefore, the two heat capacities differ only by the constant R.

We can interpret this relationship between the heat capacities in the following manner. Consider a closed system consisting of 1 mol of an ideal gas at temperature T and suppose it is heated at constant pressure until its temperature reaches $T + 1$ degrees. For a constant pressure process,

$$Q = \Delta H$$

Assuming that the heat capacity will be constant over a temperature interval of 1 degree,

$$Q = \Delta H = C_p \Delta T = C_p$$

For an ideal gas, the internal energy is a function of T only. Consequently, if the heat capacity at constant volume is assumed constant over the 1-degree temperature change, then

$$U(T + 1) - U(T) = C_v \Delta T = C_v$$

From the first Law,

$$U(T + 1) - U(T) = Q - W$$

Substituting the above expressions for ΔU and Q, the result is

$$C_v = C_p - W$$

or

$$W = C_p - C_v = R$$

Thus, R represents the energy which must be supplied over and above the amount C_v in order to account for the work of expansion in the constant-pressure process. This difference between C_p and C_v is actually observed with real gases at moderate pressures and temperatures. For example, C_p for N_2 at 20°C and 1 bar is 29.04 J/mol · K, while $C_v = 20.75$. The difference is 8.29 J/mol · K. For liquids and solids, the work of expansion is usually negligible, hence C_p is nearly equal to C_v. Because the enthalpy function is preferred to the internal energy in open-system energy balance calculations, in the remainder of this text we deal primarily with the heat capacity at constant pressure.

Heat Capacity Data   As noted earlier, the heat capacity at constant pressure is in general a funciton of both temperature and pressure. However, real gases approach ideal gas behavior at moderate pressures; hence, as with the ideal gas, the effect of pressure on the heat capacity becomes small. For most liquids and solids over moderate temperature and pressure ranges, the pressure dependence is also small. Consequently, it is conventional to report C_p data at some reference pressure

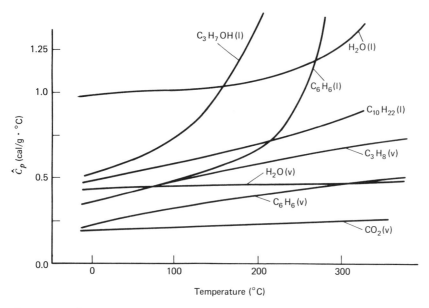

Figure 7.9 C_p versus T Curves for typical liquids and gases.

(1 atm or zero pressure are most commonly used) as a function of temperature alone and to use these data even if moderate pressure excursions from the reference are encountered. Although such data are sometimes presented in tabular form or in graphical form, such as Figure 7.9, the most compact and useful form is that of a correlated equation.

For gases, the correlation typically is expressed as a multiterm polynomial in T. For example,

$$C_p = a + bT + CT^2 + dT^3 + eT^4$$

For liquids and solids, fewer terms are usually included because for these phases there is less variation of C_p with temperature. The constants a, b, c, and d of the correlating polynomials are obtained by fitting experimentally measured heat capacity data; hence these correlations are only valid for the temperature range over which the data were taken. Since the heat capacity for each chemical species will exhibit a different temperature dependence, a separate set of correlation constants must be determined for each species. Extensive tabulation of C_p correlating coefficients for gases and liquids from the FLOWTRAN data base are given in Appendices 3 and 6.

Heat capacity correlations for complex mixtures of components such as petroleum and its derivatives or of heterogeneous solids such as coal and mineral ores must usually be determined directly from experimental measurements. Predictive equations are available which estimate heat capacities based upon several gross material properties. For instance, the heat capacity of coals and partially reacted coal, called char, can be predicted by combining the heat capacities of the aggregate components given in the proximate analysis. The component heat capacities in Btu/lb$_m$ · °F, cited in the *Coal Conversion Systems Data Book*[6] based on Kirov's work, are

For fixed carbon:

$$C_p = 0.145 + 4.70 \times 10^{-4}t - 2.63 \times 10^{-7}t^2$$
$$+ 5.25 \times 10^{-11}t^3$$

For ash:

$$C_p = 0.180 + 7.78 \times 10^{-5}t$$

For primary volatile matter:

$$C_p = 0.381 + 4.50 \times 10^{-4}t$$

For secondary volatile matter:

$$C_p = 0.699 + 3.39 \times 10^{-4}t$$

Secondary volatile matter is defined to be the portion of the volatile matter based on the first 10% of the dry-ash free coal. The primary volatile matter is the remaining volatile matter. For the moisture component, the heat capacity of liquid water is used. The heat capacity of a coal or char is calculated by taking a weight fraction average of the above component heat capacities.

Example 7.16   Calculate the heat capacity of a coal with proximate analysis (wt %) 54% fixed carbon, 21% volatile matter, 5% ash, and 20% moisture, at 100°F.

Solution   The heat capacity calculation requires determination of primary and secondary volatile fractions. The secondary volatile matter is equal to 10% of the dry ash-free coal (d.a.f.):

$$0.1(0.54 + 0.21) = 0.075$$

The primary volatile matter is thus

$$0.21 - 0.075 = 0.135$$

The heat capacity of the fixed carbon component is

$$C_p = 0.145 + 4.70 \times 10^{-2} - 2.63 \times 10^{-3} + 5.25 \times 10^{-5}$$
$$= 0.189 \text{ Btu/lb}_m \cdot °F$$

The volatile matter heat capacity consists of two portions: that of the primary matter,

$$C_p = 0.381 + 4.5 \times 10^{-2} = 0.426 \text{ Btu/lb}_m \cdot °F,$$

---

[6] *Coal Conversion Systems Technical Data Book,* U.S. DOE Report HCP/T2286-01, 1978.

and that of the secondary volatile matter,

$$C_p = 0.699 + 3.39 \times 10^{-2} = 0.733 \text{ Btu/lb}_m \cdot {}^\circ\text{F},$$

The heat capacity of ash is

$$C_p = 0.180 + 7.78 \times 10^{-3} = 0.188 \text{ Btu/lb}_m \cdot {}^\circ\text{F}$$

The heat capacity of liquid water at 100°F is almost identically equal to 1 Btu/lb · °F. Finally, the composite heat capacity of coal can be calculated using the weighted average of these component heat capacities:

$$C_p = \sum W_j C_{p_j}$$
$$= 0.54(0.189) + 0.135(0.426) + 0.075(0.733) + 0.05(0.188) + 0.2(1.0)$$
$$= 0.424 \text{ Btu/lb}_m \cdot {}^\circ\text{F}$$

Experimental heat capacity measurements for various coals are reported by the Institute of Gas Technology (IGT) to agree within 7% of values predicted using the above correlations.

Heat capacity functions for petroleum liquids and gases of undefined composition are also formulated as polynomials in temperature. However, in these cases the polynomial coefficients are themselves correlated in terms of mixture densities and experimentally measured batch distillation temperatures. For a detailed discussion of these estimation methods, the reader is referred to the *Technical Data Book—Petroleum Refining*.[7]

The Use of Heat Capacity Data   The purpose of heat capacity correlations is to allow us to calculate enthalpies of substances as they undergo various processes which do not involve phase changes. The typical operation involving heat capacities will therefore involve the evaluation of an integral of the form

$$H(T_2, P) - H(T_1, P) = \int_{T_1}^{T_2} C_p(T) \, dT$$

If the temperature interval $T_2 - T_1$ is large, say, 50 K or more, the temperature variation of C_p will be significant and the integral ought to be evaluated rigorously. Since the C_p correlations are polynomials, it is most convenient to work directly with the integrated form of the correlation, namely,

$$\int_{T_1}^{T_2} C_p(T) \, dT = a(T_2 - T_1) + \frac{b}{2}(T_2^2 - T_1^2) + \frac{c}{3}(T_2^3 - T_1^3) + \frac{d}{4}(T_2^4 - T_1^4)$$

If the temperature interval is less than 50 K, it is often sufficient to use an

---

[7] *Technical Data Book—Petroleum Refining*, 3rd ed., American Petroleum Institute, Washington, D.C., 1976.

average value of C_p obtained by evaluating the correlation of the midpoint of the temperature range. The integral is thus replaced by the product of an average heat capacity times the temperature change, or

$$\int_{T_1}^{T_2} C_p(T)\, dT = C_p\left(\frac{T_1 + T_2}{2}\right)(T_2 - T_1)$$

This is equivalent to calculating

$$\int_{T_1}^{T_2} C_p(T)\, dT \simeq a(T_2 - T_1) + \frac{b}{2}(T_2^2 - T_1^2) + \frac{c}{4}(T_2^2 - T_1^2)(T_2 + T_1)$$

$$+ \frac{d}{8}(T_2^2 - T_1^2)(T_2 + T_1)^2$$

The above approximate equation agrees with the exact integration in the first two terms but differs in the last two terms. Since, as evident from Figure 7.9, for gases, with the exception of H_2O and some of the diatomic species such as CO and NO, the $C_p(T)$ curves are concave, it can be shown that the second two terms will generally result in a higher enthalpy difference estimate than the integral. Conversely, for liquids, the $C_p(T)$ curves are typically convex, hence the average C_p approximation will generally lead to a lower enthalpy difference estimate than the integral. If the temperature difference is small, less than 50K, then in either case the error introduced will be negligible.

The reader should be cautioned, however, that in general it is not advisable to simplify the C_p integration by merely deleting the higher-order terms. Deletion of all but the first two terms, for instance, amounts to using a linear approximation with slope equal to that of the actual C_p curve at $T = 0$. It can be shown that with concave C_p curves, the two-term approximation will generally lead to a larger overestimation of the integral than the average C_p approximation. With convex C_p curves (liquids), the two-term approximation will lead to a larger underestimate of the integral than the average C_p approximation. If a linear approximation of the $C_p(T)$ curve is desired, a linear fit should be constructed specifically over the temperature range of interest rather than by deleting terms from the general correlation equation.

Example 7.17 Calculate the enthalpy change of 1 mol benzene vapor at 1 atm for a temperature change from 800°F to 1000°F, using (a) the full integral, (b) the average C_p assumption, and (c) the two-term approximation.

Solution

(a) The heat capacity equation for benzene in the gaseous state given in the Appendix is

$$C_p(T) = 16.392 + 4.0204 \times 10^{-2}T + 0.6925 \times 10^{-5}T^2 - 0.41142 \times 10^{-7}T^3$$

$$+ 0.23981 \times 10^{-10}T^4 \quad \text{Btu/lbmol} \cdot {}^\circ\text{F}$$

The integrated form is simply

$$\Delta H = C_p(T) \, dT = 16.392(200) + \frac{4.0204 \times 10^{-2}}{2} (1000^2 - 800^2)$$

$$+ \frac{0.69254 \times 10^{-5}}{3} (1000^3 - 800^3) - \frac{0.41142 \times 10^{-7}}{4} (1000^4 - 800^4)$$

$$\frac{0.23981 \times 10^{-10}}{5} (1000^5 - 800^5)$$

$$= 8793.7 \text{ Btu/lbmol}$$

(b) To employ the average C_p assumption, we first evaluate C_p at the average temperature of 900°F. Thus,

$$C_p(900) = 16.392 + 4.0204 \times 10^{-2}(900) + 0.69254 \times 10^{-5}(900)^2$$
$$- 0.41142 \times 10^{-7}(900)^3 + 0.23981 \times 10^{-10}(900)^4$$
$$= 43.927 \text{ Btu/lbmol} \cdot °\text{F}$$

Therefore,

$$\Delta H = (C_p)_{\text{av}} \, \Delta T = 8785.3 \text{ Btu/lbmol}$$

(c) If the two-term approximation were used, ΔH would simply be calculated from the truncated equation

$$\Delta H = 16.392(200) + \frac{4.0204 \times 10^{-2}}{2} (1000^2 - 800^2)$$

$$= 10{,}515 \text{ Btu/lbmol}$$

The two-term approximation leads to an overestimation of nearly 20%, whereas the simple average C_p approximation incurs only a 0.1% error.

Example 7.18 An oxygen stream at 1 bar and flowing at the rate of 100 kgmol/ h is to be heated from 25°C to 200°C in an insulated heat exchanger by condensing saturated steam available at 1.5 bar. Determine the steam consumption.

Solution As in Example 7.15, the heat exchanger may be viewed as two separate devices: one in which O_2 is heated, and the other in which steam is condensed, with transfer of heat but not mass between the devices. The system consisting of both devices is an open system which can be assumed to operate adiabatically. Neglecting any changes in potential and kinetic energy and assuming that there is no external work or friction loss, the energy balance becomes

$$0 = F_{O_2}^{\text{out}} \hat{H}_{O_2} (200°\text{C}, 1 \text{ bar}) + F_c \hat{H}_{\text{liq}} (1.5 \text{ bar, sat})$$

$$- F_{O_2}^{\text{in}} \hat{H}_{O_2} (25°\text{C}, 1 \text{ bar}) - F_s \hat{H}_{\text{vap}} (1.5 \text{ bar, sat})$$

From mass balances, $F_{O_2}^{\text{out}} = F_{O_2}^{\text{in}}$ and $F_c = F_s$. Therefore, the energy balance reduces to the equation

$$F_{O_2}(\hat{H}_{O_2}(200°C, 1 \text{ bar}) - \hat{H}_{O_2}(25°C, 1 \text{ bar}))$$

$$= F_s(\hat{H}_{vap}(1.5 \text{ bar, sat}) - \hat{H}_{liq}(1.5 \text{ bar, sat}))$$

The enthalpies of saturated steam and water can be evaluated from the steam tables to obtain

$$\hat{H}_{vap} - \hat{H}_{liq} = 2226.2 \text{ kJ/kg}$$

The enthalpy difference of the oxygen stream must be evaluated by recourse to a heat capacity equation:

$$\tilde{H}_{O_2}(200°C, 1 \text{ bar}) - \tilde{H}_{O_2}(25°C, 1 \text{ bar})$$

$$= \int C_{P_{O_2}} dT$$

$$= 29.88(175) - \frac{0.11384 \times 10^{-1}}{2}(473^2 - 298^2)$$

$$+ \frac{0.43378 \times 10^{-4}}{3}(473^2 - 248^3) - \frac{0.370 \times 10^{-7}}{4}(473^4 - 298^4)$$

$$+ \frac{0.1010 \times 10^{-10}}{5}(473^5 - 298^5)$$

$$= 5262 \text{ kJ/kgmol}$$

Note that the heat capacity is in molar units, and thus the enthalpy difference is per mole of O_2. Substituting the enthalpy differences into the energy balance and solving for the steam flow, we obtain

$$F_s = \frac{100 \text{ kgmol/h} \times 5262 \text{ kJ/kgmol}}{2226.2 \text{ kJ/kg}} = 236.3 \text{ kg/h}$$

The total amount of heat required is

$$Q = N_{O_2} \Delta H_{O_2} = 100 \times 5.262 \times 10^3 = 5.262 \times 10^5 \text{ kJ/h}$$

As illustrated in the preceding example, the use of C_p data to evaluate enthalpy differences does not introduce any significant complications into the calculation, provided the inlet and outlet temperatures are known. However, in several of the energy balance examples involving the steam tables, we encountered the converse situation, that is, the exit temperature or pressure was unknown but the exit enthalpy could be calculated. In these cases, we simply used the tables to determine the temperature or pressure value corresponding to the calculated enthalpy. When heat capacities are used to evaluate enthalpies, this situation does require a little more computational effort. Specifically, if the exit enthalpy is known but the temperature is not, then, since the integral is a fifth-order polynomial in temperature, the temperature can only be determined by finding the root of this polynomial. Root finding for a fifth-order polynomial requires an iterative calculation of the type discussed in Section 5.1.3.

The general problem can be posed as follows. Given ΔH, the lower temperature limit T_1, and the C_p correlation, determine T such that

$$\Delta H = \int C_p(T) \, dT$$

$$= a(T - T_1) + \frac{b}{2}(T^2 - T_1^2) + \frac{c}{3}(T^3 - T_1^3) + \frac{d}{4}(T^4 - T_1^4) + \frac{e}{5}(T^5 - T_1^5)$$

This equation can be rearranged for application of, say, Wegstein's algorithm, as follows:

$$T = T_1 + \frac{1}{a}\left[\Delta H - \frac{b}{2}(T^2 - T_1^2) - \frac{c}{3}(T^3 - T_1^3) \right.$$

$$\left. - \frac{d}{4}(T^4 - T_1^4) - \frac{e}{5}(T^5 - T_1^5) \right]$$

A good initial valve of T can be obtained by estimating an average C_p value and calculating

$$T \simeq T_1 + \frac{\Delta H}{(C_p)_{av}}$$

Usually, three or four iterations should suffice for convergence from this estimate.

Example 7.19 Suppose the heat exchanger of Example 7.18 was limited to a steam flow of 150 kg/h. Calculate the exit O_2 stream temperature.

Solution Assuming the same steam and inlet O_2 conditions, the energy balance becomes

$$100 \text{ kgmol/h} \int_{298}^{T} C_p \, dT = 150 \text{ kg/h} \times 2226.2 \text{ kJ/kg}$$

To determine T, the equation

$$29.88(T - 298) - \frac{1.138 \times 10^{-2}}{2}(T^2 - 298^2) + \frac{0.4338 \times 10^{-4}}{3}(T^3 - 298^3)$$

$$- \frac{0.37 \times 10^{-7}}{4}(T^4 - 298^4) + \frac{1.010 \times 10^{-11}}{5}(T^5 - 298^5) \text{ kJ/kgmol}$$

$$= 3339.3 \text{ kJ/kgmol}$$

must be solved. As an initial estimate, assume $C_{pav} = 25$ kJ/kgmol · K. Then, the initial temperature estimate can be calculated as

$$T^0 = 298 + \frac{3339.3}{25} = 431.6 \text{ K}$$

To use the root-finding algorithm, we first rearrange the polynomial to the form

$$T = 298 + \frac{1}{29.88}\left[3339.3 + \frac{1.138 \times 10^{-2}}{2}(T^2 - 298^2) \right.$$

$$- \frac{0.4338 \times 10^{-4}}{3}(T^3 - 298^3) + \frac{0.37 \times 10^{-7}}{4}(T^4 - 298^4)$$

$$\left. - \frac{1.01 \times 10^{-11}}{5}(T^5 - 298^5) \right]$$

We begin Wegstein's method by evaluating the right-hand side expression at T^0:

$$T^{(1)} = g(T^0) = 409.67 \text{ K}$$

Next, the right-hand side is evaluated at $T^{(1)}$:

$$g(T^{(1)}) = 409.997$$

Now the slope can be calculated:

$$\text{Slope} = \frac{g(T^{(1)}) - g(T^{(0)})}{T^{(1)} - T^{(0)}} = \frac{409.997 - 409.67}{409.67 - 431.6} = -0.01489$$

and then the damping parameter Θ,

$$\Theta = \frac{1}{1 - \text{slope}} = 0.9853$$

We calculate the next iterate via the equation

$$T^{(2)} = (1 - \Theta)T^{(1)} + \Theta g(T^{(1)})$$

$$= 0.0147(409.67) + 0.9853(409.997) = 409.992$$

Note that $g(T^{(2)}) = 409.993$. Thus, the solution can clearly be taken as 410 K.

7.3.2   Heats of Phase Transition

In the calculation of the properties of a wet steam mixture, we observed that there is a substantial difference in the properties of saturated liquid and those of saturated vapor at the same temperature and pressure. For instance, the enthalpy of saturated water at 100°C is 419.1 kJ/kg, while that of saturated vapor at 100°C is 2676.0 kJ/kg. When we perform energy balance calculations using complete thermodynamic tabulations such as are provided by the steam tables, these discontinuities in the properties are automatically accounted for within the tables. However, when enthalpies are evaluated using heat capacity equations, then it is necessary to explicitly account for the enthalpy discontinuities over phase transitions because each heat capacity correlation is only valid for a specified phase.

The enthalpy differences associated with phase transitions are called *heats of phase transition*, or latent heats. The term heat is used for these enthalpy differences because for a closed system undergoing a constant-pressure process, the enthalpy difference will from the first law be equal to the heat which must be supplied to the system to cause the phase change to occur. For instance, if 1 kg liquid water at 100°C and 1 atm pressure is heated at constant pressure until all of the water is converted to steam at 100°C, then the required heat Q will be equal to

$$Q = [H_{vap} \, (100°C, \, 1 \text{ atm}) - H_{liq} \, (100°C, \, 1 \text{ atm})](1 \text{ kg})$$

$$Q = (2676.0 - 419.1)(1) = 2256.9 \text{ kJ}$$

The enthalpy difference associated with a transition from liquid to vapor is called a *heat of vaporization*; that associated with a transition from solid to vapor, *heat of sublimation*; and that associated with a transition from solid to liquid, *heat of fusion* (or melting). These heats of transition will have different values for each species and for each type of transition for a given species. Moreover, from the phase rule for a system containing a single pure component which is present in two phases at equilibrium, the degree of freedom is 1. Thus, for a given species undergoing a transition between two specified phases, the heat of transition will be a function of only a single state variable. Typically, values of heats of transition are tabulated at a reference pressure of 1 atm (1.01325 bar). Such values are prefixed by the word *normal*. For instance, the heat of vaporization of water at the normal boiling point (the temperature corresponding to 1 atm) is 2256.9 kJ/kg. A tabulation of normal boiling point heats of vaporization for various species is given in Appendix 2. Additional tabulations can be found in standard handbooks.

Estimation Methods   In performing energy balance calculations, it is of course best to employ values of heats of transition obtained from experimental measurements conducted at the same conditions as those arising in the calculations. However, this is usually not possible either because the available value of the heat of transition is at a different temperature or because measured values are unavailable or unaccessible. In such cases, estimation methods prove useful.

Temperature Correction   Recall that for a pure substance undergoing transition between two phases, the state and hence the heat of transition is a function of a single state variable, say, temperature.

Given the heat of transition between phases α and β, $\Delta H_{\alpha\beta}$ at some temperature T, how can we calculate the heat of transition at some other temperature $T + dT$? Since enthalpy is a state function, we can write

$$d(\Delta H_{\alpha\beta}) = \left(\frac{\partial(\Delta H_{\alpha\beta})}{\partial T}\right)_P dT + \left(\frac{\partial(\Delta H_{\alpha\beta})}{\partial P}\right)_T dP$$

But since $\Delta H_{\alpha\beta}$ is a function of T only, that is,

$$\Delta H_{\alpha\beta} = \Delta H_{\alpha\beta}(T, \, P(T))$$

we have

$$d(\Delta H_{\alpha\beta}) = \left[\left(\frac{\partial(\Delta H_{\alpha\beta})}{\partial T}\right)_P + \left(\frac{\partial(\Delta H_{\alpha\beta})}{\partial P}\right)_T \left(\frac{dP}{dT}\right)\right] dT$$

Note that by definition the heat of transition is simply the difference between the enthalpies of the saturated phases, that is,

$$\Delta H_{\alpha\beta} = H_\alpha - H_\beta$$

Thus,

$$d(\Delta H_{\alpha\beta}) = \left\{ \left(\frac{\partial H_\alpha}{\partial T}\right)_P - \left(\frac{\partial H_\beta}{\partial T}\right)_P + \frac{dP}{dT}\left[\left(\frac{\partial H_\alpha}{\partial P}\right)_T - \left(\frac{\partial H_\beta}{\partial P}\right)_T\right] \right\} dT$$

$$= \left\{ C_{p\alpha} - C_{p\beta} + \frac{dP}{dT}\left[\left(\frac{\partial H_\alpha}{\partial P}\right)_T - \left(\frac{\partial H_\beta}{\partial P}\right)_T\right] \right\} dT$$

In principle, the heat of transition at a new temperature T^1 can thus be calculated by integrating the above expression to obtain

$$\Delta H_{\alpha\beta}(T^1) = \Delta H_{\alpha\beta}(T^0) + \int_{T^0}^{T^1} (C_{p\alpha} - C_{p\beta})\, dT$$

$$+ \int_{T^0}^{T^1} \left(\frac{dP}{dT}\right)\left[\left(\frac{\partial H_\alpha}{\partial P}\right)_T - \left(\frac{\partial H_\beta}{\partial P}\right)_T\right] dT$$

This calculation requires that in addition to $\Delta H_{\alpha\beta}(T^0)$, the following data are available:

The heat capacity correlations: $C_{p\alpha}(T)$ and $C_{p\beta}(T)$
The derivative of the pressure-temperature equilibrium curve: $\dfrac{dP}{dT}$

The partial enthalpies: $\left(\dfrac{\partial H_\alpha}{\partial P}\right)$ and $\left(\dfrac{\partial H_\beta}{\partial P}\right)$

While the C_p correlations are usually available, the latter functions are more difficult to obtain in general. For special cases, simplifications do result. For instance, for a transition from solid to liquid, since solid and liquid enthalpies are not sensitive to pressure,

$$\frac{\partial H_\alpha}{\partial P} = \frac{\partial H_\beta}{\partial P} \simeq 0$$

Thus, with α = liquid and β = solid,

$$\Delta H_{LS}(T^1) = \Delta H_{LS}(T^0) + \int_{T^0}^{T^1} (C_{pL} - C_{pS})\, dT$$

Similarly, for the case α = vapor and β = liquid,

$$\frac{\partial H_L}{\partial P} \approx 0$$

and if the vapor behaves like an ideal gas, then

$$\frac{\partial H_V}{\partial P} = 0$$

Consequently, the second integral can as an approximation again be dropped. The resulting approximate relation

$$\Delta H_{VL}(T^1) = \Delta H_{VL}(T^0) + \int_{T^0}^{T^1} (C_{pV} - C_{pL}) \, dT$$

will for real substances only be satisfactory over regions of low pressure and moderate temperatures. In general, the formula will overestimate the heat of vaporization since the term $(dP/dT)(\partial H_V/\partial P)_T$ has a negative value. This arises because for vapors, $(\partial H_V/\partial P)_T$ is usually nonnegative; and, from the Antoine formula,

$$\frac{dP}{dT} = \frac{BP}{(T + C)^2}$$

which is positive since B is usually positive.

Example 7.20   Given the heat of vaporization of water at 90°C, estimate the heat of vaporization of water at 110°C using the approximate and the exact formula.

Solution   First estimate ΔH_{VL} (110°C) using the approximation

$$\Delta H_{VL}\ (110°C) = \Delta H_{VL}\ (90°C) + (C_{pV} - C_{pL})\ (20°C)$$

The average heat capacities of water and steam in this range are

$$C_{pV} \approx 0.46 \text{ kcal/kg} \cdot °C \qquad \text{or} \qquad 1.92 \text{ kJ/kg} \cdot °C$$
$$C_{pL} \approx 1.00 \text{ kcal/kg} \cdot °C \qquad \text{or} \qquad 4.18 \text{ kJ/kg} \cdot °C$$

Thus,

$$\Delta H_{VL}\ (110°C) = 2282 \text{ kJ/kg} + (1.92 - 4.18)20 \text{ kJ/kg} = 2236.8 \text{ kJ/kg}$$

From the steam tables, the value should be 2230 kJ/kg. Next, we repeat the estimate by using the formula

$$\Delta H_{VL}\ (110°C) = \Delta H_{VL}\ (90°C) + (C_{pV} - C_{pL})\ (20°C)$$
$$+ \left.\frac{dP}{dT}\right|_{T=100°C} \left[\left(\frac{\partial H_V}{\partial P}\right)_{T=100°C} - \left(\frac{\partial H_L}{\partial P}\right)_{T=100°C}\right] (20°C)$$

From the steam tables,

$$\left.\frac{dP}{dT}\right|_{T=100°C} \approx \frac{1.01325 - 1.00}{100 - 99.6} = 0.0341 \text{ bar/K}$$

We approximate the partials of the enthalpy functions by using the nearest available entries from the superheated steam tables. First,

$$\left.\frac{\partial H_V}{\partial P}\right|_{T=100°C} \approx \frac{H_V\ (100°C,\ 1\ \text{bar}) - H_V\ (100°C,\ 0.5\ \text{bar})}{(1 - 0.5)\ \text{bar}}$$

$$= \frac{2676 - 2683}{0.5} = -14 \text{ kJ/kg·bar}$$

Similarly,

$$\left.\frac{\partial H_{\rm L}}{\partial P}\right|_{T=100°C} \simeq \frac{H_{\rm L}\ (75°C,\ 1\ {\rm bar})\ -\ H_{\rm L}\ (75°C,\ 0.5\ {\rm bar})}{(1\ -\ 0.5)\ {\rm bar}}$$

$$= \frac{314\ -\ 313.9}{0.5} = 0.2\ {\rm kJ/kg\cdot bar}$$

Thus,

$$\Delta H_{\rm VL}\ (110°C) = 2282\ {\rm kJ/kg} + (1.92\ -\ 4.18)(20)\ {\rm kJ/kg}$$

$$+\ (0.0341)(-14\ -\ 0.2)(20)\ {\rm kJ/kg} = 2227.2\ {\rm kJ/kg}$$

Note that the difference primarily arises because the quantity $(\partial H_{\rm V}/\partial P)_T$ is far from equal to zero.

Because of the difficulty in estimating the terms arising in the second integral term, temperature corrections to heats of vaporization are often calculated using the empirical relation due to Watson,

$$\Delta H_{\rm VL}(T) = \Delta H_{\rm VL}(T^0)\left(\frac{T^c\ -\ T}{T^c\ -\ T^0}\right)^n$$

In this correlation, T^c is the critical temperature of the species in question and n is an empirical parameter which is equal to 0.38 for most substances. Alternate values of n for various chemicals are tabulated by Yaws.[8]

Example 7.21   Repeat the estimate of the heat of vaporization of water at 110°C using Watson's correlation with $n = 0.38$.

Solution   From the steam tables, the critical temperature of water is 374.15°C. The estimation equation thus yields

$$\Delta H_{\rm VL}\ (110°C) = 2282\left(\frac{374.15\ -\ 110}{374.15\ -\ 90}\right)^{0.38} = 2219.6\ {\rm kJ/kg}$$

which is somewhat lower than the value 2230 kJ/kg given in the steam tables.

If the heat of transition for a species is unavailable at any temperature, then it is necessary to resort to various estimation procedures, a number of which are discussed by Reid et al.[9] The interested reader is invited to consult this source for details.

Use of Heats of Transition   Heats of transition must be used when it is necessary to evaluate the enthalpy of a species in a given phase relative to its state in another phase. For instance, to evaluate the enthalpy of 1 kg steam at 200°C and 1 atm

---

[8] C. L. Yaws, *Physical Properties*, Chemical Engineering, McGraw-Hill, New York, 1977.

[9] R. C. Reid, J. M. Prausnitz, and T. K. Sherwood, *The Properties of Gases and Liquids*, 3rd ed., McGraw-Hill, New York, 1977.

relative to the enthalpy of 1 kg water at $-10°C$ and 1 atm, we must account for all of the enthalpy changes corresponding to temperature differences while in the same phase and all of the enthalpy changes resulting from phase transitions. Thus,

$$H_V (200°C, 1 \text{ atm}) = H_S (-10°C, 1 \text{ atm}) + [H_S (0°C, 1 \text{ atm})$$
$$- H_S (-10°C, 1 \text{ atm})] + [H_L (0°C, 1 \text{ atm}) - H_S (0°C, 1 \text{ atm})]$$
$$+ [H_L (100°C, 1 \text{ atm}) - H_L (0°C, 1 \text{ atm})] + [H_V (100°C, 1 \text{ atm})$$
$$- H_L (100°C, 1 \text{ atm})] + [H_V (200°C, 1 \text{ atm}) - H_V (100°C, 1 \text{ atm})]$$

If heat capacity relations are used to evaluate the enthalpy changes occurring over constant phases, the above becomes

$$H_V (200°C, 1 \text{ atm}) = H_S (-10°C, 1 \text{ atm}) + \int_{-10°C}^{0°C} C_{pS}(T) \, dT + \Delta H_{LS} (0°C)$$
$$+ \int_{0°C}^{100°C} C_{pL}(T) \, dT + \Delta H_{VL} (100°C) + \int_{100°C}^{200°C} C_{pV}(T) \, dT$$

The heats of transition thus account for the discontinuous changes in the enthalpy, while the heat capacity integrals account for the continuous enthalpy changes with temperature.

Since the enthalpy is a state function, it is immaterial at what temperatures the heats of transition are applied. Hence, the above calculation could have been performed with the heat vaporization at 90°C and the heat of melting at $-10°C$, provided these values were consistent with the numerical values at the normal boiling and melting points, the integration limits of the heat capacity integrals were correctly adjusted, and pressure effects on the phase enthalpies were negligible. Thus,

$$H_V (200°C, 1 \text{ atm}) = H_S (-10°C, 1 \text{ atm}) + \Delta H_{LS} (-10°C)$$
$$+ \int_{-10°C}^{90°C} C_{pL}(T) \, dT + \Delta H_{VL} (90°C) + \int_{90°C}^{200°C} C_{pV}(T) \, dT$$

However, it is generally best to use the heats of transition at 1 atm because these are generally most accurately known.

Example 7.22 The oxygen stream of Example 7.18 is heated by exchanging heat with superheated benzene vapor available at 5.5 bar and 250°C. Determine the benzene flow rate assuming it leaves the heat exchanger as saturated liquid.

Solution Assuming that potential and kinetic energy terms can be neglected, the heat exchanger energy balance becomes

$$0 = F_{O_2}^{\text{out}} \hat{H}_{O_2} (200°C, 1 \text{ bar}) + F_B^{\text{out}} \hat{H}_L (T_{VL}, 5.5 \text{ bar})$$
$$- F_{O_2}^{\text{in}} \hat{H}_{O_2} (25°C, 1 \text{ bar}) - F_B^{\text{in}} \hat{H}_V (250°C, 5.5 \text{ bar})$$

From the mass balances,

$$F_{O_2}^{in} = F_{O_2}^{out} \quad \text{and} \quad F_B^{in} = F_B^{out}$$

Therefore,

$$F_B \left[\hat{H}_V (250°C, 5.5 \text{ bar}) - \hat{H}_L (T_{VL}, 5.5 \text{ bar}) \right] = F_{O_2} \int_{25°C}^{200°C} C_{p_{O_2}}(T) \, dT$$

To evaluate the benzene enthalpy term, it is necessary to determine the saturation temperature of benzene corresponding to a pressure of 5.5 bar. Using the Antoine equation for benzene,

$$\ln p = 9.5552 - \frac{2948.78}{T - 44.5633}$$

At 5.5 bar, $T_{VL} = 420.18$ K or 147°C. From the Appendix, the normal boiling point of benzene is 80.1°C and the heat of vaporization at the normal boiling point is 7.35263 kcal/mol. To correct the heat of vaporization to 147°C, we use Watson's formula with $T^c = 562.6$ K. Thus,

$$\Delta H_{VL} (147°C) = 7.35263 \left(\frac{562.6 - 420.2}{567.6 - 353.3} \right)^{0.38}$$

$$= 6.3516 \text{ kcal/mol}$$

The vapor heat capacity correlation for benzene is

$$C_{pB}(T) = 4.44235 - 0.280687 \times 10^{-2}T + 0.304765 \times 10^{-3}T^2$$
$$- 0.497093 \times 10^{-6}T^3 + 0.251743 \times 10^{-9}T^4$$

with C_p in cal/mol · K and T in K. Assuming negligible pressure effects on enthalpy,

$$\hat{H}_V (250°C, 5.5 \text{ bar}) - \hat{H}_L (T_{VL}, 5.5 \text{ bar}) = \Delta H_{VL} (147°C) + \int_{147°C}^{250°C} C_{pB}(T) \, dT$$

$$= 6351.6 \text{ cal/mol} + 2986.5 \text{ cal/mol} = 9338.1 \text{ cal/mol}$$

In terms of mass units,

$$\hat{H}_V - \hat{H}_L = \frac{9338.1 \text{ cal/mol}}{78 \text{ g/mol}} = 119.72 \text{ cal/g} = 119.72 \text{ kcal/kg}$$

From Example 7.18,

$$\int_{25°C}^{200°C} C_{p_{O_2}} \, dT = 5262 \text{ kJ/kgmol}$$

Thus,

$$F_B = \frac{100 \text{ kgmol/h} \times 5262 \text{ kJ/kgmol}}{119.72 \text{ kcal/kg} \times 4.184 \text{ kJ/kcal}} = 1050.0 \text{ kg/h}$$

Note that the required flow of benzene is about four times that of the steam calculated in Example 7.18 because of the much smaller heat of vaporization of benzene.

7.3.3 Enthalpy of Mixtures

In the material balance applications that were considered in earlier chapters, processes normally involved streams which consisted of mixtures of many component species. Yet, as noted at the beginning of Section 7.3, direct mixture enthalpy data are available only for selected mixtures such as solutions of salts, acids, and bases in water (see Figure 7.10, for instance). For most species only pure-component enthalpy data are available either in the form of tables, such as those for water, or else in the form of heat capacity correlations and heats of phase transition. The question to be addressed in this section is, how do we calculate mixture enthalpies given pure component information.

Ideal Mixtures   The preferred way to calculate extensive mixture properties, such as the mixture enthalpy, would be simply to sum the extensive properties of the pure components. For instance, it would be most convenient if the enthalpy of a mixture H_{mix} could be calculated as the sum of the component enthalpies H_s:

$$H_{mix} = \sum_{s=1}^{S} H_s$$

or, alternatively,

$$\tilde{H}_{mix} = \sum_{s=1}^{S} x_s \tilde{H}_s$$

analogous to the formula used to calculate the enthalpy of wet steam. A mixture which follows this property mixing rule is called an *ideal* mixture. Unfortunately, many single-phase gas or liquid mixtures do not exactly follow this type of mixing rule. For instance, from Figure 7.10, the enthalpy–concentration isotherms for $NaOH$–H_2O are curved lines which appear to approach straight lines at either concentration extreme. The ideal mixture equation for two components,

$$H_{mix} = x_{NaOH}(\tilde{H}_{NaOH} - \tilde{H}_{H_2O}) + \tilde{H}_{H_2O}$$

would require the mixture enthalpy to be a straight line of constant slope at a given temperature.

In fact, the ideal mixing rule is strictly satisfied only by low-density gas mixtures and dilute liquid mixtures. (Note again the low-concentration enthalpy curves of $NaOH$–H_2O.) For general mixtures, the above simple mixing rule must be modified to include a correction term ΔH_{mix}. This is called the *heat of mixing*:

$$\tilde{H}_{mix} = \sum_{s=1}^{S} x_s \tilde{H}_s + \Delta H_{mix}$$

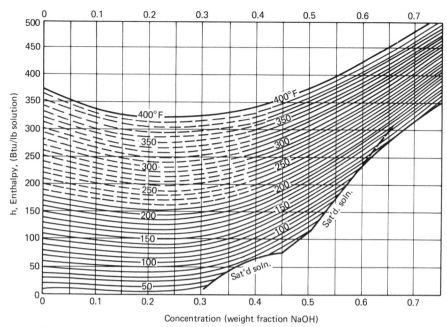

Figure 7.10 Enthalpy–concentration diagram for NaOH–H$_2$O mixtures. *Source:* W. McCabe, *Trans. AIChE.*, **31**, 129 (1935). Reproduced with permission of AIChE.

The name heat of mixing stems from the fact that if mixing is carried out in a constant-pressure and constant-temperature process, then the observed heat effect will correspond to this ΔH_{mix} correction term.

Generally, the heat of mixing will be a function of the system T and P as well as of the composition and nature of the species being mixed. While this correction is best made by using experimentally measured ΔH_{mix} values, these are not widely available. Consequently, the mixing correction is usually predicted by using empirical mixing formulas and real gas equations of state (for gas mixtures) and specialized empirical correction formulas for liquid mixtures. These considerations are treated in detail in thermodynamics texts and thus will not be discussed here. For purposes of this text, we assume that the ideal mixing rule is satisfactory for enthalpies and heat capacities. For mixtures which are neither dilute solutions nor low-density gas mixtures, this correction can lead to inaccuracies.

Species Reference Enthalpies in Mixtures The pure-component enthalpies which are summed to obtain a mixture enthalpy are always of necessity relative rather than absolute enthalpies. As noted in Section 7.2.2, care must be taken when using enthalpy values for a given species from different sources that these enthalpies all be referred to the same reference state. This caution also applies when pure component data are combined to calculate various mixture enthalpy values. However, because of the form of the energy balance equation, it is not necessary that all component enthalpies be referred to the same reference state. Instead, in the absence of chemical reactions, each species may have its own reference enthalpy value at its own reference state.

To verify this statement, consider the simplified energy balance equation

$$\frac{dQ}{dt} - \frac{dW}{dt} = \sum_{\substack{\text{outlet} \\ \text{streams } j}} N^j \tilde{H}^j - \sum_{\substack{\text{inlet} \\ \text{streams } k}} N^k \tilde{H}^k$$

Assuming that each stream is an ideal mixture, it follows that

$$N^j \tilde{H}^j = \sum_{s=1}^{S} N_s^j \hat{H}_s(T^j, P^j)$$

For each species s, $s = 1 \ldots s$, we select a reference state (T^s, P^s) and a reference enthalpy value $H_s(T^s, P^s)$. The absolute enthalpy of species s in stream j can thus be written as

$$\hat{H}_s^j = [\hat{H}_s^j - \hat{H}_s(T^s, P^s)] + \hat{H}_s(T^s, P^s) = \tilde{H}_{sr}^j + \hat{H}_s(T^s, P^s)$$

where $\tilde{H}_{sr}^j$ is defined to be the enthalpy of species s in stream j relative to the reference state (T^s, P^s). If this expression is substituted into the stream mixture enthalpy formula and the result is substituted into the energy balance equation, the following result will be obtained:

$$\frac{dQ}{dt} - \frac{dW}{dt}$$

$$= \sum_{s=1}^{S} \left\{ \sum_{\substack{\text{outlet} \\ \text{streams } j}} N_s^j [\tilde{H}_{sr}^j + \hat{H}_s(T^s, P^s)] - \sum_{\substack{\text{inlet} \\ \text{streams } k}} N_s^k [\tilde{H}_{sr}^k + \hat{H}_s(T^s, P^s)] \right\}$$

$$= \sum_{s=1}^{S} \left(\sum_{\substack{\text{outlet} \\ \text{streams } j}} N_s^j \tilde{H}_{sr}^j - \sum_{\substack{\text{inlet} \\ \text{streams } k}} N_s^k \tilde{H}_{sr}^k \right)$$

$$+ \sum_{s=1}^{S} \hat{H}_s(T^s, P^s) \left(\sum_{\substack{\text{outlet} \\ \text{streams } j}} N_s^j - \sum_{\substack{\text{inlet} \\ \text{streams } k}} N_s^k \right)$$

However,

$$\sum_{\substack{\text{outlet} \\ \text{streams } j}} N_s^j - \sum_{\substack{\text{inlet} \\ \text{streams } k}} N_s^k = 0 \qquad s = 1, \ldots, S$$

from the species material balances. Thus, the species reference enthalpies cancel from the balance equation; consequently, in the absence of chemical reactions, the choice of species reference enthalpy and state does not effect the energy balance calculations. The energy balance is always

$$\frac{dQ}{dt} - \frac{dW}{dt} = \sum_{s=1}^{S} \left(\sum_{\substack{\text{outlet} \\ \text{streams } j}} N_s^j \tilde{H}_{sr}^j - \sum_{\substack{\text{inlet} \\ \text{streams } k}} N_s^k \tilde{H}_{sr}^k \right)$$

in which each species enthalpy is a relative enthalpy. Of course, as noted earlier, for any given species, the same reference state and reference enthalpy must be used throughout. If enthalpy data for a given species are drawn from different sources, this may require corrections of the type discussed in Section 7.2.2.

Example 7.23 A stream consisting of steam at 200°C and 5 bar is let down to 48 psia and mixed adiabatically with a stream consisting of ammonia at 100°F and 48 psia to produce a composite stream at 300°F. Calculate the steam rate if the NH_3 flow is 1000 kg/h. Use the steam tables and the ammonia enthalpy data given in Table 7.1.

Table 7.1 Enthalpy Table for Pure Ammonia

Saturated Ammonia[a]

Temp. (°F)	Abs. pressure, lb./sq. in.	Volume, (cu. ft/lb)		Enthalpy, (Btu/lb)	
		Liquid	Vapor	Liquid	Vapor
−60	5.55	0.02278	44.73	−21.2	589.6
−50	7.67	.02299	33.08	−10.6	593.7
−40	10.41	.02322	24.86	0.0	597.6
−30	13.90		18.97	10.7	601.4
−20	18.30	.02369	14.68	21.4	605.0
−16	20.34		13.29	25.6	606.4
−12	22.56		12.06	30.0	607.8
−8	24.97		10.97	34.3	609.2
−4	27.59		9.991	38.6	610.5
0	30.42	.02419	9.116	42.9	611.8
4	33.47		8.333	47.2	613.0
8	36.77		7.629	51.6	614.3
12	40.31		6.996	56.0	615.5
16	44.12		6.425	60.3	616.6
20	48.21	.02474	5.910	64.7	617.8
24	52.59		5.443	69.1	618.9
28	57.28		5.021	73.5	619.9
32	62.29		4.637	77.9	621.0
36	67.63		4.289	82.3	622.0
40	73.32	.02533	3.971	86.8	623.0
50	89.19	.02564	3.294	97.9	625.2
60	107.6	.02597	2.751	109.2	627.3
70	128.8	.02632	2.312	120.5	629.1
80	153.0	.02668	1.955	132.0	630.7
90	180.6	.02707	1.661	143.5	632.0
100	211.9	.02747	1.419	155.2	633.0
110	247.0	.02790	1.217	167.0	633.7
120	286.4	.02836	1.047	179.0	634.0
125	307.8	.02860	0.973	183.9	634.0

Table 7.1   Enthalpy Table for Pure Ammonia *(Continued)*

Superheated Ammonia

v, volume, cu. ft/lb; h, enthalpy, Btu/lb; Absolute pressure, lb./sq. in. (saturation temperature, °F, in parentheses)

Temp. (°F)	24 (−9.58)		30 (−0.57)		38 (9.42)		48 (19.80)	
	v	h	v	h	v	h	v	h
Sat.	11.39	608.6	9.236	611.6	7.396	614.7	5.934	617.7
0	11.67	614.1	9.250	611.9				
10	11.96	619.7	9.492	617.8	7.407	615.0		
20	12.25	625.2	9.731	623.5	7.603	621.0	5.937	617.8
30	12.54	630.7	9.966	629.1	7.795	626.9	6.096	624.0
40	12.82	636.1	10.20	634.6	7.983	632.6	6.251	630.0
50	13.11	641.4	10.43	640.1	8.170	638.3	6.404	635.9
60	13.39	646.7	10.65	645.5	8.353	643.8	6.554	641.6
70	13.66	652.0	10.88	650.9	8.535	649.3	6.702	647.3
80	13.94	657.3	11.10	656.2	8.716	654.8	6.848	652.9
90	14.22	662.6	11.33	661.6	8.895	660.2	6.993	658.5
100	14.49	667.8	11.55	666.9	9.073	665.6	7.137	664.0
110	14.76	673.1	11.77	672.2	9.250	671.0	7.280	669.5
120	15.04	678.4	11.99	677.5	9.426	676.4	7.421	675.0
130	15.31	683.6	12.21	682.9	9.602	681.8	7.562	680.5
140	15.58	688.9	12.43	688.2	9.776	687.2	7.702	685.9
150	15.85	694.2	12.65	693.5	9.950	692.6	7.842	691.4
160	16.12	699.5	12.87	698.8	10.12	698.0	7.981	696.8
170	16.39	704.8	13.08	704.2	10.30	703.3	8.119	702.3
180	16.66	710.2	13.30	709.6	10.47	708.7	8.257	707.7
190	16.93	715.5	13.52	714.9	10.64	714.2	8.395	713.2
200	17.20	720.9	13.73	720.3	10.81	719.6	8.532	718.7
220	17.73	731.7	14.16	731.1	11.16	730.5	8.805	729.6
240	18.27	742.6	14.59	742.0	11.50	741.4	9.077	740.6
260			15.02	753.0	11.84	752.4	9.348	751.7
280					12.18	763.5	9.619	762.9
300							9.888	774.1

Solution   The flow diagram for the process is shown in Figure 7.11. For the system boundary, we choose the dotted lines which include the valve used for pressure reduction. Assuming that the process occurs adiabatically, that no work is done, and that the outlet stream is an ideal mixture, the balance equation reduces to

$$0 = F^1_{NH_3} \hat{H}_{NH_3} \,(300°F, \, 48 \text{ psia}) + F^1_{H_2O} \hat{H}_{H_2O} \,(300°F, \, 48 \text{ psia})$$
$$- F^2_{NH_3} \hat{H}_{NH_3} \,(100°F, \, 48 \text{ psia}) - F^3_{H_2O} \hat{H}_{H_2O} \,(200°C, \, 5 \text{ bar})$$

Table 7.1   Enthalpy Table for Pure Ammonia (*Continued*)

Superheated Ammonia

v, volume, cu. ft/lb; *h*, enthalpy, Btu/lb; Absolute pressure, lb./sq. in. (saturation temperature, °F, in parentheses)

Temp. (°F)	60 (30.21)		80 (44.40)		100 (56.05)		120 (66.02)	
	v	*h*	*v*	*h*	*v*	*h*	*v*	*h*
Sat.	4.805	620.5	3.655	624.0	2.952	626.5	2.476	628.4
30								
40	4.933	626.8						
50	5.060	632.9	3.712	627.7				
60	5.184	639.0	3.812	634.3	2.985	629.3		
70	5.307	644.9	3.909	640.6	3.068	636.0	2.505	631.3
80	5.428	650.7	4.005	646.7	3.149	642.6	2.576	638.3
90	5.547	656.4	4.098	652.8	3.227	649.0	2.645	645.0
100	5.665	662.1	4.190	658.7	3.304	655.2	2.712	651.6
110	5.781	667.7	4.281	664.6	3.380	661.3	2.778	658.0
120	5.897	673.3	4.371	670.4	3.454	667.3	2.842	664.2
130	6.012	678.9	4.460	676.1	3.527	673.3	2.905	670.4
140	6.126	684.4	4.548	681.8	3.600	679.2	2.967	676.5
150	6.239	689.9	4.635	687.5	3.672	685.0	3.029	682.5
160	6.352	695.5	4.722	693.2	3.743	690.8	3.089	688.4
170	6.464	701.0	4.808	698.8	3.813	696.6	3.149	694.3
180	6.576	706.5	4.893	704.4	3.883	702.3	3.209	700.2
190	6.687	712.0	4.978	710.0	3.952	708.0	3.268	706.0
200	6.798	717.5	5.063	715.6	4.021	713.7	3.326	711.8
210	6.909	723.1	5.147	721.3	4.090	719.4	3.385	717.6
220	7.019	728.6	5.231	726.9	4.158	725.1	3.442	723.1
230			5.315	732.5	4.226	730.8	3.500	729.2
240	7.238	739.7	5.398	738.1	4.294	736.5	3.557	734.9
250			5.482	743.8	4.361	742.2	3.614	740.7
260	7.457	750.9	5.565	749.4	4.428	747.9	3.671	746.5
270							3.727	752.2
280	7.675	762.1	5.730	760.7	4.562	759.4	3.783	758.0
290							3.839	763.8
300	7.892	773.3	5.894	772.1	4.695	770.8	3.895	769.6

[a] *Source: U.S. Bur. Standards Circ. 142, 1923.*

To evaluate the enthalpy functions, we use the steam tables to obtain the enthalpy of superheated steam. Note that 300°F = 148.9°C and 48 psia = 3.31 bar. Thus,

$$\hat{H}_{H_2O} (200°C, 5 \text{ bar}) = 2855 \text{ kJ/kg}$$

$$\hat{H}_{H_2O} (148.9°C, 3.31 \text{ bar}) = 2753.6 \text{ kJ/kg}$$

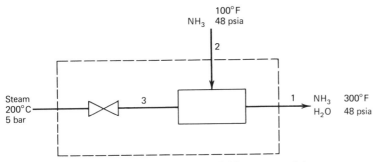

Figure 7.11 Flow diagram for Example 7.23, stream mixing.

Recall that the reference state for the steam table is water at the triple point (0.01°C and 0.00611 bar) and that at that state $\hat{U} = 0$ is chosen as the reference value.

To evaluate the enthalpy of NH_3, we use Table 7.1, which employs AE units. Thus,

$$\hat{H}_{NH_3} (100°F, 48 \text{ psia}) = 664.0 \text{ Btu/lb}_m$$

$$\hat{H}_{NH_3} (300°F, 48 \text{ psia}) = 774.1 \text{ Btu/lb}_m$$

Note that the NH_3 enthalpy table uses $-40°F$ and 10.41 psia as reference state and that the reference enthalpy of saturated liquid NH_3 at those conditions is taken to be zero. Substituting these enthalpy values into the energy balance equation and solving for the steam flow,

$$F_{H_2O} = \frac{10^3 \text{ kg/h} \times (774.1 - 664.0) \text{ Btu/lb}_m \times \dfrac{1 \text{ kcal/kg}}{1.8 \text{ Btu/lb}_m} \times 4.184 \text{ kJ/kcal}}{(2855 - 2753.6) \text{ kJ/kg}}$$

$$= 2524 \text{ kg/h}$$

The reference enthalpies of the two species are clearly different. However, it is easy to see that only the enthalpy difference for each species appears in the final equation, hence the reference enthalpies do clearly cancel out. The analysis given prior to the example assures us that this will always be the case even in more complex examples in which the cancellation is less obvious.

7.4 ANALYSIS OF NONREACTING SYSTEMS

In the preceding sections, we assembled all of the elements necessary to formulate and solve closed- and open-system energy balance problems. We developed the energy balance equations themselves and defined the energy terms that arise in these equations. Furthermore, we studied the dependence of the enthalpy and internal energy functions on the state variables of the system, considered the problems associated with interconverting between the H and U state functions and the temperature, pressure, as well as phase compositions, and finally discussed alternative data sources and approximation techniques for calculating the H and U

functions. With all of these thermodynamic considerations at hand, energy balance problems reduce simply to algebraic problems in which certain unknowns must be calculated using the available material and energy balance equations. In this section, we focus on the algebraic properties of these problems and show how the degree-of-freedom analysis developed for material balance problems can be extended to incorporate energy balances. We shall restrict our attention exclusively to open-system balances, since these complement the steady-state material balances discussed earlier and constitute the most common class of applications in chemical engineering practice.

7.4.1 The Energy Balance Variables

The general form of the open-system energy balance equation for multiple inlet and outlet streams,

$$\frac{dQ}{dt} - \frac{dW}{dt} = \sum_{\substack{\text{outlet} \\ \text{streams } j}} F^j\big(\hat{H}(w, T, P, \text{phase}) + gz + \tfrac{1}{2}v^2\big)_j$$

$$- \sum_{\substack{\text{inlet} \\ \text{streams } k}} F^k\big(\hat{H}(w, T, P, \text{phase}) + gz + \tfrac{1}{2}v^2\big)_k$$

clearly indicates that the variables which must be considered are the normal material balance stream variables, namely, the flows and compositions; the new stream variables, namely, the stream temperature, pressure, phase, elevation, and velocity; and the system variables W and Q.

In most chemical engineering applications, the stream pressures are usually fixed by considerations other than those involving energy balances, for example, chemical or physical equilibrium or flow properties. Thus, pressures are usually not treated as independent variables in energy balance calculations. Moreover, in most process applications, the potential and kinetic energy terms are negligible compared to the enthalpy, heat, and work terms and thus can for all practical purposes be excluded from consideration. Finally, for purposes of this text, we shall always assume that the stream phase or phase distribution will be known (hence, will not be an independent stream variable) and that the streams will be ideal mixtures.

With these assumptions and restrictions, the steady-state energy balance equation for a system or unit involving S species reduces to

$$\frac{dQ}{dt} - \frac{dW}{dt} = \sum_{s=1}^{S} \left[\sum_{\substack{\text{outlet} \\ \text{streams } j}} F_s^j \hat{H}_s(T^j) - \sum_{\substack{\text{inlet} \\ \text{streams } k}} F_s^k \hat{H}_s(T^k) \right]$$

In this form, it is quite evident that for a single unit or system, the stream variables will always simply be the species flows and temperature (with the phase and pressure assumed known). Moreover, associated with every unit or system around which balances are to be written will be the system variables W and Q. Counting of variables for purposes of deducing the problem degree of freedom thus is quite straightforward: every stream will require a temperature and a vector of species flows; every unit will require a W and Q.

7.4.2 Properties of the Energy Balance Equations

The first and most obvious algebraic property of the energy balance equation is that it is a *nonlinear* equation in the species flows and temperatures. If the temperatures are all known, then the equation becomes linear in the F_s^j as well as in the rates dW/dt and dQ/dt. In such cases, it can be easily used to calculate one of the unknowns, dW/dt, dQ/dt, or F_s^j, as we did in Examples 7.15, 7.18, and 7.22. However, even if the rates dW/dt and dQ/dt as well as all of the species flows are known, the calculation of one of the unknown temperatures will, because of the nonlinear dependence of H on T, require an iterative solution. We have already observed this fact with Example 7.19. In the general case in which both the species flows and the stream temperatures are unknown, the presence of an energy balance equation will always necessitate iterative solution of the balance equation set.

A second important property of the energy balance equation is that it is *homogeneous* in the rates dW/dt and dQ/dt and the species flows. Quite clearly, if a set of rates and flows

$$\frac{dW^*}{dt}, \qquad \frac{dQ^*}{dt}, \qquad F_s^{j*}$$

satisfies the energy balance equation, then any scalar multiple of these variables will also satisfy the energy balance equation. To verify this, we merely substitute in α times these variables into the equation to obtain

$$\alpha\frac{dQ^*}{dt} - \alpha\frac{dW^*}{dt} = \sum_{s=1}^{S}\left(\sum_{\substack{\text{outlets}\\j}} \alpha F_s^{j*} H_s(T^j) - \sum_{\substack{\text{inlets}\\h}} \alpha F_s^{k*} H_s(T^j)\right)$$

$$= \alpha\sum_{s=1}^{S}\left(\sum_{\substack{\text{outlets}\\j}} F_s^{j*} H_s(T^j) - \sum_{\substack{\text{inlets}\\k}} F_s^{k*} H_s(T^j)\right)$$

The right-hand side of the above equation must of course be equal to $\alpha[(dQ^*/dt) - (dW^*/dt)]$, since the original variables satisfy the balance equation. As in the material balance case, this is an important property because it allows us to arbitrarily assign a numerical value to one of the variables dW/dt, dQ/dt, or F_s^j as *basis* for the calculation if no fixed value is assigned to any of them in the problem definition. Moreover, even if a basis has been assigned in the problem statement, we can disregard it, choose our own basis for computation, and then rescale the solution to match the required basis. The reader should note again that as basis a value can be assigned to either dW/dt or dQ/dt or one of the flows.

Example 7.24 In the steam turbine system discussed in Example 7.14, suppose the exhaust steam conditions are specified to be 1 bar and a quality of 0.9. Calculate the steam feed rate required per 1 hp produced in the turbine.

Solution From Figure 7.5, the inlet and outlet steam conditions are known, the elevation and velocity terms are known, and the heat rate is specified to be

$$\frac{dQ}{dt} = 300F \text{ Btu/h}$$

The energy balance equation then becomes

$$300F - \frac{dW}{dt} = F\left[\hat{H}_2 - \hat{H}_1 + \frac{g}{g_c}(200 \text{ ft}) + \tfrac{1}{2}g_c^{-1}(1 - 10^4)\frac{\text{ft}^2}{\text{s}^2}\right]$$

where

$$\hat{H}_2 = 0.9(2675.4) + 0.1(417.5) = 2455.01 \text{ kJ/kg}$$

and

$$\hat{H}_1 = 2776.2 \text{ kJ/kg}$$

Thus,

$$300F - \frac{dW}{dt} = -138.07F \text{ Btu/h}$$

Quite clearly, the balance equation is homogeneous in F and dW/dt. We could choose a value of either F or dW/dt as basis and then could calculate the other unknown. Suppose we pick $dW/dt = 10^5$ Btu/h (work done by the system). Then,

$$F = \frac{10^5 \text{ Btu/h}}{438.07} = 228.27 \text{ lb}_m/\text{h}$$

The desired ratio is

$$\frac{F}{dW/dt} = \frac{228.27 \text{ lb/h}}{1 \times 10^5 \text{ Btu/h} \times \dfrac{778 \text{ ft} - \text{lb}_f}{\text{Btu}} \times \dfrac{1 \text{ hp} \cdot \text{s}}{550 \text{ ft} - \text{lb}_f} \times \dfrac{1 \text{ h}}{3600 \text{ s}}}$$

$$= 5.81 \text{ lb/h} \cdot \text{hp}$$

A third important property of the energy balance equation is that in addition to allowing a selection of a calculation basis, it also admits a selection of problem reference state. Consider the equation

$$\frac{dQ}{dt} - \frac{dW}{dt} = \sum_{s=1}^{S}\left(\sum_{\substack{\text{outlet}\\\text{streams } j}} F_s^j \hat{H}_s(T^j) - \sum_{\substack{\text{inlet}\\\text{streams } k}} F_s^k \hat{H}_s(T^k)\right)$$

where the species enthalpies are relative enthalpies as defined in Section 7.3.3. Suppose we select a problem reference temperature T^r (and implicitly phase and pressure P^r). We can then associate an enthalpy $H_s(T^r)$ with each species s. For each species s, add and subtract the terms

$$-F_s^j \hat{H}_s(T^r) + F_s^j \hat{H}_s(T^r) \qquad \text{for each outlet stream } j$$

and

$$-F_s^k \hat{H}_s(T^r) + F_s^k \hat{H}_s(T^r) \qquad \text{for each inlet stream } k$$

to the balance equation. Rearranging terms, the equation reduces to

$$\frac{dQ}{dt} - \frac{dW}{dt} = \sum_{s=1}^{S} \left[\sum_{\substack{\text{outlet} \\ \text{streams } j}} F_s^j (\hat{H}_s(T^j) - \hat{H}_s(T^r)) \right.$$

$$\left. - \sum_{\substack{\text{inlet} \\ \text{streams } k}} F_s^k (\hat{H}_s(T^h) - \hat{H}_s(T^r)) \right] + \sum_{s=1}^{S} \hat{H}_s(T^r) \left(\sum_{\substack{\text{outlet} \\ \text{streams } j}} F_s^j - \sum_{\substack{\text{inlet} \\ \text{streams } k}} F_s^k \right)$$

Note, however, that the third block of terms will vanish because, from the material balance equations,

$$\sum_j F_s^j - \sum_k F_s^k = 0 \qquad \text{for each } s, \; s = 1, \ldots, S$$

Thus, the modified balance equation becomes

$$\frac{dQ}{dt} - \frac{dW}{dt} = \sum_{s=1}^{S} \left[\sum_{\substack{\text{outlet} \\ \text{streams } j}} F_s^j (\hat{H}_s(T^j) - \hat{H}_s(T^r)) \right.$$

$$\left. - \sum_{\substack{\text{inlet} \\ \text{streams } k}} F_s^k (\hat{H}_s(T^k) - \hat{H}_s(T^r)) \right]$$

Observe that if T^r is set equal to one of the outlet stream temperatures T^i, then the terms corresponding to that outlet stream will vanish from the equation, that is,

$$\sum_{s=1}^{S} F_s^i (\hat{H}_s(T^i) - \hat{H}_s(T^i)) = 0$$

The balance equation will then take the form

$$\frac{dQ}{dt} - \frac{dW}{dt}$$

$$= \sum_{s=1}^{S} \left[\sum_{\substack{\text{outlet} \\ \text{streams } j \\ j \neq i}} F_s^j (\hat{H}_s(T^j) - \hat{H}_s(T^i)) - \sum_{\substack{\text{inlet} \\ \text{streams } k}} F_s^k (\hat{H}_s(T^k) - \hat{H}_s(T^i)) \right]$$

Similarly, if the problem reference state is selected to be the state of one of the input streams, then the terms associated with that stream will disappear from the balance equation.

By appropriate choice of the problem reference state, it is thus possible to delete from the balance equation the flows associated with any one of the problem streams. This can sometimes lead to a convenient simplification of the balance equation. It can be shown algebraically that the choice of T^r equal to the temperature of one of the problem streams is equivalent to using the material balance equation to eliminate the species flows of that stream from the energy balance equation.

Example 7.25   A process gas stream at 400°C is to be rapidly cooled to 200°C by direct quenching with cold liquid benzene at 20°C. If the hot stream consists of 40% C_6H_6, 30% $C_6H_5CH_3$, 10% CH_4, and 20% H_2, calculate the benzene quench rate required for a gas feed rate of 1000 kgmol/h, assuming quenching occurs adiabatically. A flow diagram for the process is shown in Figure 7.12.

Solution   Assuming that the operation is adiabatic, that the dependence of the enthalpies on pressure are negligible, and that the potential and kinetic energy effects are negligible, the energy balance equation in molar form with explicit problem reference state becomes

$$0 = N^3[\tilde{H}^3 \, (200°C) - \tilde{H}^3(T^r)] - N^2[\tilde{H}^2 \, (400°C) - \tilde{H}^2(T^r)]$$
$$- N^1[\tilde{H}^1 \, (20°C) - \tilde{H}^1(T^r)]$$

Suppose $T^r = T^3 = 200°C$ is chosen as problem reference state; then the term involving N^3 will disappear from the equation and the balance will reduce to

$$N^2[\tilde{H}^2 \, (400°C) - \tilde{H}^2 \, (200°C)] = N^1[\tilde{H}^1 \, (200°C) - \tilde{H}^1 \, (20°C)]$$

With a flow basis of 1000 kgmol/h of stream 2 and the given compositions, the left-

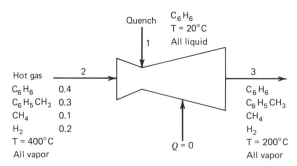

Figure 7.12   Flow diagram for Example 7.25, quench system.

hand side of the balance can be written as

$$\sum_{s=1}^{4} N_s^2 \int_{473.2}^{673.2} C_{p_s} \, dT = (200)\left(\sum_{s=1}^{4} N_s^2 a_s\right) + \frac{(673.2^2 - 473.2^2)}{2} \sum_{s=1}^{4} N_s^2 b_s$$

$$+ \frac{(673.2^3 - 473.2^3)}{3} \sum_{s=1}^{4} N_s^2 c_s + \frac{(673.2^4 - 473.2^4)}{4} \sum_{s=1}^{4} N_s^2 d_s$$

$$+ \frac{(673.2^5 - 473.2^5)}{5} \sum_{s=1}^{4} N_s^2 e_s$$

From Appendix 3, in which the heat capacity constants are given, the following table can be constructed:

s	a	$b \times 10^1$	$c \times 10^{+2}$	$d \times 10^5$	$e \times 10^{+8}$	N_s^2
C_6H_6	18.587	−0.11744	0.12751	−0.20798	0.10533	400
$C_6H_5CH_3$	31.820	−0.16165	0.14447	−0.22895	0.11357	300
CH_4	38.387	−0.73664	0.029098	−0.026385	0.0080068	100
H_2	17.639	+0.67006	−0.013149	+0.010588	−0.002918	200

Then,

$$\sum N_s^2 a_s = 2.43473 \times 10^4$$

$$\sum N_s^2 b_s = -3.5123$$

$$\sum N_s^2 c_s = 0.94625$$

$$\sum N_s^2 d_s = -1.5240 \times 10^{-3}$$

$$\sum N_s^2 e_s = 0.76420 \times 10^{-6}$$

Consequently,

$$N^2[H^2 \, (400°C) - H^2 \, (200°C)] = 2.5634 \times 10^7 \text{ kJ/h}$$

Similarly, with the normal boiling point of benzene at 353.26 K,

$$N^1(\tilde{H}_V^1 \, (200°C) - \tilde{H}_L^1 \, (20°C))$$

where

$$= N^1 \left[\int_{353.3}^{473.2} C_{pV} \, dT + \Delta H_{VL} \, (353.26°K) + \int_{293.2}^{353.3} C_{pL} \, dT \right]$$

$$\Delta H_{VL} \, (353.26 \text{ K}) = 30{,}763.4 \text{ kJ/kgmol}$$

$$C_{pL} = 59.23 + 0.2336T$$

$$\int_{353.3}^{473.6} C_{pV} \, dT = 1.3808 \times 10^4 \text{ kJ/kgmol}$$

$$\int_{293.2}^{353.3} C_{pL} \, dT = 8.0979 \times 10^3 \text{ kJ/kgmol}$$

$$N^1[\tilde{H}_V^1 \, (200°C) - \tilde{H}_L^1 \, (200°C)] = N^1(5.2669 \times 10^4) \text{ kJ/kgmol}$$

Therefore,

$$N^1 = \frac{2.5634 \times 10^7 \text{ kJ/h}}{5.2669 \times 10^4 \text{ kJ/kgmol}} = 486.7 \text{ kgmol/h}$$

Note that the problem could equally well have been solved using $T^r = T^2 = 400°C$ or $T^r = T^1 = 20°C$ or T^r set to any arbitrary temperature as problem reference state.

7.4.3 Degree-of-Freedom Analysis

On the basis of the preceding discussion, the analysis of the degree of freedom of an energy balance problem becomes quite straightforward. In general, the single-unit balance problem without chemical reaction will involve:

1. The stream variables (species flows and temperature) and the unit variables (dQ/dt and dW/dt).
2. The material balance equations.
3. One energy balance equation.
4. The problem specifications (including specifications on the temperature and possibly dQ/dt, dW/dt).
5. A basis for the computation.

The energy balance problem thus can be viewed as just a material balance problem with one additional equation, the energy balance, and several additional variables, the stream temperatures and unit dW/dt, dQ/dt variables. As can be seen from the following example, the tabulation mechanics do not involve any really new concepts.

Example 7.26 Perform a degree of freedom analysis of the problem of Example 7.25.

Solution As shown in Figure 7.12, the problem involves three streams and no energy transfer as heat or work across the system boundaries. Streams 2 and 3 each involve four species flows plus a temperature. Stream 1 involves one species flow plus a temperature. It is possible to write four material balance equations and one energy balance equation. The variables T, dQ/dt and dW/dt are all specified, as is the feed gas composition. The degree-of-freedom tabulation thus takes the familiar form shown in Figure 7.13. The degree of freedom is zero, as expected.

In addition to computing the degree of freedom of the composite material and energy balance problem, it is also useful to prepare a separate tabulation of the material balance portion of the composite problem. Such a tabulation will indicate whether the material balance problem can be solved separately from the energy balance or whether both sets of balances must be solved simultaneously. It is always preferable to separate the solution of these two balance equation types if the specifications are such that this is possible because with all flows known, the energy

Number of variables		
Stream variables	9	
T's	3	
dQ/dt, dW/dt	2	
		14
Number of balances		
Material	4	
Energy	1	
		5
Number of specifications		
Compositions	3	
Temperatures	3	
dQ/dt, $dW/dt = 0$	2	
		8
Basis		1
Degree of freedom		0

Figure 7.13   Degree-of-freedom table, Example 7.26.

balance equation becomes simpler and is less likely to require iterative solution techniques. If the material balances can be solved separately from the energy balances, the balance problem will be said to be *decoupled*. Otherwise, the problem is called a *coupled* material and energy balance problem.

Example 7.27   Analyze Examples 7.18, 7.19, and 7.25 to determine if the balances decouple.

Solution   Example 7.18 involves a heat exchanger in which two streams each consisting of a single species are contacted. The degree-of-freedom table for this problem is shown in Figure 7.14(a). The problem does have degree of freedom zero; however, the material balance problem is underspecified by one. This indicates that the balances will have to be solved in coupled form. Of course, since the material balances are trivial in this case, coupled solution is not very complicated. We can recognize that coupled solution is occurring from the fact that the energy balance equation must be used to calculate a material balance variable, namely, the steam flow.

Example 7.19 involves the same heat exchanger. However, in this case the steam flow is specified and the O_2 temperature is to be determined. From Fig. 7.14(b) it is evident that the material balance subproblem is completely specified (both stream flows are known). Thus, the energy balance can be solved separately for one of the energy balance variables, the O_2 temperature.

Finally, in the quench system of Example 7.25, the degree-of-freedom table, Figure 7.14(c), indicates that the material balance subproblem is underspecified by one. Consequently, the material balances cannot be solved independently of

	Material Balance Subproblem	Combined Balance Problem
Number of variables		
Species flows	4	4
dQ/dt, dW/dt, temperatures		6
Number of balances		
Material	2	2
Energy		1
Specifications		
Temperatures		4
dQ/dt, $dW/dt = 0$		2
Basis	1	1
Degree of freedom	1	0

Figure 7.14(a) Degree-of-freedom table for Example 7.18.

	Material Balance Subproblem	Combined Balance Problem
Number of variables		
Species flows	4	4
dQ/dt, dW/dt, temperatures		6
Number of balances		
Material	2	2
Energy		1
Specifications		
Temperatures		3
dQ/dt, $dW/dt = 0$		2
Flow	1	1
Basis	1	1
Degree of freedom	0	0

Figure 7.14(b) Degree-of-freedom table for Example 7.19.

the energy balance. In fact, the energy balance equation is used to calculate the quench liquid flow.

Although the above examples are quite elementary, they do illustrate fully the situations that can occur. Since for a given unit we can only write a single energy balance equation, only two cases can arise: the coupled case, in which the material balance problem is underspecified by one, and the decoupled case. In the coupled case, all of the primary energy balance variables (T, dQ/dt, dW/dt) are specified and the energy balance reduces to another equation relating the species

	Material Balance Subproblem	Combined Balance Problem
Number of variables		
Species flows	9	9
dQ/dt, dW/dt, temperatures		5
Number of balances		
Material	4	4
Energy		1
Specifications		
Compositions	3	3
Temperatures		3
dQ/dt, $dW/dt = 0$		2
Basis	1	1
Degree of freedom	1	0

Figure 7.14(c)   Degree-of-freedom table for Example 7.25.

flows. In the decoupled case, the material balance subproblem is fully specified, hence all flows can be determined. In the energy balances, all but one of the T, dQ/dt, and dW/dt variables will have been specified, hence the energy balance can be used by itself to determine the one unknown T, dQ/dt, or dW/dt variable.

    The above analysis can of course be extended to multiple units as well. However, we defer that discussion until after we have mastered the reacting case.

7.5   SUMMARY

In this chapter, we discussed all of the thermodynamic and algebraic issues necessary for us to be able to analyze and solve energy balances for nonreacting systems. The thermodynamic issues included the definition of the phase rule which indicates the number of intensive macroscopic properties that must be specified to fix the state of a system. It was shown that, for pure components, equations of state and pressure–temperature equilibrium curves (vapor pressure curves in the case of vapor–liquid systems) allow us to interconvert between the intensive properties P, T, and $\hat{V}$ and the phase identity. For multicomponent systems, interconversion was shown to be much more complicated. In the case of vapor–liquid systems, the pure-component properties had to be supplemented with equilibrium distribution functions or K values. Identification of phases then required that subsidiary dew and bubble point calculations be carried out. If both vapor and liquid phases were found to be present, determination of the phase compositions required a further calculation: the isothermal flash calculation. We indicated that the interconversion between the P, T, $\hat{V}$, and phase composition variables was properly the subject of thermodynamics texts. For purposes of this text, we henceforth assumed that the phase identities could always be given and that pressures would always be specified.

We next investigated several alternate ways in which thermodynamic information, specifically, the enthalpy and internal energy functions, is made available. Complete tables such as those available for steam were shown to be most convenient for use since they implicitly incorporated the P, T, $\hat{V}$ relationships. In the absence of such tabulations, we found that heat capacity correlations and heats of phase transition could be used to evaluate enthalpy differences. We noted that in general these types of calculations neglected the effects of pressure on the enthalpy but decided that in many cases this is a reasonably good approximation. Then, we considered how pure-component enthalpies could be combined to calculate the enthalpies of streams of mixed species. We concluded that for many practical cases, direct addition of pure-component enthalpies was satisfactory.

In the course of exploring these alternate ways of calculating enthalpies, we solved various simple energy balance problems involving open and closed systems. We discovered that typically the potential and kinetic energy terms of the balance equation could be neglected because they were usually small compared to the heat, work, and enthalpy terms. We analyzed the algebraic properties of the open-system energy balance equation and found that it is nonlinear and homogeneous in the species flows, dQ/dt, and dW/dt. It was shown that, in addition to picking a basis for the computations (either a species flow or dQ/dt or dW/dt), it is also possible to select a problem reference state. The choice of the state (temperature) of one of the input or output streams as problem reference was found to be useful. Finally, we extended the degree-of-freedom analysis to include the energy balance variables and equation. The separate tabulation of the material balance subproblem degree of freedom along with that of the complete energy balance problem proved useful because it allowed us to deduce when the material balances could be solved separately of the energy balances. Whenever possible, the decoupled solution was found to be intrinsically advantageous.

As a result of this chapter, we are now ready to solve most energy balance problems involving nonreacting systems. In the next chapter we seek to master the reacting case.

PROBLEMS

7.1 Use the vapor pressure equation for toluene available in Appendix 4 to answer the following questions:

    (a) If the temperature is 300°F, at what pressure will pure toluene boil?

    (b) At 300°F and 50 psia, will pure toluene exist as liquid or vapor?

    (c) At 1000 kPa and 500°K, will pure toluene exist as liquid or vapor?

7.2 A mixture of 60% methanol and 40% ethanol (molar basis) is at 345 K. Over what ranges of pressure will the mixture be all vapor, all liquid, or a two-phase (vapor/liquid) system? Assume Raoult's law is valid and use the Antoine equations for vapor pressures.

7.3 A mixture of ethanol and water is flowing in a pipe at 86°C and 0.8 bar. It appears that the mixture has a mist of liquid droplets entrained in the bulk vapor flow. Determine the likely range of the mixture composition. Assume Raoult's law is valid and use the vapor pressure equations given in Appendix 4.

7.4   **(a)**   A mixture of water and substance A consisting of 25% water (molar basis) is flowing in a pipe. If the mixture is at 133.5°C and 5 bar, what is its phase?

$$\ln p_A \text{ (bar)} = 21.79 - \frac{2000}{T \text{ (°C)} - 33.5}$$

   (b)   Suppose a substance B is added to the mixture of part (a). Beyond what composition of B will the mixture only consist of liquid?

$$\ln p_B \text{ (bar)} = 31 - \frac{3600}{T \text{ (°C)} - 13.5}$$

7.5   Given the equation of state

$$P = \frac{10.73T}{\bar{V} - 0.4125} - \frac{4.337 \times 10^5}{\bar{V}T^{1/2}(\bar{V} + 0.4125)}$$

and the vapor pressure equation

$$\ln p = 13.5633 - \frac{4253.83}{T - 41.047}$$

for ammonia, where p is pressure (psia), $\bar{V}$ is molar volume (ft³/lbmol), and T is temperature (°R), answer the following questions:

   (a)   At 50°F and 50 psia, will ammonia exist as liquid or vapor?
   (b)   At 100°F and 300 psia, will ammonia exist as liquid or vapor?
   (c)   For whichever set of conditions, (a) or (b), ammonia is a vapor, calculate the molar volume $\bar{V}$ (as initial estimate use $\bar{V} = 10.73\ T/P$).
   (d)   If the molar volume of ammonia gas at the boiling point is 250 ft³/lbmol, calculate the temperature and pressure at this boiling point.

7.6   A stream consisting of 60% 1-pentene and 40% 1-heptene is at 2 bar. Assuming Raoult's law K values, calculated using the Antoine equations given in Appendix 4, calculate **(a)** the dewpoint temperature, **(b)** the bubble point temperature, and **(c)** the vapor fraction at 90°C.

7.7   The stream of Problem 7.6 is flashed at 90°C, and the resulting liquid is flashed again at 90°C and 1 bar. Calculate the composition of the liquid obtained after the second flash. What fraction of the original heptene will appear in this liquid stream?

7.8   An equimolar mixture of toluene, *p*-xylene, and ethylbenzene is at 1 atm. Assuming K values can be calculated using Raoult's law and the Antoine equations given in Appendix 4, calculate **(a)** the dew point temperature, **(b)**

the bubble point temperature, and **(c)** the vapor fraction at 400 K and the vapor composition.

7.9 One kilogram steam goes through the following three reversible processes. Initially it is at 550°C and 40 bar. It is then expanded isothermally to state 2, which is at 10 bar. Then, it is cooled at constant volume to 5 bar (state 3). Finally, it is cooled at constant pressure to a specific volume of 0.2 m³/kg (state 4). Compute ΔH and ΔU for each step and for the entire process.

7.10 A rigid vessel having a volume of 10 m³ contains 99% water vapor and 1% liquid water (by volume) in equilibrium at 2 bar. How much heat must be transferred to just convert all the liquid to vapor?

7.11 Water is fed to a steam boiler at 40 bar and 75°C. In the boiler, this water is heated and converted to steam at 40 bar. This steam is not quite dry, its quality being 99.5%. This nearly saturated steam is next passed to a superheater where it is given 100°C superheat at constant pressure. Per 1000 kg steam, calculate the kJ required **(a)** to produce the 99.5% quality steam; **(b)** to provide the superheat.

7.12 Five kilograms wet steam at 5 bar is found to have an enthalpy of 10,000 kJ relative to liquid water at 5 bar and 50°C. Calculate the quality of the steam and the total volume occupied by the steam.

7.13 The contents of a tank containing 1 kg steam at 600°C and 150 bar are allowed to flow into an evacuated tank of equal capacity until the pressure in both tanks is the same (Figure P7.13). The process takes place isothermally.

(a) Calculate the final temperature and pressure in the combined tank system.

(b) Calculate the heat which must be supplied to allow the process to be isothermal.

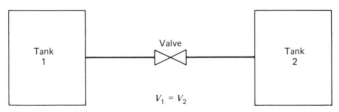

Figure P7.13

7.14 Ten kilograms steam at 500 bar is expanded at constant pressure until its volume increases to seven times its initial volume of 0.01 m³.

(a) Determine the initial and final temperature using the steam tables.

(b) Calculate the heat which must be supplied to carry out the process.

7.15 One kilogram steam at 5 bar and 400°C is compressed isothermally to 100 bar.

   (a) Prepare a plot of P vs. $\hat{V}$ for this process and integrate graphically to obtain the work performed in carrying out the compression ($W = -\int P \, dV$).

   (b) Calculate the heat which must be removed to maintain a constant temperature during this process.

7.16 Ten kilograms steam at 1 bar and a volume of 2.58 m³ are subjected to two alternate reversible compression processes both of which take the steam to the *same* final state: (I) adiabatic compression to a temperature of 600°C followed by isobaric cooling to a final volume of 0.322 m³; (II) isothermal compression directly to the final volume.

   (a) Calculate W, Q, and ΔU for each process.

   (b) Which process will require the least amount of work? Which process will cause the steam to give off the largest amount of heat?

7.17 A 1-ft³ pressure cooker on a stove contains 30% by volume liquid water and 70% by volume steam at 15 psia. Heat is transferred from the stove to the pressure cooker at the rate of 500 Btu/min. If the pressure relief valve on the cooker fails to open, how long will it take before the pressure cooker explodes? The cooker can withstand a pressure of 300 psia.

7.18 A small steam turbine which is used to generate electricity by steam expansion is fed 1000 lb/h steam at 500°F and 250 psia. The exhaust steam exits at 1 atm and has a quality of 85%. Under these conditions, the turbine is found to have a work output of 86.5 hp. Calculate the heat loss from the unit to the environment.

7.19 Steam enters a supersonic nozzle with a velocity of 10 ft/s at a pressure of 500 psia and a temperature of 1000°F. At the nozzle discharge, the pressure is found to be atmospheric and the temperature 300°F. Assuming the nozzle operates adiabatically, calculate the discharge velocity from the nozzle.

7.20 A motor which supplies 750 kW power is used to compress steam at 500 psia and 900°F adiabatically to 1000 psia and a temperature of 1140°F. Calculate the flow of steam which can be compressed assuming 100% efficiency.

7.21 Steam at 10 bar and 400°C is expanded adiabatically to 0.5 bar in a turbine which is required to generate 1500 kW power (Figure P7.21). The steam leaving the turbine is cooled by removing heat at the rate of 1.25×10^{10} J/h to produce a saturated liquid at 0.5 bar.

   (a) Calculate the flow of 10 bar steam required.

   (b) Calculate the quality of the steam leaving the turbine.

7.22 One hundred kilograms steam at 250°C and 20 bar is heated at constant pressure until 23,700 kJ have been added. The steam is then heated to 600°C at constant volume. Calculate ΔU, ΔH, Q, and W for each step and for the

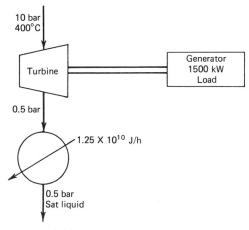

Figure P7.21

whole process (**a**) using the steam tables, and (**b**) assuming that steam is an ideal gas with $C_p = 0.5$ Btu/lb$_m$ · °R which is independent of temperature.

7.23 A hot process stream is cooled by heat exchange with boiler feed water (BFW), thus producing steam. The BFW enters at 100°C and 100 bar, the steam is saturated at 100 bar, the process stream flow is 1000 kgmol/h, its inlet molar enthalpy is 2000 kJ/kgmol, and the exit enthalpy is 800 kJ/kgmol (Figure P7.23). Calculate the required BFW flow.

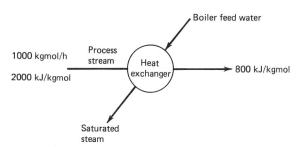

Figure P7.23

7.24 A stream of saturated water at 10 bar is available to exchange heat with a brine solution at 1 bar and 50°C. If the brine flow is twice the water flow and if the water stream can be cooled to 75°C, what is the temperature to

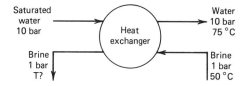

Figure P7.24

which the brine can be heated (Figure P7.24)? The brine can be assumed to have the properties of water.

7.25 An evaporator is a special type of heat exchanger in which steam is used to heat a solution to partially boil off some of the solvent. In the evaporator shown in Figure P7.25, a brine containing 1 wt % salt in water is fed at 1 bar and 50°C. The exit brine contains 2 wt % salt and is saturated liquid at 1 bar. The evaporated water is saturated steam at 1 bar. If saturated steam at 2 bar is used as heat source and its condensate is assumed to be saturated liquid at 2 bar, calculate the kilograms of 2 bar steam required per kilogram of evaporated water produced. Assume the brine has the properties of liquid water.

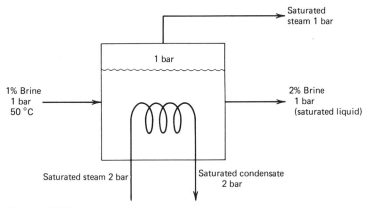

Figure P7.25

7.26 A molal heat capacity equation which has been proposed for CO_2 is

$$C_p = 9.00 + 2.71 \times 10^{-3}T - 0.256 \times 10^{-6}T^2$$

where T is expressed in °F and C_p is in Btu/lbmol · °F.

(a) Calculate the form of the equation where temperature is expressed in degrees K.

(b) Calculate the heat required to raise the temperature of 1 lb CO_2 from 60°F to 400°F.

7.27 Calculate the heat which must be supplied to raise the temperature of steam at 150°C and 1 bar to 300°C at constant pressure.

(a) Using the heat capacity equation given in Appendix 3.

(b) Using the steam tables.

7.28 Use the heat capacity equation for benzene vapor in SI units given in Appendix 3.

(a) Derive the equivalent equation expressed in terms of T (K) and C_p (cal/gmol · K).

(b) Derive the equivalent equations expressed in AE units (check against that given in Appendix 3).

(c) Verify that at 25°C (77°F) all three equations yield equivalent C_p values and that the values using the equations from parts (a) and (b) are numerically the same.

7.29 Using the C_p equation for ammonia vapor, construct a linear C_p vs. T approximation over the range 200–300°F. Compare the accuracy of your approximate and the complete C_p correlation in calculating the heat which must be supplied to raise the temperature of 10 lbmol NH_3 at constant pressure (1 bar) from 210°F to 285°F.

7.30 Using the vapor C_p equation for n-propanol, construct a linear C_p approximation over the range 100–200°C. Compare values of C_p at

(a) 150°C calculated using your approximation.

(b) 150°C using the full C_p equation.

(c) 150°C using the first two terms of the full C_p equation.

(d) The integrated mean C_p over the range 100–200°C.

7.31 A stream of pure toluene at 1 bar and 200°C flowing at 10 kgmol/h is passed through a heater rated at 100 kW. Assuming no heat losses to the surroundings, calculate the outlet temperature of the stream.

7.32 Using the enthalpy data for superheated steam at 0.0 bar pressure from the steam tables, develop a heat capacity equation for steam of the form

$$C_p = a + bT + cT^2$$

over the temperature interval 200–500°C. Plot your equation over the temperature range 50–750°C and show how it fits the steam table values which lie outside the range of validity of your equation (200–500°C). Superimpose a plot of the C_p equation for steam given in the Appendix.

7.33 Using the enthalpy data for superheated ammonia given in Table 7.1, develop two heat capacity equations: one of the form

$$C_p = a + bT + cT^2$$

and another of the form

$$C_p = a + bT + \frac{c}{T}$$

over the temperature range 30–100°F. Which gives the better predictions? How does it compare in accuracy to the C_p equation for NH_3 vapor available in the Appendix?

7.34 Calculate the enthalpy difference between water at 5 bar and 50°C and steam at 5 bar and 400°C

(a) Using the steam tables.

(b) Using the C_p equations, heat of vaporization at the normal boiling point, and Watson's correlation with $n = 0.38$.

(c) How much can the estimate obtained in (b) be improved if the steam tables are used to account for the pressure effects?

7.35 Toluene has a heat of vaporization 33,461 J/gmol at its normal boiling point of 383.8 K. Calculate its heat of vaporization at 450 K

(a) Using Watson's correlation with $n = 0.38$ and $T^c = 593.961$ K.

(b) Assuming negligible pressure effects and using the heat capacity equations.

7.36 Carbon tetrachloride has a heat of vaporization of 15867.04 Btu/lbmol at its normal boiling point of 629.46°R. Calculate its heat of vaporization at 250°F

(a) Assuming pressure effects are negligible and using the C_p equations.

(b) Using Watson's equation with $n = 0.38$.

7.37 Benzene has a heat of vaporization of 7.353 kcal/gmol at its normal boiling point of 353.26 K.

(a) Calculate the heat of vaporization at 50°C assuming pressure effects are negligible and using the C_p correlations from the Appendix.

(b) Calculate the enthalpy of benzene vapor at 200°C relative to that of the liquid at 0°C.

7.38 A mixture of ethanol and water (70% ethanol on a molar basis) fed at 2 bar and 110°C is subjected to an isothermal flash at 0.55 bar and 70°C. Calculate the required rate of heat addition to the flash unit per mole of feed. Assume $K_i = p_i/P$ can be used. For water properties, use the steam tables. For ethanol properties, use the data in the appendices. Repeat the solution assuming the flash operates at 1 bar.

7.39 Calculate the required rate of heat removal per mole of feed for the stream of Problem 7.6 if the feed is at 200°C and 2 bar and the flash is at 90°C at 2 bar.

7.40 Calculate the amount of heat which must be removed to cool the mixture of Problem 7.8 from its dew point to its bubble point.

7.41 Saturated steam at 200°C is used to boil toluene in a heat exchanger. The toluene enters at 50°C, 1 bar, and leaves as vapor at 1 bar.'The steam condensate leaving is saturated water at 200°C.

(a) If steam is supplied at 1000 kg/h and the toluene vapor is to be heated to 170°C, calculate the toluene flow which can be processed.

(b) If the steam is supplied at 1000 kg/h and the toluene flow is adjusted to 4500 kg/h, calculate the exit toluene temperature.

In either case, assume heat losses are negligible. Use the toluene data given in the Appendix.

7.42   Two liquid streams of carbon tetrachloride are to be evaporated to produce a vapor at 200°C and 1 atm. If the first feed stream is at 1 atm and 30°C with a flow rate of 1000 kg/h and the second feed stream is at 70°C and 1 atm and has a flow rate of 500 kg/h, calculate the heat which must be supplied to the evaporator.

7.43   A powdered coal at 50°C with proximate analysis 60% fixed carbon, 30% volatile matter, 5% ash, and 5% moisture is to be preheated to 300°C by entraining it in a stream of superheated steam at 10 bar and 600°C (Figure P7.43). Calculate the required ratio of steam to coal assuming the outlet pressure is 10 bar.

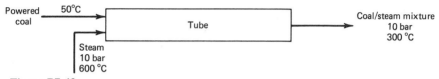

Figure P7.43

7.44   In the acetic anhydride process, 100 mol/h of a process stream consisting of 4% ketene, 10% acetic acid, 43% methane, and 43% carbon dioxide (all mol %) at 700°C is to be cooled to 400°C by direct quenching with liquid glacial acetic acid at 50°C.

    (a)   Given the data below, calculate the flow of quench acetic acid required and the composition of the cooled gas.

Gas Heat Capacities (cal/gmol · K)

	a	$b \times 10^2$	$c \times 10^5$	$d \times 10^9$
CH_4	4.750	1.2	0.303	−2.63
CO_2	6.393	1.01	−0.3405	—
CH_3COOH	8.20	4.805	−3.056	8.31
CH_2O	4.11	2.966	−1.793	4.72

$$\Delta H_{VL}(CH_3COOH, 391.4 \text{ K}) = 5.83 \text{ kcal/gmol}$$

$$\text{Mean } C_p (CH_3COOH, \text{liquid}) = 36 \text{ cal/gmol} \cdot \text{K}$$

    (b)   Suppose it was specified that 40 mol/h acetic acid was used as quench, calculate the outlet gas temperature.

CHAPTER
8

Energy Balances for Reacting Systems

To account for the generation and depletion of chemical species that take place during reaction, it was necessary, in the material balance case, to introduce a reaction rate term into the balance equation. In principle, the first law balance equation derived in Chapter 6 already takes account of the changes in the energy inventory of the system that take place during reaction. It is however convenient to rearrange the energy balance equation to explicitly formulate a term which reflects the enthalpy changes due to reaction. These modifications, the subsidiary calculations required to evaluate the reaction enthalpy differences, and the use of various forms of the resulting energy balance equation form the subject matter of this chapter.

We begin with a definition of the heat of reaction concept and review the ways of calculating heats of reaction from tabulated enthalpy data. We learn how to correct heats of reaction to account for changes in temperature, pressure, and phase by using the calculations discussed in Chapter 7. Next, the simplest form of the reacting system energy balance will be developed, followed by extensions to multiple inlet and outlet streams and multiple reactions. We also consider a special form of the energy balance equation which does not explicitly use heats of reaction. This form is often convenient for computer applications and is particularly appropriate if the reaction stoichiometry is not known or very complex. Finally, the degree-of-freedom analysis is extended to treat the reacting case as well as energy balance problems involving multiple units. The ultimate goal of the chapter is thus to carry the discussion of energy balance problems to the point at which we can solve complex multiunit problems with chemical reaction.

8.1 THE HEAT-OF-REACTION CONCEPT

In our discussion of the enthalpy of mixtures of pure components, we noted that in general mixtures did not behave ideally. Rather, if pure substances were mixed at a given temperature and pressure, then the specific enthalpy of the mixture had to be obtained by correcting the ideal mixture enthalpy by a heat-of-mixing term. Thus,

$$\Delta \tilde{H}_{mix}(T, P) = \tilde{H}_{mix}(T, P) - \sum_s x_s \tilde{H}_s(T, P)$$

Consider a reaction involving S species with stoichiometric coefficients σ_s which proceeds at some temperature and pressure T and P. Suppose we form a mixture by combining σ_s number of moles of each of the reactants s, each in some phase π_s and all at that T and P. Assume that the mixture reacts completely to form a mixture of products, each in some phase π_s and all at the same T and P. The enthalpy of the reactant mix will be given by

$$- \sum_{\text{reactants}} \sigma_s H_s(T, P, \pi_s)$$

where the sum is over all s with $\sigma_s < 0$. Similarly, the enthalpy of the product mix will be given by

$$\sum_{\text{products}} \sigma_s H_s(T, P, \pi_s)$$

where the sum is over all s with $\sigma_s > 0$. Analogous to our experience with the enthalpy of general mixtures, it is in general true that

$$- \sum_{\text{reactants}} \sigma_s H_s(T, P, \pi_s) \neq \sum_{\text{products}} \sigma_s H_s(T, P, \pi_s)$$

The difference between these two mixture enthalpies

$$\sum_{\text{products}} \sigma_s H_s(T, P, \pi_s) - \left[- \sum_{\text{reactants}} \sigma_s H_s(T, P, \pi_s) \right]$$

is defined to be the *heat of reaction* $\Delta H_R(T, P)$ of the reaction with stoichiometric coefficients σ_s at T and P and phases π_s. Thus,

$$\Delta H_R(T, P) = \sum_{s=1}^{s} \sigma_s H_s(T, P, \pi_s)$$

Again, the term heat has come into use because, if the reaction were carried out at constant T and P, then ΔH_R would represent the heat which would have to be removed from the system to maintain it at constant T. That is, for a closed system at constant pressure,

$$Q = H_2 - H_1 = \Delta H_R$$

If the heat of reaction is negative, then Q will be negative and heat will have to be removed from the system to maintain a constant T. A reaction with negative ΔH_R is said to be *exothermic*. Conversely, if the heat of reaction is positive, Q will be positive and therefore heat will have to be transferred to the system to maintain it at constant T. A reaction with positive ΔH_R is said to be *endothermic*.

Note that in general the heat of reaction will depend not only on the stoichiometry of the reaction and on the temperature and pressure but also on the phase of each product and reactant species. It is thus conventional when writing the stoichiometric reaction equation for a given reaction to not only list the species but also their phases. For instance,

$$C(s) + H_2O(g) \rightleftharpoons CO(g) + H_2(g)$$

indicates that solid carbon is to react with gaseous H_2O to form the products CO and H_2, both in gaseous form. The heat of reaction for the above reaction will be numerically different from the heat of reaction for the reaction

$$C(s) + H_2O(l) \rightleftharpoons CO(g) + H_2(g)$$

It is conventional to use the symbols, g, l, s, and aq. for gas, liquid, solid, and aqueous solution, respectively. If a solid can occur in various crystalline forms, then it is also appropriate to identify the particular form which is to be reacted.

Since the heat of reaction is an enthalpy difference, its units are those of an enthalpy term, kJ, kcal, or Btu, depending upon the system of units employed. It is conventional to report heats of reaction per mole of the reaction as expressed by the stoichiometric equation. Thus, if $\Delta H_R = 10$ kJ/mol for the reaction

$$A + B \rightarrow C$$

then the heat of reaction $\Delta H_{R'}$ for the reaction

$$2A + 2B \rightarrow 2C$$

will be

$$\Delta H_{R'} = 2\Delta H_R = 20 \text{ kJ/mol}$$

The value of the heat of reaction is thus linked to the form of the stoichiometric reaction equation.

8.2 HEAT-OF-REACTION CALCULATIONS

Since heats of reaction are functions of stoichiometry, species phase, temperature, and pressure, it would in principle be necessary to tabulate their values as functions of all of these variables. As we note in the next section, temperature, pressure, and phase tabulations can be avoided by only reporting ΔH_R values at selected reference temperature, pressure, and phase conditions. The ΔH_R at some other T, P, and phase can be calculated by updating the species enthalpies using heat capacity correlations, heats of phase transition, and enthalpy pressure corrections. Moreover, as will be discussed in subsequent sections, heat of reaction tabulations can be further compressed by reporting values for only a special set of reactions, namely, the so-called species formation reactions and the species combustion reactions. These special heats of reaction can be combined in a straightforward fashion to calculate the heats of reaction for reactions with arbitrary stoichiometry.

8.2.1 Correction of ΔH_R for T, P, and Phase

The heat of reaction for some reaction with stoichiometric coefficients σ_s at specified temperature T^0, pressure P^0, and species phases π_s^0 is by definition equal to

$$\Delta H_R(T^0,\ P^0,\ \pi^0) = \sum_{s=1}^{S} \sigma_s H_s(T^0,\ P^0,\ \pi_s^0)$$

At any other temperature, pressure, and species phases, T, P, and π_s, respectively, the heat of reaction for the same reaction will be also given by

$$\Delta H_R(T,\ P,\ \pi) = \sum_{s=1}^{S} \sigma_s H_s(T,\ P,\ \pi_s)$$

Subtracting the first equation from the second, it follows that

$$\Delta H_R(T,\ P,\ \pi) - \Delta H_R(T^0,\ P^0,\ \pi^0) = \sum_{s=1}^{S} \sigma_s[H_s(T,\ P,\ \pi_s) - H_s(T^0,\ P^0,\ \pi_s^0)]$$

Thus, given the heat of reaction at T^0, P^0, and species phases π_s^0, the heat of reaction at T, P, and species phases π_s can be calculated by simply evaluating the individual species enthalpy corrections

$$H_s(T,\ P,\ \pi_s) - H_s(T^0,\ P^0,\ \pi_s^0)$$

For instance, if $P = P^0$ and $\pi_s = \pi_s^0$ for all species s, then the temperature correction can be obtained by recourse to the species heat capacity correlations

$$\Delta H_R(T,\ P^0) = \Delta H_R(T^0,\ P^0) + \sum_{s=1}^{s} \sigma_s \int_{T^0}^{T} C_{p_s} dT$$

If $P = P^0$ and the species phases π_s^0 are all liquids while the species phases π_s are all vapors, then it is necessary to incorporate for each species s the phase transition enthalpy $\Delta H_{VL,s}$ at the phase transition temperature $T(P^0)$ corresponding to pressure P^0. Thus, for each species,

$H_s(T,\ P^0,\ \text{vapor}) - H_s(T^0,\ P^0,\ \text{liquid})$

$$= \int_{T(P^0)}^{T} C_{p_{s,V}}\, dT + \Delta H_{VL,s}(P^0) + \int_{T^0}^{T(P^0)} C_{p_{s,L}}\, dT$$

Consequently, the new heat of reaction becomes

$$\Delta H_R(T,\ P^0) = \Delta H_R(T^0,\ P^0) + \sum_{s=1}^{s} \sigma_s \left(\int_{T(P^0)}^{T} C_{p_{s,V}}\, dT \right.$$

$$\left. + \Delta H_{VL,s}(P^0) + \int_{T^0}^{T(P^0)} C_{p_{s,L}}\, dT \right)$$

Finally, if $P \neq P^0$, then in addition to the temperature correction to each of the species enthalpies, it will also be necessary to apply enthalpy pressure corrections. Consistent with our earlier discussion, we assume that the enthalpy correction $(\partial H/\partial P)_T$ will be small and can be neglected. These corrections can be evaluated using real gas equations of state or other available thermodynamic correlations which are thoroughly treated in thermodynamics texts.

In summary, heats of reaction can be corrected for changes in T, P, and phase by simply applying the corresponding corrections to each of the individual species enthalpies.

Example 8.1   Given that the heat of the reaction

$$4NH_3(g) + 5O_2(g) \rightarrow 4NO(g) + 6H_2O(l)$$

at 1 atm and 298K is -279.33 kcal/gmol, calculate the heat of reaction at 920°C, 1 atm, and with H_2O in the vapor phase.

Solution   In this case, the pressure is unchanged, the temperature is increased to 920°C, and the phase of the H_2O species is altered. Thus,

$$\Delta H_R(920°C, 1\,atm) = \Delta H_R(25°C, 1\,atm) + (-4)(\tilde{H}_{NH_3}(920°C, 1\,atm, g)$$
$$- \tilde{H}_{NH_3}(25°C, 1\,atm, g)) + (-5)(\tilde{H}_{O_2}(920°C, 1\,atm, g) - \tilde{H}_{O_2}$$
$$(25°C, 1\,atm, g)) + (+4)(\tilde{H}_{NO}(920°C, 1\,atm, g)$$
$$- \tilde{H}_{NO}(25°C, 1\,atm, g)) + (+6)(\tilde{H}_{H_2O}(920°C, 1\,atm, g)$$
$$- \tilde{H}_{H_2O}(25°C, 1\,atm, l))$$

$$= -279.33\,kcal/gmol + (-4)\int_{25°C}^{920°C} C_{p_{NH_3}} dT$$

$$+ (-5)\int_{25°C}^{920°C} C_{p_{O_2}} dT + 4\int_{25°C}^{920°C} C_{p_{NO}} dT$$

$$+ 6\left[\int_{100°C}^{920°C} C_{p_{H_2O,V}} dT + \Delta H_{VL}(100°C) + \int_{25°C}^{100°C} C_{p_{H_2O,L}} dT \right]$$

The vapor heat capacity terms amount to -3.368 kcal/gmol. The liquid heat capacity term is 8.100 kcal/gmol, and the heat of vaporization term is 6(9.6966) kcal/gmol. Thus, the heat of reaction becomes

$$\Delta H_R\,(920°C,\,1\,atm) = -279.33 - 3.368 + 8.100 + 58.180$$

$$= -216.42\,kcal/gmol$$

The major change in the heat of reaction quite obviously arises because of the phase change.

8.2.2   Heats of Formation and Their Use

In order to systematize the calculation of heats of reaction, it has proved useful to tabulate heats of reaction for a special set of reactions, namely, those reactions in which a given chemical is formed from its constituent elements. Such a tabulation is given in Appendix 7. As will be shown in the subsequent development, these special heats of reaction can be algebraically combined to calculate the heat of reaction for any reaction with arbitrary stoichiometry.

We begin this discussion with a definition of terms:

The *standard state* of a species is defined to be 25°C and 1 atm and the phase normal for that species under those conditions. The phase of water at the standard state is liquid, that of carbon dioxide is gas, and that of carbon is its graphite form, and so on. The standard state is usually indicated by a superscript zero, for example, H^0.

The *standard heat of reaction* is the heat of reaction in which all reactants and products of the reaction are at their standard states (25°C, 1 atm, and the species standard state phase).

The *formation reaction for species* s is the balanced chemical reaction in which 1 mol of species s is formed from the elements which make it up. For instance, the formation reaction for CH_3OH would be

$$C + \tfrac{1}{2}O_2 + 2H_2 \rightarrow CH_3OH$$

The *heat of formation of species* s is the standard heat of reaction of the formation reaction of species s. It is thus the reaction in which all elemental reactants and the product, species s, are in their specified standard states. For the case of methanol, the heat of formation would be the standard heat of reaction for the reaction

$$C(s) + \tfrac{1}{2}O_2(g) + 2H_2(g) \rightarrow CH_3OH(l)$$

As noted earlier, the purpose for introducing species heats of formation is to facilitate the calculation of heats of reaction for arbitrary reactions. We first motivate the calculation procedure using a simple example and then present the general relationship for calculating heats of reaction from heats of formation.

Suppose we desire to calculate the heat of reaction for the reaction (8.1)

$$CO(g) + \tfrac{1}{2}O_2(g) \rightarrow CO_2(g) \tag{8.1}$$

given the heats of formation of $CO(g)$ and $CO_2(g)$, that is, the heats of reaction of the reactions (8.2) and (8.3):

$$C(s) + O_2(g) \rightarrow CO_2(g) \qquad \Delta H_f^0 = -94{,}051.8 \text{ cal/gmol} \tag{8.2}$$

$$C(s) + \tfrac{1}{2}O_2(g) \rightarrow CO(g) \qquad \Delta H_f^0 = -26{,}415.7 \text{ cal/gmol} \tag{8.3}$$

Note that if the second reaction equation is subtracted from the first, we obtain

$$C(s) + O_2(g) - C(s) - \tfrac{1}{2}O_2(g) \rightarrow CO_2(g) - CO(g)$$

or, with our sign convention for stoichiometric coefficients,

$$CO(g) + \tfrac{1}{2}O_2(g) \rightarrow CO_2(g)$$

Since subtraction of the formation reactions results in the desired reaction equation, it seems intuitively reasonable that the heat of reaction for reaction (8.1) could be calculated by subtracting the heat of formation of reaction (8.3) from that of reaction (8.2). That is,

$$\Delta H^0_{R,1} = \Delta H^0_{R,2} - \Delta H^0_{R,3} = \Delta H^0_{f,CO_2} - \Delta H^0_{f,CO} = -67,636.1 \text{ cal/gmol}$$

Note that in reaction (8.1), CO_2 has a stoichiometric coefficient of $+1$ while CO has a stoichiometric coefficient of -1. This corresponds to the signs associated with the species heats of formation in calculating $\Delta H_{R,1}$. However, ΔH^0_f for O_2 does not seem to be involved in calculating $\Delta H_{R,1}$ even though O_2 is a reactant and has stoichiometric coefficient $-\frac{1}{2}$. The question is whether these observations can be generalized. In other words, may the heat of reaction of an arbitrary reaction be calculated as the sum of the products of the species heats of formation times their stoichiometric coefficients with the terms associated with elemental species set to zero? This is, in fact, the case.

The Heat-of-Reaction Formula   The heat of reaction of a reaction with stoichiometric coefficients σ_s is given by

$$\Delta H^0_R = \sum_{s=1}^{S} \sigma_s \, \Delta H^0_{f,s}$$

where the species s are the nonelemental species.

To verify this, note that if α_{es} are the atomic coefficients for species s, then, from the definition of the heat of reaction, the heat of formation of species s will be given by

$$\Delta H^0_{f,s} = H^0_s - \sum_{e=1}^{E} \alpha_{es} H^0_e$$

where H^0_e is the standard state enthalpy of element e. By definition, the heat of reaction of the reaction with stoichiometric coefficients σ_s is

$$\Delta H^0_R = \sum_{s=1}^{S} \sigma_s H^0_s$$

Now, substitute for H^0_s using the above equation for $\Delta H^0_{f,s}$. Then,

$$\Delta H^0_R = \sum_{s=1}^{S} \sigma_s (\Delta H^0_{f,s} + \sum_{e=1}^{E} \alpha_{es} H^0_e)$$

$$= \sum_{s=1}^{S} \sigma_s \, \Delta H^0_{f,s} + \sum_{s=1}^{S} \sigma_s \sum_{e=1}^{E} \alpha_{es} H^0_e$$

Interchanging the order of summation in the last term,

$$\Delta H^0_R = \sum_{s=1}^{S} \sigma_s \, \Delta H^0_{f,s} + \sum_{e=1}^{E} H^0_e \left(\sum_{s=1}^{S} \sigma_s \alpha_{es} \right)$$

Assuming that the stoichiometric equation with coefficients σ_s is balanced, it follows (see Section 4.4.1) that

$$\sum_{s=1}^{S} \sigma_s \alpha_{es} = 0 \qquad \text{for } e = 1, \ldots, E$$

Therefore, the second term must vanish, and we obtain

$$\Delta H_R^0 = \sum_{s=1}^{S} \sigma_s \, \Delta H_{f,s}^0$$

Thus, we have verified the general formulation for calculating heats of reaction from heats of formation of the species involved.

Example 8.2   Calculate the standard heat of reaction for the methanation reaction

$$CO(g) + 3H_2(g) \rightarrow CH_4(g) + H_2O(g)$$

Solution   From the general formula, it follows that

$$\Delta H_R^0 = \Delta H_{f,CH_4}^0 + \Delta H_{f,H_2O(g)}^0 - \Delta H_{f,CO}^0 - 3\Delta H_{f,H_2}^0$$

From the Appendix,

$$\Delta H_{f,CH_4}^0 = -17.89 \text{ kcal/gmol}$$
$$\Delta H_{f,H_2O(g)}^0 = -57.80 \text{ kcal/gmol}$$
$$\Delta H_{f,CO}^0 = -26.42 \text{ kcal/gmol}$$
$$\Delta H_{f,H_2}^0 = 0.0 \text{ kcal/gmol}$$

Thus,

$$\Delta H_R^0 = -49.27 \text{ kcal/gmol}$$

An important consequence of the fact that the terms containing the enthalpies H_e^0 vanish from the formula for ΔH_R^0 is that the values selected for the enthalpies of the elements at 25°C and 1 atm can be arbitrary. It is only necessary that the same value of H_e^0 is used for a given element e in calculating all heats of formation of species involving that element. To avoid the possibility of inconsistent element enthalpies, it has been agreed that all elements at 25°C and 1 atm in their normal phase under those conditions will have an enthalpy value of zero. Thus, H_e^0 for $O_2(g)$, $N_2(g)$, $C(s)$, $Hg(l)$, and so on, are assigned the value zero, and as a consequence their heats of formation will also be identically equal to zero.

Finally, the reader is again cautioned that heats of formation are always given for a species in a specified phase. Since the heat of formation is just a special heat of reaction, the heat of formation of a species in a phase different from the standard state phase can always be calculated using the correction formulas discussed in Section 8.2.1.

8.2.3 Heats of Combustion and Their Use

A second category of standard reactions for which heat of reaction data are reported is the set of combustion reactions. This set of reactions is typically used for organic compounds in preference to the formation reactions because it is relatively convenient to carry out combustion experiments with such compounds and to measure the heat release and hence the heat of reaction associated with the combustion reaction.

The *standard combustion reaction* used for these purposes is the reaction in which one mole of the species in question is completely oxidized to the stable oxides CO_2, H_2O, SO_2, and so on, with a stoichiometric amount of oxygen. The *heat of combustion of species s,* denoted $\Delta H^0_{c,s}$, is then simply the standard heat of reaction of the standard combustion reaction for that species. The combustion products used in the formulation of the standard combustion reaction are assumed to be in their standard state phases, that is, $CO_2(g)$, $SO_2(g)$, $H_2O(l)$, and so on. However, because in combustion it is more natural to have all products as gases, heats of combustion are sometimes also reported with $H_2O(g)$ as a combustion product. To differentiate between the two cases, the heat of combustion with $H_2O(l)$ product is sometimes called the *gross* or higher heat of combustion and the heat of combustion with $H_2O(g)$ product, the *net* or lower heat of combustion. The two differ simply by the heat of vaporization of water at 298K and 1 atm.

Since the heats of formation of the common combustion products are known and since the heat of combustion is just the heat of reaction of a specialized reaction, the heat of formation of a species can always be calculated from its heat of combustion. Given the standard combustion reaction for species A,

$$A(l) + aO_2(g) \rightarrow bCO_2(g) + cSO_2(g) + dH_2O(l)$$

it follows that

$$\Delta H^0_R = \Delta H^0_{c,A} = b\Delta H_{f,CO_2} + c\Delta H_{f,SO_2} + d\Delta H_{f,H_2O(l)} - \Delta H_{f,A}$$

Thus, the heat of combustion of A differs from the heat of formation of A by only the heats of formation associated with the combustion products.

Example 8.3 Given that the net heat of combustion of $CH_4(g)$ is -191.76 kcal/gmol, calculate its heat of formation.

Solution The standard combustion reaction for CH_4 is

$$CH_4(g) + 2O_2(g) \rightarrow CO_2(g) + 2H_2O(g)$$

Since the net heat of combustion is given, the product $H_2O(g)$ is shown in the reaction equation. The heat of combustion is then, by definition,

$$\Delta H^0_{c,CH_4} = \Delta H^0_{f,CO_2} + 2\Delta H^0_{f,H_2O(g)} - \Delta H^0_{f,CH_4} - 2\Delta H^0_{f,O_2}$$

Since

$$\Delta H^0_{f,CO_2(g)} = -94.0518 \text{ kcal/gmol}$$

$$\Delta H^0_{f,H_2O(g)} = -57.7979 \text{ kcal/gmol}$$

and $\Delta H^0_{f,O_2} = 0$, it follows that

$$\Delta H^0_{f,CH_4} = \Delta H^0_{f,CO_2} + 2\Delta H^0_{f,H_2O(g)} - \Delta H^0_{c,CH_4}$$

$$= -17.8876 \text{ kcal/gmol}$$

As a consequence of this relationship between the heats of formation and combustion, it is relatively easy to show that the heat of reaction of a reaction with stoichiometric coefficients σ_s can be calculated directly from the heats of combustion of the reactant and product species. That is,

$$\Delta H_R = -\sum_{s=1}^{s} \sigma_s \Delta H^0_{c,s}$$

Finally, since the heat of combustion is again merely a special heat of reaction, the calculations to correct for T, P and species phase differences can be carried out exactly as discussed in Section 8.2.1.

Example 8.4 Calculate the standard heat of reaction for the toluene oxidation reaction

$$C_6H_5CH_3(g) + O_2(g) \rightarrow C_6H_5CHO(g) + H_2O(g)$$

The gross heat of combustion of $C_6H_5CHO(l)$ is -841.3 kcal/gmol at 18°C. The heat of vaporization of benzaldehyde is 86.48 cal/g at its normal boiling point of 179°C. For benzaldehyde, the heat capacity of the liquid is 0.428 cal/g · °C and the heat capacity of the vapor is 31 Btu/lbmol · °F. Assume average molar heat capacities of 7.0, 8.87, and 18 for $O_2(g)$, $CO_2(g)$, and $H_2O(l)$, respectively.

Solution By definition of the heat of reaction,

$$\Delta H^0_R = \Delta H^0_{f,C_6H_5CHO(g)} + \Delta H^0_{f,H_2O(g)} - \Delta H^0_{f,C_6H_5CH_3(g)} - \Delta H^0_{f,O_2(g)}$$

To evaluate this expression, we need $\Delta H^0_{f,C_6H_5CHO(g)}$, but are given $\Delta H_{c,C_6H_5CHO(l)}$ at 18°C.

The combustion reaction with $H_2O(l)$ (gross) as product is

$$C_6H_5CHO(l) + 8O_2(g) \rightarrow 7CO_2(g) + 3H_2O(l)$$

First, we convert ΔH_c (18°C) to ΔH_c (25°C) using heat capacities:

$$\Delta H_c (25°C) = \Delta H_c (18°C) + \sum_s \sigma_s \int_{18°C}^{25°C} C_{p_s} \, dT$$

With constant average heat capacities, the second term reduces to

$$\left(\sum \sigma_s \, C_{p_s} \right) \Delta T$$

Thus, ΔH_c (25°C) $= -841.3$ kcal/gmol $+ 7\{(-8)(7.0) + (-1)(0.428)(106) + 7(8.87) + 3(18)\} \times 10^{-3}$ kcal/cal $= -841.3$ kcal/gmol $+ 0.103$ kcal/gmol $= -841.2$ kcal/gmol

Now, we can use the combustion reaction to obtain $\Delta H^0_{f,\mathrm{C_6H_5CHO(g)}}$:

$$\Delta H^0_c = -841.2 = 7\Delta H^0_{f,\mathrm{CO_2(g)}} + 3\,\Delta H^0_{f,\mathrm{H_2O(l)}} - \Delta H^0_{f,\mathrm{C_6H_5CHO(l)}}$$

or

$$\Delta H^0_{f,\mathrm{C_6H_5CHO(l)}} = -22.11 \text{ kcal/gmol}$$

But we need ΔH_f of the *gas*. Thus,

$$\Delta H^0_{f,\mathrm{C_6H_5CHO(g)}} = \Delta H^0_{f,\mathrm{C_6H_5CHO(l)}} + \Delta H_{\mathrm{VL}} \text{ (25°C)}$$

We only have ΔH_{VL} at 179°C. We must therefore convert this to 25°C. Recall that

$$\Delta H_{\mathrm{VL}} \text{ (179°C)} = H_{\mathrm{V}} \text{ (179°)} - H_{\mathrm{L}} \text{ (179°C)}$$

If pressure effects are negligible, then we can correct for temperature using heat capacities:

$$= \left(H_{\mathrm{V}} \text{ (25°)} + \int_{25°C}^{179°C} C_{p_{\mathrm{V}}} \, dT \right) - \left(H_{\mathrm{L}}(25°) + \int_{25°C}^{179°C} C_{p_{\mathrm{L}}} \, dT \right)$$

$$= \Delta H_{\mathrm{VL}} \text{ (25°C)} + \int_{25°C}^{179°C} (C_{p_{\mathrm{V}}} - C_{p_{\mathrm{L}}}) \, dT$$

If $C_{p_{\mathrm{V}}}$ and $C_{p_{\mathrm{L}}}$ are assumed to be constant over this range, then

$$\Delta H_{\mathrm{VL}} \text{ (25°C)} = 86.48 \text{ cal/g} + \left(0.428 - \frac{31}{106} \right)(154°C)$$

$$= 107.4 \text{ cal/g}$$

$$= 107.4 \text{ cal/g} \times 106 \text{ g/gmol} = 11.4 \text{ kcal/gmol}$$

Thus,

$$\Delta H^0_{f,\mathrm{C_6H_5CHO(g)}} = -22.11 + 11.4 = -10.71 \text{ kcal/gmol}$$

Finally, we can calculate ΔH_R to obtain

$$\Delta H_R \text{ (25°C)} = -10.71 + (-57.7979) - (-11.95) = -56.56 \text{ kcal/gmol}$$

8.3 ENERGY BALANCE WITH SINGLE CHEMICAL REACTION

In the previous sections, we considered the calculations that are required to evaluate the heat of reaction, that is, the enthalpy change associated with a chemical reaction. In this section, we show how this enthalpy term naturally arises in the general open-system energy balance equation in the presence of chemical reactions. We first consider the simple input–single output case, then extend the balance equation to multiple input–output streams, and finally consider an alternate form of the equation in which the heat of reaction is accounted for implicitly.

8.3.1 The Single-Input/Single-Output Case

Consider the system shown in Figure 8.1 consisting of a reactor with a single input stream, a single output stream, S species, and a single chemical reaction with stoichiometric coefficients σ_s. For simplicity, we assume that the potential and kinetic energy terms are negligible, that there is no work performed by the system, and that the streams form ideal mixtures. The energy balance equation reduces to

$$\frac{dQ}{dt} = \sum_{s=1}^{S} N_s^2 \bar{H}_s(T^2, P^2, \pi_s^2) - \sum_{s=1}^{S} N_s^1 \bar{H}_s(T^1, P^1, \pi_s^1)$$

where for completeness the pressure of the streams and phase of each species are shown explicitly.

Suppose we select a problem reference state defined by a temperature T^r, pressure P^r, and phase π_s^r for each species and we add and subtract the terms

$$\sum N_s^2 \bar{H}_s(T^r, P^r, \pi_s^r) - \sum N_s^2 \bar{H}_s(T^r, P^r, \pi_s^r)$$

and

$$\sum N_s^1 \bar{H}_s(T^r, P^r, \pi_s^r) - \sum N_s^1 \bar{H}_s(T^r, P^r, \pi_s^r)$$

to the balance equation. Grouping the terms associated with each stream, the

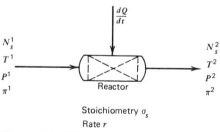

Figure 8.1 Elementary single-input/single-output reactor.

balance can be rearranged to the form

$$\frac{dQ}{dt} = \sum N_s^2 (\tilde{H}_s(T^2, P^2 \pi_s^2) - \tilde{H}_s(T^r, P^r, \pi_s^r)) - \sum N_s^1 (\tilde{H}_s(T^1, P^1, \pi_s^1)$$

$$- \tilde{H}_s(T^r, P^r, \pi_s^r)) + \sum_{s=1}^{S} \tilde{H}_s(T^r, P^r, \pi_s^r)(N_s^2 - N_s^1)$$

In the nonreacting case, $N_s^2 = N_s^1$, from material balances, and hence the last term will vanish. In the reacting case,

$$N_s^2 = N_s^1 + \sigma_s r$$

or

$$N_s^2 - N_s^1 = \sigma_s r$$

When this expression is substituted into the last term of the balance equation to eliminate $N_s^2 - N_s^1$, the last term becomes

$$\sum \tilde{H}_s(T^r, P^r, \pi_s^r)(N_s^2 - N_s^1) = r \sum \sigma_s H_s(T^r, P^r, \pi_s^r)$$

The term $\sum \sigma_s \tilde{H}_s(T^r, P^r, \pi_s^r)$ is by definition the heat of reaction of the reaction with stoichiometric coefficients σ_s at temperature T^r, pressure P^r, and species phases π_s^r. The balance equation then takes the form

$$\frac{dQ}{dt} = r \, \Delta H_R(T^r) + \sum N_s^2 (\tilde{H}_s^2 - H_s^r) - \sum N_s^1 (\tilde{H}_s^1 - H_s^r)$$

where for simplicity we have not explicitly shown P^r and the π_s^r. This is the energy balance for a system with single reaction and single input/single output stream. The first term on the right-hand side of this balance equation reflects the enthalpy change due to chemical reaction and is proportional to the rate of chemical reaction r. The other two terms reflect the so-called sensible enthalpies of the output and input streams relative to the selected problem reference state T^r, P^r, π_s^r, where $s = 1, \ldots, S$.

Note that the choice of reference state is completely arbitrary. If T^r is selected to be 298K, then from the energy balance the system can be viewed as proceeding along the following temperature path: the reactants enter at T^1, are cooled to 298K, the reaction occurs at 298K, and the products are heated to T^2. Alternatively, if T^r is selected to be equal to T^1, the system will proceed along the following path: the reactants enter at T^1, the reaction occurs at T^1, and the products are heated to T^2. Regardless of the temperature path that is taken, since the enthalpy is a state function, the value of dQ/dt calculated will only depend on the initial and final states determined by T^1, T^2, and r. Since these are the same regardless of the temperature path, dQ/dt will be the same in all cases. Consequently, T^r as well as P^r and the phases π_s^r can be selected as dictated by solution convenience.

Example 8.5 Methanol at 675°C and 1 bar is fed to an adiabatic reactor where 25% of it is dehydrogenated to formaldehyde according to the reaction

$$CH_3OH(g) \rightarrow HCHO(g) + H_2(g)$$

Calculate the temperature of the gases leaving the reactor, assuming that constant average heat capacities of 17, 12, and 7 cal/gmol · °C for CH_3OH, $HCHO$, and H_2, respectively, are acceptable over the range.

Solution As shown in Figure 8.2, this is clearly a single-input/single-output application. Using as basis 1000 mol/h CH_3OH, the material balances can easily be solved to determine the outlet flows. In particular, with a methanol conversion of 25%,

$$r = \frac{-XN_{CH_3OH}^{in}}{\sigma_{CH_3OH}} = \frac{-0.25 \times 1000 \text{ mol/h}}{-1} = 250 \text{ mol/h}$$

Then

$$N_{CH_3OH}^{out} = 1000 - r = 750 \text{ mol/h}$$
$$N_{HCHO}^{out} = 0 + r = 250 \text{ mol/h}$$
$$N_{H_2}^{out} = 0 + r = 250 \text{ mol/h}$$

Suppose that as reference state for the energy balance we select the inlet state, namely, temperature 675°C, pressure 1 bar, and gas phase for all species. The balance equation thus becomes

$$\frac{dQ}{dt} = r \, \Delta H_R \, (675°C) + \sum_s N_s^{out} \int_{675°C}^{T} c_{p_s} \, dT$$

The standard heat of reaction for this reaction is given by

$$\Delta H_R \, (25°C) = \Delta H_{f,HCHO(g)}^0 - \Delta H_{f,CH_3OH(g)}^0$$
$$= -27.70 - (-48.08) = 20.38 \text{ kcal/gmol}$$

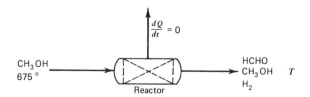

25% Conversion

Figure 8.2 Diagram for Example 8.5.

The heat of reaction at 675°C can be calculated from the equation

$$\Delta H_R\,(675°C) = \Delta H_R\,(25°C) + \sum \sigma_s \int_{25°C}^{675°C} C_{p_s}\,dT$$

Assuming constant average heat capacities, this reduces to

$$\Delta H_R\,(675°C) = 20.38 \text{ kcal/gmol} + (650)(12 + 7 - 17) \text{ cal/gmol}$$

$$= 21.68 \text{ kcal/gmol}$$

Again, assuming constant average heat capacities, the energy balance equation will become

$$0 = r(21.68) + (750 \times 17 + 250 \times 12 + 250 \times 7)(T - 675)$$

or

$$T - 675 = \frac{-250 \text{ mol/h} \times 21{,}680 \text{ cal/mol}}{250 \times 70 \text{ cal/h} \cdot °C}$$

$$T = 675 - 309.7 = 365.3°C$$

Note that the reaction is endothermic ($\Delta H_R > 0$), and consequently under adiabatic conditions the outlet temperature will be decreased.

8.3.2 Alternate Forms of the Balance Equation

The basic energy balance equation developed in the previous section can be readily extended to multiple streams and to a form which does not explicitly involve the heat of reaction. The multiple stream case can be obtained by simply repeating the previous derivation using the general form of the balance equation

$$\frac{dQ}{dt} = \sum_{s=1}^{S} \left(\sum_{\substack{\text{outlet} \\ \text{streams } j}} N_s^j \tilde{H}_s(T^j) - \sum_{\substack{\text{inlet} \\ \text{streams } k}} N_s^k H_s(T^k) \right)$$

The result will take the form

$$\frac{dQ}{dt} = r\,\Delta \tilde{H}_R(T^r) + \sum_{s=1}^{S} \left(\sum_{\substack{\text{outlet} \\ \text{streams } j}} N_s^j(\tilde{H}_s(T^j) - \tilde{H}_s(T^r)) \right.$$

$$\left. - \sum_{\substack{\text{inlet} \\ \text{streams } k}} N_s^k\,(\tilde{H}_s(T^k) - \tilde{H}_s(T^r)) \right)$$

As before, if the reference state is chosen to be equal to that of one of the problem streams, then the terms corresponding to that stream will vanish from the energy balance equation. As in the nonreacting case, a suitable choice of reference state can sometimes result in considerable simplification of the equation.

Example 8.6 Nitric oxide can be formed by the partial oxidation of NH_3 with air. In a given reactor, NH_3 fed at 25°C and preheated air at 750°C are reacted at

1 bar with 90% conversion of NH_3. If the reactor effluent may not exceed 920°C, calculate the required rate of heat removal per 1 mol NH_3 fed, assuming a feed of 2.4 mol O_2 per 1 mol NH_3.

Solution   From Figure 8.3, it is evident that the system has two input streams and a single output stream. Since the conversion is specified, the rate of reaction is easily calculated given a basis. For convenience, we select as basis 1 mol/h NH_3. Then,

$$r = \frac{0.9(1)}{4} = 0.225 \text{ mol/h}$$

The rate can now be used to calculate the reactor outlet flows. Note, however, that if $T^r = 920°C$ is selected as problem reference state, then the outlet stream term vanishes from the energy balance equation. The simplified equation becomes

$$\frac{dQ}{dt} = r\,\Delta H_R\,(920°C) + N_{O_2}^{in} \int_{750°C}^{920°C} C_{P_{O_2}}\,dT + N_{N_2}^{in} \int_{750°C}^{920°C} C_{P_{N_2}}\,dT$$

$$+ N_{NH_3}^{in} \int_{25°C}^{920°C} C_{P_{NH_3}}\,dT$$

From Example 8.1, $\Delta H_R\,(920°C) = -216.42$ kcal/gmol. Furthermore, from the O_2-to-NH_3 feed ratio and the known air composition, it follows that

$$N_{O_2}^{in} = 2.4 \text{ mol/h}$$

$$N_{N_2}^{in} = 2.4 \left(\frac{0.79}{0.21}\right) \text{ mol/h}$$

Substituting these values, we have

$$\frac{dQ}{dt} = 0.225(-216{,}420) \text{ cal/h} + 2.4 \int_{750°C}^{920°C} C_{P_{O_2}}\,dT$$

$$+ 2.4 \left(\frac{0.79}{0.21}\right) \int_{750°C}^{920°C} C_{P_{N_2}}\,dT$$

$$+ 1 \int_{25°C}^{920°C} C_{P_{NH_3}}\,dT$$

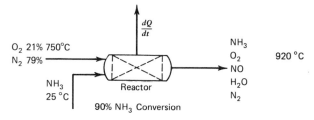

$$4\,NH_3\,(g) + 5O_2\,(g) \rightarrow 4\,NO\,(g) + 6H_2O\,(g)$$

Figure 8.3   Diagram for Example 8.6.

The O_2 and N_2 integrals can be evaluated as 15675 cal/gmol and the NH_3 integral as 10,497 cal/gmol. Therefore,

$$\frac{dQ}{dt} = -22.53 \text{ kcal/h} \qquad \text{or} \qquad -22.53 \text{ kcal/mol } NH_3$$

An alternate form of the energy balance equation which does not explicitly incorporate the heat-of-reaction term can also be obtained via a simple modification of the basic single input–single output equation. This form of the balance equation is particularly convenient for computer implementation because then a separate heat-of-reaction calculation is not required.

Consider the basic energy balance, with T^r set to the standard state, indicated by T^0:

$$\frac{dQ}{dt} = \sum N_s^2(\tilde{H}_s(T^2) - \tilde{H}_s(T^0)) - \sum N_s^1(\tilde{H}_s(T^1) - \tilde{H}_s(T^0))$$
$$+ \sum \tilde{H}_s(T^0)(N_s^2 - N_s^1)$$

Instead of eliminating the term $N_s^2 - N_s^1$, suppose we use the definition of the heat of formation,

$$\Delta H_{f,s}^0 = \tilde{H}_s^0 - \sum_{e=1}^{E} \alpha_{es} \tilde{H}_e^0$$

to eliminate $\tilde{H}_s(T^0)$ from the last term in the balance,

$$\sum \tilde{H}_s(T^0) N_s^2 - \sum \tilde{H}_s(T^0) N_s^1 = \sum N_s^2 \left(\Delta H_{f,s}^0 + \sum_e \alpha_{es} \tilde{H}_e^0 \right)$$
$$- \sum N_s^1 \left(\Delta H_{f,s}^0 + \sum_e \alpha_{es} \tilde{H}_e^0 \right)$$
$$= \sum N_s^2 \Delta H_{f,s}^0 - \sum N_s^1 \Delta H_{f,s}^0$$
$$+ \sum_e \tilde{H}_e^0 \left(\sum_{s=1}^{S} \alpha_{es}(N_s^2 - N_s^1) \right)$$

Note that the term

$$\sum_s \alpha_{es}(N_s^2 - N_s^1) = 0$$

since these are just the element balances for elements e, $e = 1, \ldots, E$. Thus, the balance reduces to

$$\frac{dQ}{dt} = \sum_s N_s^2[\Delta H_{f,s}^0 + (\tilde{H}_s(T^2) - \tilde{H}_s(T^0))]$$
$$- \sum_s N_s^1[\Delta H_{f,s}^0 + (\tilde{H}_s(T^1) - \tilde{H}_s(T^0))]$$

This corresponds to expressing the enthalpy of each species in the output stream as

$$\Delta H_{f,s}^0 + (\bar{H}_s(T^2) - \bar{H}_s(T^0))$$

and each species in the input stream as

$$\Delta H_{f,s}^0 + (\bar{H}_s(T^1) - \bar{H}_s(T^0))$$

Each of these expressions consists of the species heat of formation plus the enthalpy of species s relative to the standard state of 298K, 1 atm, and the standard state phase of that species. If the species is nonreacting, then $N_s^2 = N_s^1$ and the $\Delta H_{f,s}^0$ terms will simply cancel out. However, if the species is reacting, then $N_s^2 \neq N_s^1$ and the $\Delta H_{f,s}^0$ terms will implicitly account for the heat of reaction. This form of the balance equation is sometimes called the *total enthalpy balance*.

If we define the *total enthalpy flow* $\dot{H}^j$ for a stream j to be given by

$$\dot{H}^j = \sum_s N_s^j (\Delta H_{f,s}^0 + (\bar{H}_s(T^j) - \bar{H}_s(T^0))$$

then the total enthalpy flow balance for the multiple inlet and multiple outlet case reduces to the simple form

$$\frac{dQ}{dt} = \sum_{\substack{\text{outlets} \\ j}} \dot{H}^j - \sum_{\substack{\text{inlets} \\ k}} \dot{H}^k$$

Example 8.7 Repeat the solution of Example 8.6 using the total enthalpy balance formulation.

Solution Since none of the species undergo phase changes, we simply calculate the total enthalpy flow for each stream as follows:

$$\dot{H}^j \equiv \sum_{s=1}^S N_s^j \left(\Delta H_{f,s}^0 + \int_{T^0}^{T^j} C_{p_s} \, dT \right)$$

For the air stream, this is

$$\dot{H}_{\text{air}}^{\text{in}} = 2.4 \left(0 + \int_{298}^{1023} C_{p_{O_2}} \, dT \right) + 9.03 \left(0 + \int_{298}^{1023} C_{p_{N_2}} \, dT \right)$$

$$= 13{,}656 + 48{,}060 \text{ cal/h}$$

For the feed NH_3,

$$\dot{H}_{NH_3}^{\text{in}} = 1 \left(-10{,}920 \text{ cal/gmol} + \int_{298}^{298} C_{p_{NH_3}} \, dT \right) = -10{,}920 \text{ cal/h}$$

To evaluate the total enthalpy of the reactor effluent, we need to evaluate the material balances for the outlet flows. With $r = 0.225$, these become

NH₃ balance: $N_{NH_3}^{out} = 1 - 4(0.225) = 0.1$

NO balance: $N_{NO}^{out} = 0 + 4(0.225) = 0.9$

O₂ balance: $N_{O_2}^{out} = 2.4 - 5(0.225) = 1.275$

N₂ balance: $N_{N_2}^{out} = 9.03$

H₂O balance: $N_{H_2O}^{out} = 0 + 6(0.225) = 1.35$

The reactor effluent total enthalpy flow thus is

$$
\dot{H}^{out} = 0.1\left(-10{,}920 + \int_{298}^{1193} C_{P_{NH_3}}\, dT\right)
$$
$$
+ 0.9\left(21{,}600 + \int_{298}^{1193} C_{P_{NO}}\, dT\right)
$$
$$
+ 1.275\left(0 + \int_{298}^{1193} C_{P_{O_2}}\, dT\right)
$$
$$
+ 9.03\left(0 + \int_{298}^{1193} C_{P_{N_2}}\, dT\right)
$$
$$
+ 1.35\left(-57{,}800 + \int_{298}^{1193} C_{P_{H_2O}}\, dT\right)
$$
$$
= -41.6 + 25{,}728 + 9085 + 60{,}280 - 66{,}980 \text{ cal/h}
$$
$$
= 28.07 \text{ kcal/h}
$$

The energy balance reduces to

$$
\frac{dQ}{dt} = \dot{H}^{out} - \dot{H}_{air}^{in} - \dot{H}_{NH_3}^{in} = -22.73 \text{ kcal/h}
$$

The total enthalpy form of the balance equation is more complex for manual calculations because in effect the standard state is selected as the problem reference state. Consequently, all terms must be retained in the energy balance. However, this form of the balance is quite useful when the stoichiometry is very complex or in situations in which the species flows must be determined using element balances.

8.4 ENERGY BALANCES WITH MULTIPLE CHEMICAL REACTIONS

In our discussion of species material balances involving multiple chemical reactions, we observed that multiple reactions could be accommodated by replacing the species production rate R_s, which is equal to $\sigma_s r$ for a single reaction, by the sum $\sum_{i=1}^{R} \sigma_{si} r_i$. It is thus not surprising that the heat-of-reaction form of the energy balance equation for multiple reactions similarly will contain a term involving the rates of all of the reactions. The rearrangement of the energy balance equation to achieve this form proceeds exactly as in the single-reaction case. Consider the basic energy balance equation written in terms of a reference state denoted simply by the temperature T^r:

$$\frac{dQ}{dt} = \sum_{s=1}^{S} \tilde{H}_s(T^r)\left(\sum_{\substack{\text{outlet} \\ \text{streams } j}} N_s^j - \sum_{\substack{\text{inlet} \\ \text{streams } k}} N_s^k \right)$$

$$+ \sum_{s=1}^{S} \left[\sum_{\substack{\text{outlet} \\ \text{streams } j}} N_s^j(\tilde{H}_s(T^j) - \tilde{H}_s(T^r)) - \sum_{\substack{\text{inlet} \\ \text{streams } k}} N_s^k(\tilde{H}_s(T^k) - \tilde{H}_s(T^r)) \right]$$

In the case of multiple chemical reactions, the material balance equations give

$$\sum_{\substack{\text{outlet} \\ \text{streams } j}} N_s^j = \sum_{\substack{\text{inlet} \\ \text{streams } k}} N_s^k + \sum_{i=1}^{R} \sigma_{si} r_i$$

Consequently, the first term of the energy balance equation becomes

$$\sum_{s=1}^{S} H_s(T^r) \sum_{i=1}^{R} \sigma_{si} r_i$$

Interchanging the order of summation and using the definition of ΔH_{R_i}, this reduces to

$$\sum_{i=1}^{R} r_i \sum_{s=1}^{S} \sigma_{si} \tilde{H}_s(T^r) = \sum_{i=1}^{R} r_i \Delta H_{R_i}$$

Thus, the balance for multiple reactions is

$$\frac{dQ}{dt} = \sum_{i=1}^{R} r_i \Delta H_{R_i} +$$

$$\sum_{s=1}^{S} \left[\sum_{\substack{\text{outlet} \\ \text{streams } j}} N_s^j(\tilde{H}_s(T^j) - \tilde{H}_s(T^r)) - \sum_{\substack{\text{inlet} \\ \text{streams } k}} N_s^k(\tilde{H}_s(T^k) - \tilde{H}_s(T^r)) \right]$$

Each additional reaction only requires the presence of the product of the heat of that reaction times the rate of that reaction in the balance equation.

Example 8.8   Acetic acid is cracked in a furnace to produce the intermediate ketene via the reaction

$$CH_3COOH(g) \rightarrow CH_2CO(g) + H_2O(g)$$

The reaction

$$CH_3COOH(g) \rightarrow CH_4(g) + CO_2(g)$$

also occurs to an appreciable extent. It is desired to carry out cracking at 700°C with a conversion of 80% and a fractional yield of ketene of 0.0722. Calculate the required furnace heating rate for a furnace feed of 100 kgmol/h acetic acid. The feed is at 300°C.

Solution   As shown in Figure 8.4, this is a single input–single output system involving two chemical reactions. With a problem reference state of 700°C, the energy balance reduces to

$$\frac{dQ}{dt} = r_1 \, \Delta H_{R_1} \, (700°\text{C}) + r_2 \, \Delta H_{R_2} \, (700°\text{C}) - N_{\text{HAC}}^{\text{in}} \int_{700°\text{C}}^{300°\text{C}} C_{P\text{CH}_3\text{COOH}} \, dT$$

The standard heat of reaction of the ketene reaction is

$$\Delta H_{R_1}^0 = \Delta H_{f,\text{CH}_2\text{CO}}^0 + \Delta H_{f,\text{H}_2\text{O}}^0 - \Delta H_{f,\text{CH}_3\text{COOH}}^0$$

$$= -14.60 - 57.80 + 103.93 = 31.53 \text{ kcal/gmol}$$

The standard heat of reaction of the second reaction is

$$\Delta H_{R_2}^0 = \Delta H_{f,\text{CH}_4}^0 + \Delta H_{f,\text{CO}_2}^0 - \Delta H_{f,\text{CH}_2\text{COOH}}^0$$

$$= -17.89 - 94.05 + 103.93 = -8.01 \text{ kcal/gmol}$$

Both of these standard heats must be corrected to 700°C via the following equations:

$$\Delta H_{R_1} \, (700°\text{C}) = \Delta H_{R_1} \, (25°\text{C}) + \int_{298\text{K}}^{973\text{K}} (C_{P\text{CH}_2\text{CO}} + C_{P\text{H}_2\text{O}} - C_{P\text{CH}_3\text{COOH}}) \, dT$$

$$\Delta H_{R_2} \, (700°\text{C}) = \Delta H_{R_1} \, (25°\text{C}) + \int_{298\text{K}}^{973\text{K}} (C_{P\text{CH}_4} + C_{P\text{CO}_2} - C_{P\text{CH}_3\text{COOH}}) \, dT$$

Using the C_p relations for CH_3COOH, CH_4, CO_2, and H_2O from Appendix 3 and the following correlation for ketene,

$$C_{P\text{CH}_2\text{CO}} = 4.11 + 2.966 \times 10^{-2}T - 1.793 \times 10^{-5}T^2 + 4.22 \times 10^{-9}T^3$$

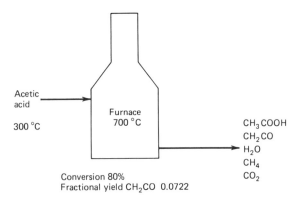

Figure 8.4   Diagram for Example 8.8, acetic acid cracking furnace.

The heats of reaction can be calculated as

$$\Delta H_{R_1} (700°C) = 31.26 \text{ kcal/gmol}$$

$$\Delta H_{R_2} (700°C) = -8.96 \text{ kcal/gmol}$$

Next, the acetic acid enthalpy difference integral is evaluated to yield

$$\int_{300°C}^{700°C} C_{P_{CH_3COOH}} \, dT = 11.55 \text{ kcal/gmol}$$

Finally, we use the acetic acid and ketene material balances

$$N_{CH_3COOH}^{out} = 100 - r_1 - r_2 \text{ kgmol/h}$$

$$N_{CH_2CO}^{out} = 0 + r_1$$

and the conversion of acetic acid, 80%, as well as the fractional yield of ketene, 0.0722, to evaluate the two rates. From the conversion,

$$r_1 + r_2 = 80 \text{ kgmol/h}$$

From the fractional yield definition,

$$0.0722 = \frac{r_1}{80} \quad \text{or} \quad r_1 = 5.776 \text{ kgmol/h}$$

Thus,

$$r_2 = 74.224 \text{ kgmol/h}$$

Now, all terms in the energy balance are known, and dQ/dt can be evaluated:

$$\frac{dQ}{dt} = 5.776(31.26) + 74.224(-8.96) + 100(11.55)$$

$$= 670.5 \times 10^3 \text{ kcal/h}$$

Quite obviously, it is this high heating rate which dictates that the reaction be carried out in a furnace.

In our discussion of material balances in the presence of multiple chemical reactions, we observed that in both the degree-of-freedom analysis and in the actual calculations it is only necessary to consider independent reactions. If the given set of reactions was dependent, then we concluded that the problem could be solved, without any loss of information, by using the largest independent subset of these reactions. Not surprisingly, the same is true in the energy balance case. As we shall confirm below, if the reactions are dependent, then we can write the energy balance equation using just the heats of reaction and reaction rates corresponding to the largest independent subset of the given set of dependent reactions.

Recall from Section 3.3.1 that a set of reactions is dependent if one or more of the vectors of stoichiometric coefficients of these reactions could be expressed as linear combinations of the remaining vectors of stoichiometric coefficients. Thus, suppose the Kth reaction of a set of K reactions is dependent. Then, following

Section 3.3.1, there must exist coefficients α_r, $r = 1, \ldots, K - 1$, such that

$$\sigma_{sK} = \sum_{r=1}^{K-1} \alpha_r \sigma_{sr}$$

Moreover, the rates of the $K - 1$ reactions can be expressed in terms of the full set of K reaction rates as follows:

$$r_i' = r_i + \alpha_i r_K$$

Let us consider the heat of reaction of the Kth reaction ΔH_{R_K}. Since the Kth reaction is dependent on the other $K - 1$ reactions, it follows that

$$\Delta H_{R_K} = \sum_{s=1}^{S} \sigma_{sK}\, \Delta H_{f,s} = \sum_{s=1}^{S} \left(\sum_{r=1}^{K-1} \alpha_r \sigma_{sr} \right) \Delta H_{f,s}$$

Interchanging the order of summation and using the definition of heat of reaction, we obtain

$$\Delta H_{R_K} = \sum_{r=1}^{K-1} \alpha_r \sum_{s=1}^{S} \sigma_{sr}\, \Delta H_{f,s} = \sum_{r=1}^{K-1} \alpha_r\, \Delta H_{R_r}$$

Evidently, the heat of reaction of the Kth dependent reaction can be calculated as a linear combination of the heats of reaction of the other $K - 1$ reactions. Now, suppose we consider the complete heat of reaction term in the energy balance equation

$$\sum_{i=1}^{K} \Delta H_{R_i} r_i$$

and separate out the term corresponding to the Kth dependent reaction

$$\Delta H_{R_K} r_k + \sum_{i=1}^{K-1} \Delta H_{R_i} r_i$$

Substituting to eliminate ΔH_R, we have

$$\sum_{i=1}^{K} \Delta H_{R_i}(\alpha_r r_K + r_i) = \sum_{i=1}^{K-1} \Delta H_{R_i} r_i'$$

Therefore, without any loss of information, we can calculate the heat-of-reaction term for the dependent set of reactions by using just the heats of reaction and the reaction rates for the largest linearly independent subset of the reactions. Once we establish such a reaction subset for material balance purposes, we merely continue to use the same subset to perform the energy balance computations.

8.5 ENERGY BALANCES WITH UNKNOWN STOICHIOMETRY

In the material balance case, we found that there was a substantial body of applications, particularly those involving fossil fuels, in which, because of the unknown species structure of the reactants themselves or because of the complexity of the reactions, the use of species material balances proved impossible or inconvenient. In such cases, element balances became the appropriate tool. In this section, we briefly consider the options available to us if it becomes necessary to solve the energy balances for such applications. We shall find that the choice is dictated in part by the available heat of reaction information.

As discussed in Section 8.3, in the presence of chemical reactions, the energy balance equation can be written in two alternate but equivalent forms: the heat-of-reaction form and the total enthalpy form. In the former case, the reaction stoichiometry must be known to calculate the reaction rate and species flows as well as to calculate the heats of reaction. In the total enthalpy form, on the other hand, the stoichiometry need not be known. In both forms, however, the heats of formation of all reactants and products must be available. If all heats of formation are known, then, in principle, two choices are open to us. We can apply the methods of Section 4.4.2 to generate a set of reactions and then use species material balances and the heat-of-reaction form of the energy balance equation. Alternatively, we can use element balances and the total enthalpy form of the energy balance equation. It appears that the choice between these two approaches will be governed by algebraic convenience. Yet, in many applications, a further consideration comes into play which forces this choice, and this is the type of heat of formation information that is available.

Recall that the standard heat of formation is by definition the heat of the reaction in which the species in question is formed from the elements. In most cases, such a reaction cannot be carried out under normal laboratory conditions, hence the heat of formation of a species cannot be directly measured. Instead, an alternate suitable reaction involving that species is carried out. Its heat of reaction is measured and the species heat of formation is back-calculated from the measured heat of reaction using the known heats of formation of the other species involved in that reaction. For instance, for organic compounds, the combustion reaction proves to be convenient for this purpose since the product heats of formation are well known. For more complex materials, specifically fossil fuels, a similar combustion experiment is carried out to measure a quantity called the gross calorific value or *high heating value* and denoted by Q_v. This quantity, which is related to the heat of combustion, is usually the only form in which heat-of-formation information is available for fossil fuels.

The *high heating value* is the heat released per unit mass of the fuel when it is reacted with oxygen to form solid residue (ash), liquid water, and the gaseous species CO_2, SO_2, and N_2, all at 77°F and 1 atm. It is usually reported as a positive quantity expressed in units of Btu/lb_m. Under the test conditions defined by ASTM specifications,[1] the experiment occurs at constant temperature and volume. For a closed system undergoing such a process, the first law yields

$$\Delta U = Q \equiv -Q_v$$

[1] D3286, Standard Method of Test for Gross Calorific Value of Solid fuel by the Isothermal-Jacket Bomb Calorimeter, *1976 Annual Book of ASTM Standards*, Part 26.

Using the definition of the enthalpy function, the standard gross heat of combustion can be related to the high heating value through the equation

$$\Delta H_c^0 = -Q_v \overline{M} + \Delta(PV)$$

where $\overline{M}$ is the average molecular weight of the fuel. Neglecting the volume contributions of the solid and liquid species and assuming that the gaseous species can be treated as ideal mixtures of ideal gases, this expression reduces to

$$\Delta H_c^0 = -Q_v \overline{M} + RT \sum_{\text{gas species } s} \sigma_s$$

where the summation term reflects the change in the number of moles of the gaseous species expressed in terms of the stoichiometric coefficients of the balanced combustion reaction for 1 mol of the fuel. Since, as discussed in Section 4.3.2, fossil fuels are characterized by an elemental analysis and, certainly in the case of coal, have unknown molecular weight, the above relation can be expressed in terms of elemental weight fractions. Given the elemental weight fractions of C, H, O, N, and S, denoted by w_C, w_H, w_O, w_N, and w_S, respectively, the balanced combustion reaction for a unit *mass* of fuel, in phase π, can be written as

$$\left[\left(\frac{w_C}{12.01}\right)C + \left(\frac{w_H}{1.008}\right)H + \left(\frac{w_O}{16.0}\right)O + \left(\frac{w_N}{14.007}\right)N + \left(\frac{w_S}{32.06}\right)S\right](\pi)$$
$$+ \left[\left(\frac{w_C}{12.01}\right) + \frac{1}{4}\left(\frac{w_H}{1.008}\right) + \left(\frac{w_S}{32.06}\right) - \frac{1}{2}\left(\frac{w_O}{16.0}\right)\right]O_2(g) \rightarrow \left(\frac{w_C}{12.01}\right)CO_2(g)$$
$$+ \frac{1}{2}\left(\frac{w_H}{1.008}\right)H_2O(l) + \frac{1}{2}\left(\frac{w_N}{14.007}\right)N_2(g) + \left(\frac{w_S}{32.06}\right)SO_2(g)$$

The heat of combustion for a unit mass of the fossil material will therefore be given by

$$\Delta \hat{H}_c^0 = -Q_v + \frac{RT}{2}\left(\frac{w_O}{16} + \frac{w_N}{14.007} - \frac{1}{2}\left(\frac{w_H}{1.008}\right)\right)$$
$$= -Q_v + 33.3w_O + 38.05w_N - 264.3w_H$$

Using this expression, the high heating value of the fossil material is corrected from constant volume to constant pressure conditions and converted to a heat of combustion per unit mass. Finally, the heat of combustion can be used to back-calculate the heat of formation of the fossil material. Specifically, using the definition of the heat of combustion, the heat of formation *per unit mass* will be given by

$$\Delta \hat{H}_f = -\Delta \hat{H}_c^0 + \frac{w_C}{12.01}\Delta H_{f,CO_2(g)}^0 + \frac{1}{2}\left(\frac{w_H}{1.008}\right)\Delta H_{f,H_2O(l)}^0 + \frac{w_S}{32.06}\Delta H_{f,SO_2(g)}^0$$

where, in AE units,

$$\Delta H_{f,CO_2(g)}^0 = -169.29 \times 10^3 \text{ Btu/lbmol}$$
$$\Delta H_{f,H_2O(l)}^0 = -122.97 \times 10^3 \text{ Btu/lbmol}$$
$$\Delta H_{f,SO_2(g)}^0 = -127.71 \times 10^3 \text{ Btu/lbmol}$$

The high heating value of a fossil fuel, particularly coal or coal char, should normally be reported as part of the characterization information of that material. If it is not available, the Institute of Gas Technology (IGT) has reported a correlation for coals and chars which predicts the high heating value as a function of the elemental analysis[2]:

$$Q_v = 14658w_C + 56878w_H + 2940w_S - 658w_{ash} - 5153(w_O + w_N)$$

where Q_v is in Btu/lb$_m$ and w_C, w_H, w_S, w_{ash}, w_O, and w_N are the weight fractions of C, H, S, ash, O, and N, respectively, all on a dry basis.

Analogous correlations are available for petroleum fractions but involve parameters in addition to the elemental analysis. The interested reader is directed to the *Technical Data Book—Petroleum Refining*.[3]

Given that the heat of formation of the fuel can be calculated in the above fashion, it is clear that energy balances involving such a substance can be solved using the total enthalpy balance form. That is, the heat of formation is simply used together with the appropriate C_p relations to determine the enthalpy of the fossil fuel containing streams relative to the elements at standard conditions. In this case, the element balances will naturally be used to perform the material balance calculations. The following example illustrates this type of solution approach.

Example 8.9 A steam-oxygen gasifier is fed 10^6 lb/h of a devolatilized char at 1700°F with ultimate analysis

78% C, 0.9% H, negligible O, 1.3% N, 0.7% S, and 19.1% ash

The char is reacted with steam fed at 1000°F and oxygen fed at 400°F to produce a synthesis gas analyzing

5% CH$_4$, 26.5% CO, 14.5% CO$_2$, 26.5% H$_2$, and 27.5% H$_2$O

all on an H$_2$S- and NH$_3$-free basis. The N and S distribution in the products is not precisely known, but it is assumed that the N and S will be reacted in the same proportion as the carbon is reacted. Hence, the C-to-N and C-to-S ratios in the feed and spent chars are assumed to be the same. The spent char is assumed to contain no H and to be dry. The steam rate is adjusted so as to feed 1.2 mol H$_2$O per 1 mol C in the feed. If the reactor is assumed to operate adiabatically at 70 bar and the temperatures of all streams leaving the reactors are the same, calculate the O$_2$ consumption and the reactor outlet temperature.

Solution From the schematic shown in Figure 8.5, it is evident that the reactor involves three gaseous species streams and two char streams. Since the elemental

---

[2] Institute of Gas Technology, Project 8979, Quarterly Report, DOE Report No. FE-2286-24, Oct. 1977.

[3] *Technical Data Book—Petroleum Refining*, 3rd ed., American Petroleum Institute, 1976, Chap. 14.

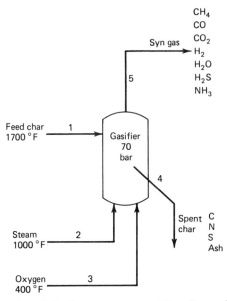

Figure 8.5   Steam-oxygen gasifier, Example 8.9.

analysis of the feed char is given, its high heating value and hence heat of formation can be calculated. The feed char heat capacities can be evaluated using Kirov's correlations (see p. 439). In order to calculate the corresponding quantities for the spent char stream, the elemental analysis of that stream must first be determined. Accordingly, let us begin with a solution of the material balances. Note that if element balances are used, the problem is correctly specified. It has 18 stream variables, six balances (five elements plus ash), the four feed char and four synthesis gas compositions, the carbon-to-nitrogen and carbon-to-sulfur ratios in stream 4, the steam-to-feed carbon ratio, and the specified char feed rate.

Using the given basis of 10^6 lb/h of stream 1, and for convenience using a separate stream (stream 6) to contain the H_2S and NH_3, the element balances are:

Sulfur balance: $0.007 \times 10^6 = 32.06 N^6_{H_2S} + F^4_S$

Nitrogen balance: $0.013 \times 10^6 = 14.007 N^6_{NH_3} + F^4_N$

Carbon balance: $0.78 \times 10^6 = 12.01\,(0.05 + 0.265 + 0.145)N^5 + F^4_C$

Hydrogen balance: $\dfrac{0.009 \times 10^6}{1.008} + 2N^2 = 2N^6_{H_2S} + 3N^6_{NH_3}$

$\qquad\qquad + [2(0.265) + 2(0.275) + 4(0.05)]N^5$

Oxygen balance: $N^2 + 2N^3 = [0.265 + 2(0.145) + 0.275]N^5$

Ash balance: $0.191 \times 10^6 = F^4_{ash}$

The specified side conditions are

$$N^2 = 1.2N_C^1 = 1.2\left(\frac{0.78 \times 10^6}{12.01}\right) = 77,935 \text{ lbmol/h}$$

$$\frac{F_N^4}{F_C^4} = \frac{F_N^1}{F_C^1} = \frac{1.3}{78}$$

$$\frac{F_S^4}{F_C^4} = \frac{F_S^1}{F_C^1} = \frac{0.7}{78}$$

The sulfur and nitrogen balances can be solved for $N_{H_2S}^6$ and $N_{NH_3}^6$, respectively, in terms of F_C^4 using the conditions on F_N^4 and F_S^4. Thus,

$$N_{H_2S}^6 = \frac{7 \times 10^3 - (0.7F_C^4/78)}{32.06}$$

$$N_{NH_3}^6 = \frac{13 \times 10^3 - (1.3F_C^4/78)}{14.007}$$

These equations can be used to reduce the hydrogen balance to

$$8.3643 \times 10^4 = 1.28N^5 - 4.1295 \times 10^{-3}F_C^4$$

This equation can be solved with the carbon balance to yield

$$N^5 = 1.2649 \times 10^5 \text{ lbmol/h}$$
$$F_C^4 = 8.1169 \times 10^4 \text{ lb/h}$$

The remaining flows can then be easily calculated with the result

$$F_N^4 = 1.353 \times 10^3 \text{ lb/h}$$
$$F_S^4 = 0.7284 \times 10^3 \text{ lb/h}$$
$$F_{ash}^4 = 1.91 \times 10^5 \text{ lb/h}$$
$$N_{NH_3}^6 = 831.5 \text{ mol/h}$$
$$N_{H_2S}^6 = 195.62 \text{ mol/h}$$
$$N^2 = 13,528 \text{ mol/h} \quad \text{or} \quad 0.208 \text{ mol } O_2/\text{mol C fed}$$

The spent char flows are equivalent to an elemental analysis of 29.6% C, 0.493% N, 0.266% S, and the rest ash. Note that in formulating the balances, we have assumed that the spent char contains no hydrogen.

For energy balance purposes, we now must calculate the char heats of formation. Using the IGT correlation, the feed char high heating value is

$$Q_v^1 = 11,773.1 \text{ Btu/lb}$$

The corresponding value for the spent char is

$$Q_v^4 = 3862.5 \text{ Btu/lb}$$

The respective heats of combustion are $-11,775.0$ and -3862.3 Btu/lb. Finally, the heats of formation can be determined.

$$\Delta \hat{H}_f^1 = 11,775 + \frac{0.78}{12.01}(-169.29 \times 10^3) + \frac{1}{2}\left(\frac{0.009}{1.008}\right)(-122.97 \times 10^3)$$

$$+ \frac{0.007}{32.06}(-127.71 \times 10^3)$$

$$= 203.4 \text{ Btu/lb}$$

$$\Delta \hat{H}_f^4 = 3862.3 + \frac{0.296}{12.01}(-169.29 \times 10^3) + \frac{0.00266}{32.06}(-127.71 \times 10^3)$$

$$= -320.6 \text{ Btu/lb}$$

The enthalpy flow of the feed char stream will be given by

$$\dot{H}^1 = 10^6 \text{ lb/h} \left(203.4 + \int_{77°F}^{1700°F} C_{p_{char}} \, dT\right) \text{ Btu/h}$$

and that of the spent char stream by

$$\dot{H}^4 = 274,251\left(-320.6 + \int_{77°F}^{T} C_{p_{char}} \, dT\right) \text{ Btu/h}$$

The char heat capacities can be calculated using Kirov's equations. The feed char is specified to be devolatilized, that is, contains no volatile matter, hence the spent char will also contain no volatile matter. Therefore, for stream 1, only the heat capacities for ash, $C_{p_{ash}}$, and fixed carbon $C_{p_{FC}}$, need to be considered.

$$\int_{77°F}^{1700°F} C_p \, dT = 0.191 \int_{77°F}^{1700°F} C_{p_{ash}} \, dT + 0.809 \int_{77°F}^{1700°F} C_{p_{FC}} \, dT$$

$$= (0.191)404.33 + (0.809)592.05 \text{ Btu/lb}$$

For stream 4, the corresponding integral is

$$\int_{77°F}^{T} C_p \, dT = 0.696 \int_{77°F}^{T} C_{p_{ash}} \, dT + 0.304 \int_{77°F}^{T} C_{p_{FC}} \, dT$$

Thus,

$$\dot{H}^1 = 7.5977 \times 10^8 \text{ Btu/h}$$

while the enthalpy flow of stream 4 is a function of the unknown reactor exit temperature T.

The complete energy balance in total enthalpy form can now be written

$$\frac{dQ}{dt} = 0 = \dot{H}^4 + \dot{H}^5 + \dot{H}^6 - \dot{H}^1 - \dot{H}^2 - \dot{H}^3$$

where $\dot{H}^4$ and $\dot{H}^1$ are given above,

$$\dot{H}^2 = N^2 \left(\Delta H^0_{f,H_2O(g)} + \int_{77°F}^{1000°F} C_{P_{H_2O(g)}} \, dT \right)$$

$$\dot{H}^3 = N^3 \left(0 + \int_{77°F}^{400°F} C_{P_{O_2(g)}} \, dT \right)$$

$$\dot{H}^6 = N^6_{H_2S} \left(\Delta H^0_{f,H_2S(g)} + \int_{77°F}^{T} C_{P_{H_2S}} \, dT \right)$$

$$+ N^6_{NH_3} \left(\Delta H^0_{f,NH_3(g)} + \int_{77°F}^{T} C_{P_{NH_3}} \, dT \right)$$

$$\dot{H}^5 = \sum_s N^5_s \left(\Delta H^0_{f,s} + \int_{77°F}^{T} C_{P_s} \, dT \right)$$

with the sum containing the species CH_4, CO, CO_2, H_2, and H_2O. The required heats of formation, $-\Delta H^0_f$ (Btu/lbmol), are summarized below:

$CH_4(g)$	32.202×10^3
$CO(g)$	47.556×10^3
$CO_2(g)$	169.29×10^3
$H_2O(g)$	104.04×10^3
$H_2S(g)$	8.676×10^3
$NH_2(g)$	19.656×10^3

When combined with the heat capacity relations for the above species drawn from Appendix 3, the following equation in temperature can be developed:

$$-6.5182 \times 10^{-9}(T^5 - 77^5) + 2.1448 \times 10^{-5}(T^4 - 77^4) - 3.9079(T^3 - 77^3)$$
$$+ 150.63(T^2 - 77^2) + 9.9490 \times 10^5(T - 77) = 1.92810 \times 10^9$$

The root of this equation calculated using any convenient root-finding method is 1688.2°F. The reactor thus operates at a temperature very close to the feed char temperature. This, however, is only coincidental. The temperature is actually governed by the relative amounts of steam and oxygen fed to the reactor.

An alternate approach to solving a problem such as Example 8.9 might be to generate a sufficient number of chemical reactions and then to use the usual species and heat of reaction balances. Thus, if the system involves S reacting species and $\rho(\alpha)$ independent element balances, then in principle $S - \rho(\alpha)$ independent reactions could be generated. However, considerable care must be exercised in the selection of the "species" which will be used to generate the independent reactions. Two basic choices can be made. Each fossil material can be treated as a species with arbitrary molecular weight and a "molecular" formula $C_aH_bO_cN_dS_eAsh_f$ in which the coefficients a through f are fixed by the elemental analysis. In this case, the heat of formation derived from the high heating value can be used to compute the necessary heats of reaction of those reactions involving the fossil pseudospecies. Alternatively, the elements comprising the fossil fuel can be treated as "species" and thus used to generate reactions. The reactions will thus typically involve a fossil fuel "pseudospecies," say, C, reacting with some reactant to produce various molecular products. This is a very natural and intuitive way of representing how the constituents of coal react. However, with this choice of pseudospecies, a difficulty arises in choosing the heats of formation for the reacting "species" present in the fuel. For instance, if in the reaction

$$C(s) + H_2O(g) \rightarrow CO(g) + H_2(g)$$

the species C(s) is carbon present in a coal or char, then the heat of formation of that species is in general not zero. The heat of formation of carbon is defined to be zero for carbon in its standard state, solid graphite. For all other states, including its combined state in fossil fuels, it need not be zero. If the fossil fuel only undergoes a single reaction, then the heat of formation for the pseudospecies can be determined for the product and reactant form of the fuel. In this way, a heat of formation for the carbon "species" can be back-calculated. In general, if multiple reactions occur, particularly if several of the constituents of the char (say, C, H, and S) all undergo reactions, then assignment of heats of formation to these pseudo species becomes quite difficult.

Example 8.10 Consider the formulation of Example 8.9 in terms of species balances.

Solution Suppose we first attempt to treat the char elements as species. The system of Figure 8.5 involves the 12 reacting species C, H, N, and S (in the chars) and O_2, H_2, CH_4, CO, CO_2, H_2O, H_2S, as well as NH_3, where for energy balance purposes the H in the char and H_2 in the gas phase are in principle different. Since there are five elements present, $12 - 5$, or seven reactions will be required to describe the changes in the relative amounts of these species. For instance, the reactions given below,

$$S^* + H_2(g) \rightarrow H_2S(g)$$
$$N^* + \tfrac{3}{2}H_2(g) \rightarrow NH_3(g)$$
$$C^* + 4H^* \rightarrow CH_4(g)$$
$$C^* + H_2O(g) \rightarrow CO(g) + H_2(g)$$
$$2C^* + O_2(g) \rightarrow 2CO(g)$$
$$C^* + 2H_2(g) \rightarrow CH_4(g)$$
$$CO(g) + H_2O(g) \rightarrow CO_2(g) + H_2(g)$$

involving the fuel species S^*, N^*, C^*, and H^*, are an independent set. To calculate the heats of reaction of these seven reactions, the heats of formation of the fuel species are required. In general, there is no simple way of predicting these quantities. In the case of coal, $\Delta H_{f,C^*} = 0$ is often assumed, and under suitable conditions the heat of reaction of the third reaction is assumed to be zero. However, these are special assumptions which must be used with great care.

Suppose, on the other hand, that the feed and spent chars are treated as separate species with pseudomolecular formulas obtained using the elemental analysis. In this case, the system will involve the feed char $C_aH_bN_cS_dAsh_e$, the spent char $C_\alpha N_\beta S_\gamma Ash_\delta$, and the normal molecular species O_2, H_2, CH_4, CO, CO_2, H_2O, H_2S, and NH_3. In this case, we could treat ash as an "element" along with C, H, O, N, and S. Thus, since there are ten species and six elements, a maximum of four independent reactions can be constructed. For instance, the four balanced chemical reactions

$$CO + \tfrac{1}{2}O_2 \rightleftharpoons CO_2$$

$$CO + H_2O \rightleftharpoons CO_2 + H_2$$

$$CO + 3H_2 \rightleftharpoons CH_4 + H_2O$$

$$C_aH_bN_cS_dAsh_e + mH_2 \longrightarrow \left(\frac{e}{\delta}\right)C_\alpha N_\beta S_\gamma Ash_\delta + \left(a - \frac{\alpha e}{\delta}\right)CH_4$$

$$+ \left(c - \frac{\beta e}{\delta}\right)NH_3 + \left(d - \frac{\gamma e}{\delta}\right)H_2S$$

where

$$m = \frac{1}{2}\left[(4a + 3c + 2d - b) - \frac{e}{\delta}(4\alpha + 3\beta + 2\gamma)\right]$$

could be generated using the methods of Section 4.4.2. The molar coefficients a through e and α through δ of the two char "species" can be obtained by assuming an arbitrary molecular weight for each (say, 100 lb/lbmol) and calculating the mole ratios using the elemental weight fractions. For instance, for the feed char,

$$a = \frac{100 \text{ lb char/mol char} \times 0.78 \text{ lb C/lb char}}{12.01 \text{ lb C/mol C}} = 6.495 \frac{\text{mol C}}{\text{mol char}}$$

and similarly

$$b = 0.893$$

$$c = 0.0928$$

$$d = 0.0218$$

For ash, an arbitrary "atomic" weight can be assumed, say, 100 lb/lbmol, and thus

$$e = 0.191$$

The empirical formula for the feed is

$$C_{6.495} \, H_{0.893} \, N_{0.0928} \, S_{0.0218} \, Ash_{0.191}$$

With known molecular formulas for all species and the generated reaction set, species material balances could clearly be written and solved for each species. Moreover, if the heats of formation per pound of the feed and spent char are calculated as before, then, using the assumed molecular weights, the molar heats of formation can be calculated. Next, the heats of reaction of all four reactions can be determined by using the heats of formation in the usual fashion. Finally, the heat of reaction form of the energy balance equation can be written and solved for the reactor exit temperature. With this approach, no additional heat of formation information beyond that employed by the total enthalpy formulation is required. Note, however, that use of this approach is nonetheless awkward because the elemental analysis of the spent char is unknown; hence its empirical formula cannot be calculated nor the stoichiometric coefficients of the fourth reaction determined. In this case, the elemental balances would first have to be solved to establish the empirical formula for the spent char, before the reactions could be generated for use in species balances. Clearly, the species balance approach is appropriate only if the elemental analysis of all fossil fuel aggregates are known in advance.

In summary, in situations in which fossil fuels are reacted, the total enthalpy form of the energy balance should normally be used. The species balance approaches should be used only if either the elemental analysis of all fossil aggregates are known or if information is available about the heats of formation of all species contained in the reaction equations.

8.6 DEGREE-OF-FREEDOM ANALYSIS

The analysis of energy balance problems involving reacting units is a straightforward extension of the analysis for nonreacting systems. In this section, we briefly examine the analysis for single-unit systems, using an example. Then, we consider the multiple-unit case using two examples. As we shall see, complications only arise when the overall balance is employed to substitute for one of the individual unit balances.

8.6.1 Single-Unit Systems

For single-unit systems, in addition to the usual material balance variables, the stream flows and reaction rates, there will be a temperature associated with every stream and a dQ/dt (as well as a dW/dt, if applicable) with the unit. As in the nonreacting case, it is expedient to check whether the material balance can be solved separately (or decoupled) from the energy balances by forming a separate material balance tabulation.

Example 8.11   Perform a degree of freedom analysis for Example 8.8.

Solution   From Figure 8.4, it is evident that the problem involves six flow variables and two reaction rates, two stream temperatures, and the dQ/dt variable. There are five possible material balance equations and a single energy balance variable. The feed rate conversion, fractional yield, and inlet and outlet temperatures are specified. Thus, the degree-of-freedom table shown in Figure 8.6 is evidence that the problem is correctly specified. Moreover, from the material balance column, it is clear that the balances can be decoupled. The material balances can be solved first, followed by the solution of the energy balances. This is in fact how the problem was solved: we determined all material balance variables in the energy balance equation and thus could use that equation to solve for dQ/dt.

	Material Balances	Combined Balances
Number of variables		
Species flows	6	6
Reaction rates	2	2
Temperatures, dQ/dt		3
Number of Balances		
Material	5	5
Energy		1
Number of specifications		
Conversion	1	1
Fractional yield	1	1
Temperatures		2
Flow	1	1
Degree of Freedom	0	0

Figure 8.6   Degree-of-freedom table for Example 8.11.

8.6.2   Multiunit Systems

The degree-of-freedom analysis of multiunit flowsheets follows directly from the multiunit analysis we performed for material balance problems alone. It is necessary to construct a table which contains the material balance and combined balance entries for each unit, and to construct a column for the entire process and, if appropriate, the material and combined balance columns for the overall balances. As before, the process column indicates whether the complete problem is correctly specified while the individual unit columns give solution sequencing information. The general strategy is to find a sequence of single-unit balance calculations in which the individual unit degree of freedom is as low as possible. When solving the single unit, the material and energy balances are solved separately, if possible. This information can of course be extracted from the material balance subproblem columns for the individual units. The degree-of-freedom analysis for a multiunit flowsheet is illustrated with the following example.

Example 8.12   Ammonia is produced via the reaction

$$N_2 + 3H_2 \rightarrow 2NH_3$$

using the process flowsheet shown in Figure 8.7. The reaction takes place in a two-stage adiabatic reactor. The conversion in the first stage is 10%, and the products from this stage are cooled back down to 425°C by mixing with fresh cold feed. The products from the second stage leave at 535°C, are first cooled by exchanging heat with the feed to the first reactor stage, and finally are refrigerated in the separator to condense out the NH_3 product together with traces of N_2 and H_2.

Calculate the refrigeration load (dQ/dt) in the separator per mole of NH_3 produced, using the following simplified heat capacity data given in calories:

Constant average molar heat capacities of gases: NH_3, 9.5; N_2, 7.0; H_2, 7.0.
Constant average molar heat capacity of NH_3 liquid: 30.0.
Heat of vaporization of NH_3: 5.581 kcal/gmol at -33.4°C, the normal boiling point.

Solution   The entire process involves just three species, NH_3, H_2, and N_2, and one reaction which takes place in each of two reactors. Note that all units are

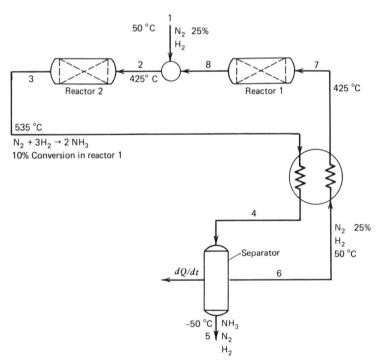

Figure 8.7   Flow diagram for Example 8.12, ammonia synthesis loop.

adiabatic except for the separator. The heat exchanger involves heat transfer between two streams. Since the individual stream material balances are trivial, it is
convenient to only list the stream variables associated with each input stream and
to drop the material balance equations from the exchanger column. In this way,
one combined column will quite obviously suffice for the heat exchanger. The
complete degree-of-freedom table is shown in Figure 8.8.

The process is correctly specified, and the solution can be initiated with reactor
1, the first reactor stage. The material balances can be solved *first*; the energy
balance, subsequently. Upon solution of the reactor 1 balances, the stream flows
and the temperature from reactor 1 will be known, hence the mixer degree of
freedom becomes zero. Note, however, that since only three flow variables have
been determined, the mixer material balances are underspecified by one, that is,
$4 - 3 = 1$. Thus, the material and the energy balances must be solved simultaneously.

After the mixer balances, the inlet flows to the second reactor stage, are known
and its degree of freedom becomes zero. Again, however, the reactor 2 material
balances are underspecified and must be solved in conjunction with the energy
balance. Next, with both streams to the heat exchanger known, the exchanger
degree of freedom becomes equal to $4 - 3 - 1 = 0$. Finally, with the input stream
flows and temperature to the separator as well as the flow of one of its output
streams known, the separator balances have degree of freedom zero. In this case,
the separator material balances can be solved separately. The complete calculations
sequence is schematically represented in Figure 8.9.

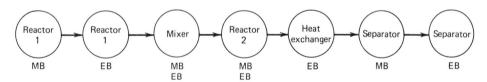

Figure 8.9 Calculation order, Example 8.12.

We begin the solution by selecting as basis 400 mol/h feed to reactor 1. From
the known composition, we thus have 100 mol/h N_2 and 300 mol/h H_2. Since the
conversion is known, the rate can immediately be determined:

$$r = \frac{0.1(100)}{1} = 10$$

The material balances then give:

    NH_3 balance: $0 + 2r = 20$

    H_2 balance: $300 - 3r = 270$

    N_2 balance: $100 - r = 90$

If 425°C is selected as reference state temperature, the energy balance takes

	Mixer		Reactor 1		Reactor 2		Heat Exchanger	Separator		Process	Overall	
	MB	Com-bined	MB	Com-bined	MB	Com-bined		MB	Com-bined		MB	Com-bined
Number of variables												
Stream	8	8	6	6	7	7	5	8	8	18	6	6
T, dQ/dt		3+1		2+1			4+1		3+1	13		2+1
Number of balances												
Material	3	3	3	3	3	3	—	3	3	12	3	3
Energy		1		1		1	1		1	5		1
Number of specifications												
Feed comp.	1	1								1	1	1
Sep. outlet comp.								1	1	1		
$dQ/dt = 0$		1		1		1	1			4		
Conversion			1	1						1		
Temperatures												
Reactor inlets				1			1			2		1
R2 outlet							1			1		1
Feed		1								1		
Sep. outlets									2	2		1
Degree of freedom	4	4	1	1	4	3	4	4	5	1	2	2
Basis			−1	−1						−1		

Figure 8.8 Degree-of-freedom table for Example 8.12.

the form

$$0 = \frac{dQ}{dt} = \Delta H_R \ (425°C)r + \sum N_s^{out} C_{p_s}(T^{out} - 425)$$

The heat of reaction at 425°C is easily calculated to be

$$\Delta H_R \ (425°C) = -25.64 \ \text{kcal/gmol}$$

Substituting all known enthalpies and flows into the energy balance, we obtain

$$0 = (10)(-25.64) \ \text{kcal} + \{20(9.5) + 90(7) + 270(7)\}(T^{out} - 425)10^{-3}$$
$$256.4 = 2.710(T^{out} - 425)$$
$$T^{out} \simeq 425 + 94.7 = 519.7°C$$

Observe that if we had not used constant heat capacities, an iterative solution would have been required for the outlet temperature.

At this point, the solution proceeds with the mixer balances. Recall that these will have to be solved in coupled form. It is nonetheless convenient to partially solve the material balances first in this case. Let N equal the mol/h of fresh feed. Then, from the species balances, it is obvious that

$$N_{NH_3}^{out} = 20 \ \text{mol/h}$$

$$N_{N_2}^{out} = 90 + \frac{N}{4} \ \text{mol/h}$$

$$N_{H_2}^{out} = 270 + \tfrac{3}{4}N \ \text{mol/h}$$

The energy balance for the nonreacting unit simplifies if the balance reference state is chosen to be that of the fresh feed stream. The result, upon substitution of the flows expressed in terms of N, is

$$0 = \sum N_j^{out}(H_j(425°C) - H_j(50°C)) - \sum N_j^{in}(H_j(T_j^{in}) - H_j(50°C))$$
$$0 = 20(9.5)(425 - 50) + \left(90 + \frac{N}{4}\right)(7)(425 - 50) + \left(270 + \frac{3}{4}N\right)(7)(425 - 50)$$

$$- \frac{N}{4}(7)(50-50) - \frac{3}{4}N(7)(50-50) - 20(9.5)(519.7 - 50)$$

$$- 90(7)(519.7 - 50) - 270(7)(519.7 - 50)$$

or $(90 \times 7 + 270 \times 7 + 20 \times 9.5)(94.7) = \left(\frac{N}{4} \times 7 + \frac{3}{4}N \times 7\right)(375)$

$$N = \frac{2710 \times 94.7}{375 \times 7} \simeq 97.8 \ \text{mol/h}$$

With N known, the input to the second reactor stage can be calculated as

$$N_{NH_3}^{in} = 20 \ \text{mol/h}$$

$$N_{N_2}^{in} = 114.45 \ \text{mol/h}$$

$$N_{H_2}^{in} = 343.35 \ \text{mol/h}$$

Proceeding with the reactor 2 balances, again coupled solution is required. The material balance equations can first be used to solve for the outlet flows expressed in terms of the reaction rate:

$$N_{NH_3}^{out} = 20 + 2r$$

$$N_{N_2}^{out} = 114.45 - r$$

$$N_{H_2}^{out} = 343.35 - 3r$$

When these expressions are substituted into the energy balance written with a reference state temperature of 425°C, the following result is obtained:

$$0 = \frac{dQ}{dt} = r\,\Delta H_R\,(425°C) + N_{NH_3}^3(9.5)(535°C - 425°C)$$

$$+ N_{N_2}^3(7)(535 - 425) + N_{H_2}^3(7)(535 - 425)$$

$$r(25.64) = 110[9.5(20 + 2r) + 7(114.45 - r) + 7(343.35 - 3r)] \times 10^{-3}$$

$$r = \frac{373.406}{26.63} = 14.0 \text{ mol/h}$$

Therefore, the flow to the heat exchanger is

$$N_{NH_3}^3 = 48 \text{ mol/h}$$

$$N_{N_2}^3 = 100.45 \text{ mol/h}$$

$$N_{H_2}^3 = 301.35 \text{ mol/h}$$

Since all stream flows are now known, the heat exchanger energy balance can be solved to yield the remaining unknown temperature. Using as reference state the state of the 50°C stream, the balance is

$$N_{N_2}^6(7)(425°C - 50°C) + N_{H_2}^6(7)(425°C - 50°C) = N_{NH_3}^3(9.5)(535°C - T)$$

$$+ N_{N_2}^3(7)(535°C - T)$$

$$+ N_{H_2}^3(7)(535°C - T)$$

or $$(535°C - T)(3.2686) = 2.8 \times 375$$

and the solution is

$$T = 214°C$$

At this point, only the separator balances remain to be solved. They can be decoupled. In fact, the material balances can simply be solved by difference. The product stream is

$$N_{NH_3} = 48 \text{ mol/h}$$

$$N_{N_2} = 0.45 \text{ mol/h}$$

$$N_{H_2} = 1.35 \text{ mol/h}$$

Using as reference state that of the product stream, the energy balance becomes

$$\frac{dQ}{dt} = \sum N_j^{out}\,(H_j(T^{out}) - H_j(-50°C))$$

$$- \sum N_j^{in}(H_j(T^{in}) - H_j(-50°C))$$

$$= 100(7)[50°C - (-50°C)]$$

$$+ 300(7)[50°C - (-50°C)]$$

$$- N_{NH_3}^{in}(H_{NH_3}(214°C) - H_{NH_3}(-50°C))$$

$$- (N_{N_2}^{in} + N_{H_2}^{in})(7)[214°C - (-50°C)]$$

or

$$\frac{dQ}{dt} = 2.8(100) - (401.8)(7)(264) \times 10^{-3} - 48\{C_{p_V}[214°C - (-33.4°C)]$$

$$+ \Delta H_{VL}(-33.4) + C_{p_L}[-33.4°C - (-50°C)]\} \approx -867 \text{ kcal/h}$$

Thus, the heat load on the separator is

$$\frac{-867 \text{ kcal/h}}{48 \text{ mol/h}} = -18 \text{ kcal/mol}$$

In the preceding example, the solution could be obtained entirely as a sequence of single-unit solutions without using the overall balances. Our experience with multiunit material balances indicates that sequences which do not employ overall balances are generally preferable because the overall balances introduce the need for more care in problem analysis. In the material balance case, we in particular noted that when overall balances are used, it is possible to exhaust independent species balances, to arrive at large dependent reaction sets, and to obtain overall reaction rate values which also are individual unit rates. These complications could easily be handled in the degree-of-freedom accounting provided that they were recognized during analysis. The same is the case when overall energy balances are used.

In writing overall energy balances, we must first of all take care to ensure that only process input and output streams and their associated temperatures appear explicitly in the equation. Secondly, it is important to realize that the heat and work rates which appear in the balance must account for all exchanges of heat and work between the system and the environment. The heat and work rates in the overall balance equation thus are the algebraic sum of the heat and work rates from all the individual units in the flowsheet. Omission of any of the individual rates from the sum will obviously lead to errors in the solution.

As a further consequence of this fact, once these net rates are calculated, the individual heat and work rates for one of the units will become dependent. For instance, if all units save one are adiabatic, then the overall heat rate will be the same as the heat rate for that one unit. When the former is calculated, the latter is also determined and no longer should be counted as an unknown. Clearly, in

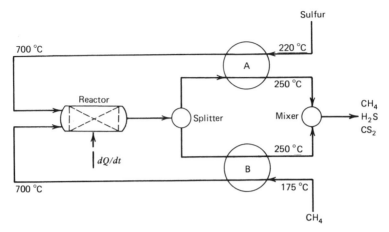

Figure 8.10   Flow diagram for Example 8.12, carbon disulfide manufacture.

problem analysis, care must be taken to recognize when individual heat and work rates have already been calculated as a result of using the overall balances so that the correct unit sequencing is obtained. The next example illustrates the use of the overall energy balance in solving a multiunit problem.

Example 8.13   Carbon disulfide is manufactured by reacting methane with sulfur vapor via the reaction

$$CH_4(g) + \tfrac{1}{2}S_8(g) \rightarrow CS_2(g) + 2H_2S(g)$$

In the process shown in Figure 8.10, complete conversion of sulfur is obtained with a feed consisting of 3 mol CH_4 per 1 mol S_8. The sulfur is fed molten at 220°C, and the products leave the process at 250°C. To carry out the reaction, the reactor feed materials must be preheated to 700°C by heat exchange with the reactor effluent.

Assume that the heat of reaction for the above reaction is 22 kcal/gmol at 250°C.

Assume heat losses in heat exchange are negligible and that the stream splitter and mixer both operate with no changes in stream temperatures. Use only the following data:

Compound	State	Average Heat Capacity (cal/gmol · °C)	ΔH_{VL} (kcal/gmol)
CH_4	g	10.0	—
H_2S	g	9.5	—
CS_2	g	7.6	—
S_8	l	7.0	2.2 at 445°C, 1 atm
S_8	g	8.0	

(a) Calculate the rate of heat addition to the reactor required per mole of carbon disulfide.

(b) Calculate the ratio in which the reactor effluent stream should be split to accomplish the preheating of the feed to 700°C.

Solution The problem involves a single chemical reaction and four species. The reactor tabulation is therefore quite straightforward. The two heat exchangers are adiabatic, and, as in the previous example, it is convenient to only count their inlet stream variables and thus implicitly account for the trivial material balances. The splitter and mixer operate with no temperature change. Thus, assuming ideal mixtures (an implicit assumption in all our work in this text), from the energy balances these two units are also adiabatic. However, these are consequences of the energy balances and not specifications. The remaining degree-of-freedom entries are straightforward. The final table is given in Figure 8.11.

The problem is correctly specified, and since the overall balances have degree of freedom 1, we should choose our basis there. Note that the material balances are completely specified; therefore, these can be done separately from the energy balances. At this point, we will have calculated the overall heat transfer rate and the reaction rate and will have used up the one independent S_8 balance that is available. Note that

$$\frac{dQ}{dt}\text{ov} = \frac{dQ}{dt}\text{A} + \frac{dQ}{dt}\text{B} + \frac{dQ}{dt}\text{react} + \frac{dQ}{dt}\text{mix} + \frac{dQ}{dt}\text{split}$$

From the problem specifications

$$\frac{dQ}{dt}\text{A} = \frac{dQ}{dt}\text{B} = 0$$

Also, as noted earlier, the no-temperature-change assumptions on the mixer and splitter imply, from the energy balance, that

$$\frac{dQ}{dt}\text{ mix} = \frac{dQ}{dt}\text{ split} = 0$$

Therefore, the overall heat transfer rate is equal to the reactor heat transfer rate:

$$\frac{dQ}{dt}\text{ reac} = \frac{dQ}{dt}\text{ ov}$$

As far as the reactor is concerned, the S_8 balance is used up, the heat and reaction rates are known, and the basis is specified. Thus, the new degree of freedom becomes

$$2 - 1(\text{basis}) - 2(\text{rates}) + 1(S_8 \text{ balance}) = 0$$

The material balances can be solved separately from the energy balance.

With the reactor outlet T known, the branch temperatures T_A and T_B will be known. For each heat exchanger, we now know one additional T, the composition

	Reactor		Splitter		Heat Ex-changer A	Heat Ex-changer B	Mixer		Process	Overall	
	MB	Com-bined	MB	Com-bined			MB	Com-bined		MB	Com-bined
Number of variables											
Stream	$5+1$	$5+1$	9	9	$4+1$	$4+1$	9	9	15	$5+1$	$5+1$
T, dQ/dt		$3+1$		$3+1$				$3+1$	15		$3+1$
Number of balances											
Material	4	4	3	3	—	—	3	3	10	4	4
Energy		1		1	1	1		1	5		1
Number of specifications											
Feed ratio	1	1							1	1	1
Splitter restrictions			2	2					1	2	
$T^A = T^{in}$				1					1		
$T^B = T^{in}$				1					1		
$\dfrac{dQ}{dt}\,A = 0$					1				1		
$\dfrac{dQ}{dt}\,B = 0$						1			1		
All temperatures 250°C					1	1		3	3	1	1
Sulfur feed T					1				1	1	1
CH$_4$ feed T						1		1	1	1	1
Reactor feed T		2			1	1		2	2		
Basis	1	1	1	1	1	1	1	1	1	1	1
Degree of freedom	2	5	4	5	4	4	6	6	0	1	0

Figure 8.11 Degree-of-freedom table for Example 8.13.

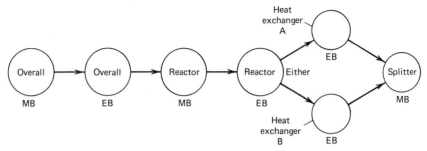

Figure 8.12 Calculation order, Example 8.13.

of the reactor product stream, and the flow of the appropriate reactor feed stream. Thus, the exchanger degree of freedom equals zero. Either exchanger can be solved.

Finally, from the splitter balance, we can calculate the flow to the other exchanger. The complete calculation sequence is shown in Figure 8.12.

We begin the solution with the overall balances and a basis of 600 mol/h CH_4. The feed ratio then implies that the S_8 feed is 200 mol/h. The material balances then simply become:

S_8 balance: $\qquad 0 = 200 - \frac{1}{2}r \rightarrow r = 400$

CH_4 balance: $\qquad N_{CH_4}^{out} = 600 - r = 200$ mol/h

CS_2 balance: $\qquad N_{CS_2}^{out} = 0 + r = 400$ mol/h

H_2S balance: $\qquad N_{H_2S}^{out} = 0 + 2r = 800$ mol/h

Note that the heat of reaction is given at 250°C and that this heat of reaction applies with all reactants in the gaseous phase, as shown in the chemical reaction equation. For convenience, suppose we select this state as problem reference state. With that choice and constant heat capacities, the energy balance equation takes the form

$$\frac{dQ}{dt} = 22 \times 10^3 \, (400) + 200(H_V \, (250°C, \, S_8) - H_L(220°C, \, S_8))$$

$$+ \, 600(10.0)(250°C - 175°C)$$

Note, however, that the enthalpy difference between S_8 vapor and liquid must be calculated, given the heat of vaporization at 445°C. This side calculation can easily be accomplished using the methods of Chapter 7 as shown below:

$$H_V(250°C, \, S_8) - H_L(220°C, \, S_8) = C_{P_L}(250 - 220) + \Delta H_{VL} \, (250°C)$$

But

$$\Delta H_{VL}(445°C) = \Delta H_{VL}(250°C) + \int_{250°C}^{445°C} C_{P_V} \, dT - \int_{250°C}^{445°C} C_{P_L} \, dT$$

or

$$\Delta H_{VL}(250°C) = 2200 - C_{P_V}(445 - 250) + C_{P_L}(445 - 250)$$

$$= 2005 \text{ cal/gmol}$$

Therefore,

$$\frac{dQ}{dt} = 22 \times 10^3(400) + 200[7(30) + 2005] + 600(10.0)(75)$$

$$= 9.693 \times 10^6 \text{ cal/h}$$

or

$$9.693 \times 10^6/400 = 2.42 \times 10^4 \text{ cal/mol } CS_2$$

At this point, we continue with the reactor material balances. Since the rate is known ($r = 400$ mol/h), we have immediately that

$$N_{CH_4}^{out} = 200 \text{ mol/h}$$

$$N_{CS_2}^{out} = 400 \text{ mol/h}$$

$$N_{H_2S}^{out} = 800 \text{ mol/h}$$

The reactor energy balance, again written using as reference 250°C and gaseous species and incorporating the calculated heating rate, reduces to

$$9.693 \times 10^6 \text{ cal/h} = 400 \, \Delta H_R (250) + [200(10) + 400(7.6) + 800(9.5)](T$$
$$- 250°C) - [600(10) + 200(8)](700°C - 250°C)$$

The solution is

$$T = 591.2°C$$

Either of the heat exchangers can now be solved. The composition of the reactor effluent is known:

$$\tfrac{2}{14}CH_4, \tfrac{4}{14}CS_2, \text{ and } \tfrac{8}{14}H_2S$$

The feed S_8 flow is known, and all four temperatures are known. Therefore, the energy balance can be used to calculate the flow of branch A. Using 220°C and liquid S_8 as reference state and noting the phase change in the S_8, the energy balance

$$200[C_{P_L}(445 - 220) + 2.2 \times 10^3 + C_{P_V}(700 - 445)]$$
$$= N^A \left[\frac{2}{14} (10) + \frac{4}{14} (7.6) + \frac{8}{14} (9.5) \right](591.2 - 250)$$

can be solved to yield

$$N^A \simeq 377.5 \text{ mol/h}$$

Finally, we can use the splitter balance to calculate the flow in branch B:

$$N^B = 1400 - 377.5 = 1022.5 \text{ mol/h}$$

Since the solution was required for a CS_2 production rate of 300 mol/min, all flows and the heat transfer rate must be rescaled. The scaled results are:

$$r = 300 \text{ mol/min}$$

$$N_{S_8}^{in} = 150 \quad \text{and} \quad N_{CH_4}^{in} = 450 \text{ mol/min}$$

$$N^B = 766.88 \text{ mol/min}$$

$$dQ/dt = 7.26 \times 10^3 \text{ kcal/min}$$

and the outlet CH_4, H_2S, and CS_2 flows are 150, 600, and 300 mol/min.

This completes the solution of the flowsheet balance problem. Note that, although all flows and temperatures satisfy the balance equations, the solution is physically unrealistic. The S_8 stream is heated from 220°C to 700°C using the reactor effluent stream which is at 591°C. It is physically impossible to heat a stream to a temperature higher than that of the heating medium. This fact is not taken into account in the energy balance equations. It is a requirement of heat transfer. To avoid such unrealistic solutions, we must always check our answers and, if need be, readjust the problem specifications. In the present instance, since the reaction is endothermic, it is clear that we need to add heat to get the reactants to 700°C and then need to remove that heat by cooling with an external medium to achieve the desired 250°C exit temperature.

8.7 SUMMARY

In this chapter, we discussed all of the thermochemistry and algebraic issues necessary to analyze and solve energy balances for multiunit reacting systems. The thermochemistry issues included the definition of the heat of reaction and its calculation using standard heats of formation and combustion. It was shown that the standard enthalpies chosen for the elements vanish from the formula for the heat of reaction. Hence, these enthalpies can be arbitrarily selected. We observed that since the heat of combustion is just the heat of reaction of a special reaction, heats of formation and combustion can be interconverted in a straightforward manner. The calculations needed to correct heats of reaction for changes in temperature, pressure, and phase were next reviewed. These calculations were shown to be obvious applications of similar calculations already discussed in Chapter 7.

We considered the open-system energy balance equation for reacting systems: first the single-input/single-output/single-reaction case, and then, extensions to multiple-input/output streams and multiple reactions. In all cases, we were always free to pick a balance reference state, and in many cases it proved convenient to select the state of one of the problem streams. The total enthalpy form of the balance equation was also developed and shown to be appropriate for cases in which the stoichiometry was inconvenient to use or was unknown. The total enthalpy form simply included the species heat of formation as part of the enthalpy of a stream. This form is often used in computer-aided balance calculations because it avoids explicit heat-of-reaction evaluations. We observed that in the presence of multiple chemical reactions, it is desirable to check that the reactions are independent and that it is only necessary to include the heats of reaction of the linearly independent reactions in the balance calculations.

Finally, the modifications needed to incorporate reacting system balances into our degree-of-freedom analysis were reviewed. In the single-unit case, these modifications merely involved maintaining a separate column for the material balance subproblem. The purpose of this supplementary tabulation was to aid in determining whether or not the balances could be decoupled. Decoupling of the material balance and energy balance calculations was desirable because it reduces algebraic complexity. The multiunit degree-of-freedom analysis with reacting systems was similarly straightforward. The only complication arose when overall energy balances were used during the solution process. The heat and work rates in the overall balance are net rates for the whole system and must be recognized as such. Moreover, once calculated, they serve to fix the corresponding rates of one of the flowsheet units, thus reducing the effective number of variables for that unit. This effect must be taken into account in the unit sequencing analysis to avoid misleading calculation paths.

With the mastery of the material in this chapter, we are ready to solve most multiunit flowsheet energy balance problems. In the next and last chapter, we complete the discussion of flowsheet balance calculations by considering manual methods for use with more complex problems and by investigating how to structure combined balance problems for solution via computer programs.

PROBLEMS

8.1 From the heat of formation data, calculate the standard heats of reaction of the following reactions:

(a) $2C_2H_4(g) + O_2(g) \rightarrow 2C_2H_4O(g)$

(b) $SO_2(g) + \frac{1}{2}O_2(g) + H_2O(l) \rightarrow H_2SO_4(l)$

(c) $C_6H_5Cl(g) + Cl_2(g) \rightarrow m\text{-}C_6H_4Cl_2(g) + HCl(g)$

(d) $CO_2(g) + 4H_2(g) \rightarrow CH_4(g) + 2H_2O(g)$

(e) $2CH_4(g) + 3O_2(g) \rightarrow 4H_2O(g) + 2CO(g)$

(f) $CH_4(g) \rightarrow 2H_2(g) + C(s)$

(g) $n\text{-}C_4H_{10}(g) \rightarrow C_2H_4(g) + C_2H_6(g)$

(h) $C_2H_5OH(l) + 3O_2(g) \rightarrow 2CO_2(g) + 3H_2O(l)$

(i) $2C_3H_8(g) \rightarrow C_3H_6(g) + C_2H_6(g) + CH_4(g)$

(j) $C(s) + H_2O(l) \rightarrow CO(g) + H_2(g)$

(k) $C(s) + CO_2(g) \rightarrow 2CO(g)$

(l) $2C(s) + H_2O(g) + H_2(g) \rightarrow CH_4(g) + CO(g)$

8.2 Calculate the heats of formation of the following compounds from the standard heat of combustion data:

(a) Benzene, $C_6H_6(l)$, $\Delta H_c^\circ = -780.98$ kcal/gmol.

(b) Diethyl ether, $C_2H_5OC_2H_5(l)$, $\Delta H_c^\circ = -2726.7$ kJ/gmol.

(c) n-Hexane (l), $\Delta H_c^\circ = -4163.1$ kJ/gmol.
(d) Carbon (graphite), $\Delta H_c^\circ = -393.51$ kJ/gmol.
(e) $H_2(g)$, $\Delta H_c^\circ = -68.317$ kcal/gmol.
(f) $H_2S(g)$, $\Delta H_c^\circ = -134.46$ kcal/gmol.
(g) Methylamine, $CH_5N(l)$, $\Delta H_c^\circ = -1071.5$ kJ/gmol.
(h) $CO_2(g)$, $\Delta H_c^\circ = 0$.

8.3 Calculate the standard heat of reaction for the reaction

$$Fe_3O_4(s) + 2CO(g) + 2H_2(g) \rightleftharpoons 3Fe(s) + 2CO_2(g) + 2H_2O(g)$$

Given:

Heat of formation of $Fe_3O_4(s)$ $= -267,000$ cal/gmol

Heat of combustion of $CO(g)$ $= -67,636$ cal/gmol

Heat for formation of $CO_2(g)$ $= -94,050$ cal/gmol

Heat of formation $H_2O(g)$ $= -57,800$ cal/gmol

8.4 Calculate the heat of reaction at 25°C for the reaction

$$CH_4(g) + Br_2(g) \rightarrow CH_3Br(g) + HBr(g)$$

The accurately determined heat of combustion of $CH_3Br(g)$ at 25°C is -162.5292 kcal/gmol with $CO_2(g)$, $H_2O(l)$, and $HBr(g)$ as combustion products.

8.5 Amyl alcohol, $C_5H_{11}CH$, is prepared industrially by chlorinating n-pentane, C_5H_{12}, to produce a chloride which is then hydrated to form the alcohol. Calculate the heat of reaction at 25°C for the overall reaction

$$C_5H_{12}(l) + Cl_2(g) + H_2O(l) \rightarrow C_5H_{11}OH(l) + 2HCl(aq)$$

from the following information (all in kcal/gmol):

ΔH_f°, $C_5H_{12}(g)$ $= -35.00$

ΔH_f°, $H_2O(l)$ $= -68.3174$

ΔH_f°, $HCl(aq)$ $= -39.56$

ΔH_f°, $CO_2(g)$ $= -94.0518$

ΔH_{VL}, C_5H_{12} (25°C) $= 6.316$

ΔH_C°, $C_5H_{11}OH(l)$ $= -786.7$

8.6 One path to aniline from nitrobenzene involves iron–acid reduction in the liquid phase

$$C_6H_5NO_2(l) + 3Fe + 6HCl(aq) \rightarrow C_6H_5NH_2(l) + 3FeCl_2(aq) + 2H_2O(l)$$

From the given data, calculate

(a) The standard heat of reaction.

Another method of manufacturing aniline from nitrobenzene involves hydrogen reduction in the vapor phase:

$$C_6H_5NO_2(g) + 3H_2(g) \rightarrow C_6H_5NH_2(g) + 2H_2O(g)$$

(b) Which of the two reaction paths will evolve the greater heat of reaction?
(c) Assuming that there was a temperature T, $(300°C > T > 25°C)$ at which the heats of reaction of the above reactions were equal, how would you find it? Is enough data given to accomplish this? If anything is missing, what is it?

Given:
Heats of formation (25°C) in kcal/gmol

$CO_2(g)$	-94.05
$HCl(aq)$	-39.56
$FeCl_2(aq)$	-99.7
$H_2O(l)$	-68.32
$H_2O(g)$	-57.80
$C_6H_5NH_2(l)$	20.28

Heat of combustion (50°C) in kcal/gmol, with final products $CO_2(g)$ $H_2O(l)$, $N_2(g)$

$C_6H_5NO_2(l)$	-739.0

Heats of vaporization in kcal/gmol (25°C)

$C_6H_5NO_2$	15.08
$C_6H_5NH_2$	17.85

Heat capacities in (cal/gmol · °C) with T in K

$H_2O(l)$ (25–90°C)	18.0
Fe	$4.13 + 0.0064T$
$FeCl_2$	$9.92 + 0.032T$
$C_6H_5NO_2(l)$	$4.16 + 0.1124T$
$C_6H_5NH_2(l)$	$32.83 + 0.0484T$

	a	b	c
$HCl(g)$	6.73	0.431×10^{-2}	0.361×10^{-6}
$H_2O(g)$	7.14	2.64×10^{-3}	0.046×10^{-6}
$CO_2(g)$	6.34	10.1×10^{-3}	-3.42×10^{-6}
$N_2(g)$	6.46	1.39×10^{-3}	-6.9×10^{-8}
$O_2(g)$	6.12	3.17×10^{-3}	-1.0×10^{-6}

where C_p (cal/gmol · °C) $= a + bT + cT^2$

8.7   Calculate the standard heat of reaction for the reaction

$$FeO(s) + 2H^+ \rightarrow H_2O(l) + Fe^{2+}$$

Given:

Heat of formation of $Fe_2O_3(s)$	$= -198,500$ cal
Heat of formation of $H_2O(l)$	$= -68,400$ cal
Heat of combustion of $FeO(s)$ (to form $Fe_2O_3(s)$)	$= -34,950$ cal

Heats of the reactions

$$\left. \begin{array}{l} \tfrac{1}{2}H_2(g) \rightarrow H^+ \\ Fe(s) + 2H^+ \rightarrow Fe^{2+} + H_2(g) \end{array} \right\} = \quad 0 \text{ cal}$$

8.8   Calculate the heat of reaction at 500°C of the reaction

$$N_2(g) + O_2(g) \rightarrow 2NO(g)$$

using the appropriate data from the appendices.

8.9   Calculate the heat of reaction at 450°C of the reaction

$$CO(g) + 3H_2(g) \rightleftharpoons CH_4(g) + H_2O(g)$$

using the appropriate data from the appendices.

8.10 Calculate the heat of reaction of the reaction

$$CO(g) + H_2O(g) \rightarrow CO_2(g) + H_2(g)$$

at 400°F. Does the heat of reaction increase or decrease with temperature?

8.11 Calculate the heats of reaction of the reactions

$$CH_3OH(l) + \tfrac{1}{2}O_2(g) \rightarrow CH_2O(g) + H_2O(g)$$
$$CH_2O(g) + \tfrac{1}{2}O_2(g) \rightarrow HCOOH(g)$$
$$CH_3OH(g) + \tfrac{1}{2}O_2(g) \rightarrow CO(g) + 2H_2O(l)$$

at 300°C and 1 atm. Which of the three reactions is most exothermic?

8.12 The temperature in a CO shift reactor can be moderated by injection of excess steam. Assuming a feed of 30% CO, 20% H_2, and 50% H_2O at 600°F and assuming 90% of the CO will be converted, determine the additional 550°F steam required (per mole of feed) to maintain the reactor outlet temperature below 850°F. The reaction stoichiometry is

$$CO(g) + H_2O(g) \rightleftharpoons CO_2(g) + H_2(g)$$

8.13 The adiabatic flame temperature of a fuel is the temperature of the oxidation products which is achieved when 1 mol of the fuel is completely burned with a stoichiometric amount of air adiabatically. The fuel and air are assumed to be supplied at 25°C, and 1 atm and air is assumed to consist of 21% O_2 and 79% N_2. The combustion products are $CO_2(g)$, $H_2O(g)$, $SO_2(g)$, and $N_2(g)$. Calculate the adiabatic flame temperatures of the following fuels:

(a) $H_2(g)$.
(b) $C_2H_6(g)$.
(c) Coal with ultimate analysis 80% C, 8% H, 2% O, 3% S, 7% ash.

8.14 In the reaction

$$2A + B \rightarrow C$$

the heat of reaction at 300K is $-10,000$ cal/gmol. The heat capacities of the substances A, B, and C, in cal/gmol · K, are:

A	$16.0 - (1.5 \times 10^3/T)$
B	$11.0 - (0.5 \times 10^3/T)$
C	$25.0 - (1.0 \times 10^3/T)$

with T expressed in K. The heat capacity equations are valid in the range

$$300 \text{ K} \leqslant T \leqslant 1000 \text{ K}$$

(a) Derive an equation for the heat of reaction as a function of temperature.
(b) Calculate the temperature at which the reaction changes from exothermic to endothermic. Use Newton's method for the iterations.
(c) Calculate the heat of reaction at 500 K assuming that substance A undergoes a change of phase at 400 K with ΔH_{VL} (400 K) = 928 cal/gmol after which its heat capacity becomes a constant 10 cal/gmol · K.

8.15 In the production of ethylene oxide by partial oxidation of ethylene, the reactions

$$2C_2H_4 + O_2 \rightarrow 2C_2H_4O$$

and

$$C_2H_4 + 3O_2 \rightarrow 2CO_2 + 2H_2O$$

take place. In a given reactor with a feed consisting of 10% C_2H_4 and the rest air, 25% conversion of C_2H_4 and an 80% fractional yield of C_2H_4O from C_2H_4 are attained when the reactor is operated at 240°C. Calculate the required heat removal rate from the reactor for a feed rate of 1000 kgmol/h if the reactor feed mixture is at 100°C. All species are in the gas phase.

8.16 Propylene can be produced by cracking of propane. Cracking is carried out in a furnace because of the large heat requirements. A feed of pure propane, C_3H_8, at 500°F and 400 psia is found to produce a product stream of the following composition at 900°F and 400 psia: 45% C_3H_8, 20% C_3H_6, 5% C_2H_4, and the rest C_2H_6, CH_4, and H_2. The ratio of C_2H_6 to CH_4 is 2:1 (all molar specifications). There is no carbon deposition observed in the furnace tubes.

(a) Calculate the complete product stream composition.
(b) Calculate the heat requirements per mole of C_3H_8 fed.

8.17 (a) Propane gas, C_3H_8, fed at 80°C is burned with air preheated to 200°C. If 50% excess air is used, calculate the adiabatic combustion temperature assuming complete combustion.
(b) Suppose the combustion of propane is carried out on the outside of tubes through which is passed saturated boiler feed water at 100 bar. What is the maximum amount of 100 bar saturated steam that could be raised per mole of propane burned? Assume that the flue gases will leave at a temperature 20°C higher than the saturated steam.

8.18 A fuel gas consisting of 91% CH_4, 8% H_2, and 1% CO is burned adiabatically with an excess of air. The fuel gas enters at 50°F and the dry air at 100°F, and the flue gases leave at 3000°F. If CO_2, H_2O, O_2, and N_2 are the only combustion products, calculate the percent excess air which was used.

8.19 A devolatilized char consisting of 80% carbon and 20% ash is burned with air in an atmospheric pressure cyclone combustor. If the char is fed at 1000°F and the air at 1200°F, calculate the air feed required per pound of char to maintain a combustor temperature of 2800°F. The cyclone is well insulated, and $CO_2(g)$ is the sole combustion product. All carbon is converted, and the ash leaves in the form of a molten slag (Figure P8.19) How is the solution changed if $CO(g)$ is assumed to be the sole combustion product? The ash fusion temperature is 2600°F, the heat of fusion is 61 Btu/lb_m, and the heat capacity of molten slag is about 0.2Btu/$lb_m \cdot$ °F.

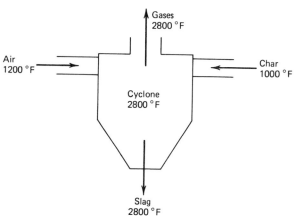

Figure P8.19

8.20 Ammonia and oxygen at 450°C and 5 bar are fed to a reactor in which 90% of the NH_3 reacts to form NO via the reaction

$$4NH_3(g) + 5O_2(g) \rightarrow 4NO(g) + 6H_2O(g)$$

Assuming that 1.5 mol O_2 and 5 mol N_2 diluent are fed per 1 mol NH_3, calculate the exit temperature of the adiabatically operated reactor. Use a constant C_p value at 1200°C to obtain an initial estimate of the outlet temperature.

8.21 Carbon monoxide is completely burned, at a pressure of 1 atm, with excess air. If reactants enter at 200°F, products leave at 1800°F, and heat losses are negligible, what percentage of excess air was used?

8.22 Sulfur dioxide may be catalytically oxidized to sulfur trioxide over a V_2O_5 catalyst at 425°C and 1 atm. Assuming this reaction is 97% complete, how much heat (in Btu/day) is liberated in a converter for a plant making 500 tons of 100% sulfuric acid per day?

8.23 A catalytic reactor is charged with a feed consisting of 5 mol H_2 to 1 mol CO_2 at 400°C and 5 bar (Figure P8.23). In the reactor, the reactions

$$CO(g) + 3H_2(g) \rightarrow CH_4(g) + H_2O(g)$$

$$CO_2(g) + H_2(g) \rightarrow CO(g) + H_2O(g)$$

take place with 90% conversion of CO_2. If the reactor operates adiabatically and the exit stream is at 400°C, calculate the composition of the exit stream.

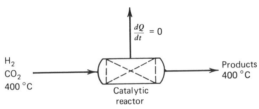

Figure P8.23

8.24 Methanol is synthesized from CO and H_2 at 50 atm and 550 K over a copper-based catalyst. With a feed of 75% H_2 and 25% CO at these conditions, a product stream consisting of 8% CH_3OH, 1% CH_4, and 20% CO and the rest H_2O and small amounts of CO_2 and H_2 is produced. The primary reaction is

$$CO(g) + 2H_2(g) \rightarrow CH_3OH(g)$$

and two secondary reactions are

$$CO(g) + 3H_2(g) \rightarrow CH_4(g) + H_2O(g)$$

$$CO(g) + H_2O(g) \rightarrow CO_2(g) + H_2(g)$$

If the feed and product streams are maintained and 50 atm and 500 K, what is heat transfer rate required to maintain isothermal conditions? Must heat be added or removed from the reactor?

8.25 Sulfur dioxide is converted to sulfur trioxide by oxidation over a V_2O_5 catalyst at atmospheric pressure. In the reactor configuration shown in Figure P8.25,

SO_2 and O_2 available at 300°C and 25°C, respectively, are preheated to 425°C and fed in stoichiometric ratio to the reactor. Conversion is 96% and, to keep the reactor temperature from rising too much, 10 kcal/g-mol of combined feed is removed by external medium. The last heat exchanger in the process is used to generate 300 psia steam by cooling down the reactor effluent to 400°C. Calculate all unknown temperatures and flows. Assume no heat losses to the environment. Given C_p (cal/gmol · °C):

	O_2	SO_2	SO_3
Avg C_p, 25–425°C	7.6	10.4	15.3
Avg C_p, 300–900°C	7.8	11.8	17.2

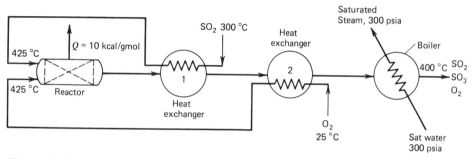

Figure P8.25

8.26 The production of ethylene oxide by partial oxidation of ethylene over a silver catalyst involves the exothermic reaction

$$2C_2H_4(g) + O_2(g) \rightarrow 2C_2H_4O(g)$$

and the secondary, even more exothermic, reaction

$$C_2H_4(g) + 3O_2(g) \rightarrow 2CO_2(g) + 2H_2O(g)$$

Temperature control is essential and is achieved by boiling a hydrocarbon heat transfer fluid on the outside of the reactor tubes. This vaporized fluid is then condensed in a heat exchanger by transferring heat to a stream of saturated liquid water at 100 bar to produce saturated steam at 100 bar. The reactor feed is 10% C_2H_4, 12% O_2, and the rest N_2 at 360°C at 10 bar. The reactor product stream is at 375°C and 10 bar and the conversion of C_2H_4 is 20% with an 85% selectivity for C_2H_4O. At the operating pressure used in this system, the heat transfer fluid boils at 350°C with a heat of vaporization of 500 Btu/lb$_m$, and has a liquid heat capacity of 0.8 Btu/lb$_m$ · °F and a vapor heat capacity of 0.4 Btu/lb$_m$ · °F. The flow of heat transfer fluid is adjusted so that it enters the reactor as liquid at 340°C and leaves as a two-phase mixture with vapor fraction of 20% (See Figure P8.26).

(a) Calculate the mass of steam raised per mole of C_2H_4O produced.

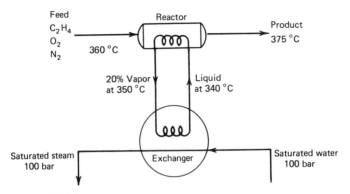

Figure P8.26

(b) Calculate the recirculation rate of the heat transfer fluid per mole of C_2H_4O produced.

(c) Why is the heat transfer fluid used instead of direct heat transfer using water?

(d) What pressure would the reactor tubes have to withstand if water were boiled at 350°C on the outside of the reactor tubes?

8.27 In the manufacture of sulfur dioxide by the direct oxidation of sulfur with pure oxygen, $S + O_2 \rightarrow SO_2$, cool sulfur dioxide must be recycled to the burner to lower the flame temperature below 1000 K, so as not to damage the burner (Figure P8.27). Per lbmol SO_2 product, calculate the lbmol SO_2 recycled and the pounds of steam produced.
Given for sulfur:

Melting point	= 113°C
Heat of fusion (at melting point)	= 0.3 kcal/gmol
Heat capacity of solid	= 5.8 cal/(gmol · K)

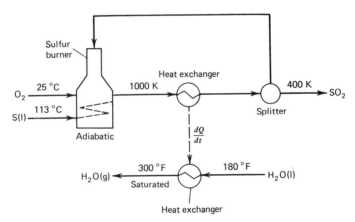

Figure P8.27

8.28 A vertical lime kiln is charged with pure limestone, $CaCO_3$, and pure coke, carbon, both at 25°C. Dry air is blown in at the bottom and provides the necessary heat for decomposition of the carbonate by burning the carbon to CO_2. Lime, CaO leaves the bottom at 950°C and contains 5% carbon and 1% $CaCO_3$. The kiln gases leave the top at 600°F and consist only of CO_2 and N_2. Assume heat losses are negligible (Figure P8.28). The two reactions are

$$CaCO_3(s) \rightarrow CaO(s) + CO_2(g)$$
$$C(s) + O_2(g) \rightarrow CO_2(g)$$

Given:

$$\Delta H_f^\circ,\ CaCO_3(s) = -289.5\ \text{kcal/gmol}$$
$$\Delta H_f^\circ,\ CaO(s)\ \ = -151.7\ \text{kcal/gmol}$$
$$\Delta H_f^\circ,\ CO_2(g)\ \ = -94.052\ \text{kcal/gmol}$$

Average Constant Molal Heat Capacities	(cal/gmol · K)
$CaO(s)$	13.7
$CO_2(g)$	12.2
$N_2(g)$	7.5
$O_2(g)$	8.0
$CaCO_3(s)$	28.0
$C(s)$	4.6

(a) Calculate the heats of reaction at 25°C.
(b) Develop a degree-of-freedom table for the problem. Can the material and energy balances be decoupled?

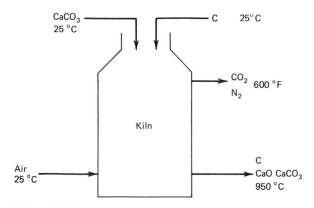

Figure P8.28

(c) Calculate the required feed ratio of $CaCO_3$ to C.

8.29 Steam at 100 psia and 700°F is to be expanded adiabatically to 10 psia in a turbine to drive both a compressor and an electric power-generating turbine. The steam leaving the expansion turbine is cooled by removing heat at the rate of 10^7 Btu/h to produce a saturated liquid at 10 psia. The generating turbine is to develop 750 kW of power and the compressor will require an equal amount of supplied power. The compressor will compress steam at 500 psia and 900°F adiabatically to 1000 psia and a temperature of 1140°F (Figure P8.29). Assume all devices operate with 100% efficient.

(a) Construct a degree-of-freedom table to show that the process is correctly specified.

(b) Deduce a calculation order sufficient to calculate all flows and rates.

(c) Calculate the quality of the 10-psia steam leaving the turbine and the flows of both streams.

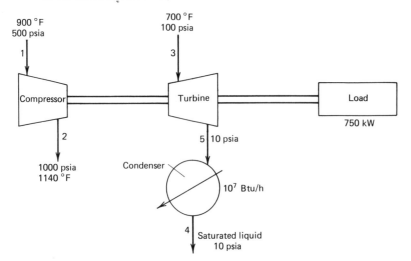

Figure P8.29

8.30 A synthesis gas is upgraded to a higher methane content in the recycle system shown in Figure P8.30. The feed gas contains a small amount of methane and analyzes 22% CO, 13% CO, and 65% H_2 (mol %) on a *methane-free basis*. The product stream (stream 6) analyzes (all mol %) CO 5%, H_2 9%, CH_4 50%, CO_2 27%, and H_2O 9%. Both feed and product are gases at 200°F. In the reactor the two reactions

$$CO + 3H_2 \rightarrow CH_4 + H_2O$$
$$CO + H_2O \rightarrow CO_2 + H_2$$

take place. The reactor effluent is cooled to 500°F in exchanger 1 and then is further cooled in exchanger 2. Part of the effluent is split off as product. The remainder is sent to a separator in which the stream is cooled to 90°F

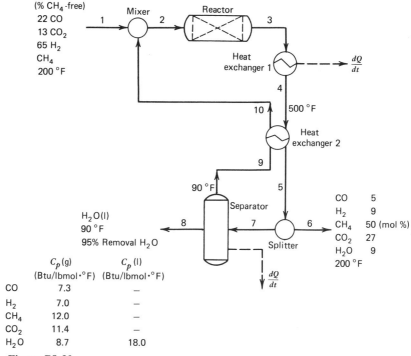

Figure P8.30

and 95% of the H₂O is separated as liquid. The remaining gas stream is reheated in exchanger 2, mixed with the fresh feed, and returned to the reactor. Assume that the mixer, the splitter, exchanger 2, and the reactor operate adiabatically. Use average C_p values as given on the flowsheet.

(a) Construct a degree-of-freedom table and show that the process is correctly specified.

(b) Determine a calculation order for the problem.

(c) Calculate the reactor outlet temperature.

8.31 A dry feed gas consisting of CO_2 and CH_4 (0.3 mol CO_2 per 1 mol CH_4) and available at 500°F is fed to a steam reformer to produce a synthesis gas with a H_2/CO ratio of 2.2. When the reformer heating rate is adjusted to 42.8 kcal/mol dry feed gas, 80% conversion of methane is obtained with a synthesis gas temperature of 1200°F. If reformer steam is supplied at 900°F (superheated), calculate the steam feed rate assuming the following four reactions take place in the reformer:

$$CH_4 + CO_2 \rightleftharpoons 2CO + 2H_2$$
$$CO_2 + H_2 \rightleftharpoons CO + H_2O$$
$$CH_4 + H_2O \rightleftharpoons CO + 3H_2$$
$$CH_4 + 2H_2O \rightleftharpoons CO_2 + 4H_2$$

The flow diagram for the reformer is shown in Figure P8.31.

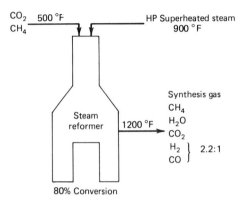

Figure P8.31

Material and Energy Balances in Process Flowsheets

In the previous chapter, we completed our study of all the essential tools necessary to formulate and solve flowsheet balance problems. In addition to mastering the structure, properties, and use of the various types of material balance equations, we achieved a comparable mastery of energy balance equations. We fully developed the open-system energy balance for the nonreacting case and for the case with multiple chemical reactions, both in the heat-of-reaction and in the total enthalpy form. In order to evaluate the enthalpy terms which arise in the balance equations, we found it necessary to study a considerable body of supporting material from the fields of thermodynamics and thermochemistry. To aid in the analysis of energy balance problems, we extended our degree-of-freedom tabulation to account for the new variables and equations. Finally, we briefly considered the use of the degree-of-freedom tabulation to determine good calculation sequences for single-unit and multiunit flowsheets.

In this chapter, which is intended to parallel Chapter 5, we examine manual and computer-aided strategies for solving the material and energy balances associated with general process flowsheets. The manual strategies will involve judicious selection of balance sets (individual unit or overall), balance types (material or energy), and iterative methods to carry out solution with complete or partial sequencing. The computer-aided approaches will consist of extensions of the sequential modular and simultaneous strategies to accommodate energy balances. We shall find that the extension of these strategies to include energy balances must address three inherent difficulties:

1. The inherent nonlinearity of the flow-enthalpy terms.
2. The need for extensive thermodynamic property data used to calculate stream enthalpy.

3. The complexity of the calculations required to interconvert between stream enthalpy, phase distribution, and the other intensive properties.

The energy balance equation essentially consists of a sum of products of stream flows and stream enthalpies. Even in the case of constant phase and constant C_p, such products are nonlinear, specifically bilinear functions of flow and temperature: with temperatures known, a linear equation in the flows is obtained; with flows known, a linear equation in the temperatures is obtained. In order to cast combined balance problems into a computationally convenient linear form, this bilinearity must be accommodated. However, as we shall discover, this can be accomplished only at the cost of either simplification of the flowsheet model or of repeated iterative calculations.

In general, the calculation of the enthalpy of a system at a known state is relatively straightforward, provided enthalpy tabulations, heat capacity correlations, and heats of transition and formation are available. In computer-aided strategies, this body of information must be introduced into the program for each species involved in the problem. This can be achieved either by requiring the user to read in the data for each application or else by developing and maintaining a numerical data base of this type of information which can be accessed by the calculation program. In either case, supplementary calculation routines must be developed which will actually retrieve the properties data and perform the stream enthalpy evaluations. Thus, programs which include energy balances calculations inherently require more data manipulation and support calculations.

Finally, we noted in Chapter 7 that the determination of one or more of the state properties of a system, particularly its phase or phase distribution, given the enthalpy of the system and the remaining state properties, was frequently a complex procedure. In fact, such interconversion between state variables often requires iterative calculations which are much more complicated than the solution of the energy balance problem itself. For instance, the calculation of the enthalpy of the outlet stream of an adiabatic reactor usually only involves summing the heat of reaction and input stream enthalpy terms. However, the determination of the outlet stream temperature, given the enthalpy, requires iterative calculation even in the case of single phase systems. A versatile balancing program must clearly be able to perform such support calculations. The inclusion of routines which implement such utility calculations further enlarges the scope and complexity of such programs. In this chapter, we focus on problem formulation and approximation strategies which address the above three issues in as simple a manner as possible. A more extensive treatment of calculations involving rigorous flowsheet models and thermodynamic property estimation programs can be found in references on process flowsheeting.[1,2]

---

[1] A. W. Westerberg, H. P. Hutchinson, R. L. Motard, and P. Winter. *Process Flowsheeting*, Cambridge University Press, Cambridge, England, 1979.
[2] C. M. Crowe, A. E. Hamielec, T. W. Hoffman, A. I. Johnson, and D. R. Woods, *Chemical Plant Simulation*, Prentice-Hall, Englewood Cliffs, N.J., 1971.

9.1 STRATEGY FOR MANUAL CALCULATIONS

The overriding consideration in performing flowsheet balance calculations by hand is to decompose the solution process into the manipulation of small equation sets consisting, if possible, of single equations. Moreover, this decomposition should be performed in a systematic fashion so that the use of redundant equations is avoided and the occurrence of dependent or inconsistent specifications is detected. The approach we have used in the material balance case and which certainly remains applicable in the combined material and energy balance case is to decompose problems into balance sets associated with individual flowsheet units. Each balance set is chosen so that at the time it is solved it has degree of freedom zero. If decomposition by units proves unsuccessful because at some point in the sequence no further degree-of-freedom zero units can be found, then we consider the use of the overall balance set as substitute for one of the unit balance sets. If even this device fails, then, as in Chapter 5, we can resort to partial solution of unit balance sets. This approach involves the carrying of unknowns from one flowsheet unit to another until a unit is reached at which all remaining unknowns can be calculated.

In combined balance problems, we have the further natural division of the unit balances into the material and the energy balance equations. The possibility exists that, depending upon the specifications, the former can be solved independently of the latter and, thus, that the unit material balances can be solved at one point in the flowsheet calculation sequence while the energy balances for that unit will be solved at a much later point in the sequence. In the extreme case, under suitable specifications, the material balances for an entire multiunit flowsheet can be solved first, followed by the flowsheet energy balances. The distinction between the unit energy and material balances thus permits an additional level of sequencing in the solution process. The degree-of-freedom table can be used to good advantage to develop such a sequence, provided that for each unit both a combined and a material balance column is generated and that both columns are updated as the sequence is developed.

While the degree of freedom-based unit sequencing strategy involves a natural utilization of the flowsheet structure, it does suffer from the disadvantage of not taking into account the relative difficulty of solving alternate unit balance sets. The degree of freedom of a balance set merely measures whether the number of its equations and specifications balances the number of unknowns involved in those equations. It gives no measure of the algebraic complexity of the set. In particular, it gives no indication of whether only individually solvable linear equations will arise. Consequently, the decision of whether one zero degree of balance set should be solved in preference to another must be made on the basis of insight into the problem, previous experience, or trial and error. Furthermore, the unit-by-unit sequencing procedure is inherently short-sighted since at each stage it only considers the current degree of freedom of the yet unsolved balance sets. It is conceivable that by first solving a balance set with nonzero degree of freedom, and thus carrying a variable, an algebraically simpler solution is achieved than if a balance set with zero degree of freedom is selected. The degree of freedom-based sequencing strategy should therefore be viewed as a guide for problem solving rather than as a rigid formula. Of course, in defense of the procedure, it is generally true that the effort involved in systematically evaluating alternative calculation sequences with respect to algebraic complexity is only worthwhile if multiple solutions are required. But in such instances, computer-aided rather than manual solution methods are clearly even more appropriate.

In the next two sections, we review the degree of freedom-based sequencing procedure and the use of iterative solution methods via a series of increasingly complex examples.

9.1.1 The Single-Unit Case

Single-unit combined balance problems in general involve the stream material balance variables (the species flow rates); the stream energy balance variables (temperature, pressure, phase distribution, bulk elevation, and bulk velocity); the unit material balance variables (the rates of any reactions); and the unit energy balance variables, dQ/dt and dW/dt. We have observed that, in most cases, the stream potential and kinetic energy components are dominated by the enthalpy term. This means that the stream elevations and bulk velocities can be neglected. Furthermore, in most cases stream pressures are selected from design considerations (heat transfer, chemical or physical equilibrium, etc.), hence are usually considered fixed for purposes of balance calculations. In most cases, each stream will consist of a single phase, and that phase will usually be known in advance. From the phase rule, such a stream with S number of species will require $S + 1$ independent intensive variables: the $S - 1$ independent mole or mass fractions, the temperature, and the pressure. For material balance purposes, the extensive total flow variable is also necessary. Thus, assuming that the pressure will always be specified and that the phase is known, each single-phase stream will have $S + 1$ independent balance variables, the S flow and composition variables, and the temperature.

For two-phase systems (vapor and liquid mixtures being the most common), the phase rule requires that $S - 2 + 2$, or S, intensive variables be specified in order to specify the state of the system. However, additional information is required to determine the amount of each phase. In the case of a single species, say, water, knowledge of one intensive variable, say, the pressure, clearly determines the temperature if both liquid and vapor phases are present. However, in order to determine how much of each phase is present, we also need to know the total flow and the quality of the steam or, alternatively, the total flow and a mixture property such as the mixture specific volume or mixture enthalpy. In the case of a two-species system, with specified T and P the mole fractions of the two phases will be fixed (recall Section 7.1.3). However, to determine the relative amounts of both phases, we need another specification, say, the bulk mixture mole fraction. Finally, to determine the flows of each phase, the total flow of the mixed stream must be known. By making use of the isothermal flash calculation analysis which we performed in Section 7.1.3, we can generalize the above discussion to conclude that, for single and multiple species systems involving either one or two phases, complete identification of a stream requires $S + 2$ variables. Hence, assuming that the pressure will always be specified and that the phases are known, there will be $S + 1$ independent stream variables associated with a stream. These will normally be the total flow, the $S - 1$ mole fractions, and the temperature. Note, however, that for a single-species, two-phase system, because P fixes T and the bulk mole fraction is unity by definition, the total flow and one additional mixture property (say, the vapor fraction or quality) will be required.

In summary, then, a single-unit balance problem with specified stream pressures and phases will involve $S + 1$ independent stream material and energy balance variables for each stream and the unit variables, the reaction, heat transfer, and work rates. The number of material balances will depend upon the balance type selected (species or element), and there will always be only one energy balance equation. Assuming that the specifications are sufficient to result in a degree of freedom equal to zero, the problem can either be solved as a coupled or as a decoupled balance problem. If there are no specifications that relate energy balance to material balance variables, then the material and energy balance subproblems can be decoupled if the material balance subproblem degree of freedom is zero. Otherwise, coupled solution will be required. Finally, the energy balance subproblem will normally require iteration if one of the stream temperatures is unknown; the material balance subproblem may require iteration only if nonlinear specifications are imposed. These situations are illustrated with the following example problems.

Example 9.1   A stream consisting of 60% n-octane and 40% n-hexadecane (mole basis) at 5 bar and 500 K is separated into vapor and liquid streams at 4 bar and 500 K. How much heat must be supplied to the flash drum per mole of feed to the unit assuming that Raoult's law is applicable?

Solution   The schematic of the system is shown in Figure 9.1. It consists of three streams (the possibly two-phase feed, the vapor outlet and the liquid outlet)

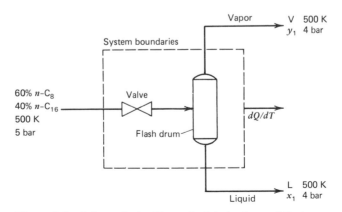

Figure 9.1   Schematic for Example 9.1, isothermal flash.

and includes the pressure reducing valve as well as the flash drum. There is no work performed by the system, the overall heat transfer rate is the primary unknown, and the system involves no reaction. The vapor and liquid outlet streams are assumed to be at equilibrium. Their mole fractions are related via equations of the form

$$K_i = \frac{y_i}{x_i} \qquad i = 1, 2$$

where the K_i will be calculated using the Raoult's law formula

$$K_i = \frac{p_i}{P}$$

The degree-of-freedom tabulation is shown in Figure 9.2. Note that the material balance subproblem has degree of freedom zero, hence can apparently be solved separately. Implicitly the K factor relations

$$\frac{p_i(T)}{P} = \frac{y_i}{x_i}$$

couple the energy balance variables T, P to the material balance variables y_i, x_i. However, since T and P are specified, that coupling is removed.

The material balance subproblem consists of the K value relations and the material balances

$$Nz_i = Vy_i + Lx_i$$

where N, V, and L are flows in mol/h. The material balance subproblem can be recognized as an isothermal flash calculation (see Section 7.1.3). Assuming a feed

	Material Balance Subproblem	Combined Problem
Number of stream variables		
Flows	6	6
T		3
Number of unit variables		
dQ/dt		1
Number of balances		
Material	2	2
Energy		1
Number of specifications		
Feed composition	1	1
K relations	2	2
Temperatures		3
Basis	1	1
Degree of freedom	0	0

Figure 9.2   Degree-of-freedom table, Example 9.1.

of 1 mol/h, the solution will be obtained by solving the equation

$$\sum_{i=1}^{S} \frac{z_i(1 - K_i)}{1 + V(K_i - 1)} = 0$$

for the unknown V. Then, the remaining mole fractions will be determined from

$$x_i = \frac{z_i}{1 + V(K_i - 1)}$$

$$y_i = K_i x_i$$

and the liquid flow by difference,

$$L = F - V = 1 - V$$

From the vapor pressure equations for n-octane (1) and n-hexadecane (2) given in Appendix 4 at 500 K,

$$\ln p_1 = 9.6316 - 3304.16/(500 - 55.2278) = 2.2027$$

$$\ln p_2 = 9.5830 - 4205.32/(500 - 119.1482) = -1.4589$$

The vapor pressures will thus be

$$p_1 = 9.0496 \text{ bar}$$

$$p_2 = 0.2325 \text{ bar}$$

To verify that both phases will be present at 500 K and 4 bar, we can determine the bubble pressure:

$$\sum_i \frac{z_i p_i}{P} = \frac{0.6(9.0496) + 0.4(0.2325)}{P} = 1$$

or

$$P = 5.523 \text{ bar}$$

and the dew pressure,

$$\sum (z_i/p_i)P = P\left(\frac{0.6}{9.0496} + \frac{0.4}{0.2325}\right) = 1$$

or

$$P = 0.5600 \text{ bar}$$

Since the specified pressure of 4 bar falls between these values, both L and V will be nonzero.

Proceeding with the flash calculation, we have

$$\frac{0.6\left(1 - \dfrac{9.0496}{4}\right)}{1 + V\left(\dfrac{9.0496}{4} - 1\right)} + \frac{0.4\left(1 - \dfrac{0.2325}{4}\right)}{1 + V\left(\dfrac{0.2325}{4} - 1\right)} = 0$$

Solving for V,

$$V = 0.3202 \text{ mol/h}$$

Then,

$$x_1 = \frac{0.6}{1 + V\left(\dfrac{9.0496}{4} - 1\right)} = 0.4273$$

$$y_1 = \frac{9.0496}{4} x_1 = 0.9667$$

$$x_2 = 1 - x_1 = 0.5727$$

$$y_2 = 1 - 0.9667 = 0.0333$$

$$L = 1 - V = 0.6798$$

With all flows known, the energy balance can now be solved to determine dQ/dt. Note, however, that to evaluate the feed enthalpy, its vapor fraction must be known. Clearly, since at 500 K the mixture bubble pressure of 5.523 bar is above the feed pressure of 5 bar, the feed will be a vapor–liquid mixture. To determine the phase distribution, another isothermal flash calculation must be performed, now at 5 bar:

$$\sum \frac{z_i(1 - K_i)}{1 + V'(K_i - 1)} = \frac{0.6\left(1 - \dfrac{9.0496}{5}\right)}{1 + V'\left(\dfrac{9.0496}{5} - 1\right)} + \frac{0.4\left(1 - \dfrac{0.2325}{5}\right)}{1 + V'\left(\dfrac{0.2325}{5} - 1\right)} = 0$$

Solving for V', we obtain

$$V' = 0.1354$$

which indicates that the feed is 13.54% vapor. As before, the phase compositions will be

$$x_1' = 0.5407$$

and

$$y_1' = \frac{9.0496}{5} x_1' = 0.9786$$

and

$$x_2' = 1 - x_1' = 0.4593$$

$$y_2' = 1 - y_1' = 0.0214$$

Assuming $dW/dt = 0$ and negligible potential and kinetic energy contributions, the open-system energy balance reduces to

$$\frac{dQ}{dt} = V\tilde{H}_V \,(500 \text{ K, 4 bar}) + L\tilde{H}_L \,(500 \text{ K, 4 bar})$$

$$- V'\tilde{H}_V \,(500 \text{ K, 5 bar}) - L'\tilde{H}_L \,(500 \text{ K, 5 bar})$$

If the state of the feed liquid is chosen as reference state, we have

$$\frac{dQ}{dt} = V\{\tilde{H}_V \,(500 \text{ K}, 4 \text{ bar}) - \tilde{H}_L \,(500 \text{ K}, 5 \text{ bar})\} + L\{\tilde{H}_L \,(500 \text{ K}, 4 \text{ bar})$$

$$- \tilde{H}_L \,(500 \text{ K}, 5 \text{ bar})\} - V'\{\tilde{H}_V \,(500 \text{ K}, 5 \text{ bar}) - \tilde{H}_L \,(500 \text{ K}, 5 \text{ bar})\}$$

Assuming negligible pressure effects, the equation reduces further to

$$\frac{dQ}{dt} = V[0.9667 \,\Delta H_{VL,1} \,(500 \text{ K}) + 0.0333 \,\Delta H_{VL,2} \,(500 \text{ K})]$$

$$- V'[0.9786 \,\Delta H_{VL,1} \,(500 \text{ K}) + 0.02136 \,\Delta H_{VL,2} \,(500 \text{ K})]$$

The heats of vaporization of the two species at 500°K can be estimated via Watson's equation using the normal boiling point, critical temperature, and heat of vaporization data for $n\text{-}C_8H_{18}$ and $n\text{-}C_{16}H_{34}$ given in Appendices 2 and 5:

$$\Delta H_{VL,1} \,(500 \text{ K}) = 34{,}940.8 \left(\frac{569.389 - 500}{569.389 - 398.828} \right)^{0.38}$$

$$= 24{,}826 \text{ J/mol}$$

$$\Delta H_{VL,2} \,(500 \text{ K}) = 51{,}142.9 \left(\frac{725 - 500}{725 - 559.956} \right)^{0.38}$$

$$= 57{,}534.3 \text{ J/mol}$$

With V and V' known, we finally obtain

$$\frac{dQ}{dt} = 4842 \text{ J/h for 1 mol/h feed}$$

The heat transfer rate is positive, indicating that heat must be transferred to the system to maintain the 500 K temperature. Essentially, this heat input is required to vaporize 0.177 mol/h octane and 0.0078 mol/h hexadecane.

The preceding example reconfirms the fact that specified values of the species flows as well as T and P are sufficient to fix the remaining stream properties, phase flows, phase compositions, and stream unit enthalpy. Alternate sets of independent variables could of course also have been selected. For instance, if the pressure and vapor fraction had been specified, the flash equation could be solved (iteratively) to determine the corresponding temperature. Note also that if more than two species had been present, the flash equation could not have been solved directly for V but instead would have required iteration. Since the balance equations are linear in the species flows, the nonlinearity is obviously introduced by the K factor relations. These are inherently quadratic in the species flows, since $y_i = K_i x_i$ is equivalent to

$$\frac{V_i}{\sum_i V_i} = K_i \frac{L_i}{\sum_i L_i}$$

where $V_i = y_i V$ and $L_i = L x_i$ are the species flow rates of the vapor and liquid streams, respectively.

In the previous example, the balances were completely decoupled. The next example illustrates coupling between the balance sets, a coupling that is introduced through the K factor relations.

Example 9.2 Suppose the pressure in the flash unit of Example 9.1 is reduced to 1 bar from a feed pressure of 4 bar but no heat is added to the process. Determine the outlet temperature and outlet stream flows assuming Raoult's law is again applicable.

Solution This problem differs from the preceding example only in the fact that the outlet temperature specification is replaced by the specification, $dQ/dt = 0$. Since the specification of one energy balance variable is replaced by another, it appears that the degree-of-freedom tabulation, Figure 9.2, is basically unchanged: the problem is correctly specified and the balances can be decoupled. This, however, is not the case because with the outlet temperature unknown, the K values cannot be calculated. Clearly, the relations

$$\frac{p_i(T)}{P} = \frac{y_i}{x_i}$$

require that T be included as a material balance variable, hence the material balance subproblem becomes underspecified by one. In the case of this *adiabatic* flash unit, the balances are coupled. Given T, the flash calculation can be performed to yield the phase flows. Given these flows, the energy balance equation can be solved to yield T. An iterative calculation will clearly be required.

The material balance subproblem can again be reduced to the solution of the flash equation:

$$\sum_i \frac{z_i(1 - K_i)}{1 + V(K_i - 1)} = \frac{0.6(1 - p_1(T))}{1 + V(p_1(T) - 1)} + \frac{0.4(1 - p_2(T))}{1 + V(p_2(T) - 1)} = 0 \quad (9.1)$$

where $K_i = p_i(T)$, because the total pressure is 1 bar. With V determined (for some assumed T), then

$$x_1 = \frac{0.6}{1 + V(p_1(T) - 1)} \qquad x_2 = 1 - x_1 \qquad (9.2)$$

$$y_1 = p_1(T)x_1 \qquad y_2 = 1 - y_1$$

The energy balance, expressed in terms of V', L' the feed vapor and liquid flows, and y'_i, x'_i, the corresponding phase compositions, takes the form

$$\frac{dQ}{dt} = V\tilde{H}_v \,(T, \, 1 \text{ bar}) + L\tilde{H}_L \,(T, \, 1 \text{ bar}) - V'\tilde{H}_v \,(500 \text{ K, } 4 \text{ bar})$$

$$- L'\tilde{H}_L \,(500 \text{ K, } 4 \text{ bar})$$

Using as reference state that of the inlet liquid and neglecting pressure effects, the energy balance reduces to

$$\frac{dQ}{dt} = V[\hat{H}_V\,(T,\ 1\ \text{bar}) - \hat{H}_L\,(500\ \text{K, 4 bar})]$$

$$+ L[\hat{H}_L\,(T,\ 1\ \text{bar}) - \hat{H}_L\,(500\ \text{K, 4 bar})]$$

$$- V'[\hat{H}_V\,(500\ \text{K, 4 bar}) - \hat{H}_L\,(500\ \text{K, 4 bar})]$$

$$= V \sum_i y_i\left[\Delta H_{VL,i}\,(1\ \text{bar}) + \int_{500\ \text{K}}^{T} C_{P_{Li}}\,dT\right] + L \sum_i x_i \int_{500\ \text{K}}^{T} C_{P_{Li}}\,dT$$

$$- V' \sum_i y_i'[H_{VL,i}\,(500\ \text{K})]$$

Using mean heat capacities $\overline{C}_{P_{Li}}$ for simplicity and recalling that $dQ/dt = 0$, the following form is obtained

$$V' \sum_i y_i'\,\Delta H_{VL,i}\,(500\ \text{K}) - V \sum_i y_i\,\Delta H_{VL,i}\,(1\ \text{bar})$$

$$= (T - 500)\left\{L \sum_i x_i\overline{C}_{P_{Li}} + V \sum_i y_i\overline{C}_{P_{Li}}\right\}$$

Solving for T and recognizing that $Nz_i = Lx_i + Vy_i$,

$$T = 500 + \frac{V' \sum_i y_i'\,H_{VL,i}\,(500\ \text{K}) - V \sum_i y_i\,H_{VL,i}\,(1\ \text{bar})}{N \sum_i z_i\overline{C}_{P_{Li}}}$$

Note that V and y_i are the only variables on the right-hand side of the above equation. For inlet conditions of 500 K and 4 bar and the given composition of octane and hexadecane, we have from Example 9.1 that

$$V = 0.3202\ \text{mol/h}$$
$$y_1 = 0.9667$$

Moreover,

$$\Delta H_{VL,1}\,(1\ \text{bar}) = 34{,}940.8\ \text{J/mol}$$
$$\Delta H_{VL,1}\,(500\ \text{K}) = 24{,}826.1\ \text{J/mol}$$
$$\Delta H_{VL,2}\,(1\ \text{bar}) = 51{,}142.9\ \text{J/mol}$$
$$\Delta H_{VL,2}\,(500\ \text{K}) = 57{,}534.3\ \text{J/mol}$$

and at 490 K, using the C_p correlations from the appendices,

$$\overline{C}_{P_{L1}} = 376.05 \quad \text{and} \quad \overline{C}_{P_{L2}} = 662.27$$

where both values are in J/mol·K.

With these parameters, the temperature equation reduces to

$$T = 516.92 - \frac{V[y_1(\Delta H_{VL,1} - \Delta H_{VL,2}) + \Delta H_{VL,2}]}{490.538} \tag{9.3}$$

Since V and y_1 are themselves functions of T through the flash equation, we have essentially reduced the problem to an implicit equation of the form

$$T = g(T)$$

which can be solved by successive substitution or Wegstein's method.

To begin the solution process, an initial estimate of T is required. This is best obtained by calculating the dew and bubble temperatures of the mixture at 1 bar and selecting an initial T within the two-phase range defined by those temperatures. In the present instance, the bubble temperature is about 417°K, and the dew temperature is 522 K. We can consequently choose a suitable temperature in that range as starting value for the iteration. For instance, the midpoint temperature $(T_d + T_b)/2$ is frequently a recommended choice.

The iteration process can proceed as follows. For given T, Eq. (9.1) can be solved for V, Eq. (9.2) for y_1 and then Eq. (9.3) for a new estimate of T. Convergence can be accelerated by applying Wegstein's method to T. Using this procedure the following sequence of iterates is obtained:

Iteration	T (K)	V	Y₁
0	490	0.6648	0.8465
1	466.20	0.5494	0.9358
2	473.45	0.5818	0.9148
3	473.73	0.5813	0.9139
4	473.73		

A plot of the iteration equations is shown in Figure 9.3. From the shape of the intersecting curves, it is evident that smooth convergence can be achieved even if only successive substitution is used. Note that further improvements in the accuracy of the solution could be obtained by updating the mean C_p values used in formulating the temperature equation.

The preceding example, which involves the classical adiabatic flash calculation, illustrates a nonlinear coupling between the material and the energy balance subproblems. In the two-species case, the flash equation can be solved directly for V and, using mean heat capacities, the energy balance equation can be partially solved for T, thereby facilitating the calculations. For more than two species, both the flash equation and the energy balance equation will require iterative solution.

As illustrated in the next example, even for two species, the sequential adiabatic flash calculations can often fail to converge, unless special safeguards are added to bound the iterations.

Example 9.3 An equimolar feed of benzene and toluene at 105°C and 1000 mm Hg is flashed adiabatically at 500 mm Hg. Determine the outlet phase flows, assuming Raoult's law is applicable.

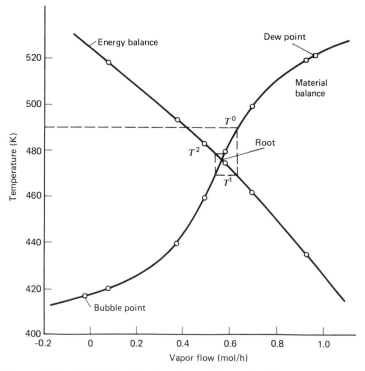

Figure 9.3 Plot of iteration functions, Example 9.2.

Solution At the inlet conditions, from Example 7.5, $V^1 = 0.4733$ and $y_1^1 = 0.6122$, where benzene is the first component. Using Watson's relation at 105°C,

$$\Delta H_{VL,1} \, (105°C) = 29,318.8 \text{ J/mol}$$

$$\Delta H_{VL,2} \, (105°C) = 33,798.8 \text{ J/mol}$$

while at 500 mm Hg,

$$\Delta H_{VL,1} \, (500 \text{ mm Hg}) = 31457.6 \text{ J/mol}$$

$$\Delta H_{VL,2} \, (500 \text{ mm Hg}) = 34293.4 \text{ J/mol}$$

At 500 mm Hg, the bubble and dew temperatures of the mixture can be calculated to be about 78.8 and 85.4°C, respectively. The liquid heat capacities at a mean temperature of 90°C are 146.09 and 175.34 J/mol · K, for benzene and toluene, respectively.

The terms arising in the temperature equation derived in Example 9.2 are

$$N \sum_i z_i \overline{C}_{P_{Li}} = 160.715 \text{ J/h} \cdot \text{K}$$

$$V' \sum y_i' \Delta H_{VL,i} \, (105°C) = 1.4699 \times 10^4 \text{ J/h}$$

Therefore, the temperature formula for this case becomes

$$T = 469.61K$$

$$- \frac{V[y_1(\Delta H_{VL,1} \, (500 \text{ mm}) - \Delta H_{VL,2} \, (500 \text{ mm})) + \Delta H_{VL,2} \, (500 \text{ mm})]}{160.715}$$

Suppose as initial estimate we use the average of the dew and bubble temperatures,

$$T^0 = \frac{78.8 + 85.4}{2} = 82.1°C = 355.25 \text{ K}$$

Following the calculation sequence of the previous example, the isothermal flash equation is first solved for V, and then y_1 is calculated. We obtain

$$V = 0.4924 \qquad \text{and} \qquad y_1 = 0.6174$$

Next, the temperature equation is evaluated with the result $T^1 = 369.91$ K. However, at this temperature, the flash equation yields $V = -99.9$, a clearly infeasible solution. As can be seen from a plot of the equations, Figure 9.4, the successive substitution calculation is very unstable in this case, because of the large difference in the relative slopes of the material and energy balance iteration equations. A temperature estimate either above or below the solution will successively generate a vapor fraction (using the material balance equation), a temperature (using the energy balance), and a subsequent vapor fraction which is considerably outside the feasible range, $0 \le V \le 1$. To achieve convergence, the temperature values generated must be restricted to be within the mixture dew and bubble point temperatures. Thus, if at any stage $T^n > T_d$, we set

$$T^n = \frac{T^{n-1} + T_d}{2}$$

Alternatively, if $T^n < T_b$, we set

$$T^n = \frac{T^{n-1} + T_b}{2}$$

For instance, in the present case, since 369.91 K is greater than the dew point, we should set

$$T^1 = \frac{355.25 + (85.4 + 273.15)}{2} = 356.9 \text{ K}$$

At this temperature, $V = 0.7330$ and $y^1 = 0.5607$. The temperature equation then yields $T^2 = 320.45$ K. Since this temperature is below the bubble point (78.8 + 273.15 = 351.95 K), the temperature is reset to

$$T^2 = \frac{351.95 + T^1}{2} = 354.43 \text{ K}$$

Convergence of the iterations can be enhanced if such bounded successive substitution is augmented by Wegstein's method. With Wegstein's method and the above

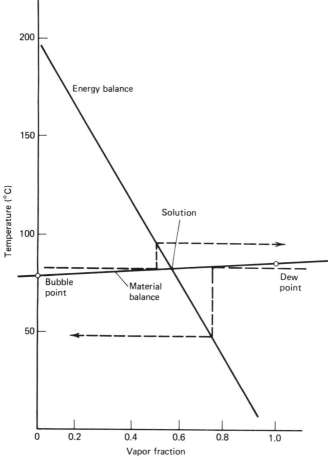

Figure 9.4 Plot of iteration functions, Example 9.3.

temperature bounding, the following iteration sequence is obtained:

	Iteration				
	0	**1**	**2**	**3**	**4**
T (K)	355.25	356.9	355.72	355.73	355.733

At the final temperature,

$$V = 0.5603$$
$$y_1 = 0.6015$$
$$x_1 = 0.3707$$
$$L = 0.4397$$

Note that in the previous example the converged temperature is only about 0.5 K from the initially guessed temperature. Yet repeated iterations were required because in this case the material balance equation was very sensitive to temperature, that is, it gave large change in V for small change in T. Situations such as these can be recognized by small differences between the mixture dew and bubble temperatures (6.6°K in Example 9.3) and are called *narrow boiling* mixtures. Mixtures with large differences between the dew and bubble temperatures, called *wide boiling* mixtures, usually will converge well with the method of Examples 9.2 and 9.3 because the magnitude of the slope of the material balance curve will be equal to or greater than the slope of the energy balance curve (see Figure 9.3). Recall from Chapter 5 that the convergence of calculations of the form $x = f(x)$ will be governed by the slope of $f(x)$ in the vicinity of the solutions. In the case of Examples 9.2 and 9.3, the implicit iteration $T = g(T)$ involves the energy balance and the isothermal flash calculation. The slope of the overall $f(T)$ can be shown to involve

$$\frac{dg}{dT} = \frac{\text{slope energy balance}}{\text{slope material balance}}$$

If $|(dg/dT)| > 1$, successive substitution will diverge (Example 9.3). If $|(dg/dT)| < 1$, convergence is likely (Example 9.2). The ratio of the energy and material balance slopes thus controls the convergence properties. The use of Wegstein's method helps to mitigate convergence problems but does not totally eliminate them. In very difficult cases, simultaneous numerical solution of the flash and energy balance equations may be the only choice.

9.1.2 The Multiunit Case

The analysis and solution of multiunit material and energy balance problems differ from those of the single-unit case in that a calculation sequence must be determined according to which the balance sets will be solved. In determining that sequence, we first consider the individual unit balance sets: either the complete combined set or the material balance subproblem or the energy balance subproblem. If none of these individual unit sets has degree of freedom zero, then we consider the use of the overall balances, either combined or separately. Finally, if even this option fails to yield a degree of freedom zero balance set, we select the unit with lowest degree of freedom and proceed with a partial solution carrying the corresponding number of unknowns. In any of these cases, iterative calculations may be required to solve any of the individual balance sets or to determine values of the carried unknown variables.

As noted in Section 8.6.2, the main conceptual difficulties in the above procedure arise when overall balances are used to replace one of the individual unit balances. First of all, we have to recognize that the reaction, heat, and work rates determined from overall balances are simply algebraic sums over all units in the flowsheet of the corresponding single-unit rates. Hence, the calculated value of an overall rate will fix the value of the corresponding rate of one of the units and thus will serve to reduce the degree of freedom of that unit. Secondly, the use of overall balances may reduce the number of independent species balances available for some subsequent single-unit solution (see Section 3.2.3). Of course, in all cases, once the overall balances are used, one of the individual unit balance sets will become dependent and must be excluded from the solution sequence. Finally, in constructing overall balances, it is important to recall that:

1. The overall set of reactions should be reduced to the largest independent subset.
2. The overall species balances should include species balances on all chemical species involved in the selected set of independent reactions, even if those species do not explicitly appear in the process input or output streams.
3. The overall energy balances should include all process input and output streams, all independent reactions, and all heat and work rates that may be specified.

These considerations are illustrated in the following series of examples.

Example 9.4   A process stream consisting of benzene vapor at 500°C and 2 bar is to be cooled to 200°C by producing steam at 50 bar, using the two-exchanger system shown in Figure 9.5. Boiler feed water at 75°C and 50 bar is first heated to saturation in exchanger 2. Saturated water is partially boiled in exchanger 1. The vapor–liquid mixture from exchanger 1 is mixed with the saturated water from heat exchanger 2 and the phases are separated in the knockout drum. If the flow of water through exchanger 1 is adjusted to be 12 times that through exchanger 2, determine the quality of steam produced in exchanger 1 and the mass of steam produced per mole of process stream.

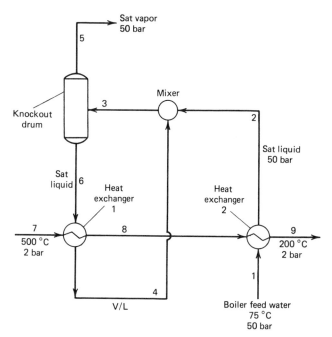

Figure 9.5   Flow diagram for Example 9.4, steam boiler loop.

Solution   This is a multiunit system in which each stream only involves a single species and in which there are no chemical reactions. Since the normal boiling point of benzene is 80°C, the process stream will be in the vapor phase throughout. From the problem statement, streams 1, 2, and 6 will be liquids, stream 5 will be vapor, and streams 3 and 4 will be vapor–liquid mixtures. Assuming that the pressure of all water streams will remain at 50 bar and the pressure of the benzene stream will be 2 bar, then for each single species stream, two variables will be necessary to define that stream: the total flow and the temperature. Recall, however, that in the case of single-species, two-phase streams, the temperature is fixed by the pressure specification. Hence, the two independent variables must be total flow and vapor fraction (or quality). The temperatures of streams 3 and 4 can therefore not be independent variables but must be replaced by the vapor fractions. Consequently, in the degree-of-freedom table shown in Figure 9.6, these two variables are listed explicitly. The knockout drum is simply a device in which a saturated vapor liquid is separated into a saturated vapor and a saturated liquid stream. Hence, the temperatures of the two output streams are known. All units are assumed to operate adiabatically, which means all have $dQ/dt = 0$.

When viewed from an overall balance point of view, the system appears as one composite heat exchanger in which the boiler feed water is converted to saturated steam while the process stream is cooled from 500°C to 200°C. The degree-of-freedom column for the overall balances can thus be assembled as if it were a heat exchanger. Note that in doing so we are implicitly making use of the simple material balance equation $F^1 = F^5$. Also, since $dQ/dt = 0$ for all units, $dQ/dt = 0$ for the overall balances since it is just the sum of the individual unit dQ/dt values. The degree-of-freedom table for this example is shown in Figure 9.6.

From the table, it is clear that the overall balance set should be solved first. As a consequence of these calculations, the flow of streams 1 and 9 will be known. Therefore, the degree of freedom of exchanger 2 becomes zero. From the exchanger 2 solution, the temperature of stream 8 becomes available. With the flow of stream 1 known, the flow of stream 6 can be calculated since the recycle ratio is specified. Therefore, the heat exchanger 1 degree of freedom is reduced to zero. Finally, either the knockout drum or the mixer balances can be solved, since either has degree of freedom zero. In either case, coupled solution is required. The degree-of-freedom updating is summarized below:

	Heat Exchanger 1	Heat Exchanger 2	Knockout Drum MB	Knockout Drum Combined	Mixer MB	Mixer Combined
	3	2	3	2	3	2
Overall	−2	−2	−1	−1	−1	−1
Heat Exchanger 2	−1					
Heat Exchanger 1			−1	−1	−1	−1

	Heat Exchanger 1	Heat Exchanger 2	Knockout Drum		Mixer		Process	Overall Combined
			MB	Combined	MB	Combined		
Number of variables								
Flow	2	2	3	3	3	3	5	2
Vapor fraction	1		1	1	2	2	2	
T	3	4		2		1	7	4
dQ/dt	1	1		1		1	4	1
							$\overline{18}$	
Number of balances								
Material			1	1	1	1	2	
Energy	1	1		1		1	4	1
Number of specifications								
$dQ/dt = 0$	1	1		1		1	4	1
Temperatures								
Stream 7	1						1	1
Stream 9		1					1	1
Stream 2		1					1	
Stream 1		1				1	1	
Stream 5				1			1	1
Stream 6	1			1			1	
Recycle ratio					1	1	1	1
							$\overline{17}$	
Degree of freedom	3	2	3	2	3	2	1	1
Basis							$\dfrac{-1}{0}$	

Figure 9.6 Degree-of-freedom table, Example 9.4.

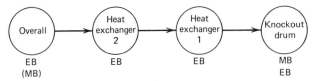

Figure 9.7 Calculation order, Example 9.4.

The calculation order is summarized in Figure 9.7.

Note that, since the implicit material balance $F^1 = F^5$ is used with the overall balances, both independent sets of material balances available in this problem are in effect being used. The balance $F^1 = F^5$ is simply the sum of the knockout drum material balance,

$$F^3 = F^5 + F^6$$

and the mixer balance,

$$F^3 = F^4 + F^2 = F^6 + F^1$$

Choosing as basis 100 mol/h of stream 7, the overall energy balance reduces to

$$\frac{dQ}{dt} = 0 = F^1[\hat{H}_V \text{ (sat, 50 bar)} - \hat{H}_L \text{ (75°C, 50 bar)}]$$

$$+ N^7[\tilde{H}_V \text{ (200°C, 2 bar)} - \tilde{H}_V \text{ (500°C, 2 bar)}]$$

Therefore,

$$F^1 = \frac{100 \text{ m/h} \int_{200°C}^{500°C} C_{p_V} \, dT}{(2794.2 - 317.9) \text{ kJ/kg}} = \frac{100 \times 4.8266 \times 10^4 \text{ J/h}}{2.4763 \times 10^6 \text{ J/kg}} = 1.949 \text{ kg/h}$$

where the water enthalpies are obtained from the steam tables and the C_p integral is evaluated using the vapor heat capacity equation for benzene.

The heat exchanger 2 energy balance is similarly

$$\frac{dQ}{dt} = F^1[\hat{H}_L \text{ (sat, 50 bar)} - \hat{H}_L \text{ (75°C, 50 bar)}]$$

$$+ N^7[\tilde{H}_V \text{ (200°C, 2 bar)} - \tilde{H}_V \text{ (T, 2 bar)}]$$

This reduces to

$$\int_{200°C}^{T} C_{p_V} \, dt = \frac{1.949(1154.5 - 317.9) \text{ kJ/h}}{100 \text{ mol/h}} = 16.31 \text{ kJ/mol}$$

An initial estimate of T can be obtained by using the average C_p of benzene vapor from the overall balances. Thus,

$$\bar{C}_p = \frac{\int_{200°C}^{500°C} C_p \, dT}{500 - 200} = \frac{4.8266 \times 10^4}{300} = 160.9 \text{ J/mol} \cdot \text{K}$$

and then,

$$160.9 \text{ J/mol} \cdot \text{K} \ (T - 200) = 16.31 \times 10^3 \text{ J/mol}$$
$$T = 200 + 101.3 = 301.3°\text{C} = 574.5 \text{ K}$$

This estimate can be improved by iteration using the C_p equation

$$T = 473.15 + \frac{1}{18.5868} \left\{ 1.6306 \times 10^4 + \frac{0.117439}{2} \times 10^{-1}(T^2 - 473.15^2) \right.$$

$$- \frac{0.127514}{3} \times 10^{-2}(T^3 - 473.15^3) + \frac{0.207984}{4} \times 10^{-5}(T^4 - 473.15^4)$$

$$\left. - \frac{0.105329}{5} \times 10^{-8}(T^5 - 473.15^5) \right\}$$

The successive iterates obtained using Wegstein's method are tabulated below:

			Iteration			
	0	**1**	**2**	**3**	**4**	**5**
T (K)	574.5	674.98	586.02	586.58	586.65	586.61

The intermediate process stream temperature is thus 313.46°C.
Continuing with the exchanger 1 balances, we have

$$0 = \frac{dQ}{dt} = N^7[\hat{H}_V \ (313.46°\text{C}, 2 \text{ bar}) - \hat{H}_V \ (500°\text{C}, 2 \text{ bar})]$$

$$+ 12F^1[\hat{H}_{\text{mix}} \ (\text{sat}, 50 \text{ bar}) - \hat{H}_L \ (\text{sat}, 50 \text{ bar})]$$

$$H_{\text{mix}} \ (\text{sat}, 50 \text{ bar}) = H_L \ (\text{sat}, 50 \text{ bar}) + \frac{100 \int_{586.61°\text{C}}^{773.15°\text{C}} C_{p_V} \ dT}{12(1.949)}$$

$$= 1154.5 \text{ kJ/kg} + \frac{100 \times 31.96}{12(1.949)} = 1291.15 \text{ kJ/kg}$$

The quality of the steam can then be determined from

$$1291.15 = X(2794.2) + (1 - X)1154.5$$
$$X = 0.08334$$

Finally, the knockout drum balances yield

$$F^3 = F^6 + F^5 = 12F^1 + F^1 = 13(1.949)$$

$$\frac{dQ}{dt} = 0 = F^5\hat{H}_V \ (\text{sat}, 50 \text{ bar}) + F^6\hat{H}_L \ (\text{sat}, 50 \text{ bar}) - F^3\hat{H}_{\text{mix}} \ (\text{sat}, 50 \text{ bar})$$

or

$$13(1.949)\hat{H}_{\text{mix}} = 1(1.949)\hat{H}_V\ (\text{sat}) + 12(1.949)\hat{H}_L\ (\text{sat})$$

$$\hat{H}_{\text{mix}} = \frac{1}{13}\hat{H}_V\ (\text{sat}) + \frac{12}{13}\hat{H}_L\ (\text{sat})$$

$$= X\hat{H}_V\ (\text{sat}) + (1 - X)\hat{H}_L\ (\text{sat})$$

Clearly,

$$X = \frac{1}{13} = 0.07692$$

Therefore, the flow of stream 3 is 25.337 kg/h and its vapor fraction or quality is 0.07692.

In all cases, the overall balances (energy or material) will always be equal to the sum of the corresponding individual unit balances. Assuming that the individual unit balances will always be linearly independent, the overall balance can always be substituted for one of the individual unit balances and the resulting balance sets will continue to be linearly independent. The sets of individual unit balances will be linearly independent in all cases save one, namely, the case of the internally recycling or *pump-around stream*. Consider the system shown in Figure 9.8(a). Suppose the internal streams 3 and 4 contain a species k which neither enters nor leaves the system through streams 1, 2, 5, and 6. If in unit 1, species k is involved in R reactions with stoichiometric coefficients σ_{ki}, then the species k material balance for unit 1 will be

$$N_k^3 = N_k^4 + \sum_{i=1}^{R} \sigma_{ki} r_i$$

Similarly if, in unit 2, species k is further involved in R' reactions with stoichiometric coefficients σ'_{kj} and rates r'_j, then the unit 2 species k balance will be

$$N_k^4 = N_k^3 + \sum_{j=1}^{R'} \sigma'_{kj} r'_j$$

Since species k neither enters nor leaves the system, the species k overall material balance is simply

$$0 = \sum \sigma'_{kj} r'_j + \sum \sigma_{ki} r_i$$

Thus, the overall balance reduces to an equation relating the reaction rates. Clearly, the overall balance can be substituted for one of the unit balances and linear independence will be maintained. However, suppose species k is nonreacting, then, since $r_i = 0$ and $r'_j = 0$ for all i and j, the two unit balances both take the form

$$N_k^4 = N_k^3$$

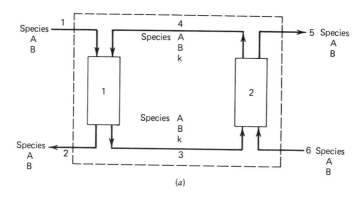

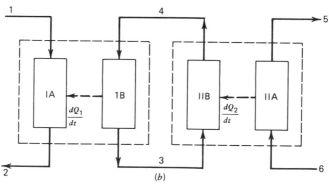

Figure 9.8 Internally recirculating flows: **(a)** totally recycled species; **(b)** recirculating heat transfer fluid.

and the species k overall material balance vanishes. Vanishing of the overall balance indicates that any nonreacting internally recycling species will give rise to a dependence between the unit material balances for that species. In fact, since the overall material balance for such a species will always vanish regardless of the number of units, M, through which the species passes, it is clear that the number of independent individual unit balances involving that species will always be one less than M. Hence, in developing the degree-of-freedom table for each totally recycled nonreacting species, the total number of independent material balances for that species entered in the process column must be reduced by one. Moreover, in developing the calculation sequence, account must be taken of the fact that the last unit to be solved involving that species will have one less independent material balance equation.

If the totally recycled species is part of a recirculating heat transfer fluid, then the energy balance for that heat transfer loop will similarly vanish. Consider the system of Figure 9.8(b) in which stream 1 is heated by transferring heat from stream 4 and stream 6 is cooled by transferring heat to stream 3. Clearly, $F^3 = F^4$, and if there is no loss of heat to the environment, the energy balance around unit IB reduces to

$$\frac{dQ_1}{dt} = F^3[\hat{H}(T^3) - \hat{H}(T^4)]$$

The energy balance around unit IIB similarly takes the form

$$\frac{dQ_2}{dt} = F^3[\hat{H}(T^4) - \hat{H}(T^3)]$$

The overall balance around the closed system consisting of IB, IIB, and the linking streams collapses to

$$\frac{dQ}{dt} = 0 = \frac{dQ_1}{dt} + \frac{dQ_2}{dt}$$

The two unit energy balances are dependent. On the other hand, the energy balances around the unit consisting of IA and IB take the form

$$\frac{dQ_I}{dt} = F^1[\hat{H}(T^2) - \hat{H}(T^1)] + F^3[\hat{H}(T^3) - \hat{H}(T^4)]$$

while that around the heat exchanger consisting of IIA and IIB is

$$\frac{dQ_{II}}{dt} = F^5[\hat{H}(T^5) - \hat{H}(T^6)] + F^3[\hat{H}(T^4) - \hat{H}(T^3)]$$

The overall balance for the entire system will be

$$\frac{dQ}{dt} = \frac{dQ_I}{dt} + \frac{dQ_{II}}{dt} = F^5[\hat{H}(T^5) - \hat{H}(T^6)] + F^1[\hat{H}(T^2) - \hat{H}(T^1)]$$

Because of the presence of the terms involving streams 1 and 5, the individual unit I and unit II balances clearly are independent and the overall balance does not vanish. The general conclusion that we can draw is that in the presence of an internally totally recirculating heat transfer fluid, the normal heat exchanger energy balances will remain independent. However, the individual unit energy balances involving only the heat transfer fluid will be dependent. The number of such independent energy balances will be one less than the number of units through which the fluid is circulated. These considerations are illustrated with the following example.

Example 9.5 Steam at 60 bar is used as heat transfer medium between a hot benzene vapor stream at 500°C and a cold methane gas stream at 100°C via the system shown in Figure 9.9. The methane is to be heated to 260°C and the benzene is to be cooled to 300°C. If the condensate leaving exchanger 1 is saturated water at 60 bar, the steam leaving exchanger 2 has a quality of 10% at 60 bar, and the benzene is fed at 200 mol/h, calculate the flow of methane which can be processed and the water circulation rates. Assume all units operate adiabatically and all water flows are at 60 bar.

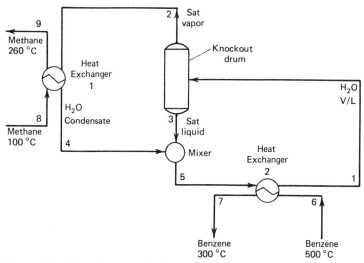

Figure 9.9 Flow diagram for Example 9.5, recirculating heat transfer system.

Solution As in Example 9.4, all streams only involve a single species. Stream 1 is the only two-phase stream. The water clearly forms an internally totally recirculating loop. Since the exchanger 1 and exchanger 2 material balances are trivial ($F^2 = F^4$ and $F^1 = F^5$), it is evident that the remaining two water material balances, that for the drum and that associated with the mixer, will be dependent. To verify this, note that the former

$$F^1 = F^2 + F^3$$

is identical to the latter,

$$F^5 = F^4 + F^3$$

if F^1 is replaced by F^5 and F^2 by F^4. The energy balances around each of the four units will, however, be independent.

Note that since the water is entirely recirculated, for overall balances, the system simply reduces to a heat exchanger in which heat is transferred adiabatically between the benzene and methane streams. The degree-of-freedom table for the system is shown in Figure 9.10. Note that only two temperatures are listed for the knockout drum and for heat exchanger 2 because the temperature is not an independent variable for stream 1. Instead, the vapor fraction is used. From the table, it is evident that the problem is correctly specified and that solution can be initiated with the overall balances. Solution of the overall energy balance will yield the flow of stream 8. This will reduce the exchanger 1 degree of freedom to zero. The exchanger 1 energy balance will serve to calculate the flow of stream 2 (and hence of stream 4 also). The knockout drum degree of freedom will be reduced to zero. The knockout drum material and energy balances will have to be solved in combined form and will yield the flow of streams 3 and 1. The mixer material balance is of course dependent, increasing its degree of freedom to 3. However,

	Heat Exchanger 1	Heat Exchanger 2	Knockout Drum MB	Knockout Drum Combined	Mixer MB	Mixer Combined	Process	Overall
Number of variables								
Flows	2	2	3	3	3	3	5	2
Vapor fraction		1	1	1			1	
T	4	3		2		3	8	4
dQ/dt	1	1		1		1	$\underline{4}$	1
							18	
Number of balances								
Material	1		1	1	1	1	1	
Energy	1	1		1		1	4	1
Number of specifications								
$dQ/dt = 0$	1	1		1		1	4	1
Vapor fraction	1	1	1	1			1	1
Temperatures								
Stream 6		1					1	1
Stream 7		1					1	1
Stream 8	1						1	1
Stream 9	1						1	1
Stream 2	1			1			1	
Stream 3				1		1	1	
Stream 4	1					1	1	1
Flow 6		1					1	
Degree of freedom	1	1	2	1	2	2	0	0

Figure 9.10   Degree-of-freedom table, Example 9.5.

with the flows of streams 3, 4, and 5 (since $F^5 = F^1$) known, it is in fact reduced to zero. Thus, either the mixer or the exchanger 2 energy balance can be solved to determine the temperature of stream 5. The calculation sequence is given in Figure 9.11.

Using the specified basis of 200 mol/h benzene, the overall energy balance can be written as

$$\frac{dQ}{dt} = 0 = N^8[\bar{H}_{CH_4} (260°C) - \bar{H}_{CH_4} (100°C)]$$

$$+ 200[\bar{H}_{C_6H_6} (300°C) - \bar{H}_{C_6H_6} (500°C)]$$

or

$$N^8 = \frac{200 \int_{300°C}^{500°C} C_{pC_6H_6} \, dT}{\int_{100°C}^{260°C} C_{pCH_4} \, dT}$$

$$= \frac{(3.4034 \times 10^4 \text{ J/mol}) \times 200 \text{ mol/h}}{6.9838 \times 10^3 \text{ J/mol}}$$

$$N^8 = 974.66 \text{ mol/h}$$

where the C_p integrals are calculated using the vapor heat capacity equations from the Appendix.

The solution continues with the exchanger 1 energy balance:

$$\frac{dQ}{dt} = 0 = N^8[\bar{H}_{CH_4} (260°C) - \bar{H}_{CH_4} (100°C)]$$

$$+ F^2[\hat{H}_L \text{ (sat, 60 bar)} - \hat{H}_V \text{ (sat, 60 bar)}]$$

Using water enthalpies from the steam tables, we can solve directly for F^2:

$$F^2 = \frac{974.66 \text{ mol/h } (6.9838 \times 10^3 \text{ J/mol})}{(2785.0 - 1213.7) \text{ J/g}} = 4332.0 \text{ g/h}$$

$$= 4.332 \text{ kg/h}$$

Next, the knockout drum balances are solved. The material balance is simply

$$F^1 = F^2 + F^3$$

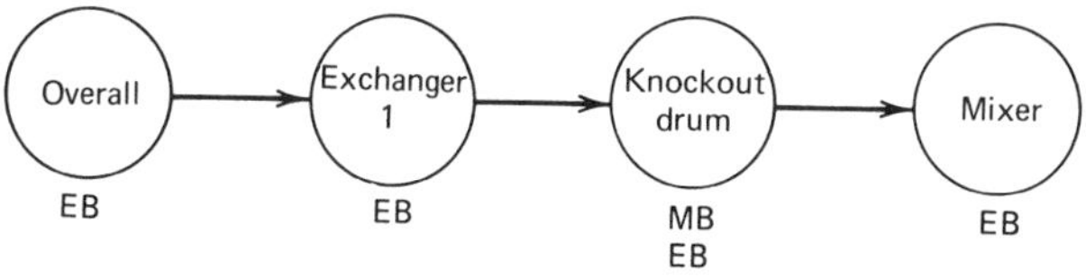

Figure 9.11 Calculation order, Example 9.5.

Using the state of stream 3 as reference, the energy balance will be given by

$$\frac{dQ}{dt} = F^2[\hat{H}_V \text{ (sat, 60 bar)} - \hat{H}_L \text{ (sat, 60 bar)}]$$
$$- F^1[\hat{H}_{mix} \text{ (60 bar)} - \hat{H}_L \text{ (sat, 60 bar)}]$$

Since the vapor fraction of stream 1 is known, the mixture enthalpy can be calculated from

$$\hat{H}_{mix} \text{ (sat, 60 bar)} = 0.1\hat{H}_V \text{ (sat, 60 bar)} + 0.9\hat{H}_L \text{ (sat, 60 bar)}$$

Therefore,

$$F^1 = \frac{F^2(2785.0 - 1213.7)}{0.1(2785.0 - 1213.7)} = 10F^2 = 43.32 \text{ kg/h}$$

The liquid stream will therefore be equal to

$$F^3 = F^1 - F^2 = 9(4.332) = 38.99 \text{ kg/h}$$

The solution is completed with the mixer energy balance:

$$\frac{dQ}{dt} = 0 = F^5\hat{H}_L \text{ } (T^5, \text{ 60 bar}) - F^3\hat{H}_L \text{ (sat, 60 bar)} - F^4\hat{H}_L \text{ (sat, 60 bar)}$$

or

$$\hat{H}_L \text{ } (T^5, \text{ 60 bar}) = \frac{9}{10}\hat{H}_L \text{ (sat, 60 bar)} + \frac{1}{10}\hat{H}_L \text{ (sat, 60 bar)}$$
$$= \hat{H}_L \text{ (sat, 60 bar)}$$

Obviously, T^5 will be the saturation temperature at 60 bar.

It is interesting to note that the above system of indirect heat transfer between streams 6 and 8 offers considerable flexibility in operation. By simply externally adding or removing saturated steam and saturated water from the closed loop, fluctuations in the flow or temperatures of either of the process streams can easily be accommodated. Using direct heat transfer, any fluctuations in one process stream will immediately be transmitted to the other stream as fluctuations in the heat transfer rate.

As the final example of this section, we consider a flowsheet in which sequencing with partial solution is necessary. The example, in addition, illustrates most of the typical situations in solving flowsheet balance problems: decoupled balances, coupled material and energy balances, iterative solution of the energy balance equation, as well as the use of overall balances.

Example 9.6 In a modification of the ammonia process flowsheet of Example 8.12, shown in Figure 9.12, three adiabatic reaction stages are used to produce ammonia via the reaction

$$N_2 + 3H_3 \rightleftharpoons 2NH_3$$

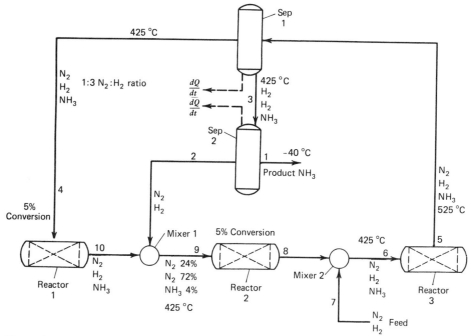

Figure 9.12 Flow diagram for Example 9.6, multistage ammonia synthesis loop.

The conversions of N_2 in the first two stages are held to 5%, and the stage inlet temperatures are maintained at 425°C. In the second and third stages, this is accomplished by mixing a colder side stream with the product stream of the previous reactor stage. The inlet temperature of the first reactor stage is adjusted to 425°C as part of the separator 1 operation. The outlet of the third reactor stage is limited to 525°C, and the liquid product NH_3 stream is refrigerated to −40°C. Assuming that the N_2 and H_2 are in stoichiometric proportions in stream 4, that the composition of stream 9 is 24% N_2, 72% H_2, and 4% NH_3 (all mol %), and that 0.2 mol of stream 7 is fed per 1 mol of stream 8, calculate all stream conditions and the heat duties on the two separators. Assume that mixing occurs adiabatically and that pressure effects can be neglected.

Solution With the exception of the product stream, all streams will be gaseous mixtures with two or more species. In each case, therefore, the stream temperature and species flows will serve as independent variables. The pressures are assumed to be specified. However, since pressure effects will be neglected in the energy balance equations, the precise values of the stream pressures need not be known. Assembly of the degree-of-freedom table is therefore quite routine. Note that, for overall balance purposes, the system appears as a nonadiabatic reactor involving a single chemical reaction with unknown conversion and a single input as well as a single output stream. In the table shown in Figure 9.13, we have grouped the variables into two sets: the variables associated with streams: flows and tempera-

	Reactor Stage 1		Reactor Stage 2		Reactor Stage 3		Mixer 1		Mixer 2		Separator 1		Separator 2		Process	Overall	
	MB	Com-bined	MB	Com-bined	MB	Com-bined	MB	Com-bined	MB	Com-bined	MB	Com-bined	MB	Com-bined		MB	Com-bined
Number of variables																	
Streams: flows,	6	8	6	8	6	8	8	11	8	11	9	12	6	9	36	3	5
T's	1	2	1	2	1	2		1		1		1		1	10	1	2
Units: dQ/dt, r's																	
Number of balances																	
Material	3	3	3	3	3	3	3	3	3	3	3	3	3	3	21	3	3
Energy		1		1		1		1		1		1		1	7		1
Number of specifications																	
Comp 9			2	2			2	2							2		
Ratio 4	1	1													1		
Ratio 7:8	1	1							1	1	1	1			2		
Conversions	1	1	1	1		1		1		1					5		
$dQ/dt = 0$				1		1		1		1		1		1			1
Temperatures																	
Stream 1												1		1	1		1
Stream 3								1				1		1	1		
Stream 4		1										1			1		
Stream 5						1									1		
Stream 6						1				1					1		
Stream 9				1											1		
Degree of freedom	2	2	1	1	4	3	3	4	4	5	5	5	3	4	1	1	2
Basis																	

Figure 9.13 Degree-of-freedom table, Example 9.6.

ture; and the variables associated with units: reaction rates and dQ/dt. (In general, the dW/dt's would also be added; but, since in the present problem they are all equal to zero, they are omitted.) From the table, it is evident that the problem is correctly specified once a basis is chosen.

The basis could either be chosen with the overall material balances or with the second reactor stage. We shall choose the latter because it allows us to also solve the energy balance for that unit. Thus, we first solve the stage 2 material balances, followed by the stage 2 energy balance, and obtain the flows and the temperature of stream 8 as well as the total flow of stream 9. As shown in Figure 9.14, the degree of freedom of mixers 1 and 2 are reduced to 3 and 1, respectively. However, with stream 8 known, the flow of stream 7 can be determined from the specified flow ratio. Hence, the overall material balances will have zero degree of freedom and can be solved to yield the flow of stream 1 and composition of stream 7. Figure 9.14 then indicates that mixer 2 can be solved in decoupled form, first the material and then the energy balances. With stream 6 calculated, reactor 3 can be solved in coupled form (degree of freedom of the material balance subproblem is 1). With the flows of stream 5 known, the degree of freedom of separator 1 becomes 2 both for the material and the combined subproblems. Since, with all stream temperatures known, the energy balances will only serve to calculate dQ/dt for that unit, we proceed by carrying two variables in the separator 1 material balances. Carrying two variables leaves stream 4 unaffected but reduces stream 3 from three to two unknowns. As a result, as shown in Figure 9.14, the degree of freedom of the separator 2 material balances is reduced to 1. This indicates that one of the two unknowns carried over from separator 1 will be determined. Hence, streams 4, 3 and 2 will all be expressed in terms of one unknown only. As a result, the mixer 1 and reactor 1 material balance subproblems are reduced to degree of freedom 1. Choosing the former, we can express stream 10 in terms of the carried unknown. At this point, all independent material balances will have been exhausted, since we have used the overall material balances. However, the unused conversion specification of reactor 1 remains available for the determination of the single carried unknown in terms of which streams 4 and 10 are expressed. With that variable determined, the reactor 1 energy balance can be solved for the reactor 1 outlet temperature. From Figure 9.14, the mixer 1 energy balance can next be solved, since the combined degree of freedom equals zero. With the temperature of stream 2 known, the separator 2 energy balance can be solved for the dQ/dt associated with that unit. The complete calculation sequence is thus as shown in Figure 9.15.

We begin the problem solution with the reactor stage 2 material balances, choosing as basis 100 mol/h of stream 9. From the conversion,

$$r = \frac{24(0.05)}{1} = 1.2$$

Therefore, the material balances yield

$$N^8 = (N_{N_2}^8, N_{H_2}^8, N_{NH_3}^8) = (22.8, 68.4, 6.4) \text{ mol/h}$$

	Reactor Stage 1		Reactor Stage 2		Reactor Stage 3		Mixer 1		Mixer 2		Separator 1		Separator 2		Overall	
	MB	Com‑bined	MB	Com‑bined	MB	Com‑bined	MB	Com‑bined	MB	Com‑bined	MB	Com‑bined	MB	Com‑bined	MB	Com‑bined
	2	2	1	1	4	3	3	4	4	5	5	5	3	4	1	2
Reactor 2 MB					−3	−3 (flows 6)										
Reactor 2 EB																
Overall MB															−1	−1 (flow 7)
Mixer 2 MB									−3	−3 (flows 8)						
Mixer 2 EB									−1	−1 (T)					−1	(T stream 7)
Reactor 3 MB/EB					−3	−3 (flows 5)			−1	(comp 7)						
Sep 1 MB											−3	−3				
Sep 2 MB							−1	−1 (carry 1)			−1		−1	−1 (carry 2)		
Mixer 1 MB	−1	−1 (carry 1)														
Reactor 1 conversion							−1									
Reactor 1 EB	−1						−1	−1 (T stream 10)								
Mixer 1 EB											−1					
Sep 2 EB													−1	(T stream 2)		
Sep 1 EB															−1	(flow 1)

Figure 9.14 Degree-of-freedom updating table, Example 9.6.

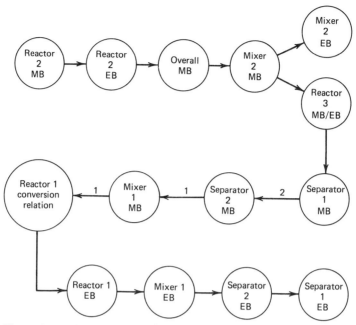

Figure 9.15 Calculation order for Example 9.6.

The energy balance requires the calculation of the heat of reaction. Using the state of the input stream as reference, the heat of reaction at 425°C is equal to $-26{,}907.1$ cal/mol.

The energy balance expressed in terms of the same reference state is

$$\frac{dQ}{dt} = 0 = r\Delta H_R\,(425°C) + \sum N_s^8 \int_{425°C}^{T} C_{p_s}\, dT$$

$$= 1.2(-26{,}907.1) + \int_{425°C}^{T} (22.8\,C_{p_{N_2}} + 68.4\,C_{p_{H_2}} + 6.4\,C_{p_{NH_3}})\, dT$$

Using average C_p values as initial estimate, we obtain

$$T^0 = \frac{26{,}907.1 \times 1.2}{(22.8\times7 + 68.4\times7 + 6.4\times9)} + 425°C = 471.4°C$$

Using the C_p functions, we can establish the iteration formula:

$$T = 698.15 + \frac{26{,}907.1 \times 1.2}{490.773} - \frac{1}{490.773}[0.5591(T^2 - 698.15^2) - 7.0155$$
$$\times 10^{-4}(T^3 - 698.15^3) + 4.3718 \times 10^{-7}(T^4 - 698.15^4) - 1.000$$
$$\times 10^{-10}(T^5 - 698.15^5)]$$

The iteration sequence obtained using Wegstein's method is listed below.

	Iteration Number			
	0	**1**	**2**	**3**
Temperature (°C)	471.4	461.44	465.77	465.73

Since stream 8 has been calculated, the specified flow ratio can be used to calculate stream 7:

$$N^7 = 0.2 N^8 = 19.52 \text{ mol/h}$$

The overall material balances are next solved, to yield $r = 4.88$ mol/h and $N^1 = 9.76$ mol/h. Consequently,

$$N^7_{N_2} = 4.88 \text{ mol/h}$$

$$N^7_{H_2} = 14.64 \text{ mol/h}$$

The mixer 2 material balances simply involve adding streams 7 and 8 to obtain the stream 6 flows,

$$N^6 = (27.68, 83.04, 6.4) \text{ mol/h}$$

Next, the mixer 2 energy balance is solved to obtain T^7. Using as reference state that of stream 6,

$$\frac{dQ}{dt} = 0 = -\sum N^8_s \int_{425°C}^{465.73°C} C_{p_s} \, dT - \sum N^7_s \int_{425°C}^{T} C_{p_s} \, dT$$

The value of the N^8 term is known from the reactor stage 2 energy balance calculations, hence the above equation reduces to

$$\int_{T}^{425°C} (4.88 C_{p_{N_2}} + 14.64 C_{p_{H_2}}) \, dT = 26{,}907.1 \times 1.2$$

Average C_p values will yield the initial estimate $T = 188.7°C$. This estimate can be improved using C_p values at the estimated mean temperature to obtain $T = 192.2°C$.

We continue with the stage 3 combined material and energy balances. Using as reference state that of the inlet stream, the energy balance is

$$\frac{dQ}{dt} = 0 = r \, \Delta H_R \, (425°C) + \int_{425°C}^{525°C} (N^5_{N_2} C_{p_{N_2}} + N^7_{H_2} C_{p_{H_2}} + N^5_{NH_3} C_{p_{NH_3}}) \, dT$$

Eliminating the species flows using the material balances, we obtain

$$0 = r(-26{,}907.1) + (27.68 - r)(741.7) + (83.04 - 3r)(702.13)$$
$$+ (6.4 + 2r)(1182.42)$$

Solving for r,

$$r = 3.154 \text{ mol/h}$$

Therefore,

$$N^5 = (24.525, 73.575, 12.71) \text{ mol/h}$$

Continuing with the separator 1 material balances, we merely express the two output streams in terms of two unknowns. Thus,

$$N^4 = \begin{cases} N^4_{N_2} \\ 3N^4_{N_2} \\ N^4_{NH_3} \end{cases} \quad \text{and} \quad \begin{aligned} N^3_{N_2} &= 24.525 - N^4_{N_2} \\ N^3_{H_2} &= 73.575 - 3N^4_{N_2} \\ N^3_{NH_3} &= 12.71 - N^4_{NH_3} \end{aligned}$$

The separator 2 material balances next allow $N^4_{NH_3}$ to be determined. The NH_3 balance yields

$$12.71 - N^4_{NH_3} = N^1_{NH_3} = 9.76$$

Therefore,

$$N^4_{NH_3} = 2.95$$

The remaining balances simply give

$$N^2_{N_2} = 24.525 - N^4_{N_2}$$

$$N^2_{H_2} = 73.575 - 3N^4_{N_2}$$

Finally, the mixer 1 material balances allow the stream 10 flows to be expressed in terms of $N^4_{N_2}$:

$$N^{10}_{N_2} = N^4_{N_2} - 0.525$$

$$N^{10}_{H_2} = 3N^4_{N_2} - 1.575$$

$$N^{10}_{NH_3} = 4.0$$

All material balances have now been exhausted. However, the conversion specification for reactor stage 1 can yet be used to determine the unknown $N^4_{N_2}$:

$$0.05 = \frac{N^4_{N_2} - N^{10}_{N_2}}{N^4_{N_2}} = \frac{0.525}{N^4_{N_2}}$$

or

$$N^4_{N_2} = 10.5 \text{ mol/h}$$

Therefore, the rate of reactor stage 1 will be

$$r = \frac{0.05(10.5)}{1} = 0.525$$

We now commence with the series of energy balance calculations. First, we use the reactor 1 energy balance to determine the temperature of stream 10. With the inlet stream conditions as reference state, the balance is

$$\frac{dQ}{dt} = 0 = -26{,}907.1(0.525) + \int_{425°C}^{T}(N_{N_2}^{10}C_{p_{N_2}} + N_{H_2}^{10}C_{p_{H_2}} + N_{NH_3}^{10}C_{p_{NH_3}})\, dT$$

Using average C_p values of 7, 7, and 11 cal/mol · K as initial estimate, we obtain

$$T^0 = \frac{26{,}907.1(0.525)}{9.975 \times 7 + 29.925 \times 7 + 4 \times 11} + 425°C$$

$$= 468.7°C$$

The iteration formula to be used for improving this estimate is

$$T = 698.15 + \frac{1}{278.35}[26{,}907.1(0.525) - 0.08137(T^2 - 698.15^2) + 8.8340$$

$$\times 10^{-5}(T^3 - 698.15^3) - 7.0686 \times 10^{-8}(T^4 - 698.15^4) + 1.9997$$

$$\times 10^{-11}(T^5 - 698.15^5)]$$

Successive iterates obtained using Wegstein's Method are shown below:

	Iteration Number			
	0	1	2	3
Temperature (°C)	468.7	466.51	466.90	466.90

With the stream 10 temperature known, the mixer 1 energy balance can be solved to obtain the stream 2 temperature. The energy balance with the stream 9 state as reference is

$$\frac{dQ}{dt} = 0 = -\sum N_s^2 \int_{425°C}^{T} C_{p_s}\, dT - \sum N_s^{10}\int_{425°C}^{466.9°C} C_{p_s}\, dT$$

The stream 10 term is just equal to the heat of reaction term in the reactor 1 energy balance. Thus, the equation reduces to

$$26{,}907.1(0.525) = \int_{T}^{425°C}(14.025\,C_{p_{N_2}} + 42.075\,C_{p_{H_2}})\, dT$$

The initial temperature estimate obtained using average C_p's of 7.0 cal/mol · K is

$$T = 389.03°C$$

The C_p values at the average temperature of 407°C, $C_{p_{H_2}} = 7.040$ cal/mol · K and $C_{p_{N_2}} = 7.312$ cal/mol · K, can be used to refine that estimate. The result is

$$T = 389.57°C$$

With this calculation, all stream flows and temperatures have been determined. To conclude the problem solution, we use the separator 2 and separator 1 energy balances to compute the heat transfer rates. The separator 2 energy balance with reference state corresponding to that of stream 3 is

$$\frac{dQ}{dt} = N^2(\bar{H}(T^2) - \bar{H}\ (425°C)) + N^1(\bar{H}_L\ (-40°C)$$

$$- \bar{H}_V\ (425°C))$$

$$= 0.525(-26907.1) - 9.76\left(\int_{-33.42°C}^{+425°C} C_{P_V}\ dT + \Delta H_{VL}\ (-33.42°C)\right.$$

$$\left. + \int_{-40°C}^{-33.42°C} C_{P_L}\ dT\right)$$

The N^2 term can be evaluated using the mixer 1 energy balance result. The enthalpy of NH_3 vapor at $425°C$ relative to that of the liquid at $-40°C$ is evaluated using the heat of vaporization of NH_3 at its normal boiling point of $-33.42°C$ and the vapor and liquid heat capacity equations. Thus,

$$\frac{dQ}{dt} = 0.525(-26907.1) - 9.76(4.499 \times 10^3 + 5581.03 + 526.84)$$

$$= -117.65\ \text{kcal/h}$$

The separator 1 energy balance with reference state corresponding to stream 3 is

$$\frac{dQ}{dt} = -\sum N_s^5 \int_{425°C}^{525°C} C_{P_s}\ dT$$

$$= -(24.525(741.7) + 73.575(702.13) + 12.71(1182.42))$$

$$= -84.878\ \text{kcal/h}$$

Evidently, this process has the potential of allowing large quantities of heat to be recovered for other useful purposes such as the generation of steam.

9.2 STRATEGY FOR MACHINE COMPUTATIONS

The solution of even modestly sized combined balance problems, such as Example 9.6 of the previous section, is certainly a tedious process but it is one that readily lends itself to computer implementation. In this section, we study the modifications which must be made to the two basic computer-oriented material balance solution strategies which were discussed in Section 5.2. It will become apparent that to employ the sequential modular strategy it is only necessary to add to each elementary material balance module the energy balance equation appropriate to that module. Recycle calculations can be performed as before by simply including the tear stream temperature or enthalpy flow as additional tear variable. However, the need to calculate stream enthalpy given its temperature, and vice versa, requires that these supplementary calculations be performed as part of the balancing program and that the extensive data required for this purpose be made available to the program. We discuss how this can be accomplished in a very rudimentary fashion.

We shall find that, in the presence of the energy balance equations, the simultaneous strategy requires the solution of sets of nonlinear equations. The nonlinearities arise both because of the nonlinear dependence of the enthalpy function on temperature and because of the bilinear enthalpy flow products which comprise the energy balance equation. Iterative solution methods will clearly be required.

As in Chapter 5, the discussion in this chapter is focused on the key ideas and techniques involved in the solution strategies and does not delve into the details of the actual coding required to implement the calculations. The reader interested in such details is referred to publications describing actual programs (J. D. Seader et al.,[3] M. K. Sood et al.,[4] or Gorczynski et al.,[5] for instance).

9.2.1 The Sequential Solution Strategy

The essential elements of the sequential modular flowsheet calculation strategy are:

1. The aggregation of the balance equations into modules corresponding to common flowsheet operations.
2. The formulation of the module balance equations so that output streams are calculated given the input streams and selected module parameters.
3. The solution of flowsheet balances by sequentially executing individual modules following the direction of the stream flows.
4. The determination of recycle streams by iterating on selected tear stream vectors.

In order to maintain this calculation strategy while also performing energy balances, it is in principle only necessary to expand the stream vector definition to include the stream temperature and phase fraction (or enthalpy) and to modify the modules so that each expanded output stream vector associated with a given module is determined from the module material and energy balance equations. As will become apparent, these modifications can be readily performed with our set of elementary modules, because the normal material balance specifications defined for these modules ensure that the material and energy balances are decoupled. As a consequence, the energy balance can be directly used to compute the selected output stream energy balance variables.

Consider a multiple inlet, multiple outlet unit with specified inlet flows and inlet stream enthalpies. In the absence of significant potential and kinetic energy

---

[3] J. D. Seader, W. D. Seider, and A. C. Pauls, *FLOWTRAN Simulation—An Introduction*, CACHE, Cambridge, MA, 1977.

[4] M. K. Sood, S. M. Clark, R. Khanna, and G. V. Reklaitis, *Linear Balancing Program, Users Manual*, Purdue Research Foundation, W. Lafayette, IN, 1980.

[5] E. W. Gorczynski, H. P. Hutchison, and A. R. M. Wajih, Development of a Modularly Organized Equation Oriented Process Simulator, *Computers and Chemical Engineering*, **3**, 353, (1979).

differences, the energy balance equation for such a system will take the form

$$\sum_{\substack{\text{outlet} \\ \text{streams } j}} N^j \bar{H}^j = \frac{dQ}{dt} - \frac{dW}{dt} + \sum_{\substack{\text{inlet} \\ \text{streams } i}} N^i \bar{H}^i$$

Alternatively, if we denote

$$\dot{H}^k \equiv N^k \bar{H}^k$$

to be the enthalpy flow of stream k, the balance reduces to the simpler form

$$\sum_{\substack{\text{outlet} \\ \text{streams } j}} \dot{H}^j = \frac{dQ}{dt} - \frac{dW}{dt} + \sum_{\substack{\text{inlet} \\ \text{streams } i}} \dot{H}^i$$

If the quantity $dQ/dt - dW/dt$ is treated as a fixed module parameter, then the above equation can be viewed as a balance on the conserved "species" $\dot{H}$. Thus, the enthalpy flow of each stream can be carried along as if it were another species flow. Of course, each module will only have a single enthalpy flow balance equation, hence the balance equation can only be used to calculate the enthalpy flow of one output stream. Clearly, as in the material balance case, additional module parameters must be defined to distribute the output enthalpy over the several output streams. However, if this is done, then, once the process input stream enthalpy flows are calculated, all other enthalpy flows can be determined in a sequential fashion without any need to deal with stream temperatures or phase fractions.

Difficulties do arise, however, if alternate specifications are desired, for instance, stream temperatures or phase fractions. If these more natural physical specifications are introduced, then the problem of interconversion between enthalpy, temperature, and phase fractions is resurrected and provisions must be made in the balance program to perform such interconversions. Finally, in some instances such as the adiabatic flash unit of Example 9.2, the material and energy balances do not decouple. In these cases, it is necessary to either treat the coupling through explicit constraint specifications or else to develop special modules which carry out the possibly iterative solution of the coupled balance set. We consider these matters in more detail in the next sections.

The Elementary Combined Balance Modules   In the material balance case, we found that the four elementary modules—mixer, separator, splitter, and stoichiometric reactor—were sufficient to model any flowsheet. In the energy balance case, we find it convenient to expand this set by the addition of an elementary heat/work transfer module. Let us consider each module and augment its balance set with the appropriate form of the enthalpy flow balance equation. In all cases, we assume that the stream pressures are always known.
Mixer   The stream mixer simply sums several input streams into a single output stream. The material balance model is

$$N_s^{\text{out}} = \sum_{\substack{\text{inlet} \\ \text{streams } i}} N_s^i \qquad s = 1, \ldots, S$$

Assuming that stream mixing takes place adiabatically and without the occurrence of work, the enthalpy flow balance reduces to

$$\dot{H}^{\text{out}} = \sum_{\substack{\text{inlet} \\ \text{streams } i}} \dot{H}_s^i$$

If all input stream vectors are known, both the species flows and the enthalpy flows, then clearly the output stream vector can immediately be calculated.

Separator The component separator separates a single input stream into several output streams. The species separation is specified via species split fractions, t_s^j, such that

$$N_s^j = t_s^j N_s^{\text{in}} \qquad s = 1, \ldots, S$$

for all output streams j, where, by definition,

$$\sum_j t_s^j = 1$$

In analogous fashion, we can introduce enthalpy separation fractions θ^j which will distribute the input enthalpy flow among the several output streams. Assuming adiabatic operation and no work, the module enthalpy flow balance reduces to

$$\dot{H}^j = \theta^j \dot{H}^{\text{in}}$$

for each output stream j, where $\sum \theta^j = 1$. Since in general the enthalpy flow of the output streams will be governed by temperature, phase, and species flows, the enthalpy split fractions are truly independent module parameters. With the t_s^j and θ^j as well as the input stream vector known, each output stream vector can be calculated.

Splitter The flow splitter separates a single input stream into several output streams while preserving stream compositions. The stream splitting is specified via stream split fractions t^j such that

$$N_s^j = t^j N_s^{\text{in}} \qquad s = 1, \ldots, S$$

for all output streams j. Again, by definition,

$$\sum_j t^j = 1$$

Since a flow splitter produces output streams of the same composition, the same phase, and the same temperature, it follows that the enthalpy split fractions will be the same as the flow split fractions. To confirm this, note that if all outlet streams are at the same temperature, say, T^1, and phase, then, by definition,

$$\dot{H}^j = N^j \bar{H}^j(T^1) = N^j \sum x_s^j \bar{H}_s(T^1)$$

The outlet streams all have the same composition and, from the material balances, these are equal to the feed composition. Therefore,

$$\dot{H}^j = N^j \bar{H}^{\text{in}}(T^1)$$
$$= t^j N^{\text{in}} \bar{H}^{\text{in}}(T^1)$$

Assuming adiabatic operation and $dW/dt = 0$, the enthalpy flow balance is

$$\dot{H}^{in} = \sum_j \dot{H}^j$$

Substituting for $\dot{H}^j$, we have

$$\dot{H}^{in} = N^{in}\tilde{H}^{in}(T^1) \sum_j t^j$$
$$= N^{in}\tilde{H}^{in}(T^1)$$

which implies that the inlet and outlet temperatures are the same. The elementary splitter is thus isothermal and the enthalpy flow will be split according to the material flow split fractions, or

$$\dot{H}^j = t^j\dot{H}^{in}$$

Note again that the elementary splitter incorporates the implicit assumption that the outlet flows and compositions are identical, that is,

$$x_s^j = x_s^{in} \qquad s = 1, \ldots, S$$

and

$$T^j = T^{in} \quad \text{for all streams } j$$

Reactor   The elementary reactor is a single-input/single-output device in which a single reaction occurs according to specified stoichiometry and key species conversion X_k. The material balance model is simply

$$N_s^{out} = N_s^{in} + \left(\frac{\sigma_s}{-\sigma_k}X_k\right)N_k^{in} \qquad s = 1, \ldots, S \quad \text{and} \quad s \neq k$$
$$N_k^{out} = (1 - X_k)N_k^{in} \qquad\qquad\qquad \text{for} \quad s = k$$

where k is the key species in terms of which the conversion is defined. Assuming again adiabatic operation and $dW/dt = 0$, the enthalpy flow balance expressed in heat of reaction form becomes

$$\dot{H}^{out} = \dot{H}^{in} + \Delta H_R^r X_k N_k^{in}$$

If the heat of reaction at reference state r is known, the conversion is specified, and the inlet stream vector is known, the outlet enthalpy flow can obviously be calculated. If alternatively, the total enthalpy form is used, that is, all enthalpy flows already incorporate the species heats of formation, then the balance collapses to

$$\dot{H}^{out} = \dot{H}^{in}$$

Heat/Work Module   This module is a single-input/single-output device in which a specified heat or work rate is transferred to a stream. The enthalpy flow balance is simply

$$\dot{H}^{out} = \dot{H}^{in} + \left(\frac{dQ}{dt} - \frac{dW}{dt}\right)$$

If the heat and work rates are specified and the input enthalpy flow is known, the output enthalpy can be immediately calculated. Note that the module material balances are trivial, namely,

$$N_s^{\text{out}} = N_s^{\text{in}}$$

This set of elementary modules is clearly sufficient to represent any flowsheet operation either singly or in combination. This feature is illustrated in the next example.

Example 9.7 Consider the flowsheet for the NH_3 plant shown in Figure 9.12 and represent it in terms of the combined elementary modules.

Solution The resulting modular flowsheet is shown in Figure 9.16. Note that separators 1 and 2 must be replaced by a combination of modules since they are nonadiabatic. Separator 1 is replaced by a heat exchanger which will cool the stream followed by an adiabatic separator. Separator 2 is also replaced by a heat exchanger which cools the stream (say, to the condition of stream 2), followed by an adiabatic separator. The ammonia stream is further cooled in exchanger C (to its liquid state,

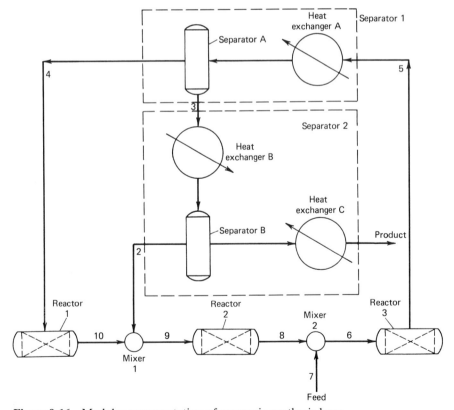

Figure 9.16 Modular representation of ammonia synthesis loop.

for instance). Note that separator 2 could equally well have been represented without exchanger C, if all of the heat required to reach states 1 and 2 is removed in exchanger B and the enthalpy split fraction of separator B is chosen so that stream 1 is a liquid. The choice of representation is dictated primarily by convenience in specifying the module parameters.

Solution Using Elementary Modules   Recall from Chapter 5 that the natural module specifications are those which allow the output streams to be calculated given the input streams. With combined modules, the natural specifications will be the following:

1. The species flows and enthalpy flows of all external process input streams.
2. The split fractions t_s^j for every species and for every output stream of all splitters and separators.
3. The enthalpy flow split fraction θ^j for every output stream of all separators.
4. The stoichiometry, conversion, and heat of reaction of all reactors.
5. The heat and work rates of all heat/work transfer modules.

In addition to these specifications, the enthalpy balances require that a reference state be selected for each unit or module. While in manual calculations we selected different reference states for each unit balance so as to simplify the solution of the individual unit balances, in computerized calculations it is preferable to choose a single reference state for use with all modules. Recall that the definition of the reference state will consist of the specification of a temperature and pressure as well as the selection of a reference phase for each species. A reference enthalpy is then assigned to each species at the reference state and these reference enthalpies will be used in the calculation of the process input stream enthalpy flows. The heats of reaction which will be supplied for all reactor modules must of course also be heats of reaction at the problem reference state. If the total enthalpy form of the balance equations is used, the reference state becomes the standard state at which the heats of formation are defined, and, in that case, heats of reaction are not required.

A problem in which all natural specifications are imposed is called an *unconstrained combined balance problem*. Such a problem will have all of the properties which we observed in the unconstrained material balance case, namely:

1. It involves only linear equations in the species and enthalpy flows.
2. It always has degree of freedom zero.
3. The order of calculations will always follow the direction of the process stream flows.
4. The number of required tear streams will always be equal to the number of essential mixers.

The number of essential mixers will of course be determined using the labeling procedure discussed in Section 5.2.1. The convergence calculations on the tear streams, the output streams of the essential mixers, will again be required except that the tear streams in the combined case must be defined to include the tear stream enthalpy flows.

Example 9.8   Identify the natural specifications for the flowsheet given in Figure 9.16 and determine the location of the convergence module.

Solution   To begin with, a reference state would have to be selected for the problem (e.g., 1 atm, 425°C, and all species in the gaseous phase). From Figure 9.16, it is next apparent that there is only one process input stream. Hence, the species flows and the enthalpy flow of stream 1 relative to the reference state must be specified. The conversion of N_2 in each reactor and the heat of reaction of the reaction

$$N_2 + 3H_2 \rightarrow 2NH_3$$

at the reference state must be supplied. The heat transfer rates of all three heat exchangers must be given and the species and enthalpy flow split fractions for separators A and B must be specified. The resulting problem will be an unconstrained combined balance problem.

To determine the number of essential mixers, we first develop the source stream list for each mixer. Mixer 1 has stream 6 as source stream, while mixer 2 has streams 9 and 7. Clearly, either mixer could be eliminated as nonessential and the remaining mixer will involve a self-loop. Consequently, either stream 6 or stream 9 could be used as tear stream. If stream 9 were used as tear, then the convergence module would be inserted as shown in Figure 9.17. An initial estimate would have to be provided of both the three species flows as well as of the enthalpy flow.

A noteworthy feature of the unconstrained problem is the total absence of concerns with phase distribution and temperature. The entire problem is solved in terms of enthalpy flows, and it is only after the solution is obtained that interconversion needs to be carried out. This convenience is however obtained at the price of a rather rigid and sometimes awkward set of specifications. For instance, it is often difficult to specify the enthalpy split of a separator. It is much more convenient to specify the outlet stream temperatures and phases. Similarly, it is often much more convenient to specify the desired outlet temperature of a heat exchanger than to specify the heat transfer rate. Such nonstandard specifications can, as in the material balance case, be accommodated by imposing them as constraint equations. Of course, for the problem to remain correctly specified, the number of constraint equations which are imposed must be exactly equal to the number of unspecified natural parameters. As before, a combined balance problem in which at least one natural specification is replaced by a constraint is called a *constrained* balance problem.

Solution of a constrained problem within the framework of the sequential modular strategy requires that a loop of iterations be imposed in addition to the

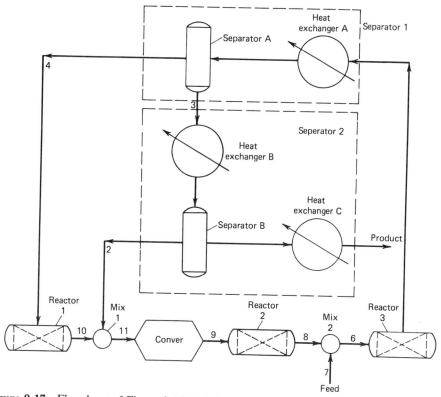

Figure 9.17 Flowsheet of Figure 9.16 with CONVER module inserted.

tear stream convergence calculations. As in the material balance case, the unspecified natural parameters are used as iteration variables whose values are adjusted until the constraint specifications are met. With each trial set of values of the unspecified natural parameters, all or part of the flowsheet tear streams must be converged. The converged stream vectors can then be used to evaluate the constraints and the resulting constraint values introduced into a suitable root-finding method in order to predict new values of the unspecified natural parameters. As in the material balance case, the calculations take the double loop form shown in Figure 5.31.

Example 9.9 Develop a sequential modular solution of the steam boiler loop of Example 9.4.

Solution As first step, a modular representation of the flowsheet is formulated as shown in Figure 9.18. Note that the two heat exchangers 1 and 2 are each separated into two elementary exchangers 1A, 1B and 2A, 2B, respectively. The knockout drum must be represented as a stream separator since the phase distribution of the two output streams are different: one is all liquid, the other is all vapor. Both the separator and the mixer are adiabatic units. The mixer clearly is

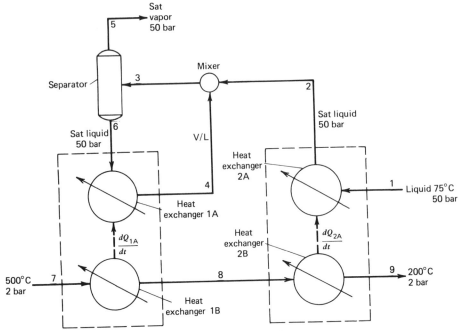

Figure 9.18   Modular representation of steam boiler loop.

an essential mixer since it is involved in a self-loop. Consequently, stream 3 should be used as tear stream and the convergence module should be inserted between the mixer and the splitter.

The natural specifications for the problem are: the species and enthalpy flows of streams 1 and 7; the heat transfer rates of all four elementary exchangers; and the flow and enthalpy split fractions of the separator-knockout drum. The natural specifications given in Example 9.4 are the species and enthalpy (temperature and pressure) of stream 7, the enthalpy (temperature and pressure) of stream 1, and the flow split for the separator (specifying the ratio of stream 1 to stream 6 is equivalent to specifying that of stream 6 to stream 3). Hence, a total of $10 - 4$, or six, natural parameters are unknown and must be supplemented by constraints. These constraints are the state of streams 2, 5, 6, and 9 or, equivalently, their molar enthalpies, as well as the requirements that

$$\frac{dQ_{1A}}{dt} = -\frac{dQ_{1B}}{dt}$$

and

$$\frac{dQ_{2A}}{dt} = -\frac{dQ_{2B}}{dt}$$

Since the number of constraints equals the number of unknown natural parameters, the problem is correctly specified. Sequential modular solution of the problem can proceed as follows.

Given an initial estimate of θ, N^1, dQ_{1A}/dt, and dQ_{2A}/dt, the heat transfer constraints are evaluated to determine dQ_{1B}/dt and dQ_{2B}/dt. Starting with a further initial guess of the tear stream variables N^3 and $\dot{H}^3$, the tear stream is converged. The four constraints

$$\dot{H}^5 = N^5 \bar{H}_V \text{ (sat, 50 bar)}$$

$$\dot{H}^6 = N^6 \bar{H}_L \text{ (sat, 50 bar)}$$

$$\dot{H}^2 = N^1 \bar{H}_L \text{ (sat, 50 bar)}$$

$$\dot{H}^9 = N^7 \bar{H}_V \text{ (200°C, 2 bar)}$$

are evaluated and the resulting values used to predict new estimates of θ, N^1, dQ_{1A}/dt, and dQ_{2A}/dt. The process is continued until convergence is attained.

In the preceding example, a problem whose solution was relatively straightforward using manual methods became rather complex to solve using the sequential modular strategy because of the rigid structure of the elementary modules. To avoid some of these complications, most sequential modular balancing programs include additional modules which admit different sets of natural specification parameters. Several typical examples are discussed in the next section.

Solution Using Supplementary Modules   The simplest supplementary modules are those which allow dQ/dt to be calculated given specifications of the outlet stream state and those which consist of commonly arising combinations of elementary modules.

Nonadiabatic Separator   In this variation of the species separator, the species split fractions and outlet stream states are specified (phase and temperature) and the module dQ/dt is computed. The material balances remain decoupled so that the outlet flows can be calculated first. Then, the outlet enthalpy flows can be evaluated since the outlet unit enthalpies are in effect specified. Finally, dQ/dt is calculated from the enthalpy flow balance. A further variation of this type of module is to allow specification of dQ/dt, to require specificaton of $J - 1$ of the J outlet stream states, and then to use the enthalpy flow balance to calculate the enthalpy flow of the remaining Jth outlet stream.

Nonadiabatic Reactor   The reactor conversion and outlet temperature and phase are specified and the module dQ/dt is computed. Alternatively, the reactor conversion and dQ/dt ($\neq 0$) is specified and the outlet stream vector is calculated. As a further variation, the reactor dQ/dt and outlet temperature and phase are specified but the conversion is not. In this case, the module balances remain linear but the material and energy balances are coupled since the reaction rate will be calculated so as to satisfy the enthalpy balance.

Rating Heat Exchanger   In this simple variation, the outlet temperature and phase are specified and the enthalpy balance is used to determine dQ/dt.

Two-Stream Exchanger   Because of the common occurrence of heat exchangers in which heat is transferred between two process streams, a combined module is often developed. Given both inlet stream vectors and the outlet state of one stream, the enthalpy flow of the second is calculated from the enthalpy flow balance assuming $dQ/dt = 0$ (alternatively, a dQ/dt could also be specified).

A further heat exchanger possibility is the case in which all four inlet and outlet states are specified as well as the flow of one stream. The enthalpy balance is then used to calculate the flow of the other stream. Although this module is also linear, it must be used with care since it violates the sequential modular structure by calculating the flow of an *input* stream. It should properly only be used if the calculated stream is a process input stream.

Single-Species Phase Separator   The single-species stream is split adiabatically and isothermally into separate phases saturated at the inlet temperature. In this module, the inlet stream vapor fraction is computed from the known enthalpy and species flows. The stream is then split according to the vapor fraction, and the saturated phase unit enthalpies are assigned to the appropriate outlet stream.

Isothermal Flash   This is a phase separation module in which the outlet liquid and vapor streams are assumed to be in equilibrium at specified temperature and pressure. The module will require equilibrium K values and will determine the outlet species flows as discussed in Example 9.1. In this case, the material balance sub-problem can be solved separately but is nonlinear and involves iteration. The energy balance is simply used to calculate dQ/dt.

General Flash   This is a phase separation module in which the outlet liquid and vapor streams are assumed to be in equilibrium at specified pressure. The unit dQ/dt is specified, but the equilibrium temperature must be calculated. The module will require equilibrium K values and will determine the outlet species flows as discussed in Example 9.2. The module balances are coupled and nonlinear and require iterative solution. Alternate specifications are of course possible. For instance, the temperature can be fixed and the pressure calculated instead.

By means of the alternate specification options permitted by an expanded library of modules, the need for constraint specifications can be considerably reduced. However, constraints that involve stream variables not incident to the module whose natural parameters they are intended to replace will nonetheless continue to require additional loops. We illustrate the simplifications made possible by using the above nonelementary modules with the next example.

Example 9.10   Develop a sequential modular solution of the steam boiler loop using the expanded module library.

Solution   The flowsheet given in Figure 9.18 can be modeled using a mixer, a single-species phase separator, and two of the two-stream exchanger modules. Exchanger 1 will be represented by the two-stream exchanger variant which will calculate the state of stream 4 given the flows of streams 6 and 7 and the states of streams 6, 7, and 8. Exchanger 2 will be represented by the two-stream exchanger variant in which the flow of stream 1 is calculated given the flow of stream 9 and the states of streams 1, 2, 8, and 9. The revised modular representation of the system will therefore coincide with the original flowsheet given in Figure 9.5.

Since the phases and saturation conditions are known for the single-species phase separator, that module is completely specified. The two heat exchanger modules each require the state (temperature) of stream 8. Hence, that unknown natural specification must be supplemented by a constraint, namely, the N^6-to-N^1 flow ratio of 12:1. The following sequential modular calculation sequence can consequently be used.

We make an initial estimate of the temperature of stream 8. Next, the exchanger 2 module is executed to give the flow of stream 1. An initial estimate is supplied of the tear stream, stream 3, and the loop consisting of the phase separator, exchanger 1, and the mixer is converged. The flow constraint ratio is then checked, a new estimate of the stream 8 temperature obtained, and the calculation sequence repeated. By using the above nonelementary modules, the constrained problem has been reduced from four constraints to one constraint.

In general, the constraints which are imposed on a sequential modular problem are of two types: *local* constraints and *global* constraints. Local constraints are relations among the stream variables of streams incident to the module whose missing natural specifications the constraints are intended to replace. Global constraints are relations which involve stream variables of streams that are not all incident to the same (underspecified) module. As shown in the previous example, local constraints can be eliminated by using modules which will admit the constraint specification as natural specifications. For instance, whereas for an elementary heat exchanger specification of the outlet temperature would have to be treated as a constraint, for the rating heat exchanger module the outlet temperature is a natural specification. However, because it is in general difficult to provide modules for all possible types of specification combinations, the occurrence of local constraints cannot be completely avoided.

From a computational point of view, local and global constraints typically require different computational strategies. If the local constraints of a module are equal in number to the number of missing natural specifications of that module, then these local constraints are always best treated by locally iterating on the unspecified natural parameters of the associated module. That is, local constraint satisfaction is enforced by iteration each time the underspecified module is encountered during the flowsheet calculation passes. Convergence calculations to satisfy local constraints thus become inner loops within the tear stream calculations. Global constraints, on the other hand, are usually more efficiently satisfied as outer loops to the tear stream calculations. In other words, the global constraints are evaluated and new values of the associated unspecified natural parameters are estimated after the tear stream calculations have converged. For instance, in Example 9.10, the tear stream was converged first and then the flow ratio constraint was converged as an outer loop. Of course, considerable judgment has to be used in deciding whether or not certain constraints should be treated via outer loops. A global constraint that involves streams separated by a path involving only a few modules, one of which is the underspecified module, might well be treated via an inner loop, that is, like a local constraint, if that path is short compared to the number of modules on the tear stream loop. Computational efficiency is the only criterion in such cases, and unfortunately the relative efficiency of alternate strategies can sometimes only be determined from the results of trial runs.

The next example illustrates the formulation of a sequential modular solution involving inner and outer constraint calculation loops.

Example 9.11   Develop a sequential modular solution of the ammonia synthesis problem as specified in Example 9.6 using the expanded module library.

Solution   From the information given in Example 9.6, all three reactors and the two mixers are adiabatic while the two separators are not. The conversions of reactor stages 1 and 2 are given, hence these two units can be modeled using elementary reactor modules. The third stage has a specified outlet temperature but unknown conversion. It can therefore be modeled using the nonadiabatic reactor version with specified dQ/dt and outlet state. The mixers can again be represented using elementary mixers. The two separators can be modeled using the nonadiabatic separator module in which the species split fractions and outlet states are specified and the unit dQ/dt is calculated. The modular representation of the flowsheet will therefore coincide with the process flowsheet given in Figure 9.12.

The missing natural specifications in this case will be:

1.   The three species split fractions for separator 1.
2.   The outlet temperature of stream 2 of separator 2.
3.   The feed temperature and composition.

The available specifications that will need to be imposed as constraints are:

1.   The temperature and two compositions of stream 9.
2.   The temperature of stream 6.
3.   The N_2-to-H_2 ratio in stream 4.
4.   The flow of stream 8.

The last specification comes about because the flow of stream 7 is chosen as basis and the ratio of stream 7 to stream 8 is fixed. The problem clearly is a constrained problem involving six constraints. Using stream 9 as tear stream, the following calculation sequence can be initiated. We begin with an initial estimate of all six missing natural specifications as well as an initial guess of the tear stream. Reactor stage 2 is executed. The mixer is executed using the estimated feed temperature and composition. The feed temperature is adjusted until the mixer outlet temperature of 425°C is met. Next, reactor stage 3 is executed, followed by separator 1 and reactor stage 1. Then, separator 2 and mixer 1 are executed. The temperature of stream 2 is readjusted and the separator 2 and mixer 1 calculations repeated until the specified mixer 1 outlet temperature of 425°C is met. This completes one iteration pass through the flowsheet. Further passes are executed until the tear stream flows have converged. At this point, the four constraints, namely, the N_2-to-H_2 ratio of stream 4, the flow of stream 7, and the two stream 9 compositions, are evaluated. Now, estimates of the feed composition and three separator 1 species split fractions are generated and another cycle of tear stream convergence calculations is initiated. The complete calculation sequence is summarized in Figure 9.19.

Note that the constraints on the temperature of stream 6 and on the temperature of stream 9 are treated as inner loops to the tear stream iteration. The former constraint is clearly a local constraint of mixer 2 and is satisfied by adjusting the inlet stream temperature. On the other hand, the stream 9 temperature constraint is satisfied by adjusting the temperature of stream 2 which is a natural parameter of the separator 2 module. The constraint is clearly not local to separator 2; how-

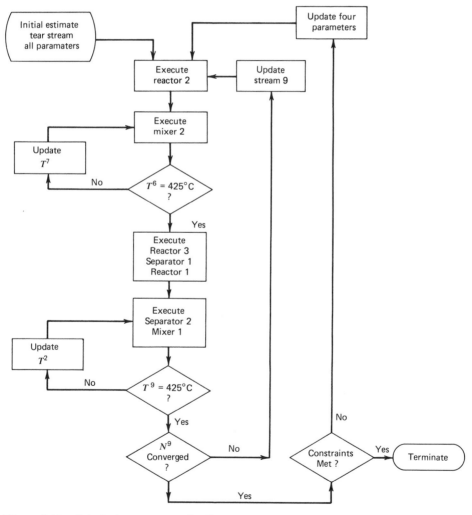

Figure 9.19 Calculation sequence for Example 9.11.

ever, because of the close proximity of the module and its associated constraint, inner loop adjustment is appropriate.

The constraint specifying the flow of stream 7, although apparently local to mixer 2, will clearly not be influenced by the composition of stream 1. Instead, it can be expected to be affected by the split fractions of separator 1. The composition of stream 9, on the other hand, will certainly be affected by the composition of stream 1 as well as by the split fractions of separator 1. Finally, the N_2-to-H_2 ratio of stream 4 is clearly a local constraint to separator 1 and thus could be treated via an inner loop. However, since separator 1 involves three unknown split fractions, it is more convenient to treat all four remaining constraints simultaneously via an outer loop.

One of the immediate consequences of introducing specification constraints and nonelementary models is the necessity for temperature–phase–enthalpy interconversions to be repeatedly carried out during the flowsheet iterations. It no longer suffices to determine the inlet stream enthalpy flows, to solve the balance problem, and then to determine the stream temperatures and phase distributions once all enthalpy flows have been calculated. Rather, interconversion calculations may be required with each module calculation. Because of the increased frequency of these thermodynamic calculations, it becomes necessary to make provisions in the flowsheet balancing program for automated enthalpy calculation and interconversion calculations. In the next section, we briefly outline the minimum requirements of such a system.

Automated Property Estimation   In order to automate balance calculations, four basic services must be provided by a property estimation support program:

1. Calculation of stream enthalpy given the state of the stream.
2. Evaluation of heats of reaction.
3. Calculation of the temperature (and phase distribution) of a stream given its composition and enthalpy.
4. Storage and retrieval of the information required to carry out services 1 through 3.

Let us consider each of these items, restricting our attention to simple gas–liquid systems and assuming that stream pressures are always specified.
Stream Enthalpy Evaluations   In order to evaluate the enthalpy of a stream given its state, we need to calculate the enthalpy of each species relative to some selected reference state and to combine the species enthalpies into a mixture enthalpy. To evaluate species enthalpies, the following is required:

1. Vapor heat capacity equation.
2. Liquid heat capacity equation.
3. Heat of vaporization.

Given these data, the evaluation of species enthalpy will, as discussed in Chapter 7, require integration of the C_p equations and may involve calculation of the heat of vaporization away from the normal boiling point. The latter calculation would probably be carried out using Watson's relation. Consequently, it will require knowledge of the normal boiling point temperature, the critical temperature, the heat of vaporization at the normal boiling point, and (possibly) the exponent of Watson's correlation. In general, the specification of the state may not include the phase distribution. Hence, in the case of vapor–liquid streams, an isothermal flash calculation may be required to determine the vapor fraction. The reader will recall that this calculation makes use of K values. In the simple Raoult's law case we have considered, vapor pressure equations will thus be needed for each species. For situations in which Raoult's law is not applicable, additional data and K value estimation procedures will be necessary.

Once the species enthalpies are calculated, these must be combined to determine a mixture enthalpy. In this text, we have assumed that all mixtures are ideal, but this need not always be the case. Thus, more complex mixing rules than a mole fraction average or introduction of heats of mixing may be necessary.

Heat of Reaction Evaluation   In order to evaluate the heat of a reaction at a given state, we need to calculate the standard heat of reaction and then to possibly correct that value to the given state. The data required for this purpose are the heats of formation of each species, the reaction stoichiometry, and the usual heat capacities and heats of vaporization used in the enthalpy updating.

Stream Temperature Calculation   Calculation of the stream temperature, given the composition, pressure, and enthalpy, requires first of all knowledge of the phase of the stream. If the stream is an ideal mixture of either gases or liquids, the coefficients of the species C_p relations can be weighted by the species mole fractions and combined to yield a temperature function. This temperature function must then be solved using a root-finding method to determine the temperature. If the stream is not known to be either a liquid or a gas in advance, then the dew and bubble point temperatures must be calculated and the enthalpies at these conditions evaluated. If the specified enthalpy falls between these enthalpies, an adiabatic flash calculation must be performed using the known stream enthalpy and composition as that of the input stream and the pressure as that of the output streams. These calculations, in addition to requiring enthalpy calculations, again make use of the K values.

Properties Data Base   The preceding review of essential property calculations indicates that for each species the following minimum information will be required:

1. Standard heat of formation.
2. Identity of the phase in the standard state (gas, liquid, or solid).
3. Critical temperature.
4. Normal boiling point temperature.
5. Heat of vaporization at the normal boiling point.
6. Watson's relation exponent.
7. The five coefficients of the vapor heat capacity equation.
8. The four coefficients of the liquid heat capacity equation.
9. The three (or more) coefficients of the vapor pressure equation.

Additional useful species data might be the species molecular weight, melting point, heat of fusion, coefficients of the solid heat capacity equation, coefficients of a real gas equation of state, and so on. If the rudimentary enthalpy and temperature calculations reviewed earlier are to be performed via special routines developed for this purpose and accessible to the balancing program, then the above set of data items must be entered into the program for each species involved in a flowsheet problem. An efficient way of avoiding tedious and repetitious manual entry of such data is to maintain the data in a permanent file stored in computer memory or on auxiliary storage devices.

In simplest form, such a properties data file can be structured as a two-dimensional array which consists of a row of entries for each chemical species and a column for each coefficient type, as shown in Figure 9.20. The rows would be numbered, with each row assigned to a particular species, and a directory would be prepared which would indicate this correspondence. Data would be stored for all species commonly encountered by the users of the balancing program. For instance, the data given in Appendices 3 through 7 of this text for some 180 species is extracted from the FLOWTRAN data base developed by the Monsanto Company. The chemicals listed presumably are all species of concern to the staff of that company in its operations. To access such a data bank, the user will simply list the directory numbers of the species involved in his problem, and, via suitable coding, the corresponding data base rows would be transferred from storage for use in the balancing program.

In summary, a rudimentary property estimation system will consist of a properties data base and of a library of calculation routines for evaluating enthalpies, heats of reaction, dew points, and bubble points as well as for executing isothermal and adiabatic flash calculations. The basic sequential modular strategy is quite flexible in accommodating modules of considerable degree of complexity. Flowsheeting systems that contain modules which go beyond simple balance calculations certainly do possess property estimation programs whose capabilities extend considerably beyond the minimum capabilities we have discussed. Typically, such systems will accommodate nonideal vapor–liquid mixtures and a variety of K value estimation methods and will contain semiempirical methods for calculating properties of complex petroleum mixtures. Despite these embellishments and extensions, the basic structure of such systems follows that of the simple prototype outlined in this section. For more details describing physical properties estimation, the reader is referred to program manuals by J. D. Seader et al.[6] and B. W. Overturf[7] and the text by Reid et al.[8]

---

[6] J. D. Seader, W. D. Seider, and A. C. Pauls, *FLOWTRAN Simulation—An Introduction*, CACHE Corporation, 1977.

[7] B. W. Overturf, *A Physical Properties Package for Computer Aided Process Design*, Purdue Research Foundation, W. Lafayette, IN, 1979.

[8] R. C. Reid, J. M. Prausnitz, and T. K. Sherwood, *Properties of Gases and Liquids*, McGraw-Hill, New York, 1977.

	1	2 3	4	5	6	7	8	9	10	11
ID	COMPONENT		FORMULA	CRITICAL TEMPERATURE (DEG R) TC	CRITICAL PRESSURE (PSIA) PC	CRITICAL VOLUME (CU.FT./MOLE) UC	CRITICAL COMP. FACTOR ZC	NORMAL BOILING (DEG R) TBP	MOLECULAR WEIGHT MW	ACENTRIC FACTOR W
1	NITROGEN		N2	2.2720E+02	4.9300E+02	1.4340E+00	2.9000E-01	1.3930E+02	2.8010E+01	2.0600E-02
2	OXYGEN		O2	2.7830E+02	7.3190E+02	1.1760E+00	2.8800E-01	1.6230E+02	3.2000E+01	2.9900E-02
3	HYDROGEN		H2	5.9760E+01	1.8810E+02	1.0410E+00	3.0500E-01	3.6700E+01	2.0160E+00	0
4	CARBON MONOXIDE		CO	2.3920E+02	5.0720E+02	1.4910E+00	2.9500E-01	1.4700E+02	2.8010E+01	-6.7000E-03
5	CARBON DIOXIDE		CO2	5.4760E+02	1.0700E+03	1.5050E+00	2.7400E-01	3.5050E+02	4.4010E+01	1.7680E-01
6	CARBONYL SULFIDE		COS	6.7500E+02	8.5240E+02	2.2400E+00	2.6000E-01	4.0120E+02	6.0070E+01	9.9000E-02
7	CARBON DISULFIDE		CS2	9.3600E+02	1.1460E+03	2.7180E+00	2.9300E-01	5.7490E+02	7.6130E+01	1.1500E-01
8	SULFUR DIOXIDE		SO2	7.7540E+02	1.1430E+03	1.9540E+00	2.6800E-01	4.7360E+02	6.4060E+01	2.4200E-01
9	HYDROGEN SULFIDE		H2S	6.7200E+02	1.2960E+03	1.5780E+00	2.8400E-01	3.8310E+02	3.4080E+01	8.6800E-02
10	AMMONIA		NH3	7.3010E+02	1.6360E+03	1.1620E+00	2.4200E-01	4.3150E+02	1.7030E+01	2.5000E-01
11	HYDROGEN CHLORIDE		HCL	5.8430E+02	1.2050E+03	1.2980E+00	2.5000E-01	3.3860E+02	3.6470E+01	1.2000E-01
12	WATER		H2O	1.1650E+03	3.2080E+03	9.1490E-01	2.3480E-01	6.7170E+02	1.8020E+01	3.4400E-01
13	METHANE		CH4	3.4310E+02	6.7200E+02	1.5860E+00	2.8800E-01	2.0100E+02	1.6040E+01	8.0000E-03
14	ETHYLENE		C2H4	5.0830E+02	7.3040E+02	2.0660E+00	2.7600E-01	3.0500E+02	2.8050E+01	9.4900E-02
15	ETHANE		C2H6	5.4970E+02	7.0830E+02	2.3710E+00	2.8500E-01	3.3220E+02	3.0070E+01	1.0640E-01

	12	13	14	15	16	17	18
ID	COMPONENT	REDLICH-KWONG CONSTANTS RKA	RKB	VAPOR PRESSURE CONSTANTS AVP	BVP	CVP	VAPOR PRESSURE TEMPERATURE LIMITS (DEG R) LTVP	UTVP
1	NITROGEN	7.7800E+04	4.2870E-01	4.7810E+00	4.6020E+02	-1.1880E+01	9.7200E+01	1.6200E+02
2	OXYGEN	8.6900E+04	4.5370E-01	4.9780E+00	5.7420E+02	-1.1610E+01	1.1340E+02	1.8000E+02
3	HYDROGEN	7.2300E+03	2.9550E-01	4.2070E+00	1.2890E+02	5.7420E+00	2.5200E+01	4.5000E+01
4	CARBON MONOXIDE	8.6800E+04	4.4350E-01	4.5260E+00	4.1450E+02	-2.3670E+01	1.1340E+02	1.9440E+02
5	CARBON DIOXIDE	3.2300E+05	4.7610E-01	8.0970E+00	2.4260E+03	-2.8800E+01	2.7720E+02	3.6720E+02
6	CARBONYL SULFIDE	6.8400E+05	7.3670E-01	4.5270E+00	4.1450E+02	-2.3670E+01	1.1340E+02	1.9440E+02
7	CARBON DISULFIDE	1.3400E+06	8.0640E-01	5.2280E+00	2.1040E+03	-5.6520E+01	4.1040E+02	6.1560E+02
8	SULFUR DIOXIDE	4.2100E+05	6.3090E-01	5.5590E+00	1.8000E+03	-6.4750E+01	3.5100E+02	5.0400E+02
9	HYDROGEN SULFIDE	4.4400E+05	4.8210E-01	5.2800E+00	1.3830E+03	-4.6910E+01	3.2220E+02	4.1400E+02
10	AMMONIA	4.3370E+05	4.1520E-01	5.6470E+00	1.6670E+03	-5.8630E+01	3.4200E+02	4.6980E+02
11	HYDROGEN CHLORIDE	3.3730E+05	4.5100E-01	5.4540E+00	1.3400E+03	-2.6010E+01	3.6000E+02	4.4670E+02
12	WATER	7.1090E+05	3.3760E-01	6.2360E+00	2.9830E+03	-3.0300E+01	4.9200E+02	1.1650E+03
13	METHANE	1.6100E+05	4.7840E-01	4.8980E+00	7.0190E+02	-1.2890E+01	1.6740E+02	2.1600E+02
14	ETHYLENE	3.9300E+05	6.4740E-01	5.0340E+00	1.0530E+03	-3.2670E+01	2.1600E+02	3.2760E+02
15	ETHANE	4.9300E+05	7.2200E-01	5.0890E+00	1.1820E+03	-3.0890E+01	2.3400E+02	3.5800E+02

Figure 9.20 Structure of a typical physical properties data base array.

(*Figure continues on following pages.*)

		19	20	21	22	23	24	25
		FREE ENERGY OF FORM. (BTU/MOLE)	HEAT OF OF FORM. (BTU/MOLE)	HEAT OF OF COMB. (BTU/MOLE)	VAPOR PHASE HEAT CAPACITY COEFFICIENTS			
ID	COMPONENT	GF	HF	HCOMB	ALPHA	BETA	GAMMA	GOUT
1	NITROGEN	-0	0	-0	6.8300E+00	5.0000E-04	0	-3.9800E+04
2	OXYGEN	-0	0	-0	7.1600E+00	5.5600E-04	0	-1.2960E+05
3	HYDROGEN	0	0	-1.2300E+05	6.4240E+00	5.8000E-04	2.4000E-08	-0
4	CARBON MONOXIDE	-5.9060E+04	-4.7550E+04	-1.2180E+05	6.7900E+00	5.4400E-04	0	-3.5640E+04
5	CARBON DIOXIDE	-1.6970E+05	-1.6930E+05	0	1.0570E+01	1.1670E-03	0	-6.6700E+05
6	CARBONYL SULFIDE	-7.1260E+04	-5.9540E+04	-2.3490E+05	5.6290E+00	1.0590E-02	-5.7130E-06	-0
7	CARBON DISULFIDE	2.8780E+04	5.0360E+04	-4.7500E+05	1.4450E+01	8.8900E-04	0	-5.8320E+05
8	SULFUR DIOXIDE	-1.2920E+05	-1.2770E+05	-0	1.1040E+01	1.0440E-03	0	-5.9160E+05
9	HYDROGEN SULFIDE	-1.4210E+04	-8.6670E+04	-2.4200E+05	7.8100E+00	1.6440E-03	0	-1.4900E+05
10	AMMONIA	-6.9480E+03	-1.9660E+04	0	7.1100E+00	3.3330E-03	0	-1.1990E+05
11	HYDROGEN CHLORIDE	-4.0990E+04	-3.9710E+04	0	6.2700E+00	6.8890E-04	-0	9.7200E+04
12	WATER	-1.0200E+05	-1.2300E+05	0	6.9600E+00	1.9240E-03	-1.4910E-07	0
13	METHANE	-2.1850E+04	-3.2200E+04	-3.7940E+05	3.3810E+00	1.0020E-02	-1.3300E-06	0
14	ETHYLENE	2.9300E+04	2.2490E+04	-5.9690E+05	2.8300E+00	1.5890E-02	-2.6900E-06	0
15	ETHANE	-1.4170E+04	-3.6420E+05	-6.6310E+05	2.2470E+00	2.1220E-02	-3.4100E-06	0

		26	27	28	29	30
		VAPOR PHASE HEAT CAPACITY TEMPERATURE LIMITS (DEG R)		LIQUID MOLAR VOLUME CONSTANTS		
ID	COMPONENT	LTCP	UTCP	VLIQ	D1	D2
1	NITROGEN	4.9200E+02	5.4000E+03	8.4800E-01	-0	-0
2	OXYGEN	4.9200E+02	5.4000E+03	4.5400E-01	-0	-0
3	HYDROGEN	4.9200E+02	5.4000E+03	4.9600E-01	-0	-0
4	CARBON MONOXIDE	4.9200E+02	4.5000E+03	5.6300E-01	-0	-0
5	CARBON DIOXIDE	4.9200E+02	2.7000E+03	7.0400E-01	-0	-0
6	CARBONYL SULFIDE	4.9200E+02	2.7000E+03	7.5600E-01	-0	-0
7	CARBON DISULFIDE	4.9200E+02	3.2400E+03	9.4300E-01	-0	-0
8	SULFUR DIOXIDE	4.9200E+02	3.6000E+03	7.2300E-01	-0	-0
9	HYDROGEN SULFIDE	4.9200E+02	3.1400E+03	6.9000E-01	-0	-0
10	AMMONIA	4.9200E+02	3.2400E+03	4.2700E-01	-0	-0
11	HYDROGEN CHLORIDE	4.9200E+02	4.9500E+03	4.8800E-01	-0	-0
12	WATER	4.9200E+02	2.7000E+03	2.8900E-01	-0	-0
13	METHANE	4.9200E+02	2.7000E+03	8.3200E-01	-0	-0
14	ETHYLENE	4.9200E+02	2.7000E+03	9.7600E-01	-0	-0
15	ETHANE	4.9200E+02	2.7000E+03	1.0880E+00	-0	-0

Figure 9.20   *(Continued)*

9.2.2   The Simultaneous Solution Strategy

Under the simultaneous solution strategy for solving material balance problems, all material balances and constraint equations are aggregated into a single composite system of equations which is solved as a block. Since the material balance equations are linear in the species flows and reaction rates, then, if all constraints are linear, the resulting system of equations will be linear. Thus, variants of row reduction methods can be used to solve the material balance problem without the need for iterative tear stream or constraint calculations. If one or more of the constraints is nonlinear, then solution of the composite system is obtained by using the Newton–Raphson method, that is, solving a series of linear subproblems. Each subproblem is generated by constructing a linear approximation to each nonlinear equation about the current best estimate of the problem solution and appending the linearization to the remaining purely linear equations.

As discussed in Section 5.2.2, the primary advantages of the simultaneous strategy are that it avoids the need for identifying tear streams as well as local and global constraints and for executing suitable sequences of constraint and tear stream iteration loops. The main disadvantages are the size of the resulting equation set and, in the presence of nonlinearities, the need to develop estimates of all stream vectors in the flowsheet to initiate the Newton–Raphson calculations. Since in problems arising in chemical engineering practice, systems with hundreds and even thousands of equations are not unusual, the memory requirements for storage of

LIQUID PHASE HEAT CAPACITY CONSTANTS / LIQUID PHASE HEAT CAPACITY TEMPERATURE LIMITS (DEG R) / HEAT OF VAPORIZATION (BTU/MOLE) / SOLUBILITY PARAMETER

ID	COMPONENT	31 ALIQ	32 BLIQ	33 CLIQ	34 DLIQ	35 LTLCP	36 UTLCP	37 HVAP	38 DELTA
1	NITROGEN	-1.4850E+01	3.0600E+01	-4.3010E-03	1.0460E-05	1.2060E+02	2.1060E+02	2.3990E+03	4.7110E+01
2	OXYGEN	-1.2520E+01	8.8660E-04	0	0	1.0190E+02	1.3100E+02	2.9340E+03	4.2440E+01
3	HYDROGEN	5.3280E+00	-2.5800E-01	8.3590E-03	-5.0160E-05	2.5200E+01	4.6800E+01	3.8880E+02	3.4400E+01
4	CARBON MONOXIDE	1.3460E+01	7.3780E+00	0	0	1.2600E+02	1.5000E+02	2.5970E+03	3.3200E+01
5	CARBON DIOXIDE	-2.0520E+03	1.4300E+01	-3.2910E-02	2.5280E-05	4.0140E+02	5.0940E+02	7.3800E+03	7.5540E+01
6	CARBONYL SULFIDE	-0	-0	-0	-0	-0	-0	2.5990E+03	1.1600E+02
7	CARBON DISULFIDE	3.2800E+01	-0	-0	-0	-0	-0	—	1.0600E+02
8	SULFUR DIOXIDE	2.0490E+01	7.6260E-05	-0	-0	4.0150E+02	5.8150E+02	1.1520E+04	6.3660E+01
9	HYDROGEN SULFIDE	9.4070E+01	-7.7320E-02	2.1530E-02	-1.9960E-05	3.4200E+02	3.8000E+02	1.0720E+04	9.3370E+01
10	AMMONIA	1.5400E+01	2.1450E-01	-4.6870E-04	3.5320E-07	2.9340E+02	3.1210E+02	1.0050E+04	1.7280E+02
11	HYDROGEN CHLORIDE	4.8520E+01	6.5010E-01	-2.2450E-03	2.5900E-06	3.5460E+02	6.7860E+02	9.9480E+03	1.1300E+02
12	WATER	1.8040E+01	-0	-0	-0	-0	-0	1.7490E+04	2.3660E+02
13	METHANE	2.4280E+01	-1.5530E-01	6.1300E-04	-6.4420E-07	1.7100E+02	2.7000E+02	3.5190E+03	6.0260E+01
14	ETHYLENE	1.5760E+01	2.5580E-02	-1.7010E-04	2.9290E-07	1.8900E+02	3.0600E+02	5.8270E+03	6.4510E+01
15	ETHANE	1.4170E+01	2.6310E-02	-1.1150E-04	1.8570E-07	1.6400E+02	4.1400E+02	6.3310E+03	6.4190E+01

PHASE OF STD. STATE / NUMBER OF ATOMS PER MOLECULE FOR COMBUSTION / MELTING POINT (DEG R) / HEAT OF FUSION (BTU/MOLE)

ID	COMPONENT	39 IFLAG	40 CA	41 HA	42 SA	43 MP	44 DELF
1	NITROGEN	1.0000E+00	0	-0	-0	1.1390E+02	3.0960E+02
2	OXYGEN	1.0000E+00	0	-0	-0	9.7920E+01	1.9080E+02
3	HYDROGEN	1.0000E+00	0	2.0000E+00	-0	2.5130E+01	5.0400E+01
4	CARBON MONOXIDE	1.0000E+00	1.0000E+00	-0	-0	2.2660E+02	3.6000E+02
5	CARBON DIOXIDE	1.0000E+00	0	-0	-0	3.8990E+02	-0
6	CARBONYL SULFIDE	1.0000E+00	1.0000E+00	-0	1.0000E+00	2.4170E+02	-0
7	CARBON DISULFIDE	1.0000E+00	1.0000E+00	-0	2.0000E+00	2.9000E+02	1.1800E+03
8	SULFUR DIOXIDE	1.0000E+00	0	-0	-0	3.5580E+02	3.1840E+03
9	HYDROGEN SULFIDE	1.0000E+00	0	2.0000E+00	1.0000E+00	3.3770E+02	1.0220E+03
10	AMMONIA	1.0000E+00	0	-0	-0	3.5170E+02	2.4320E+03
11	HYDROGEN CHLORIDE	1.0000E+00	0	-0	-0	2.8610E+02	8.5680E+02
12	WATER	1.0000E+00	0	-0	-0	4.9170E+02	2.5850E+03
13	METHANE	1.0000E+00	1.0000E+00	4.0000E+00	-0	1.6320E+02	4.0500E+02
14	ETHYLENE	1.0000E+00	2.0000E+00	4.0000E+00	-0	1.8710E+02	1.4410E+03
15	ETHANE	1.0000E+00	2.0000E+00	6.0000E+00	-0	1.6180E+02	1.2300E+03

Figure 9.20 (Continued)

the associated large coefficient arrays become a serious practical limitation, and very careful attention must be given to schemes which only store nonzero array elements. Moreover, the numerical solution itself must be carried out using specialized methods appropriate for large sparse equation systems.

The simultaneous strategy outlined above can certainly be applied to combined balance problems essentially without any modification and with the same attendant advantages and disadvantages. The key difference is that, since the combined balance problem is inherently nonlinear, its solution will in most cases require application of the Newton–Raphson or successive linearization technique. Our main concern in applying the simultaneous strategy will therefore involve the details of the linearization calculations themselves. Once the linearizations are constructed, solution of the resulting linear system will proceed as in the material balance case.

Balance Equation Linearization   Consider an open steady-state system with R reactions, S species, J output streams, and K input streams. Assume that the pressure and phase of each stream are known, that mechanical energy effects are negligible, and that each stream is an ideal mixture. The energy balance equation then takes the form

$$\frac{dQ}{dt} - \frac{dW}{dt} = \sum_{i=1}^{R} r_i \,\Delta H_{Ri} + \sum_{j=1}^{J} \sum_{s=1}^{S} N_s^j \bar{H}_s(T^j) - \sum_{k=1}^{K} \sum_{s=1}^{S} N_s^k \bar{H}_s(T^k) \quad (9.4)$$

Although this equation is linear in the variables dQ/dt, dW/dt, and r_i, the potential for nonlinearity arises because of the flow-enthalpy products. Recall that, if the material and energy balances can be decoupled, the material balance subproblem can be solved separately, perhaps using simultaneous solution, followed by solution of the energy balance equation. In this decoupled case, the product of stream flow and stream unit enthalpy can be replaced by the single enthalpy flow variable, $\dot{H}^j$, to result in a linear enthalpy flow balance equation. In general, this will not be the case, and consequently the nonlinear flow enthalpy product must be addressed directly. This means that both the species flows and the stream temperature must be treated as iteration variables and hence, following the successive linearization strategy, that a linearization must be constructed in terms of the species flows, stream temperatures, as well as the linear dQ/dt, dW/dt, and r_i variables.

In Section 5.2.2, p. 337, a linearization of a multivariable function $f(x)$ about a point x^0 was constructed using the Taylor formula expressed in terms of the partial derivatives of f evaluated at x^0. Specifically, the linearization of $f(x)$ at x^0, which we will denote by $f_L(x)$, is given by

$$f_L(x) \equiv f(x^0) + \sum_{n=1}^{N} \left(\frac{\partial f}{\partial x_n}\right)_{x=x^0} (x_n - x_n^0)$$

If we are seeking a point at which $f(x) = 0$, then it is appropriate to require that $f_L(x) = 0$. Hence, the equation $f(x) = 0$ is replaced by the linear approximation

$$f(x^0) + \sum_{n=1}^{N} \left(\frac{\partial f}{\partial x_n}\right)_{x=x^0} (x_n - x_n^0) = 0$$

In the case of our energy balance equation, the function of concern is

$$f\left(\frac{dQ}{dt}, \frac{dW}{dt}, r, N^j, N^k, T^j, T^k\right)$$

$$= \frac{dW}{dt} - \frac{dQ}{dt} + \sum r_i \Delta H_{Ri} + \sum_{j=1}^{J}\sum_{s=1}^{S} N_s^j \tilde{H}_s(T^j) - \sum_{k=1}^{K}\sum_{s=1}^{S} N_s^k \tilde{H}_s(T^k) = 0$$

Suppose we have available an estimate of the values of all of the variables, which will be denoted by superscript zero. To construct a linearization, we first need to evaluate the partial derivatives of f. These are

$$\frac{\partial f}{\partial(dQ/dt)} = -1 \qquad \frac{\partial f}{\partial(dW/dt)} = +1 \qquad \frac{\partial f}{\partial r_i} = \Delta H_{Ri}$$

$$\frac{\partial f}{\partial N_s^j} = \tilde{H}_s(T^j) \qquad \frac{\partial f}{\partial N_s^k} = -\tilde{H}_s(T^k)$$

The partial derivatives with respect to the stream temperatures are a bit more complex, namely,

$$\frac{\partial f}{\partial T^j} = \sum_{s=1}^{S} N_s^j \frac{\partial \tilde{H}_s(T^j)}{\partial T^j} = \sum_{s=1}^{S} N_s^j C_{p_s}(T^j)$$

The last equality is appropriate if the stream is of single phase, in which case, the C_p factors are the usual species heat capacity functions. The linearization of the energy balance equation will therefore take the form

$$\frac{dW^0}{dt} - \frac{dQ^0}{dt} + \sum_{i=1}^{R} r_i^0 \Delta H_{Ri} + \sum_{j=1}^{J}\left[\sum_{s=1}^{S} (N_s^j)^0 \tilde{H}_s(T^{j0})\right]$$

$$- \sum_{k=1}^{K}\left[\sum_{s=1}^{S} (N_s^k)^0 \tilde{H}_s(T^{k0})\right]$$

$$+ (-1)\left(\frac{dQ}{dt} - \frac{dQ^0}{dt}\right) + (+1)\left(\frac{dW}{dt} - \frac{dW^0}{dt}\right)$$

$$+ \sum_{i=1}^{R} \Delta H_{Ri}(r_i - r_i^0)$$

$$+ \sum_{j=1}^{J}\left[\sum_{s=1}^{S} \tilde{H}_s(T^{j0})(N_s^j - (N_s^j)^0)\right]$$

$$- \sum_{k=1}^{K}\left[\sum_{s=1}^{S} \tilde{H}_s(T^{k0})(N_s^k - (N_s^k)^0)\right]$$

$$+ \sum_{j=1}^{J}\left[\sum_{s=1}^{S} (N_s^j)^0 C_{p_s}(T^{j0})(T^j - T^{j0})\right]$$

$$- \sum_{k=1}^{K}\left[\sum_{s=1}^{S} (N_s^k)^0 C_{p_s}(T^{k0})(T^k - T^{k0})\right] = 0$$

The first two lines correspond to the value of the energy balance function at the estimated solution. The next two lines consist simply of the partial derivative terms corresponding to the linear variables. The last four lines are the partial derivative terms corresponding to the flow-enthalpy terms. Note that there is some cancellation possible of terms involving only variables evaluated at the estimated solution. The result is the equation

$$\frac{dW}{dt} - \frac{dQ}{dt} + \sum_i r_i \Delta H_{Ri} + \sum_j \sum_s (H_s^j)^0 N_s^j - \sum_k \sum_s (H_s^k)^0 N_s^k$$

$$+ \sum_j \sum_s (N_s^j)^0 C_{p_s}(T^{j^0})(T^j - T^{j^0}) - \sum_k \sum_s (N_s^k)^0 C_{p_s}(T^{k^0})(T^k - T^{k^0}) = 0 \quad (9.5)$$

which is linear in the variables dW/dt, dQ/dt, r_i, N_s^j, N_s^k, T^j, and T^k. All other terms in the equation are either constants (for instance, ΔH_{Ri}) or variables evaluated at the solution estimate. Note further that if the estimated solution and true solution become very close, then

$$T^j \simeq T^{j^0} \qquad j = 1, \ldots, J$$

and

$$T^k \simeq T^{k^0} \qquad k = 1, \ldots, K$$

The terms involving the temperatures will therefore vanish and the equation will take the form of the original energy balance equation.

Special Cases Several special cases which permit simplification of the linearized balance equation often arise in application. We briefly consider three such cases:

1. Constant C_p.
2. Specified temperature.
3. Single species streams.

If it can be assumed that the stream heat capacities are constant, then the stream enthalpy terms in the original balance equation reduce to

$$\sum_j \sum_s N_s^j H_s(T^j) = \sum_j (T^j - T^r) \sum_s C_{p_s} N_s^j$$

where T^r is the temperature of the selected problem reference state. In this case, the energy balance equation, eq. (9.4), becomes *bilinear* in stream temperature and flows. That is, with fixed flows, it is linear in temperature; with fixed temperatures, it is linear in the flows. Assuming that all species in all streams remain in their reference state phases, the linearization (9.5) reduces to

$$\frac{dW}{dt} - \frac{dQ}{dt} + \sum_i r_i \Delta H_{Ri} + \sum_j (T^{j^0} - T^r) \sum_s C_{p_s} N_s^j - \sum_k (T^{k^0} - T^r) \sum_s C_{p_s} N_s^k$$

$$+ \sum_j (T^j - T^{j^0}) \sum_s C_{p_s}(N_s^j)^0 - \sum_k (T^k - T^{k^0}) \sum_s C_{p_s}(N_s^k)^0$$

$$(9.6)$$

Whereas the more complete form, eq. (9.5), will require evaluation of stream enthalpies and temperatures each time the linearization is updated, the simplified linearization only requires heats of reaction and C_p values. Thus, while the former would in general require the use of a physical properties estimation system, the latter does not. Consequently, eq. (9.6) is appropriate for use with hand-held calculators or computers with limited storage capacity. Of course, use of constant C_p values does lead to less accurate solutions.

If any of the stream temperatures is fixed through specification, then the corresponding flow enthalpy terms become linear. Clearly, since the enthalpy term is a constant, the partial derivative with respect to temperature is zero and the corresponding C_p term in eq. (9.5) will vanish. Of course, if all the temperatures in the balance are specified, then the balance reduces to a linear equation and the balance and its linearization will be one and the same.

Finally, let us consider the case in which the streams included in the balance all involve single species. Recall from our discussion in Section 9.1.1 that, with the pressure fixed, the state of a single-species stream will be fixed by specifying either the temperature or the molar enthalpy if the stream is of single phase, or by specifying either the phase fraction (vapor fraction) or the molar mixture enthalpy if the stream has two phases. Since direct use of the stream molar enthalpy as a variable reduces the energy balance to a bilinear function in flow and enthalpy, the enthalpy is preferred to the temperature or phase fraction calculation variable. In the single-species case, the linearization thus reduces to

$$\frac{dW}{dt} - \frac{dQ}{dt} + \sum_j (H^j)^0 N^j - \sum_k (H^k)^0 N^k + \sum_j (N^j)^0 H^j$$
$$- \sum_k (N^k)^0 H^k - \sum_j (N^j H^j)^0 + \sum_k (N^k H^k)^0 = 0 \quad (9.7)$$

Note that the reaction term has been dropped since reactions involving a single species as product and reactant are of no interest. The above form of the linearization is useful in heat exchanger energy balances, even when one of the streams involves multiple species, since the stream composition will remain fixed.

Summary of Calculation Procedure   Regardless of the form of the linearization equation that is used, the simultaneous solution strategy will consist of the following steps.

Given an initial estimate of the complete solution and a specified desired accuracy $\varepsilon > 0$:

Step 1   Evaluate the coefficients of the linearization equations at the current solution estimate.

Step 2   Solve the equation set consisting of linearized and linear equations to obtain a new solution estimate.

Step 3   Check whether at the current best solution estimate the absolute value of each nonlinear equation (energy balance or nonlinear constraint) is within ε of zero. If not, continue with Step 1. Otherwise, the solution procedure terminates.

As in the material balance case, the solution estimate required to initiate the calculation sequence is best obtained by estimating values of the variables which cause the nonlinearities and by solving the remaining linear equations to calculate consistent estimates of the remaining variables. For instance, one would typically estimate the unknown temperatures (or enthalpies) and solve the linear balances to obtain estimates of the flows. This method of obtaining an initial solution estimate is usually preferable to guessing the complete set of variable values.

The next two examples illustrate the application of the simultaneous solution strategy to, first, a simple single-species case, followed by a more typical multiple-species case.

Example 9.12   Solve the steam boiler loop of Example 9.4 using the simultaneous solution strategy.

Solution   As first step, it is necessary to assemble all of the material, energy, and constraint equations appropriate to the system. These equations are the following:

Heat exchanger 1:

$$\frac{dQ}{dt} = N^7 \tilde{H}_V (T^8, 2 \text{ bar}) - N^7 \tilde{H}_V (500°C, 2 \text{ bar})$$
$$+ F^6 \hat{H}_{\text{mix}}^4 (\text{sat}, 50 \text{ bar}) - F^6 \hat{H}_L (\text{sat}, 50 \text{ bar}) = 0$$

Heat exchanger 2:

$$\frac{dQ}{dt} = F^1 \hat{H}_L (\text{sat}, 50 \text{ bar}) - F^1 \hat{H}_L (75°C, 50 \text{ bar})$$
$$+ N^7 \tilde{H}_V (200°C, 2 \text{ bar}) - N^7 \tilde{H}_V (T^8, 2 \text{ bar}) = 0$$

Mixer:

$$F^3 = F^1 + F^6$$
$$\frac{dQ}{dt} = F^3 \hat{H}_{\text{mix}}^3 (\text{sat}, 50 \text{ bar})$$
$$- F^1 \hat{H}_L (\text{sat}, 50 \text{ bar})$$
$$- F^6 \hat{H}_{\text{mix}}^4 (\text{sat}, 50 \text{ bar}) = 0$$

Knockout drum:

$$F^3 = F^5 + F^6$$
$$\frac{dQ}{dt} = F^5 \hat{H}_V (\text{sat}, 50 \text{ bar}) + F^6 \hat{H}_L (\text{sat}, 50 \text{ bar})$$
$$- F^3 \hat{H}_{\text{mix}}^3 (\text{sat}, 50 \text{ bar}) = 0$$

Specifications:

$$12F^1 = F^6$$
$$N^7 = 100 \text{ mol/h} \quad \text{(basis)}$$

The problem unknowns are F^1, F^3, F^5, F^6, T^8, and the vapor fractions of streams 4 and 3. Hence, the problem for simultaneous solution involves seven equations in seven unknowns. Since all streams involve a single species, it is convenient to use the enthalpies of streams 8, 3, and 4 as variables since this simplifies the linearized equations. Note that three of the energy balance equations are nonlinear (specifically, bilinear), since they involve products of unknown flows and enthalpies. Since N^7 is specified, the exchanger 2 balance is linear in F^1 and $\bar{H}^8$. Of course, the two material balances and the flow ratio constraint are linear equations.

The linearized equations, expressed in terms of $\bar{H}^8$, $\hat{H}^3$, and $\hat{H}^4$ as variables and following eq. (9.7), will be

Heat exchanger 1:

$$-\bar{H}_V (500°\text{C, 2 bar})100 + (\hat{H}^4)^0 F^6 - \hat{H}_L (\text{sat, 50 bar})F^6$$
$$+ 100\bar{H}^8 + (F^6)^0\hat{H}^4 - (F^6\hat{H}^4)^0 = 0$$

Heat exchanger 2:

$$\hat{H}_L (\text{sat, 50 bar})F^1 - \hat{H}_L (75°\text{C, 50 bar})F^1$$
$$+ \hat{H}_V (200°\text{C, 2 bar})100 - 100\bar{H}^8 = 0$$

Mixer:

$$F^3 - F^1 - F^6 = 0$$
$$(\hat{H}^3)^0 F^3 - \hat{H}_L (\text{sat, 50 bar})F^1 - (\hat{H}^4)^0 F^6 + (F^3)^0\hat{H}^3$$
$$- (F^6)^0\hat{H}^4 - (\hat{H}^3 F^3)^0 + (\hat{H}^4 F^6)^0 = 0$$

Knockout drum:

$$F^3 - F^5 - F^6 = 0$$
$$\hat{H}_V (\text{sat, 50 bar})F^5 + \hat{H}_L (\text{sat, 50 bar})F^6 - (\hat{H}^3)^0 F^3$$
$$- (F^3)^0\hat{H}^3 + (\hat{H}^3 F^3)^0 = 0$$

Specification:

$$12F^1 - F^6 = 0$$

If we let

$$\alpha = \hat{H}_V (\text{sat, 50 bar})$$
$$\beta = \hat{H}_L (\text{sat, 50 bar})$$
$$\gamma = \hat{H}_L (75°\text{C, 50 bar})$$
$$\delta = \bar{H}_{\text{C}_6\text{H}_6} (200°\text{C, 2 bar})$$
$$\varepsilon = \bar{H}_{\text{C}_6\text{H}_6} (500°\text{C, 2 bar})$$

then the system of linear equations can be written in detached coefficient form as shown below:

F^1	F^3	F^5	F^6	$\tilde{H}^8$	$\hat{H}^3$	$\hat{H}^4$	RHS
			$(\hat{H}^4)^0 - \beta$	100		$(F^6)^0$	$100\varepsilon + (F^6\hat{H}^4)^0$
$(\beta - \gamma)$				-100			-100δ
-1	1		-1				0
$-\beta$	$(\hat{H}^3)^0$		$-(\hat{H}^4)^0$	$-(F^6)^0$	$(F^3)^0$		$(F^3\hat{H}^3)^0 - (F^6\hat{H}^4)^0$
1		-1	-1				0
	$-(\hat{H}^3)^0$	α	β		$-(F^3)^0$		$-(F^3\hat{H}^3)^0$
12			-1				0

To begin the solution process, we need to select a reference state and to provide initial estimates of the solution. To simplify the calculations, suppose we set the enthalpy of benzene vapor at 200°C and 2 bar equal to zero and use as reference state for water the reference used in the steam tables. Thus,

$$\alpha = 2794.2 \text{ kJ/kg}$$
$$\beta = 1154.5 \text{ kJ/kg}$$
$$\gamma = 317.9 \text{ kJ/kg}$$
$$\delta = 0$$
$$\varepsilon = 48.266 \text{ kJ/gmol}$$

where ε is evaluated by integrating the vapor heat capacity equation. As initial estimate, suppose we choose

$$H^3 = 1200 \text{ kJ/kg}$$
$$\hat{H}^4 = 1350 \text{ kJ/kg}$$
$$\tilde{H}^8 = 20 \text{ kJ/gmol}$$

Then, we can solve the linear equation subset

$$(\beta - \gamma)F^1 = 100H^8$$
$$-F^1 + F^3 - F^6 = 0$$
$$F^3 - F^5 - F^6 = 0$$
$$12F^1 - F^6 = 0$$

to obtain an initial estimate of the flows. The result is

$$F^1 = F^5 = 2.389 \text{ kg/h}$$
$$F^6 = 28.670 \text{ kg/h}$$
$$F^3 = 31.060 \text{ kg/h}$$

With this initial solution estimate, all coefficients in the linearized equation array can be evaluated. The result is

F^1	F^3	F^5	F^6	$\tilde{H}^8$	$\hat{H}^3$	$\hat{H}^4$	RHS
		195.5	100			28.67	43,531.1
836.6				-100			0
-1	1		-1				0
-1154.5	1200		-1350	-28.67	31.06		-1432.5
	1	-1	-1				0
	-1200	2794.2	1154.2				$-37,272.0$
12			-1		-31.06		0

This 7×7 system of linear equations can be solved by row reduction to obtain the next solution estimate.

$$F^1 = 1.9491 \text{ kg/h} \qquad \hat{H}^3 = 1265.78 \text{ kJ/kg}$$
$$F^3 = 25.3385 \text{ kg/h} \qquad \hat{H}^4 = 1301.98 \text{ kJ/kg}$$
$$F^5 = 1.9491 \text{ kg/h} \qquad \tilde{H}^8 = 16.306 \text{ kJ/gmol}$$
$$F^6 = 23.3894 \text{ kg/h}$$

Repeating this process for two more iterations, we obtain the solution

$$F^1 = 1.9491 \text{ kg/h} \qquad \hat{H}^3 = 1280.63 \text{ kJ/kg}$$
$$F^3 = 25.3385 \text{ kg/h} \qquad \hat{H}^4 = 1291.14 \text{ kJ/kg}$$
$$F^5 = 1.9491 \text{ kg/h} \qquad \tilde{H}^8 = 16.306 \text{ kJ/gmol}$$
$$F^6 = 23.3894 \text{ kg/h}$$

Convergence is very rapid; only the wet steam enthalpies actually change. Given the three enthalpies, we can now recover the vapor fractions and the stream 8 temperature in the usual fashion. From the mixture enthalpy definitions, we have

$$1280.63 = 2794.2w^3 + (1 - w^3)1154.5$$
$$1291.14 = 2794.2w^4 + (1 - w^4)1154.5$$

or

$$w^3 = 0.07692 \qquad \text{and} \qquad w^4 = 0.08333$$

The temperature T^8 must be determined by solving the equation

$$\int_{200°C}^{T^8} C_{P_{C_6H_6}} \, dt = 16.306$$

The result is of course identical to that obtained in Example 9.4, namely, $T^8 = 574.5°K$. Recall that by properly sequencing the calculations for manual solution, no iterative solution was required, except for the calculation of T^8.

In the next example, the simultaneous solution strategy will be used with a system involving multiple species and nonlinearities arising both from the energy balance equations and through a nonlinear specification.

Example 9.13 A synthesis gas containing CO, H_2, and a small amount of CH_4, with a CO-to-H_2 ratio of 1:2.9, is to be upgraded to a higher methane content via the reaction

$$CO + 3H_2 \rightarrow CH_4 + H_2O$$

using the recycle system shown in Figure 9.21. The reactor must be operated adiabatically with a maximum outlet temperature of 1000°F to produce a product stream containing 50% CH_4 and 12% CO. The heat removal rate in heat exchanger 1 is adjusted to cool the reactor effluent stream to 500°F. The separator is operated so as to result in a recycle gas stream containing 1% H_2O and a pure liquid water stream, both at 100°F. Both the feed and the product streams are at 200°F. Assuming that the entire system operates at a constant pressure of 100 psia, calculate all flows and temperatures for a methane production rate of 100 lbmol/h.

Solution With the exception of streams 1 and 8, all streams contain the four species CO, H_2, CH_4, and H_2O. Except for stream 8 all streams are gas phase. There is a single chemical reaction, and only two of the units are nonadiabatic: the separator and heat exchanger 1. From the degree-of-freedom table shown in Figure 9.22, it is evident that the problem is correctly specified. Since, after the choice of basis, no set of balances has degree of freedom zero, manual solution will require carrying at least one unknown.

If simultaneous solution is to be used, the table indicates that the problem will

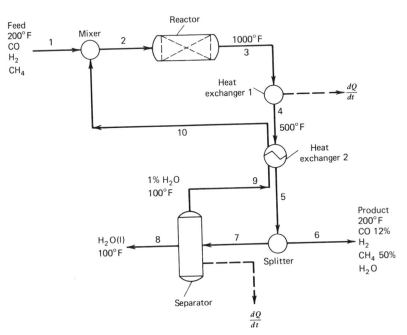

Figure 9.21 Flow diagram for Example 9.13, recycle methanation system.

	Mixer		Reactor		Heat Ex. 1	Heat Ex. 2	Splitter		Separator		Process	Overall	
	M	C	M	C			M	C	M	C		M	C
Number of stream variables													
Flow	11	11	8	8	4	8	12	12	9	9	24	8	8
T		3		2	2	4		3		3	10		3
Number of unit variables													
dQ/dt		1		1	1	1		1		1	6		1
r			1	1							1	1	1
											41		
Number of balances													
Material	4	4	4	4	—	—	4	4	4	4	16	4	4
Energy		1		1	1	1		1		1	6		1
Number of specifications													
Product Comp.									2	2	2	2	2
H_2O comp.	1	1									1	1	1
$CO:H_2$ ratio			1	1							1	1	1
Number of splitter restr.													
Comp.							3	3			3		
T								1			1		
Temperatures													
Stream 1		1									1		
Stream 3				1							1		
Stream 4					1						1		
Stream 6						1					1		
Stream 8								1			1		
Stream 9										1	1		1
$dQ/dt = 0$		1		1				1		1	4		1
Basis											1		
Degree of freedom	5	6	3	3	2	5	3	3	2	2	0	2	2

Figure 9.22 Degree-of-freedom table, Example 9.13.

involve 16 material balances, six energy balances, as well as a number of specifications. These are the feed CO-to-H_2 ratio, the stream 9 (or equivalently, stream 10) water mole fraction, the CO and CH_4 mole fractions of stream 6 and (as a result of the splitter restrictions) stream 7, as well as the remaining splitter restriction equating, say, the H_2 mole fractions of streams 6 and 7. The last specifications will lead to a nonlinear equation, while the other six will all be linear constraints. We thus will have a system of 29 equations. If the flow of CH_4 in stream 6 is used as basis, then the problem has 23 species flows, a reaction rate, two nonzero dQ/dt, and three unknown temperatures: T^2, T^5, and T^{10}. The temperature of stream 7 is equal to that of stream 6 via the splitter restriction. Actually, since $dQ/dt = 0$ for the splitter, the splitter energy balance will simply yield $T^5 = 200°F$. Thus, the problem in effect involves 28 equations in 28 unknowns. Of the 28 equations, all will be linear except for the mixer energy balance, the reactor energy balance, the heat exchanger 2 energy balance, and the splitter restriction relating the H_2 mole fraction of streams 6 and 7. The exchanger 1 and separator energy balances will be linear since their inlet and outlet stream temperatures are all known. These energy balances will serve to calculate the unit dQ/dt values.

Using as reference state 77°F, 1 atm, and all species in the gaseous phase, assuming ideal mixtures and negligible pressure effects, and assuming that $dW/dt = 0$ for all units, we can proceed to assemble the required block of balance and specification equations. In presenting the equations, we index the species in the order CO, H_2, CH_4, and H_2O. All enthalpies, except for stream 8, will be gas enthalpies.

Mixer balances:
$$N_1^2 = N_1^{10} + N_1^1$$
$$N_2^2 = N_2^{10} + N_2^1$$
$$N_3^2 = N_3^{10} + N_3^1$$
$$N_4^2 = N_4^{10}$$

$$\frac{dQ}{dt} = \sum_s N_s^2 \bar{H}_s (T^2) - \sum_s N_s^1 \bar{H}_s (200°F) - \sum_s N_s^{10} \bar{H}_s (T^{10}) = 0 \quad (9.8)$$

Reactor balances:
$$N_1^3 = N_1^2 - r$$
$$N_2^3 = N_2^2 - 3r$$
$$N_3^3 = N_3^2 + r$$
$$N_4^3 = N_4^2 + r$$

$$\frac{dQ}{dt} = r\,\Delta H_R (77°F) + \sum_s N_s^3 \bar{H}_s (1000°F) - \sum_s N_s^2 \bar{H}_s (T^2) = 0 \quad (9.9)$$

Exchanger 1 balance:
$$\frac{dQ_1}{dt} = \sum_s N_s^3 \bar{H}_s (500°F) - \sum_s N_s^3 \bar{H}_s (1000°F)$$

Exchanger 2 balance:

$$\frac{d\tilde{Q}}{dt} = \sum_s N_s^{10} \tilde{H}_s (T^{10})$$

$$+ \sum_s N_s^3 \tilde{H}_s (200°F) \qquad (9.10)$$

$$- \sum_s N_s^{10} \tilde{H}_s (100°F)$$

$$- \sum_s N_s^3 H_s (500°F) = 0$$

Splitter balances:

$$N_1^5 = N_1^6 + N_1^7$$

$$N_2^5 = N_2^6 + N_2^7$$

$$N_3^5 = N_3^6 + N_3^7 = 100 + N_3^7$$

$$N_4^5 = N_4^6 + N_4^7$$

The energy balance will not be explicitly written since it merely reduces to the result $T^5 = 200°F$.

Separator balances:

$$N_1^7 = N_1^{10}$$

$$N_2^7 = N_2^{10}$$

$$N_3^7 = N_3^{10}$$

$$N_4^7 = N_4^{10} + N^8$$

$$\frac{dQ_s}{dt} = \sum_s N_s^{10} \tilde{H}_s (100°F)$$

$$+ N^8 \tilde{H}_{H_2O} \text{ (sat liq, 100°F)}$$

$$- \sum_s N_s^7 \tilde{H}_s (200°F)$$

Constraints:

$$N_2^1 = 2.9N_1^1$$

$$N_4^{10} = 0.01(N_1^{10} + N_2^{10} + N_3^{10} + N_4^{10})$$

$$N_1^6 = 0.12(N_1^6 + N_2^6 + N_3^6 + N_4^6)$$

$$N_3^6 = 0.5(N_1^6 + N_2^6 + N_3^6 + N_4^6) \qquad \text{where} \quad N_3^6 = 100$$

$$N_1^7 = 0.12(N_1^7 + N_2^7 + N_3^7 + N_4^7)$$

$$N_3^7 = 0.5(N_1^7 + N_2^7 + N_3^7 + N_4^7)$$

$$\frac{N_2^6}{N_1^6 + N_2^6 + N_3^6 + N_4^6} = \frac{N_2^7}{N_1^7 + N_2^7 + N_3^7 + N_4^7} \qquad \text{where} \quad N_3^6 = 100 \quad (9.11)$$

As noted earlier, only eqs. (9.8) through (9.11) of this entire set are nonlinear and hence will require linearization. Equations (9.8) through (9.10) are energy balance equations involving multiple species, and thus their linearizations will take the general form of eq. (9.5).

The linearization of eq. (9.8) is

$$\sum_s (\tilde{H}_s^2)^0 N_s^2 - \sum_s \tilde{H}_s (200°F) N_s^1 - \sum_s (\tilde{H}_s^{10})^0 N_s^{10}$$

$$+ \sum_s (N_s^2)^0 C_{p_s}(T^{2,0})(T^2 - T^{2,0}) - \sum_s (N_s^{10})^0 C_{p_s}(T^{10,0})(T^{10} - T^{10,0}) = 0$$

$$(9.12)$$

The linearization of eq. (9.9), the reactor energy balance, is

$$r \, \Delta H_R(77°F) + \sum_s \tilde{H}_s(1000°F) N_s^3 - \sum_s (\tilde{H}_s^2)^0 N_s^2$$

$$- \sum_s (N_s^2)^0 C_{p_s} (T^{2,0})(T^2 - T^{2,0}) = 0 \quad (9.13)$$

Finally, the linearization of the exchanger 2 energy balance, eq. (9.10), can be written as

$$\sum_s [(\tilde{H}_s^{10})^0 - \tilde{H}_s(100°F)] N_s^{10} + \sum_s [\tilde{H}_s(200°F) - \tilde{H}_s(500°F)] N_s^3$$

$$+ \sum_s (N_s^{10})^0 C_{p_s} (T^{10,0})(T^{10} - T^{10,0}) = 0 \quad (9.14)$$

The remaining equation requiring linearization is the splitter restriction, eq. (9.11). If the denominators are cross-multiplied, eq. (9.11) can be rewritten to the form

$$f(N^6, N^7) = N_2^6(N_1^7 + N_2^7 + N_3^7 + N_4^7) - N_2^7(N_1^6 + N_2^6 + 100 + N_4^6) = 0$$

The partial derivatives are

$$\frac{\partial f}{\partial N_2^6} = \sum_s N_s^7 - N_2^7$$

$$\frac{\partial f}{\partial N_2^7} = N_2^6 - \sum_s N_s^6$$

$$\frac{\partial f}{\partial N_s^6} = -N_2^7 \qquad \text{for } s = 1, 4$$

$$\frac{\partial f}{\partial N_s^7} = N_2^6 \qquad \text{for } s = 1, 3, 4$$

The linear approximation to f is thus given by

$$f_i \equiv f^0 + \sum_i \frac{\partial f}{\partial x_i}(x_i - x_i^0)$$

$$= (N_2^6)^0 \left(\sum_s N_s^7 \right)^0 - (N_2^7)^0 \left(\sum_s N_s^6 \right)^0 + \left(\sum_s N_s^7 - N_2^7 \right)^0 [N_2^6 - (N_2^6)^0]$$

$$+ \left(N_2^6 - \sum_s N_s^6 \right)^0 [N_2^7 - (N_2^7)^0] - (N_2^7)^0 [N_1^6 - (N_1^6)^0]$$

$$- (N_2^7)^0 [N_4^6 - (N_4^6)^0] + (N_2^6)^0 \sum_{s \neq 2} [N_s^7 - (N_s^7)^0] = 0$$

After some cancellations, the following linear equation can be obtained in the unknowns N_s^6, $s = 1, 2, 4$, and N_s^7, $s = 1, 2, 3, 4$:

$$(N_2^6)^0 \sum_{s \neq 2} N_s^7 - (N_2^7)^0 (N_1^6 + N_4^6) + \left(\sum_{s \neq 2} N_2^7 \right)^0 N_2^6 - \left(\sum_{s \neq 2} N_s^6 \right)^0 N_2^7$$

$$= \left[N_2^6 \sum_s N_s^7 - N_2^7 \sum_{s \neq 3} N_s^6 \right]^0 \quad (9.15)$$

To initiate the simultaneous strategy, an estimate must be made of the solution. Note that the system of equations includes 16 linear material balances and six linear constraints relating the 23 species flows and the rate variable r. Therefore, an initial guess of two material balance variables, r and a flow, say, N_2^7, will allow all other species flows to be calculated. These flow estimates together with estimates of T^2 and T^{10} will allow dQ_1/dt and dQ_s/dt to be determined from the linear exchanger 1 and separator energy balances. In this fashion, a complete initial solution estimate can be generated from the chosen values of r, N_2^7, T^2, and T^{10} by solving the linear portion of the complete equation set.

Suppose as initial estimates we use $r = 96$ mol/h, $N_2^7 = 340$ mol/h, as well as $T^2 = 300°F$ and $T^{10} = 400°F$. Using these values, the initial stream vectors summarized in Table 9.1 can be calculated. These estimates can then be used to evaluate the coefficients of eqs. (9.12) through (9.15). The enthalpy and C_p values required for this purpose are summarized in Table 9.2. The enthalpies are all calculated as the integral of the corresponding species gas heat capacity equation from the reference 77°F to the temperature in question. Not shown are the heat of reaction, $\Delta H_R = -88.685 \times 10^3$ Btu/lbmol, and the relative enthalpy of stream 8 (liquid water at 100°F), $H^8 = 413.29$ Btu/lbmol. The full set of linear equations in which eq. (9.8) through (9.11) are replaced by linearizations (9.12) through (9.15) can now be assembled, as shown in Figure 9.23 in detached coefficient form. For clarity, the coefficients of the linear or linearized energy balances are simply denoted by X; the coefficients of the linearized splitter restriction are denoted by Y. Note that even for this small problem the array is relatively sparse and does have patterns

9.1 Initial Solution Estimate for Example 9.13

	Stream Number						
	1	2	3	6	7	8	10
Species flows (lbmol/h)							
CO	120.0	255.96	159.96	24.0	135.96	—	135.96
H_2	348.0	688.0	400.0	60.0	340.0	—	340.0
CH_4	4.0	570.49	666.49	100.0	566.49	—	566.49
H_2O	—	10.53	106.53	16.0	40.53	80.0	10.53
T (°F)	200	300	1000	200	200	100	400

9.2 Enthalpy Values at Initial Solution Estimate[a]

	H_s (100°F)	H_s (200°F)	H_s (500°F)	H_s (1000°F)	H_s (300°F)	C_{p_s} (300°F)	H_s (400°F)	C_{p_s} (400°F)
CO	0.16021	0.85813	2.9785	6.6791	1.5594	7.0348	2.2656	7.0930
H_2	0.15716	0.84952	2.9691	6.4929	1.5520	7.0582	2.2598	7.0902
CH_4	0.19888	1.0985	4.2151	10.9552	2.0630	9.9983	3.1002	10.755
H_2O	0.18486	1.6099	3.4831	7.9286	1.8114	8.2340	2.6408	8.3559

[a] H_s in 10^3 Btu/lbmol; C_{p_s} in Btu/lbmol · °F.

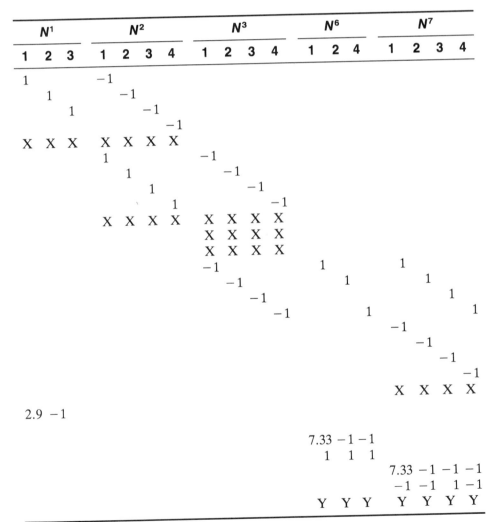

Figure 9.23   Linearized equation set in detached coefficient form.

which we exploit when performing manual unit-by-unit calculations. In simultaneous solution, these patterns are of course only indirectly used by the row reduction procedure.

This set of 28 linear equations is solved to yield the next solution estimate; the coefficients of linearizations eq. (9.12) through eq. (9.15), including the enthalpies and heat capacities of streams 2 and 10, are updated; and the process is continued until convergence is obtained. The successive values of r, T^2, T^{10}, and streams 6, 7, and 8 are shown in Table 9.3. From the table, it is evident that convergence is quite rapid. The converged stream vector values are shown in Table

N^8	N^{10}				r	T		dQ/dt		RHS	Comment
4	1	2	3	4		2	10	1	s		
	1									0	
		1								0	Mixer
			1							0	material
				1						0	
	X	X	X	X		X	X			X	Mixer energy
					−1					0	
					−3					0	Reactor
					1					0	material
					1					0	
					X	X				X	Reactor energy
							−1			0	Exchanger 1
	X	X	X	X			X			X	Exchanger 2
										0	
										0	Splitter
										100	material
										0	
	1									0	
		1								0	Separator
			1							0	material
1				1						0	
X	X	X	X	X				−1		0	Separator energy
	1	1	1	−99						0	
										0	
										100	Linear
										100	specifications
										0	
										0	
										Y	Splitter restr.

9.3 Partial Iteration Summary, Example 9.13

		Iteration Number		
	Initial Estimate	1	2	3
r (lbmol/h)	96	83.35	82.717	82.712
T^2 (°F)	300	407.69	411.67	411.69
T^{10} (°F)	400	479.75	483.01	483.03
N_1^6 (lbmol/h)	24.0	24.0	24.0	24.0
N_2^6	60.0	61.26	61.328	61.329
N_4^6	16.0	14.74	14.672	14.671
N_1^7	135.96	128.74	127.593	127.584
N_2^7	340.0	329.0	326.05	326.02
N_3^7	566.49	536.40	531.639	531.601
N_4^7	90.53	78.66	77.998	77.992
N^8	80.0	68.62	68.045	68.041

9.4. The associated unit variable values are

$$r = 82.712 \text{ lbmol/h}$$

$$\frac{dQ_1}{dt} = -6.5945 \times 10^6 \text{ Btu/h}$$

$$\frac{dQ_s}{dt} = -2.1751 \times 10^6 \text{ Btu/h}$$

In this example, it is possible to algebraically reduce the splitter condition and the reactor energy balance equation to equations in the two variables r and N_2^7 alone. The resulting reduced dimensionality equations are plotted in Figure 9.24 along with the linearization of the splitter condition. The linearization of the energy balance equation is not shown because the reactor balance is linear when expressed in terms of these two variables. From the plot, the reason for the rapid convergence

9.4 Final Stream Table, Example 9.13

	Stream Number						
	1	2	3	6	7	8	10
Species flows (lbmol/h)							
CO	106.71	234.30	151.58	24.0	127.58	—	127.58
H_2	309.47	635.49	387.35	61.329	326.02	—	326.02
CH_4	17.288	548.89	631.60	100.0	531.60	—	531.60
H_2O	—	9.951	92.663	14.671	77.992	68.041	9.951
Temperature T (°F)	200	411.69	1000	200	200	100	100

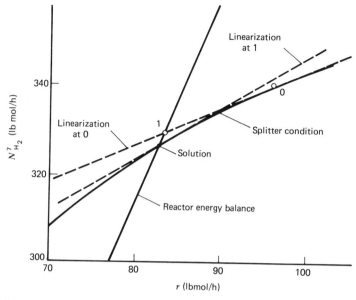

Figure 9.24   Calculation trajectory, Example 9.13.

is obvious: over the range of the variables shown, the curve is quite flat and hence the linearization proves to be a good approximation. The convergence is however sensitive to the choice of initial estimate. For instance, initial values of r and N_2^7 which are double those used in the example will lead to negative flows and poor convergence.

Implementation Considerations   In order to implement the simultaneous strategy for solving combined balance problems on a computer, the following tasks must be performed:

1. Formulation of the linear equations: balances and specifications.
2. Formulation of the linearized equations: balances and specifications.
3. Generation of an initial solution estimate.
4. Evaluation of the coefficients of the linearized equation subset.
5. Solution of the linear system.
6. Control of repeated calculation passes.

Computer implementations will essentially differ in the extent to which these tasks are automated. At the most rudimentary level, only tasks 4 through 6 will be executed via coding. In this approach, the engineer will himself write the balance and specification equations, both linear and linearized, and read into his program the coefficients of the associated array either as constants or by coding statements which will serve to evaluate the coefficients of linearized equations. For instance, in the case of Example 9.13, the engineer would define and dimension the 28×28 array and its right-hand side as shown in Figure 9.23. He would read in as numerical data the constant coefficients of the linear equations but would write a subroutine which would calculate the coefficients of the linearized equations (the X's and Y's). The engineer would also read in an initial estimate of the solution. Next, he would write a main program which would control the execution of repeated calculation passes. This program would solve the set of linear equations, check for convergence, and, if need be, update the coefficients of the equation set for subsequent solution. Such a program would presumably use a standard linear equation-solving routine available on the local computer system to solve the intermediate linear problems. Most likely, if this rudimentary approach is followed, constant heat capacities would be used, and thus the engineer would supply the heat capacity, heat of reaction, and any required heats of phase transition as part of the program data.

At a more sophisticated level, nearly all aspects of the six main tasks would be automated. Equation formulation would be automated by providing a library of modules. However, in contrast to the modules employed in the sequential modular strategy, the simultaneous modules would only serve as array entry generators. For instance, a mixer module would, given the input and output stream numbers, generate the material balance equations by simply inserting the appropriate ± 1 coefficients into the system array in the columns reserved for the species flows of those streams and the rows reserved for that mixer. The linearized energy balance equations would be generated by evaluating the coefficients of eq. (9.5) and inserting these into the appropriate columns corresponding to the stream flows and temperatures associated with the mixer.

Constraint specifications would be generated automatically in a similar fashion by allowing the user to select several types of linear and nonlinear constraints.

Evaluation of enthalpies, heat capacities, and heats of reaction would be carried out by using a physical property estimation package, such as that discussed in Section 9.2.1 under Automated Property Estimation. Generation of the complete starting estimate could be automated by requiring the user to supply a partial estimate adequate to specify the decoupled material and energy balances. The program would then be directed to solve these linear subproblems as a preparatory step to the main calculations. In such a system, an efficient sparse equation-solving method would be employed which would retain in memory a map of how the linear equation reduction was carried out during the first complete calculation pass. Since only the coefficient values and not the structure of the array would change between passes, retention of the calculation order deduced during the first pass would save considerable time in subsequent passes. Finally, a system of this type could be expanded to contain a broad module library, more extensive than that discussed in Section 9.2.1 under Solution Using Supplementary Modules, so that all types of process operations could be modeled. Needless to say, these capabilities define a large and complex program package which would require many man-years to develop. At the present time, the experimental QUASILIN system described by Hutchison and co-workers[9] is the only such system under development. Yet, this

---

[9] E. W. Gorczynski, H. P. Hutchison, and A. R. M. Wajih, Development of a Modularly Organized Equation Oriented Process Simulator, *Computers and Chemical Engineering*, **3**, 353 (1979).

is clearly the most efficient way of solving highly constrained, coupled flowsheet problems.

An extension of the decomposed material balance strategy (Ref. 4 of Chapter 5) to accommodate energy balances can be carried out. A program implementing that approach is available.[10] The interested reader is referred to the program manual for details.

9.3 SUMMARY

In this chapter, we have reviewed the manual strategy for solving complex combined material and energy balance flowsheet problems. We considered the possibility of decoupled solution, coupled solution, solution involving carrying unknown variables, and solution involving iterative calculations. We discussed the complications that can arise from the use of overall balances and considered the common case of the internally recirculating or "pump-around" stream which leads to dependence between the unit balances. With Section 9.1, we therefore developed all the tools necessary to solve any flowsheet balancing problem.

The second and larger portion of this chapter was devoted to a consideration of the adaptation of the sequential and simultaneous solution strategies to solving combined balance problems. The sequential modular strategy was found to be easily adapted to energy balances, because its modular structure allowed inclusion of modules with any degree of balance equation complexity and alternate sets of specification options. However, as in the material balance case, the treatment of constraint specifications via additional calculation loops proved awkward and inefficient. The simultaneous solution strategy, in general, requires iterations using linearized subproblems, but ultimately it is the most efficient approach for solving highly coupled and heavily constrained balance problems.

Overall, Section 9.2, together with Section 5.2, has attempted to present an overview of the problems and methods arising in computerized flowsheet calculations. This material thus provides a sound basis for further study of computer-aided process design and simulation methodology.

PROBLEMS

9.1 A flash drum is used to separate an equimolar mixture of toluene, p-xylene, and ethylbenzene fed at 5 atm and 470 K. The flash is operated at 1 atm and 400 K. Calculate the heat which must be supplied to the drum per mole of feed to the unit assuming Raoult's law is applicable.

9.2 A stream consisting of 60% n-octane and 40% n-hexadecane (mole basis)

---

[10] M. K. Sood, S. M. Clark, R. Khanna, and G. V. Reklaitis, *Linear Balancing Program,* Users Manual, Purdue Research Foundation, W. Lafayette, IN, 1980.

at 5 bar and 500 K is separated into vapor and liquid streams at 500 K. Suppose heat is supplied to the flash drum at the rate of 6 kJ/mol feed. Calculate the flash pressure, assuming Raoult's law is applicable.

9.3 Consider the flash system of Example 9.2. Suppose instead of being adiabatic, heat is added at the rate of 2 kJ/h. Calculate the flash temperature if the pressure is reduced to 2 bar in flashing.

9.4 A stream of 75% cyclohexane and 25% cyclopentane at 200°C and 5 bar is flashed adiabatically at 3.5 bar. Calculate the flash temperature and the ratio in which cyclopentane is split between the vapor and liquid streams.

9.5 A process stream consisting of 50% benzene and 50% naphthalene at 3 bar and 150°C is to be separated into benzene-rich and naphthalene-rich fractions using two flash units as shown in Figure P9.5. In the first unit, heat is added to allow the flash to proceed at 200°C and 2 bar; while in the second unit heat is removed in order to reach a 150°C temperature at 2 bar. Raoult's law may be assumed to be applicable.

(a) Construct a degree-of-freedom table to show that the problem is correctly specified.

(b) Deduce a calculation order and show that the unit material and energy balances decouple.

(c) Calculate the composition of streams 2, 4, and 5. What are the component split fractions for the process?

9.6 Consider the variation of Problem 9.5 shown in Figure P9.6. In this process, the liquid stream from the second flash unit is returned to the first flash unit, thus introducing a recycle between the two. The feed is 50% benzene and 50% naphthalene, and Raoult's law may again be employed.

(a) Construct a degree-of-freedom table to show that the problem is correctly specified.

(b) Deduce a calculation order showing that the unit material and energy

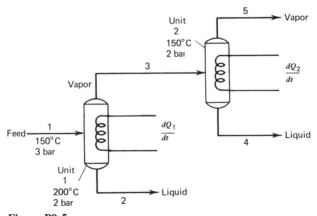

Figure P9.5

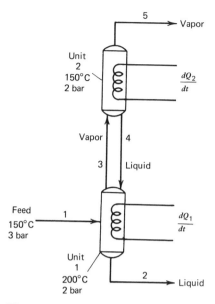

Figure P9.6

balances will decouple but that manual solution will require carrying two variables.

(c)   Formulate the equations which would have to be solved to obtain values of the carried unknown variables.

9.7   Suppose the process of Problem 9.6 is solved using the sequential modular solution strategy. Assume that the system available to you contains both the elementary and the supplementary modules.

(a)   Develop a model of the process using these modules.
(b)   Show at what point the convergence module would be inserted. Indicate how the calculations would be initiated.
(c)   Carry out three calculation cycles manually or using a suitable computer program.

9.8   Suppose the system of Problem 9.6 is solved using the simultaneous solution strategy.

(a)   Write out the complete set of equations which must be solved.
(b)   Show that the material and energy balances can be solved as separate blocks. Which of the equations introduce nonlinearities?
(c)   Show that the two isothermal flash equations can be used to condense the set of material balance equations.
(d)   Develop the set of linearized equations which will require repeated solutions. How would a good initial solution estimate be generated?
(e)   Perform one iteration of the simultaneous solution method.

9.9   The operation of a flash unit is modified as shown in Figure P9.9. The flash is carried out adiabatically, but then the overhead is cooled to its bubble temperature in a heat exchanger. The resulting liquid stream is split and a portion of it is returned to the flash drum. With a feed of 60% *n*-octane and 40% hexadecane (mole basis) at 5 bar and 500 K, the flash is carried out at 2 bar. Suppose 40% of the liquid from the heat exchanger is returned to the flash unit and Raoult's law is applicable.

(a)   Show that the process is correctly specified.
(b)   Deduce a calculation order for manual solution.
(c)   Formulate the equations which would have to be solved to obtain values of the carried unknown variables.

9.10 Suppose the system of Problem 9.9 is to be solved using the sequential modular strategy. Assuming that an expanded module library is available.

(a)   Formulate a modular representation of the flowsheet.
(b)   Locate the position of all convergence modules and indicate how the calculations would be initiated.
(c)   Carry out two calculations cycles manually or using a suitable computer program.

9.11 Consider the solution of Problem 9.9 using the simultaneous solution strategy.

(a)   Show that the simultaneous solution can be reduced to three key non-linear equations and several linear material and energy balances.
(b)   Develop a linearized form of the simultaneous equation set and find a good initial estimate of the solution.
(c)   Carry out one iteration of the linearized equations set manually.
(d)   Implement a computer program for the simultaneous solution and solve the problem to an accuracy of $\pm 1°C$ on the flash temperature.

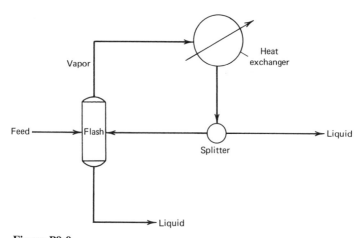

Figure P9.9

9.12 A two-stage flash system is to be designed as shown in Figure P9.12. The first stage is operated isothermally by supplying heat through suitable heating coils. The second stage is operated adiabatically with a recycle of its liquid stream to the first flash stage. The vapor from the second flash stage is cooled to its bubble point and part of it is recycled back to the second flash stage. This system essentially represents the operation of a two-stage distillation column. Suppose a feed consisting of an equimolar mixture of benzene and naphthalene at 4 bar and 180°C is flashed at 200°C and 2 bar in the first stage and at 2 bar in the second stage. The splitter is operated so as to recycle 50% of the heat exchanger effluent.

(a) Show that the problem is correctly specified.
(b) Deduce a calculation order for manual solution.
(c) Develop the equations required to carry out the calculation order deduced in part (b).
(d) Discuss the computational problems that will be encountered.

9.13 Suppose the system of Problem 9.12 is solved using the sequential modular strategy and an expanded module library.

(a) Formulate a modular representation of the flowsheet.
(b) Locate the position of all required convergence modules and suggest a procedure for obtaining a good initial estimate of the solution.
(c) Write a computer program to execute your strategy.

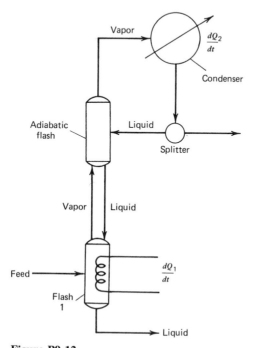

Figure P9.12

9.14 Consider the solution of Problem 9.12 using the simultaneous solution strategy.

   (a) Show that the equations can be grouped into a set of component balances, a set of flash equations, and the energy balances.

   (b) Discuss alternate strategies for solving these equations completely simultaneously or in linked blocks.

   (c) Outline a general computer program for one of the strategies developed in part (b).

   (d) Implement the computer program and use it to solve the system specified in Problem 9.12.

9.15 The generalized multistage flash system shown in Figure P9.15 can be used to model a distillation column. In this system, all stages except the first are adiabatic flash units. The feed is introduced on the kth stage and the two

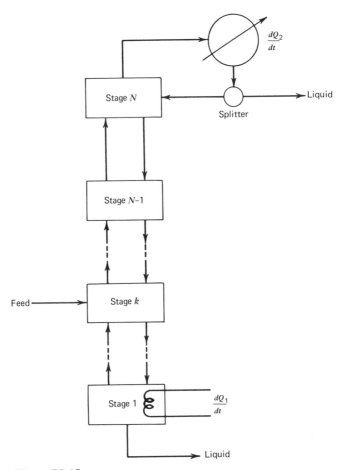

Figure P9.15

product streams are both liquids. Assume that the pressure is the same in all stages and that Raoult's law is applicable.

(a) Suppose that the feed stream conditions, the column pressure, and the splitter split fraction are specified. How many additional specifications must be imposed for the system to be correctly specified for an arbitrary number N of stages? List a reasonable set of variables which could be specified.

(b) Develop a modular representation of the system assuming an expanded module set is available and deduce the number and location of the required tear streams.

(c) Formulate a mixed solution strategy which makes use of simultaneous solution of the system species balances and adiabatic flash calculations.

9.16 Recall from Problem 7.25 that an evaporator is a device in which a solution is concentrated by boiling off some of the solvent. Normally, steam is used to supply the required heat. However, as shown in Figure P9.16(a), the solvent vapor can itself be used as a source of heat providing that its condensing temperature is raised by compressing it to a higher pressure. (If the condensing temperature is the same as the brine temperature in the evaporator, heat cannot be transferred.) The combined process is called vapor recompression evaporation. Suppose a brine containing 1% salt is concentrated to 2% (wt) by evaporation of water at 1 bar. Saturated steam at 2 bar is used as heat source, supplemented by the compressed vapor available at 160°C and 2 bar. The system can be visualized as shown in Figure P9.16(b), consisting of a tank, a compressor, and two heat exchangers. Assume the brine has the properties of liquid water.

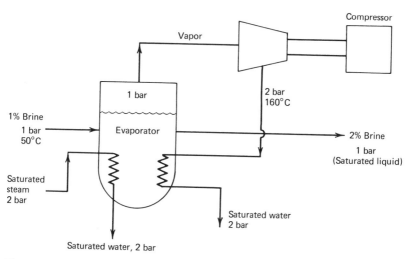

Figure P9.16(a)

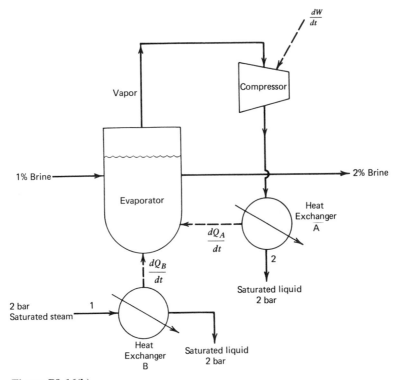

Figure P9.16(b)

(a) Construct a degree-of-freedom table to show that the process is correctly specified, assuming that compression occurs adiabatically and
there are no heat losses in transmission.

(b) Develop a calculation order for manual solution.

(c) Calculate the mass of 2 bar steam required (stream 1) per unit mass of
stream 2.

(d) If electric power for the compressor costs 10¢ per kWh, at what cost
of 2 bar steam will vapor recompression be competitive with ordinary
evaporation, assuming the additional equipment costs are neglected?

9.17 In the vapor recompression system of Problem 9.16, suppose the feed brine
can be preheated above 50°C before introducing it to the evaporator. Calculate the preheat temperature at which the recompressed vapor will itself
be adequate to supply the heating needs of the evaporator (i.e., stream 1
can be set to zero).

(a) Construct a degree-of-freedom table to show that the problem is correctly specified.

(b) Deduce a calculation order for problem solution.

(c) Calculate the feed temperature assuming the brine has properties of
liquid water.

9.18 Drinking water is to be produced at the rate of 5 million liters per day by evaporating seawater. Seawater contains 3.5 wt % salts and is available at 20°C. The evaporation is to be carried out in a two-stage forward-feed evaporator, as shown in Figure P9.18, resulting in a final brine concentration of 7 wt %. Note that the vapor produced in the first stage is used as heat source for the second stage, while saturated steam at 1 atm is used in the first stage.

The second stage is at 46°C, and the stage 1 temperature is adjusted so that the difference between the temperatures of stream 4 and stage 1 is the same as the difference between the temperatures of stream 5 and stage 2. Thus, $T_4 - T_2 = T_5 - T_3$. This is sometimes called designing the system for equal-temperature driving forces. Also, the temperature of the vapor and brine leaving each stage is the same as the stage temperature. Thus, $T_2 = T_5$, $T_6 = T_3 = 46°C$. Note that in order for boiling to occur at the different stage temperatures, the stages must be operated at different pressures. For balance calculation purposes, neglect the effects of pressure on enthalpy. Assume that seawater and brine have constant heat capacities of 1 cal/g · °C and heats of vaporization of 555 cal/g.

(a) Construct a degree-of-freedom table and show that the problem is correctly specified.

(b) Deduce a calculation order for manual solution.

(c) Calculate the steam economy of the system, that is

$$\frac{\text{Mass of vapor produced}}{\text{Mass of steam consumed}} = \frac{\text{streams 5 + 6}}{\text{stream 4}}$$

9.19 An alternate way of operating a two-stage evaporator system is to use reverse feed as shown in Figure P9.19. In such a system, the feed is introduced in the second stage and brine leaves from the first stage. Suppose the feed is seawater at 20°C (3.5 wt % salt) and the waste solution is 7 wt % brine. Stream 4 is saturated steam at 1 atm, and stage 2 is operated at 46°C. As in

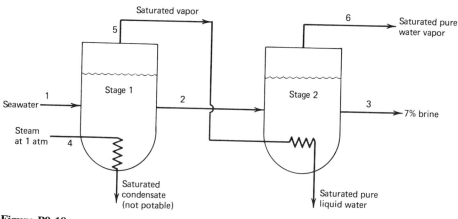

Figure P9.18

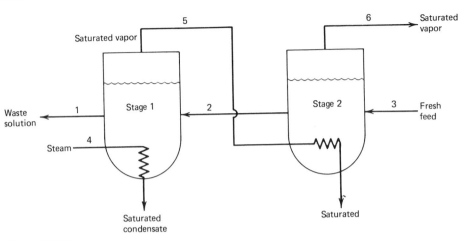

Figure P9.19

Problem 9.18, the system is designed for equal-temperature driving forces, that is,

$$T_4 - T_1 = T_5 - T_2$$

and the vapor and brine leave each stage at the stage temperature:

$$T_1 = T_5 \quad \text{and} \quad T_2 = T_6 = 46°C.$$

Assume constant solution heat capacity of 1 cal/g · °C and heat of vaporization of 555 cal/g.

(a) Construct a degree-of-freedom table, show that the problem is correctly specified, and deduce a calculation order.

(b) Assuming that 5 million liters per day of drinking water is to be produced, calculate the steam economy of the system (see Problem 9.18).

(c) Compare the systems of Figures P9.18 and P9.19. Which system is likely to have a higher steam economy? Why?

9.20 The evaporator system of Problem 9.18 can be generalized to more than two stages by adding more intermediate evaporator vessels, as shown in Figure P9.20. Using the data and specifications of Problem 9.18, consider the solution of a four-stage evaporator system for seawater desalination.

(a) Construct a degree-of-freedom table and show that the problem is correctly specified.

(b) Deduce a calculation order. How many variables must be carried?

(c) Compare the steam economy to that of a two-evaporator system (see Problem 9.18 for the definition).

9.21 Suppose the sequential modular strategy were to be used to solve a multistage evaporator system of the type shown in Figure P9.20.

(a) Outline a specialized module suitable for use with the sequential strategy which could be used to model a single evaporator stage.

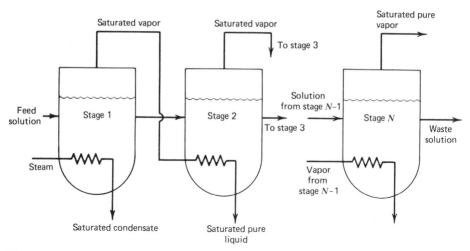

Figure P9.20

(b) What would be the natural specification parameters of the module formulated in part (a)? What physical property calculations would be required?

(c) Show how to use this module to represent the reverse-feed system shown in Figure P9.21.

9.22 Suppose the simultaneous strategy is used to solve the multistage evaporator system of Figure P9.20.

(a) Derive a reduced set of balance equations which must be solved for a system with an arbitrary number N of stages. Is linearization necessary?

(b) How is this equation set altered if a reversed-flow system must be treated?

(c) Use the data and specification of Problem 9.18 to solve the equation set of part (a) for a three-stage system.

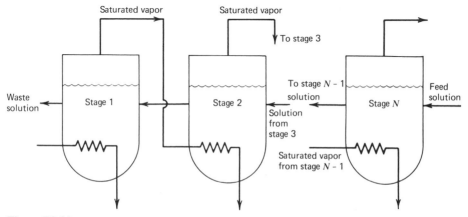

Figure P9.21

9.23 Consider the design of the two-stage evaporator system of Problem 9.19 in which the equal-temperature driving force assumption is replaced by the assumption that the heat transfer rate in the two stages is the same.

 (a) Confirm that the problem is correctly specified.
 (b) Deduce a calculation order for manual solution.
 (c) Solve the problem using the data of Problem 9.19.

9.24 Repeat the solution of Problem 9.23 assuming that seawater and brine have the properties of pure water. Use the steam tables to evaluate properties of water. What are the pressures in the two stages?

9.25 The efficiency of a multieffect evaporator system can be improved considerably by judicious exchange of heat between process streams. As shown in Figure P9.25, a portion of the vapor from the last stage can be used to preheat the cold fresh feed to within a few degrees of the last-stage temperature. (This small temperature difference ΔT is called the approach temperature.) In addition, the condensate produced at each stage can be used to further preheat the feed stream (exchangers 1B and 2B). Finally, a portion of the vapor generated at each stage can be used to preheat the feed to within ΔT of the vapor-condensing temperature. (This is referred to as a vapor bleed system.) Note that the combined condensed vapor from stage 1, stream 5, is only cooled to T_2 so that it can be combined with stream 4 at the same temperature. The combined condensate from stages 1 and 2 is only allowed to cool to T_3 so as to stay within ΔT of the temperature of the preheated feed stream leaving exchanger 3.

 Suppose this system is used to produce 5 million liters per day of drinking water from seawater available at 20°C and with 3.5 wt % salts. The waste brine is allowed to have a maximum temperature of 46°C and composition

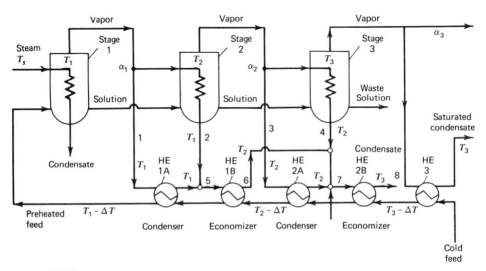

Figure P9.25

of 7 wt % salts. The steam to the first stage is saturated at 1 atm, and the design approach temperature difference, ΔT, is required to be 5°C. The system is to be designed using the equal-temperature driving force assumption (see Problem 9.18). Assume all liquids have the properties of liquid water.

(a) Show that the problem is correctly specified.
(b) Deduce a calculation order for manual solution.
(c) Calculate the vapor bleed fractions α_1, α_2, and α_3.
(d) Calculate the steam economy (see Problem 9.18).
(e) Repeat the solution assuming no preheating is carried out and compare the resulting steam economy to that obtained in part (d). *Hint:* use the fact that

$$\frac{T_s - T_3}{3} = T_s - T_1 = T_1 - T_2 = T_2 - T_3$$

9.26 The temperature in a highly exothermic catalytic reaction system can in part be controlled by depositing the catalyst on the walls of reactor tubes and passing the reactants over these surfaces while a coolant is circulated on the other side of the tube walls. In the methanation system shown in Figure P9.26, a feed consisting of 20% CO, 60% H_2, 2.5% CH_4, and the rest CO_2 at 50°C and 30 bar is mixed with a recycle stream and introduced into such a reactor. In the reactor, the reaction

$$CO(g) + 3H_2(g) \rightarrow CH_4(g) + H_2O(g)$$

takes place. The exit temperature of the gas is maintained at 500°C by circulating an organic heat transfer fluid through the reactor. The fluid boils at 370°C, it enters as a saturated liquid, and leaves containing 10% vapor.

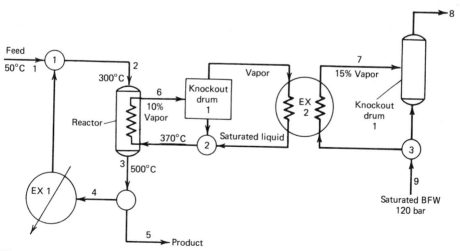

Figure P9.26

This two-phase mixture is separated in a knockout drum, the liquid is re-cycled, while the vapor is condensed in a heat exchanger. Saturated liquid water at 120 bar is boiled in the exchanger to obtain a mixture containing 15% vapor. This two-phase mixture is separated to yield saturated steam at 120 bar and saturated liquid which is mixed with makeup boiler feed water and recycled to the exchanger.

Because of heat transfer limitations, there is a limit to the rate at which heat can be removed from the gas stream in the reactor. At the conditions of the above system, suppose the transfer rate is 1000 kcal per average gram mole of gas in the reactor. The average number of gram moles in the reactor are based on the average of the input and output molar flows. The reactor inlet temperature is to be controlled to 300°C by cooling the recycle stream to a suitable temperature, and the product stream is to contain 5% CO on a H_2O- and CO_2-free basis. All units except exchanger 1 and the reactor are adiabatic.

Assume the heat transfer fluid has the following properties:

$$C_{P_L} = 0.8 \text{ cal/g} \cdot °C$$

$$C_{P_V} = 0.5 \text{ cal/g} \cdot °C$$

$$\Delta H_{VL}(370°C) = 300 \text{ cal/g}$$

For properties of the other species, use the data given in the appendix.

(a) Show that the problem is correctly specified.
(b) Deduce a calculation order for manual solution.
(c) Calculate the recycle rate, the circulation rate of heat transfer fluid, and the circulation rate of water.

9.27 Consider the system of Problem 9.26.

(a) Develop a formulation of the problem in terms of elementary modules.
(b) Select the location of any tear streams and deduce a calculation order for sequential solution.
(c) Identify the constraints which will need to be satisfied.
(d) Determine a calculation order which will include both the tear stream and the constraint iterations.

9.28 A chemical plant has the need for a large quantity of heat at 150°C as well as 1000 kW of shaft work to be delivered via a steam turbine. These needs are to be provided by burning a powdered char in a cyclonic combustor and using the hot flue gases to raise 100 bar steam via the network shown in Figure P9.28. The powdered char is dry, consists of 80 wt % carbon and the rest ash, and is fed at 100°C and 2 bar. The air is dry and preheated to 200°C. The combustor is operated adiabatically with 50% excess air and with the ash leaving as a liquid slag. The slag and flue gas will leave at the same temperature.

The flue gas is first used to heat the saturated steam to 550°C in the

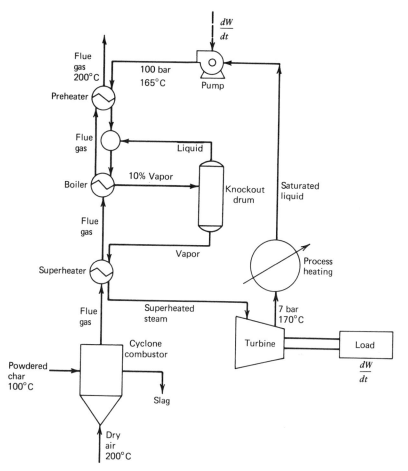

Figure P9.28

superheater, then it is used to boil saturated water to produce a 10% vapor mixture in the boiler, and finally it is used to preheat the condensate from 165°C to saturation temperature. The flue gas must leave the plant at 200°C in order to have satisfactory stack performance.

In the turbine, the superheated steam is expanded adiabatically to 7 bar and 170°C to produce the desired shaft work. After use for process heating needs, the steam is returned as saturated liquid at 7 bar and is pumped to 100 bar in a pump which operates adiabatically and essentially isothermally.

(a) Show that the process is correctly specified.
(b) Deduce a calculation sequence for manual solution.
(c) Calculate the required feed rate of char, the heat rate available for process needs, and the steam circulation rate.

Assume that the slag has a heat of fusion of 35 cal/g at 1600 K and a liquid heat capacity of 0.25 cal/g · °C.

9.29 Apply the sequential strategy to solve Problem 9.28.

(a) Develop a flowsheet formulation in terms of the expanded module library.

(b) Select the tear streams required and identify the constraint specifications.

(c) Outline the calculation sequence which must be followed to converge both the tear streams and the constraint specifications.

(d) Write a program to solve the problem as outlined in part (c).

9.30 The system of Problem 9.28 can be readily solved using the simultaneous approach.

(a) Reduce the problem equations to a set which will require iterative solution.

(b) Verify whether any of the equations are nonlinear. If so, formulate their linearizations.

(c) Carry out one iteration pass assuming as initial estimate that 1.6 kW can be generated per 1 kg char.

9.31 Conversion of a synthesis gas mixture to a synthetic natural gas can be accomplished by first reacting the gas in a shift reactor ($CO + H_2O \rightarrow CO_2 + H_2$) so as to adjust the CO-to-H_2 ratio to $1:3$ followed by catalytic reaction to produce CH_4 via the reaction

$$CO + 3H_2 \rightarrow CH_4 + H_2O$$

This reaction is strongly exothermic, consequently steps must be taken to control the temperature of the gas mixture. In the flowsheet shown in Figure P9.31, temperature control is achieved by

(a) Using a large recycle flow (six times the feed flow to the recycle loop, stream 3).

(b) Breaking the methanation reactor into three stages with injection of cold fresh feed between stages.

As a result of the recycle, it is impractical to attempt to achieve low CO concentrations in the effluent from the loop. Instead, it proves economical to use a final cleanup methanator to reduce CO levels to the 0.1% required by SNG standards. The following additional specifications are imposed on the flowsheet of Figure P9.31.

1. Feed flow (all in lbmol/h): H_2, 1.8×10^4; CO, 10^4; CH_4, 1300; CO_2, 1500; N_2, 200; H_2O, 4×10^3. Temperature, 560°F; pressure, 1000 psia.

2. The shift reactor effluent is cooled to 100°F in heat exchanger 1 and reduced to 0.1% H_2O by removing liquid H_2O in a knockout drum (separator 1).

3. The temperature rise over methanator 1 is limited to 35°F; over methanator 2, to 55°F; and over methanator 3, to 90°F.

4. The methanator 2 inlet temperature is to be controlled at 460°F and that of methanator 3, at 500°F.

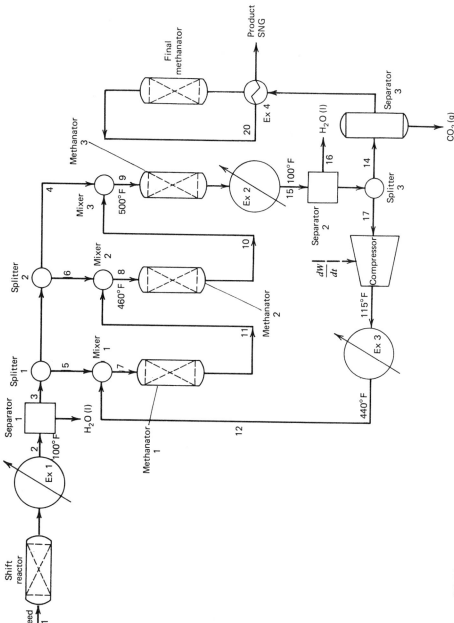

Figure P9.31

5. The effluent from methanator 3 is cooled to 100°F in exchanger 2; the condensed water is separated out in the knockout drum to result in a water mole fraction of 0.1% in the gas stream; and finally the gas stream is split for recycling.

6. Separator 3 removes 99.5% of the CO_2 to result in a feed stream to the final methanator at 100°F, 895 psia, and containing 2% CO.

7. The effluent from the final methanator is used to preheat its feed and leaves the process with a composition of 95% CH_4 and 0.1% CO.

8. The recycle stream is compressed adiabatically to 982 psia and 115°F and then is preheated to 440°F.

9. All reactors, mixers, splitters, and the compressor are adiabatic. Separator 1, separator 2, separator 3, and exchanger 4 are adiabatic but exchangers 1, 2, and 3 are not.

Use any required data from the appendices. As an initial approximation, use constant average heat capacities.

(a) Show that the problem is correctly specified.

(b) Deduce a calculation order.

(c) Determine the ratios into which the feed stream is split as well as all heat and work rates.

(d) Determine whether the effluent of methanator 3 can be used to heat stream 12 to 440°F, thus combining exchangers 2 and 3 into one.

9.32 Consider the methanation system of Problem 9.31. Assume an expanded module library is available.

(a) Formulate a flowsheet model expressed in terms of these modules.

(b) Select the tear stream locations and indicate the order of module calculations assuming the problem is unconstrained.

(c) Identify which of the specifications given in Probelm 9.31 must be treated as constraints.

(d) Formulate a sequential calculation order which includes all required constraint iterations.

(e) Write a program to carry out part (d).

9.33 Consider the system of Problem 9.31. Suppose the maximum allowable effluent temperature in the methanators is 650°F. It is possible to eliminate two of the methanators in the recycle loop and to use only one reactor? What is the minimum allowable recycle rate? Is it necessary to reduce the temperature of stream 2 to achieve feasible flows?

9.34 Consider the two-stage ammonia synthesis loop shown in Figure P9.34. In the flowsheet, a cold stoichiometric feed of N_2 and H_2 is introduced between the reactor stages. In each reactor, the synthesis reaction

$$N_2 + 3H_2 \rightleftharpoons 2NH_2$$

takes place with specified conversion. The effluent of the second reactor stage is cooled by exchanging heat with the input stream to the first reactor

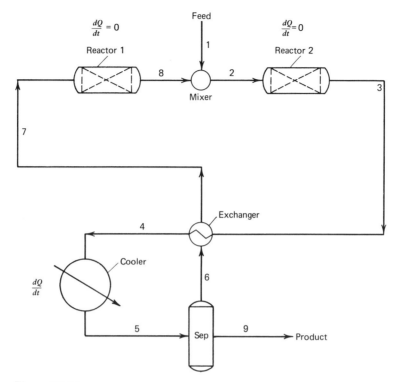

Figure P9.34

stage. After further cooling, stream 5 is separated to recover a product stream containing all of the NH_3 and some H_2 and N_2.

The following specifications are given:

1. The feed rate is 97.8 mol/h with 15% N_2 and 75% H_2; the feed temperature is 50°C.
2. The input stream to the first reactor stage is at 425°C.
3. Input stream to the separator is at 50°C and the temperatures of the outlet streams are equal to each other.
4. The recycle stream is to contain no NH_3 and 0.9945 each of the N_2 and H_2 fed to the separator.
5. The conversion of N_2 is 10% in stage 1 and 12.33% in stage 2.
6. Only the cooler is nonadiabatic.

Assume all streams are gas phase and that pressure effects can be neglected.

(a) Show that the problem is correctly specified.
(b) Show that the problem is a constrained balance problem when represented in elementary module form.
(c) Select a set of tear streams and outline a calculation procedure.
(d) Solve the problem using a suitable computer program.

9.35 An alternate process for synthesis gas methanation uses methanation catalysts that can support a higher temperature and heat exchangers to cool the
hot exit gases from each methanation stage. With these modifications, the
process of Figure P9.35 will differ from that of Figure P9.31 primarily in
that a lower recycle rate can be used, thus allowing all units on the recycle
loop to be smaller. The following specifications are given for the process of
Figure P9.35:

1. The feed flow (in lbmol/h) is H_2, 1.8×10^4; CO, 10^4; CH_4, 1300; CO_2,
    1500; N_2, 200; H_2O, 4×10^4. Temperature, 560°F; pressure, 1000 psia.
2. The shift reactor effluent is cooled to 100°F in exchanger 1 and is
    reduced to 0.1% H_2O by removing liquid H_2O in a knockout drum
    (separator 1).
3. The resulting stream 3 is split as follows: 75% is sent to mixer 1, and
    the remainder is split equally between mixers 2 and 3.
4. The inlet temperatures to methanator stages 2 and 3 are controlled to
    440°F. The outlet temperatures of all three methanation stages are
    limited to 900°F.
5. The effluent stream from methanator 3 is cooled to 720°F, then cooled
    further by exchange with the recycle stream, and finally cooled to 100°F
    in exchanger 6.
6. The water content of stream 13 is reduced to 0.1% by removing liquid
    H_2O in a knockout drum (separator 2).
7. The resulting gas stream 15 is split with a portion recycled and the
    remainder processed in separator 3, which removes 99.5% of the CO_2
    to result in a feed to the final methanator at 100°F, 943 psia, and
    containing 2% CO.
8. In the final methanator, the CO level is reduced to 0.1%, resulting in
    a CH_4 content of 95%.
9. The recycle stream is compressed adiabatically to 982 psia and 107°F
    and is preheated to 650°F by heat exchange with stream 13.
10. Assume all units operate adiabatically with the exception of heat ex-
    changers 1, 2, 3, 4, and 6.

Use any required data from the appendices.

(a) Show that the problem is correctly specified.
(b) Deduce a calculation order for manual solution.
(c) Determine the recycle ratio stream 12/stream 3 and all heat and work
     rates.
(d) How would the design be affected if all of stream 3 were sent to mixer
     1? Is there any advantage to splitting the feed?

9.36 Suppose that a sequential modular system was used to solve the balances
for Problem 9.35.

(a) Formulate a flowsheet model expressed in terms of the expanded mod-
     ule library.

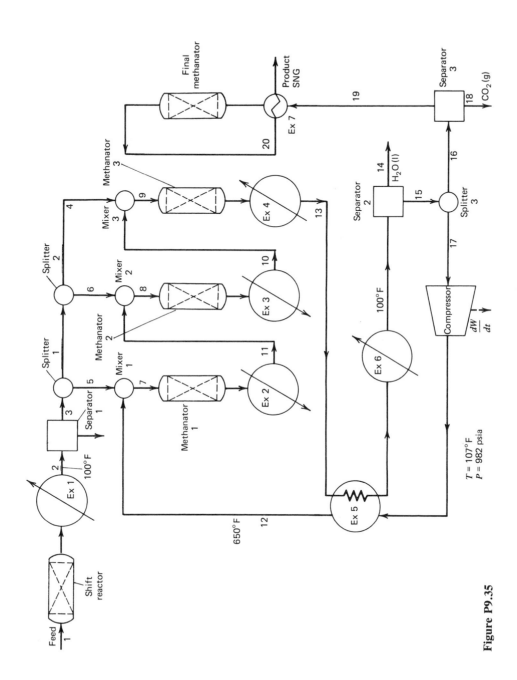

Figure P9.35

    (b) Select the tear stream locations and specify the order of the module calculations assuming the problem is unconstrained.

    (c) Formulate a calculation order which includes iterations on all problem constraint specifications.

9.37 Consider the system of Problem 9.35. Assume constant average heat capacities and negligible pressure effects.

    (a) Formulate the problem for solution using the simultaneous approach.

    (b) Identify which equations are nonlinear and will require linearization.

    (c) Formulate the linearized equation set (possibly in reduced form by eliminating some simple equations) and write a program to solve the problem which makes use of a library linear equation solving subroutine.

APPENDIX

Physical Properties Data

1 Atomic Weights and Numbers[a]

Element	Symbol	Atomic Number	Atomic Weight	Element	Symbol	Atomic Number	Atomic Weight
Actinium	Ac	89	—	Fluorine	F	9	18.9984
Aluminum	Al	13	26.9815	Francium	Fr	87	—
Americium	Am	95	—	Gadolinium	Gd	64	157.25
Antimony	Sb	51	121.75	Gallium	Ga	31	69.72
Argon	Ar	18	39.948	Germanium	Ge	32	72.59
Arsenic	As	33	74.9216	Gold	Au	79	196.967
Astatine	At	85	—	Hafnium	Hf	72	178.49
Barium	Ba	56	137.34	Helium	He	2	4.0026
Berkelium	Bk	97	—	Holmium	Ho	67	164.930
Beryllium	Be	4	9.0122	Hydrogen	H	1	1.00797
Bismuth	Bi	83	208.980	Indium	In	49	114.82
Boron	B	5	10.811	Iodine	I	53	126.9044
Bromine	Br	35	79.904	Iridium	Ir	77	192.2
Cadmium	Cd	48	112.40	Iron	Fe	26	55.847
Calcium	Ca	20	40.08	Krypton	Kr	36	83.80
Californium	Cf	98	—	Lanthanum	La	57	138.91
Carbon	C	6	12.01115	Lawrencium	Lr	103	—
Cerium	Ce	58	140.12	Lead	Pb	82	207.19
Cesium	Cs	55	132.905	Lithium	Li	3	6.939
Chlorine	Cl	17	35.453	Lutetium	Lu	71	174.97
Chromium	Cr	24	51.996	Magnesium	Mg	12	24.312
Cobalt	Co	27	58.9332	Manganese	Mn	25	54.9380
Copper	Cu	29	63.546	Mendelevium	Md	101	—
Curium	Cm	96	—	Mercury	Hg	80	200.59
Dysprosium	Dy	66	162.50	Molybdenum	Mo	42	95.94
Einsteinium	Es	99	—	Neodymium	Nd	60	144.24
Erbium	Er	68	167.26	Neon	Ne	10	20.183
Europium	Eu	63	151.96	Neptunium	Np	93	—
Fermium	Fm	100	—	Nickel	Ni	28	58.71

[a] Atomic weights apply to naturally occurring isotopic compositions and are based on an atomic mass of $^{12}C = 12$.

1 Atomic Weights and Numbers[a] (continued)

Element	Symbol	Atomic Number	Atomic Weight	Element	Symbol	Atomic Number	Atomic Weight
Niobium	Nb	41	92.906	Silicone	Si	14	28.086
Nitrogen	N	7	14.0067	Silver	Ag	47	107.868
Nobelium	No	102	—	Sodium	Na	11	22.9898
Osmium	Os	75	190.2	Strontium	Sr	38	87.62
Oxygen	O	8	15.9994	Sulfur	S	16	32.064
Palladium	Pd	46	106.4	Tantalum	Ta	73	180.948
Phosphorus	P	15	30.9738	Technetium	Tc	43	—
Platinum	Pt	78	195.09	Tellurium	Te	52	127.60
Plutonium	Pu	94	—	Terbium	Tb	65	158.924
Polonium	Po	84	—	Thallium	Tl	81	204.37
Potassium	K	19	39.102	Thorium	Th	90	232.038
Praseodymium	Pr	59	140.907	Thulium	Tm	69	168.934
Promethium	Pm	61	—	Tin	Sn	50	118.69
Protactinium	Pa	91	—	Titanium	Ti	22	47.90
Radium	Ra	88	—	Tungsten	W	74	183.85
Radon	Rn	86	—	Uranium	U	92	238.03
Rhenium	Re	75	186.2	Vanadium	V	23	50.942
Rhodium	Rh	45	102.905	Xenon	Xe	54	131.30
Rubidium	Rb	37	84.57	Ytterbium	Yb	70	173.04
Ruthenium	Ru	44	101.07	Yttrium	Y	39	88.905
Samarium	Sm	62	150.35	Zinc	Zn	30	65.37
Scandium	Sc	21	44.956	Zirconium	Zr	40	91.22
Selenium	Se	34	78.96				

2 Molecular Weights, Normal Boiling Point, Latent Heat (SI Units and AE Units)

	Name	Molecular Weight	Normal Boiling Point		Latent Heat at NBP	
			K	R	J/mol	Btu/lbmol
1	ACENAPHTHYLENE	152.180	543.161	977.690	48485.8	20859.1
2	ACETALDEHYDE	44.052	293.561	528.410	25699.1	11056.0
3	ACETIC ACID	60.052	391.661	704.990	24308.7	10457.8
4	ACETONE	58.080	329.281	592.706	29087.2	12513.6
5	ACETONITRILE	41.052	354.761	638.570	31510.7	13556.2
6	ACETYL CHLORIDE	78.500	324.081	583.346	46227.9	19887.7
7	ACETYLENE	26.036	188.401	339.122	16419.1	7063.6
8	ACRYLONITRILE	53.060	350.461	630.830	32630.1	14037.8
9	AMMONIA	17.032	239.731	431.516	23351.0	10045.8
10	ANILINE	93.116	457.291	823.124	43165.8	18570.4
11	ARGON	39.944	87.291	157.124	6527.0	2808.0
12	BENZENE	78.108	353.261	635.870	30763.4	13234.7
13	BIPHENYL	154.200	528.361	951.050	49212.9	21171.9
14	BROMINE	159.830	331.921	597.458	29413.8	12654.1
15	BROMOBENZENE	157.020	429.361	772.850	37823.6	16272.1
16	CARBON DIOXIDE	44.011	194.681	350.426	16560.9	7124.7
17	CARBON DISULFIDE	76.130	319.400	574.920	26334.4	11329.3
18	CARBON MONOXIDE	28.010	81.691	147.044	6065.3	2609.3
19	CARBON TET	153.840	349.700	629.460	36882.1	15867.0
20	CHLORINE	70.914	239.111	430.400	20410.0	8780.6
21	CHLOROACETYL-CL	112.950	382.161	687.890	57030.5	24535.1
22	CHLOROBENZENE	112.557	405.361	729.650	36449.7	15681.0
23	CHLOROFORM	119.389	334.891	602.804	29470.1	12678.3
24	CHRYSENE	228.270	721.161	1298.090	68034.6	29269.2
25	CIS-2-BUTENE	56.100	276.883	498.390	23462.9	10094.0
26	CIS-2-PENTENE	70.130	310.106	558.190	25617.7	11021.0
27	CYCLOHEXANE	84.160	353.900	637.020	29975.3	12895.7
28	CYCLOPENTANE	70.130	322.422	580.360	27246.0	11721.5
29	C3ALKYLBENZENE	120.200	438.161	788.690	38267.8	16463.2
30	C3ALKYLNAPHTHALENE	170.200	549.861	989.750	48678.5	20942.0
31	DICHLOROACETYL-CL	147.390	380.661	685.190	58136.7	25011.0
32	DICYCLOPENTADIENE	132.196	443.161	797.690	37279.4	16038.0
33	DIETHYL ETHER	74.120	307.711	553.880	26693.3	11483.7
34	DIETHYL KETONE	86.140	374.961	674.930	33178.1	14273.6
35	DIETHYLENE GLYCOL	106.120	518.800	933.840	52314.6	22506.3
36	DIMETHYL ACETAMIDE	87.120	440.161	792.290	39293.6	16904.5
37	DIMETHYL ETHER	46.068	248.321	446.978	21510.1	9253.9
38	DIMETHYL SULFIDE	62.134	310.501	558.902	26998.9	11615.2
39	DIMETHYLACETYLENE	54.090	300.200	540.360	26485.6	11394.4
40	DIMETHYLFORMAMIDE	73.094	426.161	767.090	38080.7	16382.7
41	ETHANE	30.068	184.531	332.156	14715.6	6330.8
42	ETHANOL	46.068	351.481	632.666	38577.3	16596.4
43	ETHYL ACETATE	88.100	350.261	630.470	32269.4	13882.6
44	ETHYL CHLORIDE	64.517	285.431	513.776	24685.8	10620.1
45	ETHYL FORMATE	74.080	327.500	589.500	30058.9	12931.6

2 Molecular Weights, Normal Boiling Point, Latent Heat (SI Units and AE Units) (continued)

	Name	Molecular Weight	Normal Boiling Point K	R	Latent Heat at NBP J/mol	Btu/lbmol
46	ETHYL MERCAPTAN	62.134	308.161	554.690	26778.3	11520.3
47	ETHYLACETYLENE	54.088	281.251	506.252	24726.1	10637.4
48	ETHYLAMINE	45.084	289.741	521.534	27565.2	11858.8
49	ETHYLBENZENE	106.160	409.344	736.820	35998.2	15486.8
50	ETHYLCYCLOHEXANE	112.210	404.944	728.900	34659.6	14910.9
51	ETHYLCYCLOPENTANE	98.180	376.628	677.930	32094.1	13807.2
52	ETHYLENE	28.056	169.451	305.012	13544.1	5826.8
53	ETHYLENE GLYCOL	62.070	470.600	847.080	49629.6	21351.1
54	ETHYLENE OXIDE	44.052	283.661	510.590	25526.5	10981.8
55	FLUORANTHENE	202.258	666.161	1199.090	61549.7	26479.3
56	FLUORENE	166.223	571.061	1027.910	50840.7	21872.2
57	FORMALDEHYDE	30.010	253.961	457.130	23304.0	10025.6
58	FURFURAL	96.080	434.861	782.750	43124.7	18552.7
59	HYDROGEN	2.016	20.381	36.686	1334.6	574.2
60	HYDROGEN CHLORIDE	36.465	188.127	338.629	16150.3	6948.0
61	HYDROGEN CYANIDE	27.030	298.861	537.950	26891.5	11569.0
62	HYDROGEN IODIDE	127.910	237.781	428.006	20542.7	8837.7
63	HYDROGEN SULFIDE	34.082	212.820	383.076	18678.3	8035.6
64	INDAN	118.170	450.161	810.290	39007.4	16781.4
65	INDENE	116.163	455.161	819.290	39732.0	17093.1
66	IODOBENZENE	204.020	461.491	830.684	39496.1	16991.6
67	ISO-BUTANE	58.120	261.431	470.576	21291.8	9160.0
68	ISO-BUTANOL	74.120	381.600	686.880	41253.9	17747.8
69	ISO-BUTENE	56.100	266.261	479.270	22050.6	9486.4
70	ISO-BUTYRALDEHYDE	72.110	337.161	606.890	31272.9	13453.9
71	ISO-PENTANE	72.150	301.011	541.820	24710.2	10630.6
72	ISO-PROPANOL	60.094	355.421	639.758	39874.9	17154.6
73	M-DICHLOROBENZENE	147.010	446.161	803.090	38593.9	16603.5
74	M-XYLENE	106.160	412.267	742.080	36247.6	15594.1
75	MEK	72.104	352.751	634.952	31268.6	13452.1
76	METHACRYLONITRILE	67.090	363.461	654.230	31798.3	13680.0
77	METHANE	16.042	111.671	201.008	8179.5	3518.9
78	METHANOL	32.042	337.671	607.808	35270.4	15173.7
79	METHYL ACETATE	74.080	330.411	594.740	27932.3	12016.8
80	METHYL CHLORIDE	50.491	248.941	448.094	21400.0	9206.5
81	METHYL FORMATE	60.050	304.941	548.894	28066.7	12074.6
82	METHYL IODIDE	141.950	315.661	568.190	27173.5	11690.3
83	METHYL PROPIONATE	88.100	353.100	635.580	32214.2	13858.9
84	METHYLACETYLENE	40.062	249.941	449.894	22282.1	9586.0
85	METHYLAMINE	31.058	266.711	480.080	25815.5	11106.1
86	METHYLCYCLOHEXANE	98.180	374.094	673.370	31284.9	13459.1
87	METHYLCYCLOPENTANE	84.160	344.972	620.950	29127.6	12531.0
88	METHYLSTYRENE	118.200	443.161	797.690	38478.6	16553.9
89	N-BUTANE	58.120	272.661	490.790	22416.0	9643.6
90	N-BUTANOL	74.120	390.891	703.604	43803.9	18844.9

2 Molecular Weights, Normal Boiling Point, Latent Heat (SI Units and AE Units) (continued)

	Name	Molecular Weight	Normal Boiling Point K	Normal Boiling Point R	Latent Heat at NBP J/mol	Latent Heat at NBP Btu/lbmol
91	N-BUTYLBENZENE	134.212	456.467	821.641	39243.5	16882.9
92	N-BUTYLCYCLOHEXANE	140.260	454.108	817.395	38492.9	16560.1
93	N-BUTYRIC ACID	88.104	436.431	785.576	45140.9	19420.1
94	N-DECANE	142.280	447.283	805.110	39938.8	17182.1
95	N-DODECANE	170.328	489.439	880.990	43685.1	18793.8
96	N-HEPTANE	100.200	371.589	668.860	32128.7	13822.1
97	N-HEXADECANE	226.430	559.956	1007.920	51142.9	22002.2
98	N-HEXANE	86.172	341.901	615.422	28851.3	12412.1
99	N-NONANE	128.250	423.961	763.130	37537.9	16149.2
100	N-OCTANE	114.220	398.828	717.890	34940.8	15031.9
101	N-PENTADECANE	212.410	543.789	978.820	50070.2	21540.7
102	N-PENTANE	72.150	309.233	556.620	25937.3	11158.5
103	N-PROPANOL	60.094	370.371	666.668	41221.7	17734.0
104	N-PROPYL ACETATE	102.130	374.711	674.480	34213.2	14718.9
105	N-PROPYLBENZENE	120.186	432.378	778.281	38241.8	16452.0
106	N-PROPYLCYCLOHEXAN	126.234	429.885	773.793	36063.8	15515.0
107	N-PROPYLCYCLOPENTA	112.208	404.111	727.400	34107.8	14673.5
108	N-TETRADECANE	198.380	526.733	948.120	48051.1	20672.1
109	N-TRIDECANE	184.350	508.600	915.480	46495.9	20003.0
110	N-UNDECANE	156.300	469.050	844.290	42150.3	18133.5
111	NAPHTHALENE	128.164	491.116	884.009	42180.9	18146.7
112	NEO-PENTANE	72.146	282.664	508.795	22752.4	9788.3
113	NEON	20.183	27.300	49.140	1738.8	748.1
114	NITRIC OXIDE	30.010	121.400	218.520	13778.0	5927.4
115	NITROGEN	28.016	77.361	139.250	5577.5	2399.5
116	NITROGEN DIOXIDE	46.010	294.500	530.100	38117.0	16398.3
117	NITROUS OXIDE	44.020	183.700	330.660	15594.8	6709.1
118	O-CRESOL	108.134	464.165	835.497	45192.3	19442.2
119	O-DICHLOROBENZENE	147.010	452.161	813.890	39937.0	17181.3
120	O-XYLENE	106.160	417.572	751.630	36617.0	15753.0
121	OXYGEN	32.000	90.181	162.326	6820.5	2934.2
122	P-DICHLOROBENZENE	147.010	447.061	804.710	39193.4	16861.4
123	P-XYLENE	106.160	411.511	740.720	35928.9	15457.0
124	PHENANTHRENE	178.220	612.761	1102.970	55728.7	23975.1
125	PHENOL	94.108	455.000	819.000	45693.0	19657.6
126	PHOSGENE	98.930	280.721	505.298	24402.8	10498.3
127	PROPADIENE	40.060	238.661	429.590	19092.8	8213.9
128	PROPANE	44.094	231.091	415.964	18773.1	8076.4
129	PROPIONIC ACID	74.078	414.151	745.472	41833.9	17997.4
130	PROPYL FORMATE	88.100	354.500	638.100	32400.7	13939.1
131	PROPYLENE	42.080	225.461	405.830	18372.6	7904.1
132	PYRENE	202.258	635.161	1143.290	58685.2	25247.0
133	STYRENE	104.144	418.301	752.942	36901.1	15875.2
134	SULFUR DIOXIDE	64.066	263.145	473.661	24915.7	10719.0
135	SULFUR TRIOXIDE	80.066	317.911	572.240	41797.1	17981.5

2 Molecular Weights, Normal Boiling Point, Latent Heat (SI Units and AE Units) (continued)

	Name	Molecular Weight	Normal Boiling Point		Latent Heat at NBP	
			K	R	J/mol	Btu/lbmol
136	T-BUTYL ALCOHOL	74.120	355.581	640.046	39037.5	16794.3
137	TETRAETHENE GLYCOL	194.230	592.100	1065.780	73005.5	31407.7
138	THIOPHENE	84.130	357.361	643.250	34206.4	14715.9
139	TOLUENE	92.134	383.786	690.815	33460.6	14395.1
140	TRANS-2-BUTENE	56.100	274.039	493.270	22991.8	9891.3
141	TRANS-2-PENTENE	70.130	309.517	557.130	25456.4	10951.6
142	TRICHLOROACETYL-CL	181.830	391.161	704.090	60524.5	26038.2
143	TRICHLOROETHYLENE	131.400	360.061	648.110	31448.3	13529.4
144	TRIETHYLENE GLYCOL	150.170	560.400	1008.720	60807.5	26160.0
145	TRIMETHYLAMINE	59.110	276.031	496.856	22950.7	9873.6
146	VINYLACETYLENE	52.072	278.711	501.680	24535.4	10555.4
147	VINYL CHLORIDE	62.501	259.461	467.030	20801.2	8948.9
148	WATER	18.016	373.161	671.690	40656.2	17490.7
149	1-BUTENE	56.100	266.900	480.420	22080.6	9499.3
150	1-ET-2-ME-BENZENE	120.186	438.161	788.690	38867.2	16721.0
151	1-HEPTENE	98.180	366.806	660.250	31735.2	13652.8
152	1-HEXENE	84.160	336.644	605.960	28207.4	12135.1
153	1-ME-4-ETH-NAPHTHA	182.350	548.161	986.690	51463.2	22140.0
154	1-METHYLINDENE	130.190	473.161	851.690	41368.8	17797.3
155	1-METHYLNAPHTHALEN	142.190	517.803	932.046	45481.6	19566.6
156	1-PENTENE	70.130	303.128	545.630	25412.2	10932.6
157	1-PHENYLINDENE	192.240	595.161	1071.290	54288.8	23355.6
158	1-PHENYLNAPHTHALEN	204.260	598.161	1076.690	55494.9	23874.5
159	1,1-DICHLOROETHANE	98.966	330.441	594.794	28976.9	12466.2
160	1,2-BUTADIENE	54.090	284.011	511.220	22849.5	9830.1
161	1,2-DICHLOROETHANE	98.966	356.631	641.936	31617.2	13602.0
162	1,2-DIME-3-ETHBZ	134.212	467.071	840.728	41790.0	17978.5
163	1,2,3-TRIME-INDENE	158.229	505.161	909.290	44997.3	19358.3
164	1,2,4-TRI-CL-BZ	181.455	486.641	875.954	43598.6	18756.5
165	1,3-BUTADIENE	54.090	268.750	483.750	22472.5	9667.9
166	112TRI-CL-ETHANE	133.415	386.931	696.476	34231.5	14726.7
167	2-ETHYLFLUORENE	194.279	582.161	1047.890	53803.2	23146.7
168	2-ME-1-BUTENE	70.130	304.322	547.780	24937.6	10728.4
169	2-ME-2-BUTENE	70.130	311.728	561.110	25817.6	11107.0
170	2-METHYLINDENE	130.190	481.161	866.090	42317.4	18205.4
171	2-METHYLNAPHTHALEN	142.190	514.213	925.584	45202.0	19446.4
172	2-METHYLPENTANE	86.170	333.433	600.180	28021.4	12055.1
173	2,2-DIMETHYLBUTANE	86.170	322.900	581.220	26394.5	11355.2
174	2,3-DIMETHYLBUTANE	86.170	331.150	596.070	27404.5	11789.7
175	2,3,5-TRIME-NAPHTH	182.350	558.161	1004.690	52402.0	22543.9
176	2,7-DIMETHYLNAPHTH	156.216	536.161	965.090	48772.1	20982.3
177	3-ME-1-BUTENE	70.130	293.222	527.800	23173.8	9969.6
178	3-METHYLPENTANE	86.170	336.444	605.600	28294.3	12172.5

Source: FLOWTRAN Data Base. These tabulations were prepared by Dr. D. R. Schneider of the Monsanto Company using the Physical Properties package of the FLOWTRAN system. The tabulations are reproduced with the permission of the Monsanto Company.

3 Ideal Gas Heat Capacities

$$C_p = a + bT + cT^2 + dT^3 + eT^4$$

TABLE 3A   SI Units (J, K)

	Name	Ideal Gas Heat Capacity in J/mol-K when Temperature in K				
		a	b	c	d	e
1	ACENAPHTHYLENE	-4.23583D+01	7.31681D-01	-1.90463D-04	-3.22290D-07	1.94553D-10
2	ACETALDEHYDE	2.45377D+01	7.60130D-02	1.36254D-04	-1.99942D-07	7.59551D-11
3	ACETIC ACID	6.89949D+00	2.57068D-01	-1.91771D-04	7.57676D-08	-1.23175D-11
4	ACETONE	2.31317D+01	1.62824D-01	8.01548D-05	-1.60497D-07	5.81406D-11
5	ACETONITRILE	2.44795D+01	8.91399D-02	3.50089D-05	-8.43081D-08	3.40092D-11
6	ACETYL CHLORIDE	2.64542D+01	1.83863D-01	-1.13740D-04	1.92063D-08	6.51935D-12
7	ACETYLENE	2.18212D+01	9.20580D-02	-6.52231D-05	1.81959D-08	0.0
8	ACRYLONITRILE	1.07482D+01	2.20682D-01	-1.57773D-04	4.84219D-08	-1.22867D-12
9	AMMONIA	2.75500D+01	2.56278D-02	9.90042D-06	-6.68639D-09	0.0
10	ANILINE	-2.25594D+00	3.07834D-01	2.41742D-04	-5.37543D-07	2.35694D-10
11	ARGON	2.07723D+01	0.0	0.0	0.0	0.0
12	BENZENE	1.85868D+01	-1.17439D-02	1.27514D-03	-2.07984D-06	1.05329D-09
13	BIPHENYL	-8.31984D+01	1.02502D+00	-7.22383D-04	1.42495D-07	3.97897D-11
14	BROMINE	3.36874D+01	1.02992D-02	-8.90254D-06	2.67924D-09	0.0
15	BROMOBENZENE	7.49152D-01	3.47418D-01	-1.06396D-05	-2.47154D-07	1.26969D-10
16	CARBON DIOXIDE	1.90223D+01	7.96291D-02	-7.37067D-05	3.74572D-08	-8.13304D-12
17	CARBON DISULFIDE	3.30999D+01	1.06167D-02	2.75934D-04	-3.42168D-07	1.30288D-10
18	CARBON MONOXIDE	2.90063D+01	2.49235D-03	-1.86440D-05	4.79892D-08	-2.87266D-11
19	CARBON TET	8.97631D+00	4.20036D-01	-7.51639D-04	6.27332D-07	-1.99811D-10
20	CHLORINE	2.85463D+01	2.38795D-02	-2.13631D-05	6.47263D-09	0.0
21	CHLOROACETYL-CL	2.03658D+01	3.05402D-01	-3.86004D-04	2.65736D-07	-7.54466D-11
22	CHLOROBENZENE	-2.77932D+00	3.57786D-01	-2.31480D-05	-2.40458D-07	1.25736D-10
23	CHLOROFORM	3.18924D+01	1.44743D-01	-1.11583D-04	3.07210D-08	0.0
24	CHRYSENE	-8.13533D+01	1.09414D+00	-1.66278D-04	-6.42055D-07	3.54798D-10
25	CIS-2-BUTENE	2.62404D+01	7.09338D-02	5.71217D-04	-8.38061D-07	3.70594D-10
26	CIS-2-PENTENE	1.11806D+01	2.65029D-01	2.58923D-04	-4.92368D-07	1.96524D-10
27	CYCLOHEXANE	7.04449D+00	1.30074D-01	1.08205D-03	-1.54513D-06	6.51190D-10
28	CYCLOPENTANE	-1.91083D+00	1.30865D-01	8.53142D-04	-1.30863D-06	5.87916D-10
29	C3ALKYLBENZENE	2.97401D+01	7.04306D-01	-7.62473D-04	8.82742D-07	-3.66217D-10
30	C3ALKYLNAPHTHALENE	6.44864D+01	7.84466D-01	-5.33115D-04	6.39615D-07	-2.84210D-10
31	DICHLOROACETYL-CL	1.42774D+01	4.26942D-01	-6.58268D-04	5.12266D-07	-1.57413D-10
32	DICYCLOPENTADIENE	-3.39777D+01	8.06270D-01	-3.41494D-04	-1.67598D-07	1.39284D-10
33	DIETHYL ETHER	4.67637D+01	1.00949D-01	5.69049D-04	-7.74108D-07	3.03364D-10
34	DIETHYL KETONE	2.67448D+01	2.45037D-01	3.35630D-04	-5.81979D-07	2.41068D-10
35	DIETHYLENE GLYCOL	4.46173D+01	1.44518D-01	6.82006D-04	-9.75239D-07	3.91107D-10
36	DIMETHYL ACETAMIDE	2.57501D+01	2.82595D-01	1.70219D-04	-4.03253D-07	1.76476D-10
37	DIMETHYL ETHER	4.89030D+01	-8.81692D-03	4.24399D-04	-5.12852D-07	1.93182D-10
38	DIMETHYL SULFIDE	3.34317D+01	1.13437D-01	1.39993D-04	-2.41217D-07	1.02266D-10
39	DIMETHYLACETYLENE	1.32380D+01	2.60623D-01	-1.66393D-04	7.84123D-08	-2.11896D-11
40	DIMETHYLFORMAMIDE	2.76199D+01	2.21170D-01	1.12943D-04	-2.86931D-07	1.26309D-10
41	ETHANE	3.38339D+01	-1.55175D-02	3.76892D-04	-4.11770D-07	1.38890D-10
42	ETHANOL	1.76907D+01	1.49532D-01	8.94815D-05	-1.97384D-07	8.31747D-11
43	ETHYL ACETATE	4.67922D+01	1.72055D-01	3.25949D-04	-5.20729D-07	2.11886D-10
44	ETHYL CHLORIDE	9.29967D+00	2.08358D-01	-1.00742D-04	1.66708D-08	0.0
45	ETHYL FORMATE	3.81020D+01	1.62629D-01	2.10144D-04	-3.82964D-07	1.62160D-10

3 Ideal Gas Heat Capacities (continued)

$$C_p = a + bT + cT^2 + dT^3 + eT^4$$

TABLE 3A   SI Units (J, K)

	Name	Ideal Gas Heat Capacity in J/mol-K when Temperature in K				
		a	b	c	d	e
46	ETHYL MERCAPTAN	2.15857D+01	1.76887D-01	1.30300D-05	-1.19652D-07	5.62012D-11
47	ETHYLACETYLENE	7.70702D+00	3.05514D-01	-2.21103D-04	9.09705D-08	-1.64356D-11
48	ETHYLAMINE	2.75175D+01	6.26593D-02	3.91900D-04	-5.27782D-07	2.06496D-10
49	ETHYLBENZENE	4.49950D+01	-4.58883D-02	1.83228D-03	-2.91974D-06	1.46346D-09
50	ETHYLCYCLOHEXANE	1.87681D+01	2.17024D-01	1.43654D-03	-2.29923D-06	1.08485D-09
51	ETHYLCYCLOPENTANE	-2.90029D+02	2.72203D+00	-6.38788D-03	7.80748D-06	-3.64212D-09
52	ETHYLENE	1.68346D+01	5.15193D-02	2.16352D-04	-3.45618D-07	1.58794D-10
53	ETHYLENE GLYCOL	3.58417D+01	1.08695D-01	2.90598D-04	-4.52216D-07	1.86584D-10
54	ETHYLENE OXIDE	1.79573D+01	3.43445D-02	3.51051D-04	-4.78345D-07	1.90011D-10
55	FLUORANTHENE	-6.16536D+01	1.03453D+00	-4.26744D-04	-2.51405D-07	1.94895D-10
56	FLUORENE	-2.49539D+01	6.97984D-01	-3.24458D-04	-5.28857D-07	2.70246D-10
57	FORMALDEHYDE	3.28011D+01	-3.78277D-03	4.71752D-05	-3.60606D-08	8.85123D-12
58	FURFURAL	2.52110D+01	2.21301D-01	1.30942D-04	-3.37155D-07	1.52277D-10
59	HYDROGEN	1.76386D+01	6.70055D-02	-1.31485D-04	1.05883D-07	-2.91803D-11
60	HYDROGEN CHLORIDE	3.03088D+01	-7.60900D-03	1.32608D-05	-4.33363D-09	0.0
61	HYDROGEN CYANIDE	2.08414D+01	6.79258D-02	-6.94727D-05	4.01373D-08	-9.26673D-12
62	HYDROGEN IODIDE	3.02697D+01	-1.03319D-02	2.59460D-05	-1.59529D-08	3.18621D-12
63	HYDROGEN SULFIDE	3.45234D+01	-1.76481D-02	6.76664D-05	-5.32454D-08	1.40695D-11
64	INDAN	-2.28836D+01	5.40798D-01	1.03708D-04	-5.39048D-07	2.58526D-10
65	INDENE	-1.68550D+01	5.17234D-01	-8.03445D-06	-3.75556D-07	1.91544D-10
66	IODOBENZENE	2.91636D+00	3.40319D-01	-7.86309D-07	-2.53567D-07	1.28569D-10
67	ISO-BUTANE	5.29035D+01	-1.07178D-01	1.38040D-03	-2.06667D-06	1.00888D-09
68	ISO-BUTANOL	5.61796D+01	-1.63172D-01	1.26597D-03	-1.46956D-06	5.48114D-10
69	ISO-BUTENE	1.82591D+01	2.29492D-01	1.01034D-04	-2.90919D-07	1.40589D-10
70	ISO-BUTYRALDEHYDE	2.86774D+01	1.84132D-01	2.76595D-04	-4.63760D-07	1.90207D-10
71	ISO-PENTANE	4.19952D+01	7.44763D-02	1.02577D-03	-1.61216D-06	7.79396D-10
72	ISO-PROPANOL	8.51074D+00	2.81060D-01	1.92642D-05	-2.26450D-07	1.13597D-10
73	M-DICHLOROBENZENE	-9.90706D+00	4.95077D-01	-3.31770D-04	3.99513D-08	1.15442D-09
74	M-XYLENE	4.64153D+01	-4.69727D-03	1.53345D-03	-2.37511D-06	1.33437D-09
75	MEK	4.00296D+01	1.77012D-01	2.01659D-04	-3.37579D-07	1.33437D-10
76	METHACRYLONITRILE	3.57055D+01	1.66377D-01	7.99044D-05	-1.96546D-07	8.38347D-11
77	METHANE	3.83870D+01	-7.36639D-02	2.90981D-04	-2.63849D-07	8.00679D-11
78	METHANOL	3.44925D+01	-2.91887D-02	2.86844D-04	-3.12501D-07	1.09833D-10
79	METHYL ACETATE	4.80207D+01	1.12261D-01	2.66821D-04	-4.03380D-07	1.61493D-10
80	METHYL CHLORIDE	1.94308D+01	5.96368D-02	7.06089D-05	-1.17502D-07	4.67507D-11
81	METHYL FORMATE	1.92655D+01	1.47245D-01	1.03166D-04	-2.50371D-07	1.14584D-10
82	METHYL IODIDE	1.66024D+01	1.04367D-01	-3.60895D-05	-1.84273D-08	1.38847D-11
83	METHYL PROPIONATE	4.67921D+01	1.72055D-01	3.25949D-04	-5.20729D-07	2.11886D-10
84	METHYLACETYLENE	1.36761D+01	1.94142D-01	-1.39014D-04	5.77909D-08	-1.06635D-11
85	METHYLAMINE	1.25367D+01	1.51044D-01	-6.88093D-05	1.23450D-08	0.0
86	METHYLCYCLOHEXANE	8.14527D+00	2.27701D-01	1.12022D-03	-1.77270D-06	8.07157D-10
87	METHYLCYCLOPENTANE	9.34709D+00	1.58082D-01	1.00675D-03	-1.57819D-06	7.22705D-10
88	METHYLSTYRENE	9.38426D+01	1.58742D-01	7.43415D-04	-9.37993D-07	4.29737D-10
89	N-BUTANE	6.67088D+01	-1.85523D-01	1.52844D-03	-2.18792D-06	1.04577D-09
90	N-BUTANOL	-2.06995D+00	4.29749D-01	-1.69148D-04	-7.37440D-08	6.14903D-11

3   Ideal Gas Heat Capacities (continued)

$$C_p = a + bT + cT^2 + dT^3 + eT^4$$

TABLE 3A   SI Units (J, K)

	Name	Ideal Gas Heat Capacity in J/mol-K when Temperature in K				
		a	b	c	d	e
91	N-BUTYLBENZENE	-4.87468D+01	9.34403D-01	-6.94393D-04	2.67041D-07	-4.17986D-11
92	N-BUTYLCYCLOHEXANE	-6.42985D+01	1.09466D+00	-6.65463D-04	1.71855D-07	-9.42921D-12
93	N-BUTYRIC ACID	3.04306D+01	2.44352D-01	2.16458D-04	-4.44626D-07	1.91743D-10
94	N-DECANE	6.52974D+01	3.56879D-01	1.24104D-03	-2.08987D-06	9.91165D-10
95	N-DODECANE	7.82111D+01	4.20932D-01	1.51657D-03	-2.56729D-06	1.22846D-09
96	N-HEPTANE	4.49544D+01	2.67290D-01	8.15496D-04	-1.36933D-06	6.38953D-10
97	N-HEXADECANE	1.01669D+02	5.74850D-01	1.96889D-03	-3.36510D-06	1.61483D-09
98	N-HEXANE	4.27147D+01	1.99102D-01	7.89486D-04	-1.27867D-06	5.91511D-10
99	N-NONANE	6.12326D+01	3.03023D-01	1.17647D-03	-1.95748D-06	9.28605D-10
100	N-OCTANE	5.17608D+01	2.95555D-01	9.66806D-04	-1.62822D-06	7.68379D-10
101	N-PENTADECANE	9.69035D+01	5.26624D-01	1.88708D-03	-3.20860D-06	1.53970D-09
102	N-PENTANE	8.31454D+01	-2.41925D-01	1.94653D-03	-2.80749D-06	1.35276D-09
103	N-PROPANOL	3.00764D+01	1.46604D-01	2.57793D-04	-4.03837D-07	1.51597D-10
104	N-PROPYL ACETATE	4.52996D+01	2.32266D-01	3.85041D-04	-5.38403D-07	2.62455D-10
105	N-PROPYLBENZENE	-4.61396D+01	8.29738D-01	-6.20486D-04	2.39055D-07	-3.72820D-11
106	N-PROPYLCYCLOHEXAN	-6.60979D+01	1.01662D+00	-6.44339D-04	1.85395D-07	-1.61900D-11
107	N-PROPYLCYCLOPENTA	-6.80697D+01	9.26244D-01	-6.73129D-04	2.69887D-07	-4.71546D-11
108	N-TETRADECANE	9.16493D+01	4.82659D-01	1.79251D-03	-3.03677D-06	1.45816D-09
109	N-TRIDECANE	8.19117D+01	4.78009D-01	1.57199D-03	-2.68952D-06	1.28710D-09
110	N-UNDECANE	7.29883D+01	3.79990D-01	1.40107D-03	-2.35145D-06	1.11780D-09
111	NAPHTHALENE	-5.38889D+01	7.53160D-01	-4.33264D-04	-5.52149D-09	6.75071D-11
112	NEO-PENTANE	5.13986D+01	-2.52905D-02	1.44716D-03	-2.24820D-06	1.10083D-09
113	NEON	2.07723D+01	0.0	0.0	0.0	0.0
114	NITRIC OXIDE	2.97657D+01	9.76049D-04	6.09872D-06	-3.58809D-09	5.85308D-13
115	NITROGEN	2.94119D+01	-3.00681D-03	5.45064D-06	5.13186D-09	-4.25308D-12
116	NITROGEN DIOXIDE	2.51165D+01	4.39956D-02	-9.61717D-06	-1.21653D-08	5.44943D-12
117	NITROUS OXIDE	2.05437D+01	8.12993D-02	-8.28968D-05	4.98739D-08	-1.12709D-11
118	O-CRESOL	1.29178D+00	3.61412D-01	2.48838D-04	-5.91848D-07	2.63135D-10
119	O-DICHLOROBENZENE	-9.90706D+00	4.95077D-01	-3.31770D-04	3.99513D-08	3.23324D-11
120	O-XYLENE	4.22847D+01	1.11693D-01	1.09933D-03	-1.76777D-06	8.56376D-10
121	OXYGEN	2.98832D+01	-1.13842D-02	4.33779D-05	-3.70082D-08	1.01006D-11
122	P-DICHLOROBENZENE	-9.90706D+00	4.95077D-01	-3.31770D-04	3.99513D-08	3.23324D-11
123	P-XYLENE	5.59538D+01	-7.30386D-02	1.68891D-03	-2.53135D-06	1.21448D-09
124	PHENANTHRENE	-4.68426D+01	7.81941D-01	-9.67493D-08	-6.03735D-07	3.08038D-10
125	PHENOL	-3.61498D+01	5.66519D-01	-4.11357D-04	9.39030D-08	1.80687D-11
126	PHOSGENE	2.21276D+01	2.11087D-01	-3.49699D-04	2.86091D-07	-9.13495D-11
127	PROPADIENE	1.11478D+01	1.76641D-01	-2.49984D-05	-1.22238D-07	7.96623D-11
128	PROPANE	4.72659D+01	-1.31469D-01	1.17000D-03	-1.69695D-06	8.18910D-10
129	PROPIONIC ACID	3.22170D+01	1.82665D-01	1.60153D-04	-3.29308D-07	1.41924D-10
130	PROPYL FORMATE	4.67921D+01	1.72055D-01	3.25949D-04	-5.20729D-07	2.11886D-10
131	PROPYLENE	2.43657D+01	7.12795D-02	3.38448D-04	-5.15275D-07	2.30475D-10
132	PYRENE	-7.09537D+01	1.02320D+00	-3.41758D-04	-3.62008D-07	2.39540D-10
133	STYRENE	-3.24471D+01	6.31898D-01	-3.98206D-04	4.65312D-08	3.64421D-11
134	SULFUR DIOXIDE	2.57725D+01	5.78938D-02	-3.80844D-05	8.60626D-09	0.0
135	SULFUR TRIOXIDE	1.55070D+01	1.45719D-01	-1.13253D-04	3.24046D-08	0.0

3 Ideal Gas Heat Capacities (continued)

$$C_p = a + bT + cT^2 + dT^3 + eT^4$$

TABLE 3A   SI Units (J, K)

Name	Ideal Gas Heat Capacity in J/mol-K when Temperature in K				
	a	b	c	d	e
136 T-BUTYL ALCOHOL	3.04792D+00	4.22960D-01	-1.57342D-04	-8.88132D-08	6.77124D-11
137 TETRAETHENE GLYCOL	7.77268D+01	3.67280D-01	9.86510D-04	-1.52386D-06	6.23029D-10
138 THIOPHENE	-3.35167D+01	4.71989D-01	-4.43715D-04	2.01864D-07	-3.11709D-11
139 TOLUENE	3.18200D+01	-1.61654D-02	1.44465D-03	-2.28948D-06	1.13573D-09
140 TRANS-2-BUTENE	3.49574D+01	1.06178D-01	3.88818D-04	-5.79041D-07	2.47347D-10
141 TRANS-2-PENTENE	2.75152D+01	2.15429D-01	3.46773D-04	-6.17649D-07	2.77835D-10
142 TRICHLOROACETYL-CL	8.18900D+00	5.48481D-01	-9.30532D-04	7.58795D-07	-2.39379D-10
143 TRICHLOROETHYLENE	2.06617D+01	2.94520D-01	-3.89705D-04	2.64566D-07	-7.22648D-11
144 TRIETHYLENE GLYCOL	6.15213D+01	2.57110D-01	8.29146D-04	-1.24370D-06	5.04866D-10
145 TRIMETHYLAMINE	-7.21214D+00	3.95019D-01	-2.23939D-04	-1.56727D-07	6.32532D-11
146 VINYLACETYLENE	3.29267D+01	1.14721D-01	7.98332D-05	-1.56727D-07	6.32532D-11
147 VINYL CHLORIDE	4.98798D+00	2.08934D-01	-1.72199D-04	6.80458D-08	-7.87428D-12
148 WATER	3.40471D+01	-9.65064D-03	3.29983D-05	-2.04467D-08	4.30228D-12
149 1-BUTENE	9.23232D+00	2.54744D-01	7.30443D-05	-2.65873D-07	1.24165D-10
150 1-ET-2-ME-BENZENE	-1.95616D+01	7.24058D-01	-4.70818D-04	1.46400D-07	-1.61041D-11
151 1-HEPTENE	2.65593D+01	3.95331D-01	2.91165D-04	-6.63027D-07	3.02972D-10
152 1-HEXENE	1.97547D+01	3.67049D-01	1.39914D-04	-4.04223D-07	1.73590D-10
153 1-ME-4-ETH-NAPHTHA	-3.32280D+01	7.16826D-01	2.70672D-04	-8.82466D-07	4.06741D-10
154 1-METHYLINDENE	-1.92015D+01	5.82742D-01	3.99649D-05	-4.83219D-07	2.38784D-10
155 1-METHYLNAPHTHALEN	-2.76113D+01	5.82115D-01	1.77289D-04	-6.66549D-07	3.11404D-10
156 1-PENTENE	1.45912D+01	3.25234D-01	3.17391D-05	-2.05695D-07	7.50234D-11
157 1-PHENYLINDENE	-5.11735D+01	8.92518D-01	-6.02522D-05	-6.02431D-07	3.15825D-10
158 1-PHENYLNAPHTHALEN	-5.79301D+01	8.47170D-01	1.92018D-04	-8.97733D-07	4.27089D-10
159 1,1-DICHLOROETHANE	1.76541D+01	2.38989D-01	-1.54594D-04	3.80267D-08	4.44247D-11
160 1,2-BUTADIENE	1.72814D+01	2.24295D-01	-1.89688D-05	-1.02998D-07	4.09951D-12
161 1,2-DICHLOROETHANE	2.67689D+01	2.12604D-01	-1.35500D-04	3.01151D-08	4.11215D-10
162 1,2-DIME-3-ETHBZ	-7.76813D+00	4.92779D-01	1.43321D-04	-7.05581D-07	3.35677D-10
163 1,2,3-TRIME-INDENE	-2.36187D+01	7.10688D-01	-6.40391D-04	3.20361D-07	-6.10707D-11
164 1,2,4-TRI-CL-BZ	-1.70348D+01	6.32368D-01	-2.85818D-06	-3.68547D-07	2.46459D-10
165 1,3-BUTADIENE	-5.63223D+00	3.12606D-01	-3.97122D-04	2.28305D-07	-5.24231D-11
166 112TRI-CL-ETHANE	1.21138D+01	3.57330D-01	3.48354D-05	-7.42896D-07	3.73867D-10
167 2-ETHYLFLUORENE	-5.39141D+01	9.13378D-01	2.65709D-04	-6.37392D-07	3.33393D-10
168 2-ME-1-BUTENE	1.85660D+01	2.80716D-01	2.53292D-04	-4.99083D-07	2.24537D-10
169 2-ME-2-BUTENE	1.95382D+01	2.49456D-01	3.42657D-05	-4.90222D-07	2.43351D-10
170 2-METHYLINDENE	-2.45876D+01	5.99261D-01	1.77289D-04	-6.66549D-07	3.11404D-10
171 2-METHYLNAPHTHALEN	-2.76113D+01	5.82115D-01	1.16766D-03	-1.90141D-06	9.36742D-10
172 2-METHYLPENTANE	4.86356D+01	1.16230D-01	6.52261D-04	-1.03412D-06	4.20348D-10
173 2,2-DIMETHYLBUTANE	3.15997D+01	2.55900D-01	6.66233D-04	-1.20240D-06	5.81274D-10
174 2,3-DIMETHYLBUTANE	2.56594D+01	2.77678D-01	2.57469D-04	-8.72581D-07	4.03928D-10
175 2,3,5-TRIME-NAPHTH	-3.43677D+01	7.24621D-01	2.16798D-04	-7.69003D-07	3.57471D-10
176 2,7-DIMETHYLNAPHTH	-3.10170D+01	6.53611D-01	-8.01119D-04	7.13030D-07	-2.87208D-10
177 3-ME-1-BUTENE	-1.50488D+01	6.31168D-01	8.23668D-04	-1.32253D-06	6.11324D-10
178 3-METHYLPENTANE	4.38881D+01	1.88198D-01			

Source: FLOWTRAN Data Base. These tabulations were prepared by Dr. D. R. Schneider of the Monsanto Company using the Physical Properties package of the FLOWTRAN system. The tabulations are reproduced with the permission of the Monsanto Company.

3 Ideal Gas Heat Capacities (continued)

$$C_p = a + bT + cT^2 + dT^3 + eT^4$$

Table 3B   AE Units (Btu, F)

	Name	Ideal Gas Heat Capacity in Btu/lbmol-R when Temperature in F				
		a	b	c	d	e
1	ACENAPHTHYLENE	3.04808D+01	7.75851D-02	-2.66483D-05	-5.06356D-09	4.42952D-12
2	ACETALDEHYDE	1.19092D+01	1.48112D-02	9.43915D-07	-5.01431D-09	1.72932D-12
3	ACETIC ACID	1.46392D+01	2.29877D-02	-1.02200D-05	2.58945D-09	-2.80441D-13
4	ACETONE	1.61362D+01	2.34006D-02	-1.47939D-05	-4.14355D-09	1.32372D-12
5	ACETONITRILE	1.15361D+01	1.23210D-02	-1.20045D-06	-2.03139D-09	7.74310D-13
6	ACETYL CHLORIDE	1.58551D+01	1.72566D-02	-7.11664D-06	1.06002D-09	1.48430D-13
7	ACETYLENE	9.89000D+00	8.27300D-03	-3.78300D-06	7.45700D-10	0.0
8	ACRYLONITRILE	1.37706D+01	1.98497D-02	-8.93743D-06	1.93298D-09	-2.79738D-14
9	AMMONIA	8.27650D+00	3.90060D-03	3.52450D-07	-2.74020D-10	0.0
10	ANILINE	2.01175D+01	4.53892D-02	-5.74305D-06	-1.21627D-08	5.36620D-12
11	ARGON	4.96470D+00	0.0	0.0	0.0	0.0
12	BENZENE	1.63928D+01	4.02037D-02	6.92540D-06	-4.11420D-08	2.39810D-11
13	BIPHENYL	3.20257D+01	9.11669D-02	2.28860D-06	7.50540D-09	9.05919D-13
14	BROMINE	8.55200D+00	8.33400D-04	-5.05300D-07	1.09800D-10	0.0
15	BROMOBENZENE	2.03633D+01	4.01115D-02	-1.10877D-05	-4.81355D-09	2.89080D-12
16	CARBON DIOXIDE	8.39860D+00	6.47577D-03	-3.55503D-06	1.19459D-09	-1.85170D-13
17	CARBON DISULFIDE	1.16304D+01	1.23864D-02	4.77819D-06	-8.56849D-09	2.96636D-12
18	CARBON MONOXIDE	6.95601D+00	5.91124D-05	5.07581D-07	7.64118D-10	-6.54036D-13
19	CARBON TET	1.83608D+01	1.93281D-02	-2.57605D-05	1.73446D-08	-4.54923D-12
20	CHLORINE	7.97300D+00	1.89010D-03	-1.21010D-06	2.65260D-10	0.0
21	CHLOROACETYL-CL	1.84724D+01	2.06099D-02	-1.56342D-05	7.73195D-09	-1.71774D-12
22	CHLOROBENZENE	1.99832D+01	4.08029D-02	-1.16675D-05	-4.59081D-09	2.86270D-12
23	CHLOROFORM	1.48400D+01	1.24500D-02	-6.49500D-06	1.25900D-09	0.0
24	CHRYSENE	4.25508D+01	1.20464D-01	-3.83101D-05	-1.14599D-08	8.07791D-12
25	CIS-2-BUTENE	1.65454D+01	2.96639D-02	5.47162D-06	-1.88313D-08	8.43756D-12
26	CIS-2-PENTENE	2.11241D+01	4.16978D-02	-3.05325D-06	-1.19511D-08	4.47439D-12
27	CYCLOHEXANE	2.10002D+01	5.62739D-02	1.12944D-05	-3.60617D-08	1.48261D-11
28	CYCLOPENTANE	1.62171D+01	4.64389D-02	5.94745D-06	-2.90184D-08	1.33855D-11
29	C3ALKYLBENZENE	4.13526D+01	6.15021D-02	-1.59286D-05	2.08456D-08	-8.33791D-12
30	C3ALKYLNAPHTHALENE	5.72404D+01	8.21098D-02	-1.13826D-05	1.43148D-08	-6.47080D-12
31	DICHLOROACETYL-CL	2.10897D+01	2.39632D-02	-2.41519D-05	1.44039D-08	-3.58392D-12
32	DICYCLOPENTADIENE	3.52418D+01	8.07763D-02	-3.05424D-05	-1.03767D-09	3.17117D-12
33	DIETHYL ETHER	2.34349D+01	3.45690D-02	6.98538D-06	-1.90248D-08	6.90689D-12
34	DIETHYL KETONE	2.45080D+01	4.23114D-02	-1.17334D-06	-1.37588D-08	5.48855D-12
35	DIETHYLENE GLYCOL	2.66305D+01	4.35657D-02	6.48378D-06	-2.35943D-08	8.90458D-12
36	DIMETHYL ACETAMIDE	2.46302D+01	4.01524D-02	-5.13911D-06	-9.13829D-09	4.01795D-12
37	DIMETHYL ETHER	1.59200D+01	1.59968D-02	7.89936D-06	-1.29305D-08	4.39830D-12
38	DIMETHYL SULFIDE	1.62399D+01	1.91945D-02	-3.53452D-07	-5.60440D-09	2.32836D-12
39	DIMETHYLACETYLENE	1.67682D+01	2.51710D-02	-8.45452D-06	2.32643D-09	-4.82437D-13
40	DIMETHYLFORMAMIDE	2.08472D+01	3.06900D-02	-4.23841D-06	-6.47135D-09	2.87576D-12
41	ETHANE	1.15161D+01	1.40309D-02	8.54034D-06	-1.10608D-08	3.16220D-12
42	ETHANOL	1.40485D+01	2.15315D-02	-2.15344D-06	-4.60726D-09	1.89369D-12
43	ETHYL ACETATE	2.49082D+01	3.32973D-02	7.31671D-07	-1.24703D-08	4.82415D-12
44	ETHYL CHLORIDE	1.34360D+01	2.12670D-02	-6.48930D-06	6.83200D-10	0.0
45	ETHYL FORMATE	2.09487D+01	2.73312D-02	-1.46055D-06	-8.90613D-09	3.69201D-12

3 Ideal Gas Heat Capacities (continued)

$$C_p = a + bT + cT^2 + dT^3 + eT^4$$

Table 3B   AE Units (Btu, F)

Name	Ideal Gas Heat Capacity in Btu/lbmol-R when Temperature in F				
	a	b	c	d	e
46 ETHYL MERCAPTAN	1.57394D+01	2.17597D-02	-4.17864D-06	-2.55083D-09	1.27957D-12
47 ETHYLACETYLENE	1.73883D+01	2.77898D-02	-1.16434D-05	3.04010D-09	-3.74201D-13
48 ETHYLAMINE	1.46188D+01	2.30134D-02	5.04254D-06	-1.29850D-08	4.70142D-12
49 ETHYLBENZENE	2.63783D+01	5.52627D-02	1.23968D-05	-5.83920D-08	3.33196D-11
50 ETHYLCYCLOHEXANE	3.20737D+01	7.61055D-02	7.34381D-06	-4.88120D-08	2.46996D-11
51 ETHYLCYCLOPENTANE	2.46301D+01	9.88319D-02	-1.35109D-04	1.67497D-07	-8.29225D-11
52 ETHYLENE	9.32602D+00	1.39393D-02	1.01083D-06	-7.51655D-09	3.61537D-12
53 ETHYLENE GLYCOL	1.81198D+01	2.40430D-02	1.26557D-06	-1.07217D-08	4.24809D-12
54 ETHYLENE OXIDE	1.01490D+01	1.76219D-02	4.34725D-06	-1.16491D-08	4.32609D-12
55 FLUORANTHENE	4.09534D+01	1.03618D-01	-4.00622D-05	-2.14423D-09	4.43731D-12
56 FLUORENE	3.47567D+01	8.11108D-02	-2.23269D-05	-1.03604D-08	6.15287D-12
57 FORMALDEHYDE	8.20953D+00	1.83852D-03	1.69753D-06	-1.10729D-09	2.01522D-13
58 FURFURAL	2.03865D+01	3.08530D-02	-4.99951D-06	-7.44255D-09	3.46699D-12
59 HYDROGEN	5.64782D+00	2.47265D-03	-4.55763D-06	3.11770D-09	-6.64368D-13
60 HYDROGEN CHLORIDE	6.96900D+00	-2.23600D-04	7.33300D-07	-1.77600D-10	0.0
61 HYDROGEN CYANIDE	8.19459D+00	5.26853D-03	-3.12395D-06	1.25697D-09	-2.10982D-13
62 HYDROGEN IODIDE	6.94818D+00	1.45783D-06	1.10437D-06	-5.20396D-10	7.25425D-14
63 HYDROGEN SULFIDE	8.03119D+00	9.86863D-04	2.38854D-06	-1.59311D-09	3.20329D-13
64 INDAN	2.72722D+01	6.71243D-02	-1.53515D-05	-1.12686D-08	5.88603D-12
65 INDENE	2.61158D+01	6.00721D-02	-1.62882D-05	-7.37248D-09	4.36100D-12
66 IODOBENZENE	2.05776D+01	3.96846D-02	-1.06771D-05	-5.00940D-09	2.92722D-12
67 ISO-BUTANE	2.04185D+01	3.46229D-02	1.41562D-06	-4.24613D-08	2.29699D-11
68 ISO-BUTANOL	1.79080D+01	3.08604D-02	2.61568D-05	-3.72799D-08	1.24793D-11
69 ISO-BUTENE	1.89309D+01	3.10101D-02	-4.93001D-06	-6.03695D-09	3.20089D-12
70 ISO-BUTYRALDEHYDE	2.07512D+01	3.28421D-02	-3.15237D-07	-1.10432D-08	4.33057D-12
71 ISO-PENTANE	2.49464D+01	4.44673D-02	7.05488D-06	-3.34417D-08	1.77450D-11
72 ISO-PROPANOL	1.87031D+01	3.37480D-02	-8.09768D-06	-4.52487D-09	2.58633D-12
73 M-DICHLOROBENZENE	2.28701D+01	4.45610D-02	-2.12826D-05	2.99079D-09	7.36133D-13
74 M-XYLENE	2.64279D+01	5.18815D-02	1.22123D-05	-4.90095D-08	2.62834D-11
75 MEK	2.23064D+01	2.95904D-02	-3.50639D-07	-8.24858D-09	3.03805D-12
76 METHACRYLONITRILE	1.92370D+01	2.31463D-02	-2.79353D-06	-4.54530D-09	1.90872D-12
77 METHANE	8.24522D+00	3.80633D-03	8.86474D-06	-7.46115D-09	1.82296D-12
78 METHANOL	9.80108D+00	8.43064D-03	6.66919D-06	-8.20898D-09	2.50064D-12
79 METHYL ACETATE	2.10465D+01	2.39507D-02	1.54732D-06	-9.77074D-09	3.67682D-12
80 METHYL CHLORIDE	8.96440D+00	1.00682D-02	-8.25134D-08	-2.85836D-09	1.06440D-12
81 METHYL FORMATE	1.43196D+01	2.10572D-02	-3.23184D-06	-5.46385D-09	2.60882D-12
82 METHYL IODIDE	9.71643D+00	1.10546D-02	-3.30286D-06	-1.73936D-10	3.16122D-13
83 METHYL PROPIONATE	2.49082D+01	3.32973D-02	7.31671D-07	-1.24703D-08	4.82415D-12
84 METHYLACETYLENE	1.31706D+01	1.77578D-02	-7.29642D-06	1.92198D-09	-2.42782D-13
85 METHYLAMINE	1.11920D+01	1.57100D-02	-4.37820D-06	5.05920D-10	0.0
86 METHYLCYCLOHEXANE	2.70695D+01	6.72929D-02	5.75055D-06	-3.88588D-08	1.83771D-11
87 METHYLCYCLOPENTANE	2.20273D+01	5.46597D-02	5.93519D-06	-3.44229D-08	1.64543D-11
88 METHYLSTYRENE	4.04085D+01	5.09283D-02	1.42337D-06	-2.04508D-08	9.78411D-12
89 N-BUTANE	2.07978D+01	3.14329D-02	1.92851D-05	-4.58865D-08	2.38097D-11
90 N-BUTANOL	2.28677D+01	4.42195D-02	-1.48703D-05	-4.48027D-10	1.39999D-12

3 Ideal Gas Heat Capacities (continued)

$$C_p = a + bT + cT^2 + dT^3 + eT^4$$

Table 3B   AE Units (Btu, F)

	Name	Ideal Gas Heat Capacity in Btu/lbmol-R when Temperature in F				
		a	b	c	d	e
91	N-BUTYLBENZENE	3.55780D+01	8.35466D-02	-3.73383D-05	9.19402D-09	-9.51656D-13
92	N-BUTYLCYCLOHEXANE	4.17471D+01	1.04601D-01	-3.96493D-05	6.64818D-09	-2.14681D-13
93	N-BUTYRIC ACID	2.39862D+01	3.72705D-02	-3.62564D-06	-1.01947D-08	4.36554D-12
94	N-DECANE	4.94214D+01	8.60271D-02	2.04970D-06	-4.41541D-08	2.25665D-11
95	N-DODECANE	5.90528D+01	1.02914D-01	2.24320D-06	-5.37856D-08	2.79691D-11
96	N-HEPTANE	3.49684D+01	6.08752D-02	1.21334D-06	-2.93693D-08	1.45475D-11
97	N-HEXADECANE	7.83212D+01	1.36719D-01	1.67437D-06	-7.03070D-08	3.67659D-11
98	N-HEXANE	3.01785D+01	5.19926D-02	3.04880D-06	-2.76400D-08	1.34673D-11
99	N-NONANE	4.46198D+01	7.73834D-02	2.96337D-06	-4.13472D-08	2.11422D-11
100	N-OCTANE	3.97799D+01	6.93090D-02	1.47993D-06	-3.45609D-08	1.74942D-11
101	N-PENTADECANE	7.35102D+01	1.28169D-01	2.31556D-06	-6.70384D-08	3.50554D-11
102	N-PENTANE	2.56463D+01	3.89176D-02	2.39729D-06	-5.84261D-08	3.07992D-11
103	N-PROPANOL	1.87115D+01	2.78876D-02	8.58527D-07	-9.78512D-09	3.67918D-12
104	N-PROPYL ACETATE	2.87305D+01	4.26900D-02	-9.98980D-08	-1.51759D-08	5.97548D-12
105	N-PROPYLBENZENE	3.08581D+01	7.39743D-02	-3.33376D-05	8.23617D-09	-8.48825D-13
106	N-PROPYLCYCLOHEXAN	3.69302D+01	9.59631D-02	-3.75210D-05	6.92005D-09	-3.68608D-13
107	N-PROPYLCYCLOPENTA	3.07991D+01	8.39319D-02	-3.57634D-05	9.08644D-09	-1.07360D-12
108	N-TETRADECANE	6.86980D+01	1.19660D-01	2.69646D-06	-6.34101D-08	3.31988D-11
109	N-TRIDECANE	6.38579D+01	1.11595D-01	1.11641D-06	-5.63399D-08	2.93043D-11
110	N-UNDECANE	5.42521D+01	9.42737D-02	2.72729D-06	-4.95726D-08	2.54498D-11
111	NAPHTHALENE	2.63831D+01	7.10763D-02	-3.03242D-05	2.59973D-09	1.53698D-12
112	NEO-PENTANE	2.54676D+01	4.61177D-02	1.14723D-05	-4.60514D-08	2.50634D-11
113	NEON	4.96470D+00	0.0	0.0	0.0	0.0
114	NITRIC OXIDE	7.25511D+00	4.55165D-04	2.64002D-07	-1.22544D-10	1.33261D-14
115	NITROGEN	6.94716D+00	6.60948D-05	5.69339D-07	3.22686D-11	-9.68326D-14
116	NITROGEN DIOXIDE	8.49549D+00	4.92175D-03	-1.23965D-06	-2.70430D-10	1.24071D-13
117	NITROUS OXIDE	8.76717D+00	6.36910D-03	-3.62181D-06	1.57209D-09	-2.56612D-13
118	O-CRESOL	2.41579D+01	5.18167D-02	-7.49652D-06	-1.32395D-08	5.99096D-12
119	O-DICHLOROBENZENE	2.28701D+01	4.45610D-02	-2.12826D-05	2.99079D-09	7.36133D-13
120	O-XYLENE	2.78925D+01	5.10358D-02	5.90863D-06	-3.65965D-08	1.94977D-11
121	OXYGEN	6.98650D+00	5.58110D-04	1.39992D-06	-1.09383D-09	2.29966D-13
122	P-DICHLOROBENZENE	2.28701D+01	4.45610D-02	-2.12826D-05	2.99079D-09	7.36133D-13
123	P-XYLENE	2.63986D+01	4.98221D-02	1.65837D-05	-5.28984D-08	2.76508D-11
124	PHENANTHRENE	3.44389D+01	9.08612D-02	-2.52355D-05	-1.18469D-08	7.01331D-12
125	PHENOL	1.99182D+01	4.99252D-02	-2.45162D-05	4.60471D-09	4.11381D-13
126	PHOSGENE	1.37677D+01	1.09368D-02	-1.22648D-05	7.90040D-09	-2.07981D-12
127	PROPADIENE	1.26505D+01	1.92883D-02	-6.45283D-06	-1.67465D-09	1.81372D-12
128	PROPANE	1.55868D+01	2.50495D-02	1.40426D-06	-3.52626D-08	1.86447D-11
129	PROPIONIC ACID	2.01788D+01	2.78161D-02	-2.70004D-06	-7.55436D-09	3.23127D-12
130	PROPYL FORMATE	2.49082D+01	3.32973D-02	7.31671D-07	-1.24703D-08	4.82415D-12
131	PROPYLENE	1.36327D+01	2.10700D-02	2.49845D-06	-1.14686D-08	5.24739D-12
132	PYRENE	3.89686D+01	1.05399D-01	-3.87550D-05	-4.80804D-09	5.45376D-12
133	STYRENE	2.48287D+01	5.84300D-02	-2.56930D-05	3.43249D-09	8.29702D-13
134	SULFUR DIOXIDE	9.13400D+00	5.32800D-03	-2.32300D-06	3.52700D-10	0.0
135	SULFUR TRIOXIDE	1.09640D+01	1.25100D-02	-6.52300D-06	1.32800D-09	0.0

3 Ideal Gas Heat Capacities (continued)

$$C_p = a + bT + cT^2 + dT^3 + eT^4$$

Table 3B   AE Units (Btu, F)

	Name	Ideal Gas Heat Capacity in Btu/lbmol-R when Temperature in F				
		a	b	c	d	e
136	T-BUTYL ALCOHOL	2.38069D+01	4.37823D-02	-1.46714D-05	-8.05119D-10	1.54165D-12
137	TETRAETHENE GLYCOL	5.09384D+01	8.15943D-02	4.63562D-06	-3.63689D-08	1.41849D-11
138	THIOPHENE	1.46531D+01	3.75480D-02	-2.22231D-05	6.96786D-09	-7.09689D-13
139	TOLUENE	2.11772D+01	4.63955D-02	9.96137D-06	-4.62826D-08	2.58579D-11
140	TRANS-2-BUTENE	1.88427D+01	2.76126D-02	3.09745D-05	-1.33756D-08	5.63152D-12
141	TRANS-2-PENTENE	2.29541D+01	3.85343D-02	-1.30602D-05	-1.36815D-08	6.32565D-12
142	TRICHLOROACETYL-CL	2.37071D+01	2.73164D-02	-3.26695D-05	2.10758D-08	-5.45009D-12
143	TRICHLOROETHYLENE	1.78198D+01	1.89116D-02	-1.58816D-05	7.81721D-09	-1.64530D-12
144	TRIETHYLENE GLYCOL	3.88832D+01	6.25265D-02	5.44950D-06	-2.98342D-08	1.14946D-11
145	TRIMETHYLAMINE	1.90953D+01	3.85644D-02	-1.36958D-05	2.03571D-09	-1.28213D-14
146	VINYLACETYLENE	1.55565D+01	1.71349D-02	-1.14250D-06	-3.77504D-09	1.44013D-12
147	VINYL CHLORIDE	1.15234D+01	1.77625D-02	-9.08433D-06	2.45900D-09	-1.79279D-13
148	WATER	7.98574D+00	4.63319D-04	1.40284D-06	-6.57839D-10	9.79529D-14
149	1-BUTENE	1.79614D+01	3.29702D-02	-6.05339D-06	-5.69809D-09	2.82694D-12
150	1-ET-2-ME-BENZENE	3.27457D+01	6.78723D-02	-2.69221D-05	5.32557D-09	-3.66653D-13
151	1-HEPTENE	3.26842D+01	5.76943D-02	-7.24699D-06	-1.44889D-08	6.89795D-12
152	1-HEXENE	2.78728D+01	4.92603D-02	-7.51275D-06	-9.29887D-09	3.95224D-12
153	1-ME-4-ETH-NAPHTHA	3.69298D+01	9.42101D-02	-1.81649D-05	-1.91378D-08	9.26054D-12
154	1-METHYLINDENE	2.99208D+01	6.96464D-02	-1.74683D-05	-9.80710D-09	5.43655D-12
155	1-METHYLNAPHTHALEN	2.93571D+01	7.47559D-02	-1.56030D-05	-1.42802D-08	7.08994D-12
156	1-PENTENE	2.30904D+01	4.06575D-02	-7.11790D-06	-5.28908D-09	1.70811D-12
157	1-PHENYLINDENE	3.92284D+01	1.01567D-01	-2.93745D-05	-1.14675D-08	7.19060D-12
158	1-PHENYLNAPHTHALEN	3.77155D+01	1.05967D-01	-2.42425D-05	-1.89117D-08	9.72382D-12
159	1,1-DICHLOROETHANE	1.65480D+01	2.22370D-02	-9.25490D-06	1.55840D-09	0.0
160	1,2-BUTADIENE	1.71598D+01	2.62130D-02	-5.93782D-06	-2.36130D-09	1.01145D-12
161	1,2-DICHLOROETHANE	1.73864D+01	1.98592D-02	-8.17516D-06	1.40578D-09	9.33361D-14
162	1,2-DIME-3-ETHBZ	3.21869D+01	7.62389D-02	-7.31724D-06	-2.19231D-08	9.36241D-12
163	1,2,3-TRIME-INDENE	3.74987D+01	8.87251D-02	-1.96139D-05	-1.48637D-08	7.64259D-12
164	1,2,4-TRI-CL-BZ	2.57569D+01	4.83191D-02	-3.08977D-05	1.05724D-08	-1.39044D-12
165	1,3-BUTADIENE	1.64729D+01	3.39203D-02	-1.39251D-05	-4.78631D-09	5.61130D-12
166	112TRI-CL-ETHANE	1.93707D+01	2.59822D-02	-1.79052D-05	7.16179D-09	-1.19355D-12
167	2-ETHYLFLUORENE	4.08286D+01	1.07650D-01	-2.86231D-05	-1.47942D-08	8.51209D-12
168	2-ME-1-BUTENE	2.35143D+01	4.16842D-02	-6.79796D-06	-1.21648D-08	7.59058D-12
169	2-ME-2-BUTENE	2.20851D+01	3.93216D-02	-3.03950D-06	-1.10536D-08	5.11218D-12
170	2-METHYLINDENE	2.95297D+01	7.13118D-02	-1.81527D-05	-9.90292D-09	5.54052D-12
171	2-METHYLNAPHTHALEN	2.93571D+01	7.47559D-02	-1.56030D-05	-1.42802D-08	7.08994D-12
172	2-METHYLPENTANE	3.03022D+01	5.35118D-02	5.71688D-06	-3.87087D-08	2.13274D-11
173	2,2-DIMETHYLBUTANE	2.96492D+01	5.50671D-02	1.80621D-06	-2.47831D-08	9.57033D-12
174	2,3-DIMETHYLBUTANE	2.92702D+01	5.59583D-02	-2.02828D-06	-2.49428D-08	1.32342D-11
175	2,3,5-TRIME-NAPHTH	3.69639D+01	9.45817D-02	-1.86614D-05	-1.88505D-08	9.19650D-12
176	2,7-DIMETHYLNAPHTH	3.31618D+01	8.46746D-02	-1.71489D-05	-1.65505D-08	3.13877D-12
177	3-ME-1-BUTENE	2.49862D+01	4.54600D-02	-2.70900D-05	1.71980D-08	-6.53906D-12
178	3-METHYLPENTANE	3.01717D+01	5.18987D-02	3.66338D-06	-2.86081D-08	1.39184D-11

Source: FLOWTRAN Data Base. These tabulations were prepared by Dr. D. R. Schneider of the Monsanto Company using the Physical Properties package of the FLOWTRAN system. The tabulations are reproduced with the permission of the Monsanto Company.

4 Antoine Equations for Pure Component Vapor Pressures (SI Units, kPa, K)

$$\ln P = A - B/(T + C)$$

	Name	Antoine Constants: $\ln(P^*(KPa)) = A - B/(T(K) + C)$ A	B	C
1	ACENAPHTYLENE	14.2943	4470.92	-81.1694
2	ACETALDEHYDE	15.1206	2845.25	-22.0670
3	ACETIC ACID	15.8667	4097.86	-27.4937
4	ACETONE	14.7171	2975.95	-34.5228
5	ACETONITRILE	14.8766	3366.49	-26.6513
6	ACETYL CHLORIDE	13.7462	2452.06	-55.2968
7	ACETYLENE	14.8321	1836.66	-8.4521
8	ACRYLONITRILE	14.2095	3033.10	-34.9326
9	AMMONIA	15.4940	2363.24	-22.6207
10	ANILINE	15.0205	4103.52	-62.7983
11	ARGON	13.9153	832.78	2.3608
12	BENZENE	14.1603	2948.78	-44.5633
13	BIPHENYL	14.4481	4415.36	-79.1919
14	BROMINE	14.7812	3090.86	-27.9733
15	BROMOBENZENE	14.2978	3650.77	-52.4382
16	CARBON DIOXIDE	15.3768	1956.25	-2.1117
17	CARBON DISULFIDE	15.2388	3549.90	15.1796
18	CARBON MONOXIDE	13.8722	769.93	1.6369
19	CARBON TET	14.6247	3394.46	-10.2163
20	CHLORINE	14.1372	2055.15	-23.3117
21	CHLOROACETYL-CL	14.4140	3068.09	-65.9202
22	CHLOROBENZENE	14.3050	3457.17	-48.5524
23	CHLOROFORM	14.5014	2938.55	-36.9972
24	CHRYSENE	14.5890	5915.26	-127.8828
25	CIS-2-BUTENE	13.8005	2209.76	-36.2080
26	CIS-2-PENTENE	13.7943	2451.28	-42.9501
27	CYCLOHEXANE	13.7865	2794.58	-49.1081
28	CYCLOPENTANE	13.8440	2590.03	-41.6716
29	C3ALKYLBENZENE	14.1013	3634.72	-54.9383
30	C3ALKYLNAPHTHALENE	14.4603	4553.65	-87.2011
31	DICHLOROACETYL-CL	14.8731	3393.39	-50.2758
32	DICYCLOPENTADIENE	14.0053	3599.69	-59.0556
33	DIETHYL ETHER	14.1675	2563.73	-39.3707
34	DIETHYL KETONE	14.3864	3128.36	-54.4122
35	DIETHYLENE GLYCOL	17.6738	6034.08	-53.2122
36	DIMETHYL ACETAMIDE	13.1131	2885.88	-100.3239
37	DIMETHYL ETHER	14.3448	2176.84	-24.6733
38	DIMETHYL SULFIDE	14.2291	2649.49	-34.8672
39	DIMETHYLACETYLENE	14.1712	2490.41	-39.4293
40	DIMETHYLFORMAMIDE	13.7554	3147.69	-81.3422
41	ETHANE	13.8797	1582.18	-13.7622
42	ETHANOL	16.1952	3423.53	-55.7152
43	ETHYL ACETATE	14.5813	3022.25	-47.8833
44	ETHYL CHLORIDE	14.2656	2458.21	-30.6994
45	ETHYL FORMATE	14.4017	2758.61	-45.7813

4 Antoine Equations for Pure Component Vapor Pressures (SI Units, kPa, K) (continued)

$$\ln P = A - B/(T + C)$$

Name	Antoine Constants: $\ln(P^*(KPa)) = A - B/(T(K) + C)$		
	A	B	C
46 ETHYL MERCAPTAN	14.2423	2616.51	-36.3192
47 ETHYLACETYLENE	14.0641	2279.52	-39.9148
48 ETHYLAMINE	14.4758	2407.60	-45.7539
49 ETHYLBENZENE	13.9698	3257.17	-61.0096
50 ETHYLCYCLOHEXANE	13.8180	3195.03	-57.6333
51 ETHYLCYCLOPENTANE	13.8290	2983.20	-52.7198
52 ETHYLENE	13.8182	1427.22	-14.3080
53 ETHYLENE GLYCOL	16.1847	4493.79	-82.1026
54 ETHYLENE OXIDE	14.5116	2478.12	-33.1582
55 FLUORANTHENE	14.4374	5438.77	-112.2578
56 FLUORENE	16.2019	6462.59	-13.1696
57 FORMALDEHYDE	14.3483	2161.33	-31.9756
58 FURFURAL	16.7802	5365.88	5.6186
59 HYDROGEN	12.7844	232.32	8.0800
60 HYDROGEN CHLORIDE	14.7081	1802.24	-9.6678
61 HYDROGEN CYANIDE	15.4856	3151.53	-8.8383
62 HYDROGEN IODIDE	14.3749	2133.52	-19.6195
63 HYDROGEN SULFIDE	14.5513	1964.37	-15.2417
64 INDAN	14.2465	3789.85	-56.8344
65 INDENE	14.4233	3994.97	-49.2317
66 IODOBENZENE	14.1448	3807.98	-61.9621
67 ISO-BUTANE	13.8137	2150.23	-27.6228
68 ISO-BUTANOL	15.4994	3246.51	-82.6994
69 ISO-BUTENE	13.9102	2196.49	-29.8630
70 ISO-BUTYRALDEHYDE	14.8510	2998.05	-44.6533
71 ISO-PENTANE	13.6106	2345.09	-40.2128
72 ISO-PROPANOL	15.6491	3109.34	-73.5459
73 M-DICHLOROBENZENE	19.6415	8584.16	123.8591
74 M-XYLENE	14.1146	3360.81	-58.3463
75 MEK	14.2173	2831.82	-57.3831
76 METHACRYLONITRILE	14.0891	2951.91	-51.7254
77 METHANE	13.5840	968.13	-3.7200
78 METHANOL	16.4948	3593.39	-35.2249
79 METHYL ACETATE	14.7074	2917.70	-41.3724
80 METHYL CHLORIDE	14.3114	2170.02	-25.2701
81 METHYL FORMATE	14.7233	2726.05	-35.3556
82 METHYL IODIDE	14.1329	2688.89	-33.1607
83 METHYL PROPIONATE	14.5705	3034.71	-48.3091
84 METHYLACETYLENE	13.8153	1927.99	-40.3166
85 METHYLAMINE	14.8909	2342.65	-38.7081
86 METHYLCYCLOHEXANE	13.7630	2965.76	-49.7775
87 METHYLCYCLOPENTANE	13.8064	2742.47	-46.5148
88 METHYLSTYRENE	14.2319	3687.83	-59.6117
89 N-BUTANE	13.9836	2292.44	-27.8623
90 N-BUTANOL	14.6961	2902.96	-102.9116

4 Antoine Equations for Pure Component Vapor Pressures (SI Units, kPa, K) (continued)

$$\ln P = A - B/(T + C)$$

	Name	Antoine Constants:	$\ln(P^*(KPa)) = A - B/(T(K) + C)$	
		A	B	C
91	N-BUTYLBENZENE	14.0579	3630.48	-71.8524
92	N-BUTYLCYCLOHEXANE	13.8938	3538.87	-72.5651
93	N-BUTYRIC ACID	15.8498	4096.07	-71.7697
94	N-DECANE	13.9899	3452.22	-78.8993
95	N-DODECANE	14.0587	3744.01	-92.8311
96	N-HEPTANE	13.9008	2932.72	-55.6356
97	N-HEXADECANE	14.1586	4205.32	-119.1482
98	N-HEXANE	14.0568	2825.42	-42.7089
99	N-NONANE	13.9548	3290.56	-71.5056
100	N-OCTANE	14.2368	3304.16	-55.2278
101	N-PENTADECANE	14.1466	4111.28	-112.2870
102	N-PENTANE	13.9778	2554.60	-36.2529
103	N-PROPANOL	15.2175	3008.31	-86.4909
104	N-PROPYL ACETATE	14.4147	3109.06	-57.6803
105	N-PROPYLBENZENE	13.9908	3433.51	-66.0278
106	N-PROPYLCYCLOHEXAN	13.8648	3377.56	-64.5934
107	N-PROPYLCYCLOPENTA	13.8962	3196.71	-59.5494
108	N-TETRADECANE	14.1267	4001.93	-105.7783
109	N-TRIDECANE	14.1069	3886.50	-98.9840
110	N-UNDECANE	14.7851	4152.92	-60.4933
111	NAPHTHALENE	13.7520	3701.48	-85.8319
112	NEO-PENTANE	13.7626	2304.46	-30.6944
113	NEON	13.4710	264.73	2.8276
114	NITRIC OXIDE	16.9196	1319.11	-14.1427
115	NITROGEN	13.4477	658.22	-2.8540
116	NITROGEN DIOXIDE	21.9837	6615.36	86.8780
117	NITROUS OXIDE	14.2447	1547.56	-23.9090
118	O-CRESOL	14.2678	3552.74	-95.9752
119	O-DICHLOROBENZENE	14.3011	3776.97	-63.6069
120	O-XYLENE	14.1257	3412.02	-58.6824
121	OXYGEN	13.6835	780.26	-4.1758
122	P-DICHLOROBENZENE	15.0839	4318.47	-35.3413
123	P-XYLENE	14.0891	3351.69	-57.6000
124	PHENANTHRENE	13.5159	4397.22	-119.2937
125	PHENOL	15.2767	4027.98	-76.7014
126	PHOSGENE	14.5141	2525.43	-26.1643
127	PROPADIENE	10.9901	1017.78	-78.9447
128	PROPANE	13.7097	1872.82	-25.1011
129	PROPIONIC ACID	15.4276	3761.14	-66.0009
130	PROPYL FORMATE	14.2663	2903.27	-53.5214
131	PROPYLENE	13.8782	1875.25	-22.9101
132	PYRENE	14.4694	5203.08	-106.9889
133	STYRENE	14.3284	3516.43	-56.1529
134	SULFUR DIOXIDE	14.9404	2385.00	-32.2139
135	SULFUR TRIOXIDE	13.8467	1777.66	-125.1972

4 Antoine Equations for Pure Component Vapor Pressures (SI Units, kPa, K) (continued)

$$\ln P = A - B/(T + C)$$

Name	Antoine Constants: A	$\ln(P^*(KPa)) = A - B/(T(K) + C)$ B	C
136 T-BUTYL ALCOHOL	14.4678	2490.44	-102.7144
137 TETRAETHENE GLYCOL	18.5419	8215.28	-11.3276
138 THIOPHENE	14.0097	2869.30	-51.7869
139 TOLUENE	14.2515	3242.38	-47.1806
140 TRANS-2-BUTENE	13.8064	2213.56	-33.1109
141 TRANS-2-PENTENE	13.8847	2495.40	-40.2073
142 TRICHLOROACETYL-CL	14.0716	3195.54	-53.3269
143 TRICHLOROETHYLENE	14.3346	3131.21	-37.7666
144 TRIETHYLENE GLYCOL	17.8140	6589.03	-57.0789
145 TRIMETHYLAMINE	13.8650	2239.10	-33.8347
146 VINYLACETYLENE	13.7861	2155.46	-43.4786
147 VINYL CHLORIDE	13.6163	2027.80	-33.5344
148 WATER	16.5362	3985.44	-38.9974
149 1-BUTENE	13.8817	2189.45	-30.5161
150 1-ET-2-ME-BENZENE	14.2225	3633.94	-59.9427
151 1-HEPTENE	13.8747	2895.90	-53.9388
152 1-HEXENE	13.7987	2657.34	-47.1749
153 1-ME-4-ETH-NAPHTHA	14.6221	4574.49	-90.9617
154 1-METHYLINDENE	14.2498	3939.22	-64.3544
155 1-METHYLNAPHTHALEN	14.2604	4265.00	-75.4697
156 1-PENTENE	13.7564	2409.11	-39.4834
157 1-PHENYLINDENE	14.4020	4872.88	-97.1067
158 1-PHENYLNAPHTHALEN	14.9544	5351.04	-81.5154
159 1,1-DICHLOROETHANE	13.8796	2607.81	-48.9442
160 1,2-BUTADIENE	14.4754	2580.48	-22.2012
161 1,2-DICHLOROETHANE	14.3572	3069.08	-42.3468
162 1,2-DIME-3-ETHBZ	14.2290	3799.04	-71.7621
163 1,2,3-TRIME-INDENE	14.3343	4165.42	-76.5467
164 1,2,4-TRI-CL-BZ	14.8831	4452.49	-52.7852
165 1,3-BUTADIENE	14.0719	2280.96	-27.5956
166 112TRI-CL-ETHANE	15.3614	3865.83	-26.6188
167 2-ETHYLFLUORENE	14.5052	4789.44	-97.7600
168 2-ME-1-BUTENE	13.8288	2435.13	-39.9270
169 2-ME-2-BUTENE	13.9011	2517.60	-40.5046
170 2-METHYLINDENE	14.2838	4013.05	-66.1650
171 2-METHYLNAPHTHALEN	14.2625	4238.97	-74.6666
172 2-METHYLPENTANE	13.7266	2611.47	-46.7064
173 2,2-DIMETHYLBUTANE	13.5436	2492.32	-43.6494
174 2,3-DIMETHYLBUTANE	13.6654	2595.99	-44.1960
175 2,3,5-TRIME-NAPHTH	14.6112	4648.00	-93.0922
176 2,7-DIMETHYLNAPHTH	14.7849	4734.17	-71.3275
177 3-ME-1-BUTENE	13.7142	2338.98	-36.0646
178 3-METHYLPENTANE	13.7492	2650.29	-46.1716

Source: FLOWTRAN Data Base. These tabulations were prepared by Dr. D. R. Schneider of the Monsanto Company using the Physical Properties package of the FLOWTRAN system. The tabulations are reproduced with the permission of the Monsanto Company.

5 Critical Temperature and Pressure in SI Units (kPa, K) and AE Units (psia, °R)

	Name	Critical Temperature K	R	Critical Pressure KPa	psia
1	ACENAPHTHYLENE	796.917	1434.450	3220.95	467.160
2	ACETALDEHYDE	461.161	830.090	5552.63	805.341
3	ACETIC ACID	594.761	1070.570	5785.68	839.142
4	ACETONE	509.461	917.030	4782.56	693.651
5	ACETONITRILE	547.861	986.150	4833.22	700.999
6	ACETYL CHLORIDE	507.921	914.258	5741.60	832.749
7	ACETYLENE	308.341	555.014	6138.29	890.284
8	ACRYLONITRILE	519.161	934.490	3536.25	512.890
9	AMMONIA	405.661	730.190	11402.14	1653.741
10	ANILINE	698.761	1257.770	5299.32	768.601
11	ARGON	150.651	271.172	4863.62	705.408
12	BENZENE	562.611	1012.700	4924.41	714.226
13	BIPHENYL	788.861	1419.950	3840.23	556.978
14	BROMINE	584.161	1051.490	10335.19	1498.992
15	BROMOBENZENE	670.161	1206.290	4519.11	655.442
16	CARBON DIOXIDE	304.201	547.562	7380.54	1070.457
17	CARBON DISULFIDE	552.000	993.600	7903.38	1146.288
18	CARBON MONOXIDE	132.951	239.312	3498.65	507.437
19	CARBON TET	556.400	1001.520	4559.64	661.320
20	CHLORINE	417.161	750.890	7710.86	1118.366
21	CHLOROACETYL-CL	585.991	1054.784	5106.90	740.693
22	CHLOROBENZENE	632.361	1138.250	4523.11	656.022
23	CHLOROFORM	536.561	965.810	5471.57	793.584
24	CHRYSENE	993.611	1788.500	2388.27	346.390
25	CIS-2-BUTENE	433.161	779.690	4205.80	610.000
26	CIS-2-PENTENE	478.111	860.600	3530.12	512.000
27	CYCLOHEXANE	552.961	995.330	4078.25	591.500
28	CYCLOPENTANE	511.761	921.170	4514.00	654.700
29	C3ALKYLBENZENE	653.161	1175.690	3141.11	455.580
30	C3ALKYLNAPHTHALENE	774.861	1394.750	2583.81	374.750
31	DICHLOROACETYL-CL	577.291	1039.124	4609.39	668.536
32	DICYCLOPENTADIENE	660.372	1188.670	3060.65	443.910
33	DIETHYL ETHER	466.761	840.170	3607.19	523.178
34	DIETHYL KETONE	560.961	1009.730	3738.90	542.282
35	DIETHYLENE GLYCOL	681.040	1225.872	4605.44	667.963
36	DIMETHYL ACETAMIDE	656.672	1182.010	4024.68	583.730
37	DIMETHYL ETHER	400.061	720.110	5268.92	764.192
38	DIMETHYL SULFIDE	503.061	905.510	5532.37	802.402
39	DIMETHYLACETYLENE	488.700	879.660	5084.50	737.445
40	DIMETHYLFORMAMIDE	596.900	1074.420	4709.80	683.099
41	ETHANE	305.561	550.010	4894.02	709.817
42	ETHANOL	516.261	929.270	6379.44	925.260
43	ETHYL ACETATE	523.261	941.870	3830.10	555.509
44	ETHYL CHLORIDE	460.361	828.650	5268.92	764.192
45	ETHYL FORMATE	508.500	915.300	4691.37	680.425

5 Critical Temperature and Pressure in SI Units (kPa, K) and AE Units (psia, °R) (continued)

	Name	Critical Temperature		Critical Pressure	
		K	R	KPa	psia
46	ETHYL MERCAPTAN	499.161	898.490	5491.83	796.523
47	ETHYLACETYLENE	463.722	834.700	4710.64	683.220
48	ETHYLAMINE	456.161	821.090	5623.56	815.628
49	ETHYLBENZENE	619.722	1115.500	3723.17	540.000
50	ETHYLCYCLOHEXANE	602.611	1084.700	3129.53	453.900
51	ETHYLCYCLOPENTANE	569.461	1025.030	3397.74	492.800
52	ETHYLENE	283.061	509.510	5116.94	742.150
53	ETHYLENE GLYCOL	645.200	1161.360	7528.48	1091.913
54	ETHYLENE OXIDE	468.161	842.690	7194.10	1043.416
55	FLUORANTHENE	936.589	1685.860	2606.56	378.050
56	FLUORENE	822.259	1480.067	2993.50	434.171
57	FORMALDEHYDE	415.161	747.290	6788.80	984.632
58	FURFURAL	657.122	1182.820	4924.44	714.230
59	HYDROGEN	33.191	59.744	1315.23	190.758
60	HYDROGEN CHLORIDE	324.561	584.210	8263.08	1198.459
61	HYDROGEN CYANIDE	456.661	821.990	4954.81	718.634
62	HYDROGEN IODIDE	423.161	761.690	8207.35	1190.376
63	HYDROGEN SULFIDE	373.561	672.410	9007.82	1306.474
64	INDAN	681.056	1225.900	3630.71	526.590
65	INDENE	691.944	1245.500	3817.14	553.630
66	IODOBENZENE	721.161	1298.090	4521.24	655.751
67	ISO-BUTANE	408.141	734.654	3647.71	529.056
68	ISO-BUTANOL	547.700	985.860	4295.16	622.960
69	ISO-BUTENE	417.889	752.200	3997.58	579.800
70	ISO-BUTYRALDEHYDE	505.383	909.690	4198.91	609.000
71	ISO-PENTANE	461.000	829.800	3330.17	483.000
72	ISO-PROPANOL	508.331	914.996	4764.32	691.006
73	M-DICHLOROBENZENE	683.961	1231.130	3880.76	562.857
74	M-XYLENE	619.222	1114.600	3516.33	510.000
75	MEK	535.661	964.190	4154.34	602.536
76	METHACRYLONITRILE	554.567	998.220	3883.82	563.300
77	METHANE	191.061	343.910	4640.70	673.077
78	METHANOL	513.161	923.690	7954.04	1153.636
79	METHYL ACETATE	506.861	912.350	4694.56	680.889
80	METHYL CHLORIDE	416.261	749.270	6677.34	968.466
81	METHYL FORMATE	487.161	876.890	6003.53	870.738
82	METHYL IODIDE	528.161	950.690	7315.69	1061.051
83	METHYL PROPIONATE	530.600	955.080	3982.09	577.553
84	METHYLACETYLENE	402.391	724.304	5627.50	816.200
85	METHYLAMINE	430.061	774.110	7457.55	1081.626
86	METHYLCYCLOHEXANE	572.333	1030.200	3477.72	504.400
87	METHYLCYCLOPENTANE	532.772	958.990	3785.22	549.000
88	METHYLSTYRENE	662.561	1192.610	3445.03	499.660
89	N-BUTANE	425.172	765.310	3796.94	550.700
90	N-BUTANOL	562.901	1013.222	4415.76	640.452

5 Critical Temperature and Pressure in SI Units (kPa, K) and AE Units (psia, °R) (continued)

	Name	Critical Temperature		Critical Pressure	
		K	R	KPa	psia
91	N-BUTYLBENZENE	660.461	1188.830	2886.76	418.689
92	N-BUTYLCYCLOHEXANE	645.761	1162.370	2439.94	353.884
93	N-BUTYRIC ACID	628.161	1130.690	4468.45	648.094
94	N-DECANE	618.889	1114.000	2109.80	306.000
95	N-DODECANE	660.161	1188.290	1803.60	261.590
96	N-HEPTANE	540.172	972.310	2736.80	396.940
97	N-HEXADECANE	725.000	1305.000	1378.95	200.000
98	N-HEXANE	507.861	914.150	3033.68	439.998
99	N-NONANE	595.000	1071.000	2282.16	331.000
100	N-OCTANE	569.389	1024.900	2496.59	362.100
101	N-PENTADECANE	710.000	1278.000	1516.85	220.000
102	N-PENTANE	469.778	845.600	3374.98	489.500
103	N-PROPANOL	536.861	966.350	5082.28	737.122
104	N-PROPYL ACETATE	549.361	988.850	3334.61	483.645
105	N-PROPYLBENZENE	638.311	1148.960	3199.86	464.100
106	N-PROPYLCYCLOHEXAN	619.161	1114.490	2546.20	369.295
107	N-PROPYLCYCLOPENTA	590.261	1062.470	2802.84	406.518
108	N-TETRADECANE	695.000	1251.000	1585.79	230.000
109	N-TRIDECANE	677.222	1219.000	1723.69	250.000
110	N-UNDECANE	640.000	1152.000	1944.32	282.000
111	NAPHTHALENE	748.361	1347.050	3971.95	576.083
112	NEO-PENTANE	433.761	780.770	3198.84	463.953
113	NEON	44.500	80.100	2725.65	395.322
114	NITRIC OXIDE	180.000	324.000	6484.82	940.544
115	NITROGEN	126.271	227.288	3398.45	492.904
116	NITROGEN DIOXIDE	431.000	775.800	10132.54	1469.600
117	NITROUS OXIDE	309.700	557.460	7265.03	1053.703
118	O-CRESOL	697.561	1255.610	5005.47	725.982
119	O-DICHLOROBENZENE	697.261	1255.070	4104.69	595.335
120	O-XYLENE	632.222	1138.000	3654.22	530.000
121	OXYGEN	154.781	278.606	5080.45	736.857
122	P-DICHLOROBENZENE	684.761	1232.570	3906.09	566.531
123	P-XYLENE	618.222	1112.800	3447.38	500.000
124	PHENANTHRENE	878.782	1581.808	2898.58	420.404
125	PHENOL	695.061	1251.110	6130.18	889.108
126	PHOSGENE	455.161	819.290	5674.22	822.976
127	PROPADIENE	400.967	721.740	5151.42	747.150
128	PROPANE	369.971	665.948	4256.68	617.379
129	PROPIONIC ACID	612.661	1102.790	5370.24	778.888
130	PROPYL FORMATE	538.100	968.580	4063.15	589.310
131	PROPYLENE	365.111	657.200	4598.73	666.990
132	PYRENE	892.050	1605.690	2606.56	378.050
133	STYRENE	636.861	1146.350	3854.63	559.067
134	SULFUR DIOXIDE	430.661	775.190	7893.24	1144.818
135	SULFUR TRIOXIDE	490.861	883.550	8251.94	1196.842

5 Critical Temperature and Pressure in SI Units (kPa, K) and AE Units (psia, °R) (continued)

	Name	Critical Temperature		Critical Pressure	
		K	R	KPa	psia
136	T-BUTYL ALCOHOL	506.661	911.990	3971.95	576.083
137	TETRAETHENE GLYCOL	795.750	1432.350	2099.05	304.442
138	THIOPHENE	590.161	1062.290	4863.62	705.408
139	TOLUENE	593.961	1069.130	4053.01	587.840
140	TRANS-2-BUTENE	428.161	770.690	4102.38	595.000
141	TRANS-2-PENTENE	476.089	856.960	3505.16	508.380
142	TRICHLOROACETYL-CL	589.691	1061.444	4099.32	594.556
143	TRICHLOROETHYLENE	544.161	979.490	5015.60	727.452
144	TRIETHYLENE GLYCOL	712.320	1282.176	3316.38	481.000
145	TRIMETHYLAMINE	433.261	779.870	4073.28	590.779
146	VINYLACETYLENE	456.251	821.252	4859.56	704.820
147	VINYL CHLORIDE	431.561	776.810	5339.84	774.479
148	WATER	647.301	1165.142	22109.19	3206.667
149	1-BUTENE	419.611	755.300	4019.64	583.000
150	1-ET-2-ME-BENZENE	651.111	1172.000	3040.59	441.000
151	1-HEPTENE	535.500	963.900	2842.02	412.200
152	1-HEXENE	511.111	920.000	3252.26	471.700
153	1-ME-4-ETH-NAPHTHA	774.128	1393.430	2817.13	408.590
154	1-METHYLINDENE	703.478	1266.260	3327.75	482.650
155	1-METHYLNAPHTHALEN	769.161	1384.490	3568.39	517.551
156	1-PENTENE	473.889	853.000	4040.33	586.000
157	1-PHENYLINDENE	843.667	1518.600	2695.92	391.010
158	1-PHENYLNAPHTHALEN	840.083	1512.150	2632.35	381.790
159	1,1-DICHLOROETHANE	523.161	941.690	5066.27	734.800
160	1,2-BUTADIENE	450.861	811.550	3985.65	578.070
161	1,2-DICHLOROETHANE	561.561	1010.810	5370.24	778.888
162	1,2-DIME-3-ETHBZ	680.061	1224.110	3127.75	453.642
163	1,2,3-TRIME-INDENE	720.217	1296.390	2648.00	384.060
164	1,2,4-TRI-CL-BZ	734.961	1322.930	3986.36	578.172
165	1,3-BUTADIENE	425.000	765.000	4329.91	628.000
166	112TRI-CL-ETHANE	612.161	1101.890	4834.69	701.212
167	2-ETHYLFLUORENE	811.061	1459.910	2464.88	357.500
168	2-ME-1-BUTENE	472.222	850.000	3546.66	514.400
169	2-ME-2-BUTENE	483.333	870.000	3637.67	527.600
170	2-METHYLINDENE	714.567	1286.220	3354.64	486.550
171	2-METHYLNAPHTHALEN	761.861	1371.350	3502.92	508.056
172	2-METHYLPENTANE	498.056	896.500	3034.38	440.100
173	2,2-DIMETHYLBUTANE	489.389	880.900	3106.09	450.500
174	2,3-DIMETHYLBUTANE	500.278	900.500	3139.87	455.400
175	2,3,5-TRIME-NAPHTH	788.250	1418.850	2817.13	408.590
176	2,7-DIMETHYLNAPHTH	778.161	1400.690	3222.41	467.371
177	3-ME-1-BUTENE	461.667	831.000	3495.64	507.000
178	3-METHYLPENTANE	504.333	907.800	3124.01	453.100

Source: FLOWTRAN Data Base. These tabulations were prepared by Dr. D. R. Schneider of the Monsanto Company using the Physical Properties package of the FLOWTRAN system. The tabulations are reproduced with the permission of the Monsanto Company.

6 Liquid Heat Capacities (SI Units J, mol, K)

$$C_p = a + bT + cT^2 + dT^3$$

	Name	Liquid Heat Capacity in J/mol-K when Temperature in K			
		a	b	c	d
1	ACENAPHTHYLENE	-5.67880D+01	1.31606D+00	-1.71665D-03	1.09288D-06
2	ACETALDEHYDE	1.68842D+01	8.10208D-01	-3.08085D-03	4.42590D-06
3	ACETIC ACID	-3.60814D+01	6.04681D-01	-3.93957D-04	-5.61602D-07
4	ACETONE	1.68022D+01	8.48409D-01	-2.64114D-03	3.39139D-06
5	ACETONITRILE	1.86216D+01	7.30150D-01	-2.33009D-03	2.78250D-06
6	ACETYL CHLORIDE	2.17323D+01	1.52480D+00	-5.44935D-03	6.98035D-06
7	ACETYLENE	1.21476D+01	1.11988D+00	-6.78213D-03	1.42930D-05
8	ACRYLONITRILE	1.06528D+01	9.77905D-01	-3.10778D-03	3.82102D-06
9	AMMONIA	2.01494D+01	8.45765D-01	-4.06745D-03	6.60687D-06
10	ANILINE	-1.36683D+01	9.31971D-01	-1.60401D-03	1.36715D-06
11	ARGON	-2.49300D+01	1.41664D+00	-2.86902D-03	-4.27496D-05
12	BENZENE	-7.27329D+00	7.70541D-01	-1.64818D-03	1.89794D-06
13	BIPHENYL	-8.65504D+01	1.53701D+00	-2.04329D-03	1.28288D-06
14	BROMINE	2.11979D+01	5.17990D-01	-1.75921D-03	1.95266D-06
15	BROMOBENZENE	-9.93411D+00	8.95727D-01	-1.68176D-03	1.47666D-06
16	CARBON DIOXIDE	1.10417D+01	1.15955D+00	-7.23130D-03	1.55019D-05
17	CARBON DISULFIDE	1.74151D+01	5.54537D-01	-1.72346D-03	2.07575D-06
18	CARBON MONOXIDE	1.49673D+01	2.14397D+00	-3.24703D-02	1.58042D-04
19	CARBON TET	1.22846D+01	1.09475D+00	-3.18255D-03	3.42524D-06
20	CHLORINE	1.54120D+01	7.23104D-01	-3.39726D-03	5.26236D-06
21	CHLOROACETYL-CL	2.68099D+00	1.70293D+00	-5.18598D-03	5.61119D-06
22	CHLOROBENZENE	-1.15494D+01	9.39618D-01	-1.89850D-03	1.79189D-06
23	CHLOROFORM	2.38419D+01	7.55531D-01	-2.40701D-03	2.84262D-06
24	CHRYSENE	-1.35514D+02	1.95374D+00	-2.04051D-03	9.41273D-07
25	CIS-2-BUTENE	1.59891D+01	8.59587D-01	-3.12594D-03	4.92443D-06
26	CIS-2-PENTENE	1.35240D+00	9.72143D-01	-2.73959D-03	3.71880D-06
27	CYCLOHEXANE	-1.12493D+01	8.41499D-01	-1.58331D-03	1.96493D-06
28	CYCLOPENTANE	-1.77539D+01	8.42309D-01	-2.00426D-03	2.64122D-06
29	C3ALKYLBENZENE	3.75668D+01	1.15081D+00	-2.08788D-03	2.00092D-06
30	C3ALKYLNAPHTHALENE	8.32380D+01	1.17658D+00	-1.49715D-03	1.26197D-06
31	DICHLOROACETYL-CL	-1.24709D+01	1.98455D+00	-6.07564D-03	6.56868D-06
32	DICYCLOPENTADIENE	-4.35702D+01	1.37958D+00	-2.11522D-03	1.69221D-06
33	DIETHYL ETHER	3.93869D+01	8.83221D-01	-2.83840D-03	4.18044D-06
34	DIETHYL KETONE	1.95218D+01	9.41566D-01	-2.19928D-03	2.53543D-06
35	DIETHYLENE GLYCOL	2.65129D+01	1.26205D+00	-2.69832D-03	2.56288D-06
36	DIMETHYL ACETAMIDE	1.76879D+01	8.99258D-01	-1.75431D-03	1.64643D-06
37	DIMETHYL ETHER	3.90853D+01	8.07754D-01	-3.70090D-03	6.32701D-06
38	DIMETHYL SULFIDE	2.31559D+01	7.63756D-01	-2.47469D-03	3.21702D-06
39	DIMETHYLACETYLENE	4.61307D+00	9.17069D-01	-2.87254D-03	3.70322D-06
40	DIMETHYLFORMAMIDE	2.69989D+01	9.57414D-01	-2.39927D-03	2.58304D-06
41	ETHANE	2.06881D+01	9.48588D-01	-5.98221D-03	1.31546D-05
42	ETHANOL	-3.25137D+02	4.13787D+00	-1.40307D-02	1.70354D-05
43	ETHYL ACETATE	4.29049D+01	9.34378D-01	-2.63999D-03	3.34258D-06
44	ETHYL CHLORIDE	1.79781D-01	9.04849D-01	-3.14990D-03	4.36155D-06
45	ETHYL FORMATE	3.20116D+01	8.85757D-01	-2.67612D-03	3.44613D-06

6 Liquid Heat Capacities (SI Units J, mol, K) (continued)

$$C_p = a + bT + cT^2 + dT^3$$

	Name	Liquid Heat Capacity in J/mol-K when Temperature in K			
		a	b	c	d
46	ETHYL MERCAPTAN	1.17652D+01	8.26110D-01	-2.61295D-03	3.35734D-06
47	ETHYLACETYLENE	-1.83773D+00	9.86905D-01	-3.18071D-03	4.26588D-06
48	ETHYLAMINE	2.23571D+01	8.73964D-01	-3.21587D-03	4.79241D-06
49	ETHYLBENZENE	4.31428D+00	9.00174D-01	-1.45005D-03	1.43360D-06
50	ETHYLCYCLOHEXANE	-9.49573D+00	1.06190D+00	-1.53490D-03	1.57500D-06
51	ETHYLCYCLOPENTANE	-2.31040D+02	2.61156D+00	-5.57507D-03	4.90973D-06
52	ETHYLENE	3.44364D+00	1.08420D+00	-7.13595D-03	1.65631D-05
53	ETHYLENE GLYCOL	3.10224D+01	1.10034D+00	-2.84571D-03	2.88921D-06
54	ETHYLENE OXIDE	7.41259D+00	7.42687D-01	-2.71320D-03	3.90092D-06
55	FLUORANTHENE	-8.37633D+01	1.67596D+00	-1.87334D-03	9.46409D-07
56	FLUORENE	-4.56808D+01	1.33975D+00	-1.64621D-03	1.00531D-06
57	FORMALDEHYDE	2.50990D+01	7.93671D-01	-3.82691D-03	6.10492D-06
58	FURFURAL	2.14163D+01	8.86185D-01	-1.93931D-03	1.85001D-06
59	HYDROGEN	5.88663D+01	-2.30694D-01	-8.04213D-02	1.37776D-03
60	HYDROGEN CHLORIDE	1.77227D+01	9.04261D-01	-5.64496D-03	1.13383D-05
61	HYDROGEN CYANIDE	1.68791D+01	8.50946D-01	-3.53136D-03	5.04830D-06
62	HYDROGEN IODIDE	1.67440D+01	6.72052D-01	-3.22257D-03	4.96454D-06
63	HYDROGEN SULFIDE	2.18238D+01	7.74223D-01	-4.20204D-03	7.38677D-06
64	INDAN	-3.71075D+01	1.14329D+00	-1.72505D-03	1.40023D-06
65	INDENE	-2.91550D+01	1.09168D+00	-1.71677D-03	1.38016D-06
66	IODOBENZENE	-1.06028D+01	8.50951D-01	-1.45253D-03	1.15835D-06
67	ISO-BUTANE	3.87062D+01	7.46648D-01	-2.89544D-03	5.19378D-06
68	ISO-BUTANOL	5.15292D+01	9.09017D-01	-2.75500D-03	3.69657D-06
69	ISO-BUTENE	8.45979D+00	1.00655D+00	-3.65636D-03	5.66411D-06
70	ISO-BUTYRALDEHYDE	2.53636D+01	9.66816D-01	-2.87136D-03	3.75743D-06
71	ISO-PENTANE	2.81135D+01	8.68714D-01	-2.50761D-03	3.76751D-06
72	ISO-PROPANOL	9.14963D+00	1.44133D+00	-4.60994D-03	5.88221D-06
73	M-DICHLOROBENZENE	-1.69119D+01	1.00777D+00	-1.85017D-03	1.52520D-06
74	M-XYLENE	1.40673D+01	8.70264D-01	-1.47733D-03	1.51193D-06
75	MEK	3.40321D+01	8.64140D-01	-2.40262D-03	2.94373D-06
76	METHACRYLONITRILE	2.92563D+01	8.12281D-01	-2.28011D-03	2.65682D-06
77	METHANE	-5.70709D+00	1.02562D+00	-1.66566D-03	-1.97507D-05
78	METHANOL	-2.58250D+02	3.35820D+00	-1.16388D-02	1.40516D-05
79	METHYL ACETATE	3.92701D+01	7.96701D-01	-2.47205D-03	3.23224D-06
80	METHYL CHLORIDE	7.99608D+00	7.98500D-01	-3.50758D-03	5.51501D-06
81	METHYL FORMATE	1.20769D+01	8.63546D-01	-2.87339D-03	3.83072D-06
82	METHYL IODIDE	5.22773D+00	6.85948D-01	-2.25547D-03	2.73291D-06
83	METHYL PROPIONATE	4.18960D+01	9.10092D-01	-2.50667D-03	3.12247D-06
84	METHYLACETYLENE	5.88685D+00	1.01171D+00	-4.23564D-03	6.74784D-06
85	METHYLAMINE	7.96131D+00	9.72440D-01	-3.92527D-03	5.93717D-06
86	METHYLCYCLOHEXANE	-1.36334D+01	9.73700D-01	-1.60323D-03	1.79689D-06
87	METHYLCYCLOPENTANE	-7.85540D+00	8.99222D-01	-1.87221D-03	2.34232D-06
88	METHYLSTYRENE	7.60375D+01	8.32891D-01	-1.37884D-03	1.44784D-06
89	N-BUTANE	5.18583D+01	6.56571D-01	-2.53079D-03	4.49879D-06
90	N-BUTANOL	-5.10376D-01	1.44697D+00	-3.83339D-03	4.28849D-06

6 Liquid Heat Capacities (SI Units J, mol, K) (continued)

$$C_p = a + bT + cT^2 + dT^3$$

Name	Liquid Heat Capacity in J/mol-K when Temperature in K			
	a	b	c	d
91 N-BUTYLBENZENE	-4.92918D+01	1.49204D+00	-2.39414D-03	1.94687D-06
92 N-BUTYLCYCLOHEXANE	-6.50109D+01	1.68625D+00	-2.51655D-03	2.06802D-06
93 N-BUTYRIC ACID	2.95304D+01	1.07457D+00	-2.48619D-03	2.53277D-06
94 N-DECANE	4.36906D+01	1.31857D+00	-2.03714D-03	2.00524D-06
95 N-DODECANE	4.29158D+01	1.51927D+00	-2.03948D-03	1.77265D-06
96 N-HEPTANE	3.38820D+01	1.07063D+00	-2.25615D-03	2.72250D-06
97 N-HEXADECANE	2.80584D+01	2.04324D+00	-2.42726D-03	1.75525D-06
98 N-HEXANE	3.14210D+01	9.76058D-01	-2.35368D-03	3.09273D-06
99 N-NONANE	4.34796D+01	1.21415D+00	-2.03819D-03	2.15001D-06
100 N-OCTANE	3.82405D+01	1.14275D+00	-2.13030D-03	2.39204D-06
101 N-PENTADECANE	3.32402D+01	1.92411D+00	-2.36868D-03	1.79436D-06
102 N-PENTANE	6.54961D+01	6.28628D-01	-1.89880D-03	3.18651D-06
103 N-PROPANOL	-4.88104D+02	5.78632D+00	-1.88720D-02	2.20035D-05
104 N-PROPYL ACETATE	4.10176D+01	9.93938D-01	-2.44653D-03	2.91793D-06
105 N-PROPYLBENZENE	-4.72371D+01	1.40532D+00	-2.43735D-03	2.10281D-06
106 N-PROPYLCYCLOHEXAN	-6.82123D+01	1.61015D+00	-2.57929D-03	2.24600D-06
107 N-PROPYLCYCLOPENTA	-7.05452D+01	1.53129D+00	-2.73757D-03	2.55774D-06
108 N-TETRADECANE	3.78050D+01	1.78056D+00	-2.23443D-03	1.76082D-06
109 N-TRIDECANE	4.02131D+01	1.67374D+00	-2.20936D-03	1.83719D-06
110 N-UNDECANE	4.51594D+01	1.41653D+00	-2.03836D-03	1.89124D-06
111 NAPHTHALENE	-6.32700D+01	1.24334D+00	-1.76851D-03	1.21897D-06
112 NEO-PENTANE	3.59838D+01	8.14020D-01	-2.52756D-03	4.22057D-06
113 NEON	3.30532D+02	-2.34614D+01	5.74438D-01	-4.24012D-03
114 NITRIC OXIDE	3.36324D+01	2.90498D+00	-3.26583D-02	1.20828D-04
115 NITROGEN	1.47141D+01	2.20257D+00	-3.52146D-02	1.79960D-04
116 NITROGEN DIOXIDE	1.69925D+01	1.71499D+00	-7.83962D-03	1.20017D-05
117 NITROUS OXIDE	8.58935D+00	1.05171D+00	-6.39280D-03	1.33260D-05
118 O-CRESOL	-8.59146D+00	1.03316D+00	-1.74432D-03	1.47766D-06
119 O-DICHLOROBENZENE	-1.66105D+01	1.00462D+00	-1.81232D-03	1.46252D-06
120 O-XYLENE	1.48871D+01	9.03295D-01	-1.55098D-03	1.51201D-06
121 OXYGEN	1.10501D+03	-3.33636D+01	3.50211D-01	-1.21262D-03
122 P-DICHLOROBENZENE	-1.63179D+01	1.01438D+00	-1.86814D-03	1.54201D-06
123 P-XYLENE	2.20553D+01	8.11839D-01	-1.36670D-03	1.44216D-06
124 PHENANTHRENE	-7.65516D+01	1.47204D+00	-1.67258D-03	9.13405D-07
125 PHENOL	-3.61614D+01	1.15354D+00	-2.12291D-03	1.74183D-06
126 PHOSGENE	1.35843D+01	9.04210D-01	-3.41229D-03	4.65981D-06
127 PROPADIENE	-2.57486D+00	9.04297D-01	-3.68515D-03	5.83225D-06
128 PROPANE	3.37507D+01	7.46408D-01	-3.64966D-03	7.10670D-06
129 PROPIONIC ACID	3.17072D+01	9.30795D-01	-2.33049D-03	2.45740D-06
130 PROPYL FORMATE	4.10551D+01	8.90979D-01	-2.39549D-03	2.93524D-06
131 PROPYLENE	1.22887D+01	9.18751D-01	-4.34735D-03	7.94316D-06
132 PYRENE	-9.33646D+01	1.70047D+00	-1.94567D-03	1.03340D-06
133 STYRENE	-3.80191D+01	1.19721D+00	-2.19565D-03	1.93312D-06
134 SULFUR DIOXIDE	1.92884D+01	8.45429D-01	-3.72748D-03	5.65365D-06
135 SULFUR TRIOXIDE	1.62291D+01	1.37462D+00	-5.17738D-03	6.88634D-06

6   Liquid Heat Capacities (SI Units J, mol, K) (continued)

$$C_p = a + bT + cT^2 + dT^3$$

	Name	Liquid Heat Capacity in J/mol-K when Temperature in K			
		a	b	c	d
136	T-BUTYL ALCOHOL	4.88045D+00	1.54948D+00	-4.66330D-03	5.85875D-06
137	TETRAETHENE GLYCOL	1.12812D+00	1.90413D+00	-3.04611D-03	2.26240D-06
138	THIOPHENE	-3.93905D+00	1.04007D+00	-2.38173D-03	2.34822D-06
139	TOLUENE	1.80826D+00	8.12223D-01	-1.51267D-03	1.63001D-06
140	TRANS-2-BUTENE	2.52418D+01	8.87012D-01	-3.30491D-03	5.18957D-06
141	TRANS-2-PENTENE	1.69958D+01	9.34177D-01	-2.72023D-03	3.74189D-06
142	TRICHLOROACETYL-CL	-2.32210D+01	2.14161D+00	-6.34230D-03	6.61378D-06
143	TRICHLOROETHYLENE	1.71931D+01	9.31709D-01	-2.74646D-03	3.08927D-06
144	TRIETHYLENE GLYCOL	7.99807D+00	1.76246D+00	-3.52085D-03	3.09510D-06
145	TRIMETHYLAMINE	-1.62116D+01	1.13684D+00	-3.67488D-03	5.27623D-06
146	VINYLACETYLENE	2.32080D+01	8.21960D-01	-3.04898D-03	4.36824D-06
147	VINYL CHLORIDE	-7.94162D+00	8.86957D-01	-3.33590D-03	4.82845D-06
148	WATER	1.82964D+01	4.72118D-01	-1.33878D-03	1.31424D-06
149	1-BUTENE	-7.03548D-01	1.02601D+00	-3.63907D-03	5.58528D-06
150	1-ET-2-ME-BENZENE	-2.14660D+01	1.28851D+00	-2.22048D-03	1.91681D-06
151	1-HEPTENE	2.00903D+01	1.14211D+00	-2.55737D-03	3.01294D-06
152	1-HEXENE	1.11256D+01	1.05516D+00	-2.59179D-03	3.20045D-06
153	1-ME-4-ETH-NAPHTHA	-5.45979D+01	1.49414D+00	-1.84755D-03	1.23156D-06
154	1-METHYLINDENE	-3.21486D+01	1.18921D+00	-1.74392D-03	1.35295D-06
155	1-METHYLNAPHTHALEN	-4.88822D+01	1.20998D+00	-1.53729D-03	1.03385D-06
156	1-PENTENE	5.45455D+00	1.02076D+00	-2.93262D-03	3.93623D-06
157	1-PHENYLINDENE	-7.59414D+01	1.59722D+00	-1.82973D-03	1.04060D-06
158	1-PHENYLNAPHTHALEN	-9.05449D+01	1.64726D+00	-1.84168D-03	1.04703D-06
159	1,1-DICHLOROETHANE	1.01731D+01	8.73659D-01	-2.60121D-03	3.11481D-06
160	1,2-BUTADIENE	6.25816D+00	9.13303D-01	-3.10142D-03	4.38633D-06
161	1,2-DICHLOROETHANE	1.99518D+01	8.14846D-01	-2.29913D-03	2.56998D-06
162	1,2-DIME-3-ETHBZ	-2.31556D+01	1.20294D+00	-1.70706D-03	1.43219D-06
163	1,2,3-TRIME-INDENE	-3.79693D+01	1.40735D+00	-1.87484D-03	1.38180D-06
164	1,2,4-TRI-CL-BZ	-1.73071D+01	1.10113D+00	-1.91718D-03	1.42793D-06
165	1,3-BUTADIENE	-1.61777D+01	1.08511D+00	-3.68261D-03	5.41041D-06
166	112TRI-CL-ETHANE	6.30739D+00	8.91575D-01	-2.15051D-03	2.08369D-06
167	2-ETHYLFLUORENE	-7.76821D+01	1.68493D+00	-1.97725D-03	1.19039D-06
168	2-ME-1-BUTENE	6.85913D+00	1.00156D+00	-2.83730D-03	3.82462D-06
169	2-ME-2-BUTENE	9.02049D+00	9.50674D-01	-2.68985D-03	3.60341D-06
170	2-METHYLINDENE	-3.81172D+01	1.20843D+00	-1.73213D-03	1.30948D-06
171	2-METHYLNAPHTHALEN	-4.79704D+01	1.21302D+00	-1.56000D-03	1.06587D-06
172	2-METHYLPENTANE	3.36802D+01	9.38653D-01	-2.24842D-03	3.06594D-06
173	2,2-DIMETHYLBUTANE	2.00732D+01	9.87609D-01	-2.39907D-03	3.26302D-06
174	2,3-DIMETHYLBUTANE	1.28551D+01	1.02940D+00	-2.41758D-03	3.13353D-06
175	2,3,5-TRIME-NAPHTH	-5.74309D+01	1.50154D+00	-1.82573D-03	1.18242D-06
176	2,7-DIMETHYLNAPHTH	-5.27406D+01	1.34672D+00	-1.66099D-03	1.09319D-06
177	3-ME-1-BUTENE	-2.38042D+01	1.26801D+00	-3.55766D-03	4.48758D-06
178	3-METHYLPENTANE	3.17843D+01	9.58431D-01	-2.31456D-03	3.07474D-06

Source: FLOWTRAN Data Base. These tabulations were prepared by Dr. D. R. Schneider of the Monsanto Company using the Physical Properties package of the FLOWTRAN system. The tabulations are reproduced with the permission of the Monsanto Company.

7 Standard Heats of Formation (kcal/gmol)

Formula	Name	ΔH_f^o	Formula	Name	ΔH_f^o
Ar	Argon	0.0	CH_2Cl_2	Dichloromethane	-22.80
Br_2	Bromine	0.0	CH_2O	Formaldehyde	-27.70
ClNO	Nitrosyl chloride	12.57	CH_2O_2	Formic acid	-90.49
Cl_2	Chlorine	0.0	CH_3Br	Methyl bromide	-9.00
D_2	Deuterium	0.0	CH_3Cl	Methyl chloride	-20.63
D_2O	Deuterium oxide	-59.57	CH_3F	Methyl fluoride	-55.90
F_2	Fluorine	0.0	CH_3I	Methyl iodide	3.34
F_3N	Nitrogen trifluoride	-29.78	CH_3NO_2	Nitromethane	-17.86
F_6S	Sulfur hexafluoride	-291.8	CH_4	Methane	-17.89
HBr	Hydrogen bromide	-8.66	CH_4O	Methanol	-48.08
HCl	Hydrogen chloride	-22.06	CH_4S	Methyl mercaptan	-5.49
HF	Hydrogen fluoride	-64.60	CH_5N	Methyl amine	-5.5
HI	Hydrogen iodide	6.30	CH_6N_2	Methyl hydrazine	20.4
H_2	Hydrogen	0.0	$C_2Cl_2F_4$	1,2-Dichloro-1,1,2,2-	-214.6
H_2O	Water	-57.80	$C_2Cl_3F_3$	1,2,2-Trichloro-1,1,2	-178.10
H_2S	Hydrogen sulfide	-4.82	C_2Cl_4	Tetrachloroethylene	-2.9
H_3N	Ammonia	-10.92	C_2F_4	Perfluoroethene	-157.40
H_4N_2	Hydrazine	22.75	C_2F_6	Perfluoroethane	-321.00
He(4)	Helium-4	0.0	C_2N_2	Cyanogen	73.84
I_2	Iodine	0.0	C_2HCl_3	Trichloroethylene	-1.40
Kr	Krypton	0.0	C_2H_2	Acetylene	54.19
NO	Nitric oxide	21.60	$C_2H_2F_2$	1,1-Difluoroethylene	-82.50
NO_2	Nitrogen dioxide	8.09	C_2H_2O	Ketene	-14.60
N_2	Nitrogen	0.0	C_2H_3Cl	Vinyl chloride	8.40
N_2O	Nitrous oxide	19.49	C_2H_3ClO	Acetyl chloride	-58.30
Ne	Neon	0.0	$C_2H_3Cl_3$	1,1,2-Trichloroethane	-33.10
O_2	Oxygen	0.0	$C_2H_3F_3$	1,1,1-Trifluoroethane	-178.20
O_2S	Sulfur dioxide	-70.95	C_2H_3N	Acetonitrile	21.00
O_3	Ozone	34.1	C_2H_3NO	Methyl isocyanate	-21.50
O_3S	Sulfur trioxide	-94.47	C_2H_4	Ethylene	12.50
Xe	Xenon	0.0	$C_2H_4Cl_2$	1,1-Dichloroethane	-31.05
$CBrF_3$	Trifluorobromomethane	-155.1	$C_2H_4Cl_2$	1,2-Dichloroethane	-31.00
$CClF_3$	Chlorotrifluoromethane	-166.0	$C_2H_4F_2$	1,1-Difluoroethane	-118.00
CCl_2F_2	Dichlorodifluoromethane	-115.0	C_2H_4O	Acetaldehyde	-39.76
CCl_2O	Phosgene	-52.80	C_2H_4O	Ethylene oxide	-12.58
CCl_3F	Trichlorofluoromethane	-68.0	$C_2H_4O_2$	Acetic acid	-103.93
CCl_4	Carbon tetrachloride	-24.00	$C_2H_4O_2$	Methyl formate	-83.60
CF_4	Carbon tetrafluoride	-223.0	C_2H_5Br	Ethyl bromide	-15.30
CO	Carbon monoxide	-26.42	C_2H_5Cl	Ethyl chloride	-26.70
COS	Carbonyl sulfide	-33.08	C_2H_5F	Ethyl fluoride	-62.50
CO_2	Carbon dioxide	-94.05	C_2H_5N	Ethylene imine	29.50
CS_2	Carbon disulfide	27.98	C_2H_6	Ethane	-20.24
$CHClF_2$	Chlorodifluoromethane	-119.9	C_2H_6O	Dimethyl ether	-43.99
$CHCl_2F$	Dichloromonofluorometh	-71.4	C_2H_6O	Ethanol	-56.12
$CHCl_3$	Chloroform	-24.2	$C_2H_6O_2$	Ethylene glycol	-93.05
CHN	Hydrogen cyanide	31.20	C_2H_6S	Ethyl mercaptan	-11.02
CH_2Br_2	Dibromomethane	-1.0	C_2H_6S	Dimethyl sulfide	-8.97

Source: R. C. Reid, J. M. Prausnitz, and T. K. Sherwood, *The Properties of Gases and Liquids*, McGraw-Hill Book Company, New York, 1977. Used with permission.

7 Standard Heats of Formation (kcal/gmol) (continued)

Formula	Name	ΔH_f^0	Formula	Name	ΔH_f^0
C_2H_7N	Ethyl amine	-11.00	C_4H_8O	N-butyraldehyde	-49.00
C_2H_7N	Dimethyl amine	-4.50	C_4H_8O	Isobutyraldehyde	-51.56
C_2H_7NO	Monoethanolamine	-48.18	C_4H_8O	Methyl ethyl ketone	-56.97
C_3H_3N	Acrylonitrile	44.20	C_4H_8O	Tetrahydrofuran	-44.03
C_3H_4	Propadiene	45.92	C_4H_8O	Vinyl ethyl ether	-33.50
C_3H_4	Methyl acetylene	44.32	$C_4H_8O_2$	N-butyric acid	-113.73
C_3H_4O	Acrolein	-16.94	$C_4H_8O_2$	1,4-Dioxane	-75.30
$C_3H_4O_2$	Acrylic acid	-80.36	$C_4H_8O_2$	Ethyl acetate	-105.86
C_3H_5Cl	Allyl chloride	-0.15	$C_4H_8O_2$	Isobutyric acid	-115.66
$C_3H_5Cl_3$	1,2,3-Trichloropropan	-44.40	C_4H_9Cl	1-Chlorobutane	-35.20
C_3H_5N	Propionitrile	12.10	C_4H_9Cl	2-Chlorobutane	-38.60
C_3H_6	Cyclopropane	12.74	C_4H_9Cl	tert-Butyl chloride	-43.80
C_3H_6	Propylene	4.88	C_4H_9N	Pyrrolidine	-0.86
$C_3H_6Cl_2$	1,2-Dichloropropane	-39.6	C_4H_{10}	N-butane	-30.15
C_3H_6O	Acetone	-52.00	C_4H_{10}	Isobutane	-32.15
C_3H_6O	Allyl alcohol	-31.55	$C_4H_{10}O$	N-butanol	-65.65
C_3H_6O	Propionaldehyde	-45.90	$C_4H_{10}O$	2-Butanol	-69.94
C_3H_6O	Propylene oxide	-22.17	$C_4H_{10}O$	Isobutanol	-67.69
$C_3H_6O_2$	Propionic acid	108.78	$C_4H_{10}O$	tert-Butanol	-74.67
$C_3H_6O_2$	Ethyl formate	-88.74	$C_4H_{10}O$	Ethyl ether	-60.28
$C_3H_6O_2$	Methyl acetate	-97.86	$C_4H_{10}O_3$	Diethylene glycol	-136.5
C_3H_7Cl	Propyl chloride	-31.10	$C_4H_{10}S$	Diethyl sulfide	-19.95
C_3H_7Cl	Isopropyl chloride	-35.00	$C_4H_{10}S_2$	Diethyl disulfide	-17.84
C_3H_8	Propane	-24.82	$C_4H_{11}N$	N-butyl amine	-22.00
C_3H_8O	1-Propanol	-61.28	$C_4H_{11}N$	Diethyl amine	-17.30
C_3H_8O	Isopropyl alcohol	-65.11	C_5H_5N	Pyridine	33.50
C_3H_8O	Methyl ethyl ether	-51.73	C_5H_8	Cyclopentene	7.87
$C_3H_8O_2$	1,2-Propanediol	101.33	C_5H_8	1,2-Pentadiene	34.80
$C_3H_8O_2$	1,3-Propanediol	-97.71	C_5H_8	1-trans-3-Pentadiene	18.60
$C_3H_8O_3$	Glycerol	139.8	C_5H_8	1,4-Pentadiene	25.20
C_3H_8S	Methyl ethyl sulfide	-14.25	C_5H_8	1-Pentyne	34.50
C_3H_9N	N-propyl amine	-17.30	C_5H_8	2-Methyl-1,3-butadien	18.10
C_3H_9N	Isopropyl amine	-20.02	C_5H_8	3-Methyl-1,2-butadien	31.00
C_3H_9N	Trimethyl amine	-5.70	C_5H_8O	Cyclopentanone	-46.04
C_4H_4	Vinylacetylene	72.80	C_5H_{10}	Cyclopentane	-18.46
C_4H_4O	Furan	-8.29	C_5H_{10}	1-Pentene	-5.00
C_4H_4S	Thiophene	27.66	C_5H_{10}	cis-2-Pentene	-6.71
C_4H_5N	Pyrrole	25.88	C_5H_{10}	trans-2-Pentene	-7.59
C_4H_6	1-Butyne	39.48	C_5H_{10}	2-Methyl-1-butene	-8.68
C_4H_6	2-Butyne	34.97	C_5H_{10}	2-Methyl-2-butene	-10.17
C_4H_6	1,2-Butadiene	38.77	C_5H_{10}	3-Methyl-1-butene	-6.92
C_4H_6	1,3-Butadiene	26.33	$C_5H_{10}O$	Valeraldehyde	-54.45
$C_4H_6O_2$	Vinyl acetate	-75.5	$C_5H_{10}O$	Methyl n-propyl ketone	-61.82
$C_4H_6O_3$	Acetic anhydride	137.60	$C_5H_{10}O$	Diethyl ketone	-61.82
C_4H_7N	Butyronitrile	8.14	$C_5H_{10}O_2$	N-valeric acid	-117.20
C_4H_8	1-Butene	-0.03	$C_5H_{10}O_2$	N-propyl acetate	-111.31
C_4H_8	cis-2-Butene	-1.67	$C_5H_{10}O_2$	Ethyl propionate	-112.3
C_4H_8	trans-2-Butene	-2.67	$C_5H_{11}N$	Piperidine	-11.71
C_4H_8	Cyclobutane	6.37	C_5H_{12}	N-pentane	-35.00
C_4H_8	Isobutylene	-4.04	C_5H_{12}	2-Methyl butane	-36.92

7 Standard Heats of Formation (kcal/gmol) (continued)

Formula	Name	ΔH_f^0	Formula	Name	ΔH_f^0
C_5H_{12}	2,2-Dimethyl propane	-39.67	C_7F_{14}	Perfluoromethylcycloh	-692.2
$C_5H_{12}O$	1-Pentanol	-71.4	C_7F_{16}	Perfluoro-n-heptane	-808.9
$C_5H_{12}O$	2-Methyl-1-butanol	-72.3	C_7H_5N	Benzonitrile	52.30
$C_5H_{12}O$	3-Methyl-1-butanol	-72.2	C_7H_6O	Benzaldehyde	-8.79
$C_5H_{12}O$	2-Methyl-2-butanol	-78.8	$C_7H_6O_2$	Benzoic acid	-69.36
$C_5H_{12}O$	2,2-Dimethyl-1-propan	-70.00	C_7H_8	Toluene	11.95
C_6F_6	Perfluorobenzene	-228.64	C_7H_8O	Benzyl alcohol	-22.47
$C_6H_4Cl_2$	O-dichlorobenzene	7.16	C_7H_8O	O-cresol	-30.74
$C_6H_4Cl_2$	M-dichlorobenzene	6.32	C_7H_8O	M-cresol	-31.63
$C_6H_4Cl_2$	P-dichlorobenzene	8.50	C_7H_8O	P-cresol	-29.97
C_6H_5Br	Bromobenzene	25.10	C_7H_9N	2,3-Dimethylpyridine	16.31
C_6H_5Cl	Chlorobenzene	12.39	C_7H_9N	2,5-Dimethylpyridine	15.87
C_6H_5F	Fluorobenzene	-27.86	C_7H_9N	3,4-Dimethylpyridine	16.73
C_6H_5I	Iodobenzene	38.85	C_7H_9N	3,5-Dimethylpyridine	17.39
C_6H_6	Benzene	19.82	C_7H_9N	Methylphenylamine	20.4
C_6H_6O	Phenol	-23.03	C_7H_9N	O-toluidine	
C_6H_7N	Aniline	20.76	C_7H_{14}	Cycloheptane	-28.52
C_6H_7N	4-Methyl pyridine	24.43	C_7H_{14}	1,1-Dimethylcyclopent	-33.05
C_6H_{10}	1,5-Hexadiene	20.0	C_7H_{14}	cis-1,2-Dimethylcyclo	-30.96
C_6H_{10}	Cyclohexene	-1.28	C_7H_{14}	trans-1,2-Dimethylcyc	-32.67
$C_6H_{10}O$	Cyclohexanone	-55.00	C_7H_{14}	Ethylcyclopentane	-30.37
C_6H_{12}	Cyclohexane	-29.43	C_7H_{14}	Methylcyclohexane	-36.99
C_6H_{12}	Methylcyclopentane	-25.50	C_7H_{14}	1-Heptene	-14.89
C_6H_{12}	1-Hexene	-9.96	C_7H_{14}	2,3,3-Trimethyl-1-but	-20.67
C_6H_{12}	cis-2-Hexene	-12.51	C_7H_{16}	N-heptane	-44.88
C_6H_{12}	trans-2-Hexene	-12.88	C_7H_{16}	2-Methylhexane	-46.59
C_6H_{12}	cis-3-Hexene	-11.38	C_7H_{16}	3-Methylhexane	-45.96
C_6H_{12}	trans-3-Hexene	-13.01	C_7H_{16}	2,2-Dimethylpentane	-49.27
C_6H_{12}	2-Methyl-2-pentene	-14.28	C_7H_{16}	2,3-Dimethylpentane	-47.62
C_6H_{12}	3-Methyl-cis-2-pentene	-13.80	C_7H_{16}	2,4-Dimethylpentane	-48.28
C_6H_{12}	3-Methyl-trans-2-pentene	-14.02	C_7H_{16}	3,3-Dimethylpentane	-48.17
C_6H_{12}	4-Methyl-cis-2-pentene	-12.03	C_7H_{16}	3-Ethylpentane	-45.33
C_6H_{12}	4-Methyl-trans-2-pentene	-12.99	C_7H_{16}	2,2,3-Trimethylbutane	-48.95
C_6H_{12}	2,3-Dimethyl-1-butene	-13.32	$C_7H_{16}O$	1-Heptanol	-79.3
C_6H_{12}	2,3-Dimethyl-2-butene	-14.15	$C_8H_4O_3$	Phthalic anhydride	-88.8
C_6H_{12}	3,3-Dimethyl-1-butene	-10.31	C_8H_8	Styrene	35.22
$C_6H_{12}O$	Cyclohexanol	-70.40	C_8H_8O	Methyl phenyl ketone	-20.76
$C_6H_{12}O$	Methyl isobutyl keton	-67.84	$C_8H_8O_2$	Methyl benzoate	-60.68
$C_6H_{12}O_2$	N-butyl acetate	-116.26	C_8H_{10}	O-xylene	4.54
$C_6H_{12}O_2$	Isobutyl acetate	-118.34	C_8H_{10}	M-xylene	4.12
C_6H_{14}	N-hexane	-39.96	C_8H_{10}	P-xylene	4.29
C_6H_{14}	2-Methyl pentane	-41.66	C_8H_{10}	Ethylbenzene	7.12
C_6H_{14}	3-Methyl pentane	-41.02	$C_8H_{10}O$	O-ethylphenol	-34.82
C_6H_{14}	2,2-Dimethyl butane	-44.35	$C_8H_{10}O$	M-ethylphenol	-35.01
C_6H_{14}	2,3-Dimethyl butane	-42.49	$C_8H_{10}O$	P-ethylphenol	-34.55
$C_6H_{14}O$	1-Hexanol	-75.9	$C_8H_{10}O$	2,3-Xylenol	-37.58
$C_6H_{14}O$	Diisopropyl ether	-76.20	$C_8H_{10}O$	2,4-Xylenol	-38.88
$C_6H_{15}N$	Triethylamine	-23.80	$C_8H_{10}O$	2,5-Xylenol	-38.58
			$C_8H_{10}O$	2,6-Xylenol	-38.68
			$C_8H_{10}O$	3,4-Xylenol	-37.38

7　Standard Heats of Formation (kcal/gmol) (continued)

Formula	Name	ΔH_f^o	Formula	Name	ΔH_f^o
$C_8H_{10}O$	3,5-Xylenol	−38.57	C_9H_{20}	3,3-Diethylpentane	−55.44
$C_8H_{11}N$	n,n-Dimethylaniline	20.10	C_9H_{20}	2,2,3,3-Tetramethylpe	−56.70
C_8H_{16}	1,1-Dimethylcyclohexa	−43.26	C_9H_{20}	2,2,3,4-Tetramethylpe	−56.64
C_8H_{16}	cis-1,2-Dimethylcyclo	−41.15	C_9H_{20}	2,2,4,4-Tetramethylpe	−57.83
C_8H_{16}	trans-1,2-Dimethylcyc	−43.02	C_9H_{20}	2,3,3,4-Tetramethylpe	−56.46
C_8H_{16}	cis-1,3-Dimethylcyclo	−44.16	$C_{10}H_8$	Naphthalene	36.08
C_8H_{16}	trans-1,3-Dimethylcyc	−42.20	$C_{10}H_{12}$	1,2,3,4-Tetrahydronap	6.6
C_8H_{16}	cis-1,4-Dimethylcyclo	−42.22	$C_{10}H_{14}$	N-butylbenzene	−3.30
C_8H_{16}	trans-1,4-Dimethylcyc	−44.12	$C_{10}H_{14}$	isobutylbenzene	−5.15
C_8H_{16}	Ethylcyclohexane	−41.05	$C_{10}H_{14}$	sec-Butylbenzene	−4.17
C_8H_{16}	N-propylcyclopentane	−35.39	$C_{10}H_{14}$	tert-Butylbenzene	−5.42
C_8H_{16}	1-Octene	−19.82	$C_{10}H_{14}$	1-Methyl-3-isopropylb	−7.00
C_8H_{16}	trans-2-Octene	−22.59	$C_{10}H_{14}$	1,4-Diethylbenzene	−5.32
C_8H_{18}	N-octane	−49.82	$C_{10}H_{14}$	1,2,4,5-Tetramethylbe	−10.82
C_8H_{18}	2-Methylheptane	−51.50	$C_{10}H_{18}$	cis-Decalin	−40.38
C_8H_{18}	3-Methylheptane	−50.82	$C_{10}H_{18}$	trans-Decalin	−43.57
C_8H_{18}	4-Methylheptane	−50.69	$C_{10}H_{20}$	N-butylcyclohexane	−50.95
C_8H_{18}	2,2-Dimethylhexane	−53.71	$C_{10}H_{20}$	1-Decene	−29.67
C_8H_{18}	2,3-Dimethylhexane	−51.13	$C_{10}H_{22}$	N-decane	−59.67
C_8H_{18}	2,4-Dimethylhexane	−52.44	$C_{10}H_{22}$	3,3,5-Trimethylheptan	−61.80
C_8H_{18}	2,5-Dimethylhexane	−53.21	$C_{10}H_{22}O$	1-Decanol	−96.0
C_8H_{18}	3,3-Dimethylhexane	−52.61	$C_{11}H_{10}$	1-Methylnaphthalene	27.93
C_8H_{18}	3,4-Dimethylhexane	−50.91	$C_{11}H_{10}$	2-Methylnaphthalene	27.75
C_8H_{18}	3-Ethylhexane	−50.40	$C_{11}H_{22}$	N-hexylcyclopentane	−50.07
C_8H_{18}	2,2,3-Trimethylpentan	−52.61	$C_{11}H_{22}$	1-Undecene	−34.60
C_8H_{18}	2,2,4-Trimethylpentan	−53.57	$C_{11}H_{24}$	N-Undecane	−64.60
C_8H_{18}	2,3,3-Trimethylpentan	−51.73	$C_{12}H_{10}$	Diphenyl	43.52
C_8H_{18}	2,3,4-Trimethylpentan	−51.97	$C_{12}H_{10}O$	Diphenyl ether	11.94
C_8H_{18}	2-Methyl-3-ethylpenta	−50.48	$C_{12}H_{24}$	N-heptylcyclopentane	−55.00
C_8H_{18}	3-Methyl-3-ethylpenta	−51.38	$C_{12}H_{24}$	1-Dodecene	−39.52
$C_8H_{18}O$	1-Octanol	−86.0	$C_{12}H_{26}$	N-dodecane	−69.52
$C_8H_{18}O$	2-Ethylhexanol	−87.31	$C_{12}H_{26}O$	Dodecanol	−105.84
$C_8H_{18}O$	Butyl ether	−79.80	$C_{13}H_{26}$	N-octylcyclopentane	−59.92
C_9H_{12}	N-propylbenzene	1.87	$C_{13}H_{26}$	1-Tridecene	−44.45
C_9H_{12}	Isopropylbenzene	0.94	$C_{13}H_{28}$	N-tridecane	−74.45
C_9H_{12}	1-Methyl-2-ethylben-zene	0.29	$C_{14}H_{10}$	Anthracene	53.7
			$C_{14}H_{10}$	Phenanthrene	48.4
C_9H_{12}	1-Methyl-3-ethylben-zene	−0.46	$C_{14}H_{28}$	N-nonylcyclopentane	−64.85
			$C_{14}H_{28}$	1-Tetradecene	−49.36
C_9H_{12}	1-Methyl-4-ethylben-zene	−0.49	$C_{14}H_{30}$	N-tetradecane	−79.38
			$C_{15}H_{30}$	N-decylcyclopentane	−69.78
C_9H_{12}	1,2,3-Trimethylbenzene	−2.29	$C_{15}H_{30}$	1-Pentadecene	−54.31
C_9H_{12}	1,2,4-Trimethylbenzene	−3.33	$C_{15}H_{32}$	N-pentadecane	−84.31
C_9H_{12}	1,3,5-Trimethylbenzene	−3.84	$C_{16}H_{32}$	1-Hexadecene	−59.23
C_9H_{18}	N-propylcyclohexane	−46.20	$C_{16}H_{34}$	N-hexadecane	−89.23
C_9H_{18}	1-Nonene	−24.74	$C_{17}H_{34}$	N-dodecylcyclopentane	−80.28
C_9H_{20}	N-nonane	−54.74	$C_{17}H_{36}O$	Heptadecanol	−130.47
C_9H_{20}	2,2,3-Trimethylhexane	−57.65	$C_{17}H_{36}$	N-heptadecane	−94.15
C_9H_{20}	2,2,4-Trimethylhexane	−58.13	$C_{18}H_{36}$	1-octadecene	−69.08
C_9H_{20}	2,2,5-Trimethylhexane	−60.71	$C_{18}H_{36}$	N-tridecylcyclopentan	−84.55

7 Standard Heats of Formation (kcal/gmol) (continued)

Formula	Name	ΔH_f^0	Formula	Name	ΔH_f^0
$C_{18}H_{38}$	N-octadecane	-99.08	$C_{20}H_{40}$	N-pentadecylcyclopent	-94.41
$C_{18}H_{38}O$	1-Octadecanol	-135.39	$C_{20}H_{42}$	N-eicosane	-108.93
$C_{19}H_{38}$	N-tetradecylcyclopent	-89.48	$C_{20}H_{42}O$	1-Eicosanol	-145.25
$C_{19}H_{40}$	N-nonadecane	-104.00	$C_{21}H_{42}$	N-hexadecylcyclopenta	-99.33

8A Properties of Saturated Steam (SI Units): Temperature Table

		$\hat{V}$ (m³/kg)		$\hat{U}$ (kJ/kg)		$\hat{H}$ (kJ/kg)		
T (°C)	P (bar)	Water	Steam	Water	Steam	Water	Evapo-ration	Steam
0.01	0.00611	0.001000	206.2	zero	2375.6	$+0.0$	2501.6	2501.6
2	0.00705	0.001000	179.9	8.4	2378.3	8.4	2496.8	2505.2
4	0.00813	0.001000	157.3	16.8	2381.1	16.8	2492.1	2508.9
6	0.00935	0.001000	137.8	25.2	2383.8	25.2	2487.4	2512.6
8	0.01072	0.001000	121.0	33.6	2386.6	33.6	2482.6	2516.2
10	0.01227	0.001000	106.4	42.0	2389.3	42.0	2477.9	2519.9
12	0.01401	0.001000	93.8	50.4	2392.1	50.4	2473.2	2523.6
14	0.01597	0.001001	82.9	58.8	2394.8	58.8	2468.5	2527.2
16	0.01817	0.001001	73.4	67.1	2397.6	67.1	2463.8	2530.9
18	0.02062	0.001001	65.1	75.5	2400.3	75.5	2459.0	2534.5
20	0.0234	0.001002	57.8	83.9	2403.0	83.9	2454.3	2538.2
22	0.0264	0.001002	51.5	92.2	2405.8	92.2	2449.6	2541.9
24	0.0298	0.001003	45.9	100.6	2408.5	100.6	2444.9	2545.5
25	0.0317	0.001003	43.4	104.8	2409.9	104.8	2442.5	2547.3
26	0.0336	0.001003	41.0	108.9	2411.2	108.9	2440.2	2549.1
28	0.0378	0.001004	36.7	117.3	2414.0	117.3	2435.4	2552.7
30	0.0424	0.001004	32.9	125.7	2416.7	125.7	2430.7	2556.4
32	0.0475	0.001005	29.6	134.0	2419.4	134.0	2425.9	2560.0
34	0.0532	0.001006	26.6	142.4	2422.1	142.4	2421.2	2563.6
36	0.0594	0.001006	24.0	150.7	2424.8	150.7	2416.4	2567.2
38	0.0662	0.001007	21.6	159.1	2427.5	159.1	2411.7	2570.8
40	0.0738	0.001008	19.55	167.4	2430.2	167.5	2406.9	2574.4
42	0.0820	0.001009	17.69	175.8	2432.9	175.8	2402.1	2577.9
44	0.0910	0.001009	16.04	184.2	2435.6	184.2	2397.3	2581.5
46	0.1009	0.001010	14.56	192.5	2438.3	192.5	2392.5	2585.1
48	0.1116	0.001011	13.23	200.9	2440.9	200.9	2387.7	2588.6
50	0.1234	0.001012	12.05	209.2	2443.6	209.3	2382.9	2592.2
52	0.1361	0.001013	10.98	217.7	2446	217.7	2377	2595
54	0.1500	0.001014	10.02	226.0	2449	226.0	2373	2599
56	0.1651	0.001015	9.158	234.4	2451	234.4	2368	2602
58	0.1815	0.001016	8.380	242.8	2454	242.8	2363	2606
60	0.1992	0.001017	7.678	251.1	2456	251.1	2358	2609
62	0.2184	0.001018	7.043	259.5	2459	259.5	2353	2613
64	0.2391	0.001019	6.468	267.9	2461	267.9	2348	2616
66	0.2615	0.001020	5.947	276.2	2464	276.2	2343	2619
68	0.2856	0.001022	5.475	284.6	2467	284.6	2338	2623

8A   Properties of Saturated Steam (SI Units): Temperature Table (continued)

		$\hat{V}$ (m³/kg)		$\hat{U}$ (kJ/kg)		$\hat{H}$ (kJ/kg)		
T (°C)	P (bar)	Water	Steam	Water	Steam	Water	Evapo-ration	Steam
70	0.3117	0.001023	5.045	293.0	2469	293.0	2333	2626
72	0.3396	0.001024	4.655	301.4	2472	301.4	2329	2630
74	0.3696	0.001025	4.299	309.8	2474	309.8	2323	2633
76	0.4019	0.001026	3.975	318.2	2476	318.2	2318	2636
78	0.4365	0.001028	3.679	326.4	2479	326.4	2313	2639
80	0.4736	0.001029	3.408	334.8	2482	334.9	2308	2643
82	0.5133	0.001030	3.161	343.2	2484	343.3	2303	2646
84	0.5558	0.001032	2.934	351.6	2487	351.7	2298	2650
86	0.6011	0.001033	2.727	360.0	2489	360.1	2293	2653
88	0.6495	0.001034	2.536	368.4	2491	368.5	2288	2656
90	0.7011	0.001036	2.361	376.9	2493	377.0	2282	2659
92	0.7560	0.001037	2.200	385.3	2496	385.4	2277	2662
94	0.8145	0.001039	2.052	393.7	2499	393.8	2272	2666
96	0.8767	0.001040	1.915	401.1	2501	402.2	2267	2669
98	0.9429	0.001042	1.789	410.6	2504	410.7	2262	2673
100	1.0131	0.001044	1.673	419.0	2507	419.1	2257	2676
102	1.0876	0.001045	1.566	427.1	2509	427.5	2251	2679

Source: R. W. Haywood, *Thermodynamic Tables in SI (Metric) Units,* Cambridge University Press, 1968. Adapted with permission. $\hat{V}$ = specific volume, $\hat{U}$ = specific internal energy, and $\hat{H}$ = specific enthalpy.

8B   Properties of Saturated Steam (SI Units): Pressure Table

		$\hat{V}$ (m³/kg)		$\hat{U}$ (kJ/kg)		$\hat{H}$ (kJ/kg)		
P (bar)	T (°C)	Water	Steam	Water	Steam	Water	Evapo-ration	Steam
0.00611	0.01	0.001000	206.2	zero	2375.6	+0.0	2501.6	2501.6
0.008	3.8	0.001000	159.7	15.8	2380.7	15.8	2492.6	2508.5
0.010	7.0	0.001000	129.2	29.3	2385.2	29.3	2485.0	2514.4
0.012	9.7	0.001000	108.7	40.6	2388.9	40.6	2478.7	2519.3
0.014	12.0	0.001000	93.9	50.3	2392.0	50.3	2473.2	2523.5
0.016	14.0	0.001001	82.8	58.9	2394.8	58.9	2468.4	2527.3
0.018	15.9	0.001001	74.0	66.5	2397.4	66.5	2464.1	2530.6
0.020	17.5	0.001001	67.0	73.5	2399.6	73.5	2460.2	2533.6
0.022	19.0	0.001002	61.2	79.8	2401.7	79.8	2456.6	2536.4
0.024	20.4	0.001002	56.4	85.7	2403.6	85.7	2453.3	2539.0
0.026	21.7	0.001002	52.3	91.1	2405.4	91.1	2450.2	2541.3
0.028	23.0	0.001002	48.7	96.2	2407.1	96.2	2447.3	2543.6
0.030	24.1	0.001003	45.7	101.0	2408.6	101.0	2444.6	2545.6
0.035	26.7	0.001003	39.5	111.8	2412.2	111.8	2438.5	2550.4
0.040	29.0	0.001004	34.8	121.4	2415.3	121.4	2433.1	2554.5
0.045	31.0	0.001005	31.1	130.0	2418.1	130.0	2428.2	2558.2
0.050	32.9	0.001005	28.2	137.8	2420.6	137.8	2423.8	2561.6

8B Properties of Saturated Steam (SI Units): Pressure Table (continued)

P (bar)	T (°C)	$\hat{V}$ (m³/kg) Water	$\hat{V}$ (m³/kg) Steam	$\hat{U}$ (kJ/kg) Water	$\hat{U}$ (kJ/kg) Steam	$\hat{H}$ (kJ/kg) Water	$\hat{H}$ (kJ/kg) Evaporation	$\hat{H}$ (kJ/kg) Steam
0.060	36.2	0.001006	23.74	151.5	2425.1	151.5	2416.0	2567.5
0.070	39.0	0.001007	20.53	163.4	2428.9	163.4	2409.2	2572.6
0.080	41.5	0.001008	18.10	173.9	2432.3	173.9	2403.2	2577.1
0.090	43.8	0.001009	16.20	183.3	2435.3	183.3	2397.9	2581.1
0.10	45.8	0.001010	14.67	191.8	2438.0	191.8	2392.9	2584.8
0.11	47.7	0.001011	13.42	199.7	2440.5	199.7	2388.4	2588.1
0.12	49.4	0.001012	12.36	206.9	2442.8	206.9	2384.3	2591.2
0.13	51.1	0.001013	11.47	213.7	2445.0	213.7	2380.4	2594.0
0.14	52.6	0.001013	10.69	220.0	2447.0	220.0	2376.7	2596.7
0.15	54.0	0.001014	10.02	226.0	2448.9	226.0	2373.2	2599.2
0.16	55.3	0.001015	9.43	231.6	2450.6	231.6	2370.0	2601.6
0.17	56.6	0.001015	8.91	236.9	2452.3	236.9	2366.9	2603.8
0.18	57.8	0.001016	8.45	242.0	2453.9	242.0	2363.9	2605.9
0.19	59.0	0.001017	8.03	246.8	2455.4	246.8	2361.1	2607.9
0.20	60.1	0.001017	7.65	251.5	2456.9	251.5	2358.4	2609.9
0.22	62.2	0.001018	7.00	260.1	2459.6	260.1	2353.3	2613.5
0.24	64.1	0.001019	6.45	268.2	2462.1	268.2	2348.6	2616.8
0.26	65.9	0.001020	5.98	275.6	2464.4	275.7	2344.2	2619.9
0.28	67.5	0.001021	5.58	282.7	2466.5	282.7	2340.0	2622.7
0.30	69.1	0.001022	5.23	289.3	2468.6	289.3	2336.1	2625.4
0.35	72.7	0.001025	4.53	304.3	2473.1	304.3	2327.2	2631.5
0.40	75.9	0.001027	3.99	317.6	2477.1	317.7	2319.2	2636.9
0.45	78.7	0.001028	3.58	329.6	2480.7	329.6	2312.0	2641.7
0.50	81.3	0.001030	3.24	340.5	2484.0	340.6	2305.4	2646.0
0.55	83.7	0.001032	2.96	350.6	2486.9	350.6	2299.3	2649.9
0.60	86.0	0.001033	2.73	359.9	2489.7	359.9	2293.6	2653.6
0.65	88.0	0.001035	2.53	368.5	2492.2	368.6	2288.3	2656.9
0.70	90.0	0.001036	2.36	376.7	2494.5	376.8	2283.3	2660.1
0.75	91.8	0.001037	2.22	384.4	2496.7	384.5	2278.6	2663.0
0.80	93.5	0.001039	2.087	391.6	2498.8	391.7	2274.1	2665.8
0.85	95.2	0.001040	1.972	398.5	2500.8	398.6	2269.8	2668.4
0.90	96.7	0.001041	1.869	405.1	2502.6	405.2	2265.6	2670.9
0.95	98.2	0.001042	1.777	411.4	2504.4	411.5	2261.7	2673.2
1.00	99.6	0.001043	1.694	417.4	2506.1	417.5	2257.9	2675.4
1.01325 (1 atm)	100.0	0.001044	1.673	419.0	2506.5	419.1	2256.9	2676.0
1.1	102.3	0.001046	1.549	428.7	2509.2	428.8	2250.8	2679.6
1.2	104.8	0.001048	1.428	439.2	2512.1	439.4	2244.1	2683.4
1.3	107.1	0.001049	1.325	449.1	2514.7	449.2	2237.8	2687.0
1.4	109.3	0.001051	1.236	458.3	2517.2	458.4	2231.9	2690.3
1.5	111.4	0.001053	1.159	467.0	2519.5	467.1	2226.2	2693.4
1.6	113.3	0.001055	1.091	475.2	2521.7	475.4	2220.9	2696.2
1.7	115.2	0.001056	1.031	483.0	2523.7	483.2	2215.7	2699.0
1.8	116.9	0.001058	0.977	490.5	2525.6	490.7	2210.8	2701.5
1.9	118.6	0.001059	0.929	497.6	2527.5	497.8	2206.1	2704.0
2.0	120.2	0.001061	0.885	504.5	2529.2	504.7	2201.6	2706.3
2.2	123.3	0.001064	0.810	517.4	2532.4	517.6	2193.0	2710.6
2.4	126.1	0.001066	0.746	529.4	2535.4	529.6	2184.9	2714.5

8B Properties of Saturated Steam (SI Units): Pressure Table (continued)

P (bar)	T (°C)	$\hat{V}$ (m³/kg) Water	$\hat{V}$ (m³/kg) Steam	$\hat{U}$ (kJ/kg) Water	$\hat{U}$ (kJ/kg) Steam	$\hat{H}$ (kJ/kg) Water	$\hat{H}$ (kJ/kg) Evaporation	$\hat{H}$ (kJ/kg) Steam
2.6	128.7	0.001069	0.693	540.6	2538.1	540.9	2177.3	2718.2
2.8	131.2	0.001071	0.646	551.1	2540.6	551.4	2170.1	2721.5
3.0	133.5	0.001074	0.606	561.1	2543.0	561.4	2163.2	2724.7
3.2	135.8	0.001076	0.570	570.6	2545.2	570.9	2156.7	2727.6
3.4	137.9	0.001078	0.538	579.6	2547.2	579.9	2150.4	2730.3
3.6	139.9	0.001080	0.510	588.1	2549.2	588.5	2144.4	2732.9
3.8	141.8	0.001082	0.485	596.4	2551.0	596.8	2138.6	2735.3
4.0	143.6	0.001084	0.462	604.2	2552.7	604.7	2133.0	2737.6
4.2	145.4	0.001086	0.442	611.8	2554.4	612.3	2127.5	2739.8
4.4	147.1	0.001088	0.423	619.1	2555.9	619.6	2122.3	2741.9
4.6	148.7	0.001089	0.405	626.2	2557.4	626.7	2117.2	2743.9
4.8	150.3	0.001091	0.389	633.0	2558.8	633.5	2112.2	2745.7
5.0	151.8	0.001093	0.375	639.6	2560.2	640.1	2107.4	2747.5
5.5	155.5	0.001097	0.342	655.2	2563.3	655.8	2095.9	2751.7
6.0	158.8	0.001101	0.315	669.8	2566.2	670.4	2085.0	2755.5
6.5	162.0	0.001105	0.292	683.4	2568.7	684.1	2074.7	2758.9
7.0	165.0	0.001108	0.273	696.3	2571.1	697.1	2064.9	2762.0
7.5	167.8	0.001112	0.2554	708.5	2573.3	709.3	2055.5	2764.8
8.0	170.4	0.001115	0.2403	720.0	2575.5	720.9	2046.5	2767.5
8.5	172.9	0.001118	0.2268	731.1	2577.1	732.0	2037.9	2769.9
9.0	175.4	0.001121	0.2148	741.6	2578.8	742.6	2029.5	2772.1
9.5	177.7	0.001124	0.2040	751.8	2580.4	752.8	2021.4	2774.2
10.0	179.9	0.001127	0.1943	761.5	2581.9	762.6	2013.6	2776.2
10.5	182.0	0.001130	0.1855	770.8	2583.3	772.0	2005.9	2778.0
11.0	184.1	0.001133	0.1774	779.9	2584.5	781.1	1998.5	2779.7
11.5	186.0	0.001136	0.1700	788.6	2585.8	789.9	1991.3	2781.3
12.0	188.0	0.001139	0.1632	797.1	2586.9	798.4	1984.3	2782.7
12.5	189.8	0.001141	0.1569	805.3	2588.0	806.7	1977.4	2784.1
13.0	191.6	0.001144	0.1511	813.2	2589.0	814.7	1970.7	2785.4
14	195.0	0.001149	0.1407	828.5	2590.8	830.1	1957.7	2787.8
15	198.3	0.001154	0.1317	842.9	2592.4	844.7	1945.2	2789.9
16	201.4	0.001159	0.1237	856.7	2593.8	858.6	1933.2	2791.7
17	204.3	0.001163	0.1166	869.9	2595.1	871.8	1921.5	2793.4
18	207.1	0.001168	0.1103	882.5	2596.3	884.6	1910.3	2794.8
19	209.8	0.001172	0.1047	894.6	2597.3	896.8	1899.3	2796.1
20	212.4	0.001177	0.0995	906.2	2598.2	908.6	1888.6	2797.2
21	214.9	0.001181	0.0949	917.5	2598.9	920.0	1878.2	2798.2
22	217.2	0.001185	0.0907	928.3	2599.6	931.0	1868.1	2799.1
23	219.6	0.001189	0.0868	938.9	2600.2	941.6	1858.2	2799.8
24	221.8	0.001193	0.0832	949.1	2600.7	951.9	1848.5	2800.4
25	223.9	0.001197	0.0799	959.0	2601.2	962.0	1839.0	2800.9
26	226.0	0.001201	0.0769	968.6	2601.5	971.7	1829.6	2801.4
27	228.1	0.001205	0.0740	978.0	2601.8	981.2	1820.5	2801.7
28	230.0	0.001209	0.0714	987.1	2602.1	990.5	1811.5	2802.0
29	232.0	0.001213	0.0689	996.0	2602.3	999.5	1802.6	2802.2
30	233.8	0.001216	0.0666	1004.7	2602.4	1008.4	1793.9	2802.3
32	237.4	0.001224	0.0624	1021.5	2602.5	1025.4	1776.9	2802.3
34	240.9	0.001231	0.0587	1037.6	2602.5	1041.8	1760.3	2802.1

8B Properties of Saturated Steam (SI Units): Pressure Table (continued)

P (bar)	T (°C)	$\hat{V}$ (m³/kg) Water	$\hat{V}$ (m³/kg) Steam	$\hat{U}$ (kJ/kg) Water	$\hat{U}$ (kJ/kg) Steam	$\hat{H}$ (kJ/kg) Water	$\hat{H}$ (kJ/kg) Evaporation	$\hat{H}$ (kJ/kg) Steam
36	244.2	0.001238	0.0554	1053.1	2602.2	1057.6	1744.2	2801.7
38	247.3	0.001245	0.0524	1068.0	2601.9	1072.7	1728.4	2801.1
40	250.3	0.001252	0.0497	1082.4	2601.3	1087.4	1712.9	2800.3
42	253.2	0.001259	0.0473	1096.3	2600.7	1101.6	1697.8	2799.4
44	256.0	0.001266	0.0451	1109.8	2599.9	1115.4	1682.9	2798.3
46	258.8	0.001272	0.0430	1122.9	2599.1	1128.8	1668.3	2797.1
48	261.4	0.001279	0.0412	1135.6	2598.1	1141.8	1653.9	2795.7
50	263.9	0.001286	0.0394	1148.0	2597.0	1154.5	1639.7	2794.2
52	266.4	0.001292	0.0378	1160.1	2595.9	1166.8	1625.7	2792.6
54	268.8	0.001299	0.0363	1171.9	2594.6	1178.9	1611.9	2790.8
56	271.1	0.001306	0.0349	1183.5	2593.3	1190.8	1598.2	2789.0
58	273.3	0.001312	0.0337	1194.7	2591.9	1202.3	1584.7	2787.0
60	275.6	0.001319	0.0324	1205.8	2590.4	1213.7	1571.3	2785.0
62	277.7	0.001325	0.0313	1216.6	2588.8	1224.8	1558.0	2782.9
64	279.8	0.001332	0.0302	1227.2	2587.2	1235.7	1544.9	2780.6
66	281.8	0.001338	0.0292	1237.6	2585.5	1246.5	1531.9	2778.3
68	283.8	0.001345	0.0283	1247.9	2583.7	1257.0	1518.9	2775.9
70	285.8	0.001351	0.0274	1258.0	2581.8	1267.4	1506.0	2773.5
72	287.7	0.001358	0.0265	1267.9	2579.9	1277.6	1493.3	2770.9
74	289.6	0.001364	0.0257	1277.6	2578.0	1287.7	1480.5	2768.3
76	291.4	0.001371	0.0249	1287.2	2575.9	1297.6	1467.9	2765.5
78	293.2	0.001378	0.0242	1296.7	2573.8	1307.4	1455.3	2762.8
80	295.0	0.001384	0.0235	1306.0	2571.7	1317.1	1442.8	2759.9
82	296.7	0.001391	0.0229	1315.2	2569.5	1326.6	1430.3	2757.0
84	298.4	0.001398	0.0222	1324.3	2567.2	1336.1	1417.9	2754.0
86	300.1	0.001404	0.0216	1333.3	2564.9	1345.4	1405.5	2750.9
88	301.7	0.001411	0.0210	1342.2	2562.6	1354.6	1393.2	2747.8
90	303.3	0.001418	0.02050	1351.0	2560.1	1363.7	1380.9	2744.6
92	304.9	0.001425	0.01996	1359.7	2557.7	1372.8	1368.6	2741.4
94	306.4	0.001432	0.01945	1368.2	2555.2	1381.7	1356.3	2738.0
96	308.0	0.001439	0.01897	1376.7	2552.6	1390.6	1344.1	2734.7
98	309.5	0.001446	0.01849	1385.2	2550.0	1399.3	1331.9	2731.2
100	311.0	0.001453	0.01804	1393.5	2547.3	1408.0	1319.7	2727.7
105	314.6	0.001470	0.01698	1414.1	2540.4	1429.5	1289.2	2718.7
110	318.0	0.001489	0.01601	1434.2	2533.2	1450.6	1258.7	2709.3
115	321.4	0.001507	0.01511	1454.0	2525.7	1471.3	1228.2	2699.5
120	324.6	0.001527	0.01428	1473.4	2517.8	1491.8	1197.4	2689.2
125	327.8	0.001547	0.01351	1492.7	2509.4	1512.0	1166.4	2678.4
130	330.8	0.001567	0.01280	1511.6	2500.6	1532.0	1135.0	2667.0
135	333.8	0.001588	0.01213	1530.4	2491.3	1551.9	1103.1	2655.0
140	336.6	0.001611	0.01150	1549.1	2481.4	1571.6	1070.7	2642.4
145	339.4	0.001634	0.01090	1567.5	2471.0	1591.3	1037.7	2629.1
150	342.1	0.001658	0.01034	1586.1	2459.9	1611.0	1004.0	2615.0
155	344.8	0.001683	0.00981	1604.6	2448.2	1630.7	969.6	2600.3
160	347.3	0.001710	0.00931	1623.2	2436.0	1650.5	934.3	2584.9
165	349.8	0.001739	0.00883	1641.8	2423.1	1670.5	898.3	2568.8
170	352.3	0.001770	0.00837	1661.6	2409.3	1691.7	859.9	2551.6
175	354.6	0.001803	0.00793	1681.8	2394.6	1713.3	820.0	2533.3

8B Properties of Saturated Steam (SI Units): Pressure Table (continued)

		$\hat{V}$ (m³/kg)		$\hat{U}$ (kJ/kg)		$\hat{H}$ (kJ/kg)		
P (bar)	T (°C)	Water	Steam	Water	Steam	Water	Evapo- ration	Steam
180	357.0	0.001840	0.00750	1701.7	2378.9	1734.8	779.1	2513.9
185	359.2	0.001881	0.00708	1721.7	2362.1	1756.5	736.6	2493.1
190	361.4	0.001926	0.00668	1742.1	2343.8	1778.7	692.0	2470.6
195	363.6	0.001977	0.00628	1763.2	2323.6	1801.8	644.2	2446.0
200	365.7	0.00204	0.00588	1785.7	2300.8	1826.5	591.9	2418.4
205	367.8	0.00211	0.00546	1810.7	2274.4	1853.9	532.5	2386.4
210	369.8	0.00220	0.00502	1840.0	2242.1	1886.3	461.3	2347.6
215	371.8	0.00234	0.00451	1878.6	2198.1	1928.9	366.2	2295.2
220	373.7	0.00267	0.00373	1952	2114	2011	185	2196
221.2 (critical point)	374.15	0.00317	0.00317	2038	2038	2108	0	2108

Source: R. W. Haywood, *Thermodynamic Tables in SI (Metric) Units,* Cambridge University Press, 1968. Adapted with permission. $\hat{V}$ = specific volume, $\hat{U}$ = specific internal energy, and $\hat{H}$ = specific enthalpy.

8C Properties of Superheated Steam (SI Units) (continued)

P(bars) ($T_{sat.}$ °C)		Sat'd Water	Sat'd Steam	Temperature (°C)							
				50	75	100	150	200	250	300	350
0.0 (—)	$\hat{H}$	—	—	2595	2642	2689	2784	2880	2978	3077	3177
	$\hat{U}$	—	—	2446	2481	2517	2589	2662	2736	2812	2890
	$\hat{V}$	—	—	—	—	—	—	—	—	—	—
0.1 (45.8)	$\hat{H}$	191.8	2584.8	2593	2640	2688	2783	2880	2977	3077	3177
	$\hat{U}$	191.8	2438.0	2444	2480	2516	2588	2661	2736	2812	2890
	$\hat{V}$	0.00101	14.7	14.8	16.0	17.2	19.5	21.8	24.2	26.5	28.7
0.5 (81.3)	$\hat{H}$	340.6	2446.0	209.3	313.9	2683	2780	2878	2976	3076	3177
	$\hat{U}$	340.6	2484.0	209.2	313.9	2512	2586	2660	2735	2811	2889
	$\hat{V}$	0.00103	3.24	0.00101	0.00103	3.41	3.89	4.35	4.83	5.29	5.75
1.0 (99.6)	$\hat{H}$	417.5	2675.4	209.3	314.0	2676	2776	2875	2975	3074	3176
	$\hat{U}$	417.5	2506.1	209.2	313.9	2507	2583	2658	2734	2811	2889
	$\hat{V}$	0.00104	1.69	0.00101	0.00103	1.69	1.94	2.17	2.40	2.64	2.87
5.0 (151.8)	$\hat{H}$	640.1	2747.5	209.7	314.3	419.4	632.2	2855	2961	3065	3168
	$\hat{U}$	639.6	2560.2	209.2	313.8	418.8	631.6	2643	2724	2803	2883
	$\hat{V}$	0.00109	0.375	0.00101	0.00103	0.00104	0.00109	0.425	0.474	0.522	0.571
10 (179.9)	$\hat{H}$	762.6	2776.2	210.1	314.7	419.7	632.5	2827	2943	3052	3159
	$\hat{U}$	761.5	2582	209.1	313.7	418.7	631.4	2621	2710	2794	2876
	$\hat{V}$	0.00113	0.194	0.00101	0.00103	0.00104	0.00109	0.206	0.233	0.258	0.282
20 (212.4)	$\hat{H}$	908.6	2797.2	211.0	315.5	420.5	633.1	852.6	2902	3025	3139
	$\hat{U}$	906.2	2598.2	209.0	313.5	418.4	630.9	850.2	2679	2774	2862
	$\hat{V}$	0.00118	0.09950	0.00101	0.00102	0.00104	0.00109	0.00116	0.111	0.125	0.139
40 (250.3)	$\hat{H}$	1087.4	2800.3	212.7	317.1	422.0	634.3	853.4	1085.8	2962	3095
	$\hat{U}$	1082.4	2601.3	208.6	313.0	417.8	630.0	848.8	1080.8	2727	2829
	$\hat{V}$	0.00125	0.04975	0.00101	0.00102	0.00104	0.00109	0.00115	0.00125	0.0588	0.0665

8C Properties of Superheated Steam (SI Units) (continued)

P(bars) (T_sat. °C)		Sat'd Water	Sat'd Steam	Temperature (°C)							
				50	75	100	150	200	250	300	350
60 (275.6)	$\hat H$	1213.7	2785.0	214.4	318.7	423.5	635.6	854.2	1085.8	2885	3046
	$\hat U$	1205.8	2590.4	208.3	312.6	417.3	629.1	847.3	1078.3	2668	2792
	$\hat V$	0.00132	0.0325	0.00101	0.00103	0.00104	0.00109	0.00115	0.00125	0.0361	0.0422
80 (295.0)	$\hat H$	1317.1	2759.9	216.1	320.3	425.0	636.8	855.1	1085.8	2787	2990
	$\hat U$	1306.0	2571.7	208.1	312.3	416.7	628.2	845.9	1075.8	2593	2750
	$\hat V$	0.00139	0.0235	0.00101	0.00102	0.00104	0.00109	0.00115	0.00124	0.0243	0.0299
100 (311.0)	$\hat H$	1408.0	2727.7	217.8	322.9	426.5	638.1	855.9	1085.8	1343.4	2926
	$\hat U$	1393.5	2547.3	207.8	311.7	416.1	627.3	844.4	1073.4	1329.4	2702
	$\hat V$	0.00145	0.0181	0.00101	0.00102	0.00104	0.00109	0.00115	0.00124	0.00140	0.0224
150 (342.1)	$\hat H$	1611.0	2615.0	222.1	326.0	430.3	641.3	858.1	1086.2	1338.2	2695
	$\hat U$	1586.1	2459.9	207.0	310.7	414.7	625.0	841.0	1067.7	1317.6	2523
	$\hat V$	0.00166	0.0103	0.00101	0.00102	0.00104	0.00108	0.00114	0.00123	0.00138	0.0115
200 (365.7)	$\hat H$	1826.5	2418.4	226.4	330.0	434.0	644.5	860.4	1086.7	1334.3	1647.1
	$\hat U$	1785.7	2300.8	206.3	309.7	413.2	622.9	837.7	1062.2	1307.1	1613.7
	$\hat V$	0.00204	0.005875	0.00100	0.00102	0.00103	0.00108	0.00114	0.00122	0.00136	0.00167
221.2(P_c) (374.15)(T_c)	$\hat H$	2108	2108	228.2	331.7	435.7	645.8	861.4	1087.0	1332.8	1635.5
	$\hat U$	2037.8	2037.8	206.0	309.2	412.8	622.0	836.3	1060.0	1302.9	1600.3
	$\hat V$	0.00317	0.00317	0.00100	0.00102	0.00103	0.00108	0.00114	0.00122	0.00135	0.00163
250 (—)	$\hat H$	—	—	230.7	334.0	437.8	647.7	862.8	1087.5	1331.1	1625.0
	$\hat U$	—	—	205.7	308.7	412.1	620.8	834.4	1057.0	1297.5	1585.0
	$\hat V$	—	—	0.00100	0.00101	0.00103	0.00108	0.00113	0.00122	0.00135	0.00160
300 (—)	$\hat H$	—	—	235.0	338.1	441.6	650.9	865.2	1088.4	1328.7	1609.9
	$\hat U$	—	—	205.0	307.7	410.8	618.7	831.3	1052.1	1288.7	1563.3
	$\hat V$	—	—	0.0009990	0.00101	0.00103	0.00107	0.00113	0.00121	0.00133	0.00155
500 (—)	$\hat H$	—	—	251.9	354.2	456.8	664.1	875.4	1093.6	1323.7	1576.3
	$\hat U$	—	—	202.4	304.0	405.8	611.0	819.7	1034.3	1259.3	1504.1
	$\hat V$	—	—	0.0009911	0.00100	0.00102	0.00106	0.00111	0.00119	0.00129	0.00144
1000 (—)	$\hat H$	—	—	293.9	394.3	495.1	698.0	903.5	1113.0	1328.7	1550.5
	$\hat U$	—	—	196.5	295.7	395.1	594.4	795.3	999.0	1207.1	1419.0
	$\hat V$	—	—	0.0009737	0.00009852	0.001000	0.00104	0.00108	0.00114	0.00122	0.00131

P(bars) (T_sat. °C)		Sat'd Water	Sat'd Steam	Temperature (°C)							
				400	450	500	550	600	650	700	750
0.0 (—)	$\hat H$			3280	3384	3497	3597	3706	3816	3929	4043
	$\hat U$			2969	3050	3132	3217	3303	3390	3480	3591
	$\hat V$			—	—	—	—	—	—	—	—
0.1 (45.8)	$\hat H$			3280	3384	3489	3596	3706	3816	3929	4043
	$\hat U$			2969	3050	3132	3217	3303	3390	3480	3571
	$\hat V$			21.1	33.3	35.7	38.0	40.3	42.6	44.8	47.2
0.5 (81.3)	$\hat H$			3279	3383	3489	3596	3705	3816	3929	4043
	$\hat U$			2969	3049	3132	3216	3302	3390	3480	3571
	$\hat V$			6.21	6.67	7.14	7.58	8.06	8.55	9.01	9.43
1.0 (99.6)	$\hat H$			3278	3382	3488	3596	3705	3816	3928	4042
	$\hat U$			2968	3049	3132	3216	3302	3390	3479	3570
	$\hat V$			3.11	3.33	3.57	3.80	4.03	4.26	4.48	4.72
5.0 (151.8)	$\hat H$			3272	3379	3484	3592	3702	3813	3926	4040
	$\hat U$			2964	3045	3128	3213	3300	3388	3477	3569
	$\hat V$			0.617	0.664	0.711	0.758	0.804	0.850	0.897	0.943

8C   Properties of Superheated Steam (SI Units) (continued)

P(bars) ($T_{sat.}$ °C)		Sat'd Water	Sat'd Steam	Temperature (°C) 400	450	500	550	600	650	700	750
10 (179.9)	$\hat{H}$ $\hat{U}$ $\hat{V}$			3264 2958 0.307	3371 3041 0.330	3478 3124 0.353	3587 3210 0.377	3697 3296 0.402	3809 3385 0.424	3923 3475 0.448	4038 3567 0.472
20 (212.4)	$\hat{H}$ $\hat{U}$ $\hat{V}$			3249 2946 0.151	3358 3031 0.163	3467 3115 0.175	3578 3202 0.188	3689 3290 0.200	3802 3379 0.211	3916 3470 0.223	4032 3562 0.235
40 (250.3)	$\hat{H}$ $\hat{U}$ $\hat{V}$			3216 2922 0.0734	3331 3011 0.0799	3445 3100 0.0864	3559 3188 0.0926	3673 3278 0.0987	3788 3368 0.105	3904 3460 0.111	4021 3554 0.117
60 (275.6)	$\hat{H}$ $\hat{U}$ $\hat{V}$			3180 2896 0.0474	3303 2991 0.0521	3422 3083 0.0566	3539 3174 0.0609	3657 3265 0.0652	3774 3357 0.0693	3892 3451 0.0735	4011 3545 0.0776
80 (295.0)	$\hat{H}$ $\hat{U}$ $\hat{V}$			3142 2867 0.0344	3274 2969 0.0382	3399 3065 0.0417	3520 3159 0.0450	3640 3252 0.0483	3759 3346 0.0515	3879 3441 0.0547	4000 3537 0.0578
100 (311.0)	$\hat{H}$ $\hat{U}$ $\hat{V}$			3100 2836 0.0264	3244 2946 0.0298	3375 3047 0.0328	3500 3144 0.0356	3623 3240 0.0383	3745 3335 0.0410	3867 3431 0.0435	3989 3528 0.0461
150 (342.1)	$\hat{H}$ $\hat{U}$ $\hat{V}$			2975 2744 0.0157	3160 2883 0.0185	3311 2999 0.0208	3448 3105 0.0229	3580 3207 0.0249	3708 3307 0.0267	3835 3407 0.0286	3962 3507 0.0304
200 (365.7)	$\hat{H}$ $\hat{U}$ $\hat{V}$			2820 2622 0.009950	3064 2810 0.0127	3241 2946 0.0148	3394 3063 0.0166	3536 3172 0.0182	3671 3278 0.0197	3804 3382 0.0211	3935 3485 0.0225
221.2(P_c) (374.15)(T_c)	$\hat{H}$ $\hat{U}$ $\hat{V}$			2733 2553 0.008157	3020 2776 0.0110	3210 2922 0.0130	3370 3045 0.0147	3516 3157 0.0162	3655 3265 0.0176	3790 3371 0.0190	3923 3476 0.0202
250 (—)	$\hat{H}$ $\hat{U}$ $\hat{V}$			2582 2432 0.006013	2954 2725 0.009174	3166 2888 0.0111	3337 3019 0.0127	3490 3137 0.0141	3633 3248 0.0143	3772 3356 0.0166	3908 3463 0.0178
300 (—)	$\hat{H}$ $\hat{U}$ $\hat{V}$			2162 2077 0.002830	2826 2623 0.006734	3085 2825 0.008680	3277 2972 0.0102	3443 3100 0.0114	3595 3218 0.0126	3740 3330 0.0136	3880 3441 0.0147
500 (—)	$\hat{H}$ $\hat{U}$ $\hat{V}$			1878 1791 0.001726	2293 2169 0.002491	2723 2529 0.003882	3021 2765 0.005112	3248 2946 0.006112	3439 3091 0.007000	3610 3224 0.007722	3771 3350 0.008418
1000 (—)	$\hat{H}$ $\hat{U}$ $\hat{V}$			1798 1653 0.001446	2051 1888 0.001628	2316 2127 0.001893	2594 2369 0.002246	2857 2591 0.002668	3105 2795 0.003106	3324 2971 0.003536	3526 3131 0.003953

Source: R. W. Haywood, *Thermodynamic Tables in SI (Metric) Units,* Cambridge University Press, 1968. Adapted with permission. Water is a liquid in the enclosed region between 50 and 350°C. $\hat{H}$ = specific enthalpy (kJ/kg), $\hat{U}$ = specific internal energy (kJ/kg), $\hat{V}$ = specific volume (m³/kg).

INDEX

INDEX